Engineering Design with SolidWorks 2003

A Competency Project Based Approach
Utilizing 3D Solid Modeling

David C. Planchard & Marie P. Planchard

Schroff Development Corporation

www.schroff.com
www.schroff-europe.com

Trademarks and Disclaimer

SolidWorks and its family of products are registered trademarks of the Dassault Systemes. Microsoft Windows and its family of products are trademarks of the Microsoft Corporation. Other software applications and parts described in this book are trademarks or registered trademarks of their respective owners.

When parts from companies are used, dimensions may be modified for illustration purposes. Every effort has been made to provide an accurate text. The authors and the manufacturers shall not be held liable for any parts developed or designed with this book or any responsibility for inaccuracies that appear in the book.

Copyright © 2003 by David C. Planchard and Marie P. Planchard

All rights reserved. This document may not be copied, photocopied, reproduced, transmitted, or translated in any form or for any purpose without the express written consent of the publisher Schroff Development Corporation.

Introduction

Engineering Design with SolidWorks is written to assist students, designers, engineers and professionals. The book is focused on providing a solid foundation in SolidWorks by utilizing projects with step-by-step instructions. Desired outcomes and usage competencies are listed for each project.

The book contains information for users to translate conceptual design ideas and sketches to manufacturing reality.

Designers and engineers work in a collaborative environment. They receive information from internal departments such as marketing and purchasing; they supply information to external sources such as vendors and customers.

Collaborative information translates into many formats such as paper drawings, electronic files, rendered images and animations. On-line intelligent catalogs guide designers to the product that meets both their geometric requirements and performance functionality.

In Project 1 through 6 the book provides step-by-step instructions on using and applying SolidWorks. Each Project contains exercises. The exercises analyze and examine usage competency.

The authors developed the industry scenarios by combining their own industry experience with the knowledge of engineers, department managers, vendors and manufacturers. These professionals are directly involved with SolidWorks everyday. Their responsibilities go far beyond the creation of just a 3D model.

The book is designed to compliment the On-line tutorials contained within SolidWorks.

About the cover

Displayed on the front cover is the Linkage assembly. Gears Educational Systems, LLC, Hanover, MA USA designed the Linkage assembly. The Air Cylinder is manufactured by SMC Corporation of America, Indianapolis, IN, USA. The Linkage assembly is a sub-assembly of the Pneumatic Test Module developed in the Project exercises. All parts are used with permission.

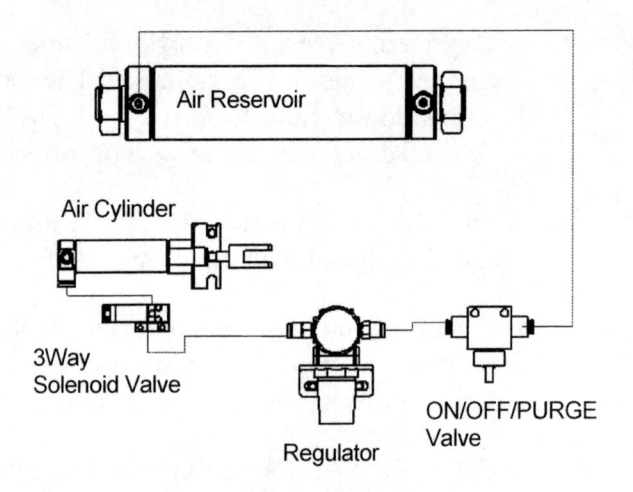

Pneumatic Components Diagram
Courtesy of SMC Corporation of America and Gears Educational Systems.

About the Authors

Marie Planchard is CAD department manager at Mass Bay College in Wellesley Hills, MA. Before developing the CAD engineering design program, she spent 13 years in industry and held a variety of High Technology management positions including Beta Test Manager for CAD software at Computervision Corporation. She has written and presented numerous technical papers on 3D modeling. She was Vice-President of the New England Pro/Users Group for 6 years, an active member of the SolidWorks Educational Advisory Board, a SolidWorks Research Partner, CSWP, SolidWorks World 2003 Presenter and coordinator for the New England SolidWorks Users Group.

David Planchard is the President of D & M Education, LLC. Before starting D & M Education LLC, he spent over 23 years in industry and academia holding various Engineering and Marketing positions. He has five U.S. and one International patent. He has published and authored numerous papers on equipment design. He is a member of the New England Pro/Users Group, New England SolidWorks Users Group and the Cisco CCNA Regional Academy Users Group.

David and Marie Planchard are Co-founders of D & M Education LLC and are active industry and education consultants. They are Co-authors of the following SDC Publication books:

- **Assembly Modeling with SolidWorks 2001Plus/2003.**
- **Engineering Design with SolidWorks 1999, 2000, 2001, 2001Plus.**
- **Drawing and Detailing with SolidWorks 2001/2001Plus.**
- **SolidWorks Tutorial 2001/2001Plus.**
- **Applications in Sheet Metal Using Pro/SHEETMETAL & Pro/ENGINEER.**
- **An Introduction to Pro/SHEETMETAL.**

Acknowledgement

The authors would like to acknowledge the following professionals for their contribution to the enhancements made to this revision and the corresponding instructors guide. Their assistance has been invaluable.

- Mike Bastoni and Mark Newby, Gears Educational Design Systems, LLC.
- Dave Kempski, Steve Hoffer and the Etech Team, SMC Corp. of America.
- Rosanne Kramer, Pierre Devaux and the SolidWorks Educational Team.
- Computer Aided Products Application Engineers: Jason Pancoast, Keith Pederson, Adam Snow, Joe St Cyr.
- Dave Pancoast and the SolidWorks Training Team.
- CSWPs: Mike J. Wilson, Devon Sowell, Scott Baugh, Paul Salvadore and Matt Lombard for their support of engineering education.
- SDC Publications: Stephen Schroff, Mary Schmidt and the SDC team.

For this 5th edition of Engineering Design with SolidWorks we realize that keeping software application books up to date is important to our customers. We value the hundreds of professors, students, designers and engineers that have provided us input to enhance our books. We value your suggestions and comments.

Please contact us with any comments, questions or suggestions on this book or any of our other SolidWorks SDC Publications.

Marie P. Planchard	David C. Planchard
Mass Bay College	D & M Education, LLC
mplanchard@massbay.edu	dplanchard@msn.com

Trademarks, Disclaimer and Copyrighted Material

SolidWorks and its family of products are registered trademarks of the SolidWorks Corporation. Microsoft Windows, Microsoft Office and its family of products are registered trademarks of the Microsoft Corporation. Pro/ENGINEER is a registered trademark of PTC. AutoCAD is a registered trademark of AutoDesk. Other software applications and parts described in this book are trademarks or registered trademarks of their respective owners.

Dimensions of parts are modified for illustration purposes. Every effort is made to provide an accurate text. The authors and the manufacturers shall not be held liable for any parts or drawings developed or designed with this book or any responsibility for inaccuracies that appear in the book. World Wide Web and company information was valid at the time of this printing.

Information in this text is provided from the ASME Engineering Drawing and Related Documentation Publications:

ASME Y14.1 1995
ASME Y14.2M-1992 (R1998)
ASME Y14.3M-1994 (R1999)
ASME Y14.5M-1994

The illustrations and part documents were recreated in SolidWorks. Note: By permission of The American Society of Mechanical Engineers, Codes and Standards, New York, NY, USA. All rights reserved.

References:

- SolidWorks Users Guide, SolidWorks Corporation, 2003.
- COSMOSXpress On-line help 2003.
- COSMOS/Works On-line help 2003.
- ASME Y14 Engineering Drawing and Related Documentation Practices.
- NBS Handbook 71, Specifications for Dry Cells and Batteries.
- Beers & Johnson, Vector Mechanics for Engineers, 6th ed. McGraw Hill, Boston, MA.
- Betoline, Wiebe, Miller, Fundamentals of Graphics Communication, Irwin, 1995.
- Earle, James, Engineering Design Graphics, Addison Wesley, 1999.
- Gradin, Hartley, Fundamentals of the Finite Element Method, Macmillan, NY 1986.
- Hibbler, R.C, Engineering Mechanics Statics and Dynamics, 8th ed, Prentice Hall, Saddle River, NJ.
- Hoelscher, Springer, Dobrovolny, Graphics for Engineers, John Wiley, 1968.
- Jensen, Cecil, Interpreting Engineering Drawings, Glencoe 2002.
- Jensen & Helsel, Engineering Drawing and Design, Glencoe, 1990.
- Lockhart & Johnson, Engineering Design Communications, Addison Wesley, 1999.
- Olivo C., Payne, Olivo, T, Basic Blueprint Reading and Sketching, Delmar 1988.
- Planchard & Planchard, Drawing and Detailing with SolidWorks, SDC Pub., Mission, KS 2002.
- Planchard & Planchard, Apps in SheetMetal Using Pro/ENGINEER, SDC Pub., Mission, KS 2000.
- Walker, James, Machining Fundamentals, Goodheart Wilcox, 1999.
- 80/20 Product Manual, 80/20, Inc., Columbia City, IN, 2002.
- GE Plastics Product Data Sheets, GE Plastics, Pittsfield, MA. 2000.
- Reid Tool Supply Product Manual, Reid Tool Supply Co., Muskegon, MI, 2002.
- Simpson Strong Tie Product Manual, Simpson Strong Tie, CA, 1998.
- Ticona Designing with Plastics – The Fundamentals, Summit, NJ, 2000.
- SMC Corporation of America, Product Manuals, Indiana, USA, 2003.
- Gears Educational Design Systems, Product Manual, Hanover, MA, USA 2003.
- Emerson-EPT Bearing Product Manuals and Gear Product Manuals, Emerson Power Transmission Corporation, Ithaca, NY, 2003.
- Alden Products, On-line catalog, Brocton, MA, 2003.
- Emhart – A Black and Decker Company, On-line catalog, Hartford, CT, 2003.
- DE-STA-CO Industries, On-line catalog, 2003.
- Boston Gear, On-line catalog, 2003.
- Enerpac a Division of Actuant, Inc., On-line catalog, 2003.

Notes:

Table of Contents

Introduction   **I-1**
About the Cover   I-2
About the Authors   I-2
Acknowledgement, Trademarks and Disclaimer   I-3
References   I-5
Table of Contents   I-7
What is SolidWorks?   I-13
Overview of Projects   I-15
Command Syntax   I-18
Note to Instructors   I-20

Project 1 - Fundamentals of Part Modeling   **1-1**
Project Objective   3
Project Situation   4
Project Overview   5
Windows 95/98/2000/NT4.0 Terminology   7
Keyboard Shortcuts   10
File Management   11
Starting a SolidWorks Session   13
System Options   17
Part Document Template and Document Properties   19
PLATE Part Overview   23
Extruded Base Feature   26
Machined Part   27
Reference Planes and Orthographic Projection   28
Create the PLATE - Extruded Base Feature   32
Create the PLATE - Modify Dimensions   38
Display Modes and View Modes   41
Fasteners   43
View Orientation   45
Create the PLATE - Holes as Cut Features   45
Create the PLATE - Fillet Feature   50
Create the PLATE - Hole Wizard Countersink Hole Feature   52
ROD Part   54
ROD Feature Overview   55
Create the ROD   56
Create the ROD - Chamfer Feature   59
Create the ROD - Extruded-Cut Feature   59
View Orientation and Named Views   61
Create the ROD - Copy/Paste Function   63
Create the ROD - Design Change with Rollback
    and Edit Definition   65
Create the ROD - Recover from Rebuild Errors   67

Create the ROD - Edit Part Color	70
GUIDE Part	71
GUIDE Feature Overview	71
Create the GUIDE Part	72
Create the GUIDE - Extruded Cut	75
Create the GUIDE – Mirror Feature	77
Create the Guide - HOLES	77
Create the GUIDE - Linear Pattern Feature	80
Project Summary	82
Project Terminology	82
Project Features	83
Questions/Exercises	85

Project 2 - Fundamentals of Assembly Modeling 2-1

Project Objective	3
Project Situation	3
Project Overview	5
Tolerance and Fit	6
Assembly Modeling Approach	6
Linear Motion and Rotational Motion	7
GUIDE-ROD Assembly	8
GUIDE-ROD Assembly - Insert Components	9
Mate Types	14
GUIDE-ROD Assembly - Mate the ROD Component	15
GUIDE-ROD Assembly - Mate the PLATE Component	18
Edit Component Dimension	22
Add a Component from the Parts Library	22
SmartMates	24
Copy the Component	26
Socket Head Cap Screw	28
Exploded View	35
Step Editing Tools and Viewing Exploded State	38
Section View	39
Analyze a Problem	40
Suppressed Components and Light Weight Components	42
Make-Buy Decision	43
CosmosXpress	53
Project Summary	70
Project Terminology	70
Project Features	72
Questions/Exercises	73

Project 3 - Fundamentals of Drawing 3-1

Project Objective	3
Project Situation	5
Project Overview	5
Drawing Template and Sheet Format	7
Title Block	14

Company Logo	16
Create the GUIDE Drawing from the GUIDE Part	21
Move Views - Named View	24
Additional Views	26
Insert Dimensions from the Part	29
Move Dimensions in the Same View	31
Partial Auxiliary View – Crop View	33
Move Dimensions to a Different View	34
Dimension Holes	36
Create Center Marks	38
Modify the Dimension Scheme	40
Clearance Fit for the ROD and GUIDE	43
Add An Additional Feature	44
General Notes	46
Part Number and Part Name	48
Notes in the Title Block	50
Exploded View and Bill of Materials	55
Associative Part, Assembly and Drawing	65
Project Summary	67
Project Terminology	67
Questions/Exercises	70
Geometric Tolerancing and Datum Feature Exercise	79

Project 4 - Extrude and Revolve Features   4-1

Project Objective	3
Project Situation	5
Project Overview	7
BATTERY	8
BATTERY Feature Overview	9
Create the Template	10
Create the BATTERY	12
Create the BATTERY - Fillet Feature	15
Create the BATTERY - Extruded Cut Feature	16
Create the Battery - Fillet Feature on the Top Face	17
Create the BATTERY - Extruded Boss Feature	18
BATTERY PLATE	23
BATTERY PLATE Feature Overview	23
Create the BATTERYPLATE	24
Create the BATTERYPLATE - Delete and Edit Features	25
Create the BATTERYPLATE - Extruded Boss Feature	26
Create the BATTERYPLATE - Edge Fillets	34
LENS	35
LENS Feature Overview	36
Create the LENS	37
Create the LENS - Shell Feature	40
Create the LENS - Extruded Boss Feature	41
Create the LENS - Hole Wizard Counterbore Hole Feature	41
Create the LENS - Boss Revolve Thin Feature	44
Create the LENS - Extruded Boss Feature	47
BULB	48

BULB Feature Overview	50
Create the BULB - Revolved Base Feature	50
Create the BULB - Revolved Boss Feature	51
Create the BULB - Revolved Cut Thin Feature	54
Create the BULB - Dome Feature	56
Create the BULB - Circular Pattern	56
Customizing Toolbars	60
Project Summary	60
Project Terminology	61
Project Features	62
Questions/Exercises	63

Project 5 - Sweep and Loft Features — 5-1

Project Objective	3
Project Situation	4
Project Overview	4
O-RING	7
O-RING Feature Overview	7
Create the O-RING	8
SWITCH	11
SWITCH - Feature Overview	11
Create the SWITCH – Loft Feature	11
Create the SWITCH - Dome Feature	16
LENSCAP	17
LENSCAP Feature Overview	18
Create the LENSCAP - Extruded Base Feature	19
Create the LENSCAP - Extruded Cut Feature	20
Create the LENSCAP - Shell Feature	21
Create the LENSCAP - Revolved Cut Thin Feature	22
Create the LENSCAP - Circular Pattern	23
Suppress Feature	27
Create the LENSCAP - Sweep Feature	27
HOUSING	34
HOUSING Feature Overview	36
Create the HOUSING - Extruded Base Feature	36
Create the HOUSING - Loft Feature	37
Create the HOUSING - Extruded Boss Feature	41
Create the HOUSING - Shell Feature - Extruded Boss Feature	42
Create the HOUSING - Draft Feature	44
Create the HOUSING - Thread with Sweep Feature	45
Create the HOUSING - Handle with Sweep Feature	49
Create the HOUSING - Extruded Cut Feature for SWITCH	53
Create the HOUSING - First Rib Feature	54
Create the HOUSING - Linear Pattern of Ribs	57
Create the HOUSING - Second Rib Feature	59
Mirror the Second Rib	62
FLASHLIGHT Assembly	64
FLASHLIGHT Assembly Overview	65
Assembly Techniques	66
Create an Assembly Template	67

LENSANDBULB Sub-assembly	69
BATTERYANDPLATE Sub-assembly	78
CAPANDLENS Sub-assembly	82
Complete the FLASHLIGHT Assembly	88
Addressing Design Issues	93
Export Files	95
Project Summary	98
Project Terminology	98
Project Features	99
Questions/Exercises	100

Project 6 - Top Down Assembly Modeling   6-1

Project Objective	3
Project Situation	5
Top Down Design Approach	6
BOX Assembly Overview	8
Layout Sketch	11
Link Values and Equations	16
Insert Component - MOTHERBOARD	19
Insert Component - POWERSUPPLY	26
Sheet Metal Overview	30
Material Thickness - Design State - Neutral Bend Line	31
Relief	34
Insert Component - CABINET	34
Create the Rip Feature & Sheet Metal Bends	37
Create the Flange Walls	40
Create a Hole and a Pattern of Holes	43
Add a Die Cutout Palette Feature	47
Add a Louver Forming Tool	50
Manufacturing Considerations	52
PEM Inserts	57
Add an Assembly Hole Feature	64
FeatureManager and the Assembly	65
Equations	70
Design Table	74
Project Summary	78
Project Terminology	78
Project Features	79
Questions/Exercises	80

Appendix

ECO Form	A1
Cursor Feedback	A2
On-line URLs for additional SolidWorks assistance	A7

Index

Notes:

What is SolidWorks?

SolidWorks is a design automation software package used to produce parts, assemblies and drawings. SolidWorks is a Windows native 3D solid modeling CAD program. SolidWorks provides easy to use, highest quality design software for engineers and designers who create 3D models and 2D drawings ranging from individual parts to assemblies with thousands of parts.

The SolidWorks Corporation, headquartered in Concord, Massachusetts, USA develops and markets innovative design solutions for the Microsoft Windows platform. More information on SolidWorks and its family of products can be found at their URL, **www.SolidWorks.com**.

In SolidWorks, you create 3D parts, assemblies and 2D drawings. The part, assembly and drawing documents are all related.

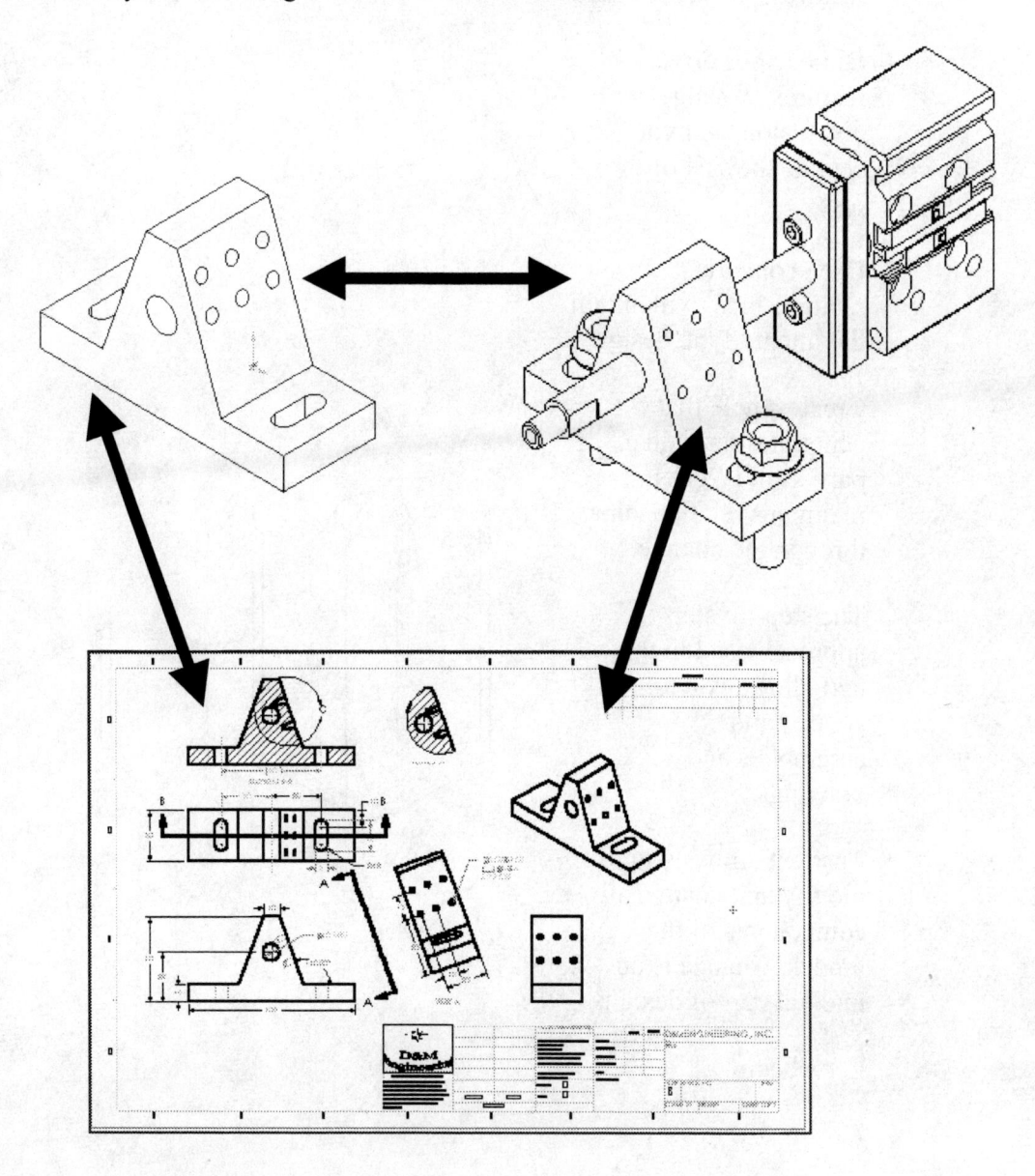

Features are the building blocks of parts. Use features to create parts, such as: Extruded-Boss, Cut, Hole, Fillet and Chamfer. Other features are sketched, such as an Extruded-Boss.

Create features by selecting edges or faces of existing features, such as a Fillet.

Dimensions drive features. Change a dimension, and you change the size of the part.

Use geometric relationships to maintain the intent of the design.

Create a hole that penetrates through a part. SolidWorks maintains relationships through the change.

The step-by-step approach used in this text allows you to create parts, assemblies and drawings.

The text allows you to modify and change all components of the model. Change is an integral part of design.

Let's begin.

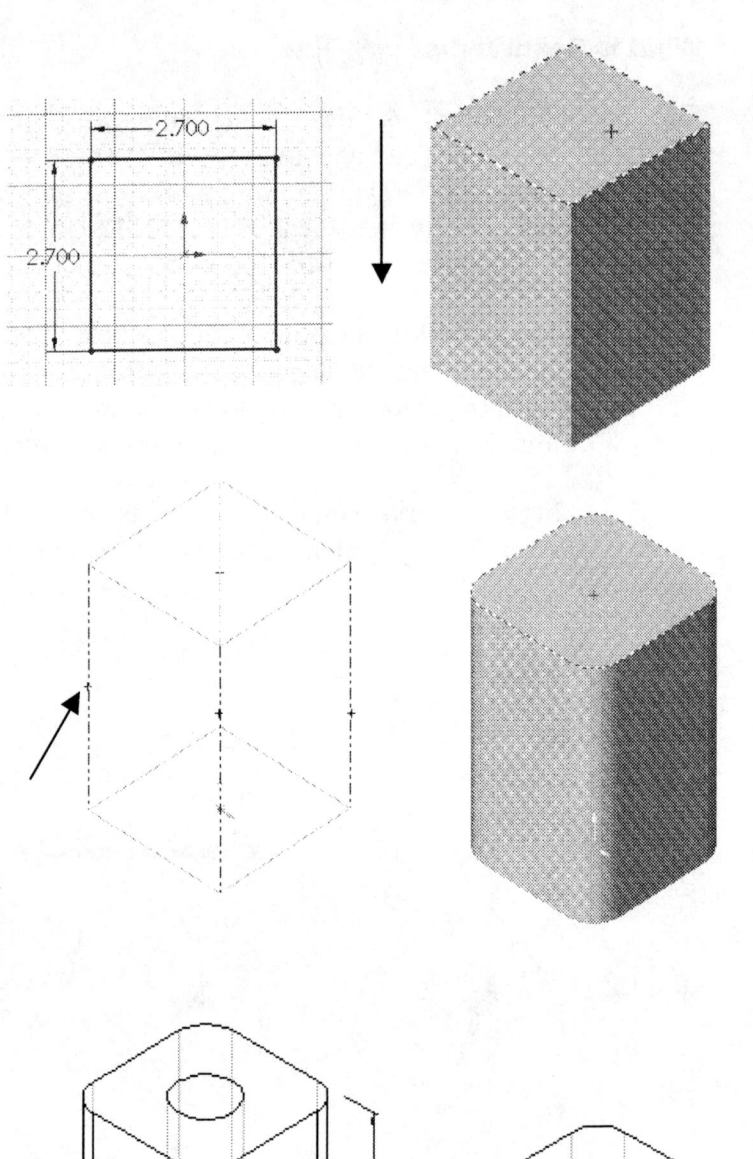

Engineering Design with SolidWorks   Introduction

Overview of Projects

Project 1: Fundamentals of Part Modeling.

Project 1 introduces the basic concepts behind SolidWorks and the SolidWorks user interface.

Create file folders and sub-folders to manage projects. Develop custom templates for part documents. Create three parts: PLATE, ROD and GUIDE.

Utilize the following features: Extruded-Base, Extruded-Boss, Extruded-Cut, Fillet, Chamfer, Hole Wizard and Linear Pattern.

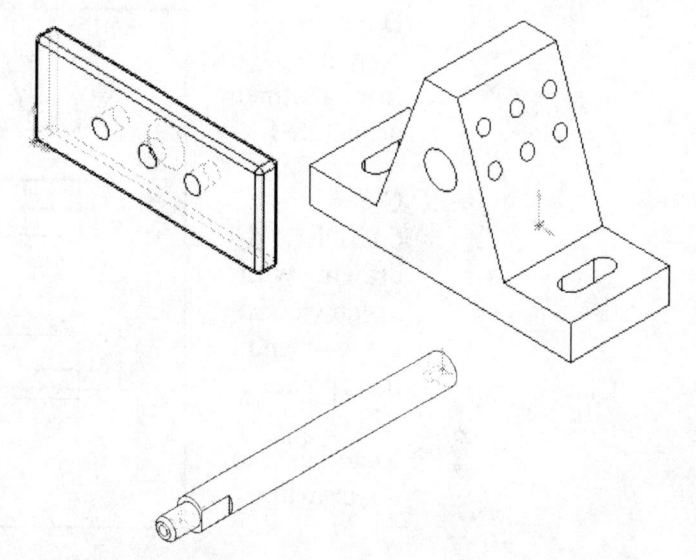

Project 2: Fundamentals of Assembly Modeling.

Project 2 introduces the fundamentals of Assembly Modeling.

Create an Assembly template. Review the FeatureManager syntax. Create two assemblies. Edit component dimensions and address tolerance and fit.

Incorporate design changes into an assembly. Obtain additional SolidWorks parts from the World Wide Web and the Feature Palette.

Apply COSMOSXpress, a software application analysis tool to view VonMises stress plot and deflection animation.

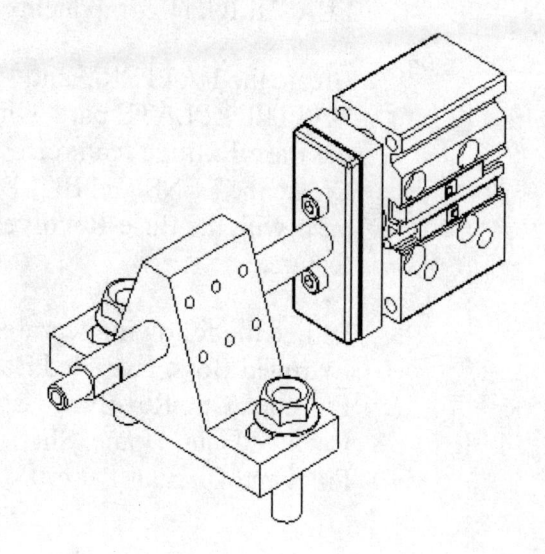

Project 3: Fundamentals of Drawing.

Project 3 covers the development of a customized drawing template.

Review the differences between the sheet and the sheet format.

Develop a company logo from a bitmap or picture file.

Create a GUIDE drawing with seven views. Develop and incorporate a Bill of Materials into the drawing.

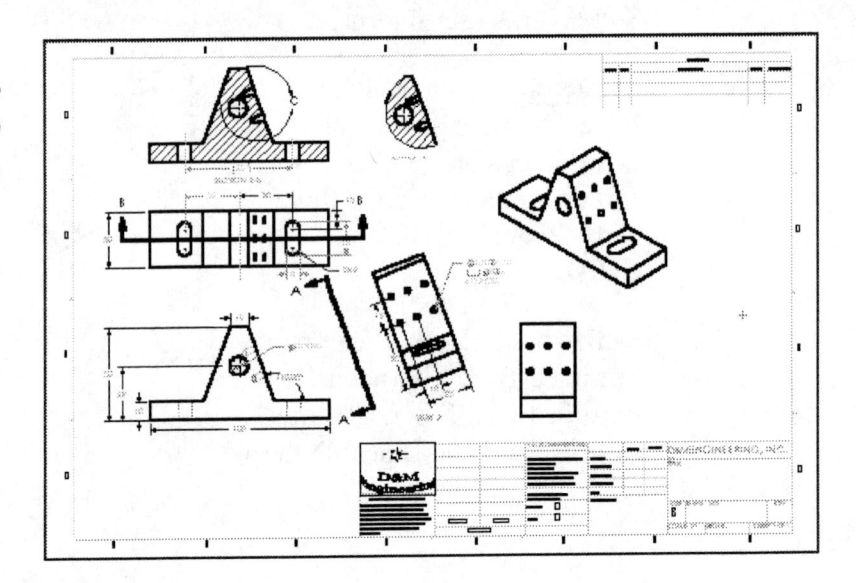

Project 4: Extrude and Revolve Features.

Project 4 focuses on the customer's design requirements. Create four key FLASHLIGHT components. Develop an English part template.

Create the BATTERY and BATTERYPLATE parts with the Base-Extrude feature. Create the LENS and BULB parts with the Base-Revolve feature.

Utilize the following features: Extruded Boss, Extruded Base, Extruded Cut, Revolved Base, Revolved Cut, Dome, Shell, Fillet and Circular Pattern.

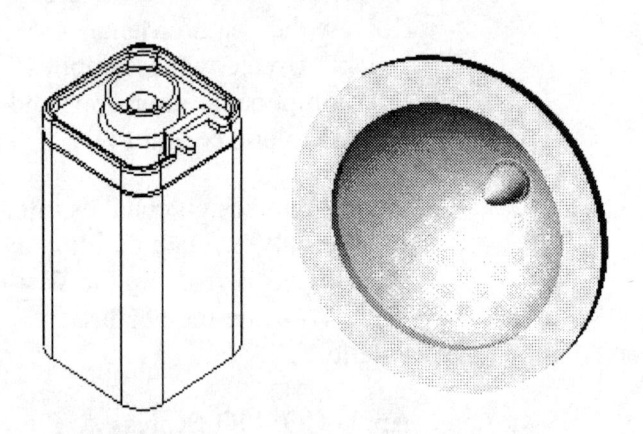

Project 5: Sweep and Loft Features.

Project 5 develops four additional components to complete the FLASHLIGHT assembly.

Utilize the following features: Sweep, Loft, Rib, Linear Pattern, Circular Pattern, Draft and Dome.

Create an English assembly template.

Insert the parts and sub-assemblies to create the final FLASHLIGHT assembly.

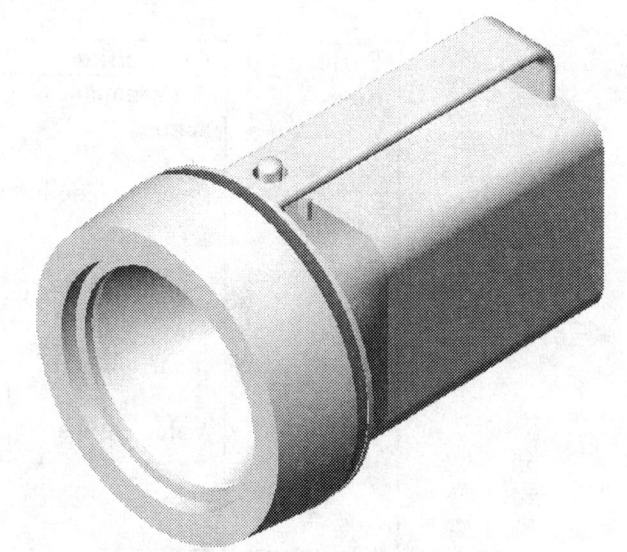

Project 6: Top Down Assembly.

Project 6 focuses on a Top Down assembly modeling approach. Develop a Layout Sketch. Create components and modify them In Context of the assembly.

Create Sheet metal features. Utilize the following features: Rip, Insert Sheet metal Bends, Edge Flange and Hem.

Utilize the Die Cut Palette Feature and Louver Form Tool.

Add IGES format part files from World Wide Web.

Replace fasteners in the assembly and redefine Mates.

Utilize equations, link values and a Design Table.

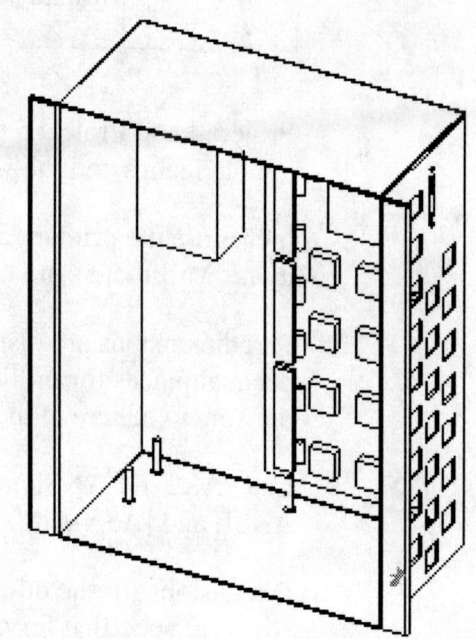

Command Syntax

The following command syntax is used throughout the text. Commands that require you to perform an action are displayed in **Bold** text.

Format:	Convention:	Example:
Bold	All commands actions.	Click **Save**. Click **Tools**, **Options** from the Main menu.
	Selected icon button.	Click **Rectangle** from the Sketch Tools toolbar.
	Selected geometry: line, circle, arc, point and text	Select the **centerpoint**. Drag the **circle** downward. Click the **arc**.
	Value entries.	Enter **3.0** for Radius. Click **60mm** from the Depth spin box.
Capitalized	Filenames, part names, assembly, component and drawing names.	The BATTERY is contained inside the FLASHLIGHT assembly.
	First letter in a feature name.	Click the **Fillet** feature. Click the **Extrude Base** feature.

Project 4 and Project 5 part dimensions and assembly dimensions are provided in both inches and millimeters.

Inches are the primary design units and millimeters are the secondary design units. Millimeters are displayed in brackets [x].

Part dimensions are displayed with three decimal places for inches and two decimal places for millimeters. Control individual decimal display in your drawing to adhere to the following drawing standards: ANSI or ISO.

Use PART-IN-ANSI and ASY-IN-ANSI for inch units. Use the [PART-MM-ANSI] and [ASY-MM-ANSI] for millimeter units.

Dimensions for the other projects in this book are provided in millimeters. Units are specified for each exercise.

This text follows the ASME Y14 Engineering Drawing and Related Documentation Practices for drawings. Display of dimensions and tolerances are as follows:

Review the ASME Y14.5 standard for Types of Decimal Dimensions.

TYPES of DECIMAL DIMENSIONS (ASME Y14.5M)			
Description:	**Example: MM**	**Description:**	**Example: INCH**
Dimension is less than 1mm. Zero precedes the decimal point.	0.9 0.95	Dimension is less than 1 inch. Zero is not used before the decimal point.	.5 .56
Dimension is a whole number. Display no decimal point. Display no zero after decimal point.	19	Express dimension to the same number of decimal places as its tolerance. Add zeros to the right of the decimal point. If the tolerance is expressed to 3 places, then the dimension contains 3 places to the right of the decimal point.	1.750
Dimension exceeds a whole number by a decimal fraction of a millimeter. Display no zero to the right of the decimal.	11.5 11.51		

TABLE 1		
TOLERANCE DISPLAY FOR INCH AND METRIC DIMENSIONS (ASME Y14.5M)		
DISPLAY:	**INCH:**	**METRIC:**
Dimensions less than 1	.5	0.5
Unilateral Tolerance	$1.417^{+.005}_{-.000}$	$36^{0}_{-0.5}$
Bilateral Tolerance	$1.417^{+.010}_{-.020}$	$36^{+0.25}_{-0.50}$
Limit Tolerance	.571 .463	14.50 11.50

Note to Instructors

Please contact the publisher www.schroff.com for additional materials that support the usage of this text in the classroom.

The physical parts and assemblies utilized in the exercises in this book are available through Gears Educational Systems, Hanover, MA (www.gearseds.com).

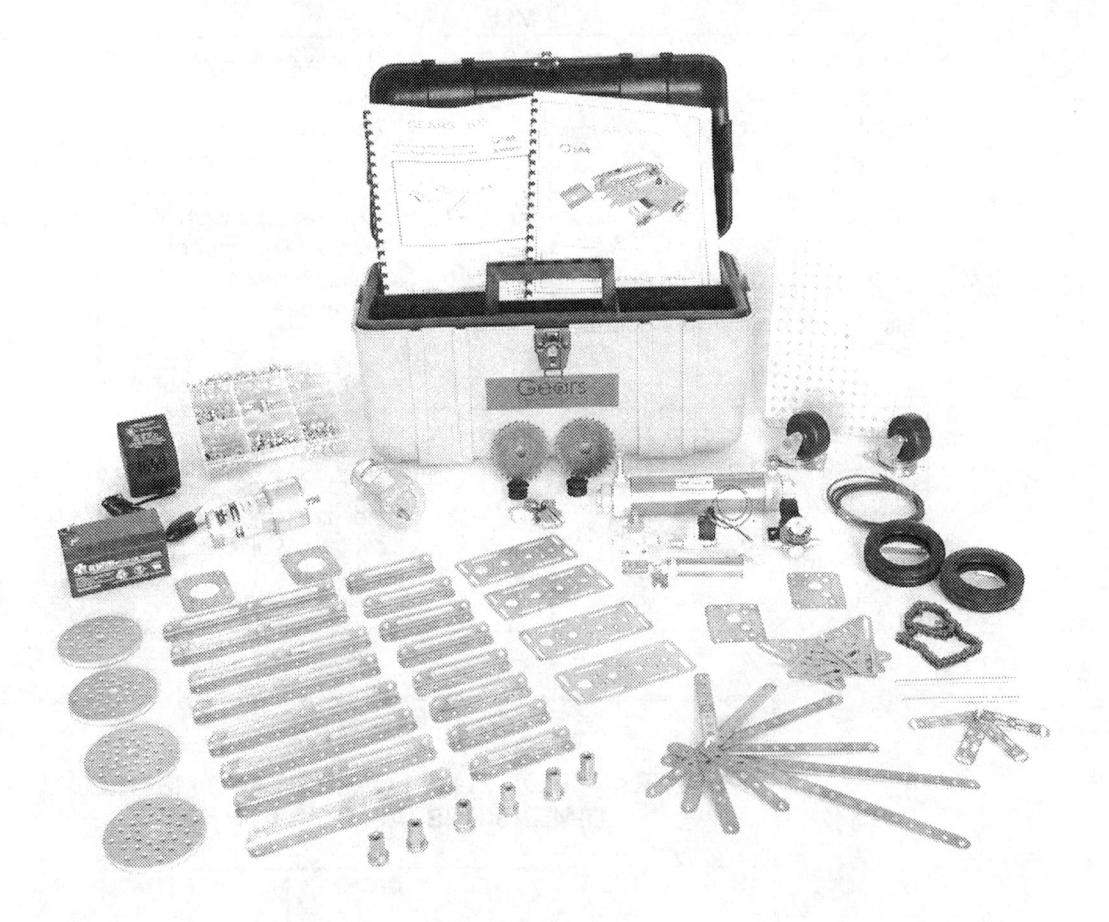

Gears Educational Systems
Hanover, MA USA

Project 1

Fundamentals of Part Modeling

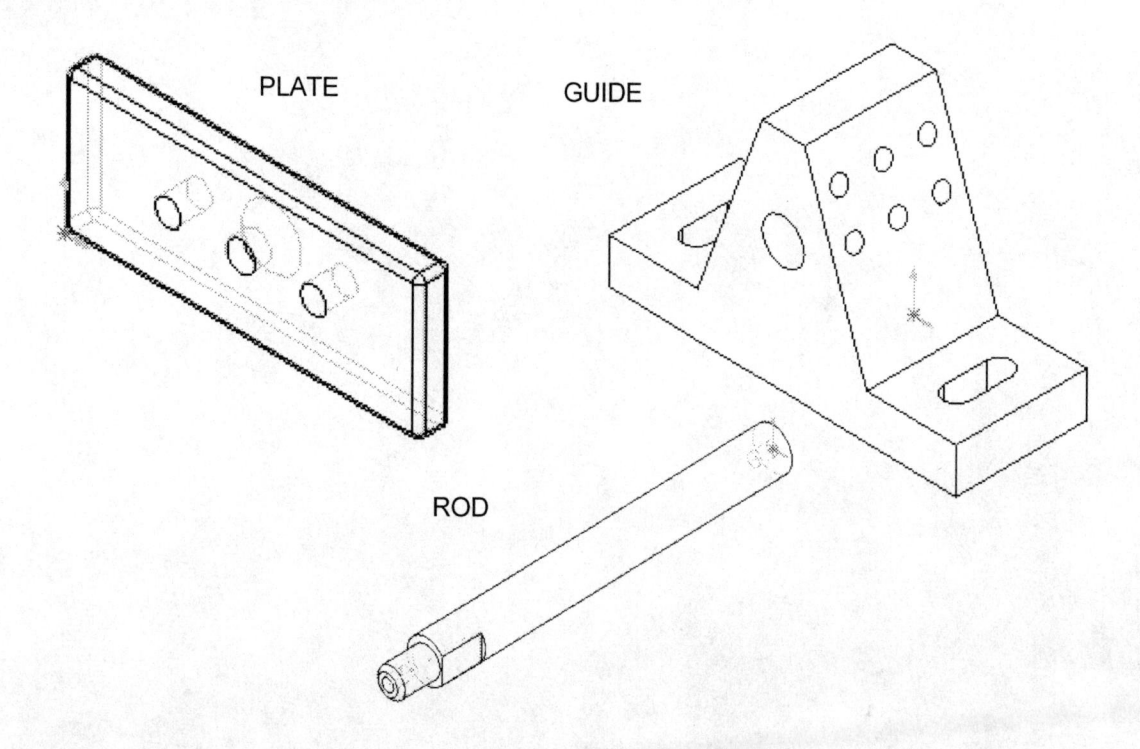

PLATE

GUIDE

ROD

Below are the desired outcomes and usage competencies based on the completion of Project 1.

Project Desired Outcomes:	Usage Competencies:
A comprehensive understanding of the SolidWorks User Interface and the customer's design intent.	Ability to start and setup a SolidWorks session.
Custom Templates and File Management folders.	Ability to use the following SolidWorks features: Extruded-Base, Extruded-Cut, Extruded-Boss, Fillet, Mirror, Chamfer, HoleWizard and Linear Pattern.
Three new parts: • ROD • GUIDE • PLATE	Ability to generate and edit SolidWorks sketches and features.

Notes:

1 Project 1 – Fundamentals of Part Modeling

Project Objective

Provide a comprehensive understanding of the SolidWorks User Interface: Menus, Keyboard shortcuts, Toolbars, System feedback, Document Properties, Templates and File Management.

Obtain the knowledge of the following SolidWorks features: Extruded-Base, Extruded-Cut, Extruded-Boss, Fillet, Mirror, Chamfer, HoleWizard and Linear Pattern.

Address the ability to generate and edit SolidWorks sketches and features.

Model three individual parts, (PLATE, GUIDE & ROD) for the requested customer GUIDE-ROD assembly.

On the completion of this project, you will be able to:

Choose the best profile for sketching.

Choose the proper sketch plane.

Set System Options.

Create three new parts: PLATE, ROD and GUIDE.

Generate a sketch.

Add Dimensions.

Add Geometric Relations.

Add Sketch Relations.

Use the following SolidWorks Features:

Extruded-Base.

Extruded-Cut.

Extruded-Boss.

Fillet.

Mirror.

HoleWizard.

Chamfer.

Project Situation

You receive a fax from a customer, Figure 1.0. The customer is retooling an existing assembly line.

You are required to design and manufacture a ROD. The ROD part is 10mm in diameter x 100mm in length.

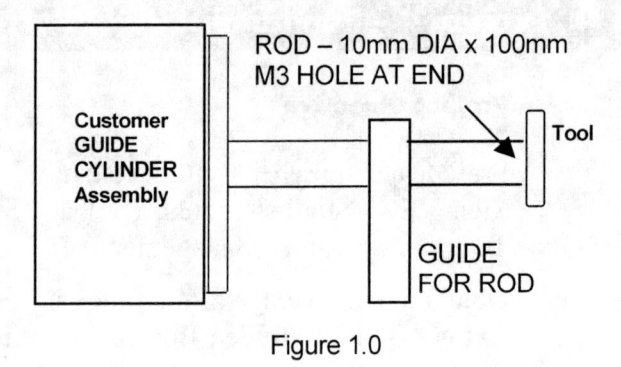

Figure 1.0

One end of the ROD connects to an existing customer GUIDE CYLINDER assembly.

The other end of the ROD connects to the customer's tool. The ROD contains a 3mm hole and a key-way to attach the tool.

The ROD requires a support GUIDE. The ROD travels through the support GUIDE. The GUIDE-ROD assembly is the finished customer product.

The GUIDE-ROD assembly is a component used in a low volume manufacturing environment.

Before you begin, investigate a few key design issues:

How will the customer use the GUIDE-ROD assembly?

How are the parts: PLATE, ROD and GUIDE used in the GUIDE-ROD assembly?

Does the GUIDE-ROD assembly affect other components?

Identify design requirements for load, structural integrity or other engineering properties.

Identify cost effective materials.

How will the parts be manufactured and what are their critical design features?

How will each part behave when modified?

You may not have access to all of the required design information. Placed in a concurrent engineering situation, you are dependent upon others and are ultimately responsible for the final design.

Dimensions for this project are in millimeters.

Design information is provided from various sources. Ask questions. Part of your learning experience is to know which questions to ask and who to ask.

The ROD part requires support. During the manufacturing operations, the ROD exhibits unwanted deflection.

The engineering group calculates working loads on test samples. Material test samples include: 8086 Aluminum, 303 Stainless steel and a machinable plastic acetal.

The engineering group recommends 303 Stainless steel for the ROD and GUIDE parts.

In the real world, there are numerous time constraints. The customer requires a quote, design sketches and a delivery schedule, YESTERDAY! If you wait for all of the required design information, you will miss the project deadline.

In a design review meeting, you create a rough concept sketch with notes, Figure 1.1a.

Your colleagues review and comment on the concept sketch.

The ROD cannot mount directly to the customer's GUIDE CYLINDER assembly without a mounting PLATE part.

The PLATE part mounts to the customer's PISTON PLATE part in the GUIDE CYLINDER assembly.

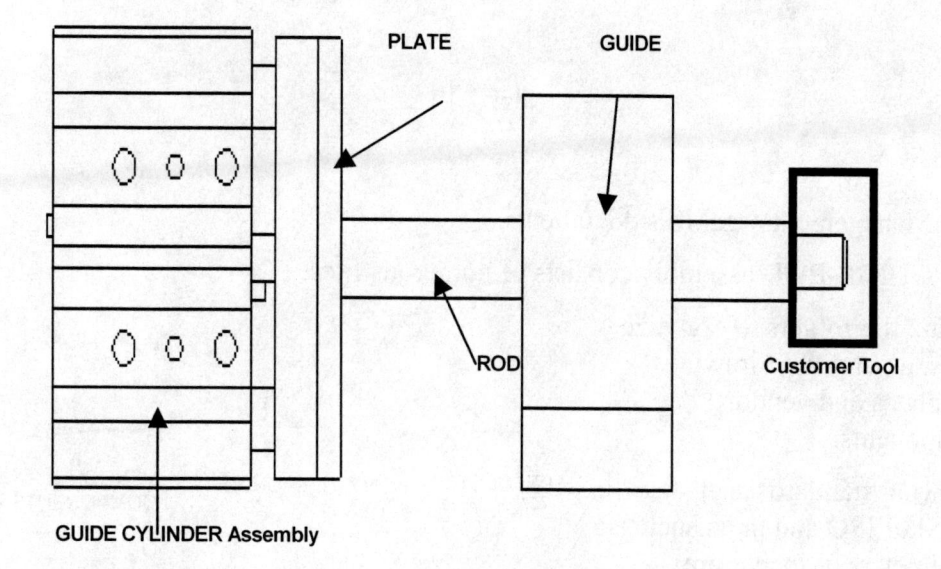

Figure 1.1a

Project Overview

A key goal in this project is to create three individual parts, (PLATE, GUIDE & ROD) for the requested GUIDE-ROD assembly.

In Project 2, the goal is to incorporate the three parts into the GUIDE-ROD assembly, Figure 1.1b.

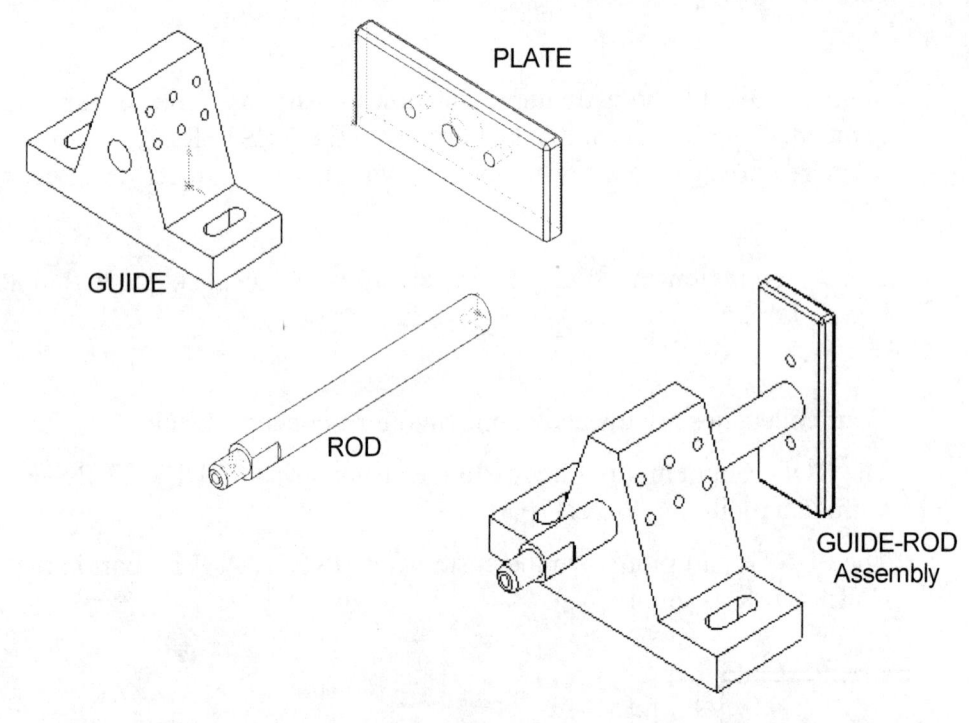

Figure 1.1b

File Management organizes documents.

The GUIDE-ROD assembly consists of numerous files.

Create file folders to organize parts, assemblies, drawings, templates and vendor components.

PROJECTS   MY-TEMPLATES   VENDOR COMPONENTS

Drawing standards such as ANSI or ISO and units such as millimeters or inches are defined in the Document Properties and are stored in the Part Template.

Plan file organization and templates before you create parts.

Create the PART-MM-ANSI Template for the metric parts required for the GUIDE-ROD assembly.

Windows 2000/NT Terminology

The mouse pointer provides an integral role in executing SolidWorks commands.

The mouse pointer executes commands, selects geometry, displays Pop-Up menus and provides information feedback.

A summary of mouse pointer terminology is displayed below:

Item:	Description:
Click	Press and release the left mouse button.
Double-click	Double press and release the left mouse button.
Click inside	Press the left mouse button. Wait a second, and then press the left mouse button inside the text box. Use this technique to modify Feature names in the FeatureManager design tree.
Drag	Point to an object, press and hold the left mouse button down. Move the mouse pointer to a new location. Release the left mouse button.
Right-click	Press and release the right mouse button. A Pop-up menu is displayed. Use the left mouse button to select a menu command.
ToolTip	Position the mouse pointer over an Icon (button). The command is displayed below the mouse pointer.
Mouse pointer feedback	Position the mouse pointer over various areas of the sketch: part, assembly or drawing. The cursor provides feedback depending on the geometry.

Fundamentals of Part Modeling **Engineering Design with SolidWorks**

Let's review various Windows terminology that describes: menus, toolbars and commands that constitute the graphical user interface in SolidWorks; Figure 1.3a and Figure 1.3b.

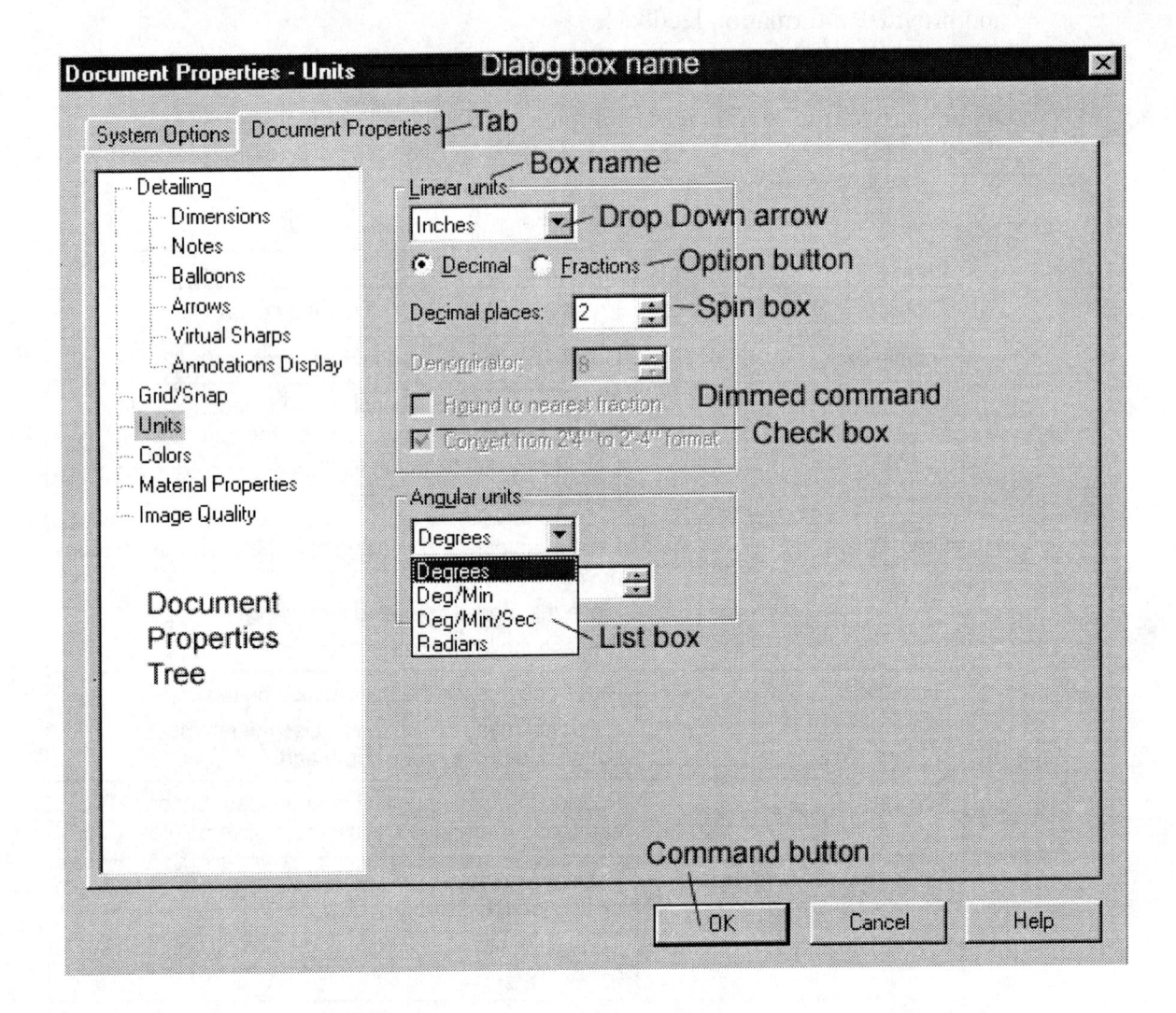

Figure 1.3a

The following Windows terminology is used throughout the text:

Item:	Description:
Dialog box name	Name of a window to enter information in order to carry out a command.
Box name	Name of a sub-window area inside the dialog box.
Check box	A square box. Click to turn on/off an option.
Spin box	A box containing up/down arrow to scroll or type by numerical increments.
Dimmed command	A menu command that is not currently available (light gray).
Tab	Dialog box sub-headings to simplify complex menus.
Option button	A small circle to activate/deactivate a single dialog box option.
List box	A box containing a list of items. Click the list drop down arrow. Click the desired option.
Text box	A box to type text.
Drop down arrow	Opens a cascading list containing additional options.
OK	Executes the command and closes the dialog box.
CANCEL	Closes the dialog box and leaves the original dialog box settings.
APPLY	Executes the command. The dialog box remains open.

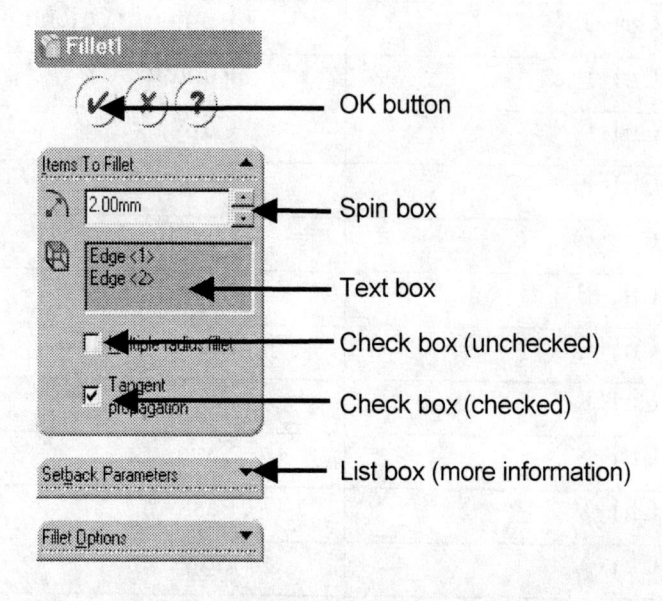

Figure 1-3b

Keyboard Shortcuts

Listed below are the pre-defined keyboard shortcuts for view options:

Shortcut Keys:	Action:
Arrow Keys	Rotate the view
Shift+Arrow Keys	Rotate the view in 90° increments
Alt+Left or Right Arrow Keys	Rotate about normal to the screen
Ctrl+Arrow Keys	Move the view
Shift+z	Zoom In
z	Zoom Out
f	Zoom to Fit
Ctrl+1	Front Orientation
Ctrl+2	Back Orientation
Ctrl+3	Left Orientation
Ctrl+4	Right Orientation
Ctrl+5	Top Orientation
Ctrl+6	Bottom Orientation
Ctrl+7	Isometric Orientation
Ctrl+B	Rebuild
Ctrl+C	Copy
Ctrl+N	New
Ctrl+O	Open
Ctrl+P	Print
Ctrl+Q	Force Rebuild
Ctrl+R	Redraw
Ctrl+S	Save
Ctrl+V	Paste
Ctrl+X	Cut
Ctrl+Z	Undo

File Management

File management organizes parts, assemblies and drawings.

Why do you need file management? In a large assembly, there could be hundreds or even thousands of parts.

To facilitate time, parts and assemblies are distributed between team members. Design changes occur frequently in the development process. How do you manage and control changes? Answer: Through file management. File management is a very important tool in the development process.

The GUIDE-ROD assembly consists of many files. Utilize file folders to organize projects, vendor parts and assemblies, templates and libraries.

File folders exist on the local hard drive, example C:\. Folders also exist on a network drive, example Z:\. The letters C:\ and Z:\ are used as examples for a local drive and a network drive respectfully.

Create a new file folder.
1) Start Windows Explorer. Click **Start** on the Windows Taskbar, **Start**. Click **Run**. Enter **Explorer** in the text box. Click **OK**.

2) Expand My Computer. Click the **Plus** icon. Double-click the **location** for the new project file folder. Example: C:\.

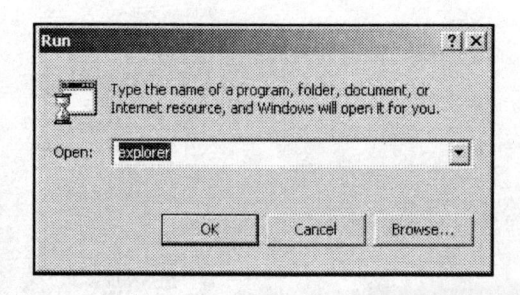

3) Click **File**, **New**, **Folder**. Enter the new file folder name. Enter **ENGDESIGN-W-SOLIDWORKS**.

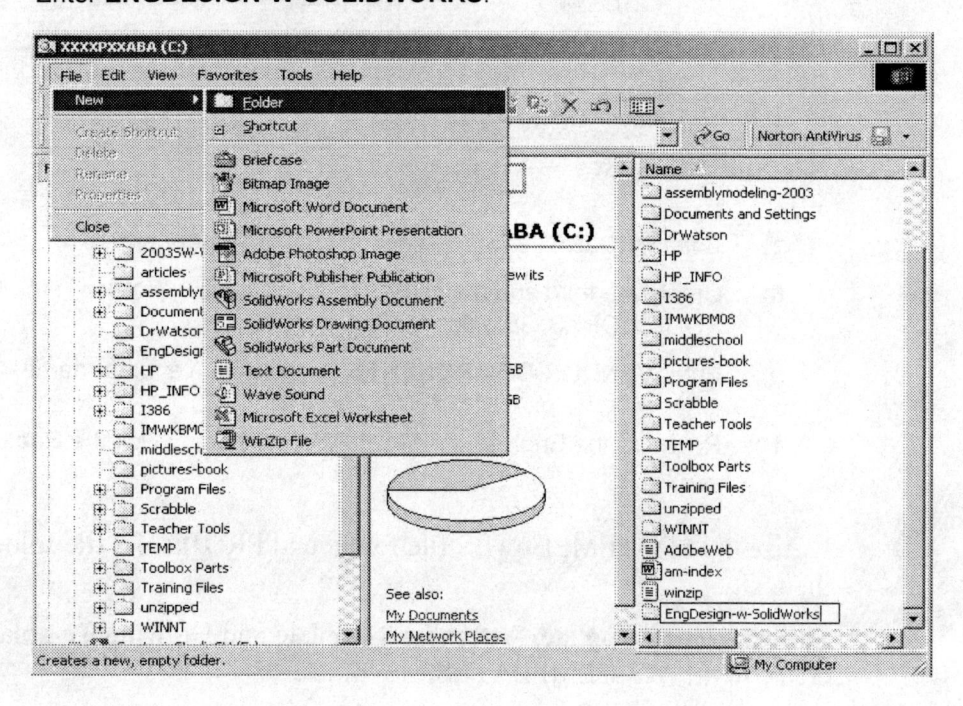

PAGE 1-11

Fundamentals of Part Modeling   **Engineering Design with SolidWorks**

Select the Microsoft Window commands from the Main toolbar, icons and with the right mouse button.

Create the file folders.

4) Double-click the **ENGDESIGN-W-SOLIDWORKS** file folder. Create the first sub-file folder. Right-click in the **ENGDESIGN-W-SOLIDWORKS** Graphics window. Right-click **New**, **Folder**. A New Folder icon is displayed.

5) Enter **MY-TEMPLATES** for the file folder name.

6) Create the second sub-file folder. Right-click in the **ENGDESIGN-W-SOLIDWORKS** Graphics window. Right-click **New**, **Folder**.

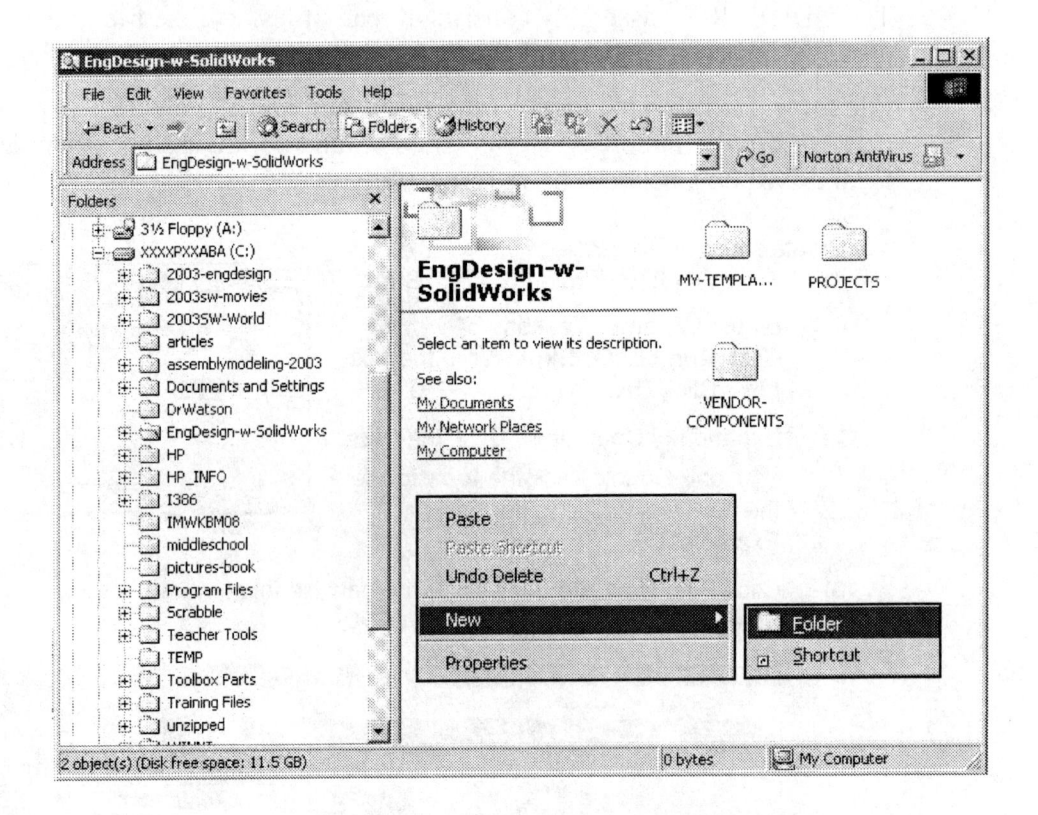

7) Enter **PROJECTS** for the second sub-file folder name.

8) Create the third sub-file folder. Right-click **New**, **Folder** in the ENGDESIGN-W-SOLIDWORKS Graphics window.

9) Enter **VENDOR-COMPONENTS** for the third file folder name.

10) Return to the EngDesign-W-SolidWorks folder. Click the **Back** icon.

Utilize the MY-TEMPLATES file folder and PROJECTS file folders throughout the text.

Store the Part Template, Assembly Template and Drawing Template that you create in the MY-TEMPLATES file folder.

Store the parts, assemblies and drawings that you create in the PROJECTS file folder.

Store the parts and assemblies that you download from the World Wide Web the VENDOR-COMPONENTS file folder.

Note: The pathname to the MY-TEMPLATES folder utilized in the text is as follows: ENGDESIGN-W-SOLIDWORKS\MY-TEMPLATES.

Start A SolidWorks Session

Set System Properties and Document Properties. Create a custom Part Template.

Start a SolidWorks session.
11) Click **Start** on the Windows Taskbar, Start.
12) Click **Programs**.
13) Click the **SolidWorks 2003** SolidWorks 2003 folder.
14) Click the **SolidWorks 2003** SolidWorks 2003 application.
15) The SolidWorks program window opens. Read the Tip of the Day dialog box. Click **Close**.

Create a new part.
16) Click **New** from the Standard toolbar. The Templates tab is the default tab. Part is the default template from the New SolidWorks Document dialog box. Click **OK**.

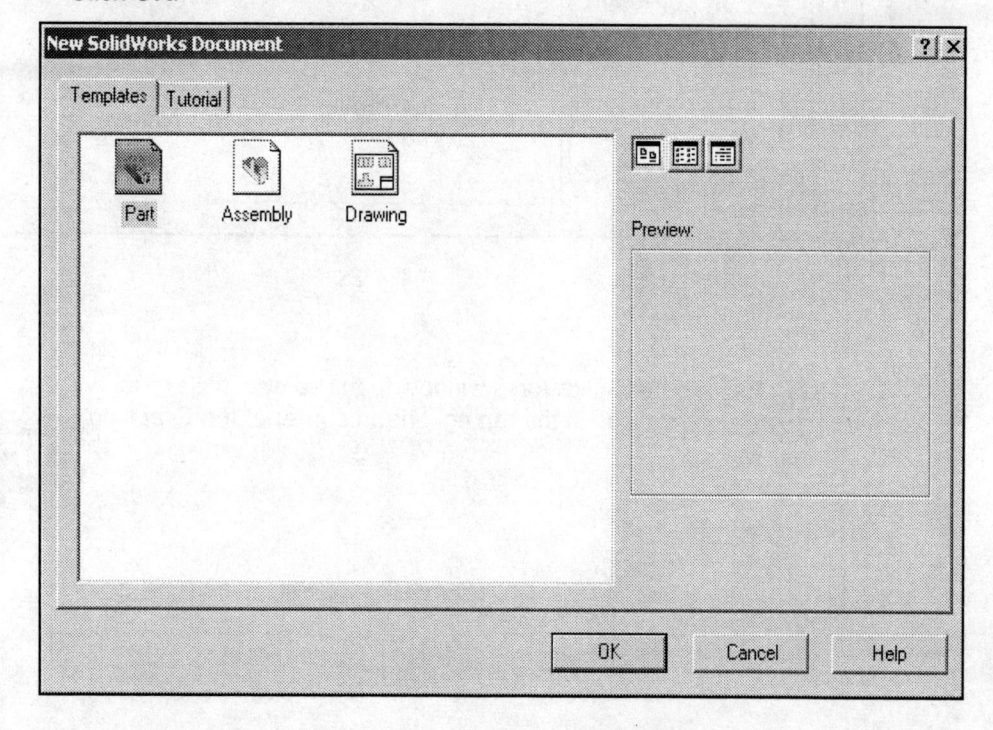

Fundamentals of Part Modeling **Engineering Design with SolidWorks**

Note: In a networked educational license installation, the user is prompted to select a working file folder for the default Templates.

Select MY-TEMPLATES for the default Templates location.

The default Part, Assembly and Drawing Templates are copied to the MY-TEMPLATES file folder.

Part1 is displayed. Part1 is the new default part window name.

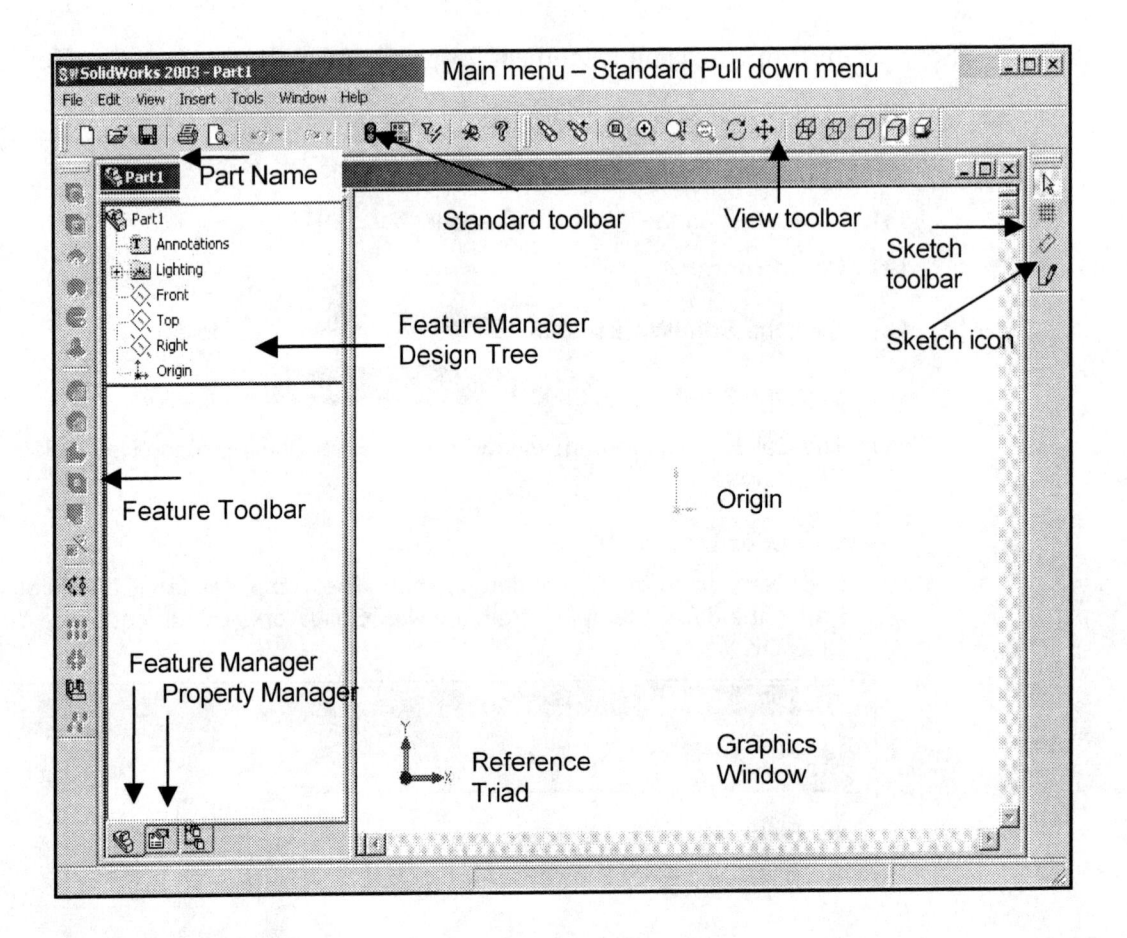

17) Expand the SolidWorks window to full screen. Click the **Maximize** icon in the top right hand corner of the Graphics window.

18) Display the toolbars. Click **View** from the Main menu. Click **Toolbars**. The system places a checkmark in front of the displayed toolbars. Click the **Standard Views**. The Standard Views toolbar is displayed below the Main menu.

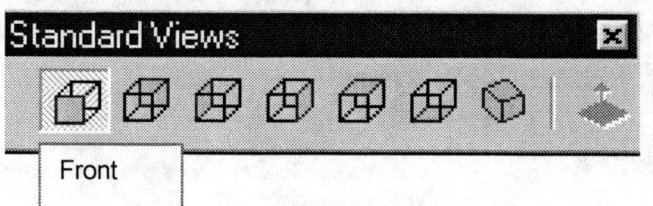

19) Position the **mouse pointer** on an individual toolbar icon to receive a ToolTip.

Enter commands from the Toolbar icons, Main Pull down menu and or Short Cut Keys. Examples:

Toolbar icon   Save icon.
Main Pull down menu   File, Save.
Short Cut Key   Ctrl + S.

The On-line Help contains step-by-step instructions for various commands. A few commands contain an AVI file.

The Help icon appears in the Feature toolbars.

Fundamentals of Part Modeling **Engineering Design with SolidWorks**

Display On-line Help for the rectangle.
20) Click **Help** from the Main menu.
21) Select SolidWorks Help Topics.

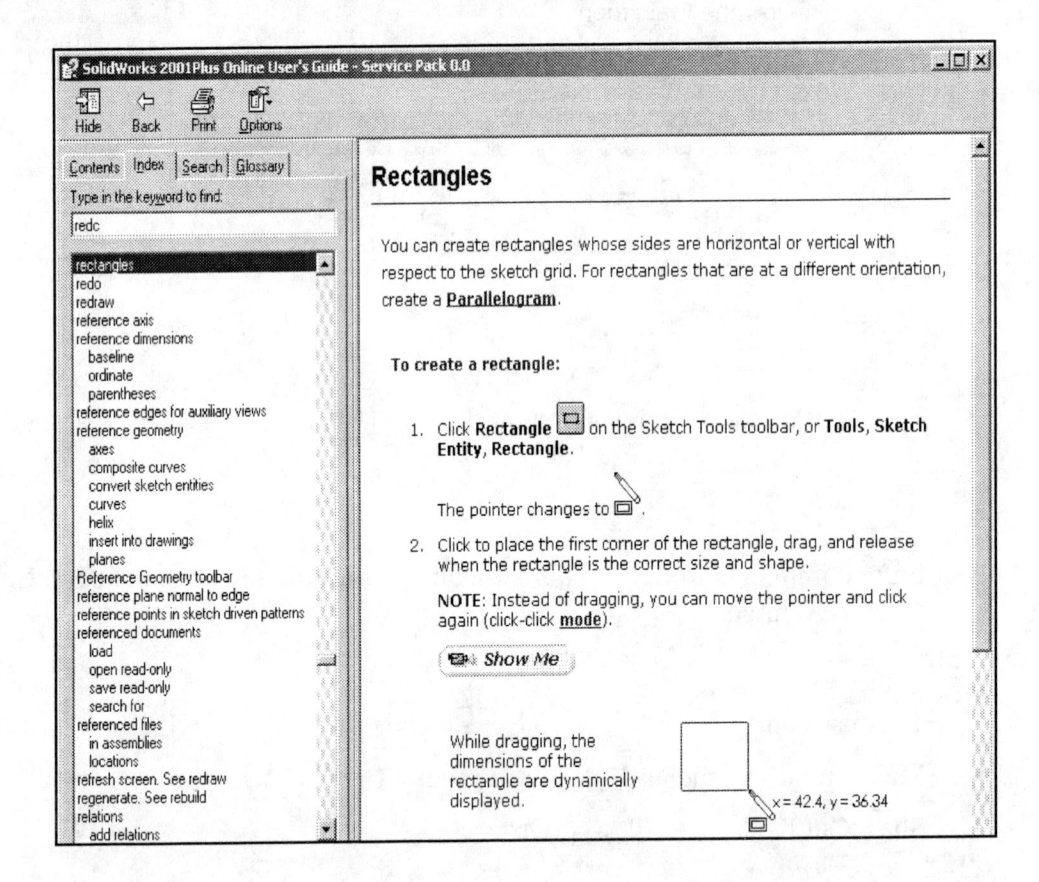

22) Click **Index**.

23) Enter **rectangle**. The description appears in the right window.

24) Play the Rectangle .AVI file. Click the **Show Me** button.

25) Close Help. Click **Close** ⊠.

The Help menu contains the SolidWorks Online Tutorial, Introducing SolidWorks and Design Portfolio documents.

These documents contain additional information on using SolidWorks.

Note: The default Toolbar icons and the mouse Right-click Pop-up menus are used throughout the text. The Standard Main Pull down menu is referred to as the Main menu.

System Options

System Options are stored in the registry of the computer. System Options are not part of the document. Changes to the System Options affect current and future documents.

Review and modify the System Options. If you work on a local drive, C:\, the System Options are stored on the computer.

If you work on a network drive, Z:\ set System Options for each SolidWorks session.

Set the System Options before you start a project. The File Locations Option contains a list of file folders.

Add the ENGDESIGN-W-SOLIDWORKS\MY-TEMPLATES folder path name to the Document Templates File Locations list.

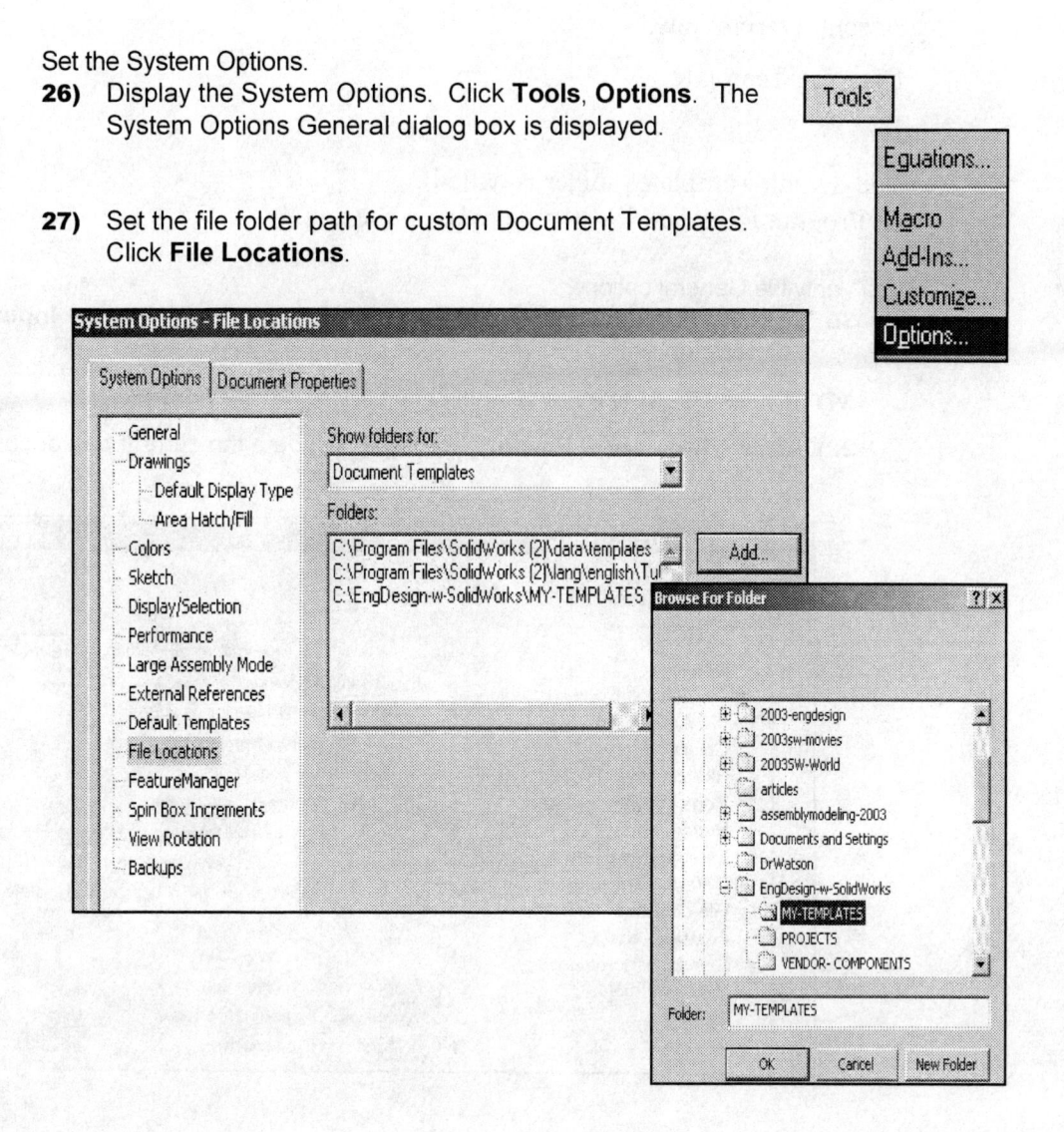

Set the System Options.
26) Display the System Options. Click **Tools**, **Options**. The System Options General dialog box is displayed.

27) Set the file folder path for custom Document Templates. Click **File Locations**.

Fundamentals of Part Modeling   **Engineering Design with SolidWorks**

28) Click the **Add** button.

29) Select the **ENGDESIGN-W-SOLIDWORKS\MY-TEMPLATES** file folder in the Browse for Folder dialog box. Click **OK**.

Each file folder listed in the Document Templates Show Folders For option produces a corresponding Tab in the New SolidWorks Document dialog box.

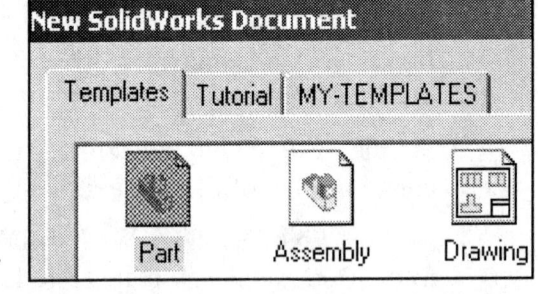

The Tab is visible when the file folder contains a SolidWorks:

Part Template.

Assembly Template.

Drawing Template.

The default Templates folder is called: C:\ProgramFiles\SolidWorks\data\Templates.

Verify the General options.

30) Click the **General** option. Verify the default options. Check the **Input dimension value**.

31) Check Use shaded face highlighting.

32) Check **Auto-show PropertyManager**. Update the current part document.

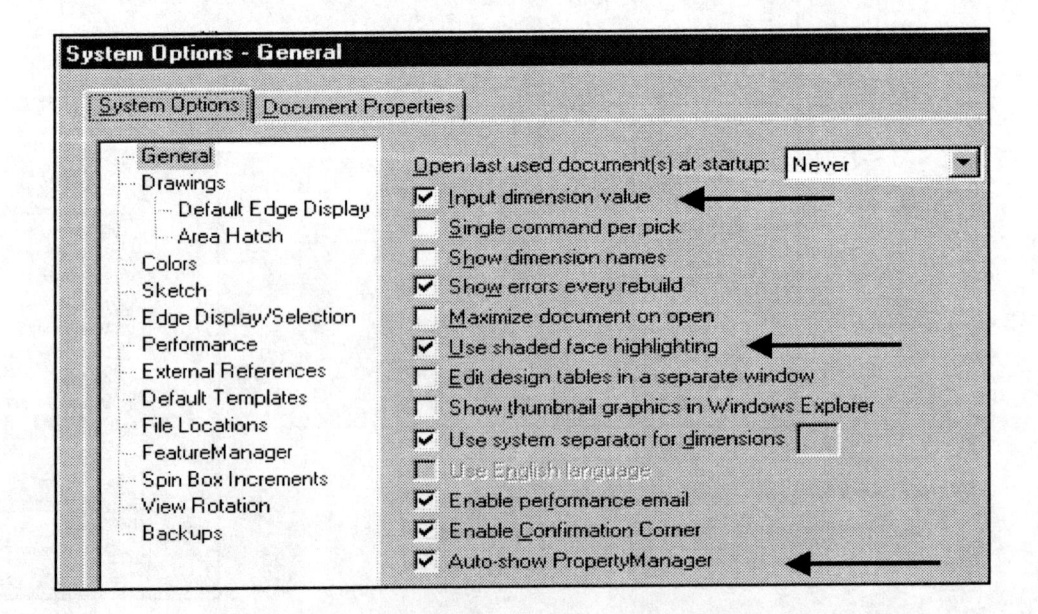

PAGE 1 - 18

Part Document Template and Document Properties

Document Templates are the foundation for parts, assemblies and drawings. The foundation of a SolidWorks Part is the Part Template.

Drawing standards such as ANSI or ISO and units such inches or millimeters are defined in the Document Properties and are stored in the Part Template.

The name of the Default Part Template is: Part.prtdot.

Create a Part Template named PART-MM-ANSI from the Default Part Template. Save the Custom Part Template in the ENGDESIGN-W-SOLIDWORK\MY-TEMPLATES file folder.

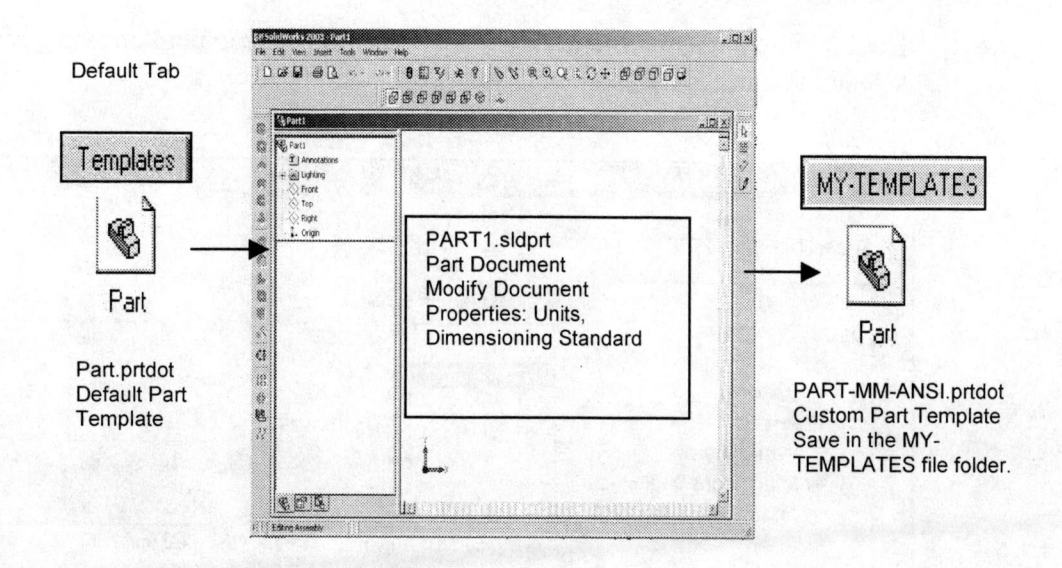

Utilize the PART-MM-ANSI Part Template for all metric parts.

Document Properties provide the ability to address: dimensioning standards, units, text style, center marks, witness lines, arrow styles, tolerance, precision and other parameters.

Document Properties apply only to the current document. The current parameters are stored with the template.

New documents that utilize the same template contain the saved parameters.

Units vary for Assembly, Part and Drawing Templates. Verify the document units.

Conserve modeling time. Set the Document Properties and create templates before starting a project.

Fundamentals of Part Modeling   **Engineering Design with SolidWorks**

Set the Document Properties.
33) Click the Document Properties tab.

34) Select **ANSI** from the Dimensioning standard: drop down list. Detailing options are available depending on the selected standard.

The Dimensioning standard options are: ANSI, ISO, DIN, JIS, BSI, GOST and GB. Obtain additional drawing options through the SolidWorks On-Line Help.

Display the parts, assemblies and drawings created in the GUIDE-ROD assembly in the ANSI standard.

Display the SMC components downloaded from the World Wide Web in the ISO standard.

Document units assist the designer by presenting a graphical drawing grid with selectable units of measurement.

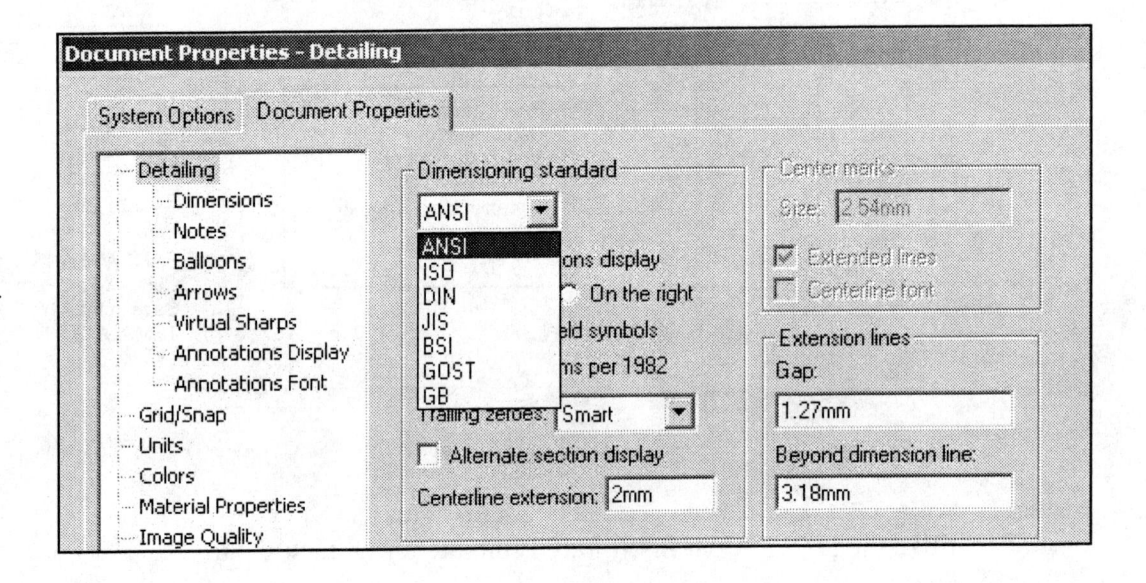

Set the document units.
35) Click the **Units** option from the Document Properties tab. The Document Properties – Units dialog box is displayed.

Engineering Design with SolidWorks **Fundamentals of Part Modeling**

36) Click the **drop down arrow** under the Linear units box.

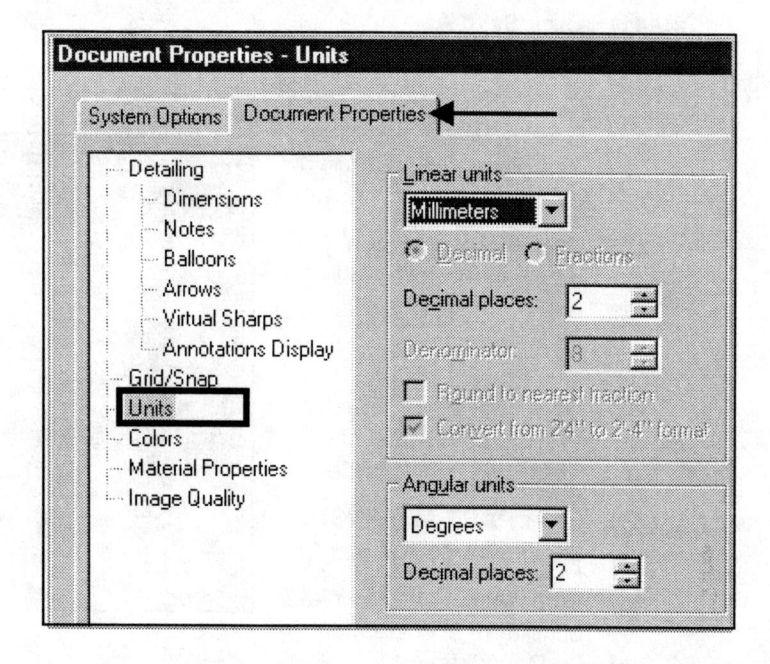

37) Select **Millimeters** from the list box.

38) Enter **2** in the Decimal places spin box.

Set the grid units.

39) Click the **Grid/Snap** option. The Document Properties – Grid/Snap dialog box is displayed. Review the default Grid/Snap properties. Click **OK**.

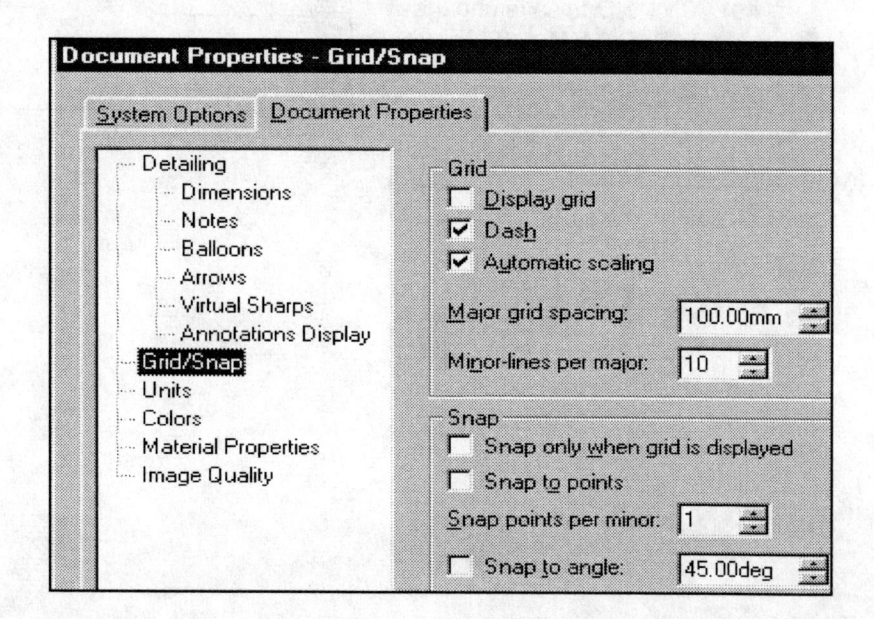

Fundamentals of Part Modeling **Engineering Design with SolidWorks**

Create the part document template.

40) Click **File** from the Main menu.

41) Click **Save As**.

42) Click **Part Templates *.prtdot** from the Save As type list box. The default Templates file folder is displayed.

43) Click the **drop down arrow** in the Save in list box.

44) Select the ENGDESIGN-W-SOLIDWORKS\ MY-TEMPLATES file folder.

45) Enter **PART-MM-ANSI** in the File name text box.

46) Enter Metric part template, units-mm, ansi standard for Description.

47) Click **Save**.

48) The PART-MM-ANSI.prtdot Part Template is displayed in the Graphics window. Close the PART-MM-ANSI Part Template. Click **File**, **Close** from the Main menu.

49) Click **NO** to save changes to PART-MM-ANSI.prtdot.

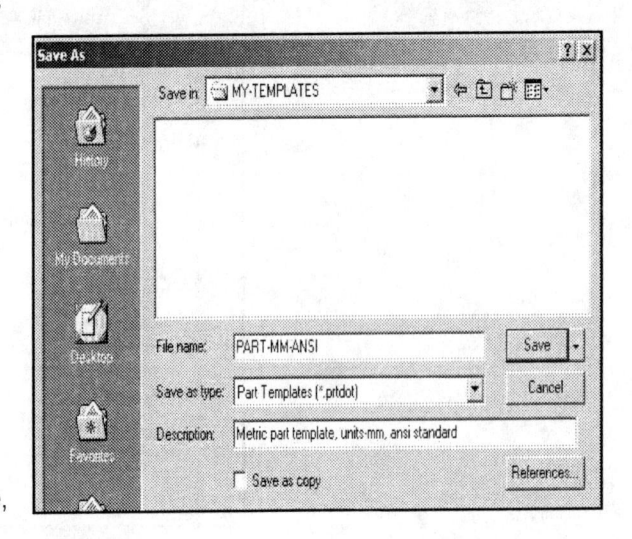

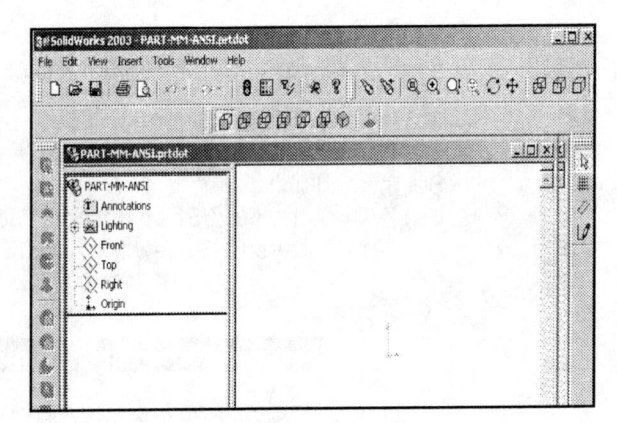

PLATE Part Overview

The PLATE part mounts to the customer's PISTON PLATE part in the GUIDE CYLINDER assembly.

The dimensions of the PISTON PLATE part are 56mm x 22mm.

The dimensions of the PLATE part are 56mm x 22mm.

Locate the 4mm mounting holes with respect to the PISTON PLATE.

Review all dimensions before designing parts.

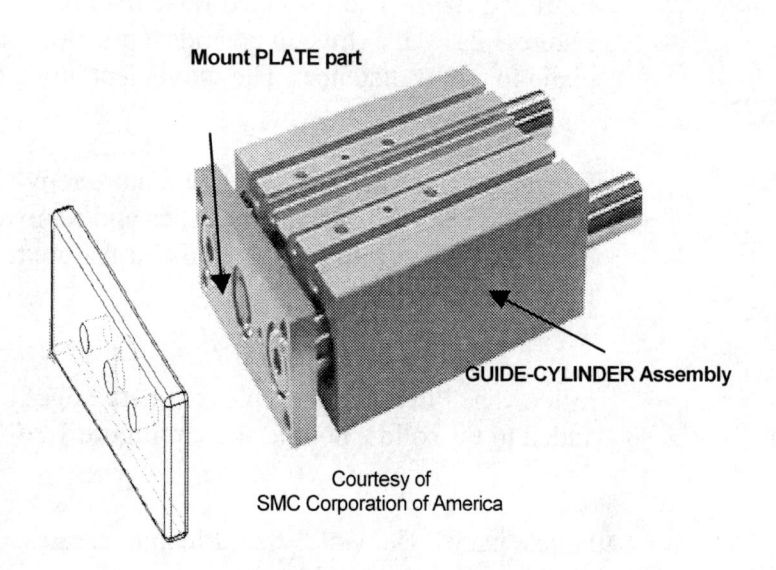

Note: The GUIDE CYLINDER assembly dimensions referenced in this project are derived from the SMC Corporation of America (www.smcusa.com).

The GUIDE CYLINDER assembly was obtained in a document format.

Start the translation of the initial design sketch into SolidWorks features.

Features are geometry building blocks.

Features add or remove material.

Features are created from sketched profiles or from edges and faces of existing geometry.

Fundamentals of Part Modeling   **Engineering Design with SolidWorks**

Use the following features to create the PLATE part:

Extruded Base. The Extruded Base feature is the first feature of the PLATE, Figure 1.2a. An extrusion extends a profile along a path normal to the profile plane for some distance. The movement along that path becomes the solid model.

Extruded Cut. The Extruded Cut feature removes material. This is the opposite of the boss. Cuts begin as 2D sketches and remove materials by extrusions, or revolutions. Utilize the Extruded Cut feature to create the two holes for the PLATE, Figure 1.2b.

Fillet. The Fillet feature removes sharp edges of the PLATE. Fillets are generally added to the solids, not the sketch, Figure 1.2c.

HoleWizard. The HoleWizard feature creates a countersink hole at the center of the PLATE, Figure 1.2d.

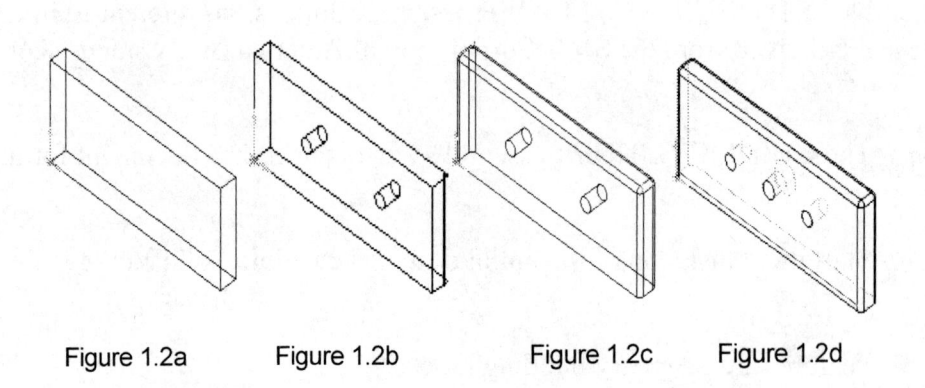

Figure 1.2a   Figure 1.2b   Figure 1.2c   Figure 1.2d

Engineering Design with SolidWorks   Fundamentals of Part Modeling

Create a new part.

50) Click **New**.

51) Click the **MY-TEMPLATES** tab.

52) Click **PART-MM-ANSI** icon.

53) Click **OK**. Part2 is displayed.

In a SolidWorks session, the first system default part filename is named: Part1.sldprt.

The system attaches the .sldprt suffix to the created parts.

The second created part in the same session, increments to the filename: Part2.sldprt.

There are numerous ways to manage a part.

Use Project Data Management (PDM) systems to control, manage and document file names and drawing revisions.

Use appropriate filenames that describe the part.

Save the empty part.

54) Click **Save**.

55) Click the **Drop down arrow** in the Save in list box.

56) Click the ENGDESIGN-W-SOLIDWORKS\PROJECTS file folder.

57) Enter **PLATE** for File name.

58) Enter **PLATE 56MM x 22MM** for Description.

59) Click **Save**.

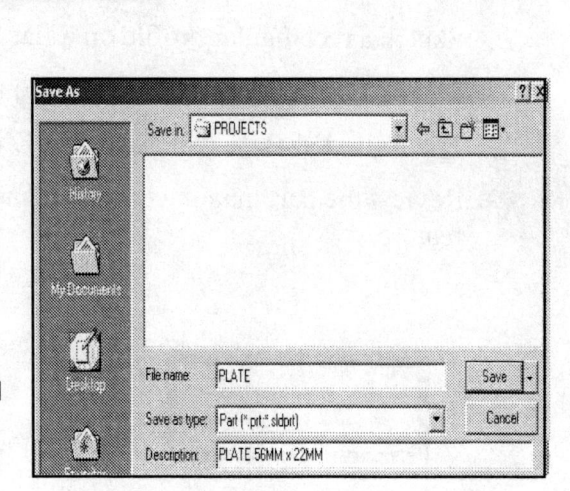

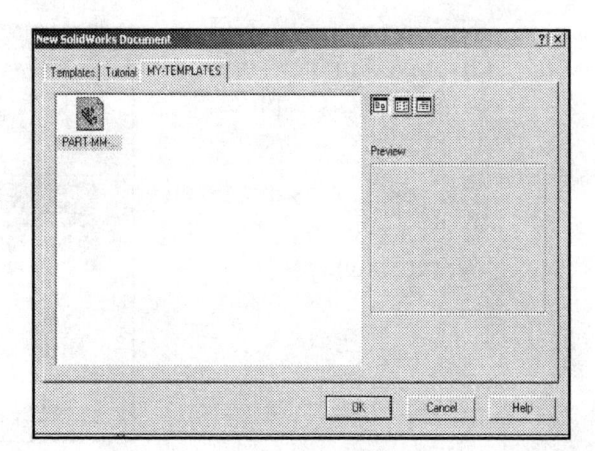

The part name PLATE is displayed in the SolidWorks Graphics window.

The part icon is displayed at the top of the FeatureManager.

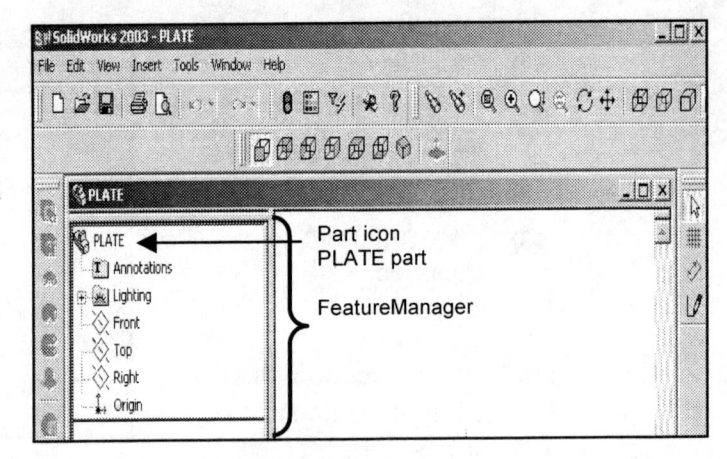

Extruded -Base Feature

What is a Base feature? The Base feature is the first feature that is created.

The Base feature is the foundation of the part. Note: Keep the Base feature simple!

The Base feature geometry for the PLATE is an extrusion. The extrusion is named Extruded-Base feature.

How do you create a solid Extruded-Base feature?

Sketch a rectangular profile on a flat 2D plane, Figure 1.4a.

Extend the profile perpendicular ($\perp$) to the Sketch plane, Figure 1.4b.

The Extruded-Base feature is the 3D block, Figure 1.4c.

Review the part manufacturing and assembly procedures before creating a Base feature.

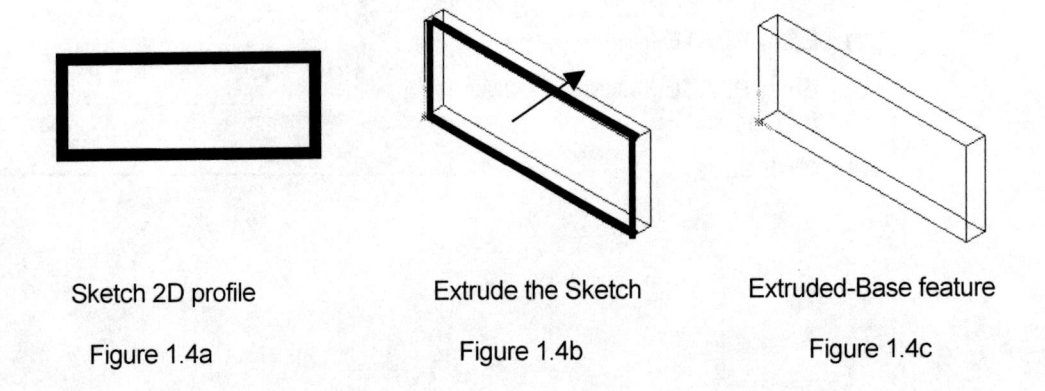

Sketch 2D profile

Figure 1.4a

Extrude the Sketch

Figure 1.4b

Extruded-Base feature

Figure 1.4c

Machined Part

In earlier conversations with manufacturing, a decision was made that the part would be machined. Your material supplier stocks raw material in rod, sheet, plate, angle and block forms.

You decide to start with a standard plate form. A standard plate form will save time and money.

Select the best profile for the extrusion. The best profile is a simple 2D rectangle.

The Extruded Boss and square internal cuts are very costly to manufacture from machined bar stock.

As a designer, review your manufacturing options.

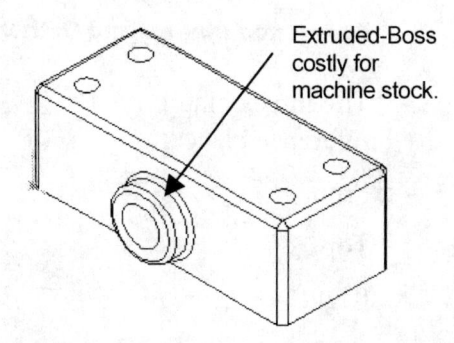

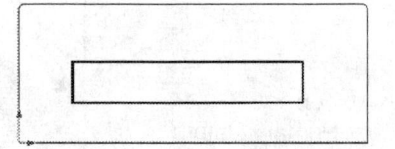

Extruded-Boss costly for machine stock.

Internal Square Cut (expensive)

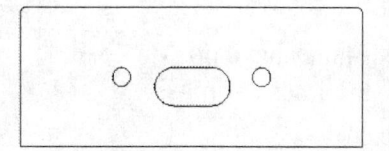

Holes and Slots (less expensive)

Machined parts require datum planes for referenced dimensions. Sketch the rectangular profile. The bottom and left edges are aligned to the reference planes. Dimension holes and cuts from the same dimension reference planes, Figure 1.5a.

In 3D modeling, dimension the second hole from the edge of the Extruded-Base feature to its center point. Avoid dimension references between the center points of the holes, Figure 1.5b.

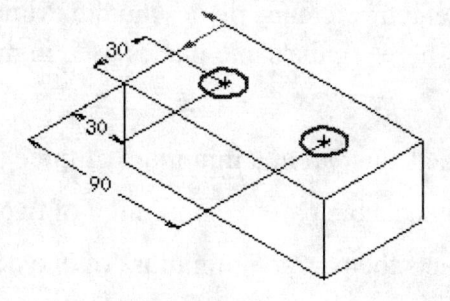

Figure 1.5a

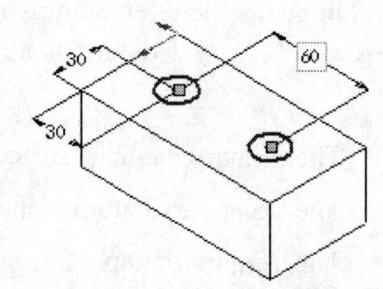

Figure 1.5b

Referencing dimensions to the edge of the Extruded-Base feature provides information on how the part is manufactured and leads to fewer part rebuilding, "regeneration" problems and calculation errors.

Add dimensions and notes on the drawing if mating parts require center-to-center distances between holes.

Reference Planes and Orthographic Projection

The three default ⊥ reference planes:

Front

Top

Right

represent infinite 2D planes in 3D space, Figure 1.6.

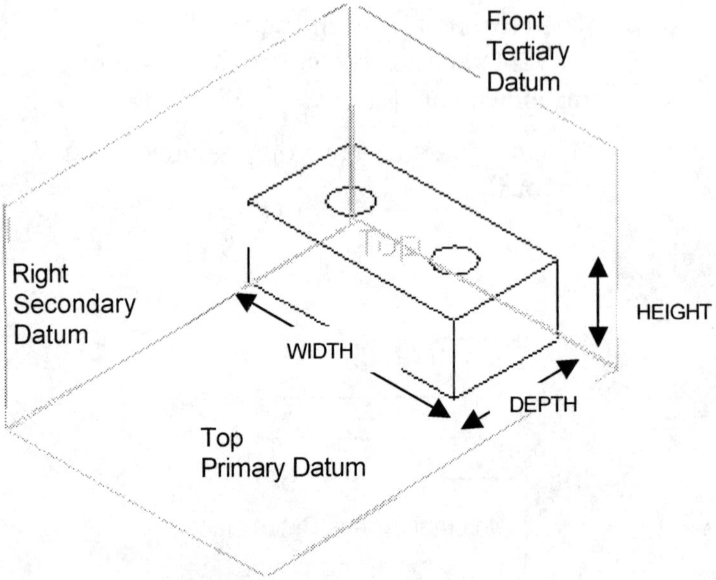

Planes have no thickness or mass.

Orthographic projection is the process of projecting views onto parallel planes with ⊥ projectors.

Figure 1.6

The default reference planes are the Front, Top and Right viewing planes.

In geometric tolerancing, the default reference planes are the Primary, Secondary and Tertiary ⊥ datum planes. These are the same planes used in manufacturing.

The Primary datum plane contacts the part at a minimum of three points.

The Secondary datum plane contacts the part at a minimum of two points.

The Tertiary datum plane contacts the part at a minimum of one point.

Orthographic Projection

In third angle Orthographic projection, the standard drawing views are Front, Top, Right and Isometric, Figure 1.7.

There are two Orthographic projection drawing systems.

The projection systems are called third angle projection and first angle projection respectively, Figure 1.8.

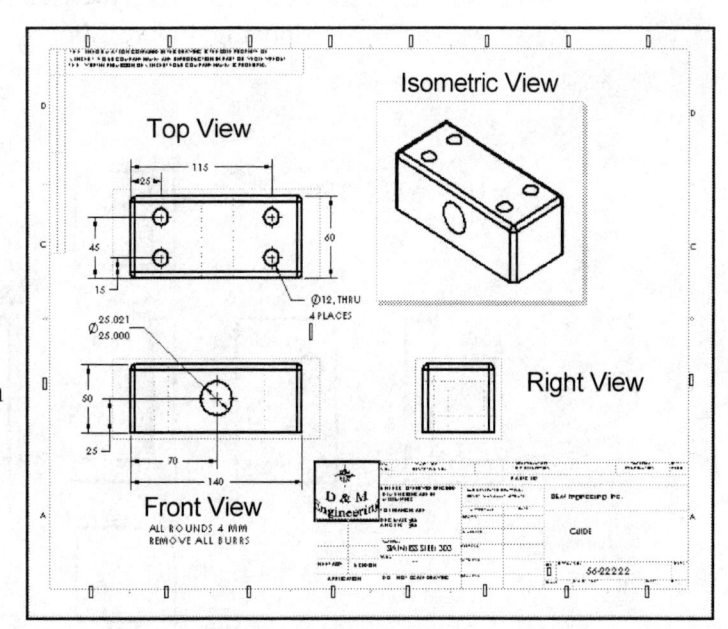

Figure 1.7

The systems are derived from positioning a 3D object in the third or first quadrant.

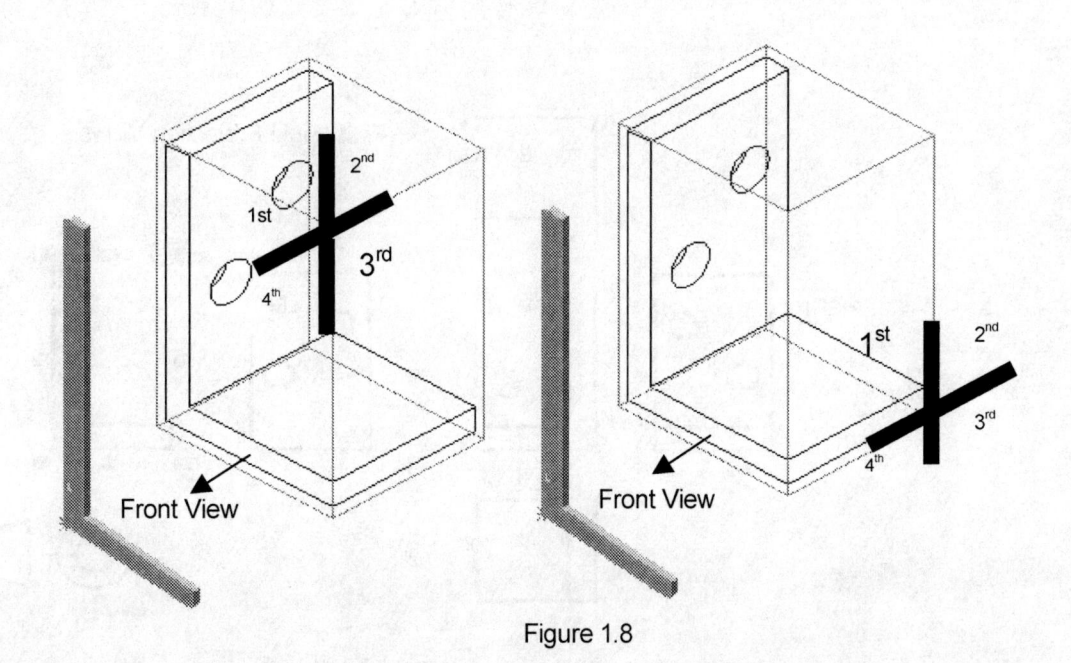

Figure 1.8

Fundamentals of Part Modeling **Engineering Design with SolidWorks**

In third angle projection, the part is positioned in the third quadrant. The 2D projection planes are located between the viewer and the part. The projected views are placed on a drawing, Figure 1.9.

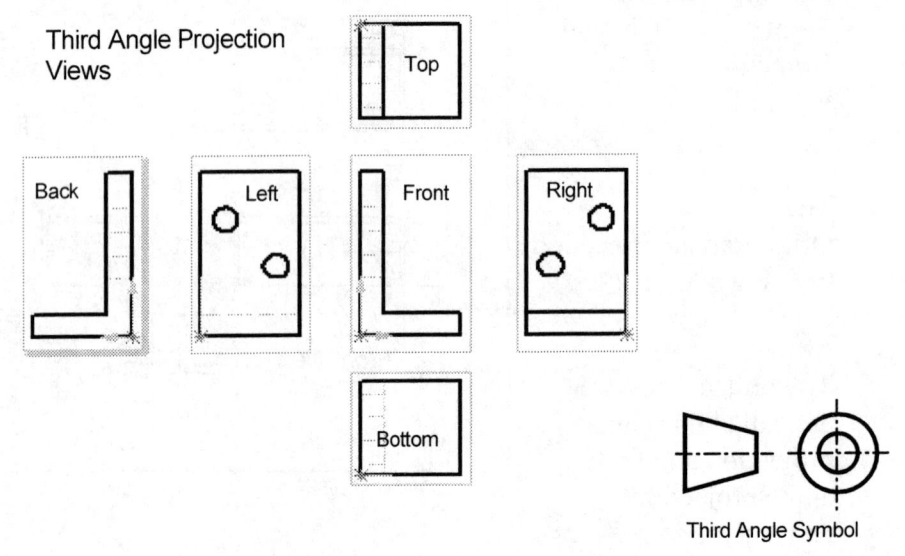

Figure 1.9

In first angle projection, the part is positioned in the first quadrant. Views are projected onto the planes located behind the part. The projected views are placed on a drawing, Figure 1.10.

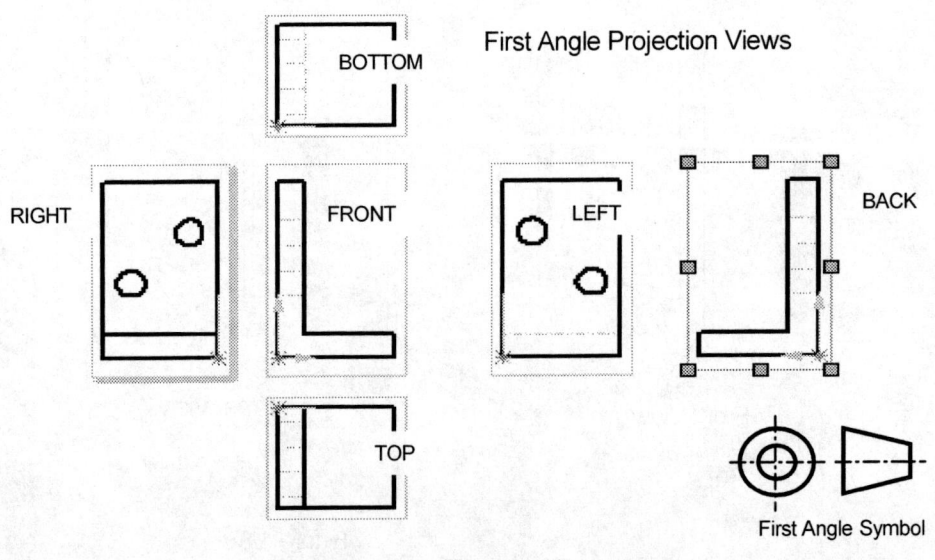

Figure 1.10

First angle projection is primarily used in Europe and Asia. Third angle projection is primarily used in the United States and is the ANSI standard. Designers should have knowledge and understanding of both systems. There are numerous multi-national companies.

Example: A part is designed in Europe, manufactured in Japan and is destined for the United States.

Note: Third angle projection is used in this text. A truncated cone symbol appears on the drawing to indicate the projection system:

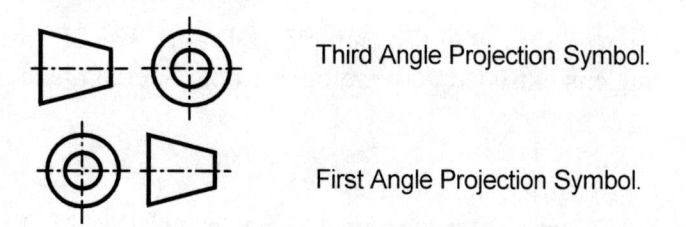

Third Angle Projection Symbol.

First Angle Projection Symbol.

The selected Sketch plane defines the orientation of the part in 3D space, Figure 1.11.

Illustrated below are three examples of utilizing the Front, Top or Right planes as Sketch planes for the Extruded Base feature. The overall dimensions are physically the same. The orientation is different.

Note: The Front sketch plane is the Default Sketch plane.

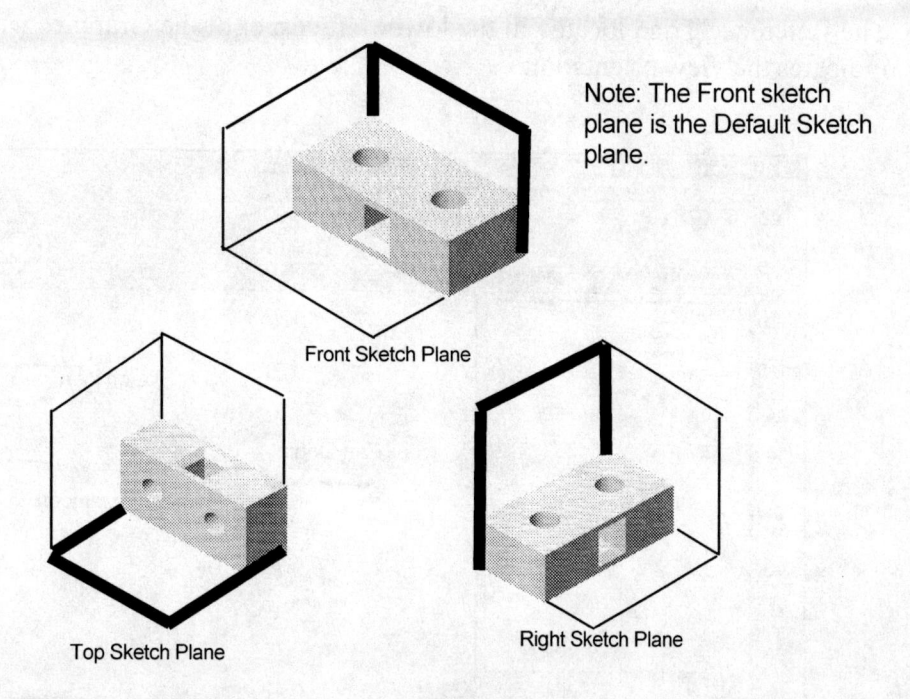

Front Sketch Plane

Top Sketch Plane

Right Sketch Plane

Figure 1.11

Before incorporating your design ideas into the Sketch plane, ask a question:

How will the part be oriented in the assembly?

Answer: A part should be oriented or aligned to assist in the final assembly.

Create the PLATE – Use the Extruded Base Feature

In SolidWorks, the name used to describe a 2D profile is Sketch. A Sketch requires a Sketch plane and a 2D profile.

The Sketch uses the default Sketch plane, Front. The 2D profile is a rectangle. The rectangle is extruded perpendicular to the Sketch plane.

Select the Sketch plane.
60) The Front plane is the default sketch plane. Click the **Front plane** from the FeatureManager. Note: Planes are flat and infinite. They are represented on the screen with visible edges. They are used as the primary sketch surface for creating Boss and Cut features.

61) Create a new Sketch. Click **Sketch** from the Sketch toolbar. The Sketch opens on the Front plane. The red Origin represents the intersection of the Front, Top and Right planes.

The Reference Triad located in the lower left corner of the Graphics window indicates the view orientation.

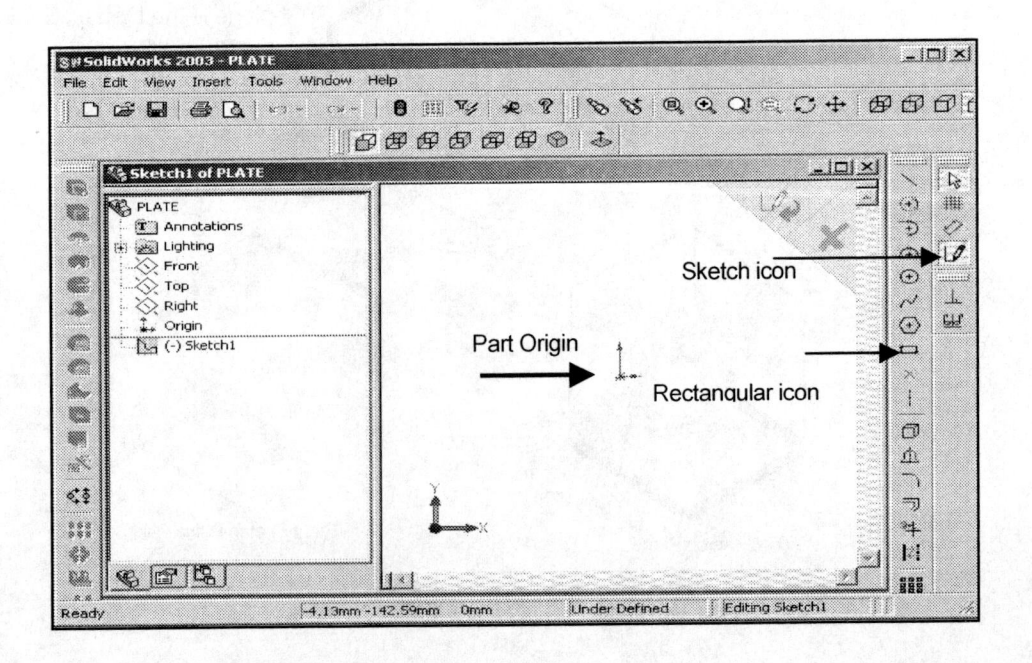

Sketch a rectangle profile.

62) Click **Rectangle** from the Sketch Tools toolbar. The mouse pointer is displayed as a pencil with a rectangle. Note: The pointer displays a pencil with an orange point, when the mouse pointer is located on the Origin.

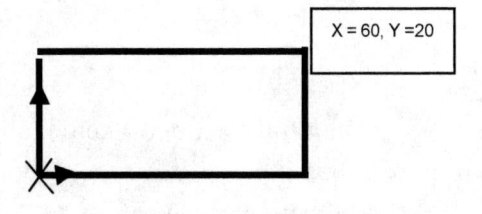

63) Create the first corner point of the rectangle. Click the **Origin**. Hold the **left mouse button** down. Create the second corner point. Drag the **mouse pointer** up and to the right. Release the **left mouse button**.

X = 60, Y = 20

64) The X-Y coordinates of the rectangle are displayed above the mouse pointer. The X-Y coordinates display different values. Define exact width and height with dimensions. The rectangle is displayed in green.

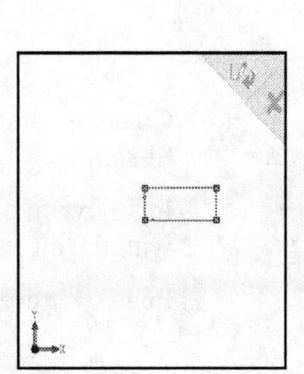

Note: If you make a mistake, select the UNDO button.

Dimensions provide location and size information. Models require dimensions for manufacturing.

Dimensions are not required to create features in SolidWorks.

SolidWorks uses color and cursor feedback to aid in the sketching process.

The current rectangular Sketch is displayed in green.

Green indicates that the sketched geometry is currently selected.

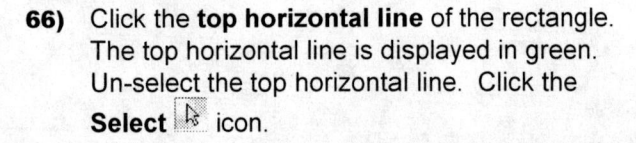

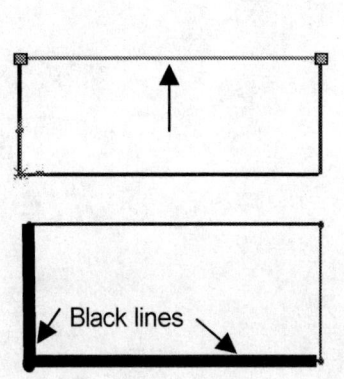

65) Select the top horizontal line. Click the **Select** icon.

66) Click the **top horizontal line** of the rectangle. The top horizontal line is displayed in green. Un-select the top horizontal line. Click the **Select** icon.

Black lines

The rectangular sketch is displayed in two colors: blue and black. The geometry consists of four lines and four vertices.

The vertex Origin is black.

The horizontal and vertical line positioned along the axes of the Origin is black.

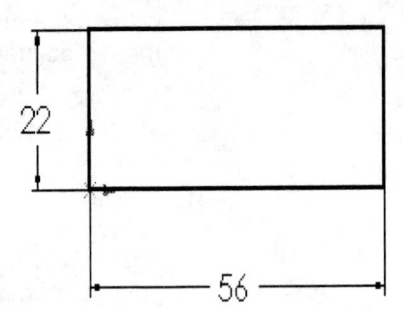

In a fully defined Sketch, all entities are displayed in black.

A fully defined Sketch has a defined position, dimensions and or relationships.

In a under defined Sketch, the entities that require position, dimensions or relationships are displayed in blue.

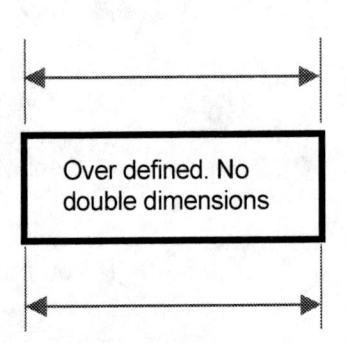

Click and drag under defined geometry to modify the Sketch.

In an over defined Sketch, there is geometry conflict with the dimensions and or relationships.

In an over defined Sketch, entities are displayed in red.

Engineering Design with SolidWorks Fundamentals of Part Modeling

Add dimensions.
67) Dimension the horizontal line. Click **Dimension** from the Sketch toolbar. The mouse pointer changes shape to the dimension symbol,

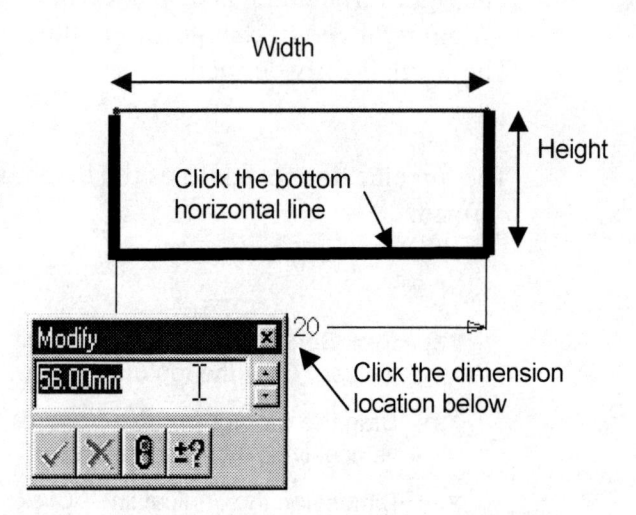

68) Click the bottom **horizontal line** of the rectangle. Drag the **mouse pointer** downward.

69) Click the **dimension text** location below the green line.

70) Enter **56** in the Modify text box. Accept the value.

71) Click the **Green Check Mark** in the Modify dialog box.

Note: Press the Enter key or click the Green Check Mark to accept the value.

The Dimension PropertyManager is displayed. Change Dimension properties such as Tolerance/Precision and Arrow is displayed in the PropertyManager.

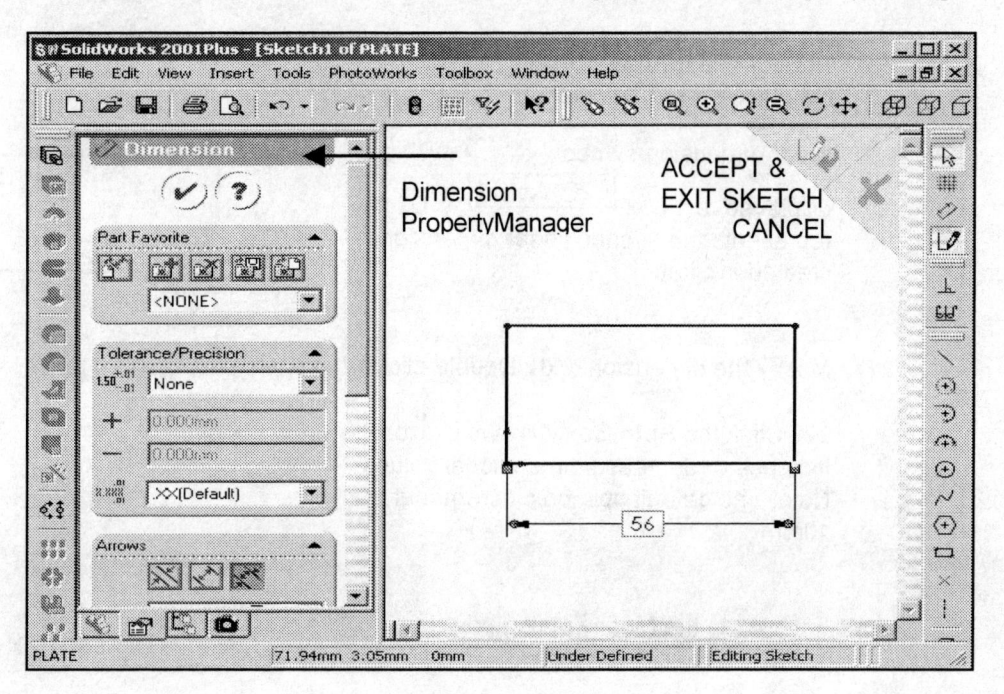

Fundamentals of Part Modeling **Engineering Design with SolidWorks**

The right vertical line is displayed in black. The bottom right vertex changes from blue to black. The width is fully defined.

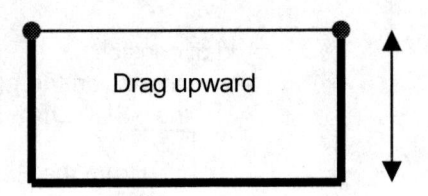

The top blue line that defines the height is under defined.

72) Click **Select** from the Sketch toolbar. Click the top **blue line**.

73) Drag the **mouse pointer** upward. The vertical lines increase in size.

74) Dimension the vertical line. Click **Dimension**. Click the **left most vertical line**. Drag the **mouse pointer** to the left. Click a position to the **left** of the Sketch. Enter **22**. The Sketch is fully defined. All lines and vertices are displayed in black. Click the **Green Check Mark** in the Modify dialog box.

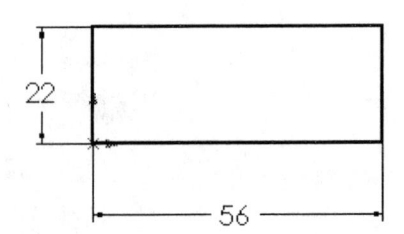

Modify the dimension values.

75) Modify the dimension values to increase or decrease the size of the Sketch. Click **Select**.

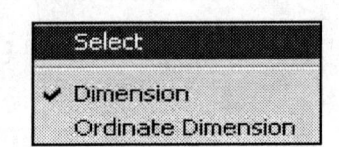

76) Position the **mouse pointer** over the dimension text. The pointer changes to a linear dimension symbol, with a displayed text box D2@Sketch1. D2 represents the second linear dimension created in Sketch1.

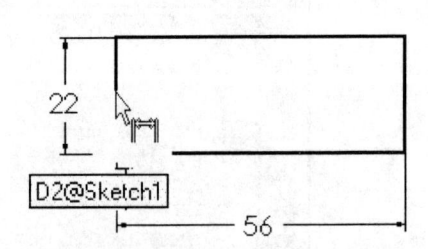

77) Modify the dimension text. Double-click **22**. Click the **Spin Box Arrows** to increase or decrease dimensional values. Note: The default spin box increments are 10mm.

78) Enter **42** in the Modify dialog box.

Note: The Check ✓, saves the current value and exits the Modify dialog box.

Restore ✗, restores the original value and exits.

Rebuild 🔄, rebuilds the model with the current value.

Reset ±?, modifies the spin box increment.

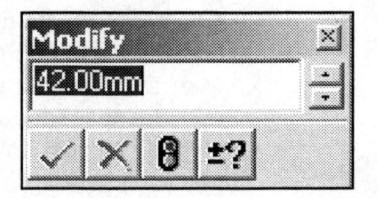

79) Return to the original vertical dimension. Enter **22**. Accept the value. Click the **Green Check Mark** ✓ in the Modify dialog box.

Note: Click Undo ↺ from the Standard toolbar to reverse changes to the sketch.

Extrude the Sketch.

80) Extruding the Sketch adds depth to the rectangle profile. Click **Extruded Boss/Base** in the Features toolbar. The Extrude PropertyManager is displayed. The extruded Sketch is previewed in an Isometric view. The preview displays the direction of the feature creation.

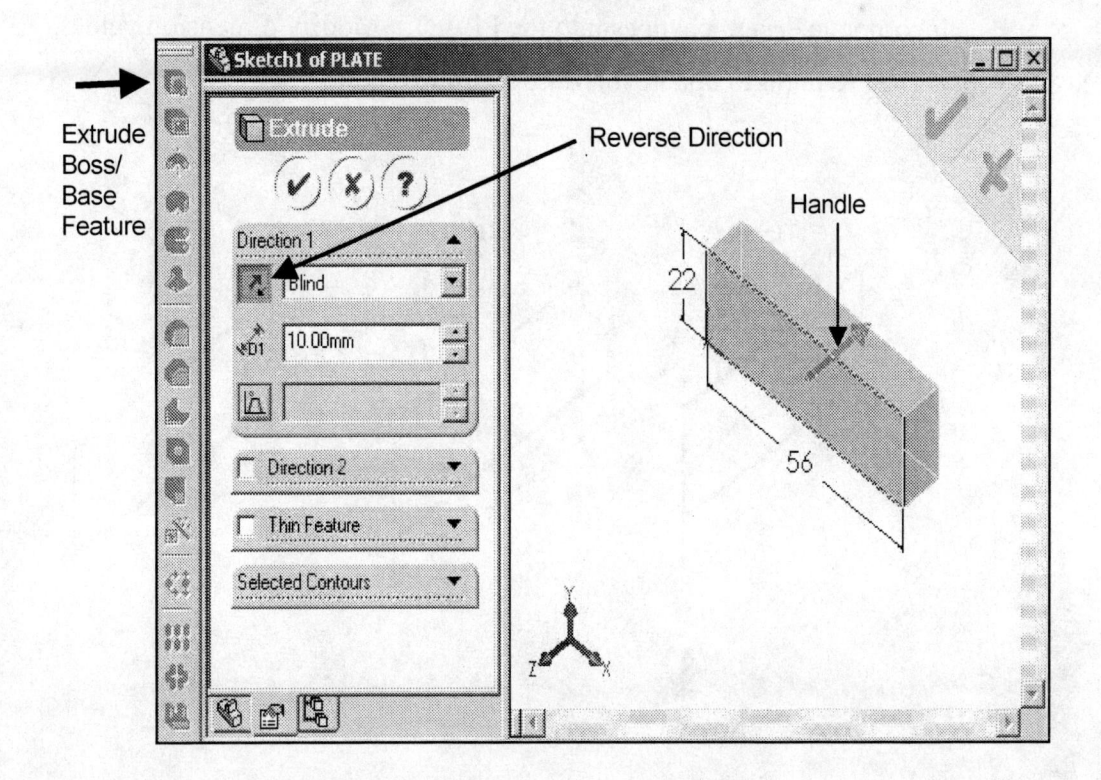

PAGE 1 - 37

Fundamentals of Part Modeling   **Engineering Design with SolidWorks**

81) Reverse the direction of the extruded depth. Click the **Reverse Direction** check box. Specify the depth of the extrusion. Blind is the default Type option. Drag the **Handle** in the extrusion direction and review the mouse pointer feedback for depth. The depth of the extrusion increases respectively with the depth spin box value. The Origin is displayed in the lower left corner.

82) Enter **10** in the Depth spin box. Click **OK** ✔. Note: The Green Check Mark represents the OK command. Note: The Tool tip for the Green Check Mark displays OK, Apply, Close Accept and Exit depending on the command.

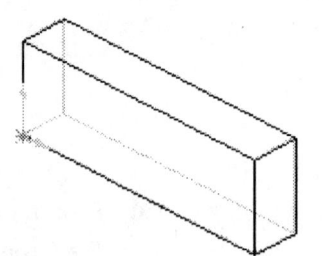

Note: If you exit the Sketch before selecting Extrude Boss/Base, Sketch1 is displayed in the FeatureManager.

Click Sketch1 from the FeatureManager. Click Extrude Boss/Base to create the feature.

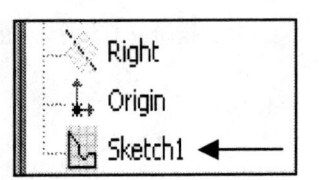

Create the PLATE - Modify The Dimensions

Incorporate design changes into the PLATE. Modify dimension values. Utilize Rebuild to update the Base Extrude feature.

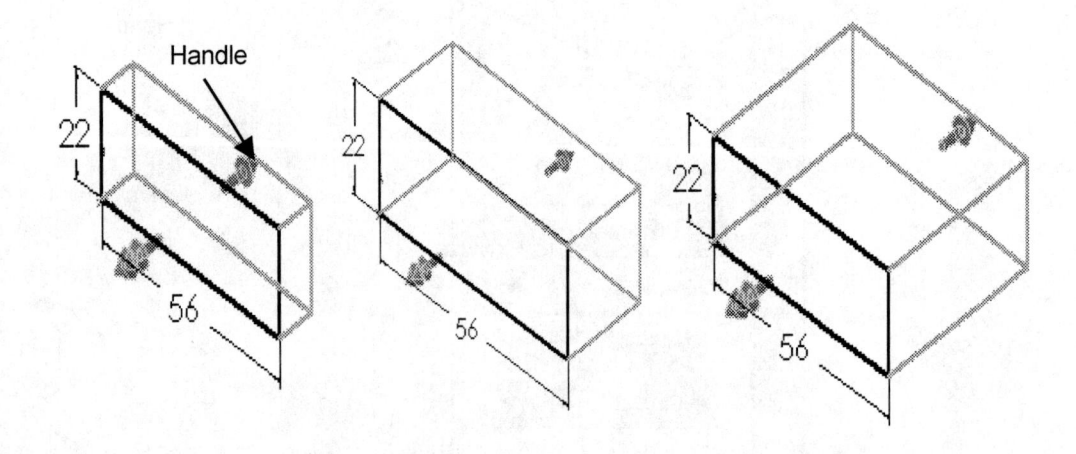

Engineering Design with SolidWorks **Fundamentals of Part Modeling**

Modify the PLATE.
83) Double-click on the **Extrude1** front face. Change the width dimension.

84) Double-click on **10**. The Dimension PropertyManager is displayed.

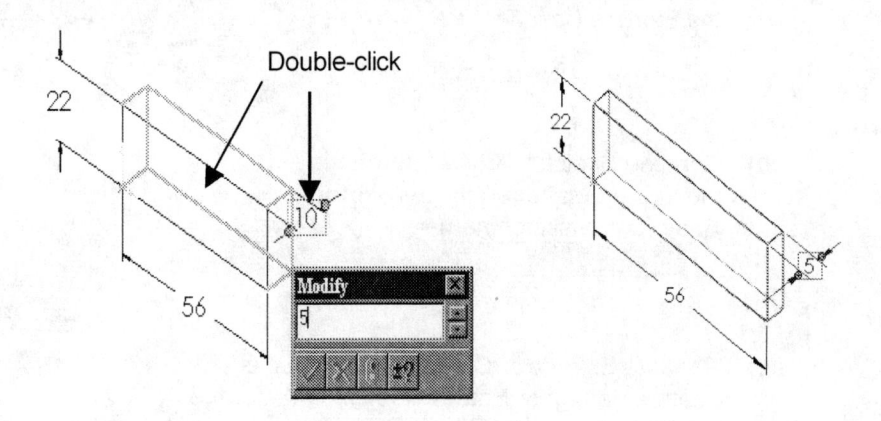

85) Enter **5** in the Modify dialog box. Click the **Green Check Mark** in the Modify dialog box.

86) Update the feature dimension. Click **Rebuild** from the Standard toolbar.

87) Fit the model to the screen. Press the **f** key.

Incorporate the machining and assembly process into the PLATE design.

Dimensions are referenced from the three datum planes with the machined Origin in the lower left hand corner of the PLATE.

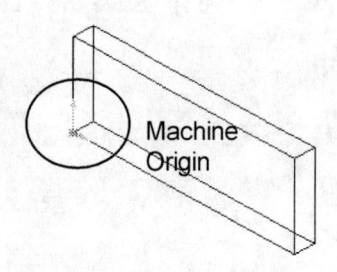

Maintain the Origin on the front lower left hand corner of the Base-Extrude feature and co-planar with the Front plane.

An Extruded-Base feature is named Base-Extrude in the FeatureManager design tree.

The Plus Sign icon indicates that additional feature information is available.

Fundamentals of Part Modeling   **Engineering Design with SolidWorks**

88) Display the FeatureManager. Click the **FeatureManager**.

89) Click the **Plus Sign** of the Extrude1 feature. Sketch1 is the name of the Sketch used to create the Extrude Base feature.

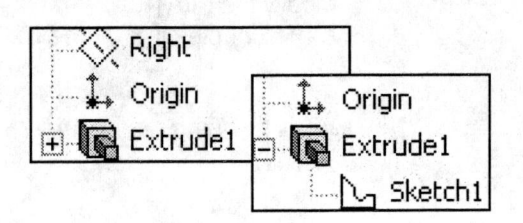

90) Rename Sketch1. Click **Sketch1** in the FeatureManager. Enter **Sketch-Base** for the sketch name.

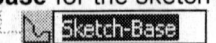

91) Rename Extrude1. Click **Extrude1** in the FeatureManager. Enter **Base-Extrude** for the feature name.

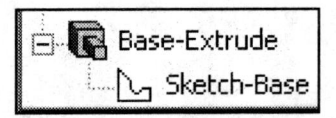

92) The Minus Sign icon indicates that the feature information is expanded. Click the **Minus Sign** to collapse the Base-Extrude feature.

93) Save the PLATE. Click **Save**.

Display Modes and View Modes

The Display modes specify various model appearances in the Graphics window. The View modes manipulate the model in the Graphics window.

Display the GUIDE.

Click the following icons from the View Display toolbar to assist in model visualization.

View various Display modes:

94) Click **Wireframe**.

95) Click Hidden Lines Visible.

96) Click Hidden Lines Removed.

97) Click **Shaded**.

Fundamentals of Part Modeling   **Engineering Design with SolidWorks**

Note: The Hidden Lines Removed option may take longer to display than the Shaded option depending on your computer configuration and the size of the part and file.

The View modes manipulate the model in the Graphics windows.

Display the various View modes.

98) Click **Zoom to Fit** to display the full size of the part in the current window.

99) Click **Zoom to Area**. Select two **opposite corners** of a rectangle to define the boundary of the view. The defined view fits to the current window.

100) Click **Zoom In/Out**. Drag upward to **zoom in**. Drag downward to **zoom out**. Press the lower case **z** key to zoom out. Press the upper case **Z** key to zoom in.

101) **Right-click** in the Graphics window. Click **Select**.

Click the **front edge**. Click **Zoom to Selection**. The selected geometry fills the current window.

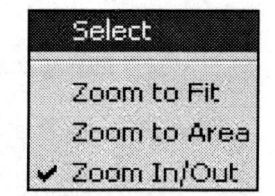

102) Click **Rotate**. Drag the **mouse pointer** to rotate about the screen center. Use the computer keyboard **arrow keys** to rotate in 15-degree increments.

103) Click **Pan**. Drag the **mouse pointer** up, down, left, or right. The model scrolls in the direction of the mouse.

104) Right-click in the **Graphics window** area to display the zoom options. Click **Zoom to Fit**.

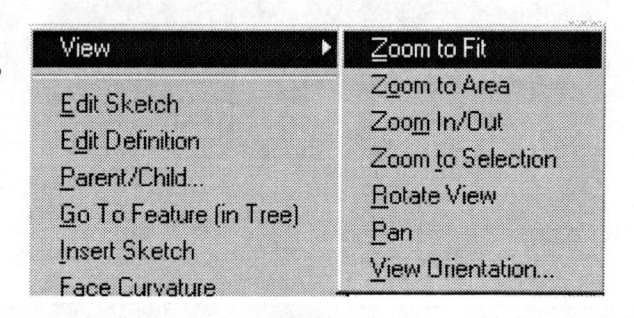

Note: View modes remain active until deactivated from the View toolbar or unchecked from the pop-up menu.

105) Return to the standard Isometric view. Click **Isometric** from the Standards View toolbar.

PAGE 1 - 42

Fasteners

Screws, bolts and fasteners are used to joint parts together. Use standard available fasteners whenever possible. This will decrease product cost and reduce component purchase lead times.

The American National Standard Institute (ANSI) and the International Standardization Organization (ISO) provide standards on various hardware components.

The SolidWorks Library contains a variety of standard fasteners to choose from.

Below are a few general selection and design guidelines that are utilized in this text:

Use standard industry fasteners where applicable.

Reuse the same fastener types where applicable. Dissimilar screws and bolts require different tools for assembly, additional part numbers and increase inventory storage and cost.

Decide on the fastener type before creating holes. Dissimilar fastener types require different geometry.

Create notes on all fasteners. Notes will assist in the development of a Parts list and Bill of Materials.

Use caution when positioning holes. Do not position holes too close to an edge. Stay one radius head width at a minimum from an edge or between holes. Review manufacturer's recommended specifications.

Design for service support. Insure that the model can be serviced in the field and or on the production floor.

Fundamentals of Part Modeling

Use standard M4x8 Socket head cap screws in this exercise.

M4 represents a metric screw.

4mm represents the major outside diameter.

8mm represents the overall length of the thread.

Determine the dimensions for the mounting holes from the drawing of the GUIDE CYLINDER assembly.

The customer will attach the PLATE to the GUIDE CYLINDER assembly with the M4 Socket head cap screws.

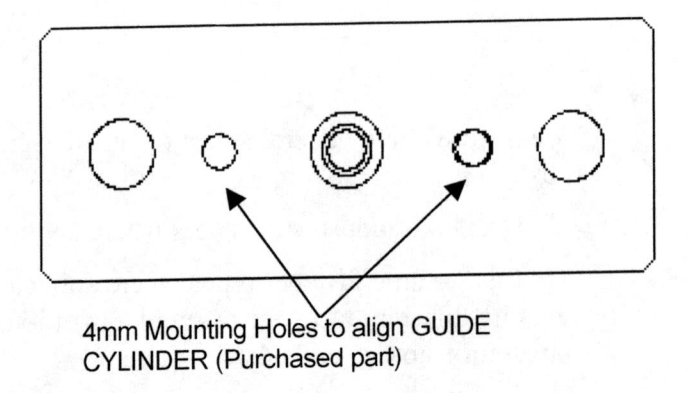

4mm Mounting Holes to align GUIDE CYLINDER (Purchased part)

View Orientation

The view orientation defines the preset position of the model in the Graphics window.

The Standard View toolbar displays eight view options: Front, Back, Left, Right, Top, Bottom, Isometric and Normal To.

The Isometric view displays the part in 3D with two equal projection angles.

The Normal To view displays the part ⊥ to the selected plane.

Different View orientation is selected for sketches, parts and assemblies.

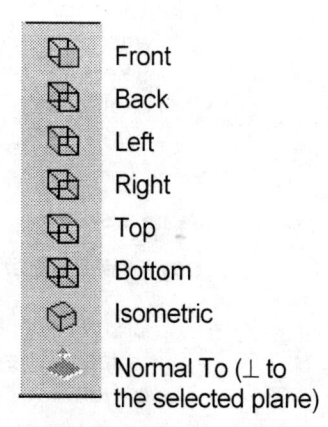

- Front
- Back
- Left
- Right
- Top
- Bottom
- Isometric
- Normal To (⊥ to the selected plane)

Create the PLATE Part – Use Holes as Cut Features

Holes remove material. There are numerous ways to create holes. Use the Extruded Cut feature in this example.

An Extruded Cut feature removes material ⊥ to the Sketch for a specified depth.

The PLATE part requires mounting holes. The holes are machined through the front face of the PLATE.

A critical feature of the PLATE is the location of the holes. The holes are dependent on the corresponding holes in the GUIDE CYLINDER assembly.

The machinist references all dimensions from the reference datum planes to product the PLATE holes.

Note: Reference all dimensions from the Right datum plane and the Top datum plane.

Select the front face of the PLATE.
106) Position the mouse pointer on the **front face** of the PLATE. The pointer displays various items depending on geometry: Line, Face, Point or Vertex. Example:

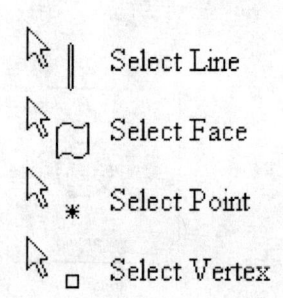

- Select Line
- Select Face
- Select Point
- Select Vertex

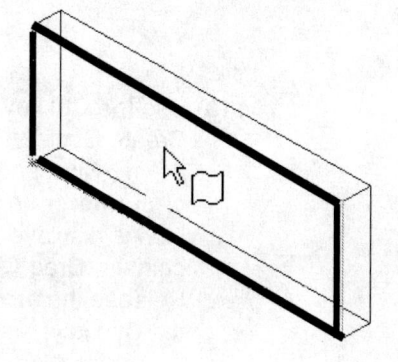

Fundamentals of Part Modeling **Engineering Design with SolidWorks**

Select the Sketch plane.
107) Click the **front face** of the Extrude1 feature. The boundary is highlighted with dash lines when a face of a feature is selected. Note: Select the Sketch plane before creating a Sketch.

Create a Sketch.
108) Click **Sketch**. Display the Front view.

Click **Front** from the Standard view toolbar. The Origin is located in the lower left corner.

109) Display the reference plane location with respect to the Sketch plane. Click the **Top** and **Right** plane from the FeatureManager design tree. The Top plane is displayed as the bottom horizontal line. The Right plane is displays at the left vertical line.

110) Create Circle1. Click **Circle** from the Sketch Tools toolbar. Create the first circle point. Click the **center point**. Create the second circle point. Drag the **mouse pointer** to the right, until R is approximately 3. Release the **mouse button**.

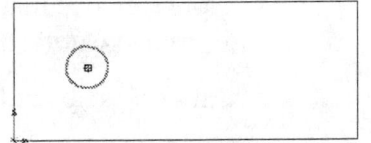

The circle is displayed in green. The circumference and the center point of the circle are currently selected. The circle sketch icon is display on the mouse pointer.

Copy the Sketch geometry.
111) Right-click in the **Graphics window**.

112) Click **Select** from the Pop-up menu. The center point and circle circumference is displayed in blue. Blue indicates an undefined sketch.

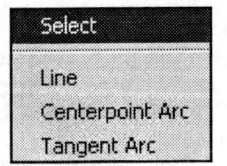

113) Use the Ctrl key to copy Circle1. Create Circle2. Hold the **Ctrl** key down. Click and drag the **circumference** of Circle1. The Circle Icon is displayed on the mouse pointer. Drag **Circle2** to the right. Release the **mouse button**. Release the **Ctrl** key.

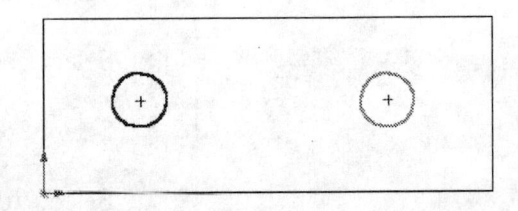

PAGE 1 - 46

Geometric relations are dimensional relationships between one or more creations. The holes may appear aligned and equal, but they are not!

Note: Add Relations is used to create a geometric relationship such as parallel or collinear between sketch elements.

Add geometric relations.

114) Circle2 is the selected system entity, named Arc2. Hold the **Ctrl** key down. Click the **Circle1** circumference. Circle1 and Circle 2 are displayed in green. Arc1 and Arc2 are displayed in the Selected Entities text box in the PropertyManager. Release the left mouse **button**. Release the **Ctrl** key.

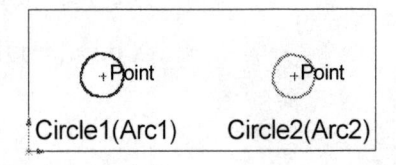

115) Click the **Equal** button. Both circles have the same diameter. The Equal radius/length is displayed in the Existing Relations text box. Click **inside** the Graphcis window to apply the relation.

Note: The SolidWorks default name for curve geometry is an Arc#. There is an Arc# for each circle.

Note: Click the circumference of the two circles. If you selected the center point and not the circumference, right-click in the Selected Entities window and click Clear Selections.

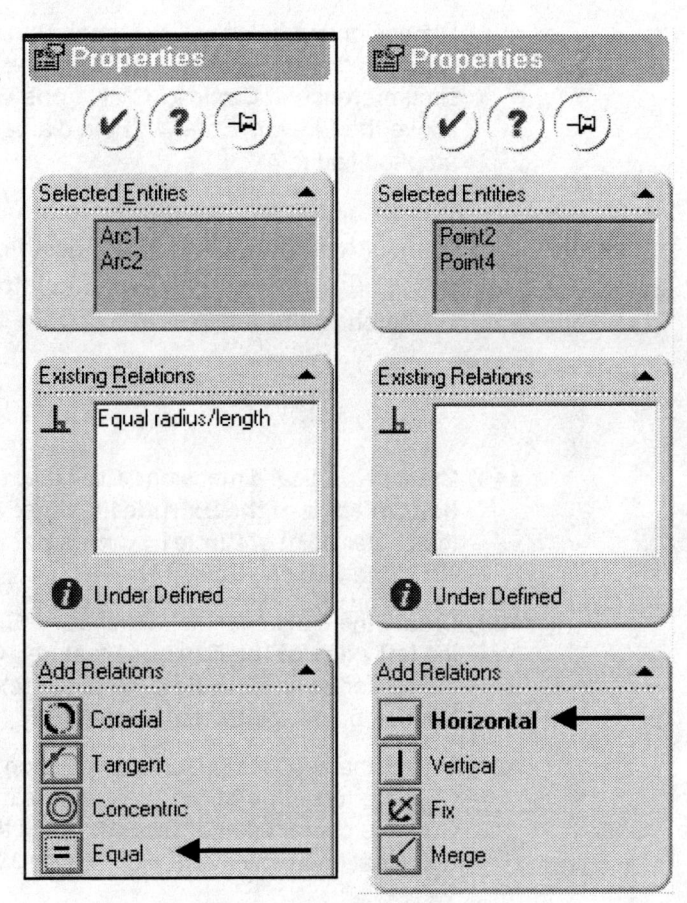

All geometry is removed from the Selected Entites text box.

If you create or delete the geometry in a different order, the system selected entity names will be different.

Fundamentals of Part Modeling **Engineering Design with SolidWorks**

Align the center points of Circle1 and Circle2.
116) Hold the **Ctrl** key down. Click the **two center points**.

117) Click the **Horizontal** button for the horizontal point alignment. Release the **Ctrl** key. Click **OK** ✔.

The two center points are system entities named Point2 and Point4. The default name for any point is Point#. Each circle requires two points:

- A center point.
- A point on the diameter.

Add dimensions.

118) Dimension the holes. Click **Dimension**. Create a diameter dimension. Click the circumference of **Circle1**. Click a **position** above the Sketch. Enter **4**. Both diameters are modified to 4.

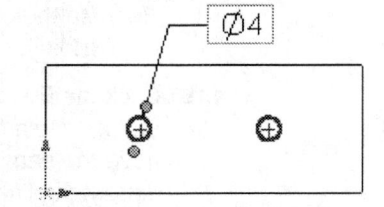

Note: If the Sketch is too large in the Graphics window, you will not be able to view the dimension. Press the f key to fit the Sketch to the screen.

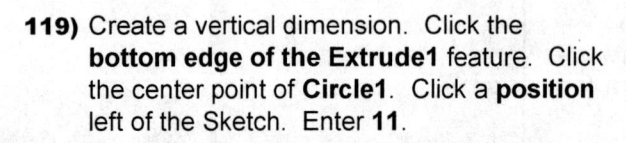

119) Create a vertical dimension. Click the **bottom edge of the Extrude1** feature. Click the center point of **Circle1**. Click a **position** left of the Sketch. Enter **11**.

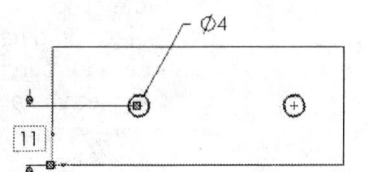

120) Create the first horizontal dimension. Click the **left edge of the Extrude1** feature. Click the center point of **Circle1**. Click the **text** location below the Sketch. Enter **16.5**.

121) Create the second horizontal dimension. Click the **left edge of the Extrude1** feature. Click the center point of **Circle1**. Click the **text** location below the Sketch. Enter **39.5**.

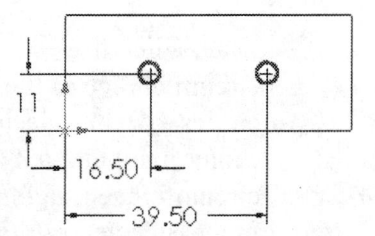

122) Flip the dimension arrows to the inside. Click the **16.50** dimension text. Click the **arrow head green dot** to display the arrows inside the extension lines.

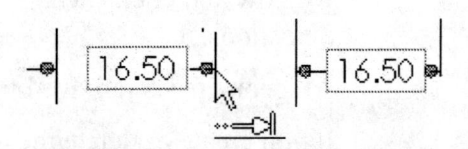

Extrude the Sketch.

123) Display the Isometric view. Click **Isometric**.

124) Click **Extruded-Cut** on the Features toolbar. The Extruded-Cut PropertyManager is displayed. Determine the depth of the hole. The direction arrow points into the Extruded-Base feature.

125) Click the **drop down arrow** from the Type list box.

126) Click the **Through All** option.

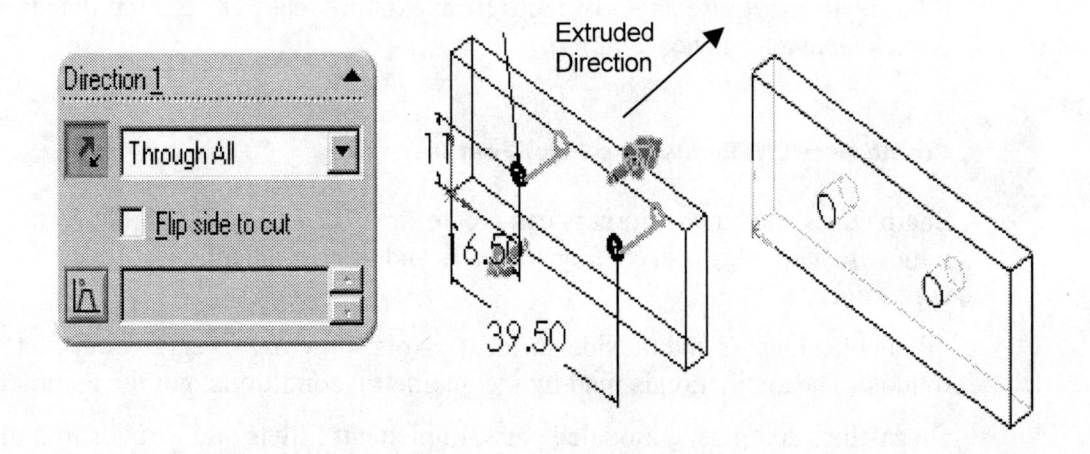

127) Create the two holes. Click **OK**.

Rename the Cut-Extrude1 feature.
128) Rename the Cut-Extrude1 feature to the Mounting Holes feature. Click the feature **Cut-Extrude1** from the FeatureManager.

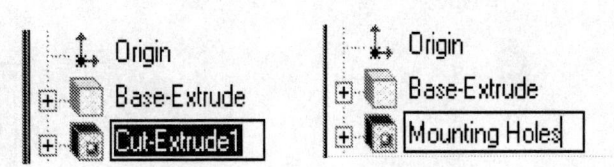

129) Enter **Mounting Holes** for the new feature name.

130) Click the **PLATE icon** at the top of the FeatureManager.

131) Save the PLATE. Click **Save**.

An Extrude Boss/Base adds material. An Extrude Cut removes material. Extrude features require the following:

> Sketch Plane.
>
> Sketch.
>
> Geometric Relations.
>
> Dimensions.

Fillet features require edges or faces from existing features. Sketch Planes and Sketch Geometry is not required.

Create the PLATE – Use the Fillet Feature

Sharp edges and square corners can create stress in a part. The Fillet feature removes sharp edges, strengthens corners and or cosmetically improves appearance.

Fillets blend inside and outside surfaces. Note: Filleting refers to both filets and rounds. The distinction is mad by the geometric conditions, not the command.

On castings and plastic modeled parts, implement Fillets and Rounds into the initial design.

If you are uncertain of the exact radius value, input a small test radius of 1mm.

It takes less time for a supplier to modify an existing dimension than to create one.

Create an Edge Fillet.
132) Display Hidden Lines to select edges to fillet. Click **Hidden Lines Visible**.

133) Fit the model to the Graphics window. Press the **f** key.

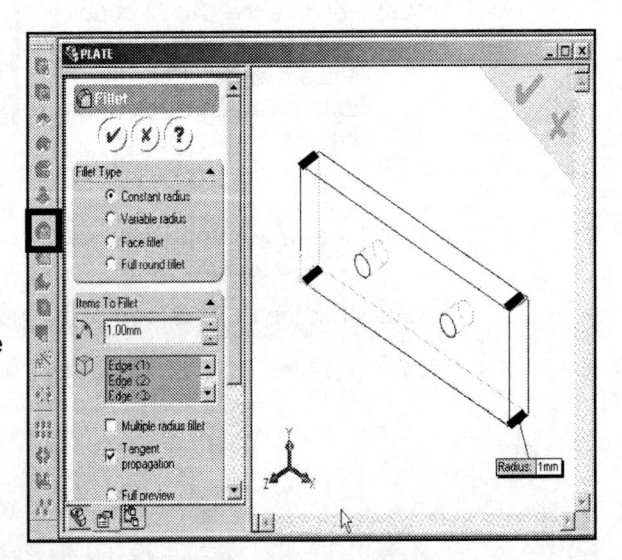

Engineering Design with SolidWorks   **Fundamentals of Part Modeling**

134) Create a Fillet/Round feature. Click **Fillet**. The Fillet feature PropertyManager is displayed. Enter **1** in the Fillet Radius list box. Click the **4 small corner edges**. Each edge is added to the Items to Fillet list.

135) Display the edge Fillet. Click **OK**.

136) Rename Fillet1 to Small Edge Fillet.

137) Save the PLATE. Click **Save**.

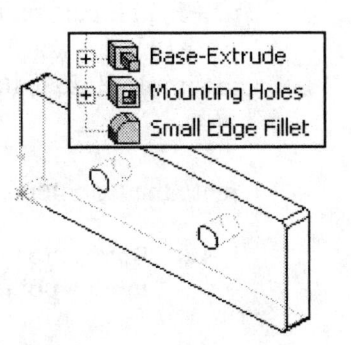

Minimize the number of Fillet Radius sizes created in the FeatureManager. Combine Fillets and Rounds that have a common radius.

Select all Fillet/Round edges. Add them to the Items to Fillet list in the Fillet/Round feature dialog box.

In the previous step, you selected single edges to Fillet. Select tangent edges in the next step.

Create a Tangent Edge Fillet.

138) Click **Fillet**. The Tangent Propagation check box is checked.

139) Click the **front top horizontal edge**. All tangent edges are selected.

140) Click the back top horizontal edge.

141) Click **OK**.

142) Rename Fillet2 to Front-Back Edge Fillet.

143) Save the PLATE part. Click **Save**.

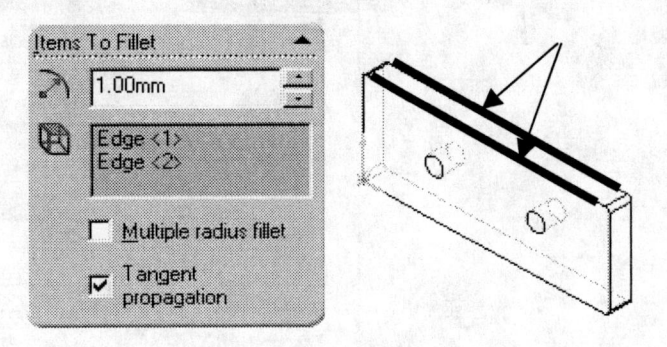

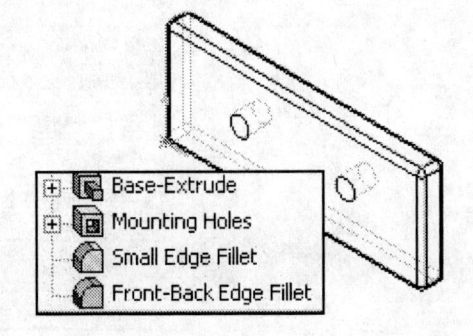

Create the PLATE – Use the HoleWizard Countersink Hole Feature

The PLATE part requires a Countersink Hole feature. Use the HoleWizard. The HoleWizard creates complex and simple Hole features.

Create the Countersink Hole.

    144) Rotate the PLATE part. Click **Rotate**. Drag the **mouse pointer** to rotate the part.

    145) Click **Rotate** to deactivate.

    146) Select the Sketch plane. Click the **back face**.

    147)

    148) View the Sketch plane for the HoleWizard. Click **Back**. The Origin is displayed at the lower right edge.

Note: Dimension the Countersink Hole to the Origin not to the Fillet edge.

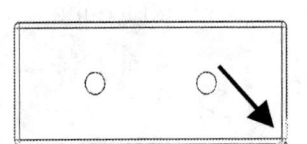

    149) Click **HoleWizard**. The Hole Definition dialog box is displayed. Click the **Countersink** tab.

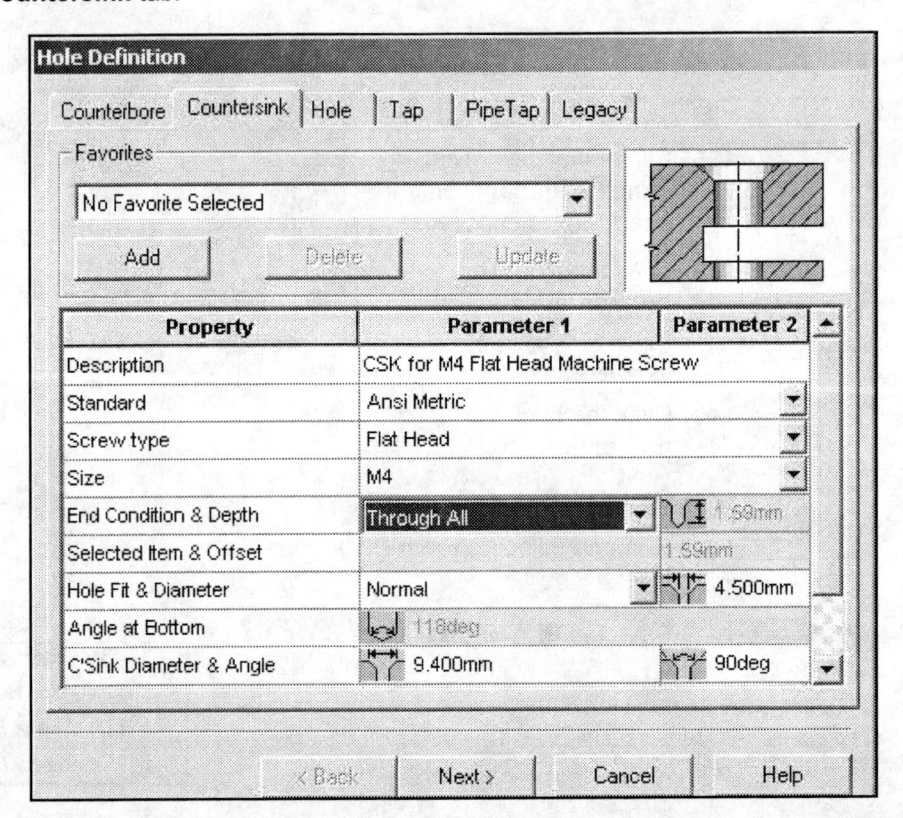

Property	Parameter 1	Parameter 2
Description	CSK for M4 Flat Head Machine Screw	
Standard	Ansi Metric	
Screw type	Flat Head	
Size	M4	
End Condition & Depth	Through All	1.59mm
Selected Item & Offset		1.59mm
Hole Fit & Diameter	Normal	4.500mm
Angle at Bottom	118deg	
C'Sink Diameter & Angle	9.400mm	90deg

Engineering Design with SolidWorks — Fundamentals of Part Modeling

150) Enter **ANSI Metric** for the Standard.

151) Enter **Flat Head** for Screw type.

152) Enter **M4** for Size.

153) Select **Through All** from the drop down list for End Condition & Depth. Accept the other default values.

154) Click **Next** from the Hole Placement dialog box. The Hole Placement dialog box is displayed. Note: DO NOT SELECT THE FINISH BUTTON AT THIS TIME in the Hole Placement Dialog box. The Countersink hole is displayed in orange.

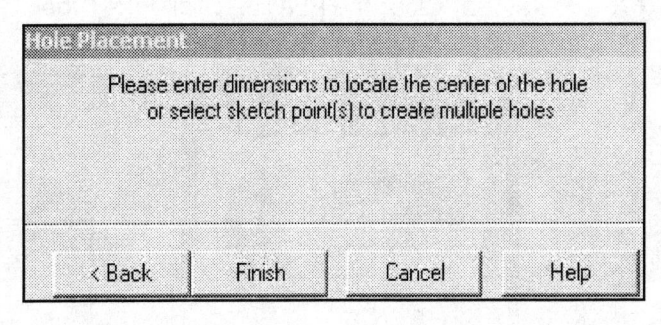

Orange is a preview color. The Sketch Point tool is automatically selected. No other holes are required.

Dimension the hole relative to the Origin.

155) Create the horizontal dimension. Click **Dimension**. Click the **center point** of the countersink hole. Click the **Origin**. Click a **position** below the bottom horizontal edge. Enter **56/2**. The dimension value 28 is calculated. Click the **Green Check Mark**.

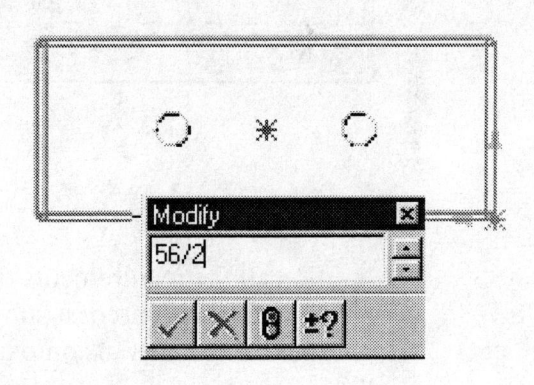

156) Create the vertical dimension. Click the **center point of the countersink hole**. Click the **Origin**. Click a **position** to the right of the profile. Enter **11**. Click the **Green Check Mark**.

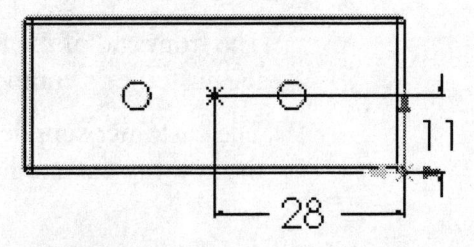

157) Click **Finish** from the Hole Placement dialog box.

158) Click **Isometric** from the Standards View toolbar.

The Countersink Hole is named CSK for M4 Flat Head Screw 1 in the FeatureManager.

The PLATE is complete.

159) Save the PLATE. Click **Save**.

160) Close the PLATE. Click **File**, **Close**.

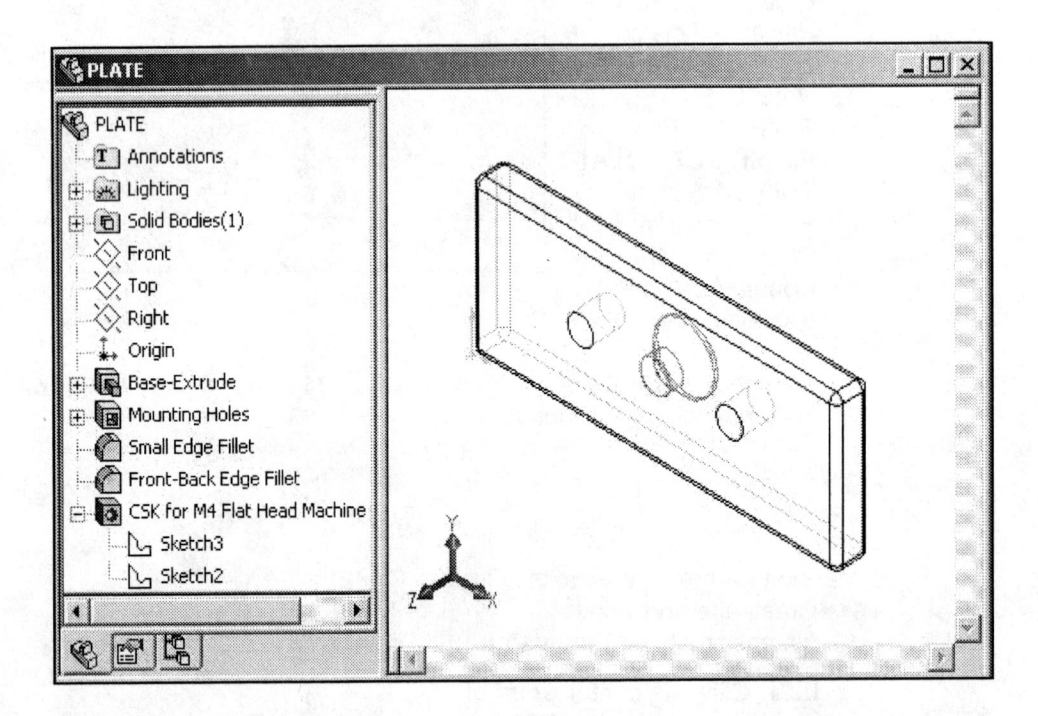

ROD

Recall the requirements of the customer. The ROD is part of a sub-assembly that positions materials onto a conveyor belt.

The back end of the ROD is fastened to the PLATE.

The front end of the ROD is mounted to the customer's components.

The customer supplies dimensions for the keyway cut and hole.

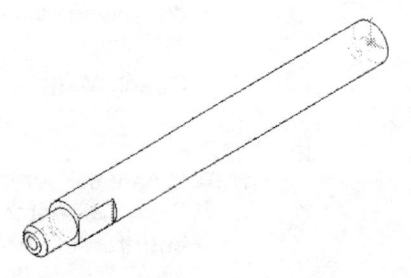

ROD Feature Overview

Features are geometry building blocks. Features add or remove material.

Features are created from sketched profiles or from edges and faces of existing geometry.

Use the following features to create the ROD part:

Extruded Base: Create the Extruded Base feature on the Front plane with a circular sketched profile, Figure 1.13a.

Extruded Cut: Create the Extruded Cut feature by sketching a circle on the front face to form a hole, Figure 1.13b.

Chamfer: Create the Chamfer feature from the circular edge of the front face, Figure 1.13b.

Extruded Cut: Create the Extruded Cut feature from a converted edge of the Extruded Base feature to form a Keyway, Figure 1.13c.

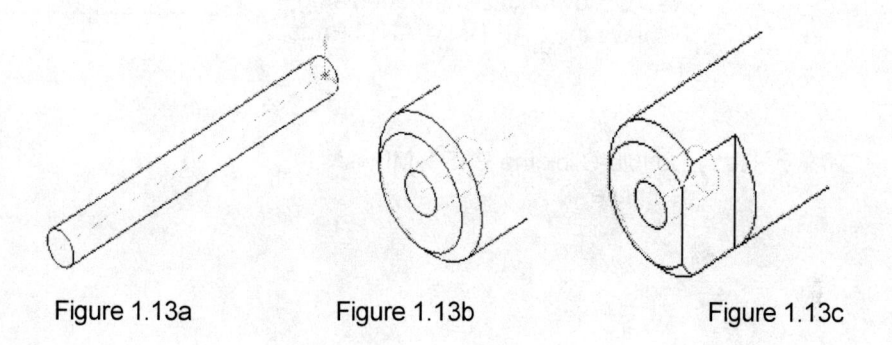

Figure 1.13a   Figure 1.13b   Figure 1.13c

Extruded Cut: Create the back hole with the Copy/Paste function. Copy the front hole to the back face, Figure 1.13d.

Extruded Cut: Add a new Extruded Cut to the front face, Figure 1.13e. Redefine the Chamfer and Keyway Cut, Figure 1.13f.

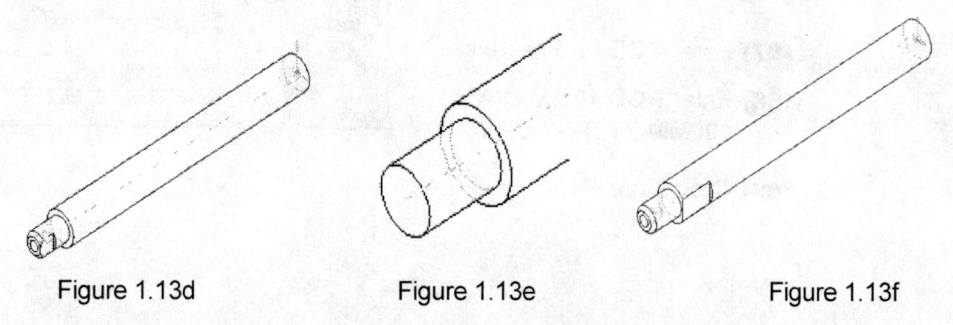

Figure 1.13d   Figure 1.13e   Figure 1.13f

Create the ROD Part

The geometry of the Base feature is a cylindrical extrusion. The Extruded Base feature is the foundation for the ROD.

What is the shape of the sketched 2D profile? Answer: A circle.

What is the Sketch plane?

Before you answer the question, remember how the ROD is positioned in the assembly. Answer: The Front plane is the Sketch plane.

Create a new part.
161) Click **New**.

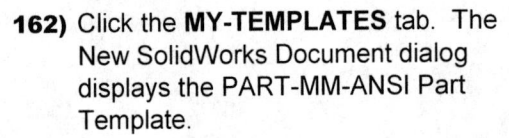

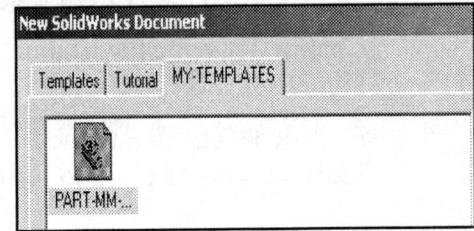

162) Click the **MY-TEMPLATES** tab. The New SolidWorks Document dialog displays the PART-MM-ANSI Part Template.

163) Double-Click the PART-MM-ANSI template.

164) Save the part. Click **Save**.

165) Click the **drop down arrow** in the Save in list box.

166) Select ENGDESIGN-W-SOLIDWORKS\PROJECTS

167) Enter **ROD** for File name.

168) Enter **ROD 10MM DIA x 100MM** for Description.

169) Click **Save**.

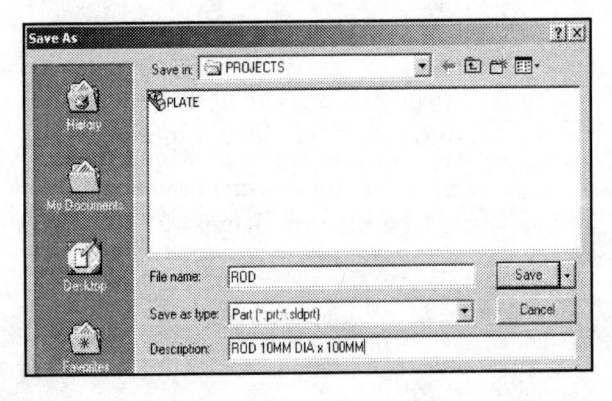

Create a Sketch.

170) Select the Sketch plane. The Front plane is the default Sketch plane. Click the **Front plane** from the FeatureManager. Open a Sketch. Click **Sketch** from the Sketch toolbar.

171) Create a circle. Click **Circle** from the Sketch Tools toolbar. Create the first point. Click the **Origin**. Create the second point. Drag the **mouse pointer** to the right. Click a **position** to the right of the Origin.

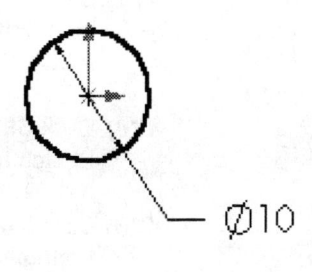

172) Add the dimension. Click **Dimension**. Click the **circumference** of the circle. Click a **position** off the profile. Enter **10**. Click the **Green Check Mark**.

Extrude the Sketch.

173) Click **Extruded Boss/Base** on the Features toolbar. Blind is the default Type option. Enter **100** in the depth text box.

174) Display the Extruded-Base feature. Click **OK**.

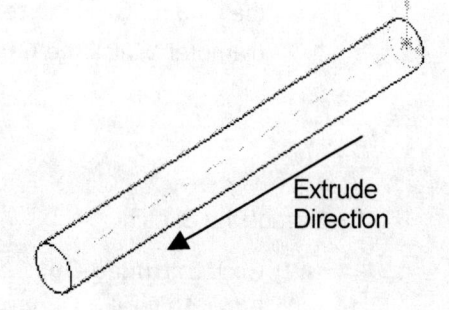

175) View the Extruded-Base feature. Click **Zoom to Fit**.

176) Rename Extrude1 to Base Extrude.

Fundamentals of Part Modeling **Engineering Design with SolidWorks**

Create an Extruded Cut.
177) Select a Sketch plane. Click the **front circular face** of the ROD.

178) Click **Sketch** from the Sketch toolbar.

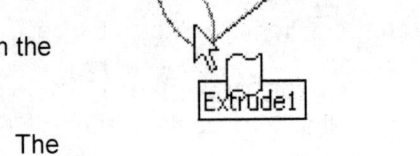

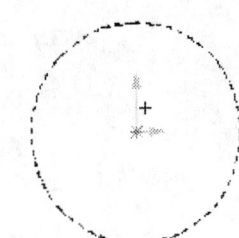

179) Click **Normal To**. The front face of the ROD is displayed.

Create a Sketch.
180) The 2D profile of the hole is a circle. Create a circle. Click **Circle**. Create the first point. Click the **Origin**. Create the second point. Drag the **mouse pointer** away from the Origin. Create the circle. Click a **position** to the right of the Origin.

181) Click **Dimension**. Click the **circumference of the circle**. Drag the **mouse pointer** directly to the right of the Origin. Click the **text location**. Enter **3** for the new diameter. Click the **Green Check Mark**.

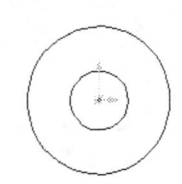

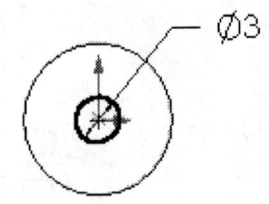

Extrude the Sketch.
182) Click **Extruded-Cut**. Blind is the default Type option. Enter **10** for the Depth. Click **OK**. The feature is named Extrude-Cut1.

183) Rename Cut-Extrude1 to Front Hole.

184) Save the ROD. Click **Save**.

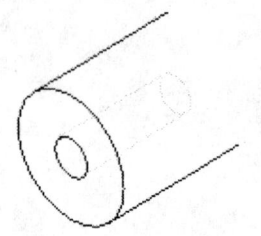

PAGE 1 - 58

Create the ROD – Chamfer Feature

The Chamfer feature removes material along an edge. The Chamfer feature assists the ROD by creating beveled edges for ease of movement in the GUIDE-ROD assembly.

Create a Chamfer feature.
185) Click on the **front outer circular edge**. Click **Chamfer**. An arrow indicates the direction in which the distance of the chamfer is measured.

186) The Chamfer feature PropertyManager is displayed. Enter **1** for Distance.

187) Accept the default **45** degree Angle.

188) Display the Chamfer feature. Click **OK**.

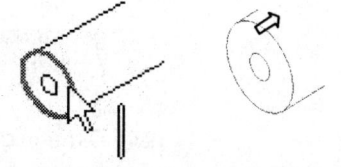

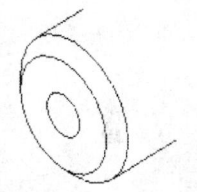

Create the ROD – Use the Extruded-Cut Feature

The Extruded-Cut feature removes material from the front of the ROD to create a keyway.

A keyway locates the orientation of the ROD into a mating part in the assembly. Utilize the Convert Entities tool in the Sketch Tools toolbar to extract existing geometry.

Create an Extruded Cut.
189) Select the Sketch plane. Click the **front circular face** of the ROD. Click **Sketch**. Click **Normal To**. The front face of the ROD is displayed.

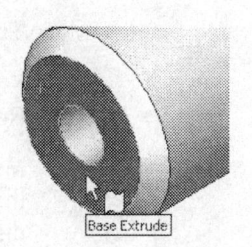

190) Sketch a Profile. Click **Select**. Click the **outside circular edge**. Click **Convert Entities** from the Sketch Tools toolbar. The system extracts the outside edge and positions it on the Sketch plane.

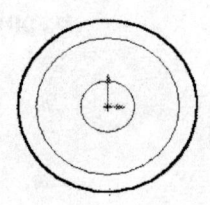

191) Click **Line**. Sketch a **line**. The end points of the line are coincident with the circumference of the circle.

192) Add a vertical relation. Click **vertical** from the PropertyManager. The relation vertical is displayed in the Existing Relations text box.

193) The Trim command deletes the sketched geometry. Click **Trim** from the Sketch Tools toolbar. Click the **left outside edge** of the circle.

194) Dimension the Sketch. Click **Dimension**. Click the vertical **line**. Click a **position** to the right of the profile. Enter **6.0**. Click the **Green Check Mark**. The end points of the vertical line are coincident with the outside edge of the circle.

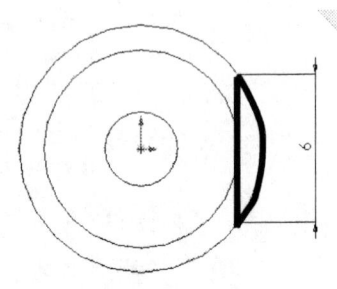

Extrude the Sketch.

195) Click **Extruded-Cut**. Enter **5** for the Depth.

196) Display the Extruded-Cut. Click **OK**. The feature is named, Cut-Extruded2.

197) Rename Cut-Extrude2 to Keyway Cut.

198) Orient the part in the Isometric view. Click **Isometric** from the Standards View toolbar.

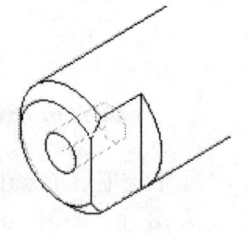

Move/Size an Extruded feature.

199) Resize the Keyway Cut feature. Click **Move/Size** from the Features toolbar. The Move/Size handles are displayed on the Keyway Cut feature.

200) Click the **Resize** handle. Drag the **Resize** handle towards the Origin.

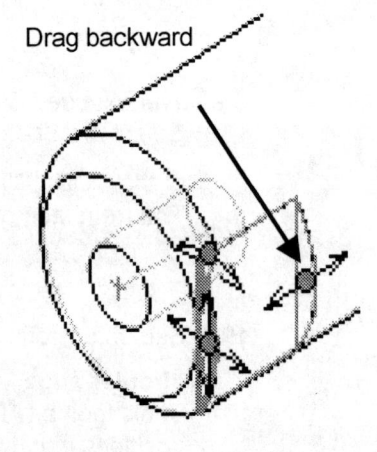

Drag backward

201) Release the **mouse button** when 15 is displayed on the pointer.

202) Deactivate the Move/Size command. Click **Move/Size**.

203) Save the ROD. Click **Save**.

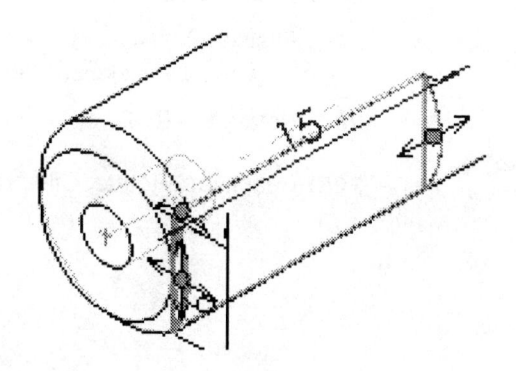

View Orientation and Named Views

The View Orientation defines the preset position of the ROD in the Graphics window. It is helpful to display different views when creating and editing features.

The View Orientation options provide the ability to:

Create a named view.

Select standard views: Normal To, Front, Back, Left, Right, Top, Bottom and Isometric.

Select the Trimetric and Diametric views.

Redefine the standard views or return them to the system default setting.

Drag Split Bars to display multiple view orientations in the same Graphics window.

Example: The mouse pointer displays ⇳ when positioned over the horizontal Split bar.

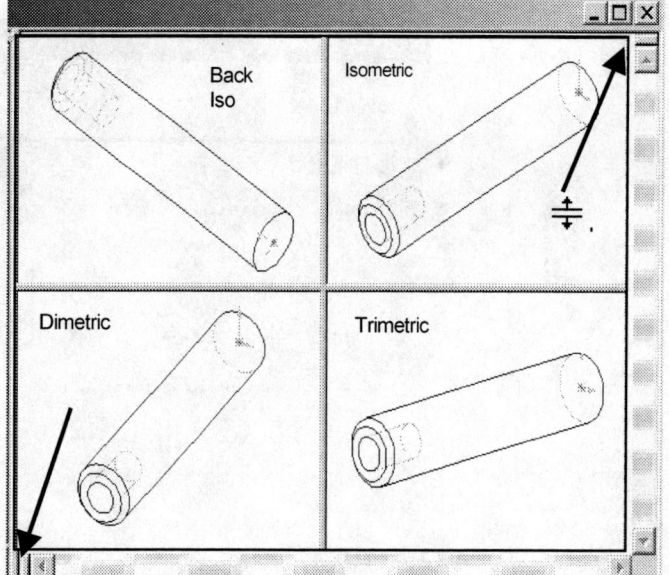

Fundamentals of Part Modeling **Engineering Design with SolidWorks**

Create a new view.

204) Position the ROD. Click **Rotate**.
View the back circular face.

205) Rotate the **ROD**.

206) Deactivate Rotate. Click **Rotate**.

207) Create a new View Orientation. Click **View Orientation** from the View toolbar.

208) Click the **Push Pin** from the View toolbar to maintain the displayed menu.

209) Click the **New View** in the Orientation menu. Enter Name text box.

210) Click **OK**. **Close** the View Orientation dialog box.

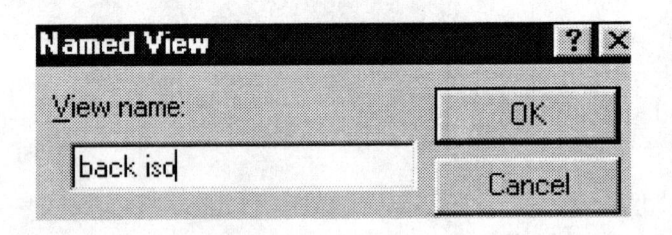

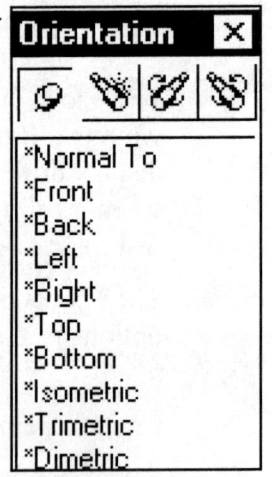

Engineering Design with SolidWorks Fundamentals of Part Modeling

Create the ROD – Use the Copy/Paste Function

Simple sketch features and some applied features can be copied and then pasted onto a planar face.

Multi-sketch features such as sweeps and lofts cannot be copied.

The ROD requires an additional hole on the back face.

Copy the Front Hole feature to the back face.

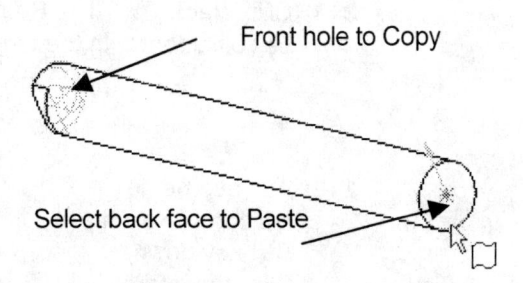

211) Click the **Front Hole** from the FeatureManager.

212) Click **Edit**, **Copy** from the Main menu. Click the **back face**.

213) View the mouse pointer for face feedback. Click **Edit**, **Paste**.

Note: The Copy Confirmation dialog box appears automatically. The box states that there are external constraints in the feature being copied.

External constraints are the dimensions used to place the Front Hole.

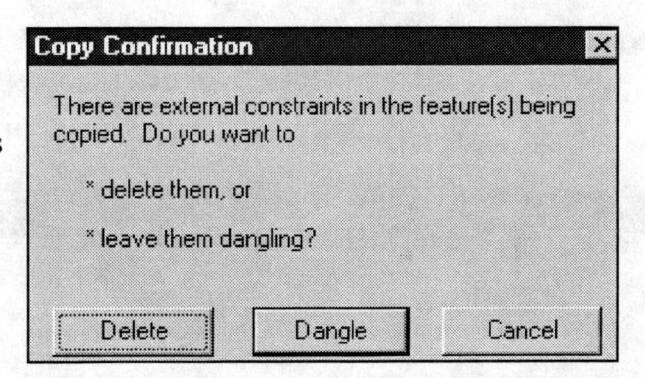

214) Delete the old dimensions. Click **Delete** from the Copy Confirmation dialog box. A copy of the Front Hole feature is placed on the back face of the ROD.

215) Rename the **Cut-Extrude3** feature to **Back Hole**.

PAGE 1 - 63

Fundamentals of Part Modeling   **Engineering Design with SolidWorks**

216) Locate the hole in the center of the back face. Right-click the **Back Hole** feature from the FeatureManager.

217) Click **Edit Sketch**. Display the Back view.

218) Click **Back**. The Back Hole's center point must be coincident with the Origin.

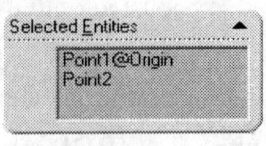

219) Add geometric relations. Hold the **Ctrl** key down. Click the **center point** of the 3mm circle. Click the **Origin** of the ROD. Click **Coincident**. Release the **Ctrl** key. Click **OK**.

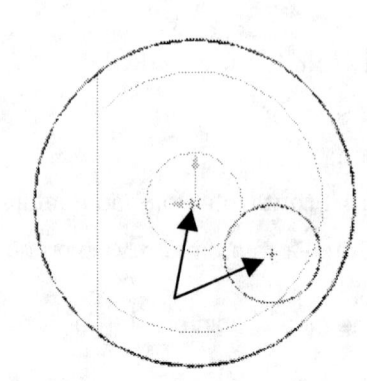

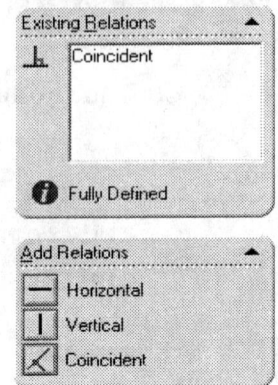

The ROD Back Hole is aligned to the PLATE M4 Countersunk Hole.

220) Modify the diameter dimension of the ROD. Double-click the diameter dimension **3**. Enter **4**. Click **Rebuild** from the Modify dialog box. Click the **Green Check Mark**.

221) Close the Sketch. Click **Sketch**.

222) Click Isometric.

223) Close the View Orientation dialog box. Click **Close**.

224) Save the ROD. Click **Save**.

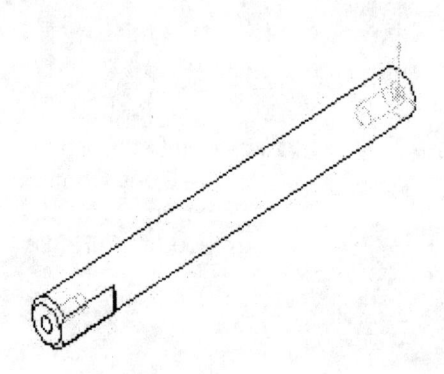

Create the ROD – Perform Design Changes with Rollback and Edit Definition

You are finished for the day. The phone rings. The customer voices concern with the GUIDE-ROD assembly. The customer provided incorrect dimensions for the mating assembly to the ROD.

The ROD must fit into a 7mm hole. The ROD fastens to the customer's assembly with a 4mm Socket head cap screw.

You agree to make the changes at no additional cost since the ROD has not been machined. You confirm the customer's change in writing. The customer agrees but wants to view a copy of the GUIDE-ROD assembly design by tomorrow!

You are required to implement the design change and to incorporate it into the existing part. You begin the design change for the ROD.

The Rollback and Edit Definition functions are used to implement the design change.

The Rollback function allows a feature to be redefined in any state or order. The Edit Definition function allows feature parameters to be redefined.

Implement the design change. First, add the new Extruded Cut feature to the front face of the ROD.

Second, edit the Chamfer feature to include the new edge.

Third, redefine a new Sketch Plane for the Keyway Cut feature.

In this procedure you are exposed to rebuild errors. Information is provided to correct these errors.

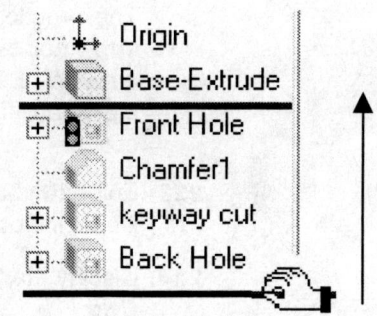

Create the Extruded Cut.
225) Position the Rollback bar. Place the **mouse pointer** over the yellow Rollback bar at the bottom of the FeatureManager design tree. The mouse pointer displays a symbol of a hand.

226) Drag the **Rollback** bar upward to below the Base-Extrude feature. The Base-Extrude feature is displayed.

Sketch the Profile.
227) Select the Sketch plane. Click the **front face** of the Extruded-Base. Click **Sketch** from the Sketch toolbar.

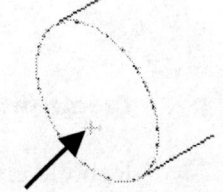

228) Display the **Front** view. Click **Front**.

Fundamentals of Part Modeling   **Engineering Design with SolidWorks**

229) Create a circle. Click **Circle**. Create the first point. Click the **Origin** of the Extruded-Base feature. The mouse pointer displays a pencil with a point. Create the second point. Drag the **mouse pointer** away from the Origin until r is approximately 3. Click the **left mouse button**.

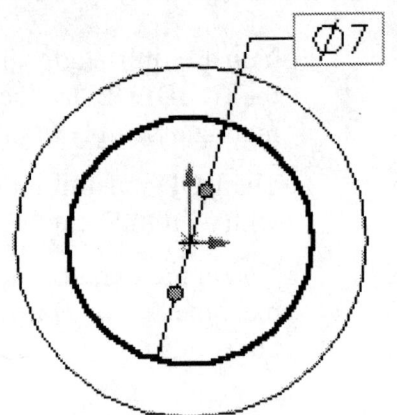

230) Add a dimension. Click **Dimension**. Click the **circumference**. Click a **position** on the profile. Enter **7**. Click the **Green Check Mark**.

Extrude the Sketch.
231) Click **Extruded-Cut**.

232) Click the **Flip side to cut** check box. The direction arrow points outward. Blind is the default Type option.

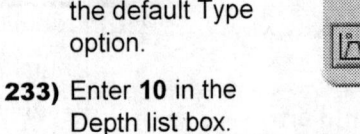

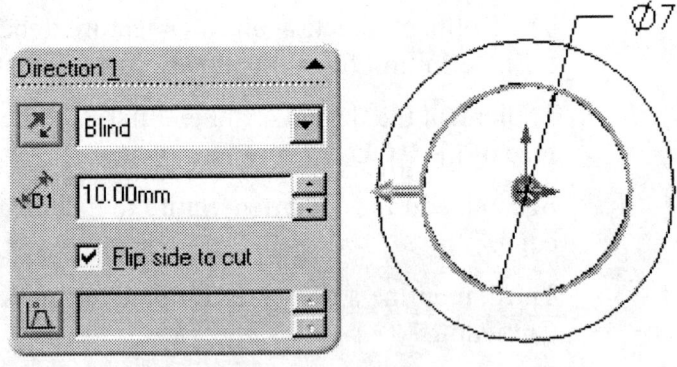

233) Enter **10** in the Depth list box.

234) Display the Extruded-Boss feature. Click **OK**.

235) Rename Cut-Extrude4 to Front Cut.

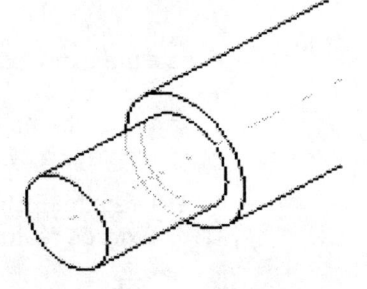

Create the ROD - Recover from Rebuild Errors

Rebuild errors can occur when using the Rollback function. A common error occurs when an edge or face is missing. Redefine the edge for the Chamfer feature. Redefine the face from the Keyway Sketch plane.

PAGE 1 - 66

Engineering Design with SolidWorks   **Fundamentals of Part Modeling**

Redefine the Chamfer feature.

236) Drag the **Rollback** bar downward below the Chamfer1 feature. A red explanation point is displayed next to the name of the Chamfer feature. A red arrow is displayed next to the ROD. Red indicates a model Rebuild error.

237) The ROD Rebuild Error dialog box is displayed. Click **Close**. The original edge from the Extruded-Base feature was deleted when the Front Cut feature was created.

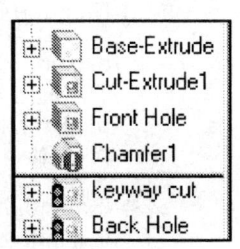

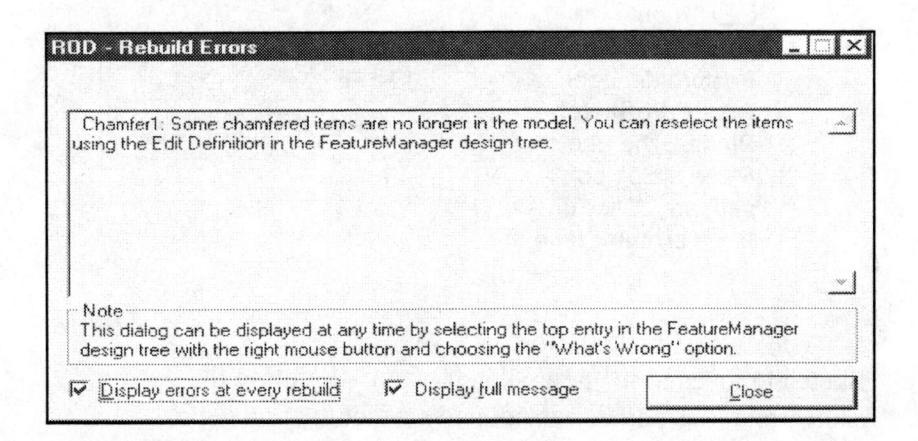

238) Create the Chamfer feature on the Front Cut edge. Right-click **Chamfer1** in the FeatureManager. Click **Edit Definition** from the Pop Up menu.

239) The Caution dialog box displays the message, "Chamfer1 is missing 1 edge". Click **OK**.

240) The Chamfer Feature PropertyManager appears. Click the **front edge** of the Base Extrude feature. Enter **1.0**mm for Distance.

241) The item appears inside the Items to Chamfer list box. Create the Chamfer feature. Click **OK** ✓ from the Chamfer Feature dialog box.

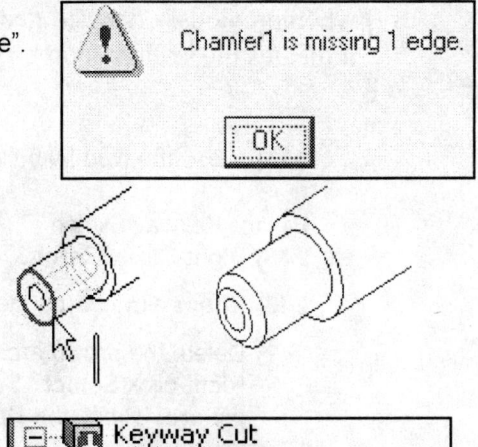

PAGE 1 - 67

Fundamentals of Part Modeling   **Engineering Design with SolidWorks**

Redefine the face for the Keyway Sketch Plane.

242) Drag the **Rollback** bar downward below the Keyway Cut feature. A red explanation point is displayed next to the name of the Keyway Cut feature. The Rod Rebuild Error box is displayed. Redefine the sketch plane and sketch relations to correct the four rebuild errors. Click **Close** from the Rod Rebuild Error Box.

243) Expand the Keyway Cut in the FeatureManager. Click the **Plus Sign**. Right-click **Sketch3** in the FeatureManager. Click **Edit Sketch Plane**. The Sketching Plane dialog box appears. Click the **large circular face** of the Front-Cut. Click **OK**.

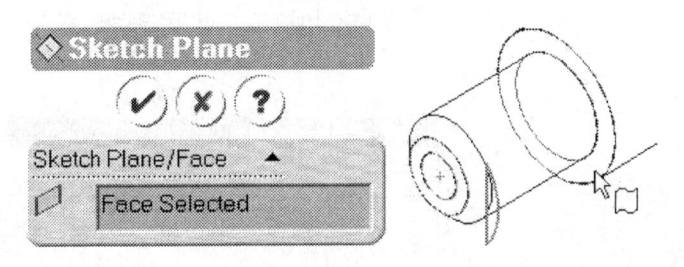

Sketch3 moves to the new Sketch plane. Redefine the sketch relations and dimensions to correct the two build errors.

Dangling sketch entities result from a dimension or relation that is unresolved. An entity become dangling when previously defined geometry is deleted. Brown indicates the sketch geometry is dangling.

244) Close the Rod Rebuild Error box. Click **Close**.

Edit the Keyway Sketch.
245) Right-Click **Sketch3**. Click **Edit Sketch**.

246) Zoom window on the Keyway-Cut.

247) Delete the brown arc. Right-click **Select**. Select the **arc**. Press the **Delete** key.

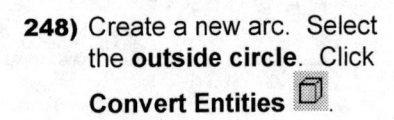

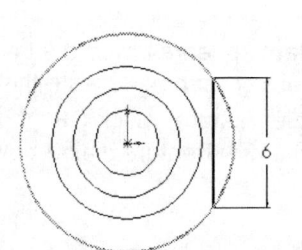

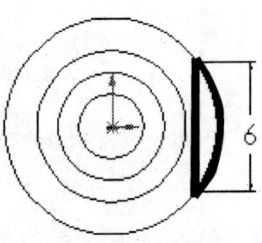

248) Create a new arc. Select the **outside circle**. Click **Convert Entities**.

PAGE 1 - 68

Engineering Design with SolidWorks   **Fundamentals of Part Modeling**

249) Delete unwanted geometry. Click **Trim** from the Sketch Tools toolbar. Click the **left outside edge** of the circle. The end points of the vertical line are coincident with the outside edge of the circle.

250) Close the Sketch. Click **Sketch**. The Keyway Cut is rebuilt.

251) Drag the **Rollback** bar downward to the bottom of the FeatureManager. All feature icons in the FeatureManager design tree are displayed in yellow. Yellow indicates dangling dimensions or relations.

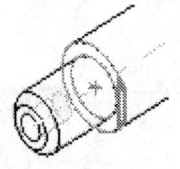

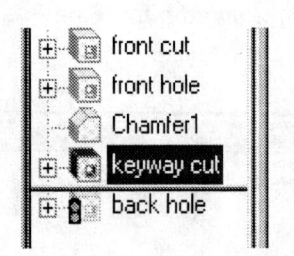

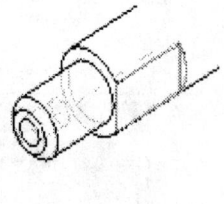

252) Display the Isometric view. Click **Isometric**.

253) Modify the depth of the back hole. Double-click **Back Hole** from the FeatureManager. Double-click Depth, 10 in the Graphics window. Enter **20** for Depth.

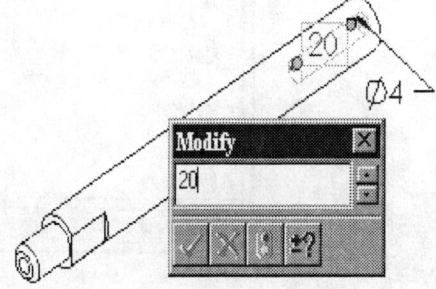

254) A Rebuild icon is displayed before the Back Hole feature. Rebuild all features. Click **Rebuild** from the Main toolbar.

255) Save the ROD. Click **Save**.

Create the ROD - Edit the Part Color

Parts are shaded tan by system default. Modify the system default color.

256) Click the **ROD Part** icon on the top of the FeatureManager design tree.

257) Click **Edit Color**. Basic colors are displayed as small squares, called swatches.

258) Click a **color swatch** from the Basic color palette.

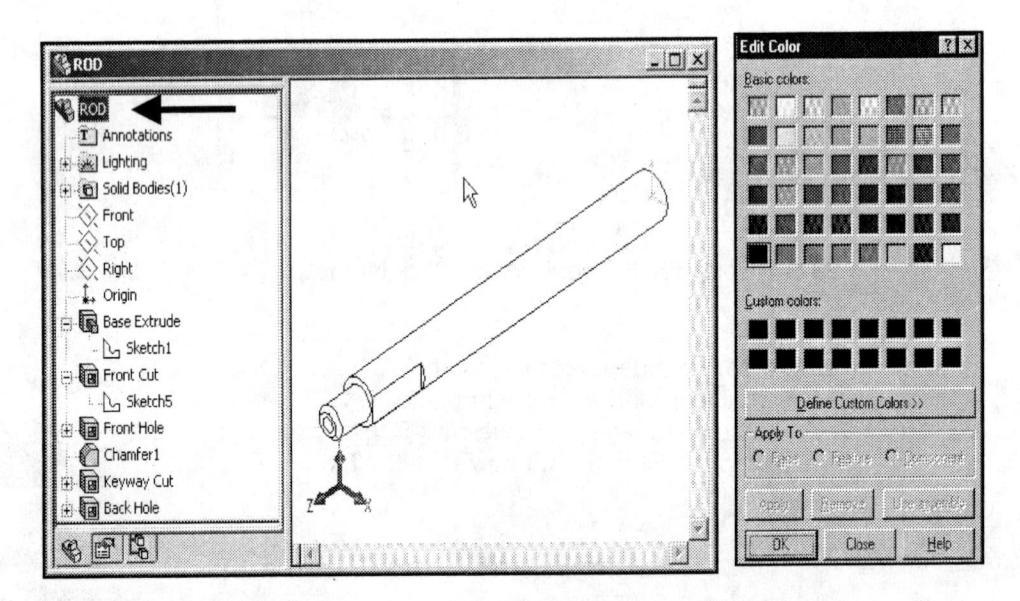

259) Set the color for the ROD. Click **OK** from the Edit Color dialog box.

260) Display the color for the ROD. Click **Shade** from the View toolbar.

261) Save the ROD. Click **Save**.

262) Close the ROD. Click **File**, **Close**.

GUIDE Part

The GUIDE part supports the ROD. The ROD moves linearly through the GUIDE.

Add slot cuts to the GUIDE to allow for positioning flexibility during field installation.

The GUIDE supports a small sensor that will be mounted on the angled right side. You do not have the information on the exact location of the sensor. The information will be provided to you during the field installation.

Create a pattern of holes on the angled right side of the GUIDE to address this issue.

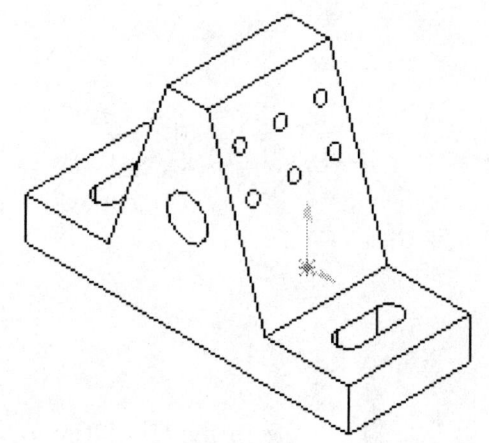

GUIDE Feature Overview

Identify the required GUIDE features:

Extruded Base: Create a symmetrical sketched profile with the Extruded Base feature, Figure 1.15a.

Extruded Cut: The Extruded Cut feature is created with sketching lines and arcs on the top face to form the slot, Figure 1.15b.

Mirror: Use the Mirror feature to create a second slot about the Right plane, Figure 1.15c.

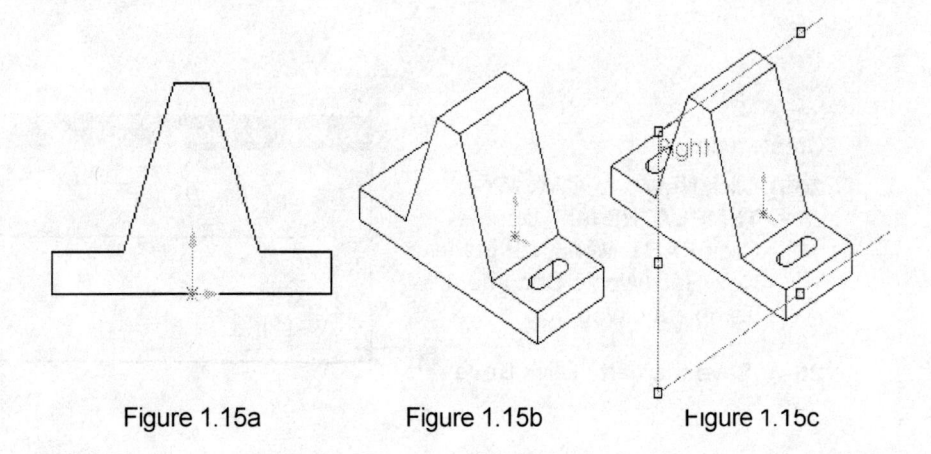

Figure 1.15a   Figure 1.15b   Figure 1.15c

Extruded Cut: Use the Extruded Cut feature to create the Guide Hole. The ROD glides through the Guide Hole, Figure 1.15d.

Fundamentals of Part Modeling   **Engineering Design with SolidWorks**

HoleWizard and Linear Pattern: Use the HoleWizard to create a M3 tapped hole. Use the Linear Pattern to create multiple instances of the M3 tapped hole, Figure 1.15e.

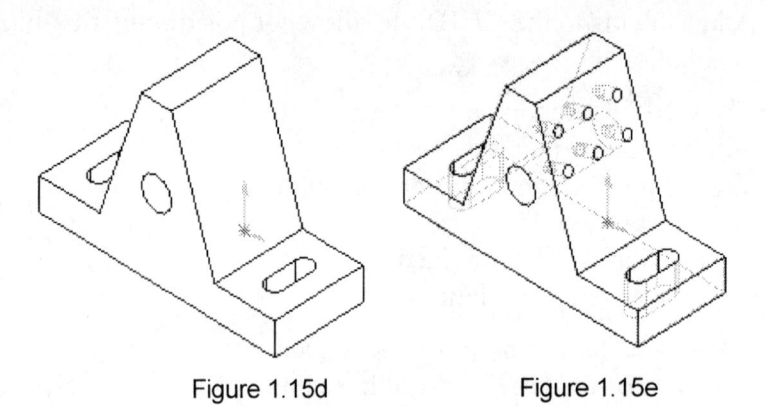

Figure 1.15d   Figure 1.15e

Let's create the GUIDE part.

Create the GUIDE Part

The geometry of the Base feature is a symmetric extrusion. The Extruded Base feature is the foundation for the GUIDE.

How do you sketch a symmetrical 2D profile? Answer: Use a sketched centerline and the Mirror Sketch tool.

What is the Sketch plane? Answer: The Front plane is the Sketch plane.

Create a new part.

263) Click **New**. Click **MY-TEMPLATES** tab. Double-Click **PART-MM-ANSI** in the New SolidWorks Document Template dialog box.

264) Save the part. Click **Save**.

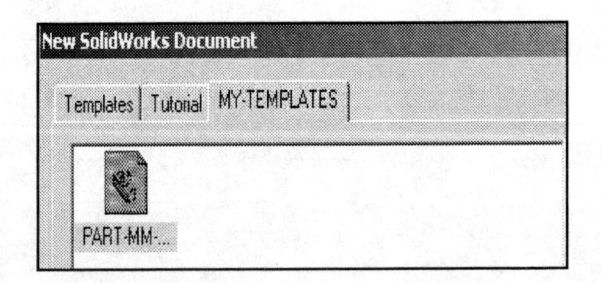

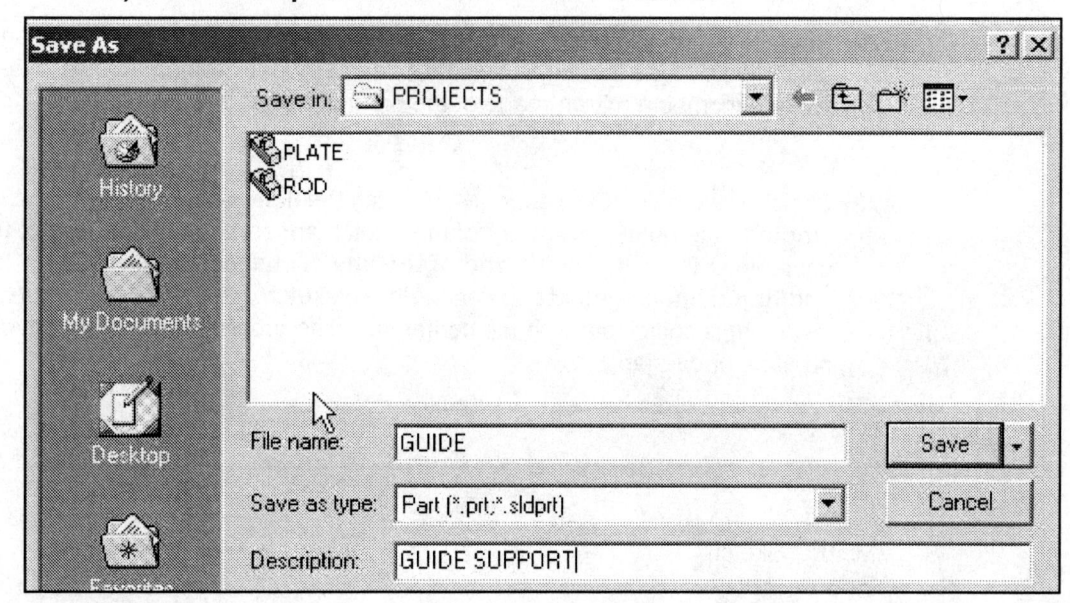

Engineering Design with SolidWorks **Fundamentals of Part Modeling**

265) Click the **drop down arrow** in the Save in list box.

266) Select ENGDESIGN-W-SOLIDWORKS\PROJECTS.

267) Enter **GUIDE** for the File name.

268) Enter **GUIDE SUPPORT** for the Description. Click **Save**.

Create the Sketch.

269) Select the Sketch plane. The Front plane is the default Sketch plane. Click the **Front** plane from the FeatureManager. Open the Sketch. Click **Sketch** from the Sketch toolbar.

270) Sketch a centerline. A centerline is a line just like a regular sketch line except it has a property that makes it exempt from the normal rules that govern sketches. When the system validates a sketch, it does not include centerlines when determining if the contour is disjoint or self-intersecting. Click **Centerline** from the Sketch Tools toolbar. Create the first point. Click the **Origin**. Create the second point. Drag the **mouse pointer** vertically upward. Click the **mouse button**.

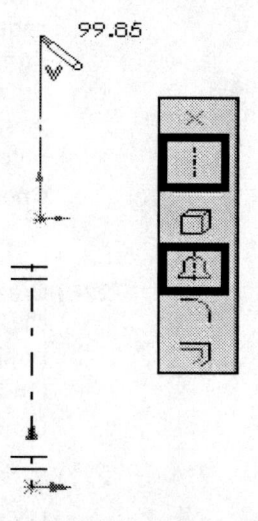

PAGE 1 - 73

Fundamentals of Part Modeling **Engineering Design with SolidWorks**

271) Activate the Mirror Sketch Tool. Click **Sketch Mirror** from the Sketch Tools toolbar. Two mirror marks appear on the centerline.

272) Note: Mirroring involves creating a centerline and using the Mirror option. The centerline acts as the mirror axis that geometry gets copied across. The copied entity becomes a mirror image of the original across the centerline.

273) Sketch the profile. Click **Line**. Sketch the first horizontal line from the **Origin** to the right. The mirror of the Sketch appears on the left side of the centerline. Double click the **end of the line** to end the Sketch. Sketch the **four additional line segments** to complete the sketch. See the below figure. The last point is coincident with the centerline. The profile is continuous; there are no gaps or overlaps.

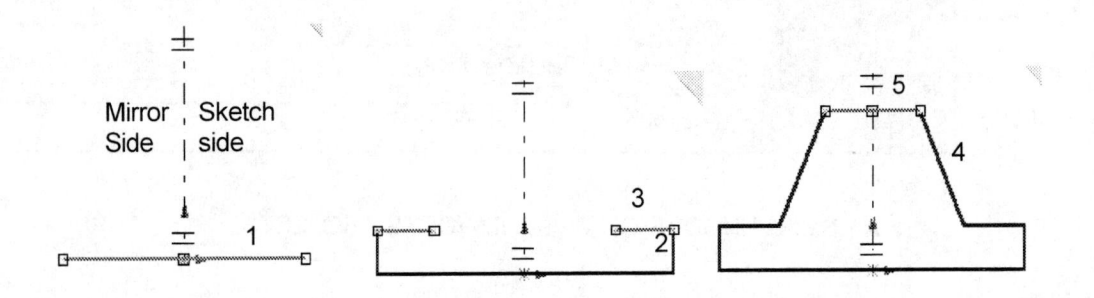

274) Deactivate the Mirror Sketch Tool. Click **Sketch Mirror**.

275) Dimension the Sketch. Click **Dimension**. Create an angular dimension. Select the **right horizontal line**. Select the **right-angled line**. Click a **position** between the two lines. Enter **110**. Click the **Green Check Mark**.

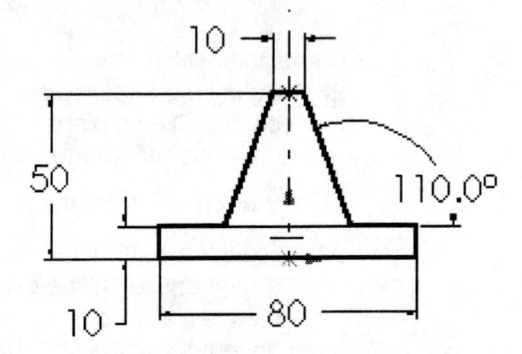

276) Create two vertical dimensions. Click the **bottom horizontal line**. Click the **top horizontal line**. Click a **position** between the two lines. Enter **50**. Click the **Green Check Mark**. Click the small **vertical line**. Click a **position** between the two lines. Enter **10**. Click the **Green Check Mark**.

277) Create two linear horizontal dimensions. Click the **bottom horizontal line**. Click a **position** below the profile. Enter **80**. Click the **Green Check Mark**. Click the **top horizontal line**. Click a **position** below the profile. Enter **10**. Click the **Green Check Mark**.

Engineering Design with SolidWorks   Fundamentals of Part Modeling

Note: Select edges instead of points to create linear dimensions. Points are removed when Fillet and Chamfer features are added.

Extrude the Sketch.

278) Click **Extruded Boss/Base** from the Features toolbar. Blind is the default Type option. Enter **30** in the depth text box. Display the Extruded-Base feature. Click **OK**.

279) Rename Extrude1 to Base Extrude.

280) Save the GUIDE. Click **Save**.

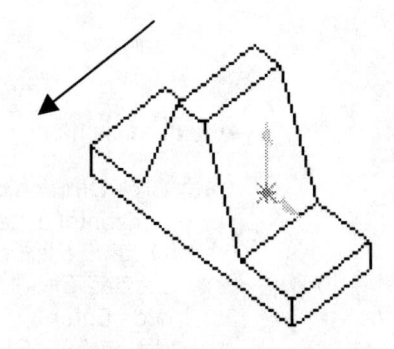

Create the GUIDE – Use The Extruded Cut Feature

281) Select the Sketch plane. Click the **top right face** of the GUIDE.

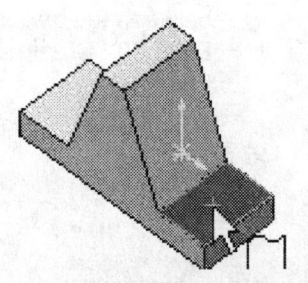

Create a Sketch.

282) Click **Sketch** from the Sketch toolbar.

283) Click **Top**. The top face of the GUIDE is displayed.

284) The 2D profile of the slot is sketched with two lines and two 180-degree Tangent Arcs. Click **Line**.

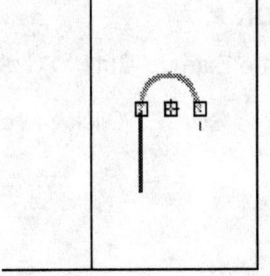

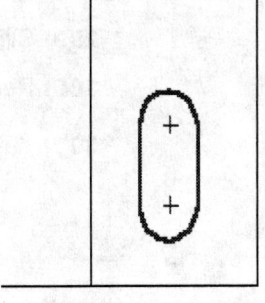

285) Sketch a short **vertical line**.

286) Click **Tangent Arc**.

287) Click the **endpoint** of the vertical line.

288) Drag the **arc** upward and to the right. **Create** a 180° arc. The three arc points are aligned horizontally. View the mouse pointer for feedback.

289) Create a second vertical line. Click **Line**.

290) Sketch a short **vertical line**.

291) Click **Tangent Arc**.

Fundamentals of Part Modeling **Engineering Design with SolidWorks**

292) Click the **endpoint** of the second vertical line.

293) Drag the **arc** downward and to the left.

294) Create a 180° arc.

295) Close the Sketch. Double-click the **first point** of the first vertical line.

Add dimensions.

296) Click **Dimension**. Create a horizontal dimension. Click the **Origin**. Click a **position** above the profile. Click the **center point** of the arc. Enter **30**. Create a vertical dimension. Click the **Origin**. Click the **center point** of the arc. Click a **position** to the right of the profile. Enter **10**. Create a dimension between the **two arc center points**. Enter **10**. Create a radial dimension. Enter **3**.

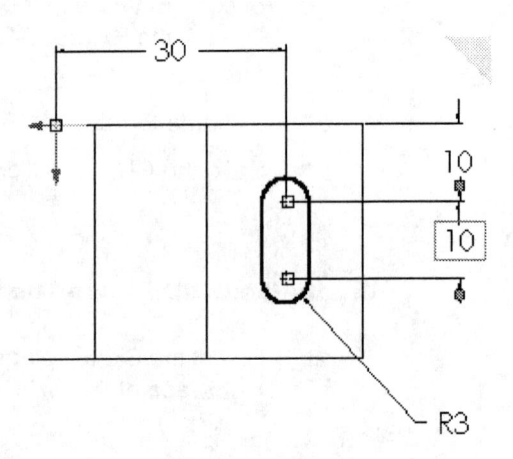

Extrude the Sketch.

297) Click **Extruded-Cut**. Blind is the default Type option.

298) Enter **Through All** for the Depth.

299) Click **OK**.

300) Rename Extrude-Cut1 to Slot Cut.

301) Save the GUIDE. Click **Save**.

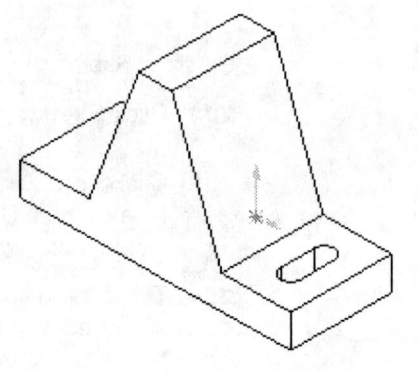

Engineering Design with SolidWorks Fundamentals of Part Modeling

Create the GUIDE – Use The Mirror Feature

The Mirror feature mirrors a select feature about a mirror plane.

Create a Mirror feature.

302) Click **Mirror** from the FeatureToolbar. Click the **Mirror Feature Property button** to display the FeatureManager. Click the **Right plane** for the Mirror plane. The Slot Cut is displayed in the Features to Mirror text box. Click the **Geometric Pattern** check box. Click **OK**. The Slot Cut is mirrored about the Right plane.

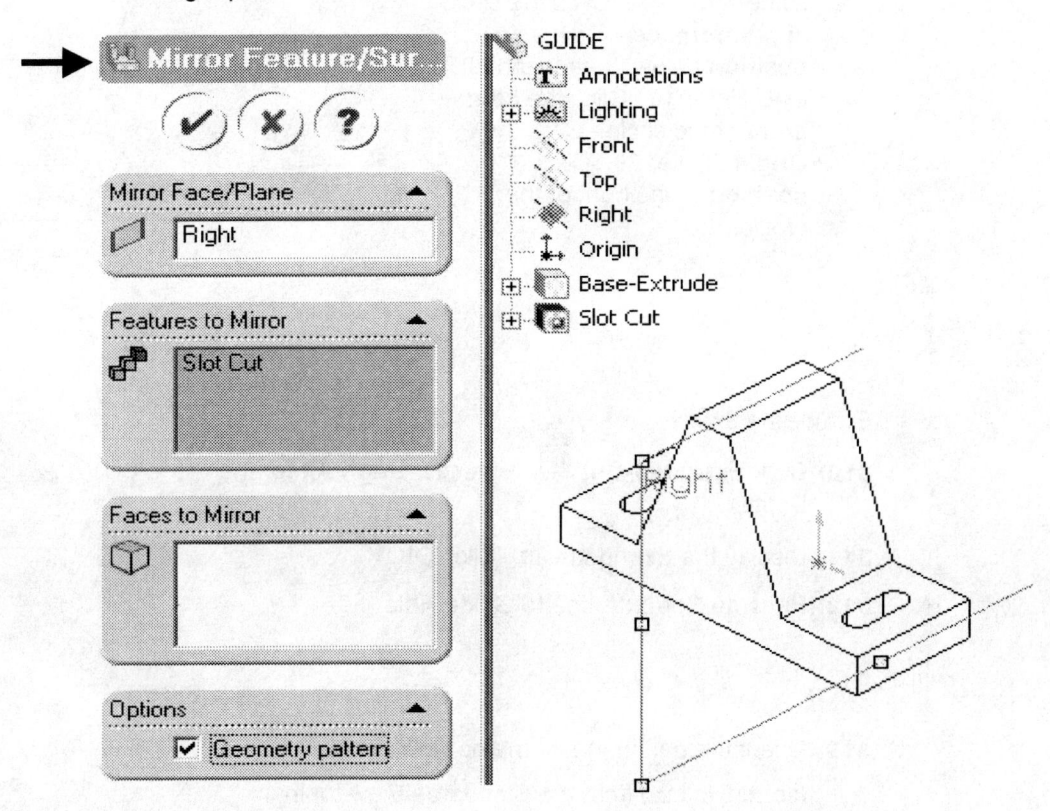

Create the GUIDE – HOLES

The Extruded-Cut feature removes material from the front of the GUIDE to create a Guide Hole. The ROD moves through the Guide Hole.

Create the tapped holes with the Hole Wizard and a Linear Pattern.

Dimension the tapped holes relative to the Guide Hole.

Create the Extruded-Cut.
303) Click **Isometric** view.

304) Select a Sketch plane. Click the **front face** of the GUIDE.

305) Click **Sketch**.

Fundamentals of Part Modeling **Engineering Design with SolidWorks**

306) Click **Normal To**. The front face of the GUIDE is displayed.

307) Click **Circle**. Sketch the **circle**.

308) Add geometric relations. Right-click **Select**. Hold the **Ctrl** key down. Click the **Origin**. Click the **center point** of the circle. Click **Vertical** from the Add Relations text box. Release the **Ctrl** key.

309) Dimension the circle. Click **Dimension**. Click the circle **circumference**. Click a **position** below the horizontal line. Enter **10**. Click the **center point** of the circle. Click the **Origin**. Enter **29**. Click a **position** to the right of the profile.

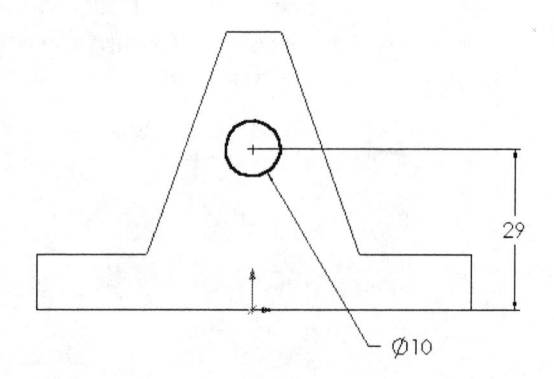

Extrude the Sketch.

310) Click **Extruded-Cut**. Enter **Through All** for the Depth.

311) Display the Extruded-Cut. Click **OK**.

312) Rename Cut-Extrude2 to Guide Hole.

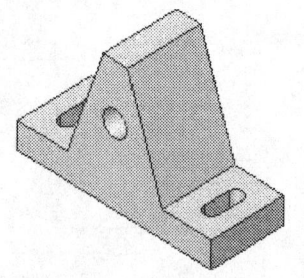

313) Orient the part in the Isometric view. Click **Isometric** from the Standards View toolbar.

314) Display the Temporary axis. Click **View**. Check **Temporary axis** from the Main menu.

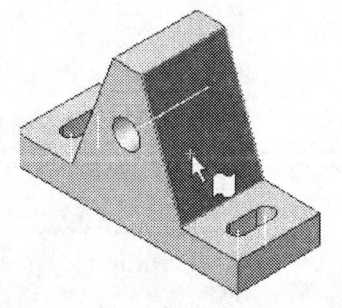

Create the Tapped Hole.

315) Select the Sketch plane. Click the **right angled face** below the Guide Hole Temporary Axis.

316) Click **HoleWizard**. The Hole Definition dialog box is displayed. Click the **Tap** tab.

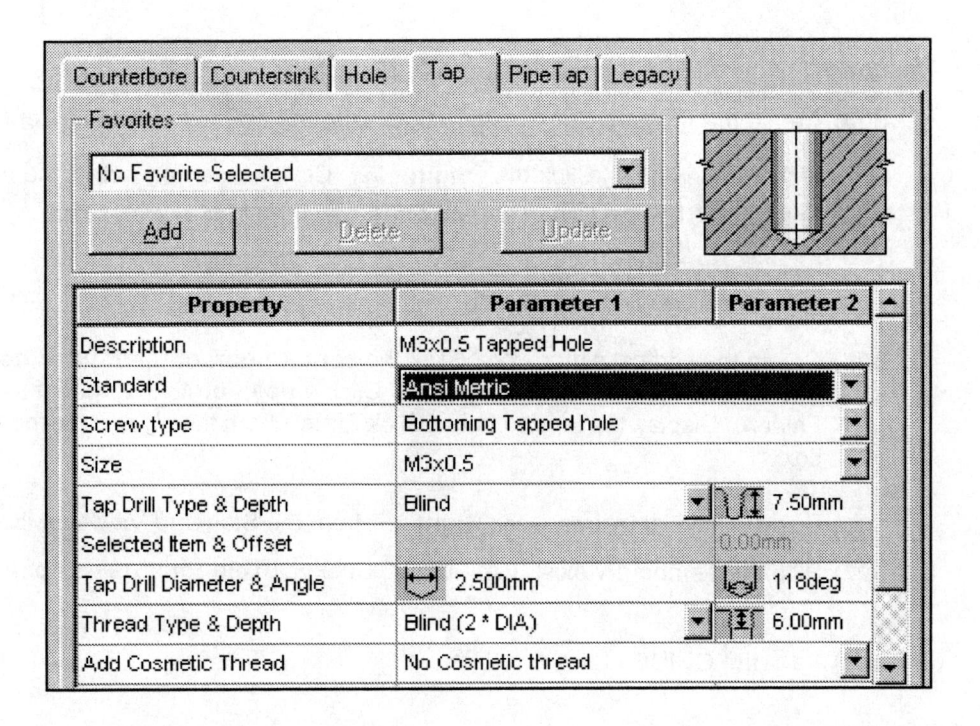

Engineering Design with SolidWorks   **Fundamentals of Part Modeling**

317) Select **ANSI Metric** for Standard.

318) Select **Bottom Tapped hole** for Screw type.

319) Select **M3x0.5** for Size.

320) Select **Blind** from the drop down list for Tap Drill Type/Depth. Accept the other default values in the table.

321) Click **Next** from the Hole Placement dialog box. The Hole Placement dialog box is displayed. Note: The orange tapped hole is displayed.

322) Do not select the Finish Button at this time. Drag the **Hole Placement dialog** box off the Graphics window to improve visibility.

323) View the Sketch plane. Click **Normal To**.

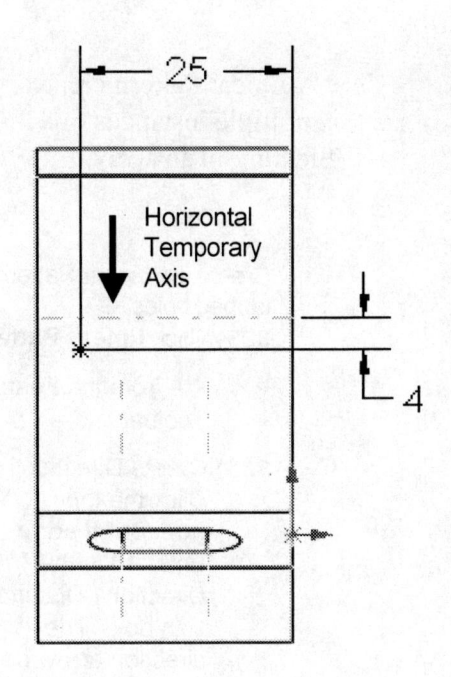

PAGE 1 - 79

Dimension the tapped hole relative to the Guide Hole.

324) Create the horizontal dimension. Click **Dimension**. Click the **center point** of the tapped hole. Click the **Origin**. Click a **position** above the profile. Enter **25**. Click the **Green Check Mark** in the Modify dialog box.

325) Create the vertical dimension. Click the **center point of the tapped hole**. Click the horizontal **Temporary axis**. Click a **position** to the right of the profile. Enter **4**. Display the tapped hole. Click **Finish** from the Hole Placement dialog box.

326) Display the Right view. Click **Right** from the Standard Views toolbar.

327) Hide the Temporary axis. Click **View**. Uncheck **Temporary axis** from the Main menu.

328) Save the GUIDE. Click **Save**.

Create the GUIDE – Use The Linear Pattern Feature

A Linear Pattern creates multiple instances of a feature in an array.

Create the Linear Pattern of tapped holes.

329) Click **Linear Pattern** from the Features Toolbar.

330) Create Direction1. Click the top **horizontal edge**. Enter **10** in the Direction1 Distance spin box. The direction arrow points to the right. Enter **3** in the Number of Instances spin box.

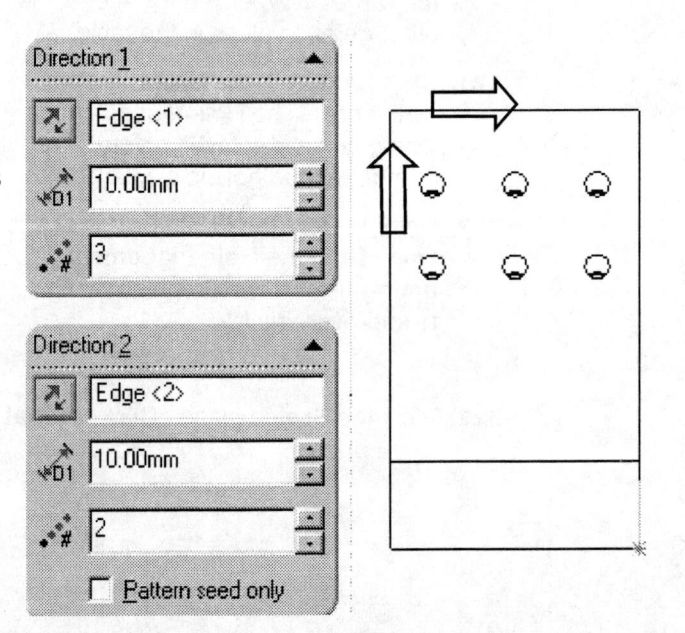

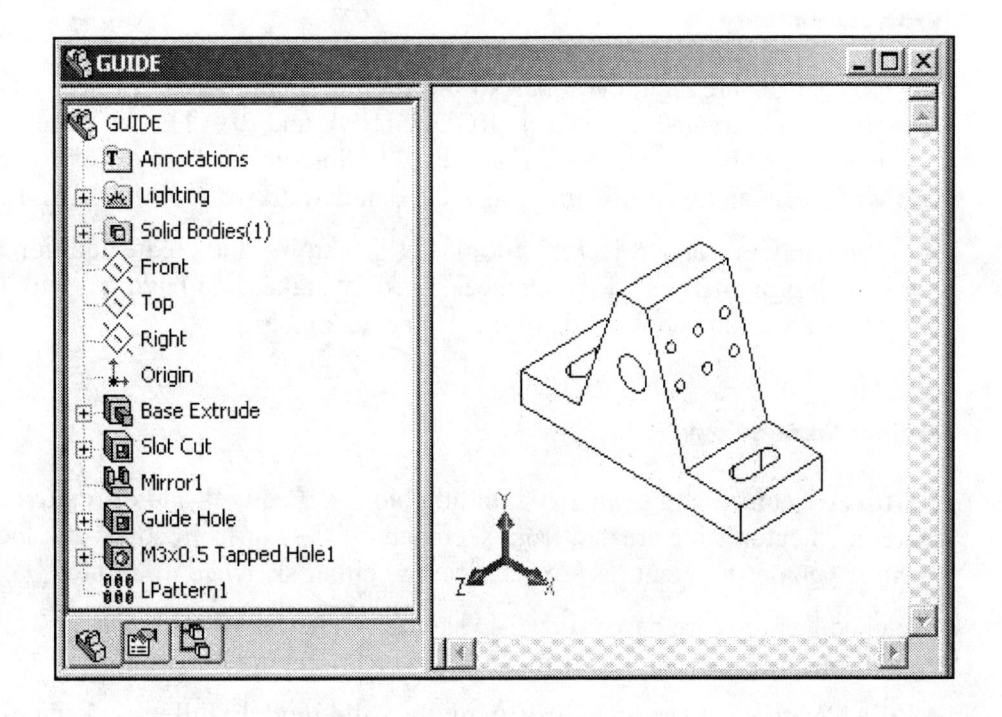

Engineering Design with SolidWorks **Fundamentals of Part Modeling**

331) Create Direction 2. Click the **left vertical edge**. The direction arrow points upward. Reverse the direction arrow if required. Click the **Reverse Direction** button on the Direction2 text box. Enter **10** for Distance in the Direction 2 Distance spin box. Enter **2** in the Number of Instances spin box.

332) Display the Pattern. Click **OK** ✔.

333) Click the **Isometric** view.

334) Save the GUIDE. Click **Save**.

Project Summary

Design changes are an integral part of the engineering process. Project 1 is complete. You created three parts: ROD, GUIDE and PLATE. In Project 2, you will incorporate the ROD, GUIDE and PLATE into an assembly. In Project 3, you will create an assembly drawing and a detailed drawing of the GUIDE.

Are you ready to start Project 2? Stop! Let's examine and create additional parts. Perform design changes. Take chances, make mistakes and have fun with the various features and commands in the project exercises.

Project Terminology

Features: Features are geometry building blocks. Features add or remove material. Features are created from sketched profiles or from edges and faces of existing geometry. Features are classified as either sketched or applied.

Sketched Features are based on a 2-D sketch. Generally that sketch is transformed into a solid by extrusion, rotation, sweeping or lofting.

Applied Features are created directly on the solid model. Fillets and chamfers are examples of this type of feature.

Base feature: The Base feature is the first feature that is created. The Base feature is the foundation of the part. Keep the Base feature simple. The Base feature geometry for the PLATE is an extrusion. The extrusion is named Extruded-Base feature.

Sketch: The name to describe a 2D profile is called a sketch. Sketches are created on flat faces and planes within the model. Typical geometry types are lines, arcs, rectangles, circles and ellipses.

Status of a Sketch: Three states are utilized in this Project:

Under Defined: There is inadequate definition of the sketch, (Blue).

Fully Defined: Has complete information, (Black).

Over Defined: Has duplicate dimensions, (Red).

Plane: To create a sketch choose a plane. Planes are flat and infinite. They are represented on the screen with visible edges. The default reference plane for this project is Front.

Centerline: A centerline is a line. When the system validates a sketch, it does not include centerlines when determining if the contour is self-intersecting.

Sketch Mirroring: Involves creating a centerline and using the Sketch Mirror option. The centerline acts as the mirror axis that geometry gets copied across. The copied entity becomes a mirror image of the original across the centerline.

Geometric Relations: To force a behavior on a sketch element to capture the customers design intent.

Add Relations: Provide a way to connect related geometry. Some common relations in this book are concentric, tangent, coincident and collinear.

Dimensions: Dimension the sketch to display the design intent.

Rebuild: After you make changes to the dimensions, you must rebuild the model to cause those changes to take affect.

Menus: Menus provide access to the commands that the SolidWorks software offers.

Toolbars: The toolbar menus provide shortcuts enabling you to quickly access the most frequently used commands.

Mouse Buttons: The left and right mouse buttons have distinct meanings in SolidWorks.

System Feedback: Feedback is provided by a symbol attached to the cursor arrow indicating your selection. As the cursor floats across the model, feedback is provided in the form of symbols, riding next to the cursor.

Copy and Paste: Simple sketched features and some applied features can be copied and then pasted onto a planar face. Multi-sketch features such as sweeps and lofts cannot be copied.

Project Features:

Extruded Boss: Use to add material by extrusions.

Extruded Cut: Use to remove material from a solid. This is the opposite of the boss. Cuts begin as a 2D sketch and remove materials by extrusions.

Fillet: Removes sharp edges of the PLATE. Fillets are generally added to the solids, not the sketch. By the nature of the faces adjacent to the selected edge, the system knows whether to create a round (removing materials) or a fillet (adding material). Some general filleting rules:

Leave cosmetic fillets until the end.

Create multiple fillets that will have the same radius in the same command.

When you need fillets of different radii, you should generally make the larger fillets first.

Fillet order is important. Fillets create faces and edges that are used to generate more fillets.

HoleWizard:. The HoleWizard is used to create specialized holes in a solid. It can create simple, tapped, counterbored and countersunk holes using a step by step procedure. Use the HoleWizard to create a countersink hole at the center of the PLATE. Use the HoleWizard to create tapped holes in the GUIDE.

Questions

1. Identify at least four key design areas that you should investigate before starting a project.

2. Why is file management important?

3. Identify the steps in starting a SolidWorks session.

4. Describe the default reference planes.

5. What is the Base feature? Provide an example.

6. Why should the Base feature be kept simple?

7. How do you recover from Rebuild errors?

8. When do you use the HoleWizard feature?

9. Describe the difference between an Extruded Base feature and an Extruded Cut feature.

10. Describe a Fillet feature. Provide an example.

11. Name the command keys that are used to Copy sketched geometry.

12. Describe the Edit feature. Provide an example.

13. Describe a Chamfer feature. Provide an example.

14. Describe the Rollback function. Provide an example.

15. Identify the type of Geometric Relations that can be added to a Sketch.

16. Describe the Move/Size feature. Provide an example.

17. Describe the procedure in creating a part document template.

Exercises

Exercise 1.1 and Exercise 1.2

A fast and economical way to manufacture parts is to utilize Steel Stock, Aluminum Stock, Cast Sections, Plastic Rod and Bars, and other stock materials. Stock materials are available in different profiles and different dimensions from tool supply companies. Create a new part. Each part contains one Extrude feature. Select Units. Utilize the Front plane for the Sketch Plane.

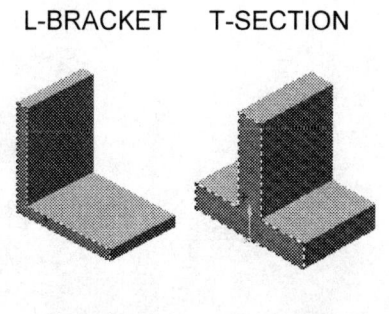

L-BRACKET   T-SECTION

Exercise 1.1   Exercise 1.2

Exercise 1.1: L-BRACKET

Ex.:	A:	B:	C:	LENGTH:	Units:
1.1a	3	3	5/8	10	IN
1.1b	4	4	3/4	10	
1.1c	6	4	1/2	25	
1.1d	8	6	3/4	25	
1.1e	12	12	3/2	25	
1.1f	7	4	3/4	30	
1.1g	75	75	10	250	MM
1.1h	75	100	10	250	
1.1i	200	100	25	500	

Exercise 1.2: T-SECTION

Ex.:	A:	B:	C:	LENGTH:	Units:
1.2a	3	3	5/8	30	IN
1.2b	4	4	3/4	30	
1.2c	6	6	1	30	
1.2d	75	75	10	250	MM
1.2e	75	100	10	250	
1.2f	200	100	25	500	

Parts manufactured from machine stock can save design time and money. Machine stock is purchased from material supply companies.

Machine stock sizes vary from supplier. Utilize the World Wide Web and determine common sizes for other machine stock. Example: 1/4in Aluminum Rod or 10mm Steel Plate.

Part information courtesy of the Reid Tool Supply Co. Muskegon, MI USA (www.reidtool.com).

Exercise 1.3: Joining Plate. Use symmetry and geometric relations to create the 6mm extruded plate. The 8.3mm thru holes are spaced 40mm apart.

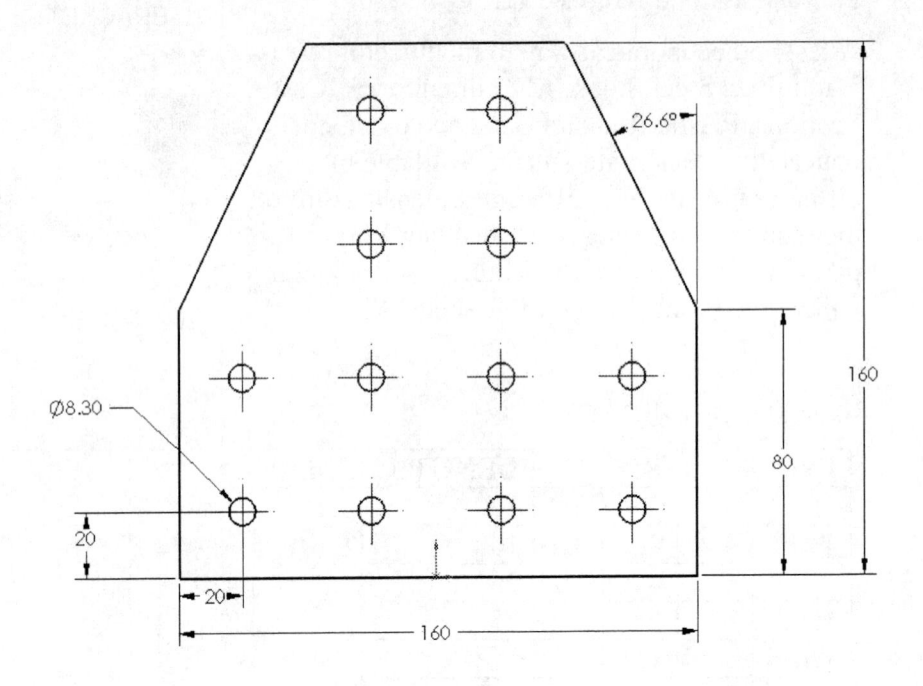

The company, 80/20 Inc., manufactures modular aluminum structural extrusions to reduce fabrication time and cost. 80/20 manufacturers both Metric and English products.

Part courtesy of 80/20, Inc. Columbia City, IN USA (www.8020.net) and its distributor, Air Inc. Franklin, MA USA (www.airinc1.com)

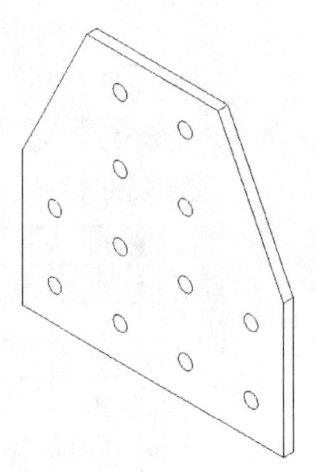

Exercise 1.4a: Part document templates. Create a Metric part document template using an ISO dimension standard.

Exercise 1.4b: Part document templates. Create an English part document template using an ANSI dimension standard. Set Drawing/Units to the appropriate values for each template.

Exercise 1.5: Linkage Assembly.

Develop the mechanical parts required for a Linkage Assembly. The Linkage assembly contains four machined parts:

FLAT BAR - 3 HOLE.

FLAT BAR - 9 HOLE.

AXLE.

SHAFT COLLAR.

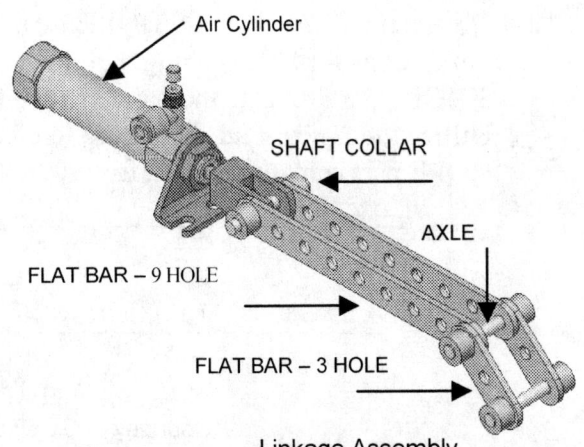

Linkage Assembly
Courtesy of
Gears Education Systems, LLC and
SMC Corporation of America.

The pneumatic air cylinder is utilized to translate the Linkage Assembly. The pneumatic air cylinder is a purchased part manufactured by SMC Corporation of America. Develop four machined parts. Provide the primary units in inches. Provide the secondary units in [millimeters].

Exercise 1.5a: FLAT BAR - 3 HOLE.

Create the FLAT BAR - 3 HOLE part. Utilize the Front Plane for the Sketch Plane. Utilize a Linear Pattern for the three holes. The FLAT BAR – 3 HOLE part is manufactured from 0.060in [1.5mm] Stainless Steel.

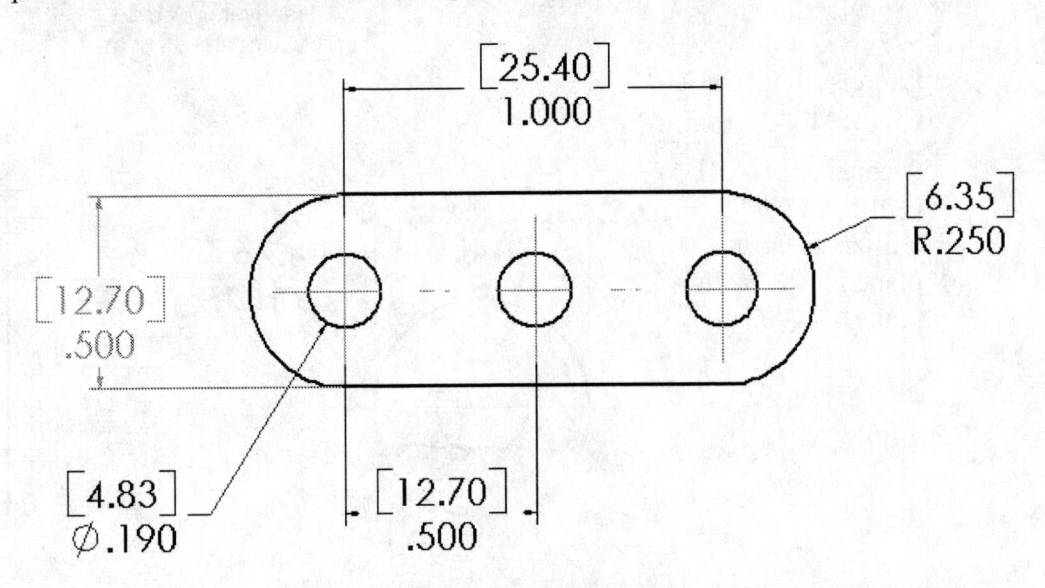

FLAT BAR, 3 HOLE
Courtesy of GEARS Educational Systems, LLC
Hanover, MA USA
www.gearseds.com

Exercise 1.5b: FLAT BAR - 9 HOLE

Design the FLAT BAR, 9 HOLE part. Manually sketch the 2D profile. The dimensions for hole spacing, height and end arcs are the same as the FLAT BAR, 3 HOLE part. Create the part. Utilize the Front Plane for the Sketch Plane. Utilize the Linear Pattern feature to create the ho e pattern. The FLAT BAR – 9 HOLE part is manufactured from 0.060in [1.5mm] Stainless Steel.

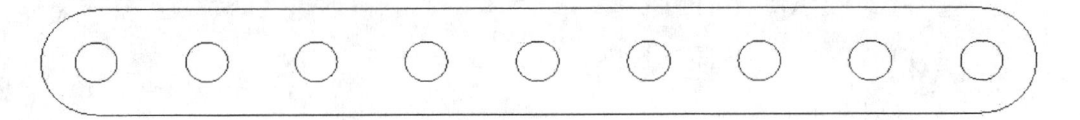

FLAT BAR, 9 HOLE
Courtesy of GEARS Educational Systems, LLC
Hanover, MA USA
www.gearseds.com

Exercise 1.5c: AXLE.

Create the AXLE part. Utilize the Front plane for the Sketch plane.

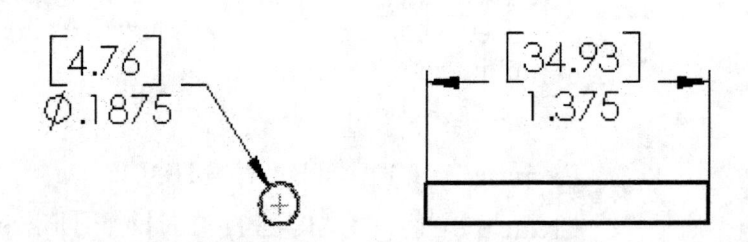

[4.76]
⌀.1875

[34.93]
1.375

AXLE
Courtesy of GEARS Educational Systems, LLC
Hanover, MA USA
www.gearseds.com

Exercise 1.5d: SHAFT COLLAR.

Create the SHAFT COLLAR part. Utilize the Front plane for the Sketch plane.

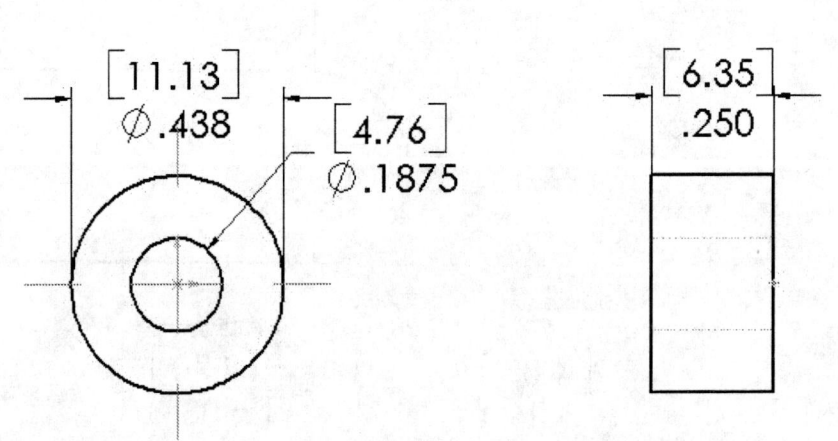

[11.13]
⌀.438

[4.76]
⌀.1875

[6.35]
.250

SHAFT COLLAR
Courtesy of GEARS Educational Systems, LLC
Hanover, MA USA
www.gearseds.com

Exercise 1.6: Industry Collaborative Exercise.

An Engineering Change Order (ECO) was issued for the PLATE. Modify the PLATE to include 4 Thru Holes.

Use the Hole Wizard and a Linear Pattern.

The 4 holes of your PLATE correspond to the 4 corner mounting holes on the Piston Plate.

The Piston Plate nominal dimensions referenced for this example are derived from SMC Corporation of America (www.smcusa.com).

Review the dimensions from the Piston Plate drawing. Identify the required dimensions to create the SolidWorks features.

Review the specification table. Identify the size mounting hole sizes for the Piston Plate for a 12mm Bore.

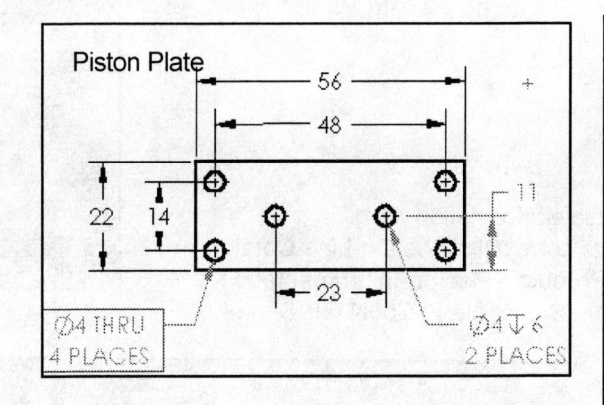

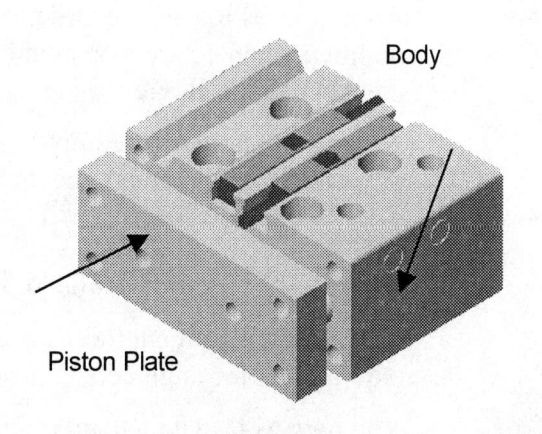

Feature		12	16
Bore Size		12 mm	16mm
Piston Rod Size		6 mm	8mm
Guide Rod Size		8 mm	10 mm
Bearing Type			
Port Size		M5 X 0.8	M5 X 0.8
Port Locations	Top	2	
	Side	2	
No of Mounting Holes	Body	4	
	Plate	4	
Mounting Thread	Body	M5 X 0.8	M5 X 0.8
	Plate	M4 X 0.7	M5 X 0.8
Locating Pin Size	Body	3mm H7	3mm H7

Fundamentals of Part Modeling **Engineering Design with SolidWorks**

An Engineering Change Order or Engineering Change Notice (ECO/ECN) is required when a part is created or modified. A blank ECO form is contained in the Appendix. Review this form.

A drawing number is required. Drawings are an integral part of the design process. You will create the drawing at the end of Project 3.

Exercise 1.7: Industry Collaborative Exercise.

On-line intelligent engineering catalogs save time. They help eliminate the requirement for engineers to design, build and test all components from scratch.

You now work with a team of engineers on a new project. The senior engineer has a question on the voltage requirements, size requirements and color scheme for a particular connector. You have been provided the directive to locate an in-line receptacle connector with the following specifications:

Component Type:	In-Line Receptacle
Color Scheme:	Red Housing-Black Adapter or Black Housing-Black Adapter
Contact Type:	Machine Gold Plated
Circuit Count:	2+2
Key Pattern:	1
Contact Size:	2#16 & 2#22

1) Obtain the Receptacle information from www.aldenproducts.com. Click the On-Line Catalog button. Click the Product Selector, Instrumentation Connectors button. Select the PL700N GP Series.

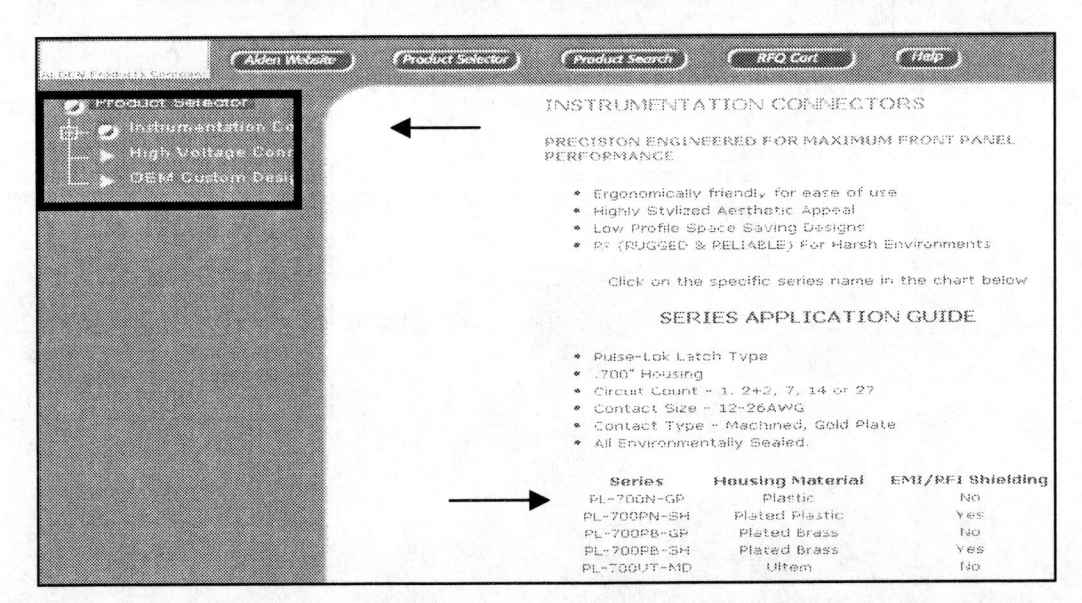

Engineering Design with SolidWorks | Fundamentals of Part Modeling

2) Over 370 part numbers are listed. The term "drill down" describes the process of finding the correct component based on your specifications. Select In-Line Receptacle for Component Type. The number of parts to select is reduced to 140.

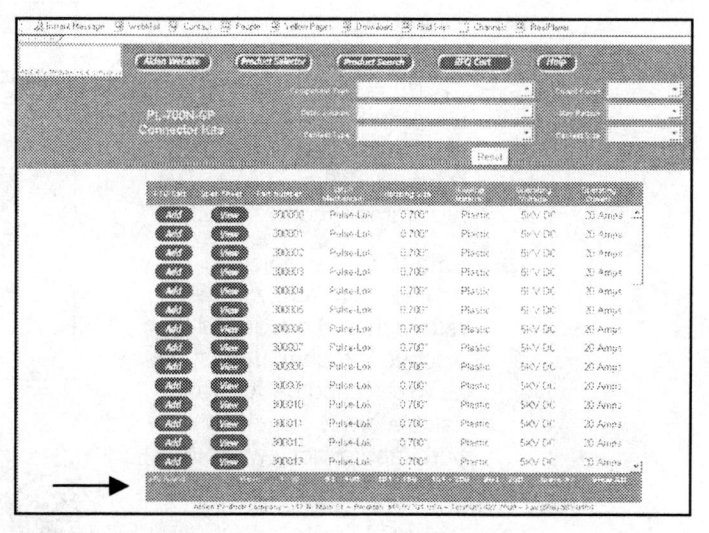

3) Select Red Housing with Black Adapter.

4) Select Machined, Gold Plated.

5) Select 2+2 for Circuit Count.

6) Select 1 for Key Pattern.

7) Select 2-#16 and 2-#22 for Contact Size. Only one part number, 300250 is listed.

8) Click the View button to display the In-line receptacle.

9) Click Layout Dimensions in the Specification column to obtain the overall size of the In-line receptacle.

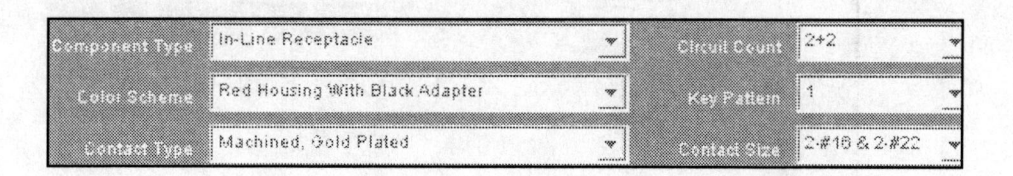

10) What is the voltage specification of this component?

11) What is the operating current specification?

12) What are the overall dimensions in millimeters?

13) Is the component available in both color schemes? What are the part numbers for the components?

PAGE 1 - 91

Fundamentals of Part Modeling **Engineering Design with SolidWorks**

14) SmartCAT®, the on line catalog content management (CCM) software utilized on the Alden Products web site was developed by i-MARK®, Hartford, CT (www.imark.com).

15) SmartCAT® enables engineers to apply in-depth research and advanced discovery techniques, such as wizard and performance calculators to specify and purchase exactly what they require.

16) SmartCAT® electronic catalogs guide engineers to the product that meets both geometric and performance characteristics.

17) The next section required Windows Media Player. Review the video that describes the latch mechanism of the connector.

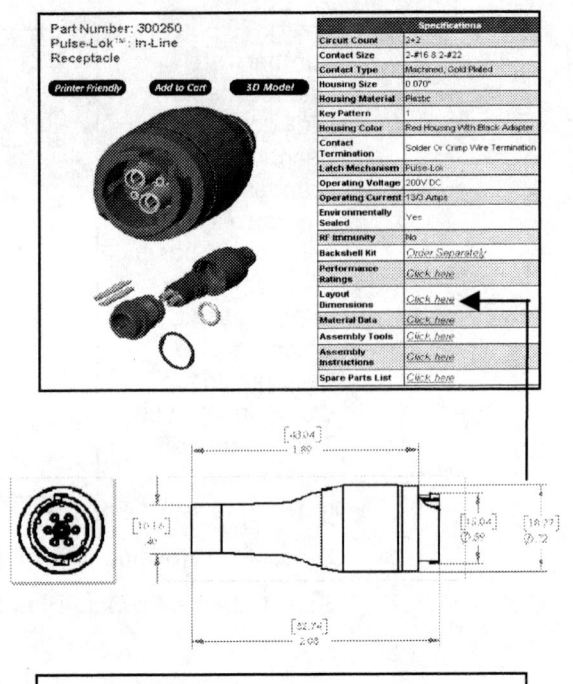

Components Courtesy of
ALDEN Products, Brockton, MA USA

Engineering Design with SolidWorks | Fundamentals of Part Modeling

18) Click the Product Selector button.

19) Expand Instrumentation Connectors. Click the + button.

20) Select PL700N-GP.

21) Click the Video Demonstration button to view the animation create with SolidWorks documents.

The video file takes a few seconds to download. The Windows Media Player opens. The movie contains animated part files from SolidWorks. Materials and voice was added to complete the demonstration.

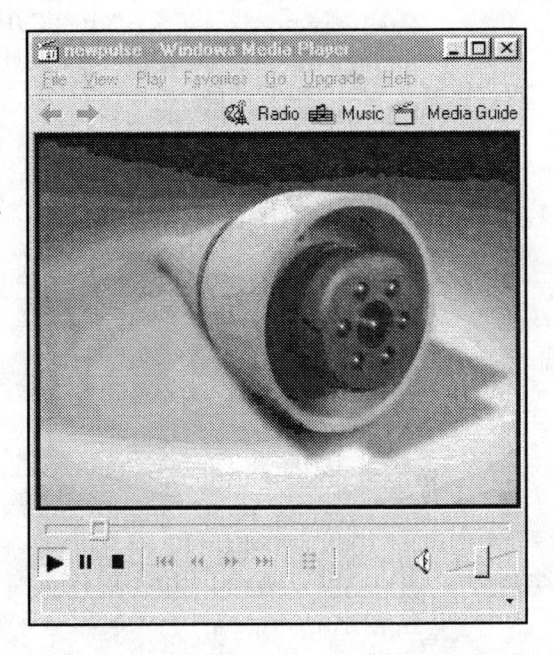

Fundamentals of Part Modeling   **Engineering Design with SolidWorks**

Exercise 1.8: Industry Collaborative Exercise.

Note: PhotoWorks Add-In is required to complete this exercise. Select Tools, Add-Ins. Select PhotoWorks. PhotoWorks toolbar and On-Line Help are added to the Main menu.

Create picture images of the GUIDE and PLATE parts. Incorporate the new images into a PowerPoint presentation.

PhotoWorks software application renders photo-realistic images of SolidWorks 2003 parts and assemblies.

PhotoWorks generates an image directly from the view in the active SolidWorks Graphics window.

Rendering effects include materials, lighting, image background, image quality and image output format. A scene contains a combination of these rendering effects.

The Render Wizard steps you through the process to apply a material to the model, select a scene to display the model and render the image to the screen.

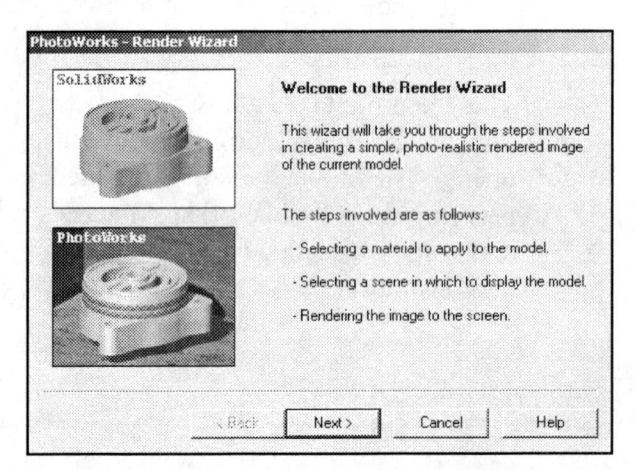

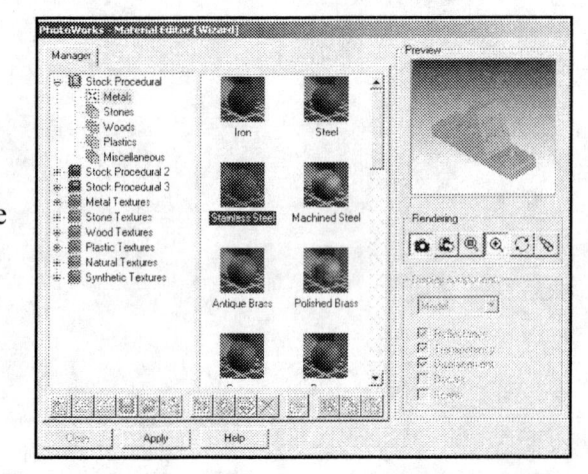

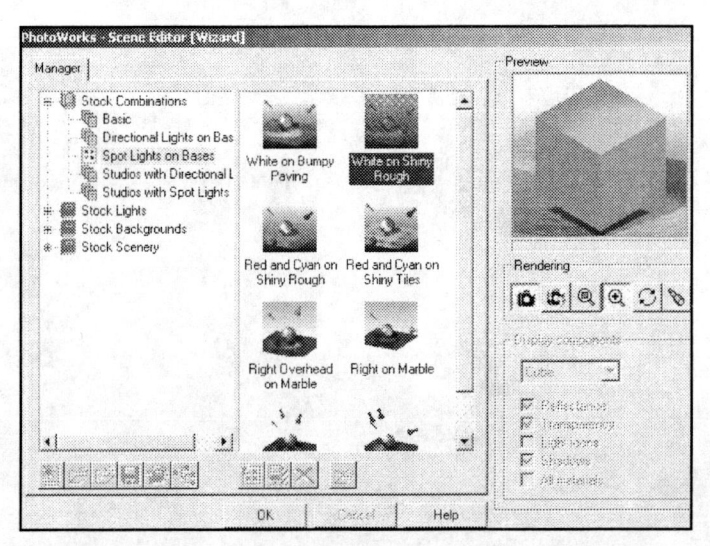

PAGE 1 - 94

Save the image in .JPG format.

Click Render Image to File.

Select JPEG for Save as Type. The default picture size is 320x240 pixels.

Insert the GUIDE.JPG file as a Picture into a new PowerPoint document.

Repeat the above procedure for the PLATE. Select a different material and scene.

Additional PhotoWorks examples are found in Help, Getting Started and Help PhotoWorks Help Topics.

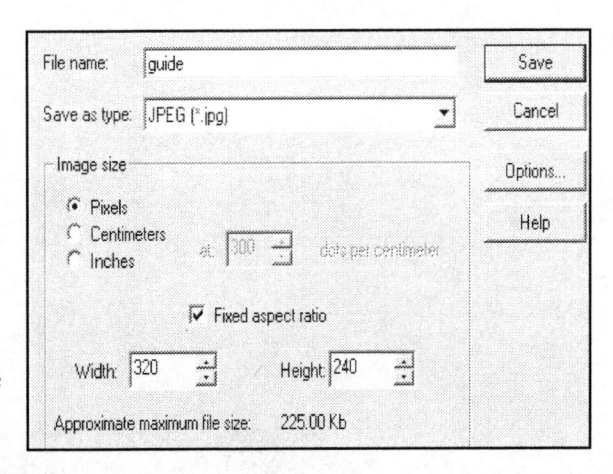

Note: Image size and file type affect file size and render time. Images below were created with 320x240 pixels and cropped. JPEG format creates a smaller file size than Bitmap for PowerPoint and web-based presentations.

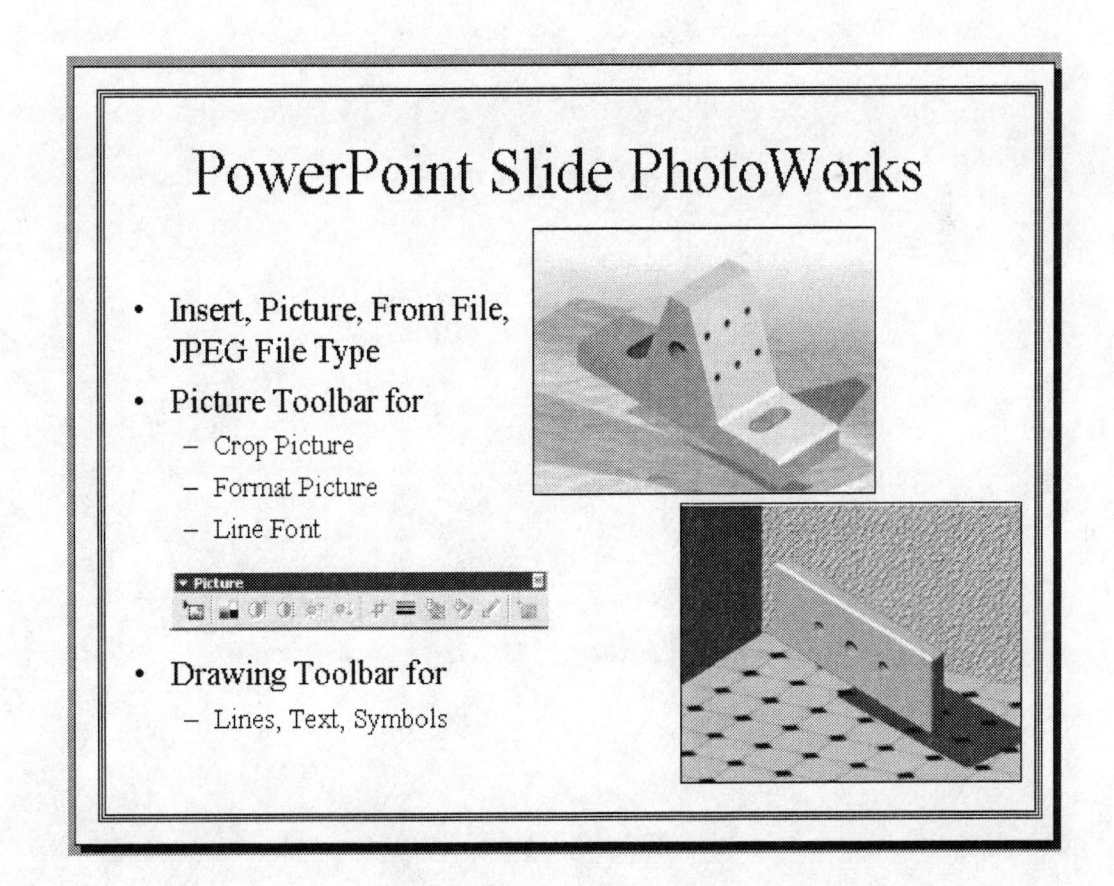

PowerPoint Slide PhotoWorks

- **Insert, Picture, From File, JPEG File Type**
- **Picture Toolbar for**
  - Crop Picture
  - Format Picture
  - Line Font
- **Drawing Toolbar for**
  - Lines, Text, Symbols

Notes:

Project 2

Fundamentals of Assembly Modeling

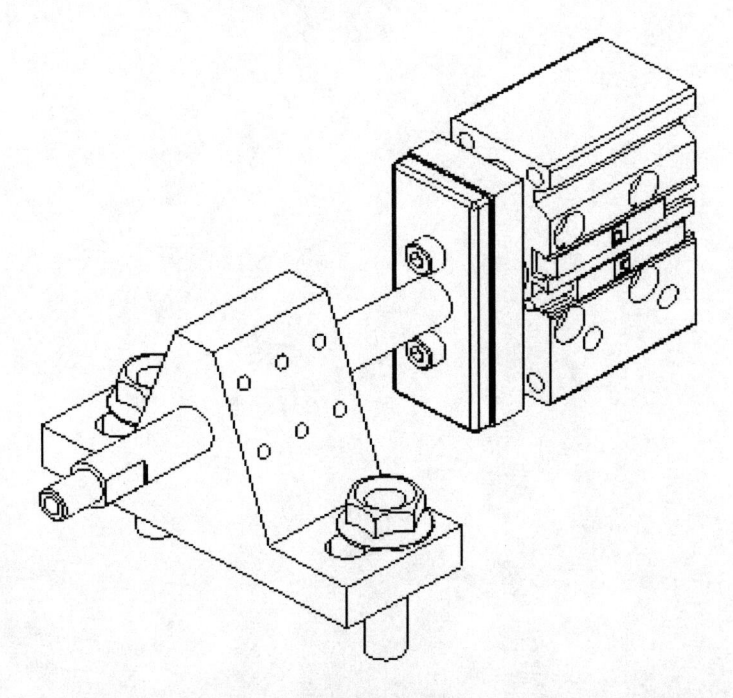

Below are the desired outcomes and usage competencies based on the completion of Project 2.

Project Desired Outcomes:	Usage Competencies:
Knowledge and experience in combining parts to create assemblies using the Bottom Up assembly modeling design technique.	Ability to insert components and create mates between components.
A comprehensive understanding of incorporating design changes into an assembly.	Ability to Modify, Edit, and Redefine assembly features.
Obtain an assembly from SMC USA.	Ability to comprehend the function of assembly creation.
Create two Assemblies: • GUIDE-ROD assembly. • CUSTOMER assembly.	Understanding of Tolerance and Fit.
Ability to use COSMOSXpress.	An introduction to Applied Finite Element Analysis.

Notes:

Project 2 – Fundamentals of Assembly Modeling

Project Objective

Create two assemblies using the Bottom-Up assembly design approach. Bottom-Up assemblies are created by adding and orienting existing parts in an assembly. Parts added to the assembly appear as components. Orient and position the components in the assembly using Mates.

Create the first assembly, GUIDE-ROD by using the ROD, GUIDE and PLATE parts. The ROD, GUIDE and PLATE parts were created in Project 1.

Create the second assembly, CUSTOMER. The CUSTOMER assembly consists of two sub-assemblies: GUIDE-ROD and GUIDE-CYLINDER. Download the GUIDE-CYLINDER assembly from SMC USA.

On completion of this project, you will be able to:

- Understand the FeatureManager Syntax.
- Insert components into an assembly.
- Set System Options.
- Rename parts.
- Create Mates.
- Incorporate design changes into an assembly.
- Add a component from the Parts Library.
- Edit component dimensions.
- Modify, Edit, and Redefine assembly features.
- Suppress features
- Address Tolerance and Fit.
- Apply COSMOSXpress.
- Use the following SolidWorks features:
    - Extruded-Cut.
    - Extruded-Boss.

- o Fillet.

- o Chamfer.

Project Situation

Perform the following steps.

Step 1: Assemble the ROD, GUIDE and PLATE into a GUIDE-ROD assembly.

Step 2: Obtain the customer's GUIDE-CYLINDER assembly from SMC USA. The assembly is obtained to insure proper fit between the GUIDE-ROD assembly and the customer's GUIDE-CYLINDER assembly.

Step 3: Create the CUSTOMER assembly. The CUSTOMER assembly combines the GUIDE-ROD assembly with the GUIDE-CYLINDER assembly.

Let's start. Review the GUIDE-ROD assembly design constraints:

- The ROD requires the ability to travel through the GUIDE.

- The ROD keyway is parallel to the right surface of the GUIDE. The top surface of the GUIDE is parallel to the work area.

- The ROD mounts to the PLATE. The GUIDE mounts to a flat work surface.

The PISTON PLATE is the front plate of the GUIDE-CYLINDER assembly. The PLATE from the GUIDE-ROD assembly mounts to the PISTON PLATE.

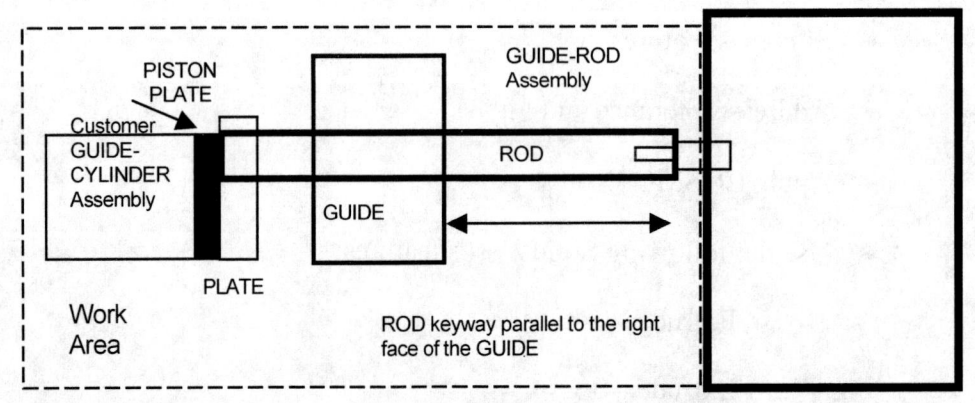

Rough Sketch of Design Situation: CUSTOMER Assembly

Create a rough sketch of the conceptual assembly.

An assembly combines two or more parts. In an assembly, parts are referred to as components.

Design constraints directly influence the assembly design process. Other considerations indirectly impact the assembly design, namely: cost, manufacturability and serviceability.

Project Overview

Translate the rough conceptual sketch into a SolidWorks assembly.

The GUIDE-ROD is the first assembly.

Determine the first component of the assembly.

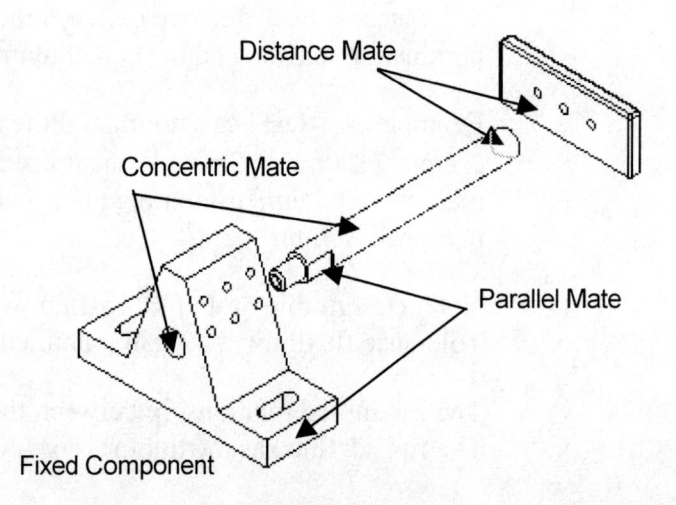

Figure 2.1

The first component is the GUIDE. The GUIDE remains stationary. The GUIDE is a fixed component.

The action of assembling components in SolidWorks is defined as Mates, Figure 2.1.

Mates are relationships between components that simulate the construction of the assembly in a manufacturing environment.

The CUSTOMER assembly combines the GUIDE-CYLINDER assembly with the GUIDE-ROD assembly, Figure 2.2.

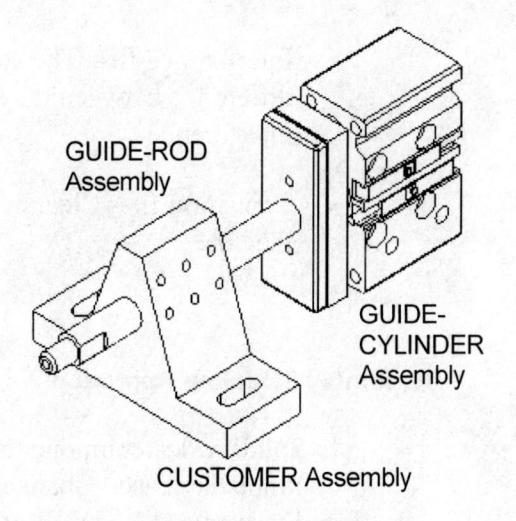

Figure 2.2

Tolerance and Fit

The ROD travels through the GUIDE in the GUIDE-ROD assembly. The shaft diameter of the ROD is 10mm. The hole diameter in the GUIDE is 10mm. A 10mm ROD cannot be inserted into a 10mm GUIDE hole without great difficulty!

Note: The 10mm dimension is the nominal dimension. The nominal dimension is approximately the size of a feature that corresponds to a common fraction or whole number.

Tolerance is the difference between the maximum and minimum variation of a nominal dimension and the actual manufactured dimension.

Example: A ROD has a nominal dimension of 100mm with a tolerance of ± 2mm, (100mm ± 2mm). This translates to a part with a possible manufactured dimension range between 98mm to 102mm. The total ROD tolerance is 4mm.

Note: Design rule of thumb: Design with the maximum permissible tolerance. Tolerance flexibility saves in manufacturing time and cost.

The assembled relationship between the ROD and the GUIDE is called the fit. The fit is defined as the tightness or looseness between two components.

This project discuss three major types of fits:

- Clearance fit - The shaft diameter is less than the hole diameter.

- Interference fit – The shaft diameter is larger than the hole diameter. The difference between the shaft diameter and the hole diameter is called interference.

- Transition fit – Clearance or interference can exist between the shaft and the hole.

Assembly Modeling Approach

In SolidWorks, components and their assembly are directly related through a common database. Changes in the components directly affect the assembly and vise a versa. SolidWorks provides two assembly-modeling techniques:

- Top Down.

- Bottom Up.

In the Top Down approach, major design requirements are translated into assemblies, sub-assemblies and components.

Note: You do not need all of the required component design details. Individual relationships are required.

Example: A computer. The inside of a computer can be divided into individual key sub-assemblies such as a: motherboard, disk drive, power supply, etc. Figure 2.3.

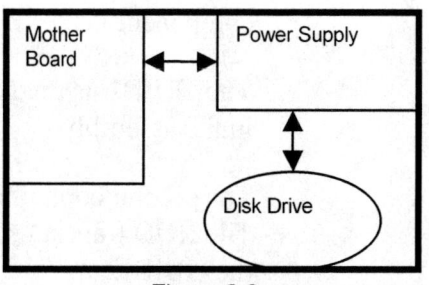

Figure 2.3

Relationships between these sub-assemblies must be maintained for proper fit.

In the Bottom Up approach, components are assembled using part dependencies and parent-child relationships. In this approach, you possess all of the required design information for the individual components.

The child is automatically modified during a parent modification. Avoid unwanted references and dependencies when establishing parent-child relationships.

Use the Bottom Up design approach for the GUIDE-ROD assembly and the CUSTOMER assembly.

Linear Motion and Rotational Motion

In dynamics, motion of an object is described in linear and rotational terms. Components possess linear motion along the x, y and z-axes and rotational motion around the x, y, and z-axes. In an assembly, each component has 6 degrees of freedom: 3 translational (linear) and 3 rotational. Mates remove degrees of freedom. All components are rigid bodies. The components do not flex or deform.

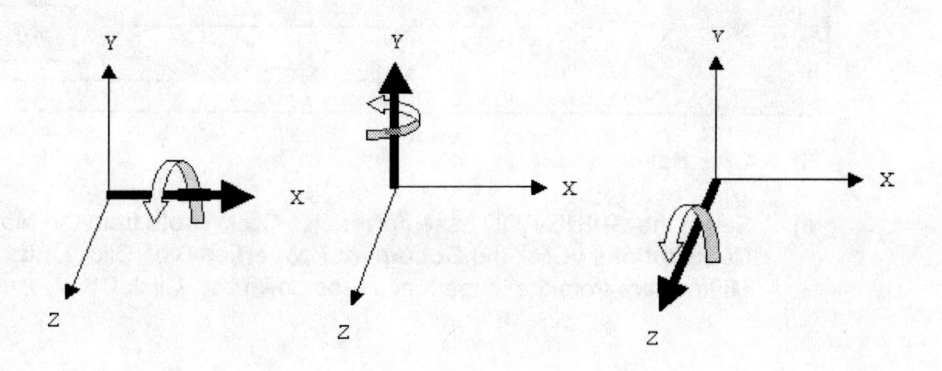

GUIDE-ROD Assembly

The GUIDE-ROD assembly consists of three components. The first component is the GUIDE.

The GUIDE is the fixed component in the assembly.

The second component is the ROD. The ROD translates linearly through the GUIDE.

The third component is the PLATE.

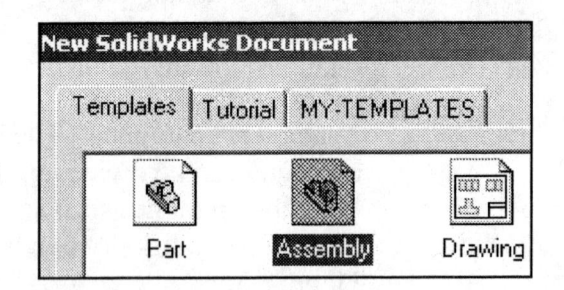

Create the GUIDE-ROD assembly.

1) Click **New** from the Standard toolbar. The New SolidWorks Documents dialog box is displayed.

2) Click **Assembly** from the Templates folder. Click **OK**. The default file name is ASSEM1.

3) Click **File** from the Main menu. Click **Save**.

4) Enter the assembly name. Enter **GUIDE-ROD**. Enter **GUIDE-ROD Assembly** for Description.

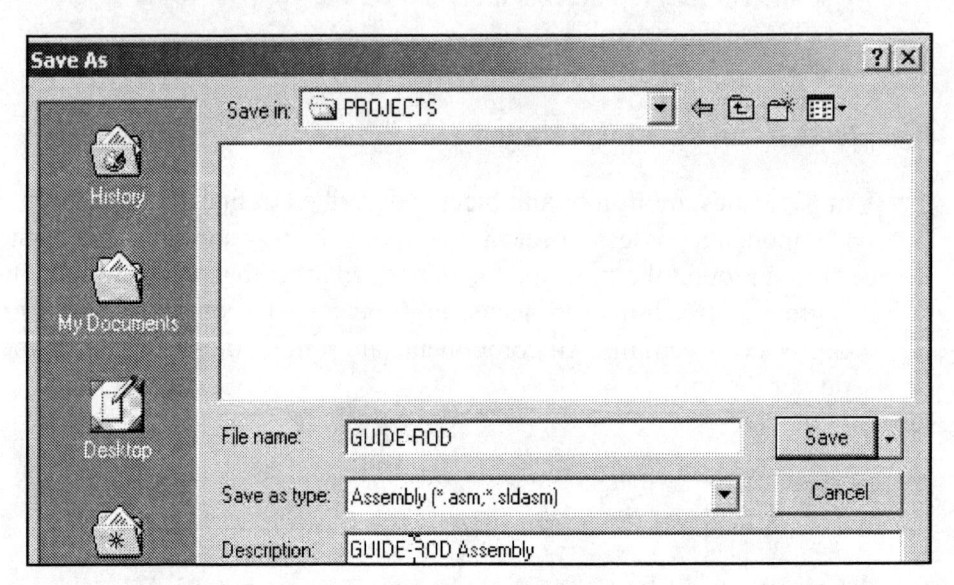

5) Click **Save**.

6) Select the GUIDE-ROD assembly units. Click **Tools** from the Main menu. Click **Options**. Click the **Document Properties** tab. Click **Units**. Click **Millimeters** from the Linear units drop down list. Click **OK**.

Engineering Design with SolidWorks   Fundamentals of Assembly Modeling

7) Open the GUIDE. Click **Open**. Double-click **GUIDE**.

8) Open the ROD. Click **Open**. Double-click **ROD**.

9) Click **Window** from the Main menu. The following components are displayed: ROD, GUIDE and GUIDE-ROD. Display the assembly and components.

10) Click **Tile Horizontally** from the Window menu. Review the windows. The Origin feature of the GUIDE-ROD assembly is displayed in the Graphics window. The FeatureManager is displayed on the left side of the three Graphics windows.

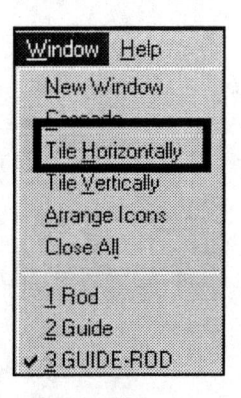

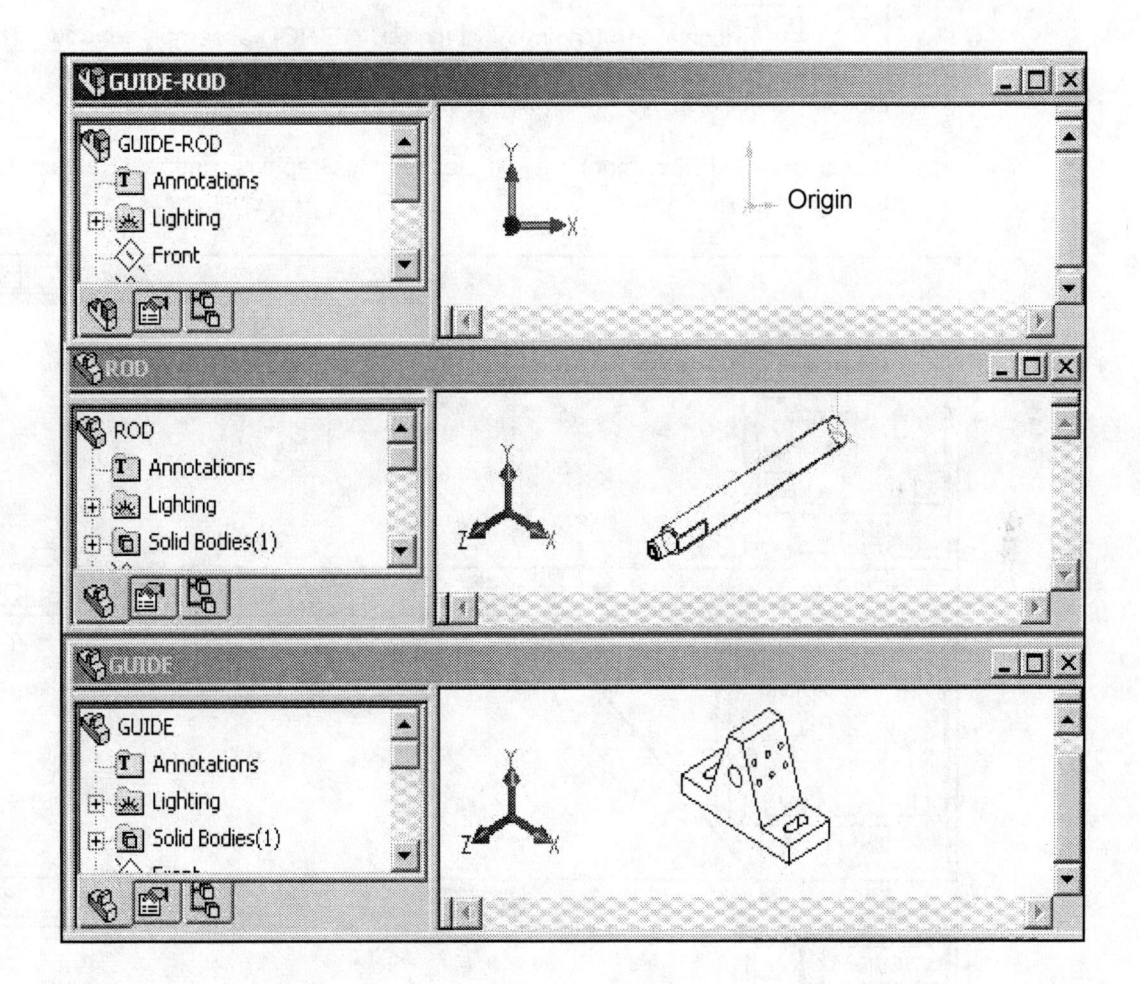

GUIDE-ROD Assembly – Insert Components

The first component is the foundation of the assembly. The GUIDE is the first component in the GUIDE-ROD assembly. The ROD is the second component in the GUID-ROD assembly. Add components to assemblies utilizing the following techniques:

Fundamentals of Assembly Modeling　　　　　　　　**Engineering Design with SolidWorks**

- Click Insert from the Main menu.
- Click Component.
- Click From File.
- Drag the components from the Windows Explorer.
- Drag the components from the Feature Palette window.
- Drag the components from the Open part files.

Add the first component. Add the GUIDE.

11) Position the GUIDE inside the assembly Graphics window. Drag the **GUIDE** GUIDE icon to the Origin of the GUIDE-ROD assembly window. The mouse pointer displays when position on the Origin.

12) Display the GUIDE component in the assembly Graphics window. Release the **mouse button**.

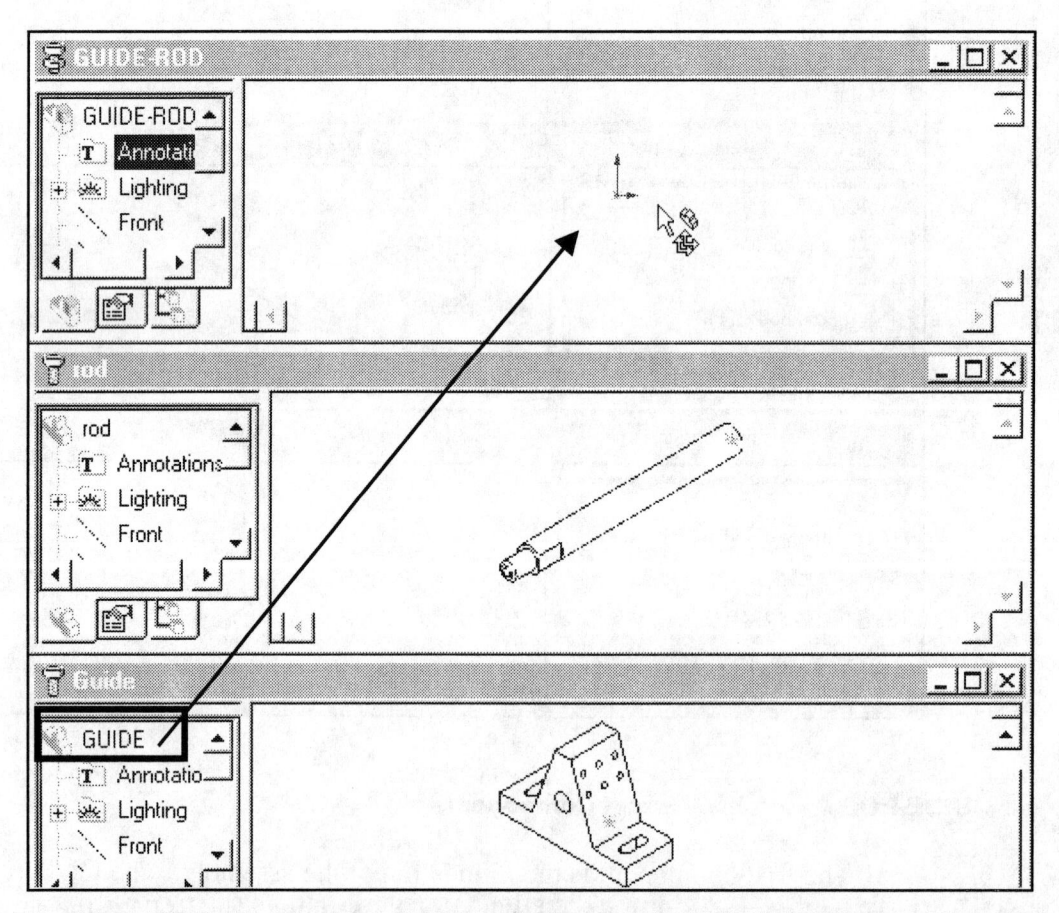

Note: The front view of the GUIDE is displayed in the GUIDE-ROD assembly. The GUIDE name is added to the GUIDE-ROD assembly FeatureManager with the symbol (f). The symbol (f) represents a fixed component.

Engineering Design with SolidWorks   **Fundamentals of Assembly Modeling**

A fixed component cannot move and is locked to the assembly Origin. To remove the fixed state, Right-click a component name in the FeatureManager. Click Float.

Add the second component. Add the ROD.

13) Drag the **ROD** 🔧 ROD icon component to the assembly window, left of the GUIDE. The mouse pointer displays ◇ when positioned inside the GUIDE-ROD Graphics window. Display the ROD component. Release the **mouse button**.

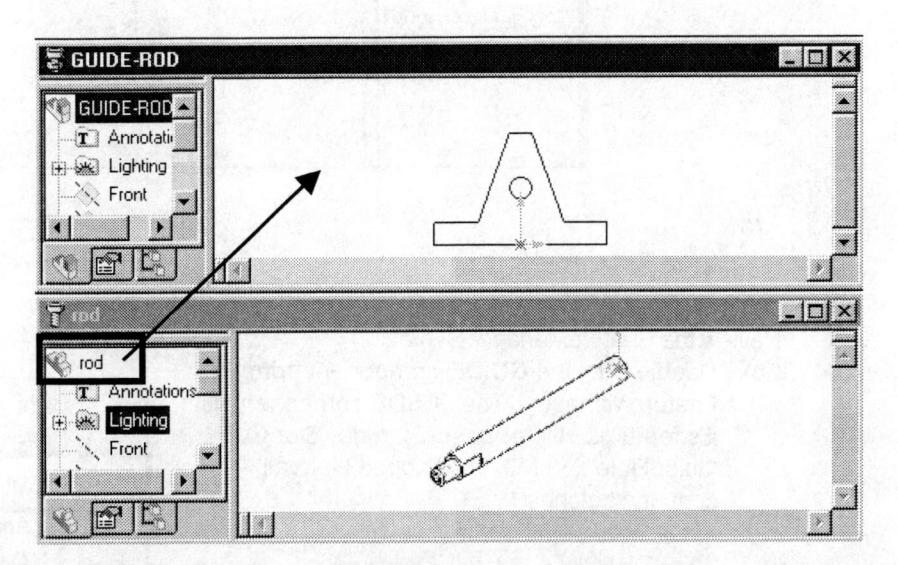

14) Close the GUIDE. Click **Close** ⊠.

15) Close the ROD. Click **Close** ⊠.

16) Display the Isometric view. Click **Isometric** 🔲 in the GUIDE-ROD window.

17) Enlarge the assembly window. Click **Maximize** 🔲 in the upper right hand corner of the GUIDE-ROD window.

18) Fit all components in the Graphics window. Press the **f** key.

19) Save GUIDE-ROD assembly. Click **Save** 💾.

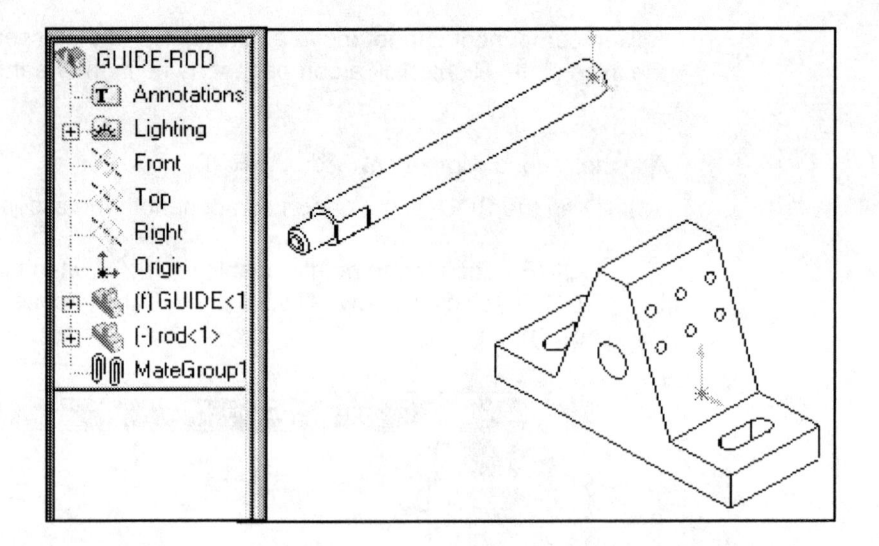

Fundamentals of Assembly Modeling　　　　**Engineering Design with SolidWorks**

Review the FeatureManager Syntax.
20) Double click the **GUIDE** component from the FeatureManager. The GUIDE component lists it's features. Note: Base-Extrude, Slot-Cut, GuideHole and M3x0.5 Tapped Hole all contain sketches.

Review the ROD syntax components in the FeatureManager.

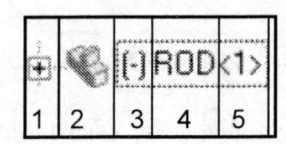

1. A Plus sign ⊞ icon indicates that additional feature information is available. A minus sign indicates that the feature list is fully expanded.

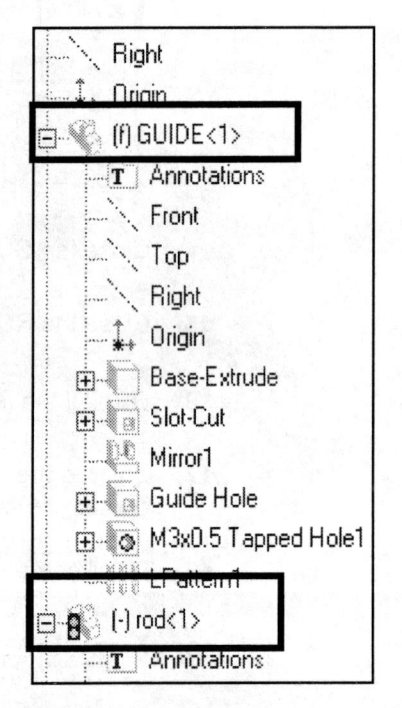

2. A component icon indicates that the ROD is a part. The assembly icon indicates that the GUIDE-ROD is an assembly. Sub-assemblies display the same icon as an assembly.

3. Column 3 identifies the Component State:

   - A minus sign (−) indicates that the component is under defined and requires additional information.

   - A plus sign (+) indicates that the component is over defined.

   - A fixed symbol (f) indicates that the component does not move.

   - A question mark (?) indicates that additional information is required.

4. ROD - Name of component.

5. The symbol <#> indicates the number of copies in the assembly. The symbol <1> indicates the original component, "ROD" in the assembly.

Move the ROD component.
21) Components located in an assembly are free to move. Move the ROD. Click the **ROD**. Click the **Move Component** from the Assembly toolbar. Click the **shaft** of the ROD in the Graphic window. The mouse pointer displays.

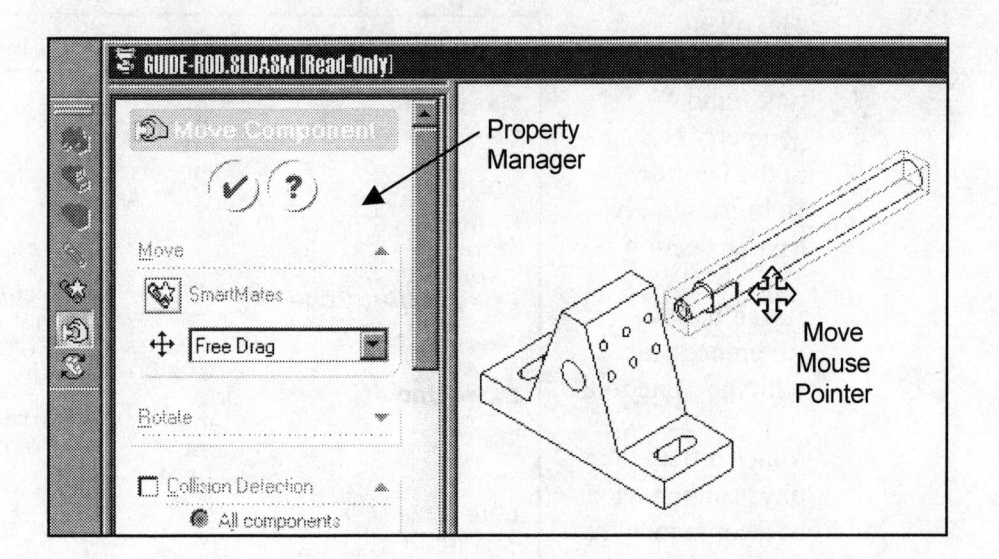

22) Position the **ROD** in back of the Guide Hole. Click **OK**. The PropertyManager is displayed on the left side of the Graphics window.

23) Fit the model to the Graphics window. Click the **f** key.

Mate Types

Establishing the correct component relationship in an assembly requires forethought on component interaction. Mates are geometric relationships that align and fit components in an assembly. Mates remove degrees of freedom from a component.

Components are assembled with various Mate types. The Mate types are:

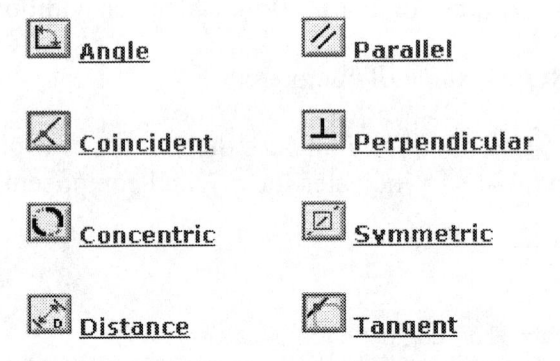

- Angle
- Parallel
- Coincident
- Perpendicular
- Concentric
- Symmetric
- Distance
- Tangent

Mates require geometry from two different components. Selected geometry includes Planar Faces, Conical faces, Linear edges, Circular/Arc edges, Vertices, Axes, Temporary axes, Planes, Points and Origins.

The rules governing Mate type valid geometry is listed in the On-line Help, Assembly Mating section.

Assemble Reference planes with the Mate tool.

Mates reflect the physical behavior of a component in an assembly. Utilize Concentric and Parallel Mates between the ROD and GUIDE.

Coincident Mate Type Combinations:

Cone
- Cone

Cylinder
- Line
- Point
- Circular/Arc Edge

Extrusion
- Point

Line
- Cylinder
- Line
- Plane
- Point

Plane
- Line
- Plane
- Point
- Circular/Arc Edge

Point
- Cam
- Cylinder
- Extrusion
- Line
- Plane
- Point
- Sphere
- Surface

Sphere
- Point

Circular/Arc Edge
- Plane
- Cylinder
- Circular/Arc Edge

Surface
- Point

The Concentric Mate utilizes the conical face of the ROD's shaft with the conical face of the GUIDE Hole.

The Parallel Mate utilizes two planar faces from the ROD keyway cut and the right face of the GUIDE.

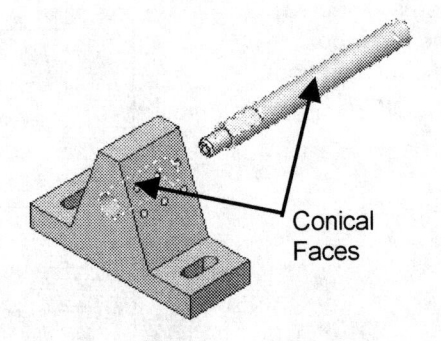

Concentric Mate – 2 Conical faces 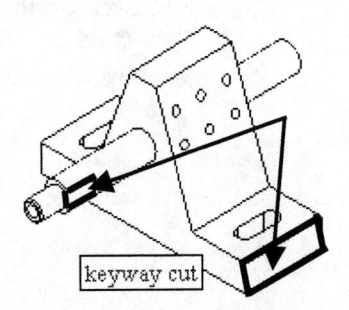   Parallel – 2 Planar faces

The Concentric Mate and the Parallel Mate allow the ROD to linearly translate through the GUIDE Hole. The ROD does not rotate.

The operation of the Mate tool is as follows.

1. Select Mate from the Assembly Toolbar.
2. Select the geometry from the first component (usually the Part).
3. Select the geometry from the second component (usually the Assembly).
4. Select the Mate Type.
5. Select Preview.
6. Select OK.

GUIDE-ROD Assembly - Mate the ROD Component

Recall the initial assembly design constraints. The ROD requires the ability to travel through the GUIDE.

The ROD keyway cut face is parallel to the right face of the GUIDE.

Smart Mate Combinations:

Mating entities	Type of mate
2 linear edges	Coincident
2 planar faces	Coincident
2 vertices	Coincident
2 conical faces, or 2 axes, or 1 conical face and 1 axis	Concentric
2 circular edges (the edges do not have to be complete circles)	Concentric (conical faces) coincident (adjacent planar faces)

Fundamentals of Assembly Modeling   **Engineering Design with SolidWorks**

Create the first Mate.

24) Click **Mate** from the Assembly toolbar. The Assembly Mating PropertyManager is displayed. Click the **conical face** of the ROD. Click the **conical face** of the GUIDE hole.

25) The faces are added to the Items Selected list. Click **Concentric** from the Mate box. Click **Closest** from the Alignment Condition text box. Click **Preview**. Click **OK** ✔.

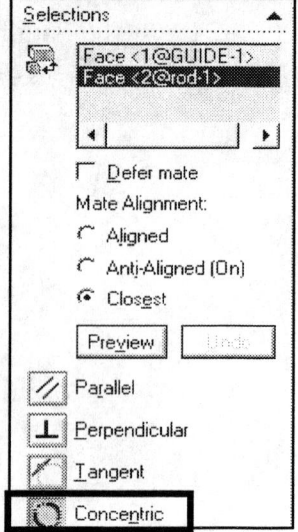

Note: When selecting faces, position the mouse pointer in the middle of the face. Do not position the pointer near the edge of the face. If the wrong face or edge is selected, click the face or edge again to remove it from the Items Selected text box.

Right-click in the Graphics window. Click Clear Selections to remove all geometry from the Items Selected text box.

Utilize the UNDO button to begin the Mate command again.

Select hidden geometry with the Select Other option.

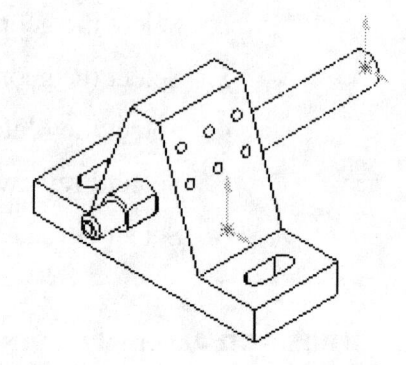

Select hidden geometry.

26) Right-click the **front face** of the Guide. Click **Select Other**. The Select Other icon is displayed from the Pop-up menu.

27) **Right-click** until the bottom face of the GUIDE is selected. Accept the bottom face. Left **click** the mouse.

The ROD is positioned concentric with the GUIDE. The ROD has the ability to move and rotate while remaining concentric to the GUIDE hole.

PAGE 2 - 16

Engineering Design with SolidWorks   Fundamentals of Assembly Modeling

Move and rotate the ROD.

28) Click **Move Component** from the Assembly toolbar.

29) Drag the **ROD** in a horizontal direction. The ROD travels linearly in the GUIDE.

30) Drag the **ROD** in a vertical direction. The ROD rotates in the GUIDE.

31) Rotate the **Rod** until the Keyway Cut is approximately parallel to the Right face of the GUIDE. Click **OK**.

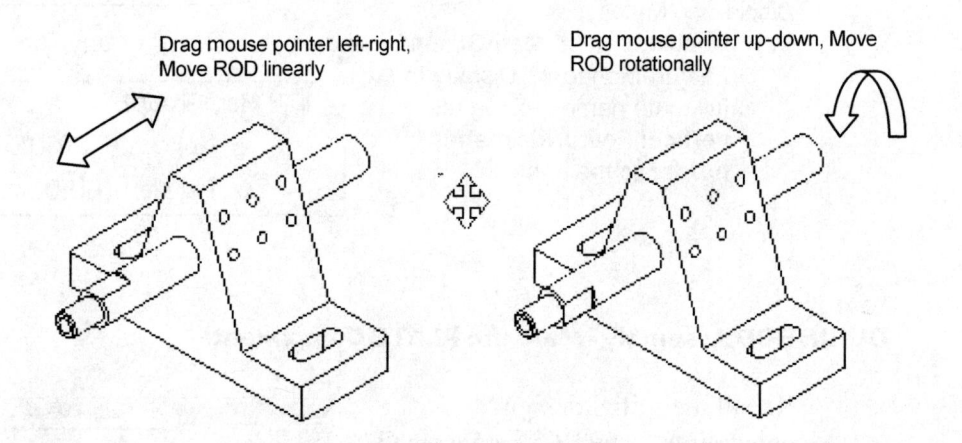

Drag mouse pointer left-right, Move ROD linearly

Drag mouse pointer up-down, Move ROD rotationally

Recall the second assembly design constraint. The flat end of the ROD must remain parallel to the top surface of the GUIDE.

32) Create the second Mate. Click **Mate** from the Assembly toolbar. Click the **Keyway Cut** of the ROD. Click the **flat right face** of the GUIDE. Both faces are added to the Items Selected list. Click **Parallel** from the Mate Types box. Click **Closest** from the Alignment Condition text box.

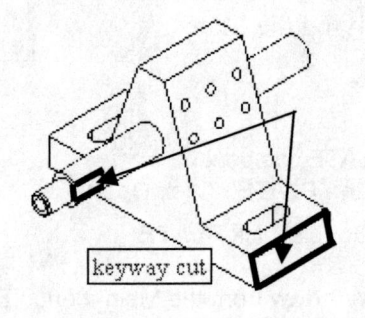

keyway cut

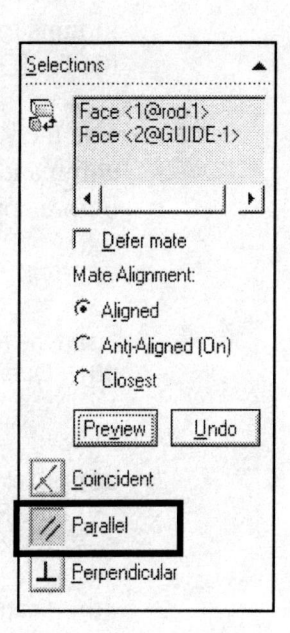

Selections
Face <1@rod-1>
Face <2@GUIDE-1>

☐ Defer mate
Mate Alignment:
● Aligned
○ Anti-Aligned (On)
○ Closest

[Preview] [Undo]

∠ Coincident
// Parallel
⊥ Perpendicular

33) Click **Preview**. Click **OK**.

Fundamentals of Assembly Modeling   **Engineering Design with SolidWorks**

34) The ROD travels linearly through the hole. Click **Move Component** from the Assembly toolbar.

35) Position the **ROD** approximately in the center of the GUIDE. The ROD remains under defined. Click **OK** ✓.

36) Hide the GUIDE component. Right-click on the **GUIDE** from the FeatureManager. Click **Hide Components**.

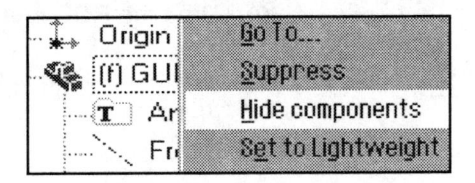

Display the Mate types.
37) Double-click on **MateGroup1** in the FeatureManager. Display the full mate names. Drag the **vertical FeatureManager border** to the right.

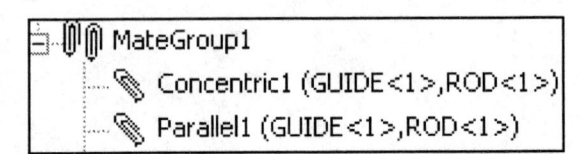

GUIDE-ROD Assembly - Mate the PLATE Component

Recall the initial design constraints. The ROD is fastened to the PLATE. The PLATE mounts to the PISTON PLATE of the GUIDE-CYLINDER assembly.

Use the Rotate Component command to position the PLATE before applying the Mates.

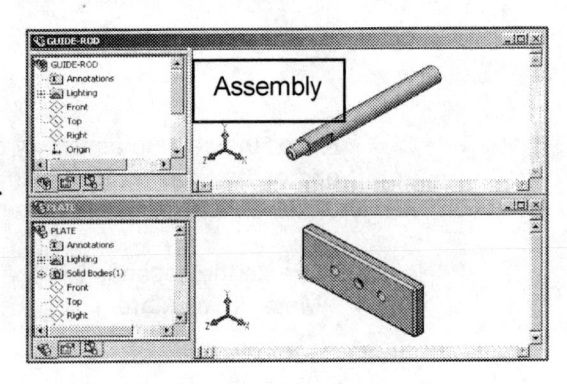

Insert the PLATE component.
38) Open the PLATE. Click **Open**. Double-click **PLATE**.

39) Click **Window** from the Main menu. Display the PLATE and the GUIDE-ROD assembly. Click **Tile Horizontally**.

40) Drag the **PLATE** icon into the GUIDE-ROD Graphics window. Release the **mouse pointer** when the PLATE is positioned behind the ROD.

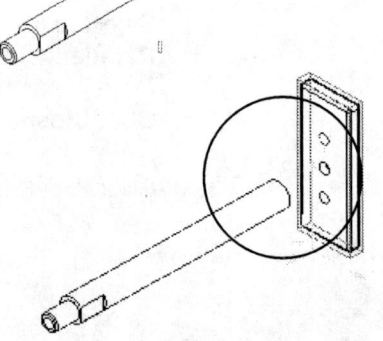

PAGE 2 - 18

Engineering Design with SolidWorks   Fundamentals of Assembly Modeling

41) **Close** the PLATE window. **Maximize** the GUIDE-ROD assembly window. Fit the model to the Graphics window. Press the **f** key.

42) Rotate the PLATE. Click **Rotate Component** from the Assembly Toolbar. Click the **PLATE**. Drag the **mouse pointer** downward until the PLATE countersunk hole and the PLATE faces backwards. Click **OK**.

43) Zoom in on the back end of the ROD and the center of the PLATE. Click **Zoom to Area**.

Selection Filters are used to select difficult individual features such as: faces, edges and points.

Activate the face Selection Filter. 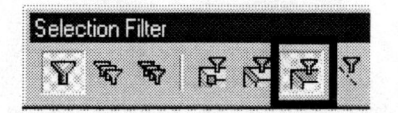
44) Click **View** from the Main menu. Click **Toolbars**. Click **Selection Filter**. Click **Filter Faces** from the Selection Filter toolbar. Note: You can only select component faces if the Face Filter is activated.

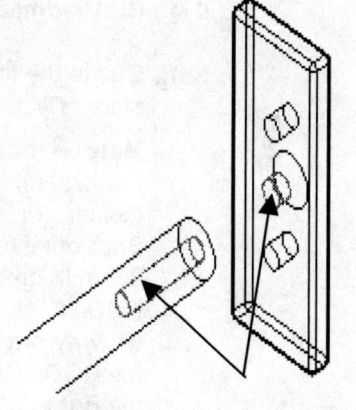

45) Create the third Mate. Click **Mate** from the Assembly toolbar. Right-click the **ROD**. Click **Select Other**. Select the hidden back hole.

The Select Other icon is displayed from the Pop-up menu. Right-click until the hidden back hole is displayed.

46) Accept the back hole. **Left-click**.

47) Click the **center conical face** from the Countersink hole of the PLATE.

48) The back hole of the ROD face and the Countersink hole of the PLATE face are added to the Items Selected list. Click **Concentric** from the Mate Types box.

49) Click **Closest** from the Alignment Condition text box.

50) Click **Preview**. Click **OK**.

51) Fit the model to the Graphics window. Press the **f** key.

Fundamentals of Assembly Modeling Engineering Design with SolidWorks

52) Create the fourth Mate. Click **Mate** from the Assembly toolbar. Press the **left arrow** key to rotate the view until the back face of the ROD is visible. Click the **back circular face** of the ROD. Press the **right arrow** key to rotate the view until the front face of the PLATE is visible. Click the **front rectangular face** of the PLATE. Click **Distance** from the Mate Types box. Enter **20**.

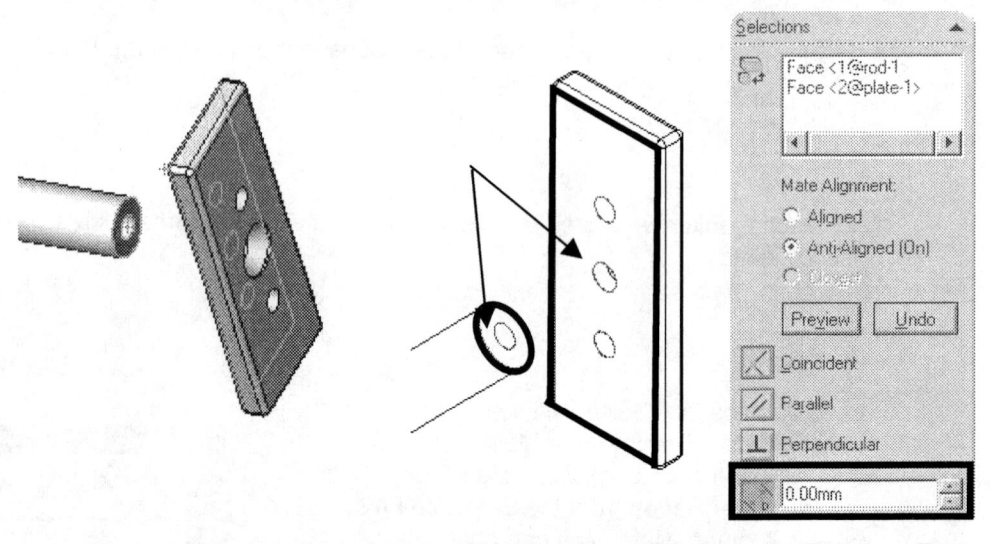

53) The two faces are 20mm apart. Click **Preview**. Enter **0** in the Distance Text box. The two faces are coincident.

54) Click **Isometric** view. Click **Preview**. Click **OK**.

55) Create the fifth Mate. Click **Mate** from the Assembly toolbar. Press **Shift** once to Zoom In on the ROD. Click the **keyway cut** of the ROD. Click the **right rectangular face** of the PLATE. Click **Parallel** from the Mate Types box.

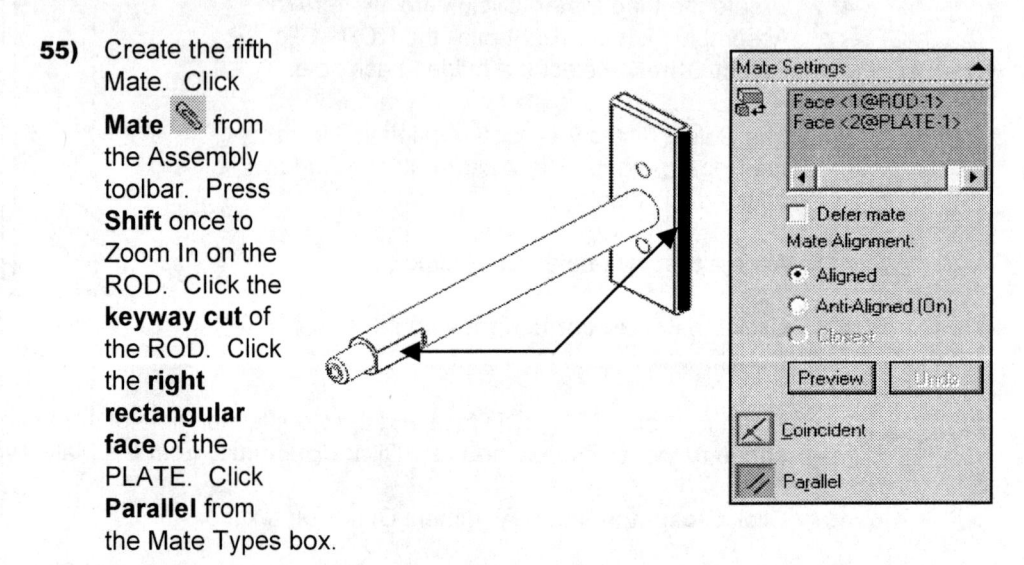

56) Click **Preview**. Click **OK**.

Engineering Design with SolidWorks | Fundamentals of Assembly Modeling

57) Display the GUIDE. Right-click **GUIDE** from the FeatureManager. Click **Show Component**.

58) Move the PLATE to the correct location. Click **PLATE** in the FeatureManager. Click **Move Component** from the Assembly Toolbar. Drag the **PLATE** backward until the PLATE clears the GUIDE. The PLATE is free to rotate about the Z axis.

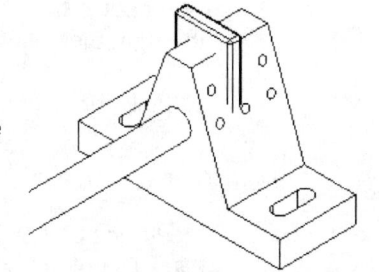

59) Click the **Collision Detection** checkbox.

60) Click the **Stop at Collision** check box from the Move Property Manager.

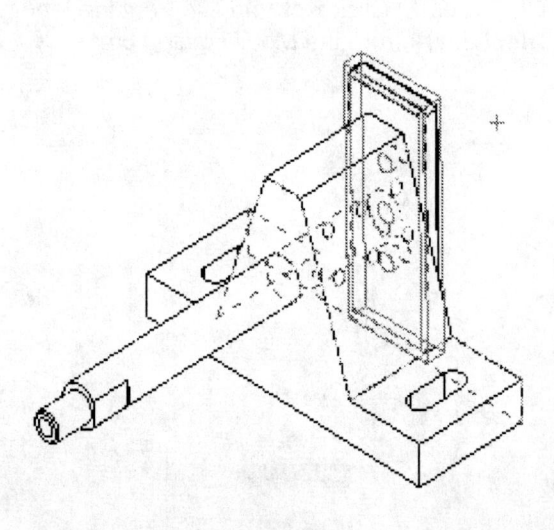

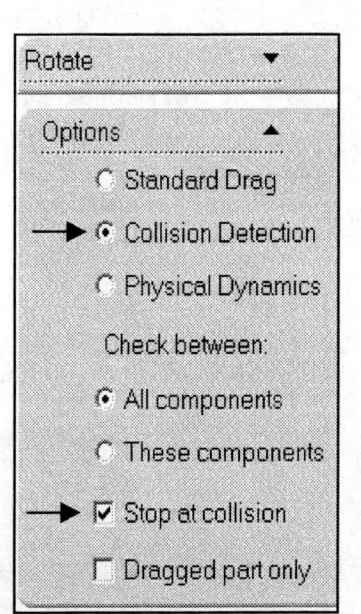

61) Slowly drag the **PLATE** forward. The GUIDE back face turns red when the PLATE front face collides with the back face of the GUIDE.

62) Drag the **PLATE** backward until the ROD is approximately halfway through the GUIDE. Click **OK**.

Display the Mate types.
63) Double-click on **MateGroup1** in the FeatureManager. Display the full mate names.

64) Reset the Filters. Click **Clear All Filters** from the Selection Filters Toolbar.

65) Save the GUIDE-ROD assembly. Click **Save**.

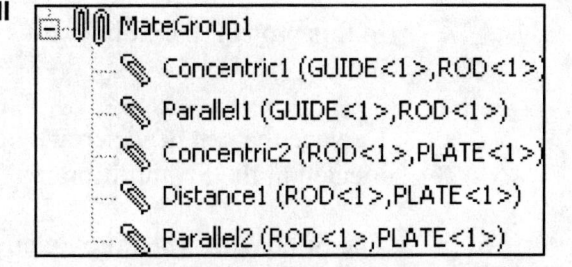

MateGroup1
- Concentric1 (GUIDE<1>,ROD<1>)
- Parallel1 (GUIDE<1>,ROD<1>)
- Concentric2 (ROD<1>,PLATE<1>)
- Distance1 (ROD<1>,PLATE<1>)
- Parallel2 (ROD<1>,PLATE<1>)

Edit Component Dimension

You realize from additional documentation that the Slot in the GUIDE is 4mm. Modify the right Slot Cut feature dimensions in the GUIDE-ROD assembly.

Rebuild the assembly. The left Mirror Slot Cut and right Slot Cut update.

Modify the Slot of the Guide.
66) Double-click on the right **Slot Cut** of the GUIDE in the Graphics window. Modify radial dimension. Double-click **3**. Enter **4**.

67) Update the GUIDE part. Click **Rebuild** from the Modify dialog box. Click the **Green Check Mark** from the Modify dialog box.

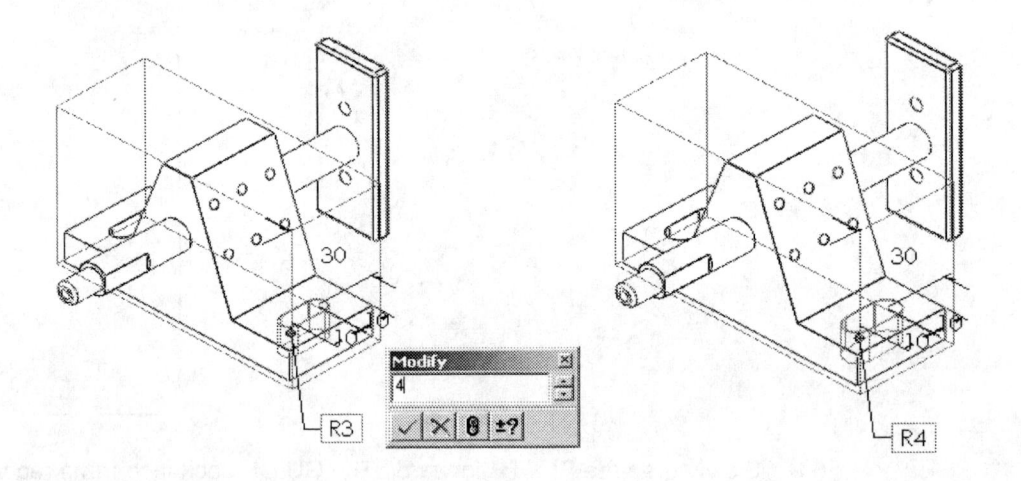

Add a Component from the Parts Library

A parts library contains components used in a design creation.

Your company issued a design policy. The policy states that you are required to only use parts that are presently in the company's parts library. The policy is designed to lower inventory cost, purchasing cost and design time.

In this project, the SolidWorks Feature Palette simulates your company's part library.

Replace the cap head screw with a hex flange bolt. The hex flange bolt is located in the FeatureLibrary.

The SolidWorks Feature Palette provides examples of common industry components for design creation.

Engineering Design with SolidWorks   Fundamentals of Assembly Modeling

Display the Feature Palette.
- **68)** Click **Tools**, **Feature Palette** from the Main menu. Double-click the **Parts folder**, (SolidWorks\data\Palette Parts).

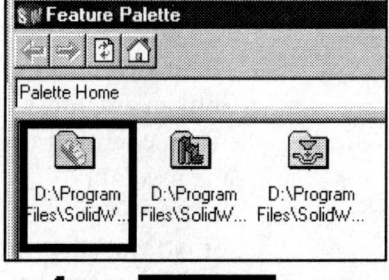

- **69)** Double-click the **hardware folder** hardware icon. The hardware Feature Library components are displayed. The flange bolt flange bolt icon represents a family of similar shaped components in various configurations.

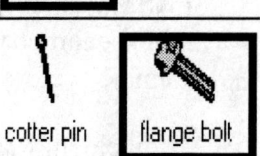

- **70)** Add a flange bolt to the assembly. Click and drag the **flange bolt** icon to the left of the GUIDE-ROD assembly into the Graphics window. Release the **mouse button**.

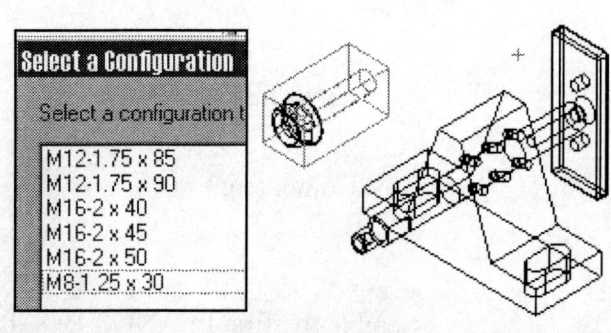

- **71)** Select a 8mm flange bolt, **M8-1.25 x 30** from the configuration list.

- **72)** Click the **OK** button. The flange bolt appears in the assembly.

- **73)** **Close** the Feature Palette.

- **74)** The flange bolt is free to travel and rotate. Assemble the flange bolt. Click **Rotate Component**.

- **75)** Click the **shaft** of the flange bolt. Drag the **mouse pointer** in a vertical direction. The flange bolt rotates.

- **76)** Click **OK**.

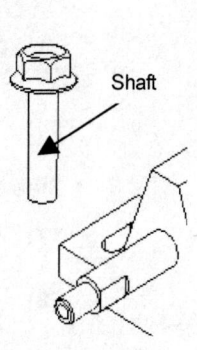

SmartMates

A SmartMate is a Mate that automatically occurs when a component is placed into an assembly. The mouse pointer displays a SmartMate feedback symbol when common geometry and relationships exist between the component and the assembly.

SmartMates are Concentric or Coincident. A Concentric SmartMate assumes that the geometry on the component has the same center as the geometry on an assembled reference.

A Coincident Planes SmartMate assumes that a plane on the component lies along a plane on the assembly.

As the component is dragged into place, the mouse pointer provides feedback such as:

- Concentric.
- Coincident.

Assemble the first FLANGE BOLT. The conical face of the FLANGE BOLT mates Concentric with the conical face of the top left slot.

The flat bottom face of the FLANGE BOLT mates Coincident with the top face of the GUIDE.

Create the first SmartMate for the FLANGE BOLT.

77) Click **SmartMates** in the Assembly toolbar.

78) Double-click the **conical face** of the FLANGE BOLT. The mouse pointer displays . The FLANGE BOLT is translucent.

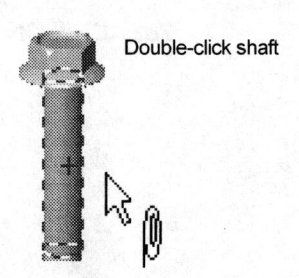

Double-click shaft

Engineering Design with SolidWorks — Fundamentals of Assembly Modeling

79) Click the **back radial face** of the right Slot. The SmartMate mouse pointer feedback displays Concentric.

80) Press the **Tab** key if the FLANGE BOLT is upside down.

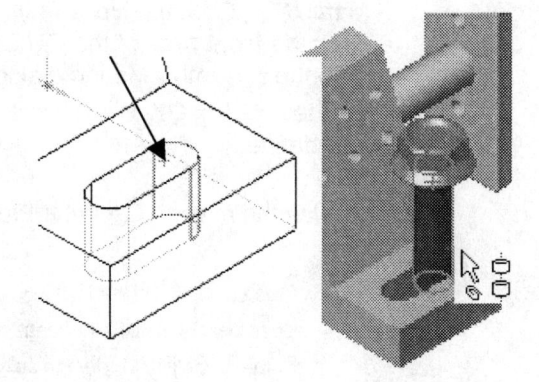

Create the second SmartMate.

81) View the flat bottom face of the Flange Bolt Head feature. Rotate the view. Press the **up arrow key** four times to display the flat bottom face of the bolt. Double-click the **flat bottom face** of the FLANGE BOLT. The mouse pointer feedback displays.

Double-click Flange Face

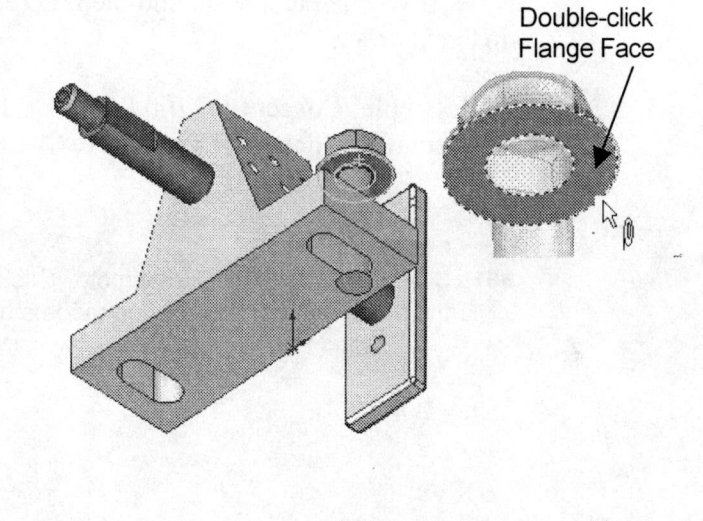

82) Click **Isometric** from the View menu. Click the **top face** of the Slot. The SmartMate mouse pointer feedback displays Coincident Planes. The mating face of the GUIDE is displayed in green. Green indicates the geometry is selected.

83) Click **OK**. The FLANGE BOLT is under defined. The FLANGE BOLT is free to rotate about its centerline.

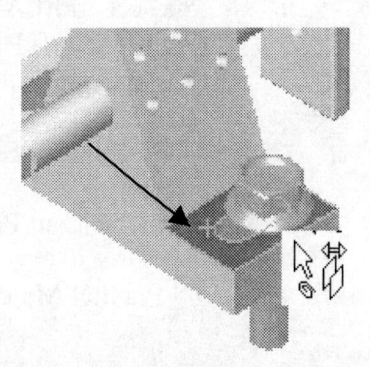

Create the third mate.

84) The front face of the FLANGE BOLT head is parallel with the front face of the GUIDE. Click **Mate**. Click the **front face** of the hex head. Click the **front face** of the GUIDE. Click **Parallel** from the Assembly Mating dialog box. Click **Preview**. Click **OK**. The FLANGE BOLT is fully defined.

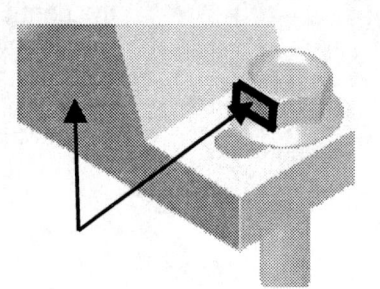

85) Display the mates. Click the **Plus sign** in MateGroup1.

- Concentric2 (GUIDE<1>,flange bolt<1>)
- Coincident1 (GUIDE<1>,flange bolt<1>)
- Parallel2 (GUIDE<1>,flange bolt<1>)

Note: If you delete a Mate and then recreate it, the Mate numbers will be in a different order.

Example: Concentric3 (GUIDE<1>, FLANGE BOLT <1>) instead of the displayed Concentric2 (GUIDE<1>, FLANGE BOLT <1>).

86) Save the GUIDE-ROD assembly. Click **Save**. Click **Yes** to the question, "Save the document and referenced models now?" The referenced models are the GUIDE and ROD.

Copy the Component

The second FLANGE BOLT is a copy of the first FLANGE BOLT. The second FLANGE BOLT requires the following mates:

- Concentric SmartMate.

- Coincident Planes SmartMate.

- Parallel Mate.

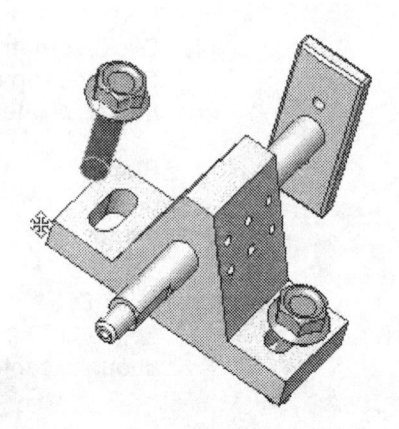

Engineering Design with SolidWorks   Fundamentals of Assembly Modeling

Copy the first FLANGE BOLT.

87) Click the **FLANGE BOLT** name in the FeatureManager, flange bolt<1> (M8-1.25x30). Hold the **Ctrl** key down. Drag the **FLANGE BOLT name** into the Graphics window. The mouse pointer displays the component icon . Release the **mouse button**. Release the **Ctrl** key. The instance number <2> appears after the second FLANGE BOLT, flange bolt<2> in the FeatureManager.

Create the first SmartMate for the second FLANGE BOLT.

88) Click **SmartMates** in the Assembly toolbar.

89) Double-click the **conical face** of the FLANG feedback displays .

90) Click the **back radial face** of the left Slot Cu feedback displays Concentric . Press upside down.

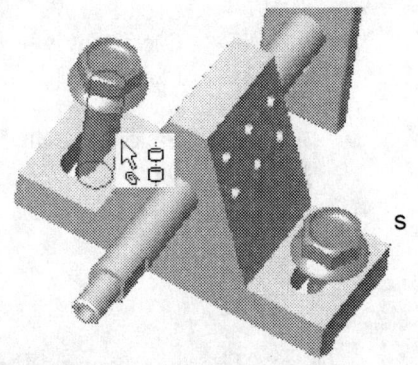

Create the second SmartMate.

91) Double-click the **flat bottom face** of the second FLANGE BOLT Head feature. The mouse pointer feedback displays .

92) Click **Isometric** from the View menu.

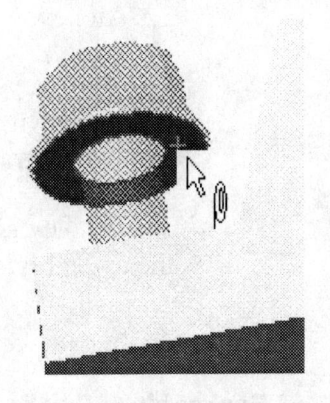

PAGE 2 - 27

93) Click the **top face** of the left Slot. The SmartMate mouse pointer feedback displays Coincident Planes. The mating face of the GUIDE is displayed in green.
Click **OK**.

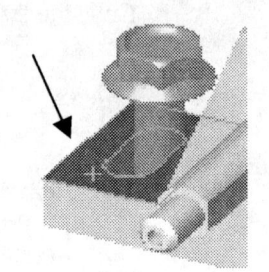

Create the third mate.

94) The front face of the second FLANGE BOLT hex head is parallel with the front face of the GUIDE. Click **Mate**. Click the **front face** of the left hex head Flange Bolt. Click the **front face** of the GUIDE. Click **Parallel** from the Mate dialog box. Click **Preview**. Click **OK**. The second FLANGE BOLT is fully defined.

95) Save the GUIDE-ROD assembly. Click **Save**.

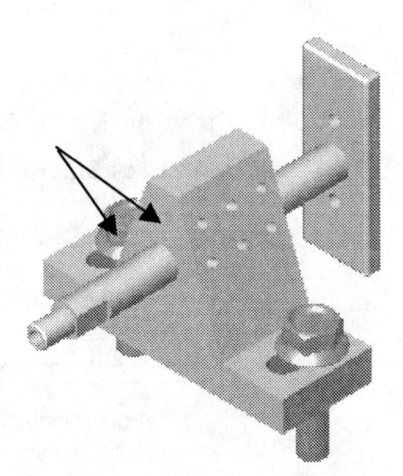

There are two additional keyboard commands that will assist you in the SmartMate process:

1. Shift key.

2. Alt key.

The Shift key is used to drag a component into the assembly. The component is viewed with a normal (opaque) appearance. Release the Shift key to display the transparent appearance.

The Alt key is used to suspend SmartMates while placing a component into the assembly.

Socket Head Cap Screw

The PLATE mounts to the PISTON PLATE of the GUIDE CYLINDER assembly with two M4x0.7 Socket Head Cap Screws.

Create a simplified version of the M4CAPSCREW based on the ANSI B 18.3.1M-1986 standard.

How do you determine the overall length of the M4CAPSCREW? Answer: The depth of the PLATE plus the required blind depth of the PISTON PLATE provided by the manufacturer.

When using fasteners to connect two plates, a design rule of thumb is to use a minimum of 75% to 85% of the second plate's blind depth.

The geometry of the Base feature is a cylindrical extrusion. The Extruded Base feature creates the head of the M4CAPSCREW. The Extruded Boss feature creates the shaft of the bolt.

The Chamfer feature is added to the 2 ends. A hex profile creates the Extruded-Cut.

Create the M4CAPSCREW.

96) Click **New**.

97) Double-click **PART-MM-ANSI** for Template.

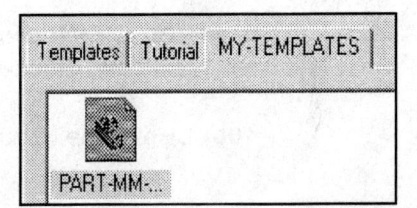

98) Save the part. Click **Save**. Select **ENGDESIGN-W-SOLIDWORKS/PROJECTS** for Save in file folder.

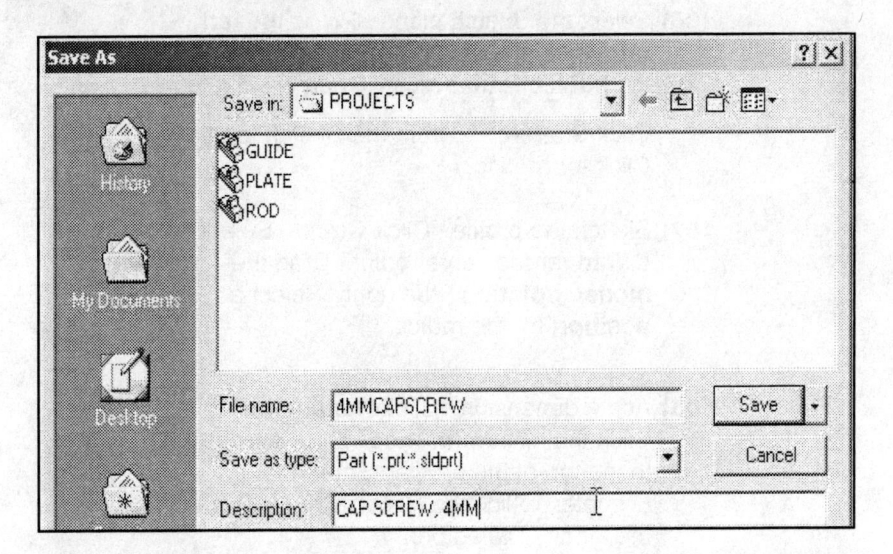

99) Enter **4MMCAPSCREW** for Filename.

100) Enter **CAP SCREW, 4MM** for Description. Click **Save**.

Fundamentals of Assembly Modeling   **Engineering Design with SolidWorks**

Create the Extrude Base feature.
101) Select the Sketch plane. Click **Front** in the FeatureManager. Open the Sketch. Click **Sketch** from the Sketch toolbar.

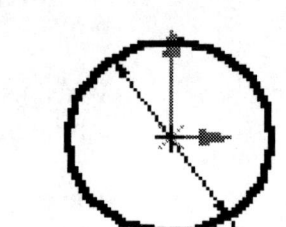

102) Sketch the profile. Click **Circle**. Select the **Origin** for the center point. Drag the **mouse pointer** to the right. Select a **position** for the radius.

103) Add a dimension. Click **Dimension**. Click the **circumference**. Click a position to the right of the profile. Enter **7** for diameter. Click the **Green Check Mark** in the Modify Dialog box.

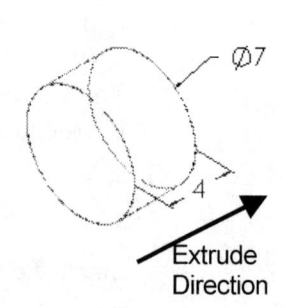

104) Extrude the Sketch. Click **Extruded Boss/Base**. Click **Reverse Direction** from the Direction1 text box. Enter **4** in the depth text box.

105) Display the Extruded-Base feature. Click **OK**.

Create the Extruded Boss feature.
106) Select the Sketch plane. Press the **left arrow** key 5 times to rotate the view. Click the **back circular face**. Open the Sketch. Click **Sketch** from the Sketch toolbar. Click **Isometric**.

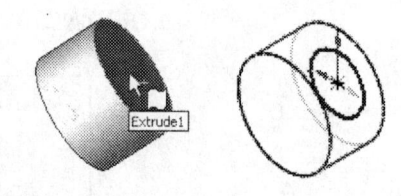

107) Sketch the profile. Click **Circle**. Select the **Origin** for the center point. Drag the **mouse pointer** to the right. Select a **position** for the radius.

108) Add a dimension. Click **Dimension**. Click the **circumference**. Click a **position** to the top of the profile. Enter **4** for diameter. Click the **Green Check Mark** in the Modify Dialog box.

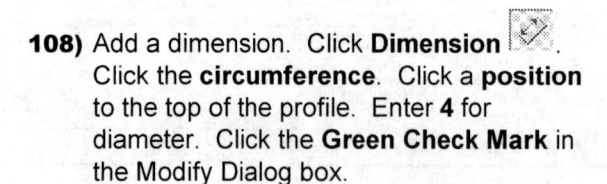

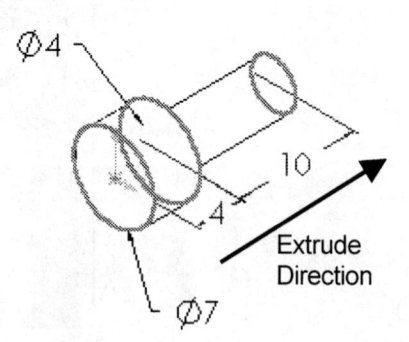

109) Extrude the Sketch. Click **Extruded Boss/Base**. Enter **10** for Depth.

110) Display the Extruded-Base feature.

111) Fit the model to the Graphics window. Press the **f** key. Click **OK**.

PAGE 2 - 30

Engineering Design with SolidWorks   **Fundamentals of Assembly Modeling**

Create a Chamfer feature.
112) Click **Chamfer**.

113) Select the **front circular edge** of the 4MMCAPSCREW.

114) Select the **back circular edge** of the 4MMCAPSCREW. Enter **0.4** in the Distance box. Accept the default **45** degree Angle.

115) Display the Chamfer feature. Click **OK**.

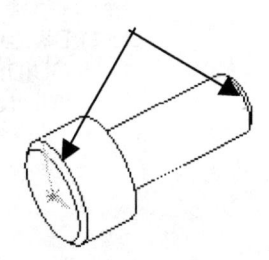

Create the Hex Head Cut.
116) Select the Sketch plane. Click the **front circular face**. Open the Sketch. Click **Sketch** from the Sketch toolbar.

117) Click the **Front** view.

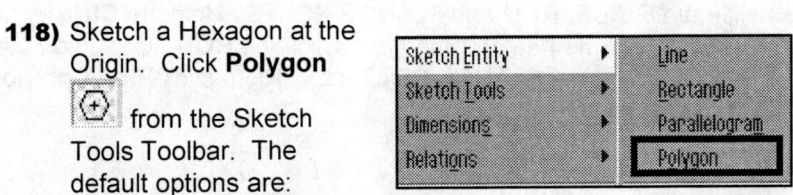

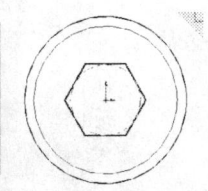

118) Sketch a Hexagon at the Origin. Click **Polygon** from the Sketch Tools Toolbar. The default options are: Inscribed circle and 6 sides. Create the first point. Click the **Origin**. Create the second point. Click a **position** to the right of the Origin. Create the **circle**. Enter **2.0** for diameter in the Circle Diameter spin box.

119) Click **Extruded-Cut**. Blind is the default Type option. Enter **4** for the Depth.

120) Click **OK**.

121) Save the 4MMCAPSCREW. Click **Save**.

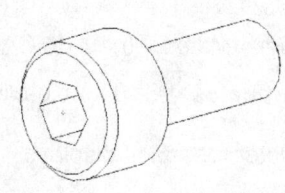

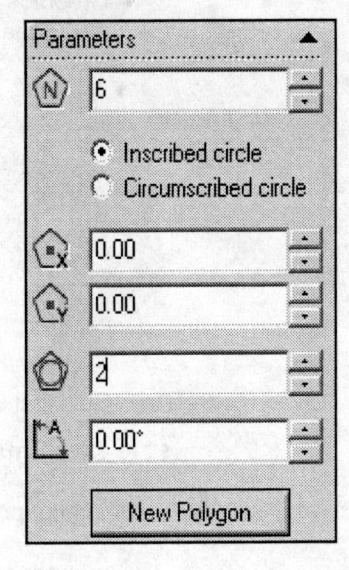

Fundamentals of Assembly Modeling **Engineering Design with SolidWorks**

Insert the first 4MMCAPSCREW.
122) Display the 4MMCAPSCREW and the GUIDE-ROD assembly. Click **Window** from the Main menu. Click **Tile Horizontally** from the Window menu.

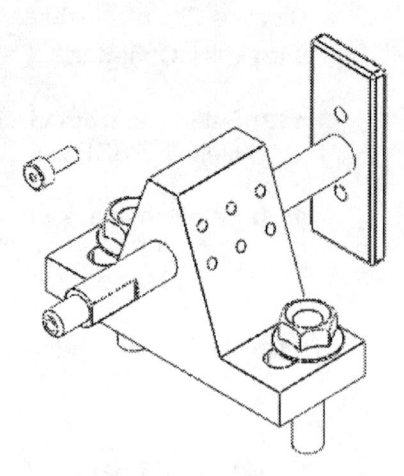

123) Drag the **4MMCAPSCREW** icon into the GUIDE-ROD Graphics window. Release the **mouse pointer** to the left of the PLATE.

124) Maximize the **GUIDE-ROD** assembly window.

125) Hide the GUIDE, ROD and FLANGE BOLTS. Hold the **Ctrl** key down. Click the **GUIDE** in the FeatureManager. Click the **ROD**. Click the **FLANGE BOLT<1>** and the **FLANGE BOLT <2>**. Right-click **Hide Components**. Release the **Ctrl** key.

126) Fit the model to the Graphics window. Press the **f** key.

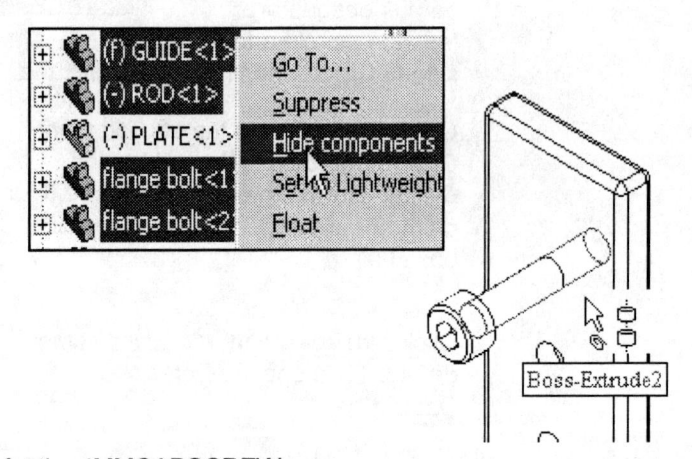

Create the first SmartMate for the 4MMCAPSCREW.
127) Click **SmartMates**. Double-click the **shaft** of the 4MMCAPSCREW. The mouse pointer feedback displays. Click the **inside face** of the Top mounting hole. The mouse pointer feedback displays Concentric.

128) Press the **Tab** key if the 4MMCAPSCREW is backward.

PAGE 2 - 32

Engineering Design with SolidWorks   Fundamentals of Assembly Modeling

Create the second SmartMate.
129) Rotate the assembly. Press the left arrow key **5 times**. Double-click the **bottom face of the head** of the 4MMCAPSCREW. The mouse pointer feedback displays.

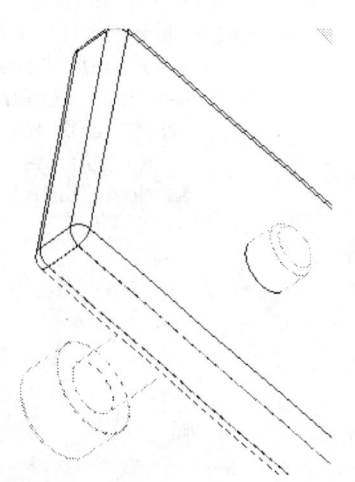

130) Click **Isometric** from the View menu.

131) Click the **front face** of the PLATE. The SmartMate mouse pointer feedback displays Coincident Planes.

132) The Mating face of the PLATE is displayed in green. Click **OK**.

Create the third mate.
133) The Right plane of the 4MMCAPSCREW is parallel with the Right plane of the PLATE. Click **Mate**.

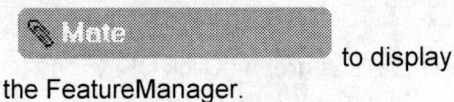

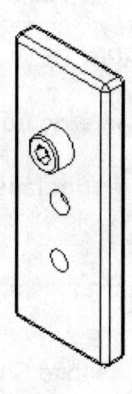

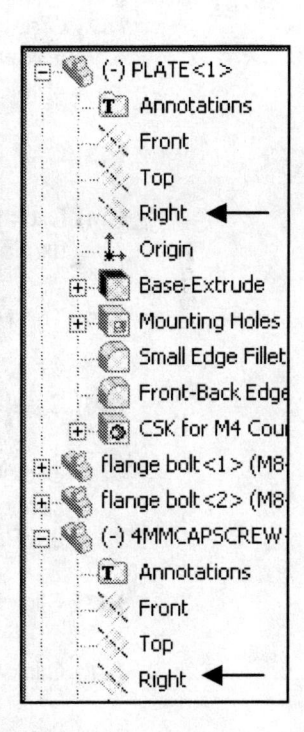

134) Click the **Mate button** to display the FeatureManager.

135) Click the **Right plane** of the 4MMCAPSCREW in the FeatureManager.

136) Click the **Right plane** of the PLATE.

137) Click **Parallel** from the Assembly Mating dialog box.

138) Click **OK**. The first 4MMCAPSCREW is fully defined.

139) Save the GUIDE-ROD assembly. Click **Save**.

PAGE 2-33

Fundamentals of Assembly Modeling   **Engineering Design with SolidWorks**

Insert the second 4MMCAPSCREW.
140) Click the **4MMCAPSCREW** from the FeatureManager. Hold the **Ctrl** key down. Drag the **4MMCAPSCREW** name from the FeatureManger into the Graphics window. Click a **position** in front of the PLATE. Release the **Ctrl** key.

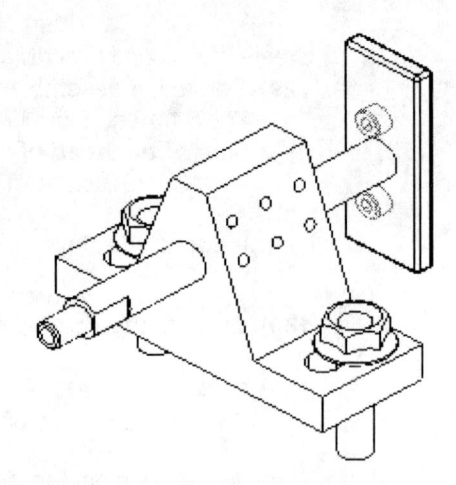

141) Create the first SmartMate for the second 4MMCAPSCREW. Click **SmartMates**. Double-click the **shaft** of the 4MMCAPSCREW. Click the **inside face** of the bottom Mounting Hole of the PLATE. The mouse pointer feedback displays Concentric. Press the **Tab key** if the 4MMCAPSCREW is backwards.

142) Create the second SmartMate for the second 4MMCAPSCREW. Press the **left arrow** key 5 times. Double-click the **bottom face** of the head of the second 4MMCAPSCREW.

143) Click **Isometric** from the View menu.

144) Click the **front face** of the PLATE. The SmartMate mouse pointer feedback displays Coincident Planes.

145) The Mating face of the PLATE is green. Click **OK**.

146) Create the third SmartMate for the second 4MMCAPSCREW. Click **Mate**. Click the **Mate** button. Click the **Right** plane of second 4MMCAPSCREW from the FeatureManager. Click the **Right plane** of the PLATE in the FeatureManager. Click **Parallel**.

147) Click **Preview**.

148) Click **OK**. The second 4MMCAPSCREW is fully defined.

149) Display the hidden components. Right-click on the **GUIDE** in the FeatureManager. Click **Show component**. Repeat for the **ROD** and the two **FLANGE BOLTS**.

150) Fit the assembly to the Graphics window. Press the **f** key.

151) Save GUIDE-ROD assembly. Click **Save**.

Engineering Design with SolidWorks   Fundamentals of Assembly Modeling

Exploded View

Exploded views assist the designer in the viewing of the design creation. You can fully explode an assembly in a single or multi step procedure.

AutoExplode explodes an assembly in a single step procedure. Explode the assembly in a multi step approach.

Explode Step1.
152) Click **Insert** from the Main menu. Click **Exploded View**. Click **Yes** to explode the light weight components. Create the first explode step.

153) Click **New Step** from the Step Editing Tools.

154) Click the **top right** edge of the GUIDE.

155) Enter **100** in the Distance spin box.

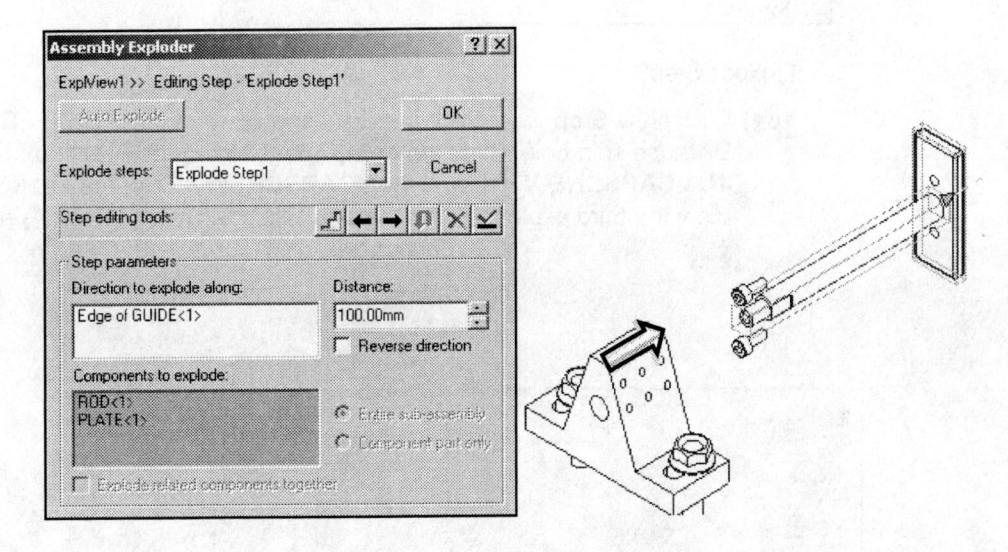

156) Click the **Component to explode** text box.

157) Click the **ROD** and the **PLATE** in the FeatureManager.

158) View the first explode step, named Explode Step1. Click the **Apply Check Mark** from the Assembly Exploder dialog box.

Fundamentals of Assembly Modeling   **Engineering Design with SolidWorks**

Explode Step2.

159) Click **New Step**. Click the **right front vertical edge** of the GUIDE. Enter **50** in the Distance spin box.

160) Click the **component to explode** text box. Click **FLANGE BOLT <1>** and **FLANGE BOLT <2>** from the FeatureManager. View the second exploded step, named Explode Step2. Click the **Apply Check Mark**.

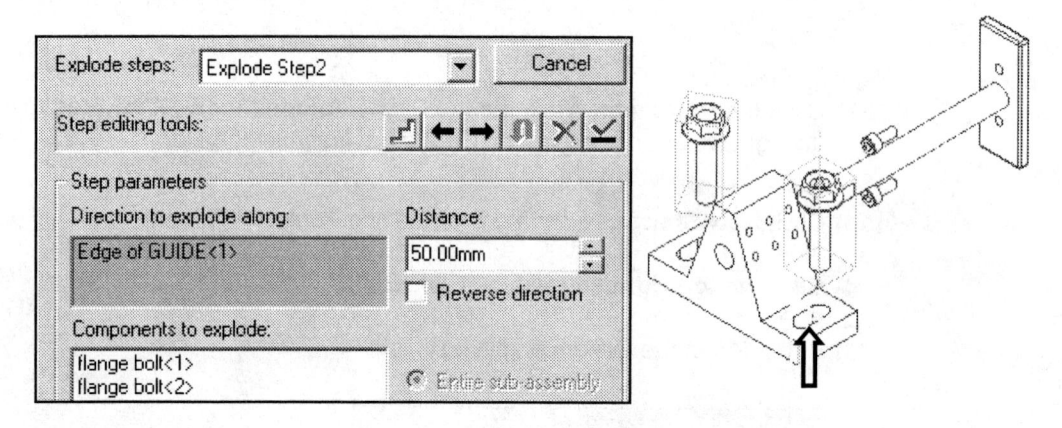

Explode Step3.

161) Click **New Step**. Click the **right top edge** of the GUIDE. Enter **70** in the Distance spin box. Click the **component to explode** text box. Click **4MMCAPSCREW <1>** and **4MMCAPSCREW <2>** in the FeatureManager. View the third exploded step, named Explode Step3. Click **Apply Check Mark**.

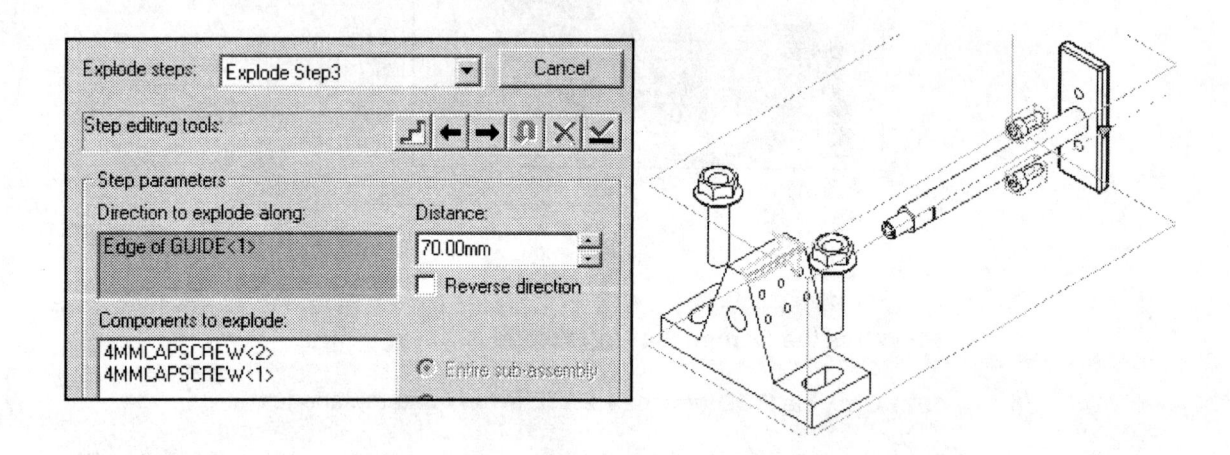

Explode Step4.

162) Click **New Step**.

163) Click the **right top edge** of the GUIDE.

164) Enter **10** in the Distance spin box.

165) Click the **Reverse direction** check box.

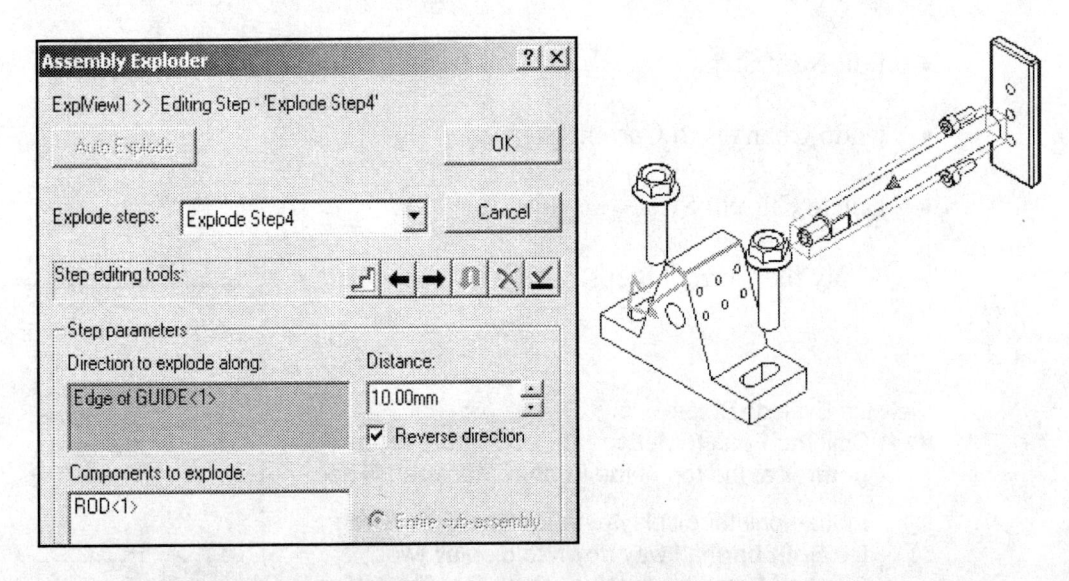

166) Click the **component to explode** text box.

167) Click **ROD** in the FeatureManager. View the forth exploded step, named Explode Step4.

168) Click **Apply Check Mark**.

169) Click **OK**.

170) Fit the Exploded view to the Graphics window. Press the **f** key.

Step Editing Tools and Viewing Exploded State

Edit steps in the Exploded view with the following Step Editing tools. From left to right the following icons represents: .

- New Step.

- Edit Previous Step.

- Edit Next Step.

- Undo Changes to Current Step.

- Delete Current Step.

- Apply the Current Steps.

View the Explode Steps.
171) Split the FeatureManager. Position the **mouse pointer** at the top of the FeatureManager. The mouse pointer displays the Split bar . Drag the **Split bar** half way down to display two FeatureManager windows. Click **Configuration Manager** to display the Exploded states in the upper window.

172) Expand Default. Click **Plus** to the left of the Default entry.

173) Expand ExpView1. Click **Plus** to the left of ExpView1 entry. The FeatureManager is displayed in the bottom window.

174) Remove the Exploded state. Right-click in the **Graphic window**.

175) Click **Collapse** from the Pop-up menu.

176) Save the GUIDE-ROD assembly. Click **Save**.

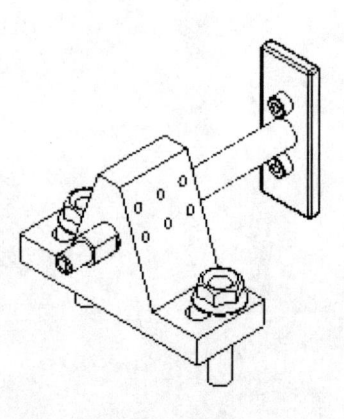

PAGE 2 - 38

Section View

Section views display the internal cross section of a component or assembly. The Section view dissects a model like a knife slicing through a stick of butter. Section views can be performed anywhere in a model. The location of the cut corresponds to the Section plane. A Section plane is a planar face or reference plane.

Section views detect potential problems before manufacturing. Use a Section to determine the interference between the 10mm Guide Hole and the Linear Pattern of 3mm Holes in the GUIDE.

Display the Section View.
177) Create the Section View on the Front plane of the GUIDE. Click **Front** under the GUIDE component in the FeatureManager. Click **View** from the Main menu. Click **Display**, **Section View**. Enter **15** for Section Position. Set the viewing direction of the Section cut. The view arrow points outward. Click **Preview** to display the section. Click **OK**.

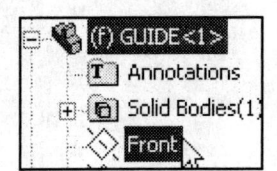

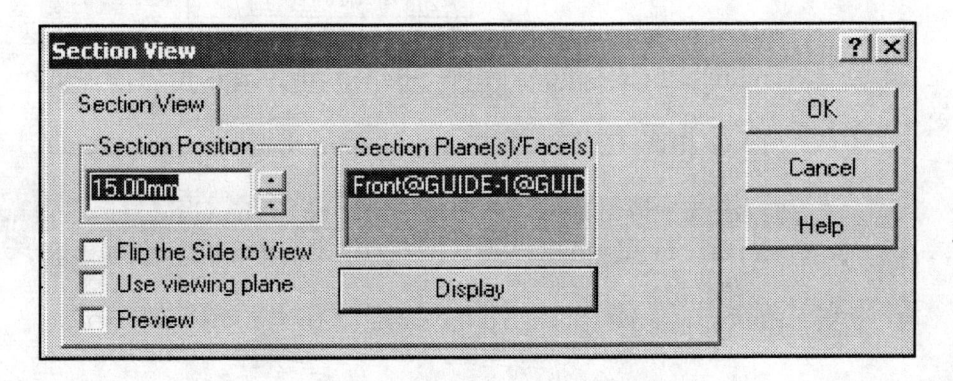

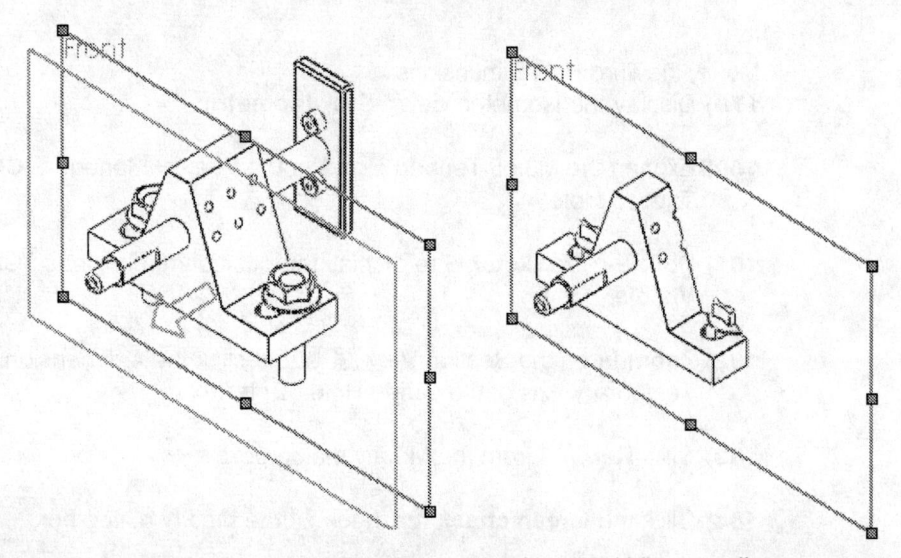

178) Display the Back view. Click **Back** from the Standard Views toolbar.

Analyze a Problem

An interference problem exists between the 10mm Guide Hole and the 3mm tapped Holes. Review your design options:

- Reposition the Guide Hole.

- Modify the size of the Guide Hole.

- Adjust the length of the 3mm Holes.

- Reposition the Linear Pattern.

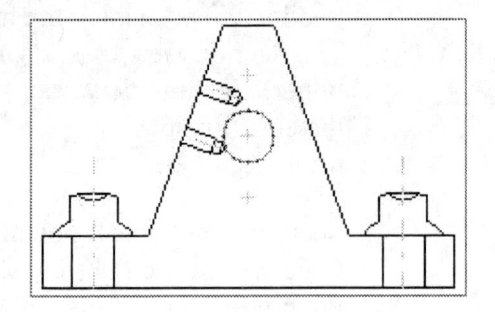

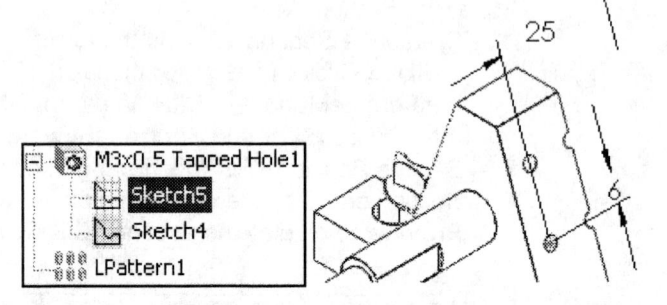

The first three options affect other components in the assembly.

The GUIDE-CYLINDER and PLATE determine the Guide Hole location.

The ROD diameter determines the size of the Guide Hole. The sensor will not properly fasten to the GUIDE if you shorten the length of the 3mm Holes.

Reposition the Linear Pattern and its first 3mm Thru Hole.

Modify the Thru Hole Dimensions.
179) Display the Isometric view. Click **Isometric**.

180) Expand the M3x.5 Tapped Hole1 in the FeatureManager. **Click** ⊞ on the M3x5 Tapped Hole1.

181) Double-click **Sketch 5** to display the position dimensions. Click **Hidden Lines Visible**.

182) Zoom in on the Section View. Double-click the **4** dimension created from the Temporary axis of the Guide Hole. Enter **6**.

183) Click **Rebuild** from the Modify dialog box.

184) Click the **Green check mark** to exit the Modify dialog box.

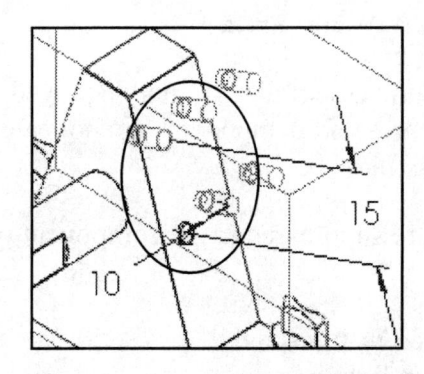

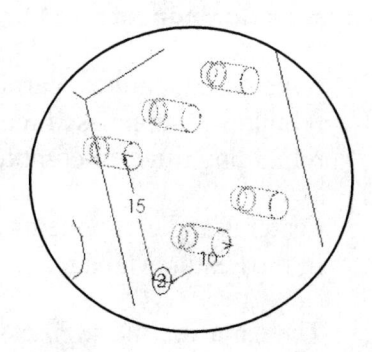

Modify the Linear Pattern Dimension.
185) Double-click the **Linear Pattern1** of the GUIDE from the FeatureManager.

186) Double-click the vertical **10**.

187) Enter **15**.

188) Click **Rebuild** from the Standards toolbar.

Display the Full View.
189) Click **View** from the Main menu.

190) Click **Display**.

191) Click **Section View**.

Suppressed Components and Light Weight Parts

Suppressed features, parts and assemblies are not displayed. During model rebuilding, suppressed features and components are not calculated. This saves rebuilding time for complex models.

Features and components are suppressed at the component or assembly level in the FeatureManager.

The names of the suppressed features and components are displayed in light gray.

A lightweight part is a part that only loads a portion of its model data into memory.

A fully resolved part loads all model data into memory. Lightweight parts are more efficient for large assemblies. They are quicker to open and Rebuild in an assembly.

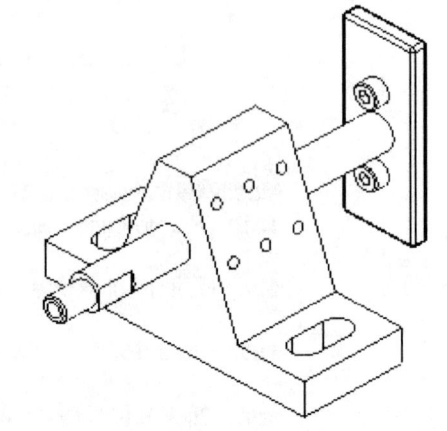

After a lightweight part is edited, the component is fully resolved into its features.

Parts are displayed in their lightweight state when an assembly is opened. Lightweight parts are loaded in the Shaded display mode.

Suppress the FLANGE BOLT.
192) Hold the **Ctrl** key down. Click **FLANGE-BOLT<1>** and **FLANGE-BOLT<2>** from the FeatureManager. Right-click **Suppress**. Release the **Ctrl** key.

193) Display the GUIDE-ROD assembly in the Shaded mode. Click **Shaded**.

194) Save the GUIDE-ROD assembly. Click **Save**.

195) Click **Yes** to the question, "Save the document and referenced models now?" The referenced models are the GUIDE, PLATE and ROD.

196) Close all parts and assemblies. Click **Window** from the Main menu.

197) Click **Close All**.

198) Open the GUIDE-ROD assembly. Click **File, Open** from the Main menu.

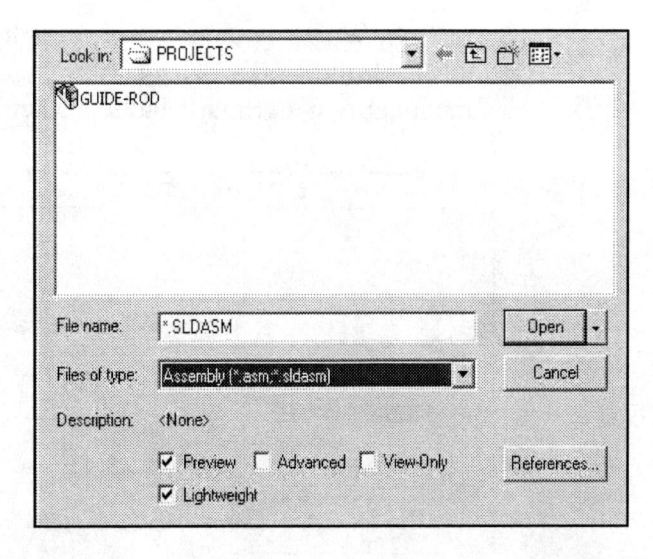

199) Select **Assembly** for Files of Type.

200) Click the **GUIDE-ROD**.

201) Select the **Lightweight** check box.

202) Click **Open**. Note: Parts are displayed in their lightweight state when an assembly is opened.

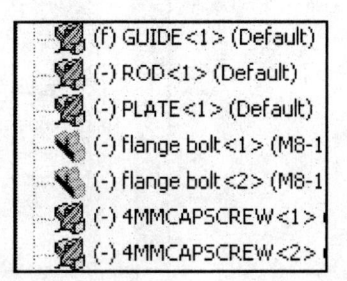

Make-Buy Decision

In a make-buy decision process, a decision is made on which parts to manufacture and which parts to purchase.

In assembly modeling, a decision is made on which parts to design and which parts to purchase.

SolidWorks contains a variety of designed parts in the FeaturePalette.

The SolidWorks Toolbox/SE is a library of feature based design automation tools for SolidWorks. The Toolbox uses the window's drag and drop functionality with SmartMates. Fasteners are displayed with full thread detail.

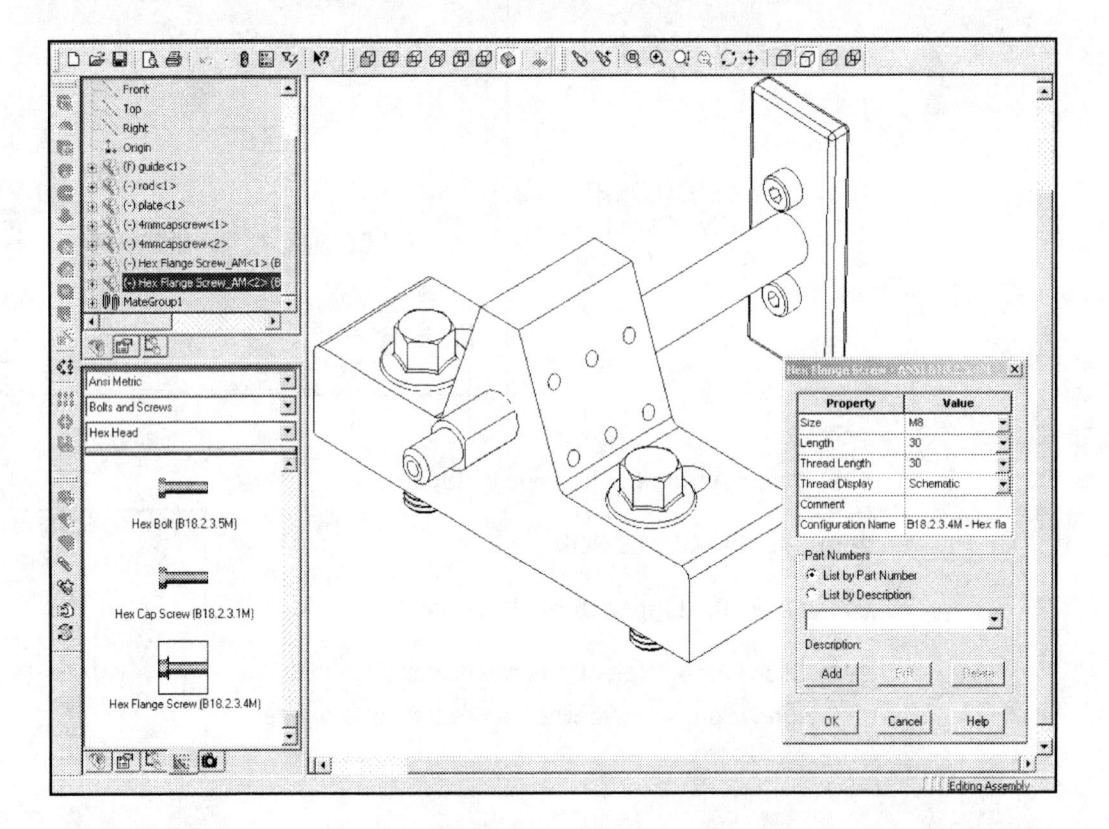

SolidWorks SmartFastener uses the HoleWizard to automatically SmartMate the corresponding fasteners in an assembly.

SolidWorks provides an Internet service, 3DPartStream. The service is utilized by vendors to share model information with their customers.

SMC USA of America sizes their components based upon customer input. The customer answers a series of questions on: working pressure, load conditions, travel, etc.

An exercise on how to obtain an SMC component via 3DPartStream is provided below:

Access to the World Wide Web is required for the following steps.

Size and download a component for the CUSTOMER assembly.

203) Obtain the GUIDE CYLINDER assembly from SMC USA. **Invoke a web** browser. Enter the following URL: **www.smcusa.com**.

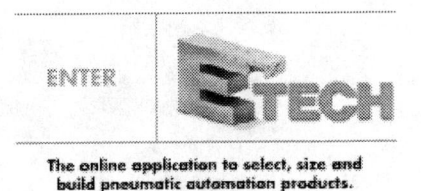

204) Select the **E-TECH** icon in the right corner of the home web page.

205) Enter your **email address** and **password**. Note: If you are a new user, click the **Register button** and enter the **requested** information.

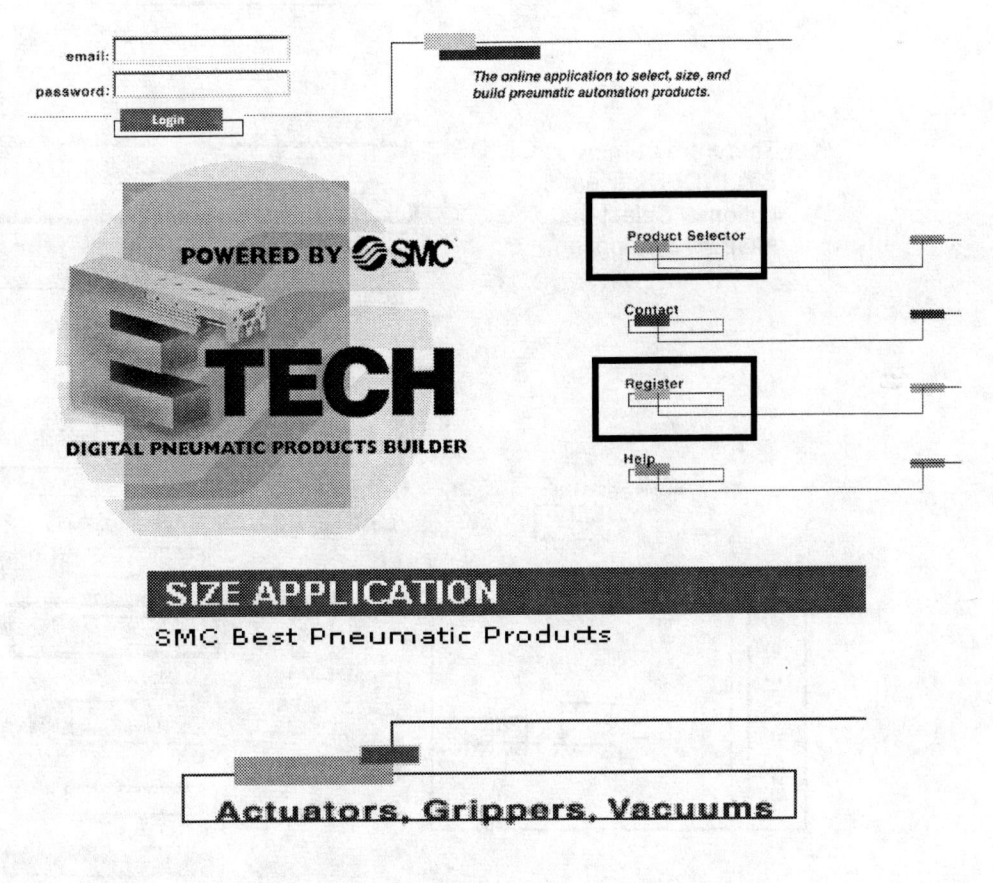

206) Select the **Product Selector** button.

207) Click **Size Applications**. Click **Actuators/Grippers/Vacuum**.

208) Click **Guided**. Read the disclaimer.

Fundamentals of Assembly Modeling　　　　　　　　　　**Engineering Design with SolidWorks**

209) Accept the disclaimer.

210) Enter the design parameters. Enter **10**mm for Stoke. Enter **0.5** MPa for Supply Pressure. Click **Size Bore and Find Product** button.

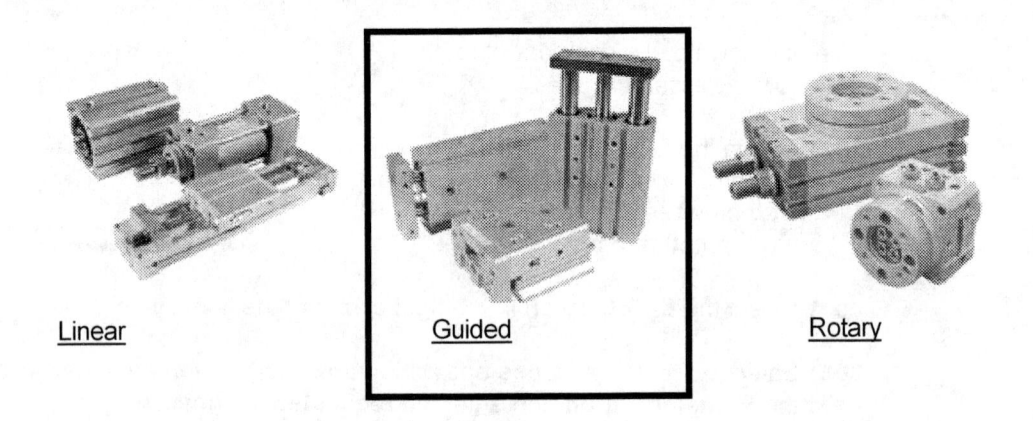

Linear　　　　　Guided　　　　　Rotary

211) There are numerous GUIDE CYLINDER assembly options. Select the **MGPM12-10** option.

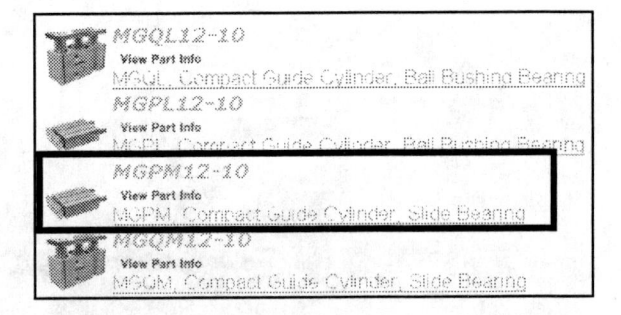

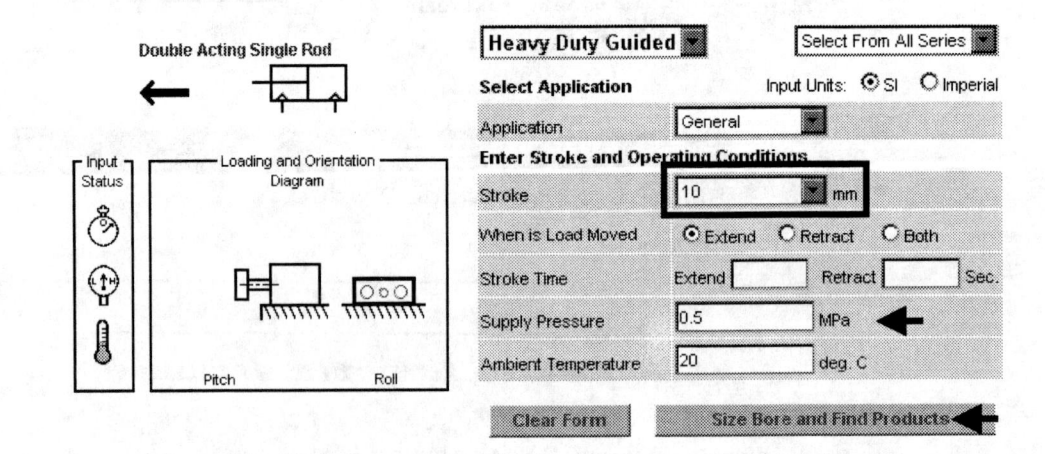

PAGE 2 - 46

Engineering Design with SolidWorks   Fundamentals of Assembly Modeling

212) The part number, description and picture are displayed.

213) There are three options: View 3D Model, View 2D Drawing and Download CAD File. Click the **Download CAD File** button.

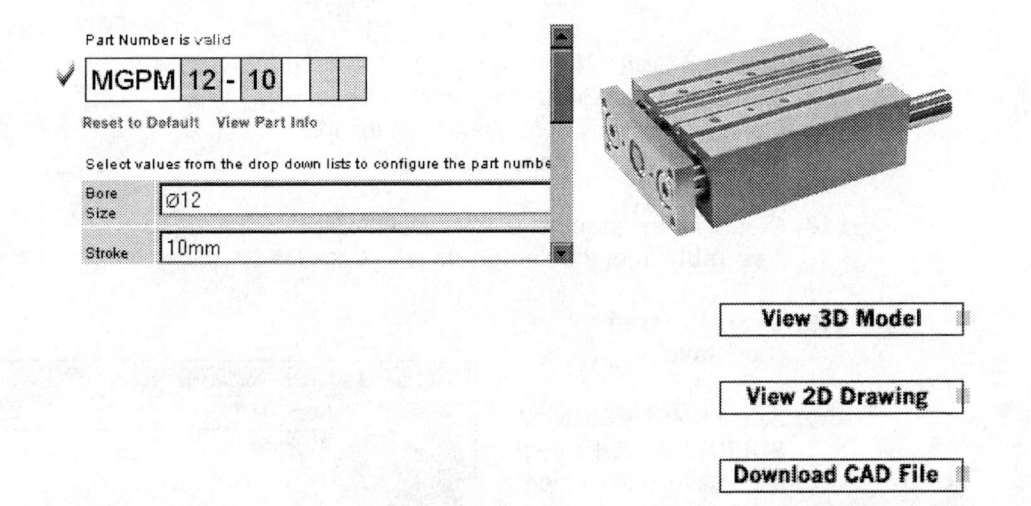

214) Display the download file formats. Click **3D Formats**. Select the **SolidWorks Assembly (.sldasm)**. Select **2003** for Version. Click the **Download Files** button.

215) The zipped SolidWorks file is downloaded to your computer. Store the downloaded files. Select **ENGDESIGN-W-SOLIDWORKS\VENDOR COMPONENTS** file folder. Unzip the file. Click the **Unzip Now** button.

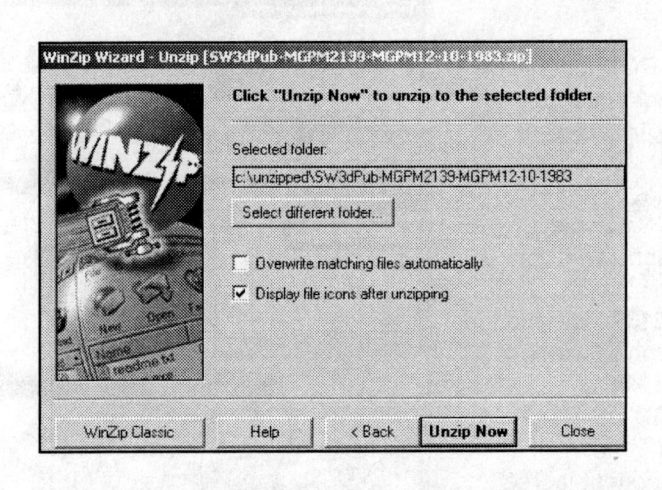

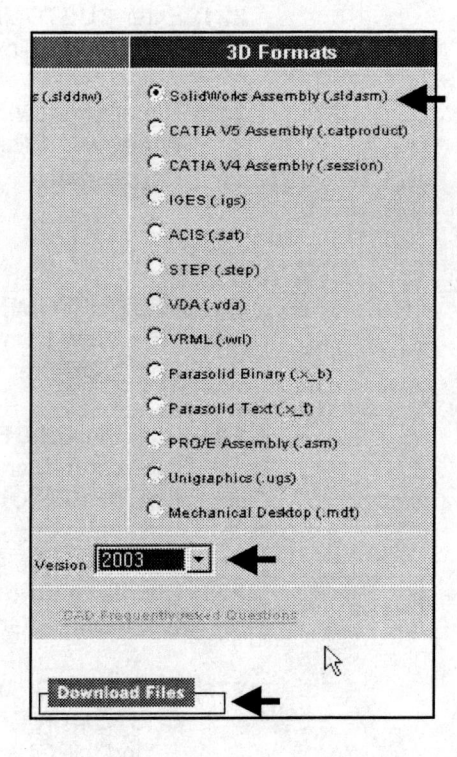

Open the MGPM2139 assembly and the GUIDE-ROD assembly.

216) Three documents are contained in the unzipped folder: MGPM2139.sldasm, MGPROD.sldprt and MGPTUBE.sldprt. Open the assembly. Double-click the **MGPM2139 assembly icon**.

PAGE 2-47

Fundamentals of Assembly Modeling   Engineering Design with SolidWorks

vendor components   Mgpm2139   MGPRod   MGPTube

MGPM2139

217) Open the GUIDE-ROD assembly. Click **Open**. Select the **GUIDE-ROD** assembly. Deactivate Planes for clarity. Click **View** from the Main toolbar. Uncheck **Planes**.

218) Create a new assembly. Click **New**. Click **Assembly** from the Templates tab. Click **OK**.

219) Save the assembly. Click **Save**.

220) Select **ENGDESIGN-W-SOLIDWORKS/Projects** for Save in file folder.

221) Enter **CUSTOMER** for File name. Click **Save**.

222) Tile all windows. Click **Windows**. Click **Tile Horizontally**.

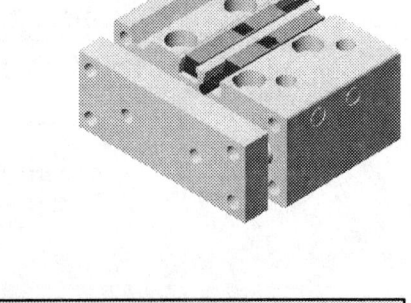

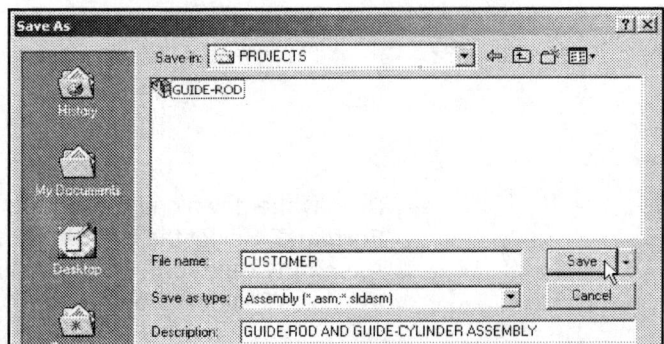

Create the CUSTOMER assembly.

223) Click **View** from the Main menu. Check **Origins**.

224) Drag the **GUIDE-ROD** icon from the FeatureManager into the CUSTOMER **Origin**.

225) Click **Isometric** in the CUSTOMER Graphics window.

226) Drag the **MGPM2139** icon into the **CUSTOMER** Graphics window behind the GUIDE-ROD assembly. The GUIDE-ROD assembly is fixed. The MGPM2139 component is free to rotate.

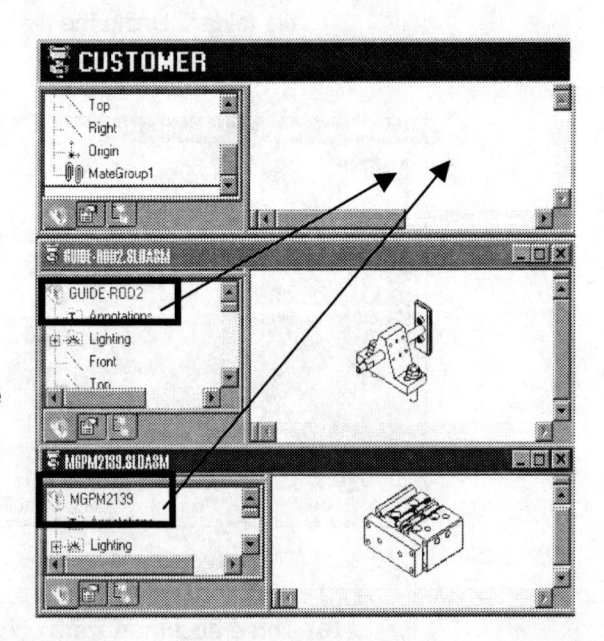

PAGE 2 - 48

Engineering Design with SolidWorks   Fundamentals of Assembly Modeling

227) Maximize the CUSTOMER Graphics window.

228) Fit the model to the Graphics window. Press the **f** key.

229) Deactivate origins and planes for clarity. Click **View** from the Main toolbar. Uncheck **Origins**. Click **View**. Uncheck **Planes**.

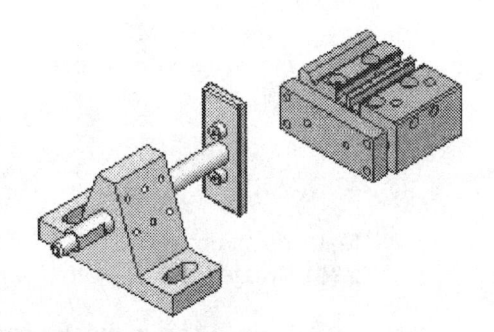

230) Hide the 4MMCAPSCREWS and ROD component. Expand the **GUIDE-ROD** in the CUSTOMER FeatureManager. Hold the **Ctrl** key down. Click the two **4MMCAPSCREWS**. Right-click **Hide components**. Release the **Ctrl** key.

231) Rotate the component. Click **MGPM2139**. Click **Rotate Component** from the Assembly toolbar. Click the **down arrow** from the Free Rotate text box.

232) Select **By Delta XYZ**.

233) Enter **90** in the Z text box.

234) Click **Apply**. Click **OK** ✔.

235) The MGPM2139 rotates about the CUSTOMER assembly Z axis. Fit the model to the Graphics window to view the rotated MGPM2139 assembly. Press the **f** key.

236) Move the MGPM2139 assembly. Click **Move Component** from the Assembly toolbar.

237) Drag the **MGPM2139 assembly** behind the GUIDE-ROD assembly. Note: The Piston Plate is behind the GUIDE-ROD assembly.

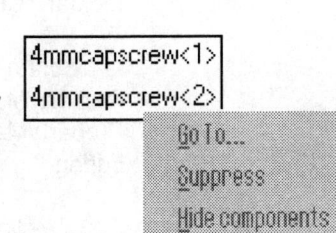

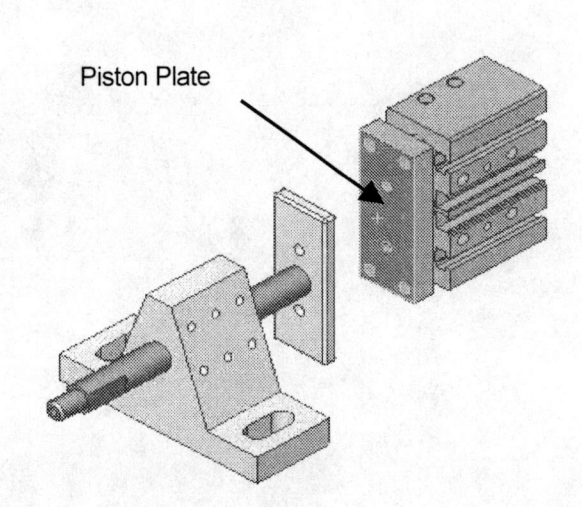

Fundamentals of Assembly Modeling   **Engineering Design with SolidWorks**

238) Zoom in on the PLATE of the GUIDE-CYLINDER assembly and the Piston Plate of the MGPM2139 assembly. Click **Zoom to Area**. Position the mouse **pointer** at the first corner. Drag the mouse **pointer** diagonally to the opposite corner.

Mate the components.

239) Create the first Mate. Select **Mate** from the Assembly toolbar. Click the **Pin** button from the Mate PropertyManager. The Mate PropertyManager remains open during the next three mates.

240) Double-click the **inside M4 hole face** of the PLATE.

241) Click the **inside M4 hole face** of the PISTON PLATE.

242) Click **Concentric** for Mate type.

243) Click **Preview**. Click **OK**.

244) Create the second Mate. Click the **front face** of the MGPM2139 PISTON PLATE. Press the **left arrow key** six times to rotate the view. Click the **back face** of the PLATE of the GUIDE-ROD assembly. Click **Coincident** for Mate type.

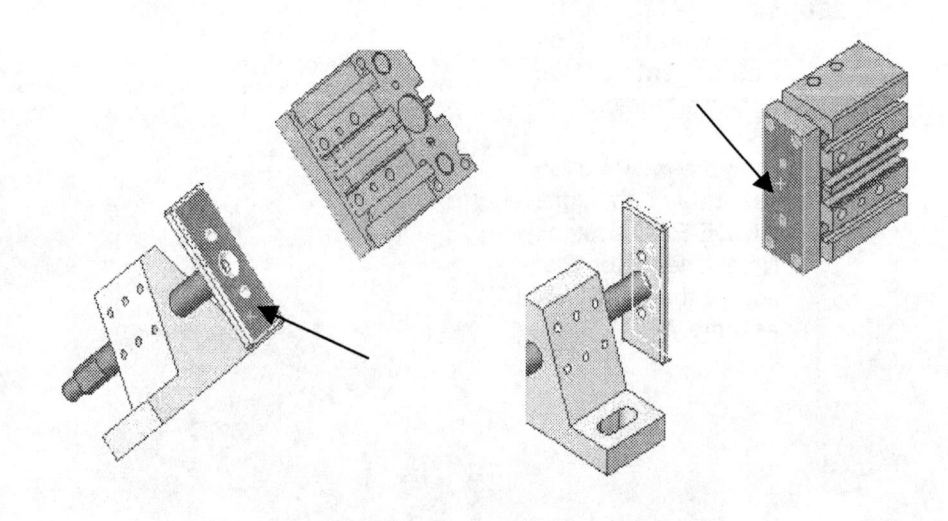

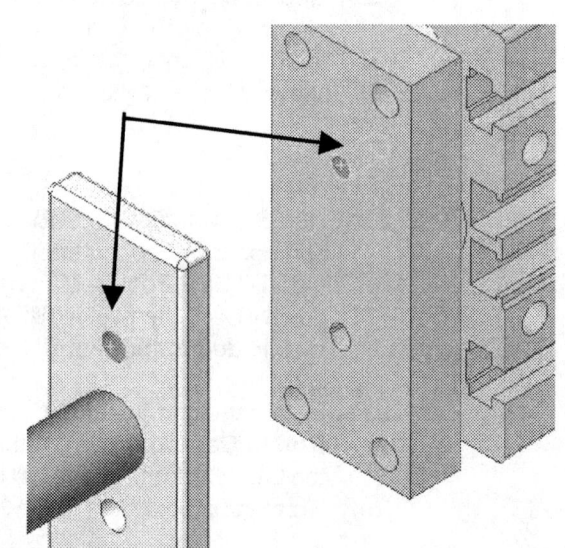

Engineering Design with SolidWorks   Fundamentals of Assembly Modeling

245) Click **Preview**. Click **OK**.

246) Create the third Mate. Click the **Mate** button to display the FeatureManager.

247) Click the **Top** plane of the PLATE and **PLANE2** of MGPM2139 from the FeatureManager.

248) Click **Parallel** for Mate type. Click **Preview**. Click **OK**.

249) Close the Mate PropertyManager. Click **OK**.

250) Display hidden components. Hold the **Ctrl** key down. Click the two **4MMCAPSCREW**s from the CUSTOMER FeatureManager. Right-click **Show components**. Release the **Ctrl** key.

251) Save the CUSTOMER assembly in the ENGDESIGN-W-SOLIDWORKS/PROJECTS file folder. Click **Save**.

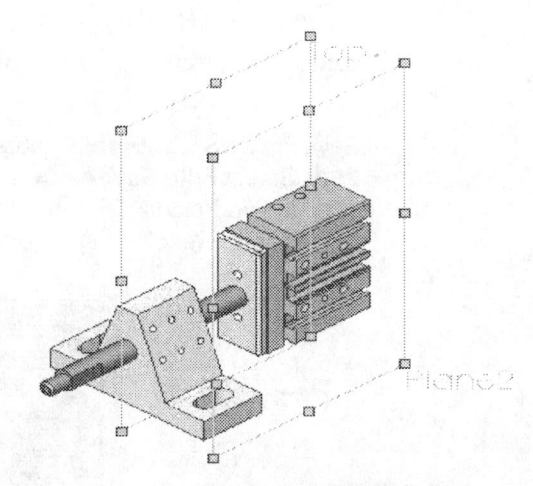

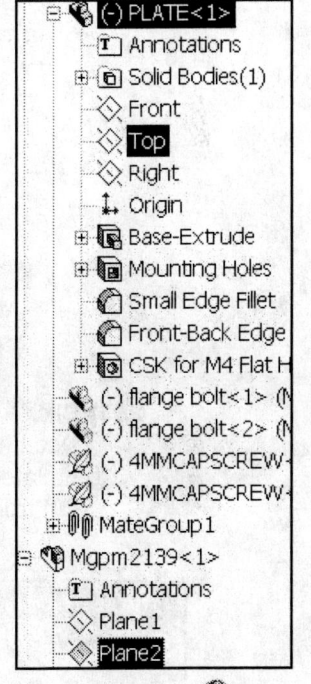

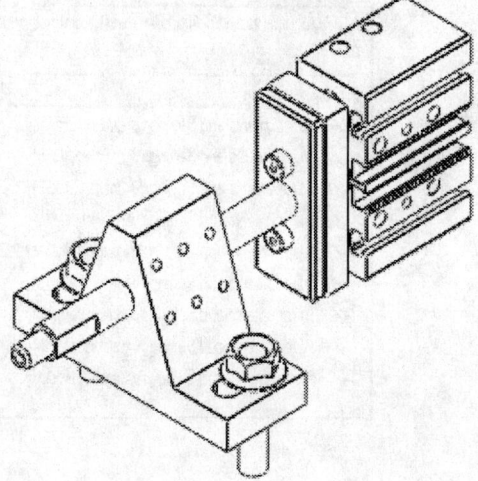

PAGE 2 - 51

Fundamentals of Assembly Modeling **Engineering Design with SolidWorks**

Send a copy of the CUSTOMER assembly to a colleague for review. Copy the CUSTOMER assembly and all of the components into a different file folder. Reference all component file locations to the new file folder.

Save the CUSTOMER assembly to a new file folder.
252) Select **File**, **SaveAs** from the Main menu. Select **My Documents** for the Save in File Folder. Click the **References** button from the Save dialog box. Click **Yes** to Resolve Lightweight Components dialog box.

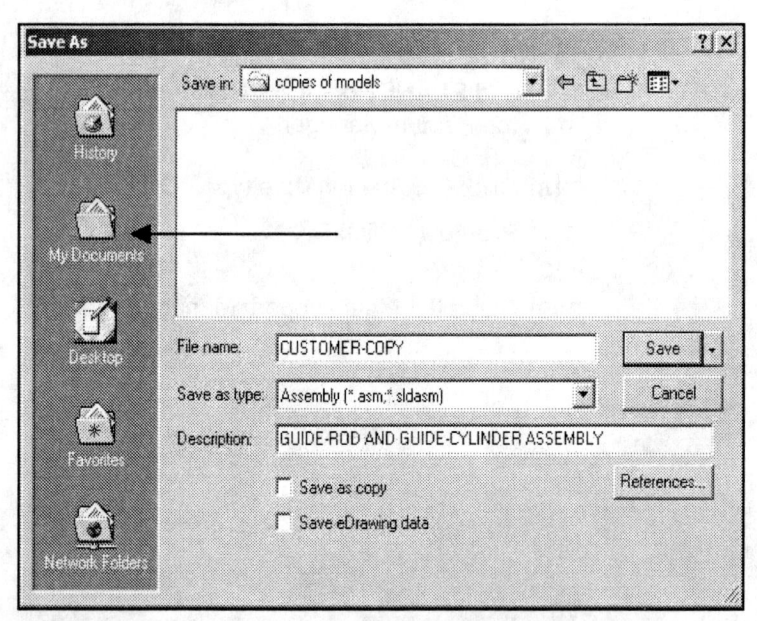

253) Click the **Select All** from the dialog box. Click the **Browse** button. Select My Documents file folder. Click **OK**. The pathnames to the components are updated and a green checkmark is displayed before each component. Click **OK** from the Edit Reference file locations dialog box.

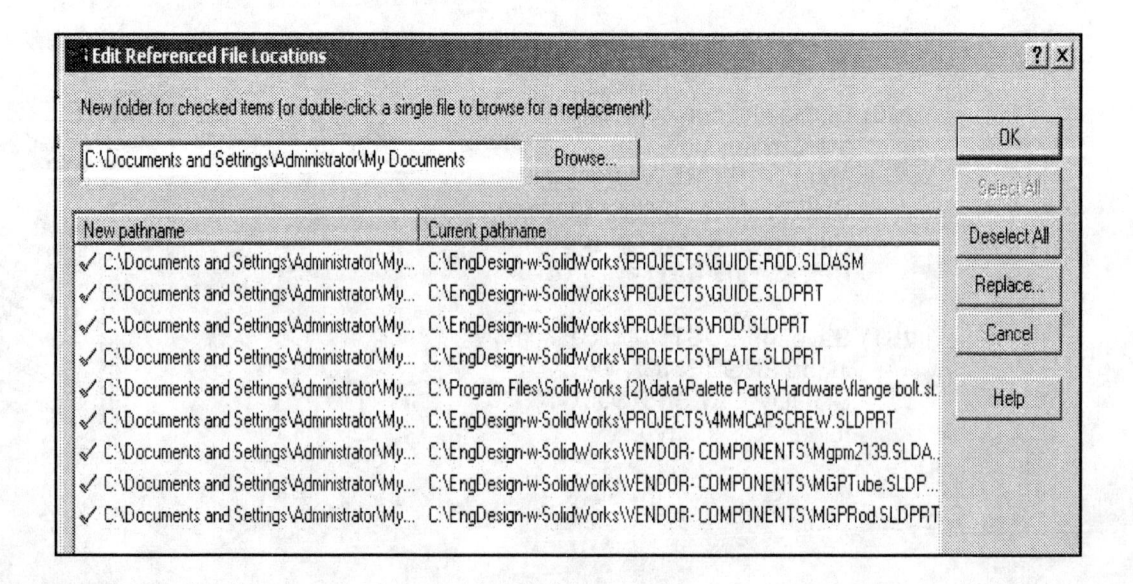

254) Save the copy of the assembly. Check the Save as copy text box. Enter **CUSTOMER-COPY** for assembly file name.

255) Click the **Save** button.

256) Click **Yes** to the question, "Save the document and the referenced models now?"

257) Close all files. Click **Windows** from the Main menu. Click **Close All**.

Will the ROD experience unwanted deflection or stress? Insure a valid GUIDE-ROD assembly by using the COSMOSXpress tool.

COSMOSXpress

COSMOSXpress is a Finite Element Analysis (FEA) tool.

COSMOSXpress calculates the displacement and stress in a part based on material, restraints and static loads.

When loads are applied to a part, the part tries to absorb its effects by developing internal forces.

Stress is the intensity of these internal forces.

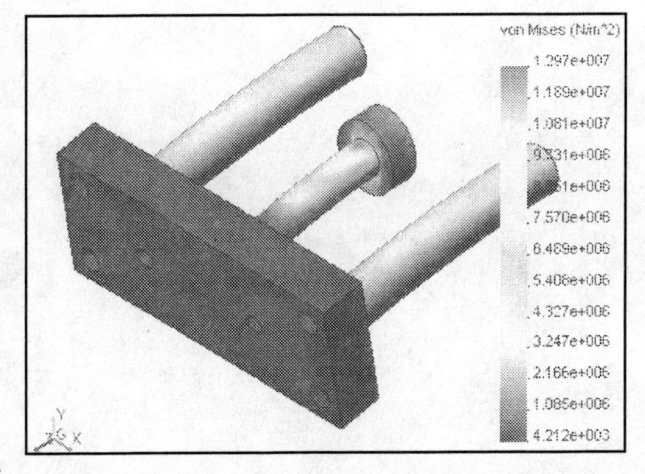

Stress is defined in terms of Force per unit Area (F/A). Different materials have different stress property levels. Mathematical equations derived from Elasticity theory and Strength of Materials are utilized to solve for displacement and stress.

These analytical equations solve for displacement and stress for simple cross sections.

Example: Bar or Beam. In complicated parts, computer based numerical methods such as Finite Element Analysis are used.

Bar   Beam

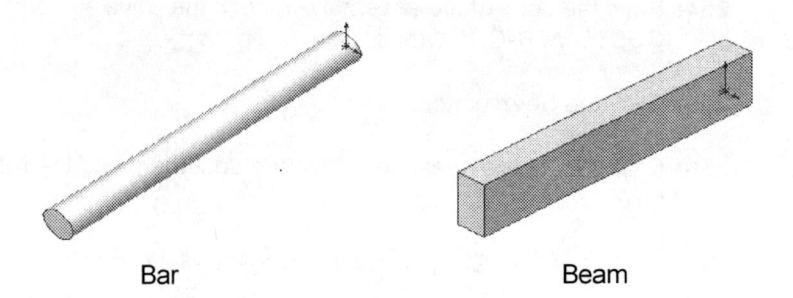

COSMOSXpress utilizes linear static analysis based on the Finite Element Method. The Finite Element Method is a numerical technique used to analyze engineering designs.

FEM divides a large complex model into numerous smaller models. A model is divided into numerous smaller segments called elements.

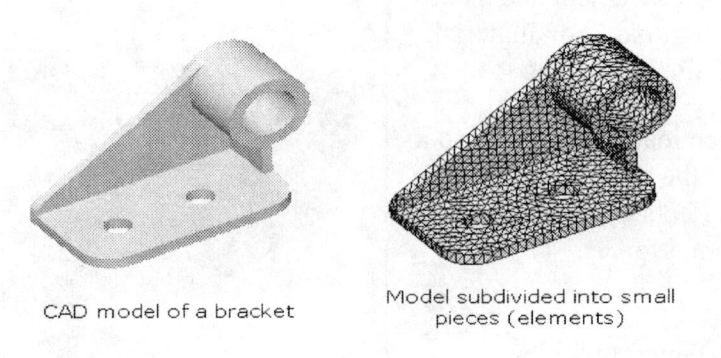

CAD model of a bracket

Model subdivided into small pieces (elements)

COSMOSXpress utilizes a tetrahedral element containing 10 nodes. Each node contains a series of equations. COSMOSXpress develops the equations governing the behavior of each element.

The equations relate displacement to material properties, restraints, "boundary conditions" and applied loads.

The COSMOSXpress Solver organizes a large set of simultaneous algebraic equations.

The Finite Element Analysis (FEA) equation is:

$[K]\{U\} = \{F\}$ where

1. $[K]$ is the structural stiffness matrix.

2. $\{U\}$ is the vector of unknown nodal displacements.

3. $\{F\}$ is the vector of nodal loads.

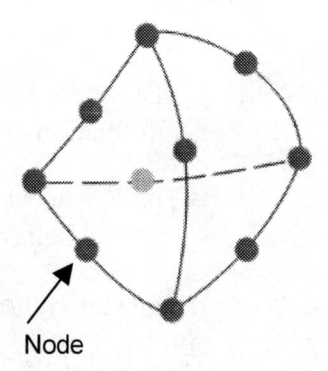

Node

Tetrahedral Element

The COSMOSXpress Solver determines the X, Y and Z displacement at each node. This displacement is utilized to calculate strain.

Strain is defined as the ratio of the change in length, δL to the original length, L. Stress is proportional to strain in a Linear Elastic Material.

The Elastic Modulus (Young's Modulus) is defined as stress divided by strain.

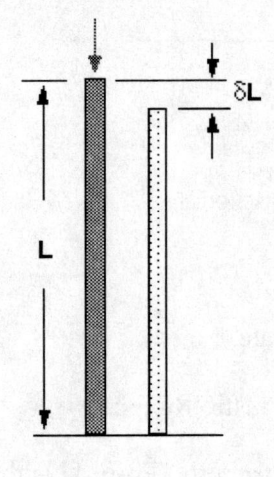

Strain = $\delta L / L$

Compression Force Applied
Original Length L
Change in Length δL

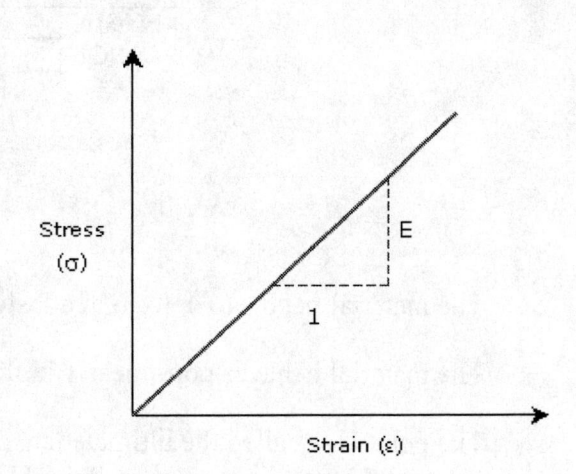

Elastic Modulus is the ratio of Stress to Strain for Linear Elastic Materials.

The COSMOSXpress Solver determines the stress for each element based on the Elastic Modulus of the material and the calculated strain.

The Stress versus Strain Plot for a Linearly Elastic Material provides information about a material.

The Elastic Modulus, E is the stress required to cause one unit of strain. The material behaves linearly in the Elastic Range.

The material remains in the Elastic Range until it reaches the elastic limit.

The point EL is the elastic limit. The material begins Plastic deformation.

The point Y is called the Yield Point.

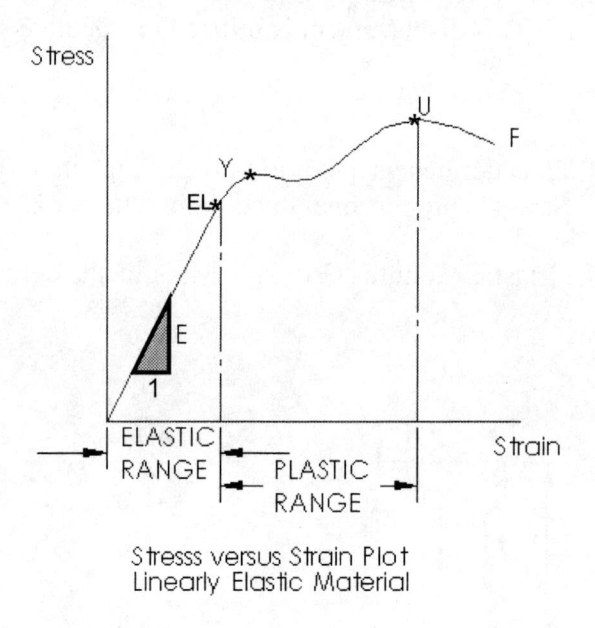

Stresss versus Strain Plot
Linearly Elastic Material

The material begins to deform at a faster rate.

The material behaves non-linearly in the Plastic Range.

The point U is called the ultimate tensile strength. Point U is the maximum value of the non-linear curve. Point U represents the maximum tensile stress a material can handle before a facture or failure. Point F represents where the material will fracture.

Designers utilize maximum and minimum stress calculations to determine if a part is safe. COSMOSXpress reports a recommended Factor of Safety during the analysis.

The COSMOSXpress Factor of Safety is a ratio between the material strength and the calculated stress.

The von Mises stress is a measure of the stress intensity required for a material to yield. The COSMOSXpress Results plot displays von Mises stress.

The COSMOSXpress design analysis wizard steps through six task tabs.

The tabs are defined as follows:

- Select the Welcome tab to set units and to store the results in a file folder.

- Select the Material tab to assign or input Material Properties to a part.

- Select the Restraint tab to apply boundary conditions to a face of a part.

- Select the Load tab to apply force or pressure to a face of a part.

- Select the Analyze tab to modify the default settings, run the analysis, automatically apply a mesh and solve a series of simultaneous equations to obtain displacement and stress results.

- Select the Results tab to view the analysis.

The Results are viewed as follows:

1. Show critical areas where the factor of safety is less than a specified value.

2. Show the stress distribution of a part.

3. Show the deformed shape of the part.

4. Generate an HTML report.

5. Generate eDrawing files for analysis.

Run the COSMOSXpress wizard: Analysis of the MGPRod part.

The MGPRod part is a component of the GUIDE-CYLINDER assembly.

Add three Restraints to the back circular faces of the MGPRod part.

The three MGPRod faces are fixed.

Apply a 100N Force on the front face of the MGPRod part. The Force is perpendicular to the Top Plane.

Analyze the MGPRod Part utilizing COSMOSXpress.

258) The MGPRod is a referenced part in the MGPM2139 assembly. Open the **MGPM2139 assembly**t.

259) Click **Open**.

260) Select **ENGDESIGN-W-SOLIDWORKS\VENDORCOMPONENTS** for Look In File Folder.

261) Double-click the **MGPM2139** assembly.

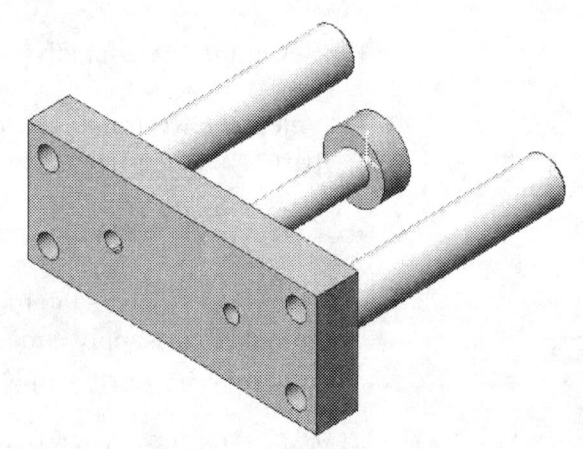

262) Open the MGPRod Part. Right-click **MGPRod** from the MGPM2139 FeatureManager.

263) Click **Open MGPRod.sldprt**.

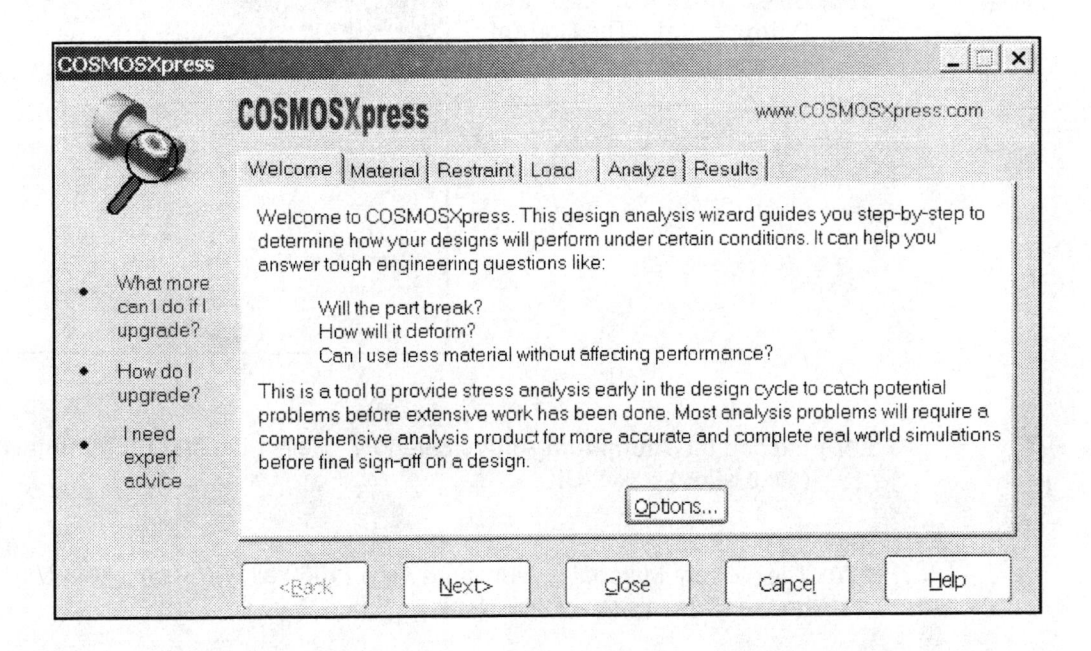

Engineering Design with SolidWorks **Fundamentals of Assembly Modeling**

264) Select the **CosmosXpress Analysis Wizard** from the Standard toolbar.

265) Select the Units and the results file folder location. Click the **Options** button from the Welcome tab. Select **SI** for Units.

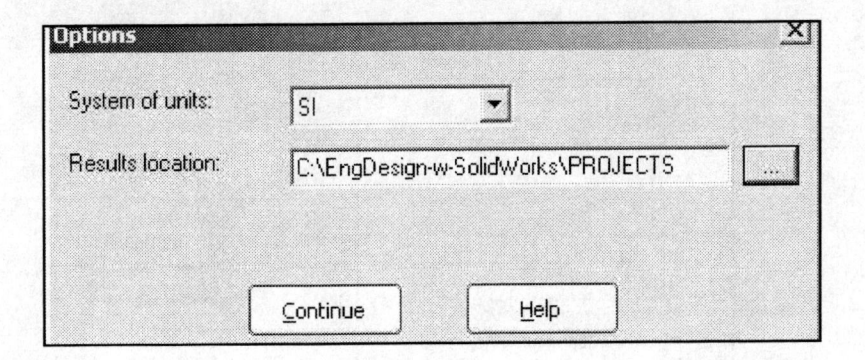

266) Select the **ENGDESIGN-W-SOLIDWORKS\PROJECTS** file folder to store the results.

267) Click **Continue**. The Material tab is selected.

Fundamentals of Assembly Modeling   **Engineering Design with SolidWorks**

268) Select a Material. Click the **Define** button. The Current Material is set to None.

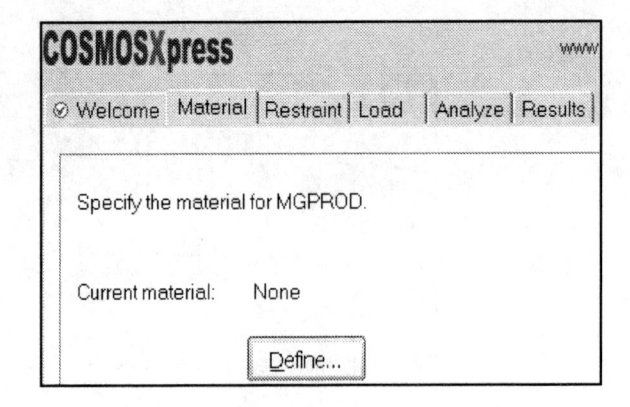

269) Expand the **Aluminum Alloys** category. Select the first **Aluminum Alloy (1060 alloy)**. Click **OK**.

270) The Current Material is Aluminum Alloy (1060 alloy). A checkmark is displayed on the Materials tab. Click **Next**.

271) The Restraint tab is selected. Click **Next**.

272) Drag the **COSMOSXpress** dialog box to the lower right corner of the Graphics window. Do not close the dialog box.

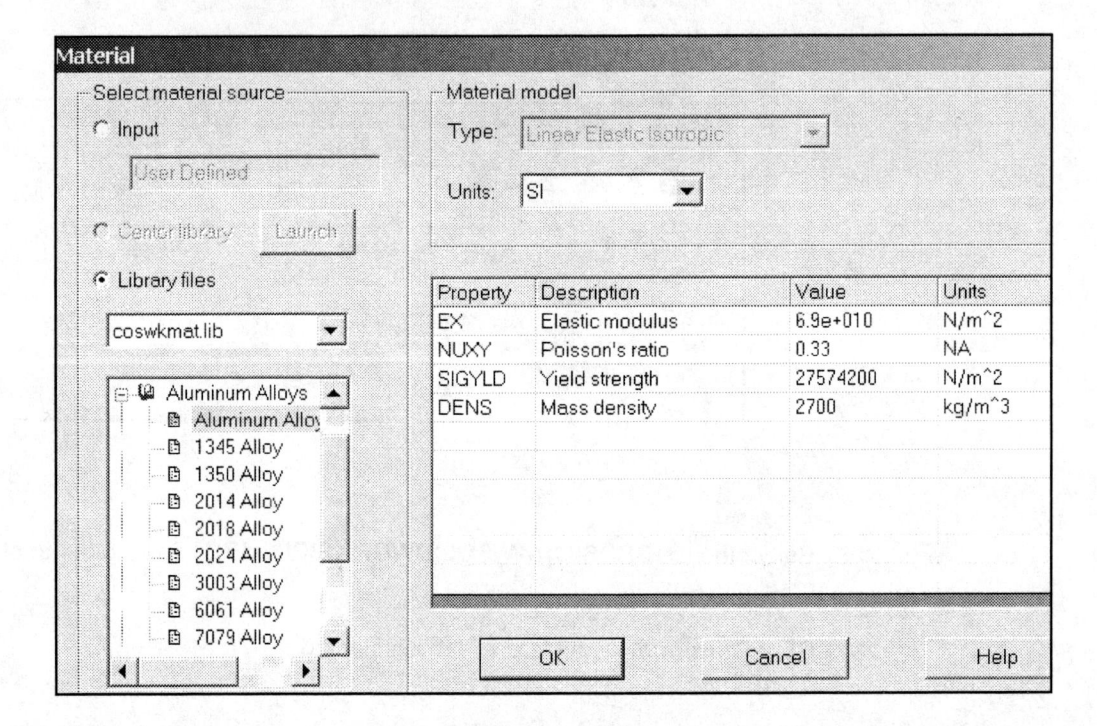

The Material dialog box lists four Material Properties for the Aluminum Alloy 1060:

- Elastic Modulus.
- Poisson's ratio.
- Yield Strength.
- Mass Density.

Note: Select the Input option from the Selected Material Source to define a new material. Input the Material Properties. COSMOSXpress utilizes Linear Elastic Isotropic materials.

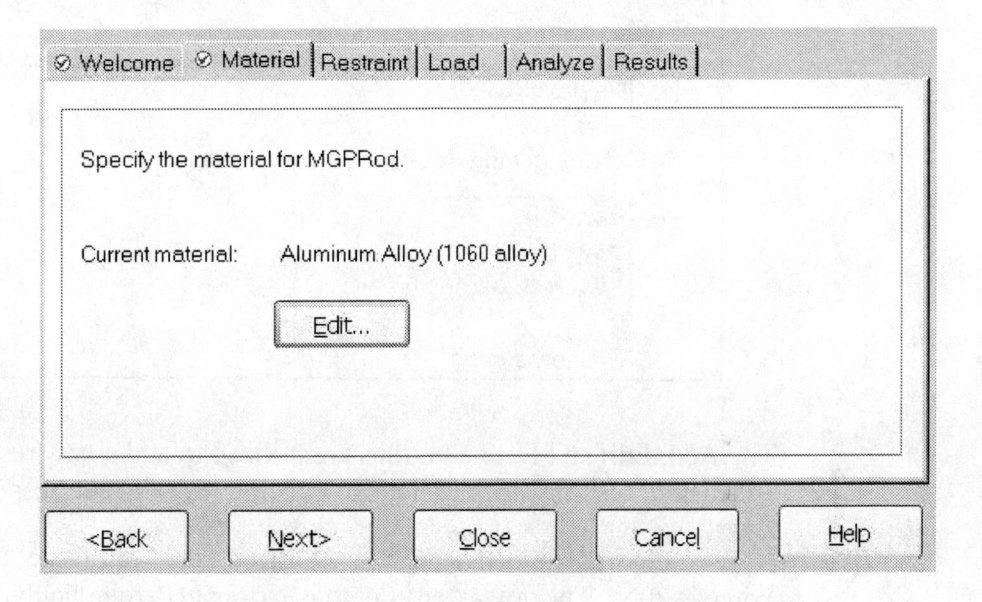

Add the Restraints to the three circular back MGPRod Part faces.

The Restraints create fixed boundary conditions.

Add Restraints.
273) Position the MGPRod part. Click a **position** inside the Graphics window. Press the **arrow keys** until the three circular back faces are displayed.

274) The current restraint set is named Restraint1. Hold the **Ctrl** key down. Click the **back middle circular face**. Small hatch symbols are displayed on the circular face.

Fundamentals of Assembly Modeling   **Engineering Design with SolidWorks**

275) Select the **back left circular face**. Select the **back right circular face**. Release the **Ctrl** key. Click **Next** from the COSMOSXpress dialog box. Face <1>, Face<2> and Face<3> are added to the restraint set. There are no additional restraint sets. Click **Next** from the COSMOSXpress dialog box.

276) A checkmark is displayed on the Restraint tab. Restraint1 is listed in the Existing Restraint Set list box. Click **Next**. The Load tab is selected.

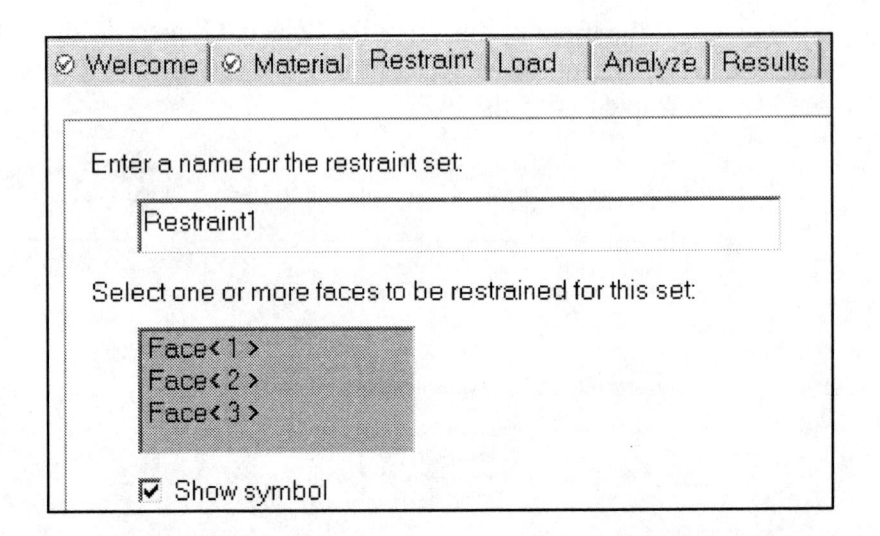

The Add, Edit and Delete buttons are utilized to add a new restraint set, edit an existing restraint set or delete an existing restraint set, respectfully.

Apply a distributed load set to the MGPRod Part front face. The applied Force is 20N. The downward Force is Normal (Perpendicular) to the Plane2 (Top) Reference Plane.

Add loads.
277) Add a Load set. Click the **Next** button. Click **Force**. Click **Next**.

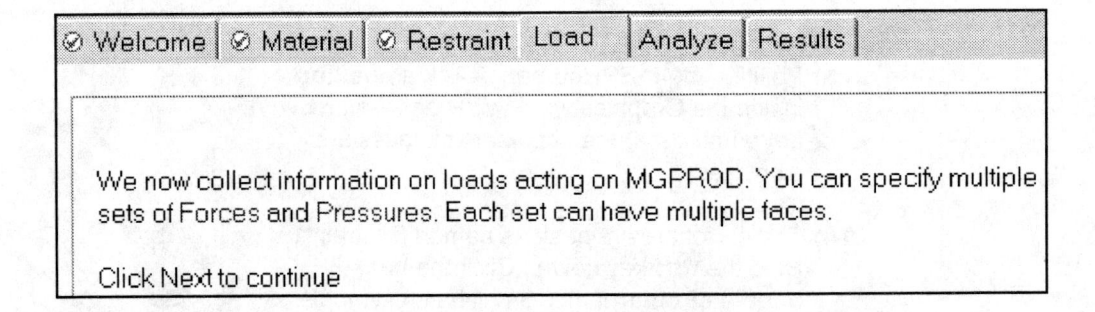

PAGE 2 - 62

Engineering Design with SolidWorks **Fundamentals of Assembly Modeling**

278) Load1 is the default name for the first load set. Click a **position** inside the Graphics window.

279) Click **Isometric**.

280) Click the MGPRod **front face**.

281) Click **Next**.

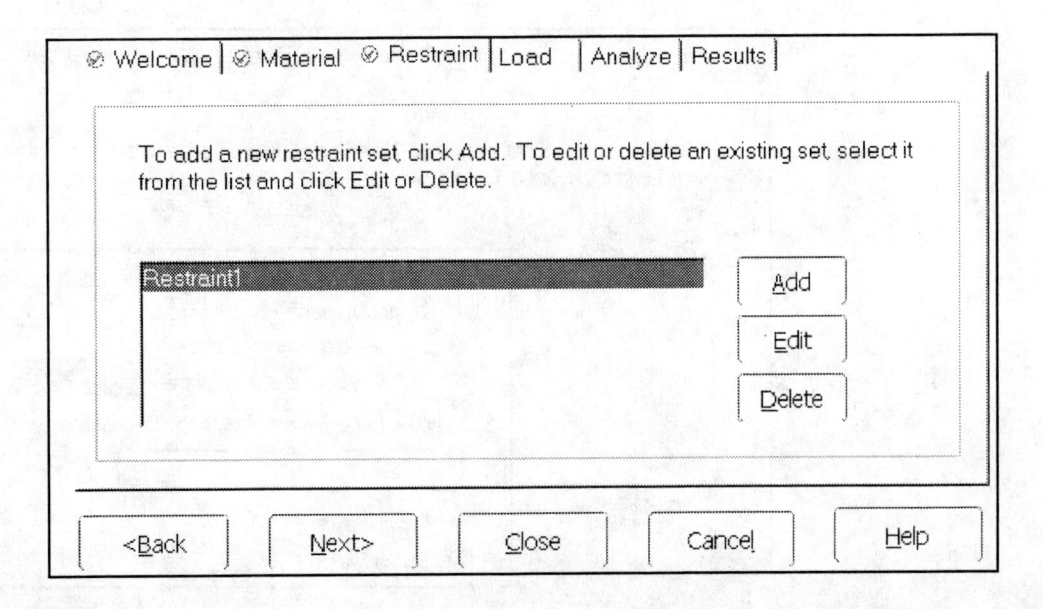

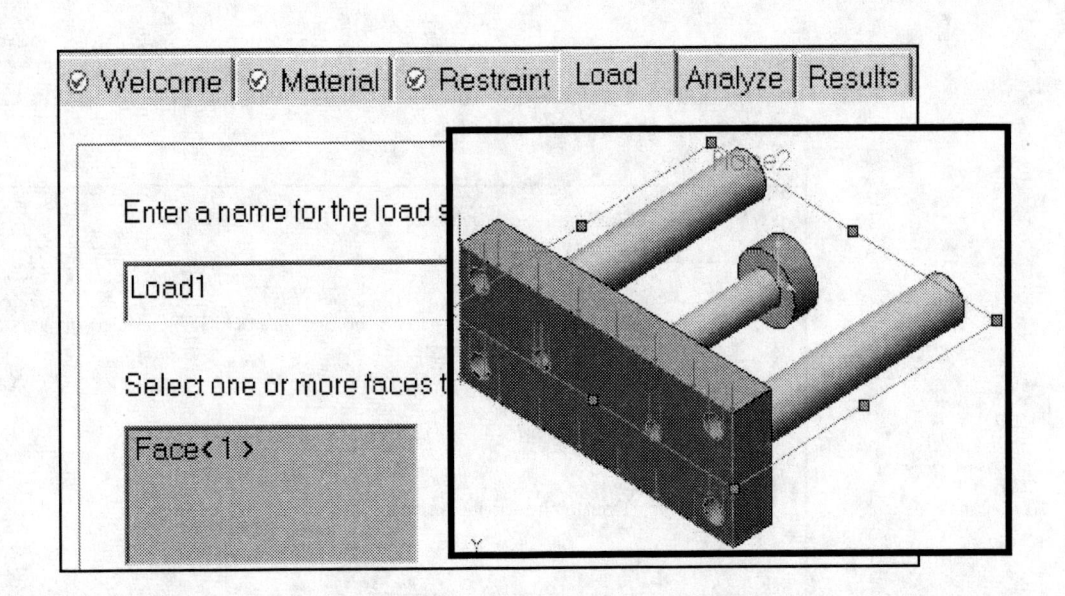

Fundamentals of Assembly Modeling **Engineering Design with SolidWorks**

282) Select the direction for Load1. Click **Normal to a reference plane**. Select **Plane2** from the FeatureManager. Enter **20N** for the Force value. Click the **Flip direction** check box. The force symbols point downward. Click **Next**.

283) Load1 is listed in the Load text box. Click the **Next** button. A checkmark is displayed on the Load tab.

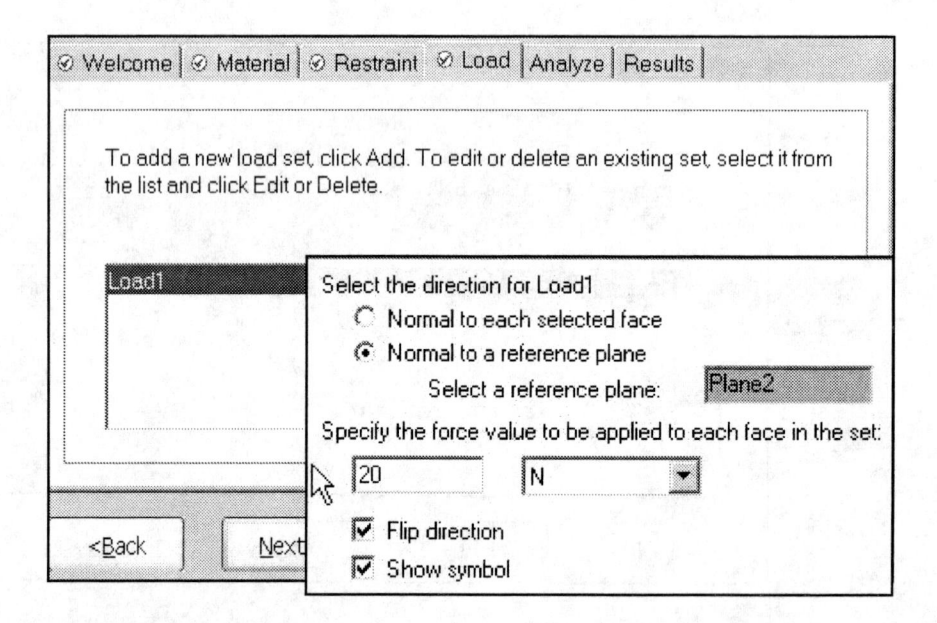

284) Click **Next**. The Analyze tab is selected.

Run the analysis.
285) The Yes button is selected to run the analysis with the default settings. Click **Next** to utilize the default settings.

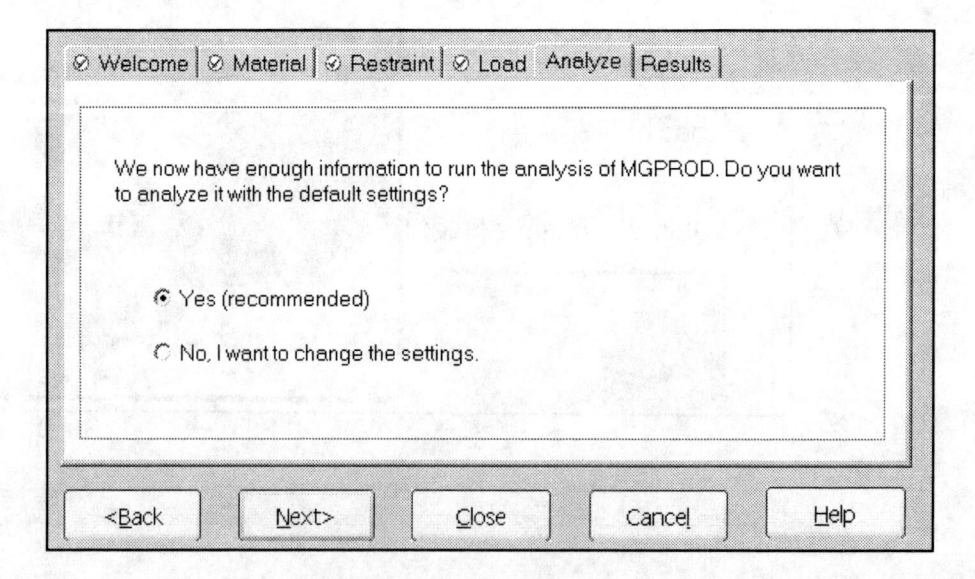

Engineering Design with SolidWorks **Fundamentals of Assembly Modeling**

286) Click the **Run** button to apply the mesh and to calculate the results. A series of simultaneous equations is solved.

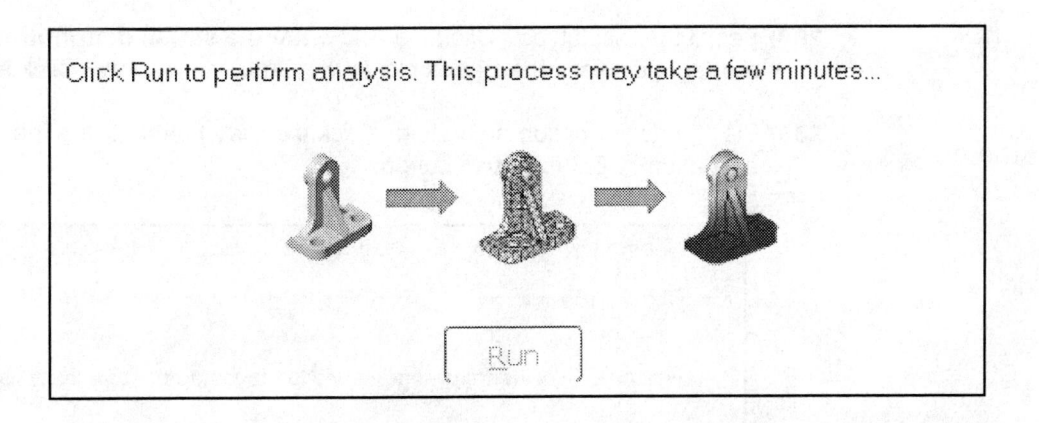

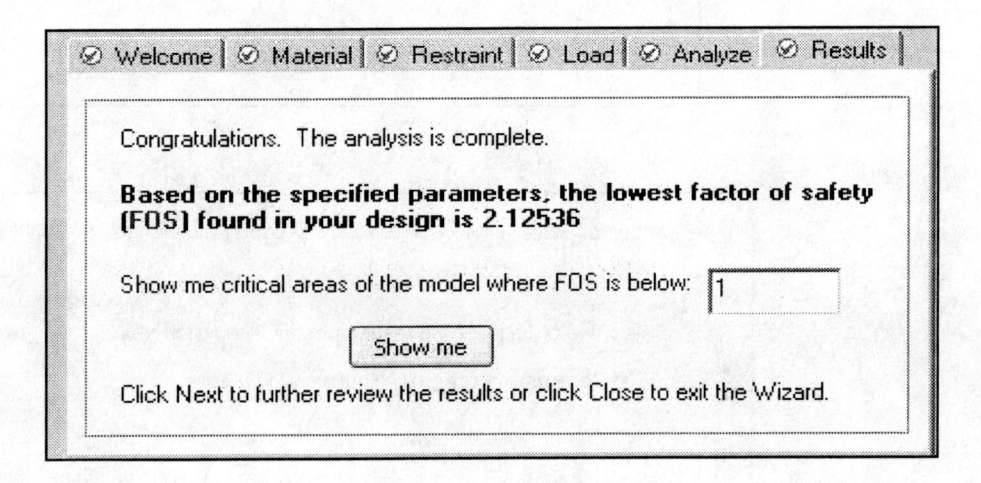

```
Click Run to perform analysis. This process may take a few minutes...

              [part] → [mesh] → [result]

                        [ Run ]
```

```
⊘ Welcome | ⊘ Material | ⊘ Restraint | ⊘ Load | ⊘ Analyze | ⊘ Results

  Congratulations. The analysis is complete.

  Based on the specified parameters, the lowest factor of safety
  (FOS) found in your design is 2.12536

  Show me critical areas of the model where FOS is below:  [ 1 ]

                        [ Show me ]

  Click Next to further review the results or click Close to exit the Wizard.
```

287) A checkmark is displayed on the Analysis tab.
The Lowest Factor of Safety (FOS) for the MGPRod Part is 2.12536. Display the Results. Click the **Show me** button. The model is displayed in blue. The model meets your lowest factor of safety. Click **Next**.

288) Select the result type. Click the **Show me the stress distribution in the model** to displace the Stress plot in the Graphics window. Click **Next**.

289) Play the Deformation animation. Click the **Play** button. Stop the Deformation animation. Click the **Stop** button.

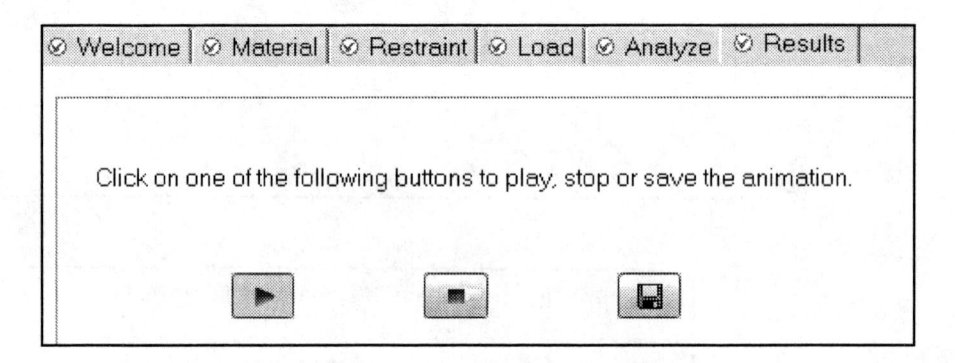

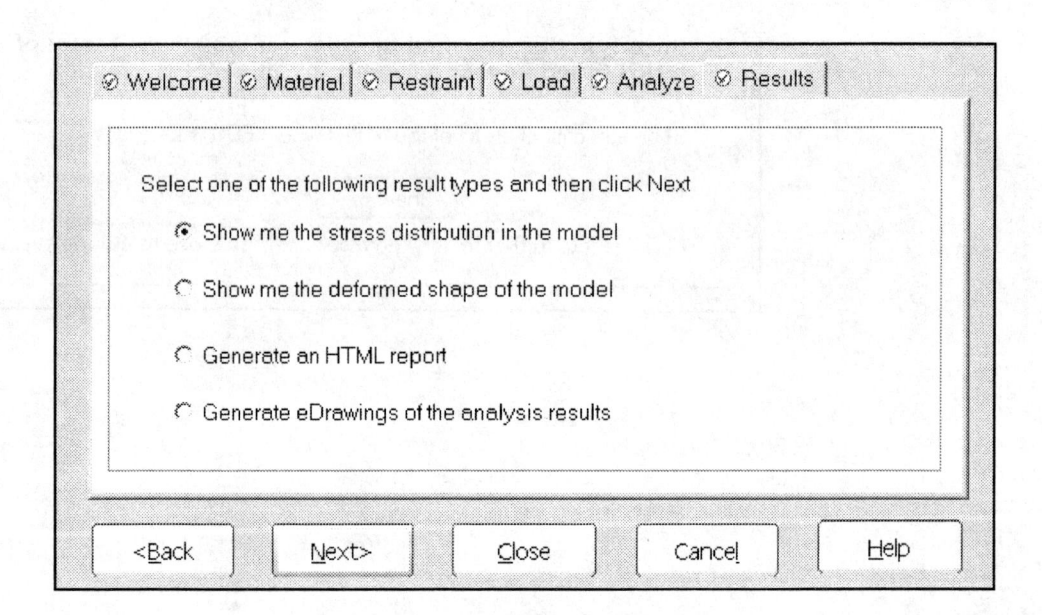

The MGPROD Part is within the Factor of Safety design criteria based on the 20N Force.

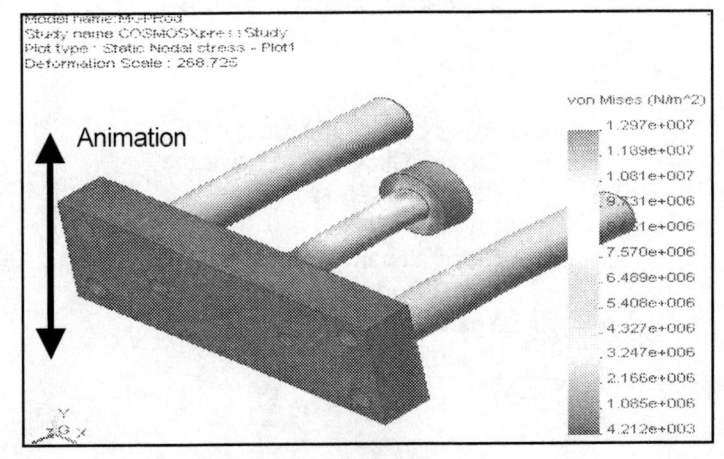

The customer requires a second design application that utilizes a 100N Load. Return to the Load tab. Modify the Force to 100N. Run the analysis and review the results.

Modify the force.
290) Click the **Load** tab. Click **Edit**. Click **Next**. Enter **100N** for Force. Click **Next**.

291) Calculate the Analyze and Results for the new Force. Click the **Analyze** tab. Click **Next**. Click **Run**.

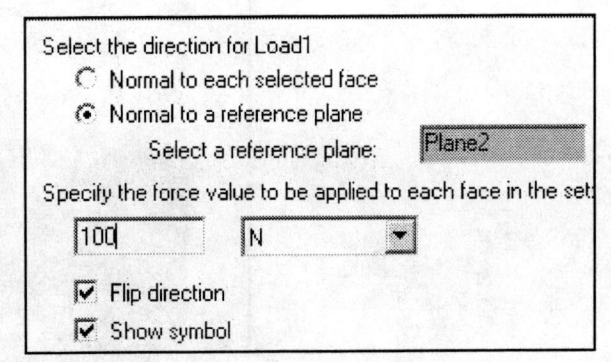

292) Review the Results. The lowest Factor of Safety is .425071. Display the critical areas of the model. Click the **Show me** button.

Congratulations. The analysis is complete.

Based on the specified parameters, the lowest factor of safety (FOS) found in your design is 0.425071

Show me critical areas of the model where FOS is below: 1

Show me

Click Next to further review the results or click Close to exit the Wizard.

Fundamentals of Assembly Modeling **Engineering Design with SolidWorks**

293) Display the critical area where the Factor of Safety is less than 1. The critical area is displayed in red.

294) Close COSMOSXpress. Click **Close**. Click **Yes** to Save the COSMOSXpress data.

295) Close all parts and assemblies. Click **Windows**, **Close All**. Click **Yes to All** to Save modified documents.

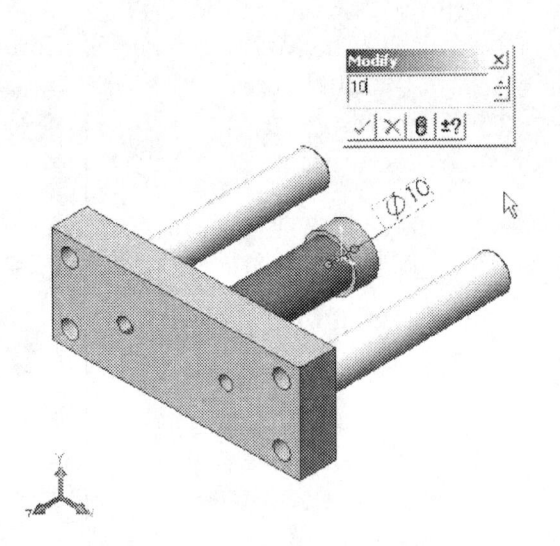

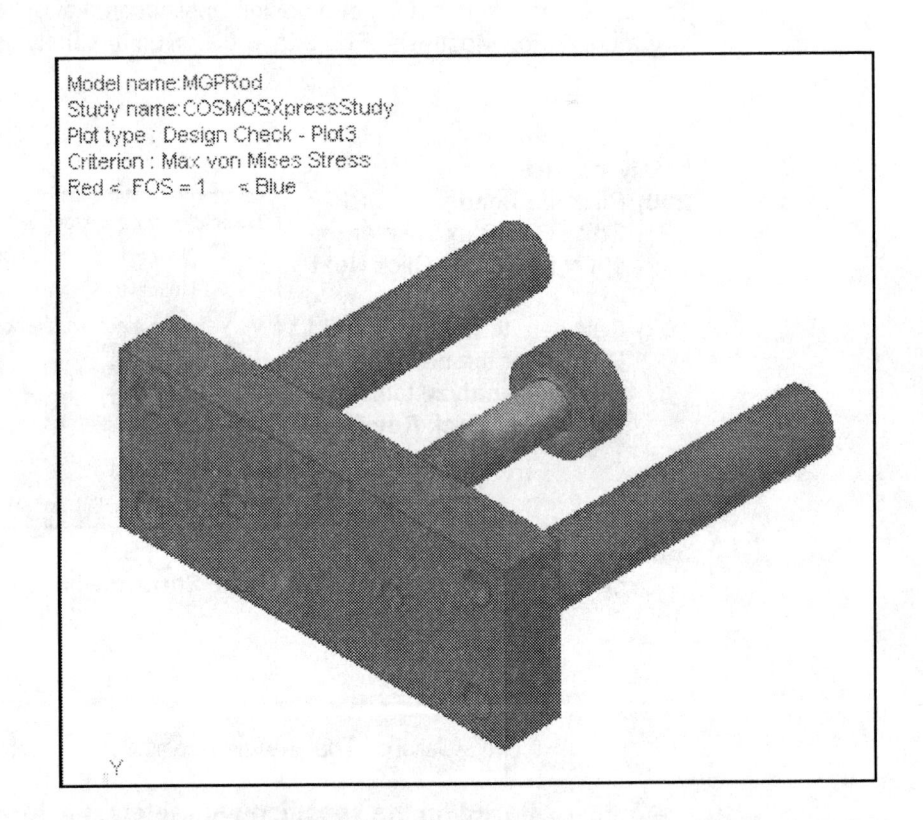

```
Model name: MGPRod
Study name: COSMOSXpressStudy
Plot type : Design Check - Plot3
Criterion : Max von Mises Stress
Red < FOS = 1   < Blue
```

Property	Description	Value	Units
EX	Elastic modulus	2.1e+011	N/m^2
NUXY	Poisson's ratio	0.28	NA
SIGYLD	Yield strength	6.20422e+008	N/m^2
DENS	Mass density	7700	kg/m^3

Material Properties for Steel

There is an area of concern near the back face of the Piston.

How do you increase the Factor of Safety? There are two suggestions:

- Modify the Material from Aluminum to Steel.

- Increase the piston diameter from 6mm to 10mm.

The MGPRod Part is a purchased part. You cannot modify this part.

Return to SMCUSA.com to locate the GUIDE-CYLINDER with a larger bore size for your customer's second application.

Here are a few tips in performing the analysis. Remember you are dealing with thousands or millions of equations. These tips are a starting point. Every analysis situation is unique.

- Utilize symmetry. If a part is symmetric about a plane, one half of the model is required for analysis. If a part is symmetric about two planes, one fourth of the model is required for analysis.

- Suppress small fillets and detailed features in the part.

- Avoid parts that have aspect ratios over 100.

- Utilize consistent units.

- Estimate an intuitive solution based on the fundamentals of stress analysis techniques.

- Factor of Safety is a guideline for the designer. The designer is responsible for the safety of the part.

Project Summary

In the assembly process, you are required to incorporate design changes to various components. Changes are based on customer requirements, materials, company standards, manufacturing concerns, etc.

COSMOSXpress provides an analysis of a part during the development process.

You created the GUIDE-ROD assembly from the following parts: GUIDE, ROD, PLATE, FLANGE BOLT and 4MMCAPSCREW.

Through the World Wide Web, you downloaded the SMC MGPM2139 assembly. The CUSTOMER assembly contains the MGPM2139 assembly and the GUIDE-ROD assembly.

Project 2 is completed. In Project 3, you will create an assembly drawing and a detailed drawing of the GUIDE.

Project Terminology

Bottom-Up assembly design approach: Components are assembled using part dependencies and parent-child relationships. In this approach, you possess all of the required design information for the individual components. Components are oriented and positioned in the assembly using Mates. Use the Bottom-Up assembly design approach for the GUIDE-ROD and the CUSTOMER assemblies.

Top-Down assembly design approach: Major design requirements are translated into assemblies, sub-assemblies and components. You do not need all of the required component design details. Individual relationships are required.

Assembly: An assembly combines two or more parts. In an assembly, parts are referred to as components.

Sub-assemblies: Are added to the assembly just like parts. Sub-assemblies behave as a single piece of geometry. When an assembly file is added to an existing assembly, it is referred to as a sub-assembly, (*.sldasm).

Editing the assembly: Individual parts can be edited while in the assembly. Changes can be made to the values of a part's dimensions while active in the assembly.

Mates: The action of assembling components in SolidWorks is defined as Mates. Mates are geometric relationships that align and fit components in an assembly. Mates remove degrees of freedom from a component. Mates require geometry from two different components. Selected geometry includes Planar Faces, Conical faces, Linear edges, Circular/Arc edges, Vertices, Axes, Temporary axes, Planes, Points and Origins.

SmartMates: A SmartMate is a Mate that automatically occurs when a component is placed into an assembly. The mouse pointer displays a SmartMate feedback symbol when common geometry and relationships exist between the component and the assembly. SmartMates are Concentric or Coincident.

A Concentric SmartMate assumes that the geometry on the component has the same center as the geometry on an assembled reference.

A Coincident Planes SmartMate assumes that a plane on the component lies along a plane on the assembly

Tolerance: The difference between the maximum and minimum variation of a nominal dimension and the actual manufactured dimension.

Fits: There are three major types of fits addressed in this Project:

- Clearance fit - The shaft diameter is less than the hole diameter.

- Interference fit – The shaft diameter is larger than the hole diameter. The difference between the shaft diameter and the hole diameter is called interference.

- Transition fit – Clearance or interference can exist between the shaft and the hole.

Sketch: The name to describe a 2D profile is called a sketch. Sketches are created on flat faces and planes within the model. Typical geometry types are lines, arcs, circles, rectangles and ellipses.

Component: Components and their assembly are directly related through a common file structure. Changes in the components directly affect the assembly and vise a versa. When a part is inserted into an assembly, it is called a component.

Hidden Geometry: Geometry that is not displayed.

Geometric Relations: To force a behavior on a sketch element to capture the customers design intent.

Add Relations: Provide a way to connect related geometry. Some common relations in this book are concentric, tangent, coincident and collinear.

Dimensions: Dimension the sketch to display the design intent.

Modify Dimension: Double-click the dimension text to invoke the modify dialog box. Enter the new value. Click the Green Check Mark. Rebuild.

Rebuild: After changes are made to the dimensions, rebuild the model to cause those changes to take affect.

Save As: Utilize the reference button to save all parts in a new assembly.

Exploded view: You can fully explode an assembly in a single or multi step procedure. AutoExplode explodes an assembly in a single step procedure.

Section view: Displays the internal cross section of a component or assembly. Section views can be performed anywhere in a model. The location of the cut corresponds to the Section plane. A Section plane is a planar face or reference plane.

Suppressed features and components: Suppressed features, parts and assemblies are not displayed. During model rebuilding, suppressed features and components are not calculated. Features and components are suppressed at the component or assembly level in the FeatureManager. The names of the suppressed features and components are displayed in light gray.

Lightweight part: A part only loads a portion of its model data into memory. A fully resolved part loads all model data into memory. Lightweight parts are more efficient for large assemblies.

COSMOSXpress: A Finite Element Analysis (FEA) tool. COSMOSXpress calculates the displacement and stress in a part based on material, restraints and static loads.

Elastic Modulus, E: The stress required to cause one unit of strain.

Project Features:

Extruded Base: Is the first feature in the 4MMCAPSCREW. The Extruded Base requires a 2-D circular sketch on the FRONT plane.

Extruded Boss: Adds material to the 4MMCAPSCREW.

Chamfer: Removes material from the front and back edges of the 4MMCAPSCREW.

Extruded Cut: Cut features are used to remove material form a solid. This is the opposite of the boss. Cuts begin as a Hexagon 2D sketch and remove materials by extrusions.

Questions

1. Describe an assembly or sub-assembly.

2. What are Mates and why are they important in assembling components?

3. Name and describe the three major types of Fits.

4. Name and describe the two assembly modeling techniques in SolidWorks.

5. Describe Dynamic motion.

6. In an assembly, each component has_____# degrees of freedom? Name them.

7. True or False. A fixed component cannot move and is locked to the Origin.

8. What is the SolidWorks Feature Palette?

9. Describe the different types of SmartMates. Utilize on line help to view the animations for different Smart Mates.

10. How are SmartMates used?

11. Describe a Section view.

12. What is a lightweight component? Describe the features and benefits.

13. What are Suppressed features and components? Provided an example.

14. True or False. COSMOSXpress calculates the displacement and stress in a part based on material, restraints and static loads.

Exercises

Exercise 2.1: Create the 3-GUIDE-ROD Assembly.

Design a BASE PLATE to fasten three GUIDE-ROD assemblies.

The BASE PLATE is the first component in the 3-GUIDE-ROD Assembly:

Modify the BASE PLATE to contain six slotted holes.

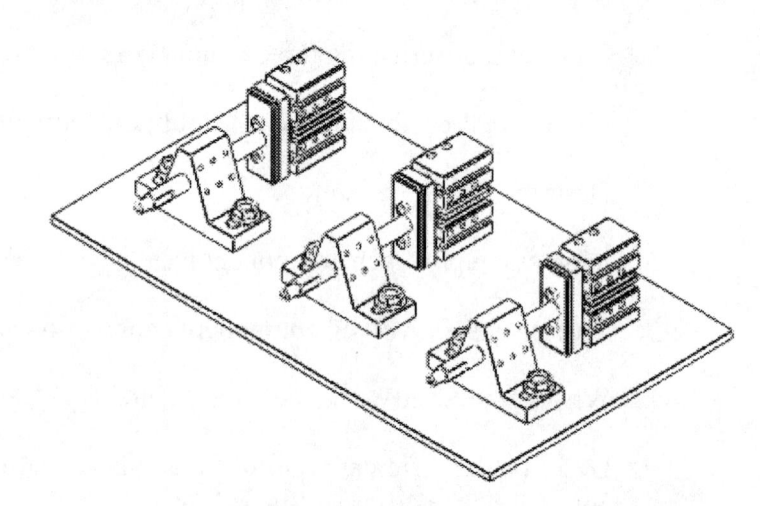

Utilize the SolidWorks/Toolbox. Click Tools, Add-Ins. Select the Toolbox checkbox. Hardware from SolidWorks/Toolbox is applied in the assembly.

Note: SolidWorks/Toolbox is required for this exercise or utilize the Flange Bolt in the FeaturePalette.

Exercise 2.2: Create the PLATE4H-GUIDECYLINDER Assembly.

The PLATE 4H part utilizes four front outside holes of the MGPM2139 SMC GUIDECYLINDER Assembly.

MGPM2139 Assembly

Manually sketch the overall dimensions of the PLATE4H part.

Dimension the 4 outside holes.

Create the new part, PLATE4H.

Create the new assembly, PLATE4H-GUIDECYLINDER.

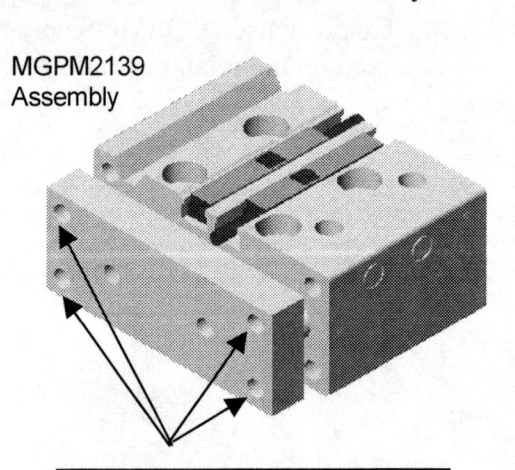

Fasten PLATE4H to the front face of the MGPM2139 SMC GUIDE CYLINDER assembly

Exercise 2.3a:
LINKAGE Assembly.

In Exercise 1.5, you created four machined parts from Gears Education Systems (www.gearseds.com).

- o FLAT BAR - 3 HOLE.
- o FLAT BAR - 9 HOLE.
- o AXLE.
- o SHAFT COLLAR.

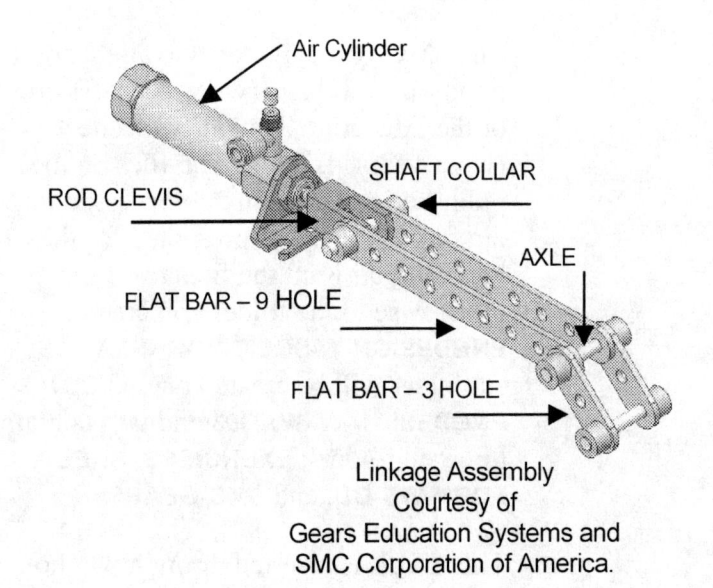

Linkage Assembly
Courtesy of
Gears Education Systems and
SMC Corporation of America.

Utilize these four parts to create the LINKAGE Assembly. The LINKAGE Assembly incorporates an SMC Air Cylinder.

The SMC Air Cylinder is the first component in the LINKAGE Assembly. When compressed air goes in to the air in let, the Piston Rod is pushed out. Without compressed air, the Piston Rod is returned by the spring.

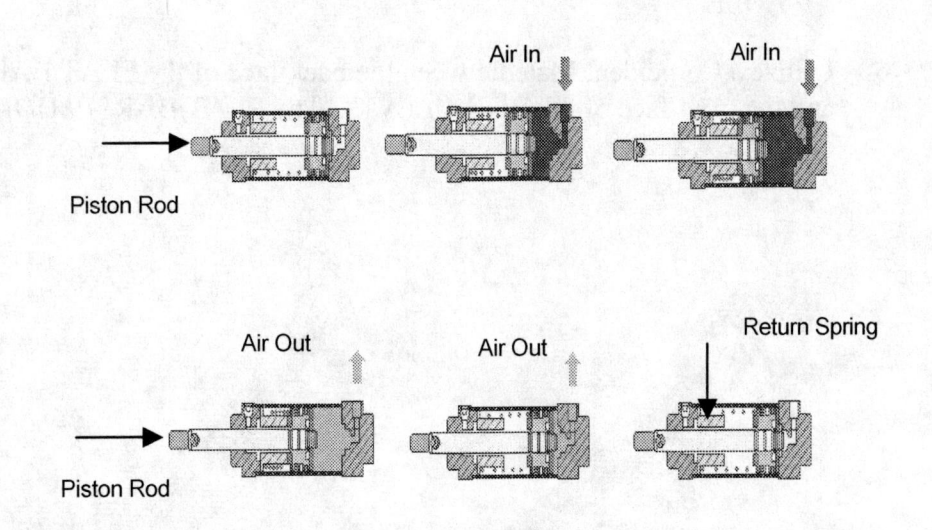

The Piston Rod linearly translates the ROD CLEVIS in the LINKAGE assembly.

Fundamentals of Assembly Modeling **Engineering Design with SolidWorks**

The SMC Air Cylinder is fixed to the LINKAGE Origin.

The SMC Air Cylinder is available on the publisher's website (www.schroff1.com) or the Education Systems web site (www.gearseds.com). The files on the publisher's website are compressed into a zip file. You will be prompted for the location you want the files to be unzipped to. Browse to the folder you created: **ENGDESIGN-W-SOLIDWORKS**. A subfolder will be created named **CD-ENGDESIGN-W-SW2003** and will contain three subfolders: **EXERCISES**, **SHEET-FORMATS-CD**, and **SMC-GEARS**.

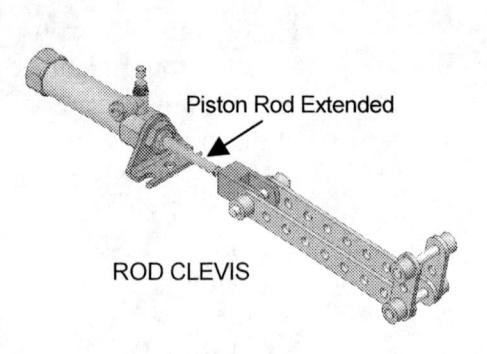

The AXLE is inserted through two holes on the ROD CLEVIS.

Utilize a Concentric Mate and a Distance Mate to center the AXLE in the ROD CLEVIS holes.

Position the FLAT BAR – 9 HOLE on the left side of the AXLE.

Utilize a Concentric Mate between the AXLE and the first hole of the FLAT BAR – 9 HOLE.

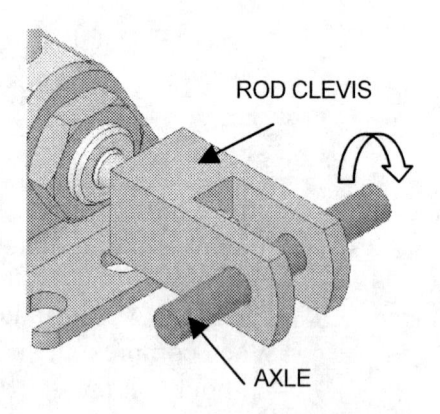

Utilize a Coincident Mate between the back face of the FLAT BAR – 9 HOLE and the front face of the ROD CLEVIS. The FLAT BAR – 9 HOLE rotates about the AXLE.

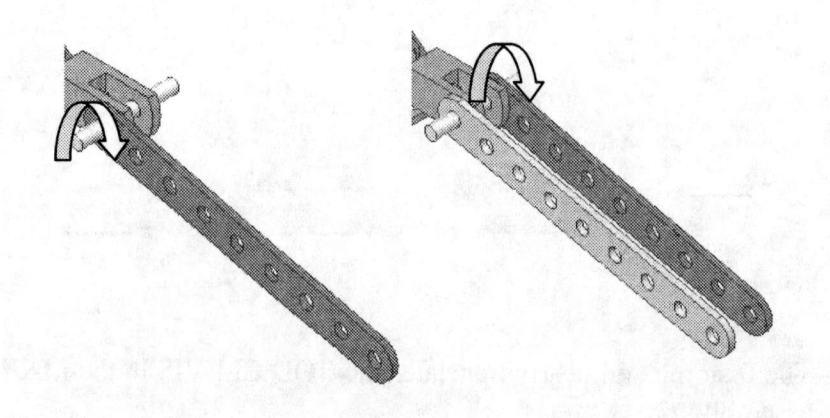

PAGE 2 - 76

Repeat the Concentric Mate and Coincident Mate for the second FLAT BAR – 9 HOLE.

The second FLAT BAR – 9 HOLE is free to rotate about the AXLE, independent from the first FLAT BAR.

Add a Parallel Mate between the two top narrow faces of the FLAT BAR – 9 HOLE. The two FLAT BAR – 9 HOLEs rotate together.

Insert the second instance of the AXLE. Mate the AXLE to the right hole of the FLAT BAR – 9 HOLE.

Add a Distance Mate to center the AXLE between the two FLAT BAR – 9 HOLEs.

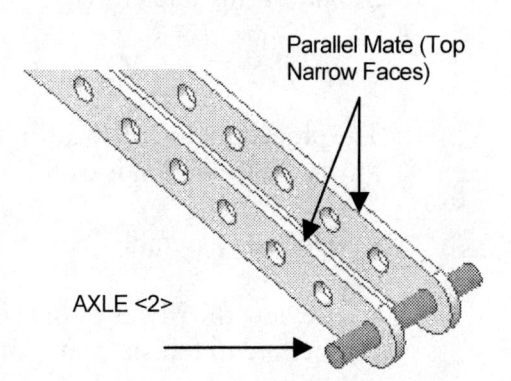

Insert the first FLAT BAR – 3 HOLE. Add Mates. The back face of the FLAT BAR – 3 HOLE is coincident front face of the FLAT BAR – 9 HOLE.

Repeat for the second FLAT BAR – 3 HOLE. The two FLAT BAR – 3 HOLEs are free to rotate about AXLE <2>.

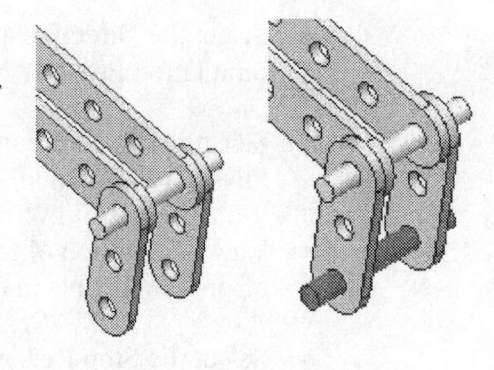

Insert the third AXLE to complete the LINKAGE Assembly.

Note: There is more than one mating technique to create the LINKAGE assembly. Utilize Mates that reflect the physical behavior of the assembly.

Exercise 2.3b: Physical Simulation for Linkage Assembly.

The Physical Simulation tools represent the effects of motors, springs and gravity on an assembly. The Physical Simulation tools are combined with Mates and Physical Dynamics to translate and rotate components in an assembly.

The Simulation Toolbar contains four simulation tools: Linear Motor, Rotary Motor, Spring and Gravity.

The procedure to create a Rotary Motor Physical Simulation is as follows:

- Open an assembly.

- Review the Mates of the components that are free to translate and rotate.

- Apply a Physical Simulation tool to a face of a part in the assembly.

- Select the Direction and Velocity from the Simulation Property Manager.

- Select the Record button to record the Physical Simulation.

- View the motion of the part with respect to other components in the assembly.

- Select the Stop Record button to complete the Physical Simulation.

- Select the Replay button to play the Physical Simulation animation.

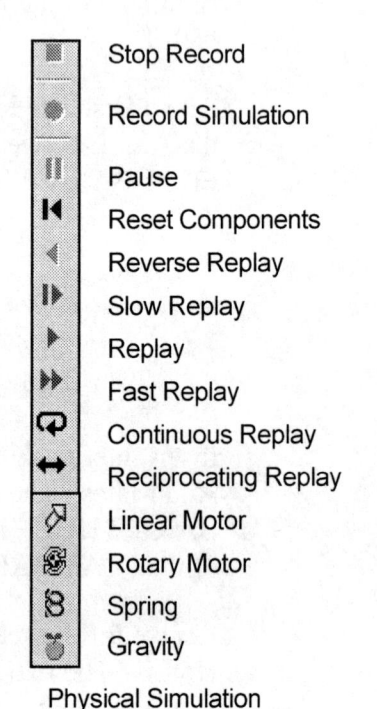

Stop Record
Record Simulation
Pause
Reset Components
Reverse Replay
Slow Replay
Replay
Fast Replay
Continuous Replay
Reciprocating Replay
Linear Motor
Rotary Motor
Spring
Gravity

Physical Simulation Toolbar

Engineering Design with SolidWorks **Fundamentals of Assembly Modeling**

The LINKAGE Assembly was created in the Exercise 2.3.

The FLATBAR – 3HOLE is free to rotate about AXLE<2>.

The FLATBAR – 9 HOLE is free to rotate about AXLE<1>.

Apply a Rotary Motor Physical Simulation tool to the front face of the FLAT BAR – 3HOLE<1>.

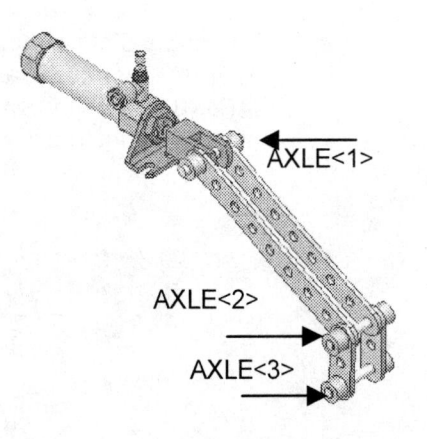

Linkage Assembly
Courtesy of
Gears Education Systems and SMC Corporation of America.

1) Rotate FLATBAR – 3HOLE. Utilize Rotate Component to position the AXLE<3> below the AXLE<2>

2) Click **Rotary Motor** from the Simulation toolbar. Click the **FLAT BAR – 3HOLE<1> front face**. A red Rotary Motor icon is displayed.

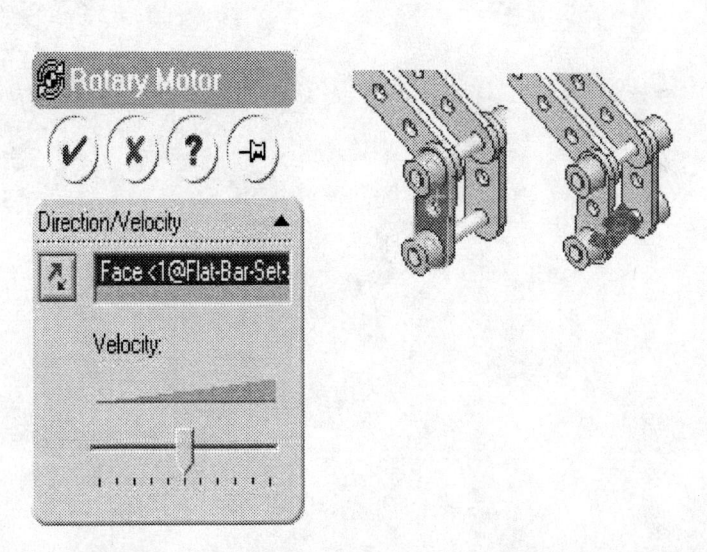

3) Click the **Direction arrow** button from the Rotary Motor Property Manager. The Direction arrow points counterclockwise. Position the **Velocity Slide bar** in the middle of the PropertyManager. Click **OK**.

Fundamentals of Assembly Modeling   **Engineering Design with SolidWorks**

4) Record the Simulation. Click **Record Simulation** ●. The FLAT BAR – 3HOLE<1> rotates in a counterclockwise direction until collision with the FLAT BAR – 9 HOLE<1> component. The FLAT BAR – 9HOLE component begins to rotate in a counterclockwise direction.

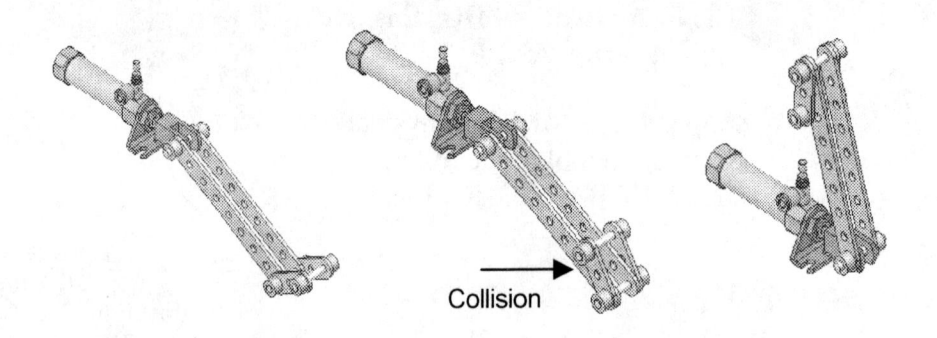

← Collision

LINKAGE Assembly Simulations

5) Stop the Simulation. Click **Stop Simulation** ■.

6) Replay the Simulation. Click **Replay**.

Exercise 2.3c: Engineering Analysis.

The AXLE created in Exercise 1.5 and assembled in Exercise 2.3 is utilized in the following exercise.

Determine the Factor of Safety, the von Mises Stress Plot and Deflection Plot utilizing COSMOSXpress.

The AXLE is fixed at both ends. A 100N load is applied along the entire cylinderical face perpendicular to the Top plane.

a) Utlize Aluminum 1060.

b) Utilize Stainless Steel.

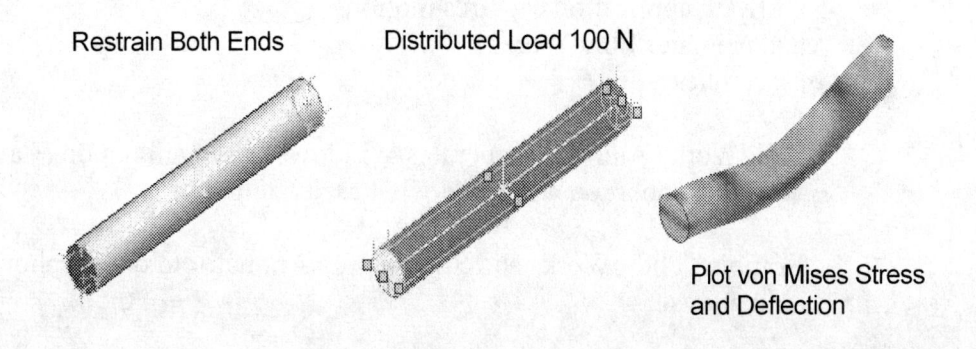

Restrain Both Ends

Distributed Load 100 N

Plot von Mises Stress and Deflection

Determine the modifications to the material, geometry, restraints and loading conditions that will increase the Factor of Safety in your design.

Exercise 2.3c: Exploded View and Animation.

a) Insert a new Exploded view for the LINKAGE Assembly. The Exploded view represents how the components will be assembled.

Note: SolidWorks Animator Add In is required for this exercise. Select Tools, Add-Ins, SolidWorks/Animator.

Create a collapsed animation of the LINKAGE assembly. Incorporate the new animation into a PowerPoint presentation.

SolidWorks Animator add-in software application captures motion and animates SolidWorks assemblies.

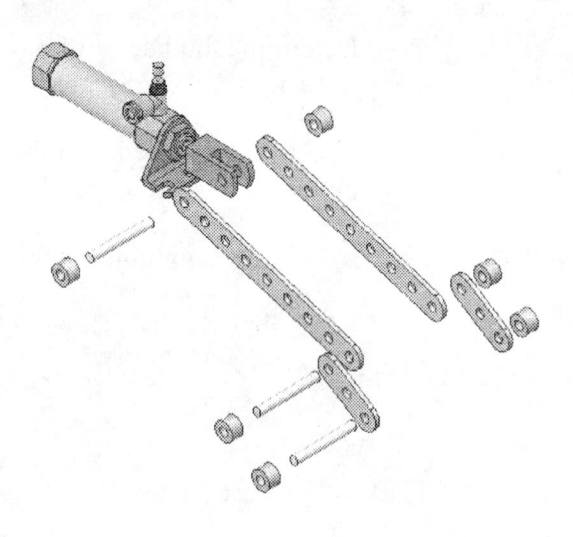

SolidWorks Animator generates Windows-based animations (.avi files) that you can support on any Windows-based computer.

Combine PhotoWorks and SolidWorks Animator to output photo-realistic animations.

Start with an exploded view of the LINKAGE assembly. Click Tools, Add Ins. Check the SolidWorks Animator check box.

Create a collapse animation of the GUIDE-ROD assembly.

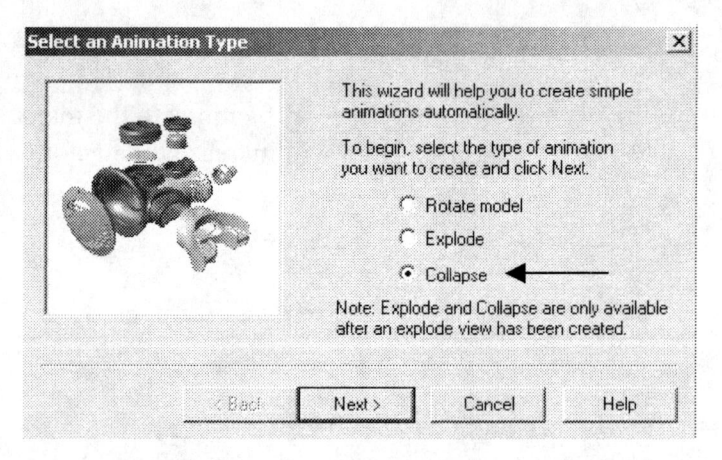

Click the Animator Wizard. Click Collapse. Click Next. Enter 20 seconds for Duration. Enter 0 for Start Time. Click Finish to view the collapse.

Record the animation and create the .AVI file. Click Record Animation to File. The Renderer is the SolidWorks screen.

Microsoft Video 1 is the default Video Compression. The time required to create the .AVI file is greater than the time required to create the SolidWorks/Animator screen file. Play the .AVI file with the Windows Media Player.

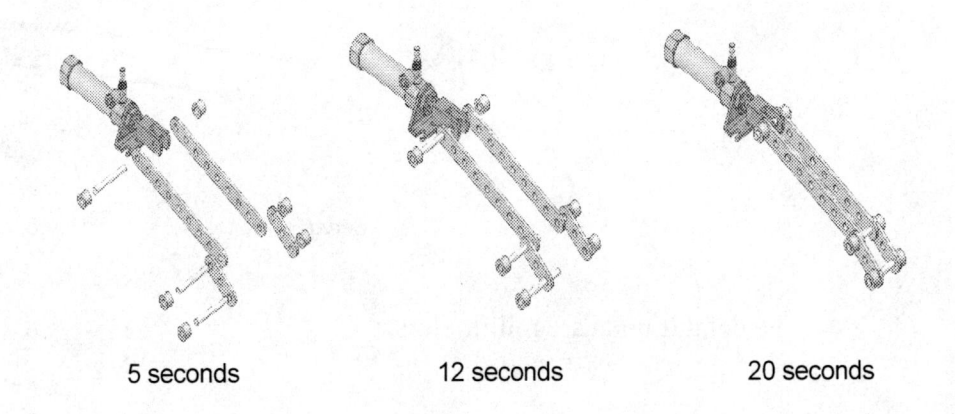

5 seconds   12 seconds   20 seconds

Exercise 2.4: Pneumatic On-Off-Purge Valve Assembly.

Create the Pneumatic On-Off-Purge-Valve Assembly.

The new assembly is comprised of the Servo Bracket Part and the SMC Purge Valve Assembly.

The SMC Purge Valve Assembly is included in the files from the publisher's website, or from Gears Educational Systems (www.gearseds.com).

The Servo Bracket Part is machined from 0.06in [1.5mm] Stainless Steel flat stock.

The default units are inches.

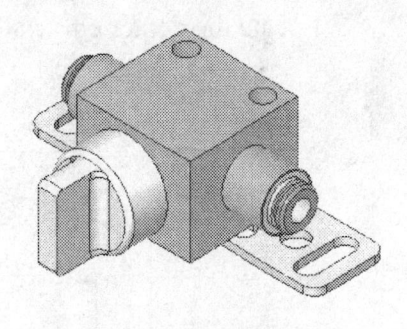

Pneumatic On-Off-Purge-Valve Assembly
Courtesy of
Gears Education Systems and SMC Corporation of America

The ⌀4.2mm [.165in] Mounting Holes fasten to the back Slot Cuts of the Servo Bracket.

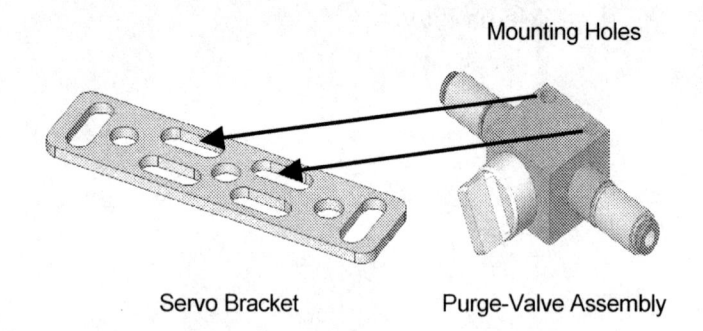

Mounting Holes

Servo Bracket   Purge-Valve Assembly

The default units are millimeters.

Engineers and designers work with components in multiple units such as inches and millimeters. Utilize Tools, Options, Document Properties, Units to check default units and precision.

The Servo Bracket Part illustration represents part dimensions, only.

Locate the center circle at the part Origin.

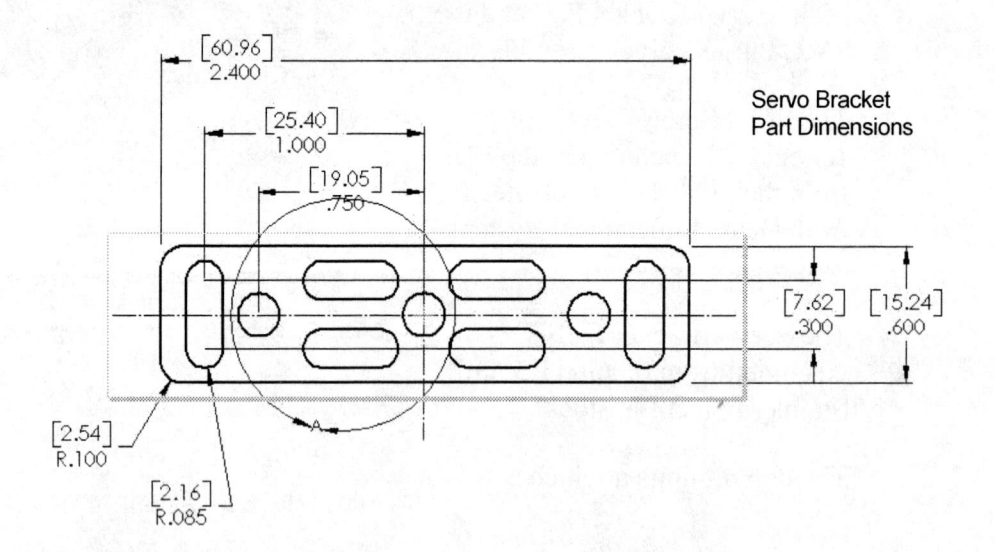

Servo Bracket
Part Dimensions

Utilize Mirror and Linear Pattern to create the features.

Mate the Right planes to center the two components.

Utilize a Distance Mate to align the Mounting Holes of the Purge Valve to the Slot Cuts of the Servo Bracket.

The Shut Off Valve knob indiates the direction of flow. The Shut Off Valve is Off when the knob is perpendicular to the direction of flow.

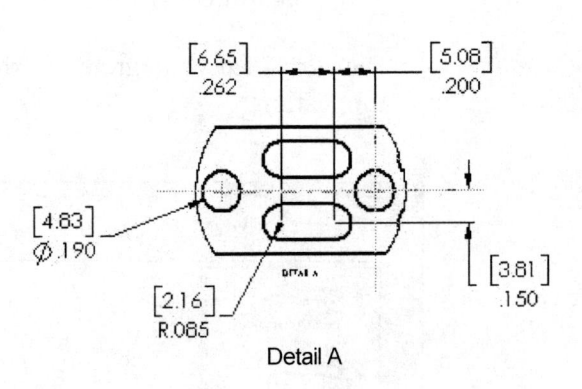

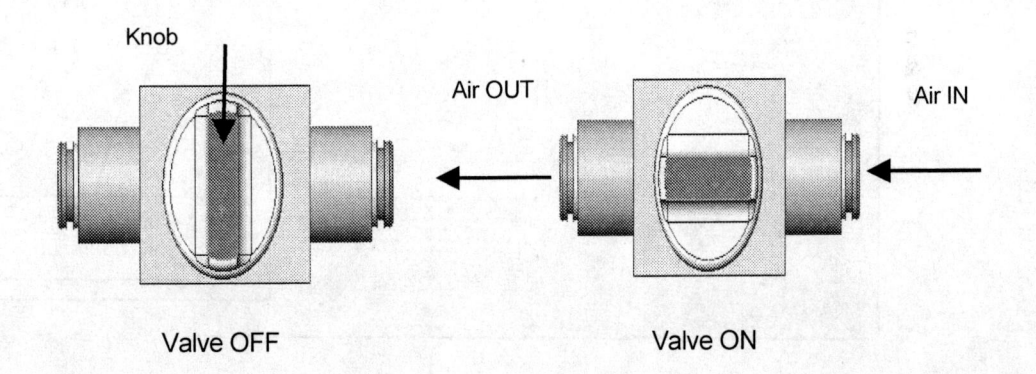

The Shut Off Valve is On when the knob is parallel to the direction of flow. Review the Mates. The Shut Off Valve Angle Mate controls the orientation of the Knob.

Exercise 2.5: Computer Aided Manufacturing.

How do you insure that the parts machined are the same as the parts modeled? CAMWorks, manufactured by TekSoft, Scottsdale, AZ USA (www.teksoft.com).

They are a fully integrated CAM (Computer Aided Manufacturing) application.

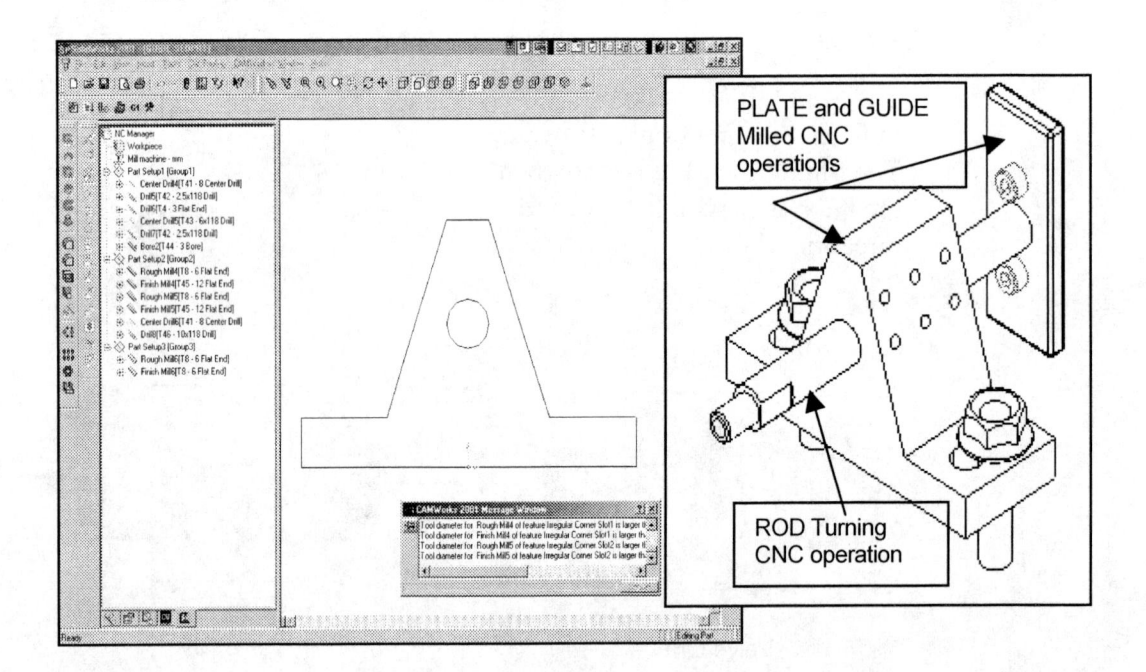

The GUIDE-ROD assembly requires Milling and Turning CNC (Computer Numerical Control) operations.

CAMWorks offers knowledge-based, feature recognition and associate machining capabilities within SolidWorks.

Research other Mill and Turning CNC Operations using the World Wide Web.

Exercise 2.6: COSMOSXpress.

Utilize COSMOSXpress to analyze the ROD. Apply a Restraint on the right face.

Apply a 50N Load on the Front Cut.

Utilize Aluminum Alloy.

Calculate the Factor of Safety.

Utilize Steel Alloy.

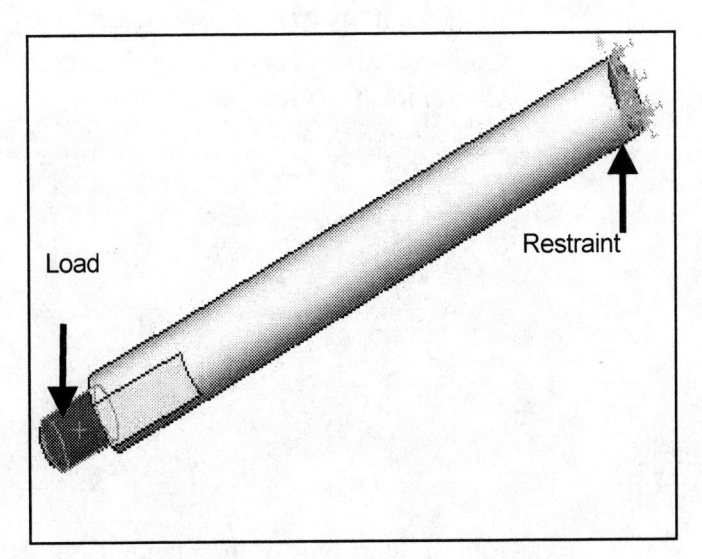

Calculate the Factor of Safety.

Exercise 2.7: Industry Collaborative Exercise.

Note: SolidWorks Animator Add In is required for this exercise. Select Tools, Add-Ins, SolidWorks/Animator.

Create a collapsed animation of the GUIDE-ROD assembly. Incorporate the new animation into a PowerPoint presentation.

SolidWorks Animator add-in software application captures motion and animates SolidWorks assemblies.

SolidWorks Animator generates Windows-based animations (.avi files) that you can support on any Windows-based computer.

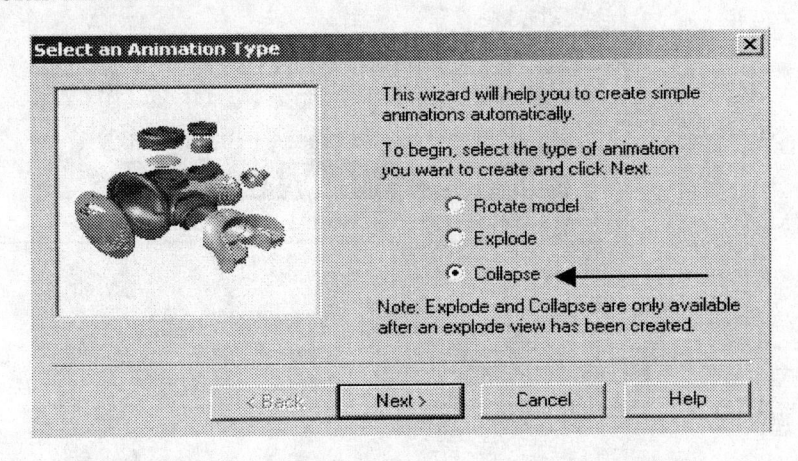

Combine PhotoWorks and SolidWorks Animator to output photo-realistic animations.

Fundamentals of Assembly Modeling **Engineering Design with SolidWorks**

Start with an exploded view of the GUIDE-ROD assembly. Click Tools, Add Ins. Check the SolidWorks Animator check box. Create a collapse animation of the GUIDE-ROD assembly. Click the Animator Wizard. Click Collapse. Click Next. Enter 20 seconds for Duration. Enter 0 for Start Time. Click Finish to view the collapse.

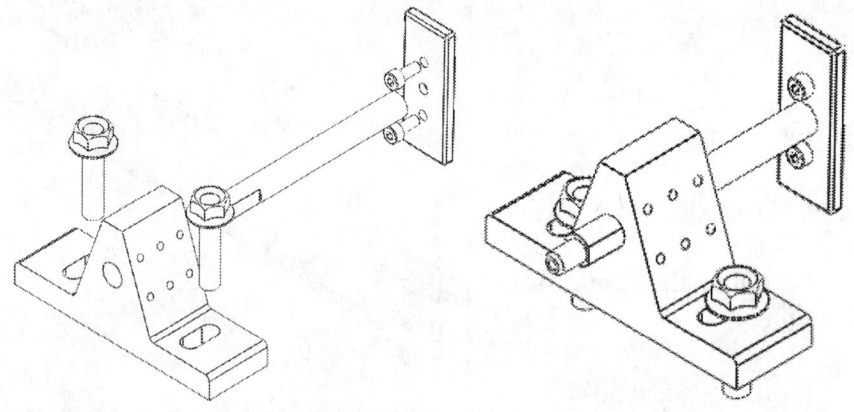

Edit the start time of the Flange Bolt. Right-click on Flange Bolt1 in the Animator FeatureManager. Click Edit Path. Enter 20 for Start time. Enter 10 for Duration.

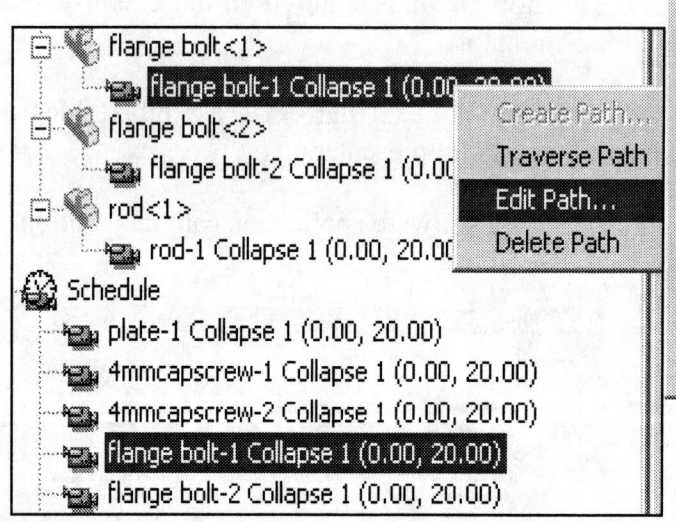

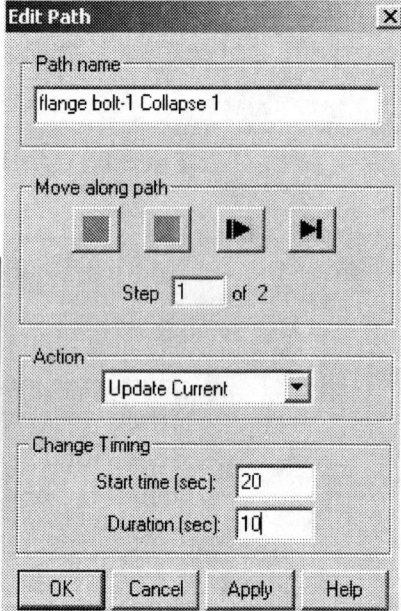

PAGE 2 - 88

Modify the start
time for the ROD
and each
4MMCAP
SCREWs.
Click Edit Path.

Enter 5 for Start
time.
Enter 15 for
Duration.

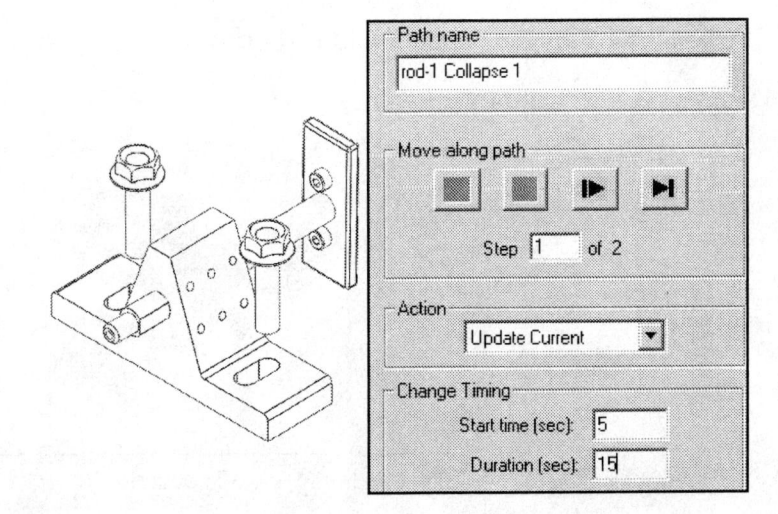

The PLATE will move first, then the ROD, then the 4MMCAP Screws and then the FLANGE BOLTS.

Rotate the model. Click the Animation Wizard.

Click Rotate the model about the Y-axis. Enter 30 for Start time. Enter 20 for Duration. Record the animation and create the .AVI file.

Click Record Animation to File . The Renderer is the SolidWorks screen.

Microsoft Video 1 is the default Video Compression. The time required to create the .AVI file is greater than the time required to create the SolidWorks/Animator screen file.
Play the .AVI file with the Windows Media Player.

Insert the .AVI file as a Movie into PowerPoint and create a presentation.

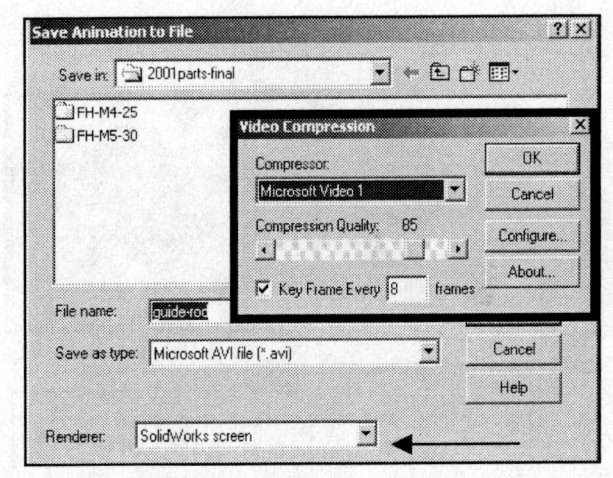

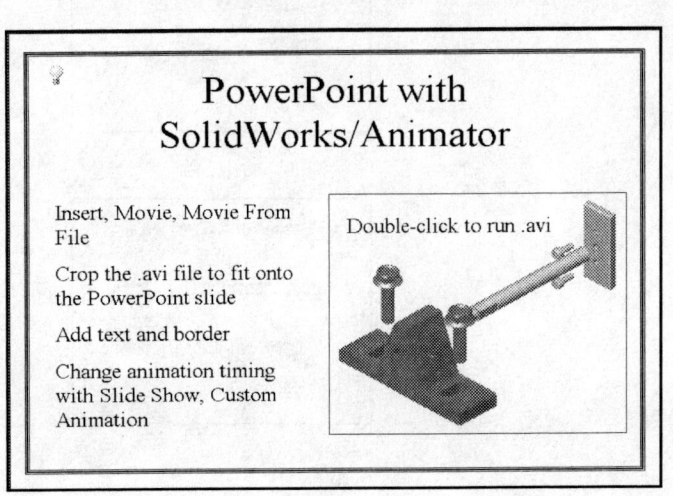

PowerPoint with SolidWorks/Animator

Insert, Movie, Movie From File

Crop the .avi file to fit onto the PowerPoint slide

Add text and border

Change animation timing with Slide Show, Custom Animation

Double-click to run .avi

Fundamentals of Assembly Modeling **Engineering Design with SolidWorks**

Play the animation in PowerPoint.

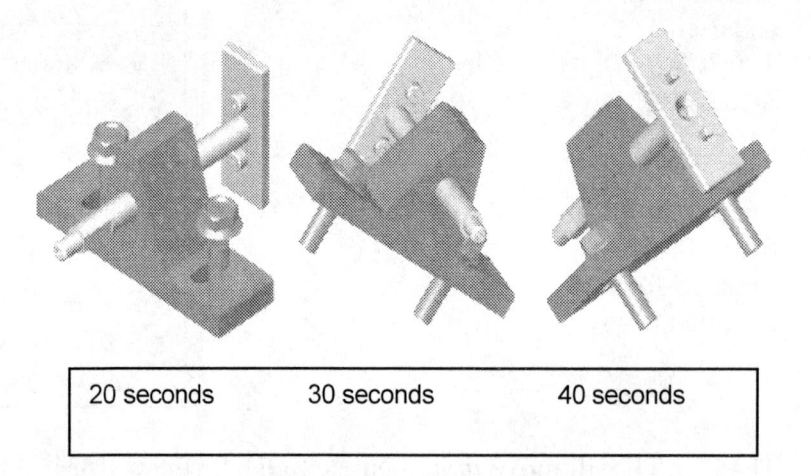

20 seconds   30 seconds   40 seconds

Note: A manufacturing procedure can be created by combining parts and assemblies from SolidWorks, PhotoWorks, SolidWorks/Animator and PowerPoint. The following example illustrates the process to create the 3 Cylinder Rack assembly.

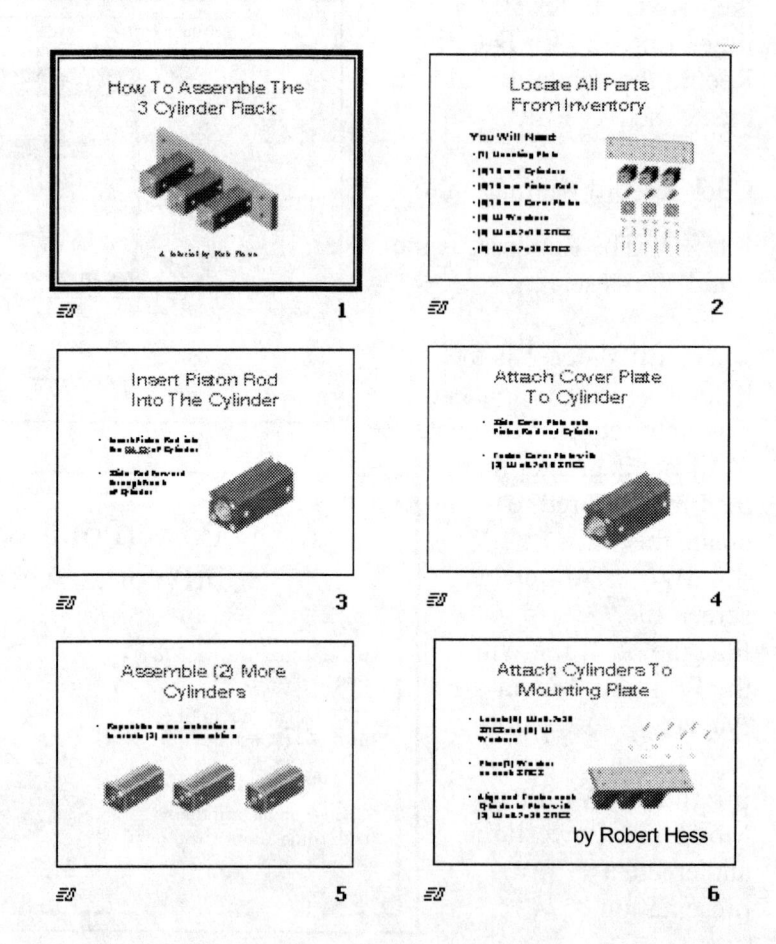

Exercise 2.8: Industry Collaborative Exercise.

In a make-buy decision process, you decide on which parts to manufacture and which parts to purchase.

In assembly modeling, you decide on which parts to design and which parts to obtain from other sources.

You now work with a team of engineers on a new project.

The senior engineer has specified a steel shaft size of 1 n diameter, [25.4mm] by 12in length, [305mm] to be used in conjunction with a mounted ball bearing.

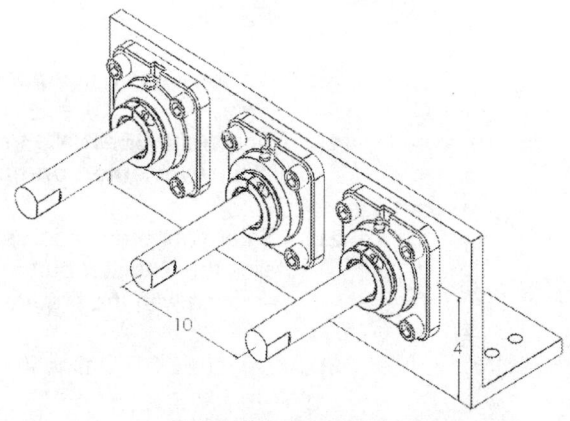

Mounted Bearing Assembly
Model and Images
Courtesy of Emerson Power Transmission Corporation,
Ithaca, NY a subsidiary of Emerson.

Three bearings are mounted to a support plate. Each shaft is separated by a minimum of 10 inches, [254mm].

The center of each shaft is located 4 inches, [100mm] about the bottom face of the support plate.

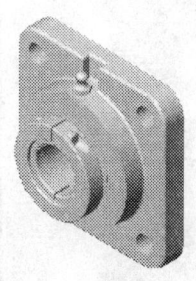

The senior engineer provides two key bearing requirements.

- The bearing contains a concentric collar.

- The bearing must be located in a 4-bolt flange block.

Time is critical. Obtain purchased components directly from the manufacturer, Emerson-EPT, website: http://www.emerson-ept.com.

Design the L-SHAPE SUPPORT PLATE to hold three bearings. Assemble the shaft to the bearing. Utilize SolidWorks Toolbox or create the fasteners for this project.

Fundamentals of Assembly Modeling   **Engineering Design with SolidWorks**

Note: Access is required to the World Wide Web with a valid email address to complete this exercise.

Size and download a component for the CUSTOMER assembly.

1) Obtain the BEARING from EMERSON-EPT. Invoke a web browser. Enter the URL: **www.emerson-ept.com**.

2) Enter your email address and password. If you are a new user to the web site, click the Register button and enter the requested information. You will be emailed with the required information.

3) Select EDGE Online in the left side of the home web page.

4) Select eCatalog.

5) Select the search criteria.

6) Click the Mounted Bearings for Product Line Search.

EDGE Online: eCatalog

EPT PART LOOKUP

Your Search Criteria

Drill Down Levels:	Mounted Bearings >> Ball Bearings >> Flange Units >> Four Bolt Mounted >> Shaft Size 1.0000 >> Locking Device Concentric Collar

Select a part for further information

EPT Part Number	Brand	Extended Part Description
VF4B-216	Browning	Browning - Standard Duty - VF4B200 Series - 4 Bolt Flange - Concentric Lock - Contact Seal
VF4B-316	Browning	Browning - Medium Duty - VF4B300 Series - 4 Bolt Flange - Concentric Lock - Contact Seal

7) Select Ball Bearings.

8) Select Flange Units.

9) Select 4 Bolt Mounted for Housing Type.

10) Select 1.0000 for Shaft Size.

11) Select Concentric Collar for Locking Device.

Engineering Design with SolidWorks — Fundamentals of Assembly Modeling

12) The Part Number and Description are displayed. Two part numbers are valid. Double-click VF4B-216 to display the part details.

13) View the 3D Image and display the file formats to download.

14) Click View 3D button.

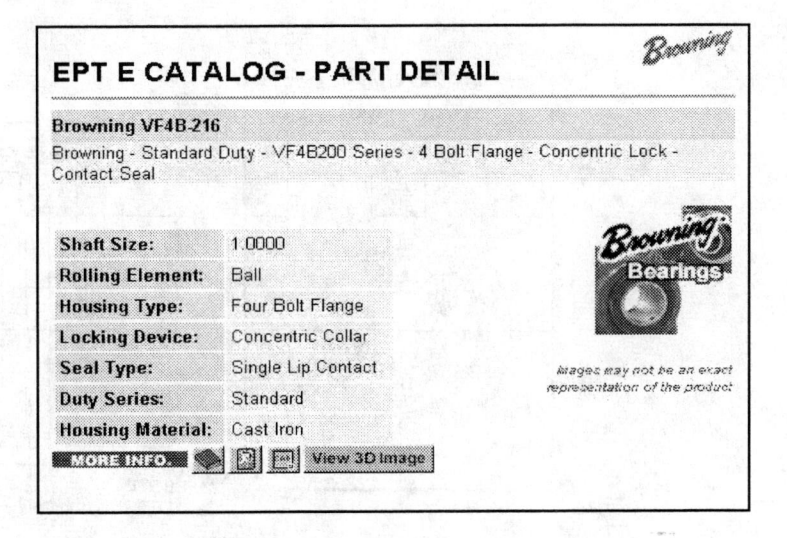

EPT E CATALOG - PART DETAIL

Browning VF4B-216
Browning - Standard Duty - VF4B200 Series - 4 Bolt Flange - Concentric Lock - Contact Seal

Shaft Size:	1.0000
Rolling Element:	Ball
Housing Type:	Four Bolt Flange
Locking Device:	Concentric Collar
Seal Type:	Single Lip Contact
Duty Series:	Standard
Housing Material:	Cast Iron

MORE INFO   View 3D Image

Images may not be an exact representation of the product

15) Select the SolidWorks Part/Assembly.

16) Select 2003 for Version.

17) Click the Download button. Save the zipped file to the ENGDESIGN-W-SOLIDWORKS\VENDOR-COMPONENTS file folder. The zipped SolidWorks file is downloaded to your computer. Unzip the file.

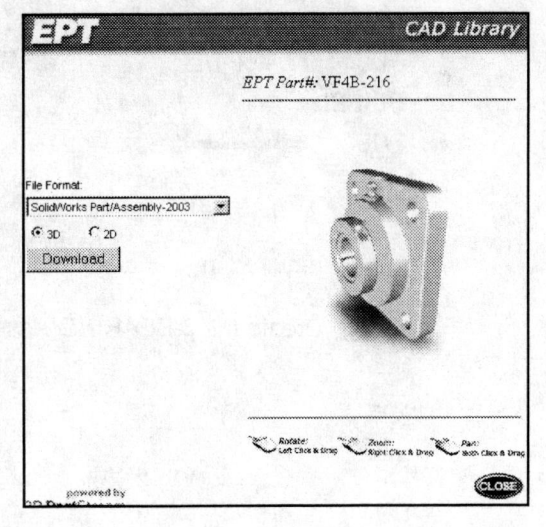

EPT — CAD Library

EPT Part#: VF4B-216

File Format:
SolidWorks Part/Assembly-2003
● 3D   ○ 2D
Download

Rotate: Left Click & Drag   Zoom: Right Click & Drag   Pan: Both Click & Drag

powered by

Fundamentals of Assembly Modeling **Engineering Design with SolidWorks**

18) Close the 3D View window.

19) Click the CLOSE button.

20) Click the Back button to obtain the specification sheet. Utilize the bolt size

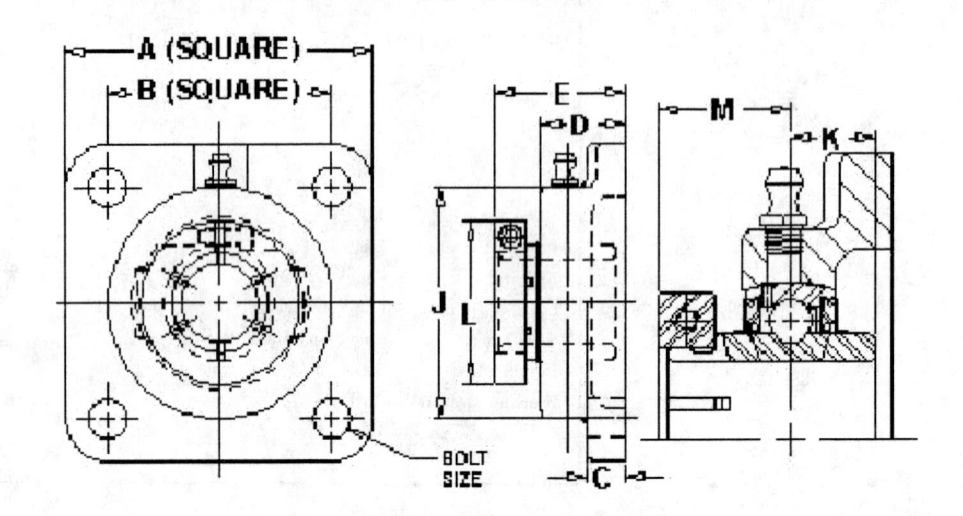

Part No	Description	Shaft Dia.	Insert No.	A	B	C
766227	VF4B-216	1.0000	VB-216	3 3/4	2 3/4	1/2

D	E	F	J	K	L	M
61/64	1 1/2	35/64	2 3/4	9/16	1 15/16	7/8

Bolt Size
7/16 ←

required in your design.

21) Create the 3-BEARING Assembly. The Shaft diameter requires a close running fit.

Engineering Design with SolidWorks — Fundamentals of Assembly Modeling

A 500lb (2224N) load is applied to the end of the 12in shaft fixed at the bearing.

A static analysis is performed utilizing COSMOSXpress.

The engineer requires a factor of safety greater than 2. Increase the shaft diameter from 1in (25.4mm) to 1.5 in.

Return to the Emerson-ept.com web site.

Model name: emerson-shaft
Study name: COSMOSXpressStudy
Plot type : Design Check - Plot3
Criterion : Max von Mises Stress
Red < FOS = 2 < Blue

LOAD
12.00
Ø1.0000

Locate a 4Bolt Flange Ball Bearing for a 1.5in shaft.

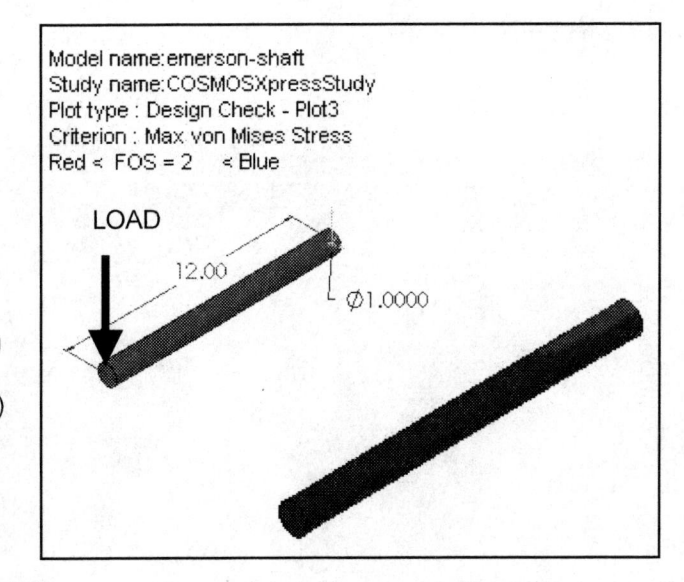

Identify the key dimensions in the new 4Bolt Flange Ball Bearing affect the overall assembly.

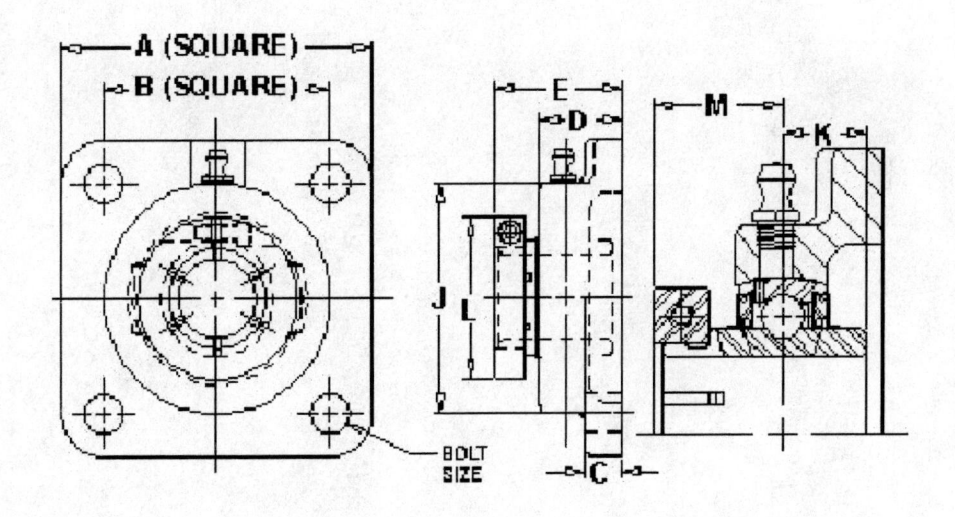

Part No	Description	Shaft Dia.	Insert No.	A	B	C
766234	VF4B-224	1.5000	VB-224	5 1/8	4	5/8
D	E	F	J	K	L	M
1 19/64	2 5/64	25/32	4 1/8	3/4	2 11/16	1 1/4

Bolt Size
1/2

Notes:

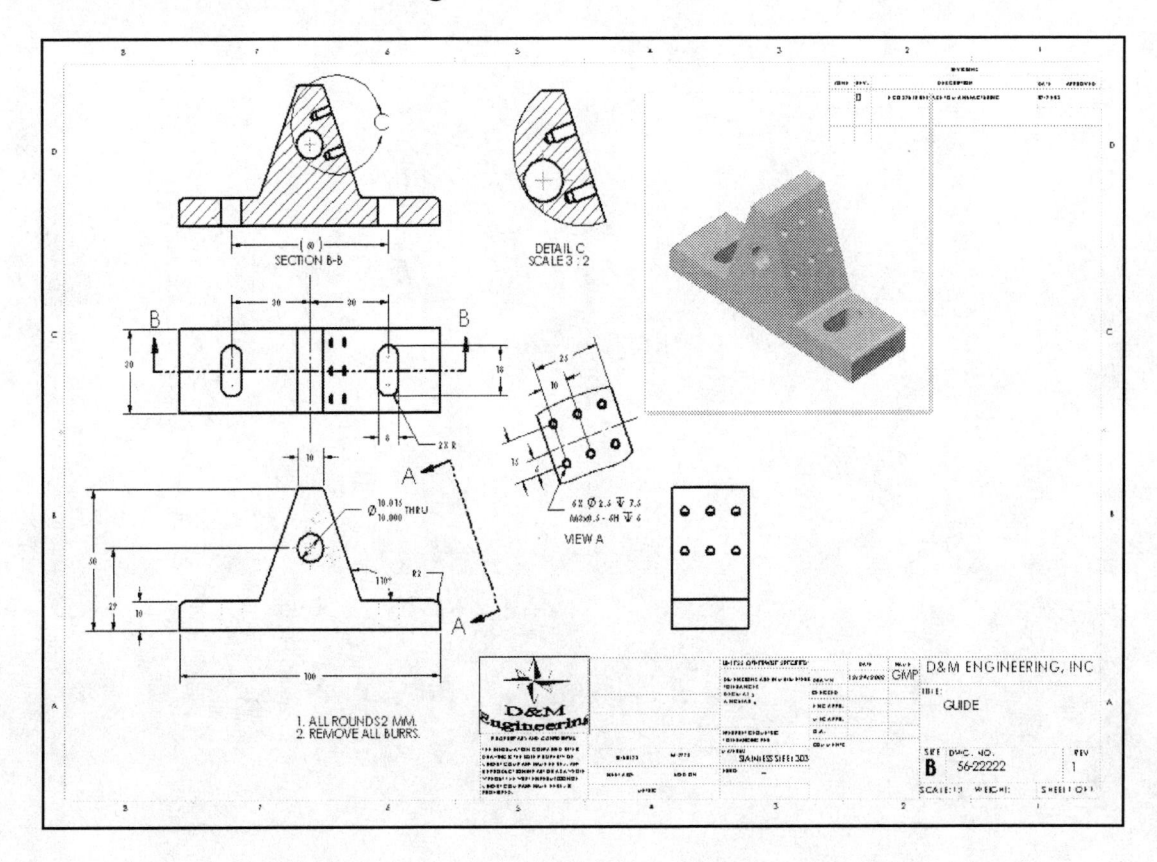

Engineering Design with SolidWorks — Fundamentals of Drawing

Project 3
Fundamentals of Drawing

Below are the desired outcomes and usage competencies based on the completion of Project 3.

Project Desired Outcomes:	Usage Competencies:
A GUIDE drawing with a customized sheet format containing a company logo, title block and sheet information.	Ability to create a custom sheet format, logo, title block and sheet information.
	An understanding of displaying Standard, Auxiliary, Detail and Section views.
	Skill to insert, create and modify dimensions.
A GUIDE-ROD assembly drawing with a Bill of Materials.	Knowledge to develop and incorporate a Bill of Materials with Customer Properties into a drawing.

Notes:

Project 3 – Fundamentals of Drawing

Project Objective

Create a GUIDE drawing with a customized sheet format containing a company logo, title block and sheet information.

Obtain an understanding of displaying the following views with the ability to insert, create and modify dimensions:

- Principle.
- Auxiliary.
- Detail.
- Section.

Create a GUIDE-ROD assembly drawing with a Bill of Materials.

Obtain knowledge to develop and incorporate a Bill of Materials with Custom Properties.

On completion of this project, you will be able to:

- Create a new Drawing Template.
- Create a new Sheet Format.
- Generate a customized sheet format.
- Produce a Bill of Materials with Custom Properties.
- Develop the following views:
  - Detailed view.
  - Section view.
  - Auxiliary view.
- Reposition views on a drawing.
- Add a Named view.

- Set the Dimension Layer.
- Add Hole Callouts.
- Move dimensions in the same view
- Move dimensions in a different view.
- Use Edit Sheet Format and Edit Sheet.
- Create Center Marks.
- Modify the dimension scheme.
- Create a parametric drawing note.
- Link notes in the title block to SolidWorks properties.
- Rename parts and drawings.

Engineering Design with SolidWorks — Fundamentals of Drawing

Project Situation

The individual parts and assembly are completed. What is the next step? You are required to create drawings for various internal departments, namely: production, purchasing, engineering, inspection and manufacturing.

Each drawing contains unique information and specific footnotes. Example: A manufacturing drawing would require information on assembly, Bill of Materials, fabrication techniques and references to other relative information.

Project Overview

Generate two drawings in this project:

- A GUIDE drawing with a customized sheet format.

- A GUIDE-ROD assembly drawing.

The GUIDE drawing contains three standard views, (principle views) and an Isometric view.

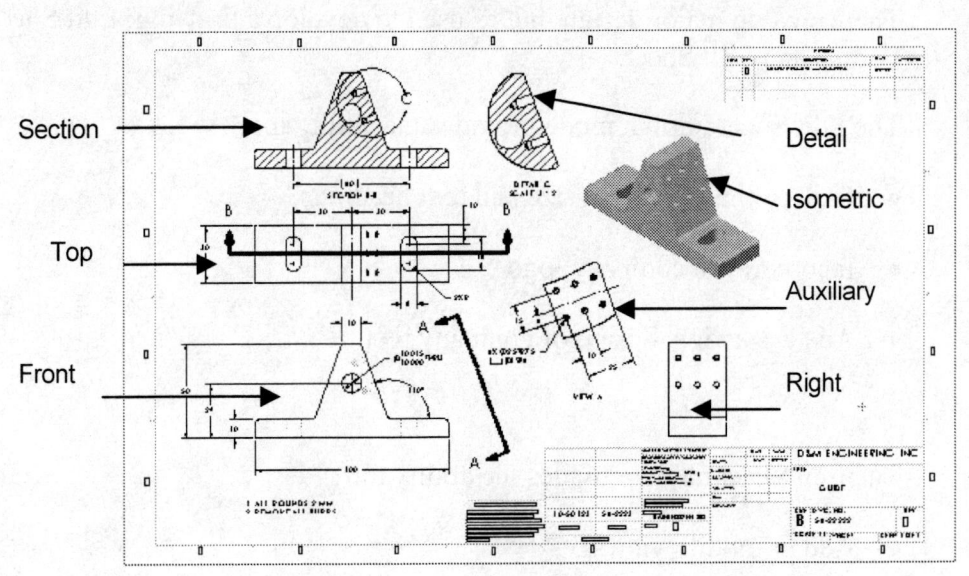

Figure 3.1

Do you remember what the three standard views are? They are: Top, Front and Right side, Figure 3.1.

Three new views are introduced: Detailed view, Section view and Auxiliary view. Orient the views to fit the drawing sheet.

Fundamentals of Drawing   **Engineering Design with SolidWorks**

Incorporate the GUIDE dimensions into the drawing.

The GUIDE-ROD assembly drawing contains an exploded view.

The drawing contains a Bill of Materials and balloon text, Figure 3.2.

Note: Microsoft EXCEL 97 or a later version is required to create the Bill of Materials.

ITEM NO.	QTY.	PART NO.	MATERIAL
1	1	GUIDE	SS 303
2	1	ROD	SS 303
3	1	PLATE	SS 303
4	2	M8-1.25 x 30	ALUMINUM
5	2	4MM CAPSCREW	ALUMINUM

Bill of Materials

Exploded View

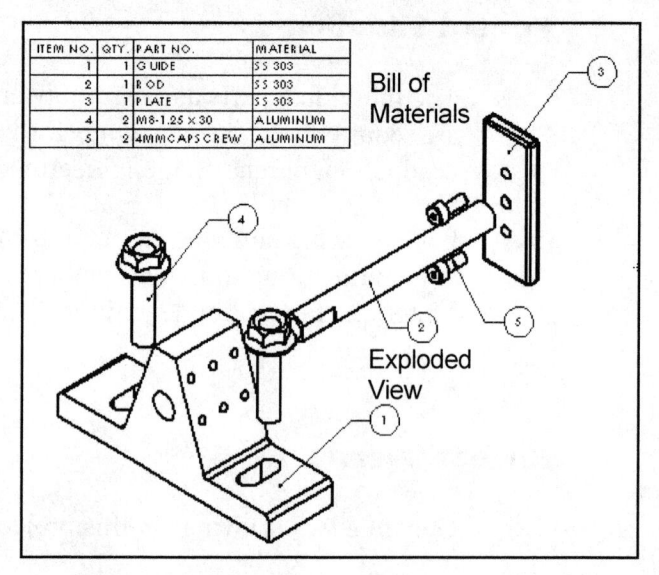

Figure 3.2

Both drawings utilize a custom sheet format containing a company logo, title block and sheet information.

There are two major design modes used to develop a drawing: Edit Sheet Format and Edit Sheet.

The Edit Sheet Format mode provides the ability to:

- Change the title block size and text headings.

- Incorporate a company logo.

- Add a drawing, design or company text.

The Edit Sheet mode provides the ability to:

- Add or modify views.

- Add or modify dimensions.

- Add or modify text.

Drawing Template and Sheet Format

The foundation of a SolidWorks drawing is the Drawing Template. Drawing size, drawing standards, company information, manufacturing and or assembly requirements, units and other properties are defined in the Drawing Template.

The Sheet Format is incorporated into the Drawing Template. The Sheet Format contains the border, title block information, revision block information, company name and or logo information, Custom Properties and SolidWorks Properties. Custom Properties and SolidWorks Properties are shared values between documents.

Utilize the B-size Drawing Template with no Sheet Format. Set Units, Font and Layers. Add the ENGDESIGN-B.slddrt Sheet Format contained in the enclosed CD.

1. Add the Company Name and Company Logo.

2. Create a new Drawing Template and Sheet Format.

3. Save the Drawing Template and Sheet Format in the MY-TEMPLATE file folder.

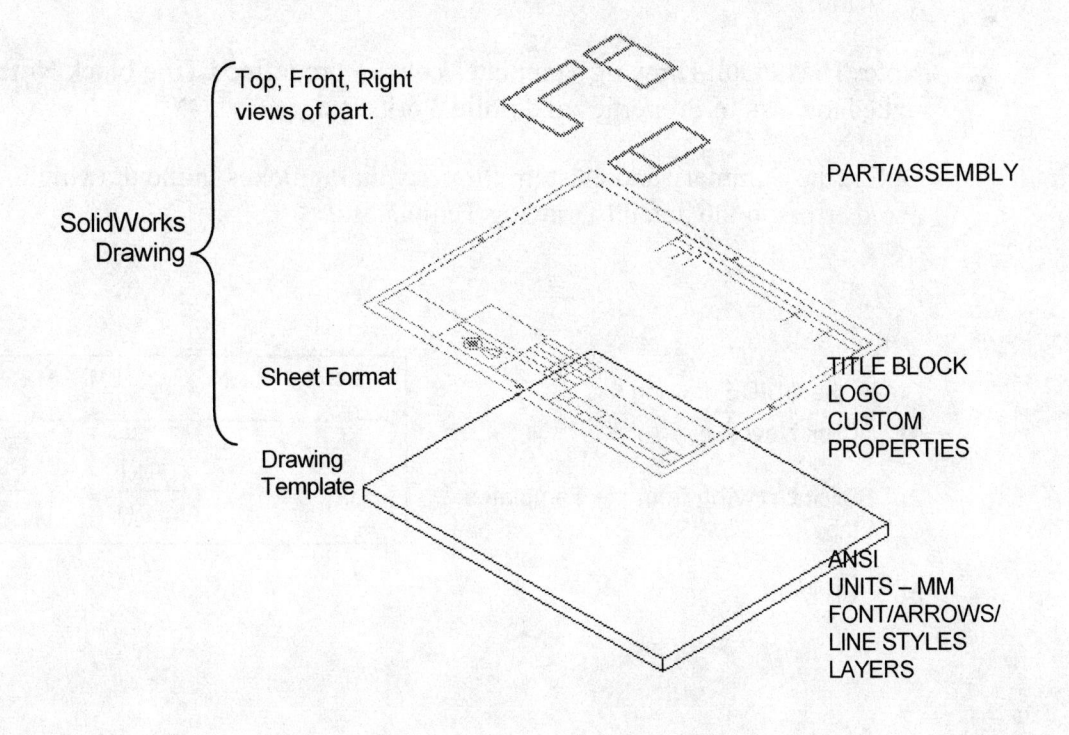

Views from the part or assembly are inserted into the SolidWorks Drawing.

There are various sheet format options in SolidWorks:

- Standard Sheet Format.

- Custom Sheet Format.

- No Sheet Format.

Note: A third Angle Projection scheme is illustrated in this project.

For non-ANSI dimension standards, the dimensioning techniques are the same, even if the displayed arrows and text size are different.

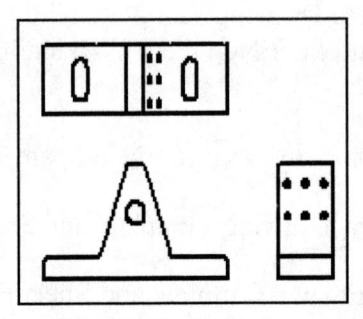

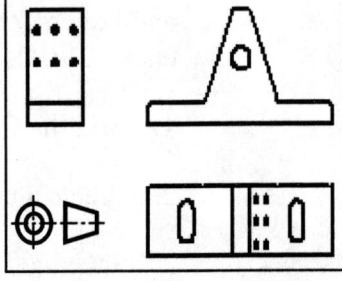

Third Angle Projection   First Angle Projection

For printers supporting millimeter paper sizes, select A3-Landscape (420mm x 297mm).

Note: The default Drawing Templates contain predefined Title block Notes linked to Custom Properties and SolidWorks Properties.

Define the Summary and Custom Property dialog boxes in the drawing to avoid errors in the default Drawing Templates.

Create the GUIDE drawing.
1) Click **New**.

2) Click **Drawing** from the Templates tab.

3) Click **OK**.

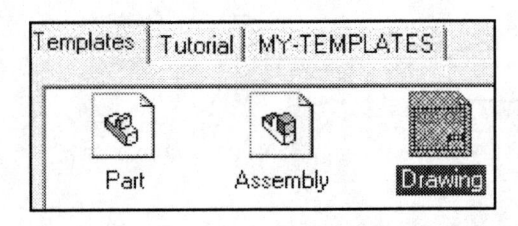

4) Select **No Sheet Format**.

5) Select **B-Landscape** from the Paper Size drop down list.

6) Click **OK**.

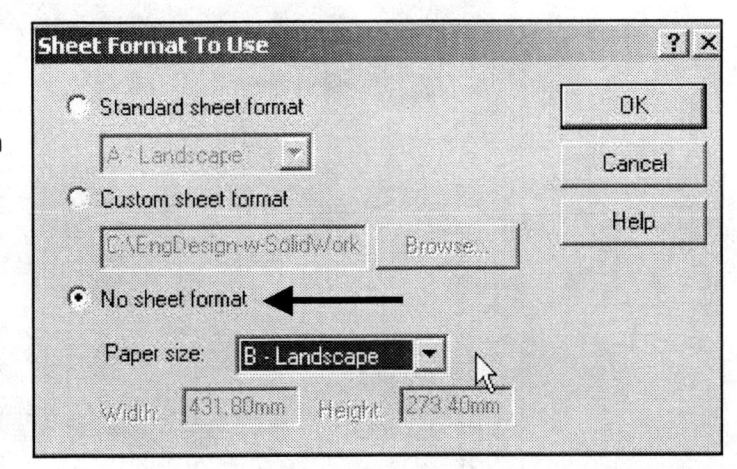

The B-Landscape paper is displayed in a new Graphics window. The sheet border defines the drawing size, 17" x 11", (431.8mm x 279.4mm).

The Annotation toobar, Drawing toolbar and Line Format toolbar are displayed on the left side of the Graphics window.

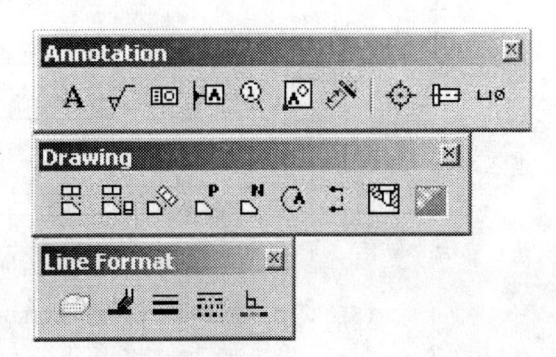

7) Right-click in the **Graphics window**.

8) Click **Properties**. The Sheet Setup Properties are displayed.

9) Select Sheet Scale **1:1**.

10) Select the **Third Angle** projection.

11) Click **OK**.

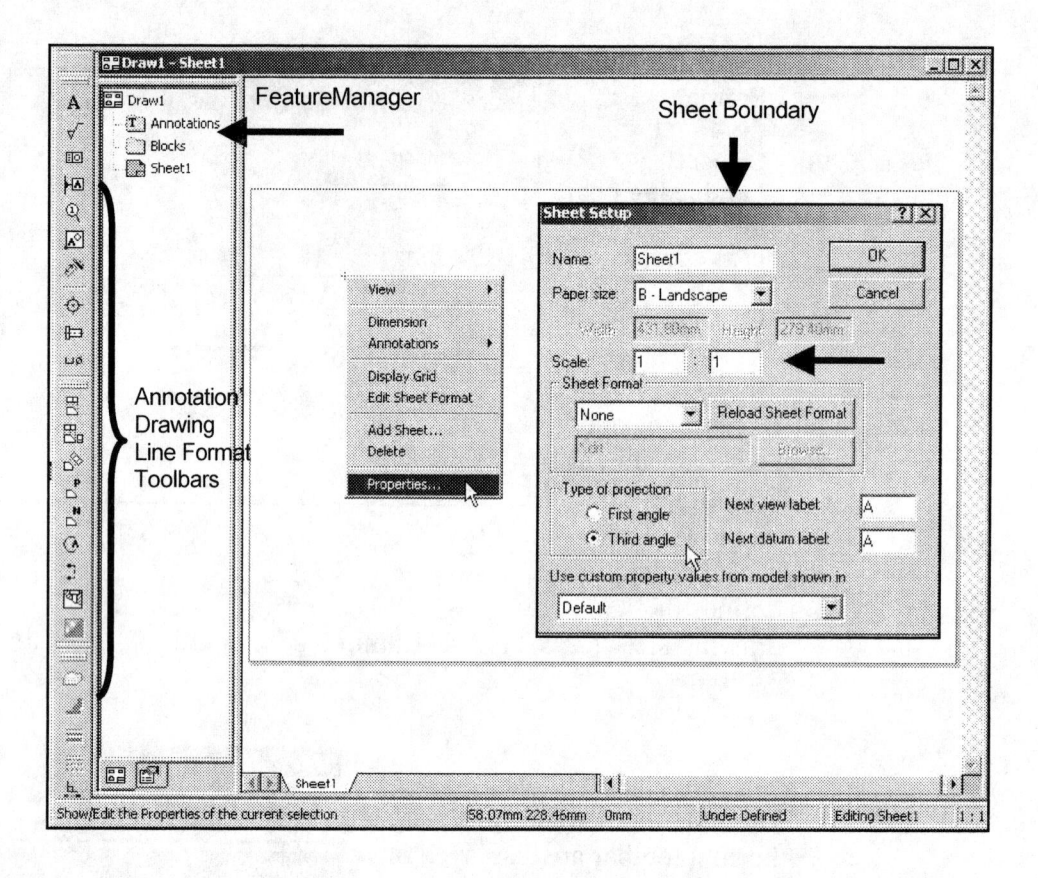

12) Set the ANSI dimensioning standard. Click **Tools**, **Options**, **Document Properties tab**.

13) Click **ANSI** for Dimensioning Standard.

14) Set the Drawing Units. Click **Units**, **Millimeters**.

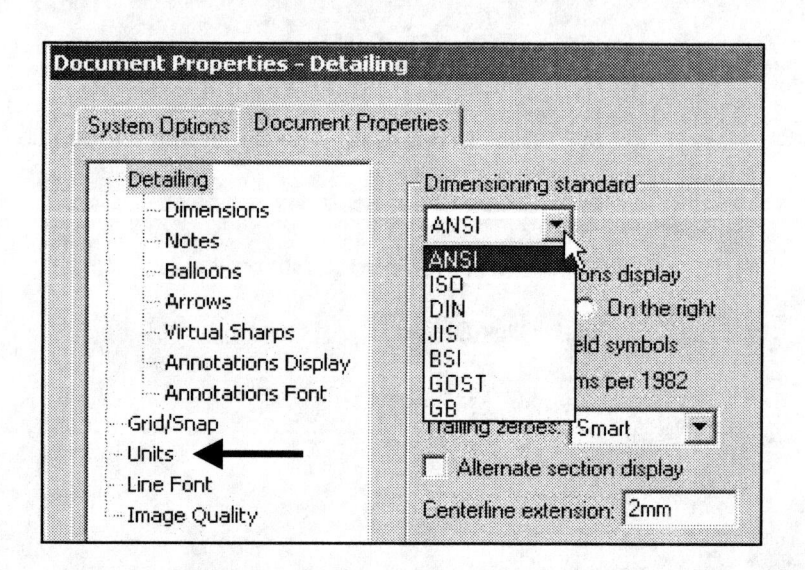

Detailing options provide the ability to address: dimensioning standards, text style, center marks, extension lines, arrow styles, tolerance and precision.

There are numerous text styles and sizes available in SolidWorks. Companies develop drawing format standards and use specific text height for Metric and English drawings. Numerous engineering drawings use the following format:

- Font: Century Gothic – All capital letters.

- Text height: .125in. or 3mm for drawings up to B Size, 17in. x 22in.

- Text height: .156in. or 5mm for drawings larger than B Size, 17in x 22in.

- Arrow heads: Solid filled with a 1:3 ratio of arrow width to arrow height.

Set the Dimension Font.

15) Change the default text height. Click **Annotations Font**. Click **Dimension**. Set the dimension text height.

16) Click the **Units** button.

17) Enter **3.0** for height.

18) Click **OK**.

19) Change the arrow height. Click the **Arrows** entry on the left side, below the Detailing entry. The Detailing - Arrows dialog box is displayed. Enter **1** for arrow Height. Enter **3** for arrow Width. Enter **6** for arrow Length.

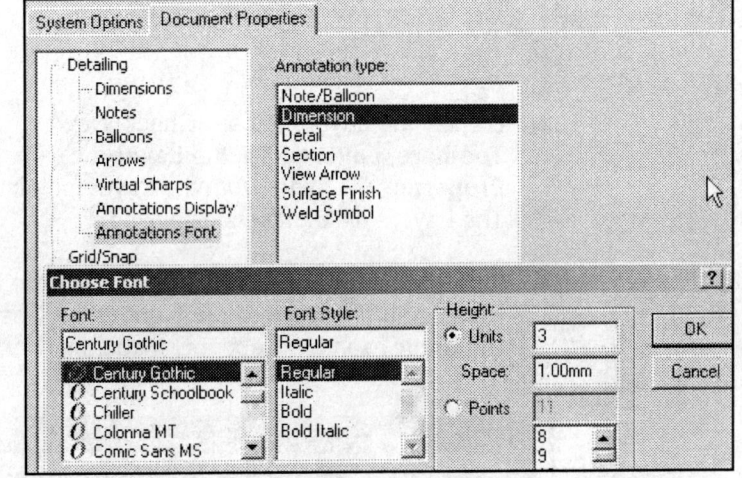

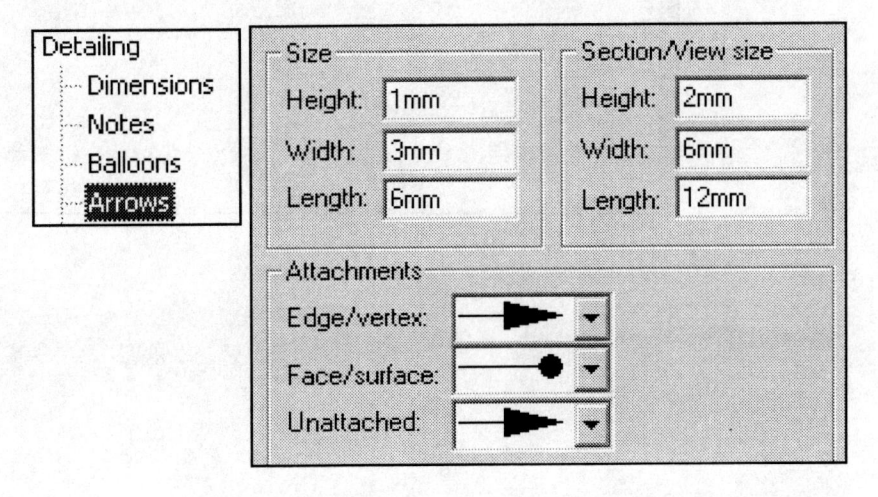

Fundamentals of Drawing　　　　　　　　　　　　　　**Engineering Design with SolidWorks**

20) Set Section/View size. Enter **2** for arrow Height. Enter **6** for arrow Width. Enter **12** for arrow Length.

21) Set the arrow style. Click the solid **filled arrow head** from the Edge/vertex list box.

22) Click **OK** from the Document Properties dialog box.

Drawing Layers contain dimensions, annotations and geometry. Create a new drawing layer to contain dimensions and notes.

Create a second drawing layer to contain hidden feature dimensions. Select the Light Bulb ♀ to turn On/Off Layers.

Create a Layer.
23) Display the Layer toolbar. Check **View**, **Toolbars**, **Layer**. Click the **Layer Properties** file folder from the Layer toolbar. The Layers dialog box is displayed.

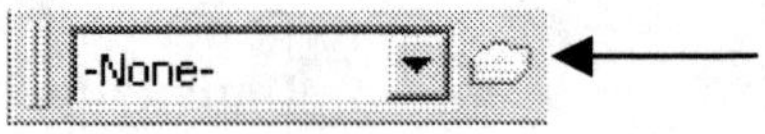

24) Create the new dimension Layer. Click the **New** button. Enter **Dims** in the Name column. **Double-click** under the Description column. Enter **Dimensions** in the Description column.

Name	Description	O	C	Style	Thickness
Dims	Dimensions	♀	■		
Notes	General Notes	♀	■		
→ Hide Dims	Hidden Insertec...	♀	■		

Buttons: OK, Cancel, Help, New, Delete, Move

PAGE 3 - 12

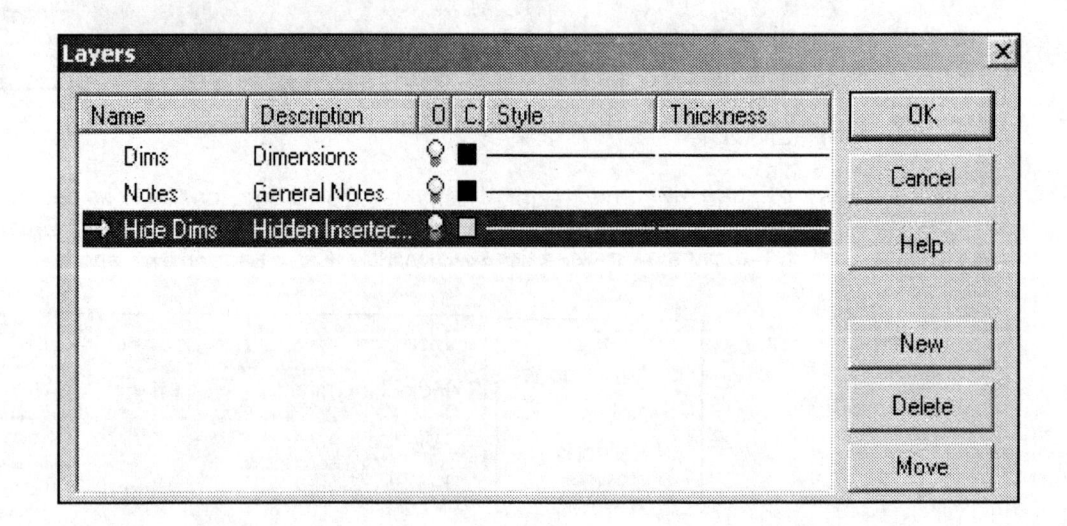

25) Create the Notes Layer. Click the **New** button. Enter **Notes** for Name.

26) Enter **General Notes** for Description.

27) Create the Hide Dims Layer. Click the **New** button. Enter **Hidden Dims** for Name. Enter **Hidden Insert Dimensions** for Description. Turn the Hidden Insert Dimension Layer off. Click **On/Off**. The light bulb is displayed in light gray.

28) Click the **Color Square**.

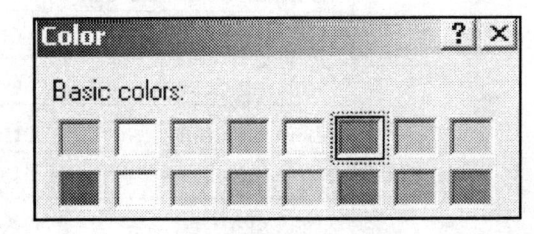

29) Select **Blue**. Click **OK**. Click **OK**.

30) Set the None Layer. Click the **Layer drop down arrow**. Click **None**.

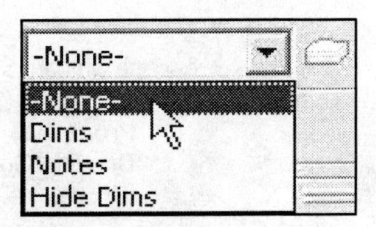

Note: The None Layer is saved with the Drawing Template.

The Drawing Template contains the drawing Size, Document Properties and Layers.

The Sheet Format contains the title block information.

Utilize the ENGDESIGN-B Sheet Format (17in. x 11in.) or the ENGDESIGN-A3 Sheet Format (420mm x 297mm) on the enclosed CD.

Fundamentals of Drawing   **Engineering Design with SolidWorks**

Title Block

The title block contains vital part or assembly information. Each company can have a unique version of a title block.

Most title blocks contain the following information:

Company Name/Logo:	CAD file name:	Part number:
Part name:	Quantity required:	Drawing number:
Drawing description:	Checked by:	Revision number:
Sheet number:	Engineering Change Orders:	Material & Finish:
Tolerance:	Drawn by:	Drawing scale:
Sheet size:	Approved by:	Revision block:

Add the Custom Sheet Format.

31) Right-click **Properties** in the Drawing Graphics window.

32) Click **Custom** from the Sheet Format drop down list.

33) Click the **Browse** button.

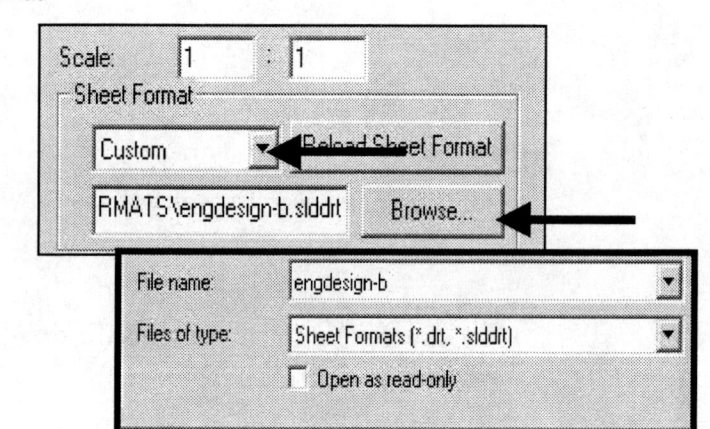

34) Select **ENGDESIGN-W-SOLIDWORKS \ CD-ENGDESIGN-W-SW2003 \SHEET-FORMATS\ ENGDESIGN-B.SLDDRT** for Sheet Format.

35) Click **Open**.

36) Click **OK** from the Sheet Setup dialog box.

Engineering Design with SolidWorks   **Fundamentals of Drawing**

The title block is located in the lower right hand corner of Sheet1.

The Drawing contains two modes:

1. Edit Sheet.

2. Edit Sheet Format.

Modify the Sheet Format text, lines or title block information in the Edit Sheet Format mode.

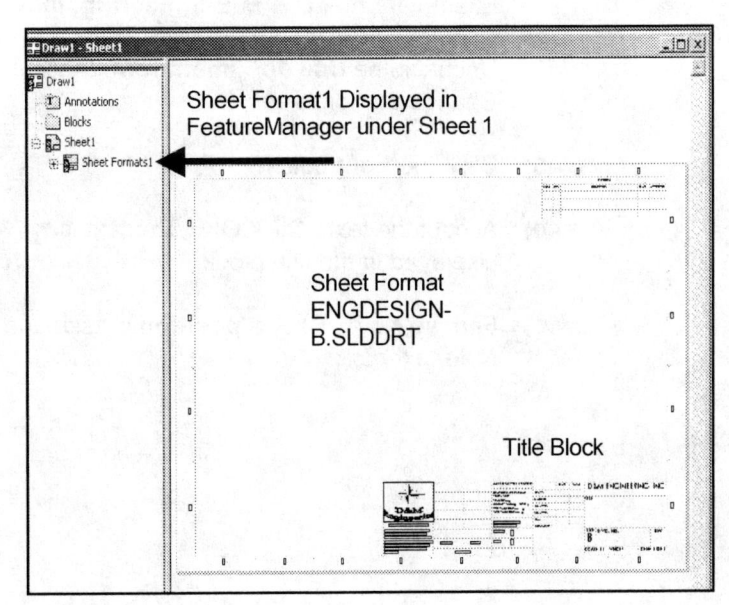

Insert views in the Edit Sheet mode.

Edit the Sheet Format - Title Block.

37) Edit the name of the company. Right-click **Edit Sheet Format** from the Pop-up menu in the Graphics window. The drawing title block lines turn blue.

38) View the right side of the title block. Click **Zoom to Area** on the sheet format title block.

39) Right-click on the **D&M ENGINEERING, INC.** text.

40) Click **Edit Text**.

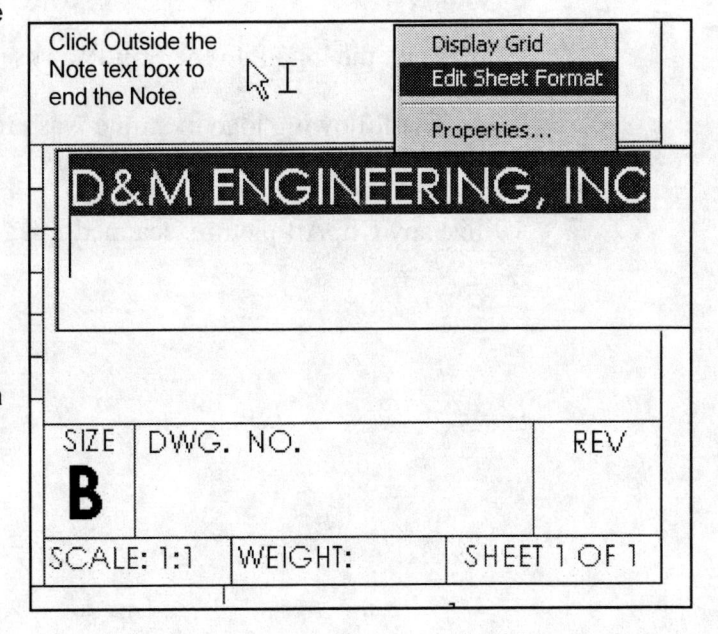

PAGE 3 - 15

Fundamentals of Drawing   **Engineering Design with SolidWorks**

41) Enter the name of the company. Click a **position** inside the Note text box. Example: Enter **D & M Engineering, Inc**.

42) Uncheck the **Use document font** check box. Change the font size.

43) Click the **Font** button.

44) Accept the text. Click **OK**. The text is displayed in the title block.

45) End the Note. Click a **position** outside the Note text box.

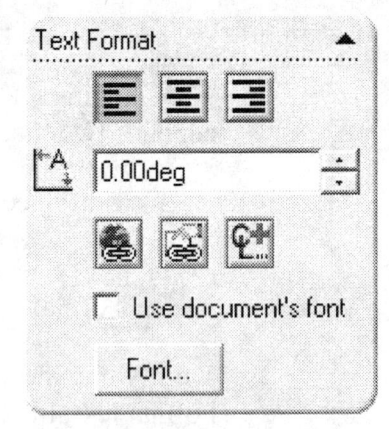

Company Logo

A company logo is normally located in the title block. Create a company logo by copying a picture file from Microsoft ClipArt using Microsoft Word.

Copy/Paste the logo into the SolidWorks drawing.

Note: The following logo example was created in Microsoft Word 2000 using the COMPASS.wmf and WordArt text.

Utilize any ClipArt picture, scanned image or bitmap.

Engineering Design with SolidWorks — Fundamentals of Drawing

Create a logo.

46) Create a New Microsoft Word Document. Click **Start**.

47) Click **MS Word**.

48) Click **New** from the Standard toolbar in MS Word.

49) Click **ClipArt** from the Draw toolbar.

Rectangle / WordArt / ClipArt

50) Double-click on the **COMPASS.wmf** file.

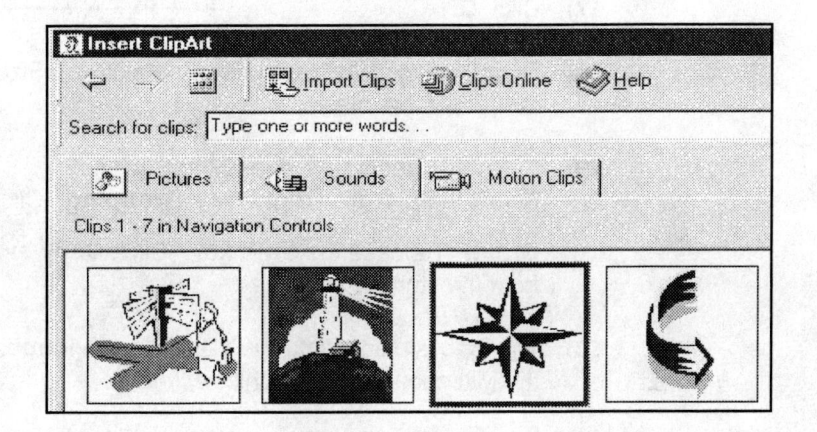

51) Insert the **compass.wmf** picture file from the ClipArt menu.

52) Click **Insert Clip**.

53) Redefine the picture layout. Display the drag handles. Right-click **Format Picture**. Click **Layout**. Click **Square**. Click **OK**.

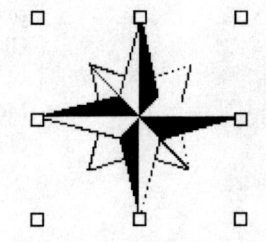

Fundamentals of Drawing   **Engineering Design with SolidWorks**

54) Add text to the logo picture. Click **Insert Word Art** from the Draw toolbar. Click a **WordArt** style. Click **OK**.

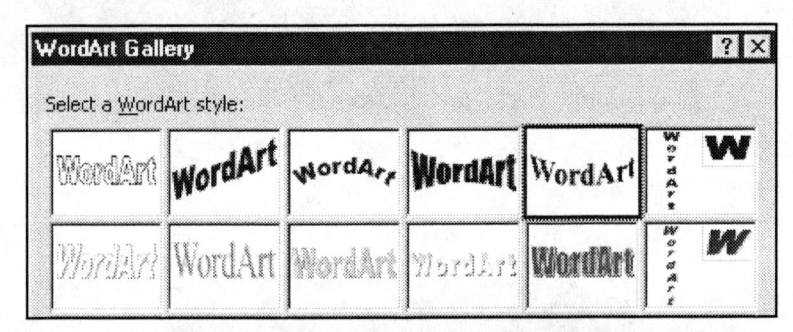

55) Enter **D & M Engineering** in the text box.

56) Click **24** from the Size drop down list.

57) Click **OK**.

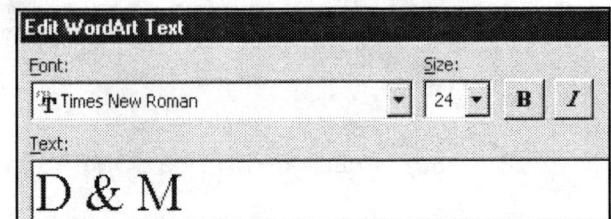

58) Drag the **Word Art text** under the picture. Size the **Word Art text** by dragging the picture handles.

59) Select the geometry. Press the **Shift** key. Select the **picture**. Right-click **Grouping**. Click **Group**. Release the **Shift** key. Click **Copy** from the standards toolbar.

60) The logo is placed into the Clipboard. **Minimize** the Microsoft Word Graphics window.

61) Click a **position** in the SolidWorks Graphics window. Zoom out if required.

62) Delete the current logo. Click the **D & M** logo. Press the **Delete** key. Click **Yes** to Confirm Delete.

63) Paste the logo into the title block. Click a **position** on the left side of the title block. Click **Edit**, **Paste**.

64) **Move** and **Size** the logo to the SolidWorks title block by dragging the picture handles.

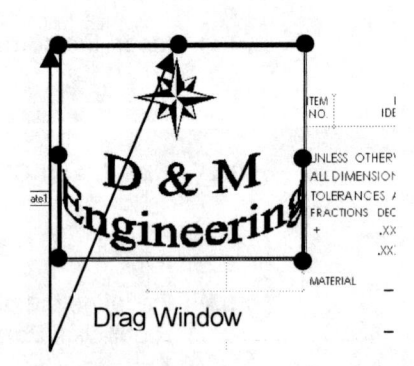

Drag Window

PAGE 3 - 18

Display Sheet1.

65) Return to Edit Sheet. Right-click in the **Graphics window**. Click **Edit Sheet**. The title block is displayed in gray.

66) Fit the Sheet Format to the Graphics window. Press the **f** key.

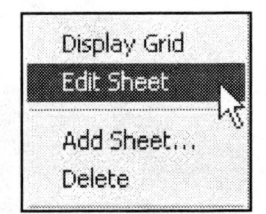

Save the Sheet Format.

67) Click **File**, **Save Sheet Format**. The Save Sheet Format dialog box appears.

Save the Sheet Format as a custom Sheet Format. Use the custom Sheet Format for the drawings in this project.

Combine the Sheet Format with the Drawing Template to create a custom Drawing Template.

68) Click the **Custom Sheet Format** button. Click the **Browse** button from the Save Sheet Format dialog box.

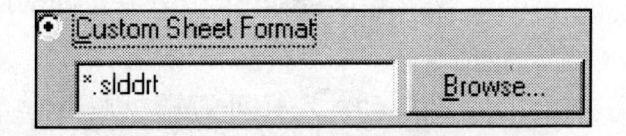

69) The default Sheet Format file folder is called data. The file extension for Sheet Format is .slddrt. Select **ENGDESIGN-W-SOLIDWORKS\MY-TEMPLATES** for Save In File Folder.

70) Enter **CUSTOM-B** for File name.

71) Click **Save** from the Save Sheet Format dialog box. Use the CUSTOM-B Sheet Format with the B size Drawing Templates.

72) Click **OK**.

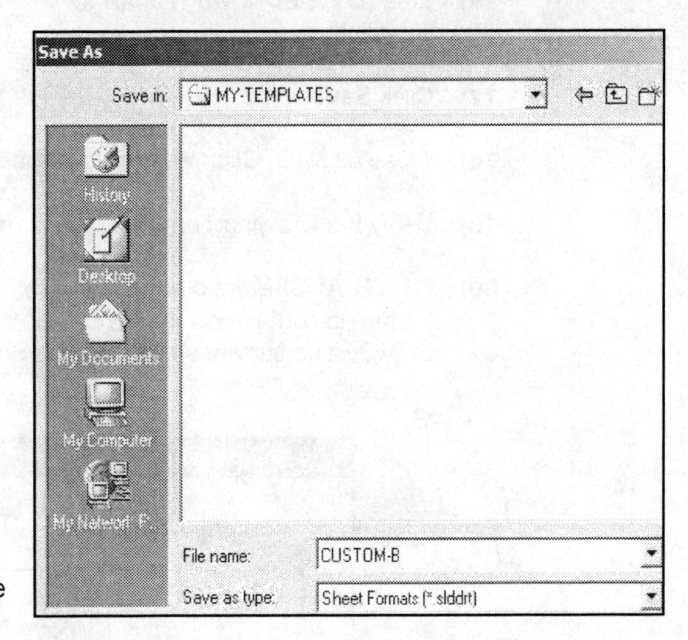

Fundamentals of Drawing   **Engineering Design with SolidWorks**

Save the Drawing Template.

73) Click **File**, **Save As**. Click **Drawing Template[*.drwdot]** from the Save as type list.

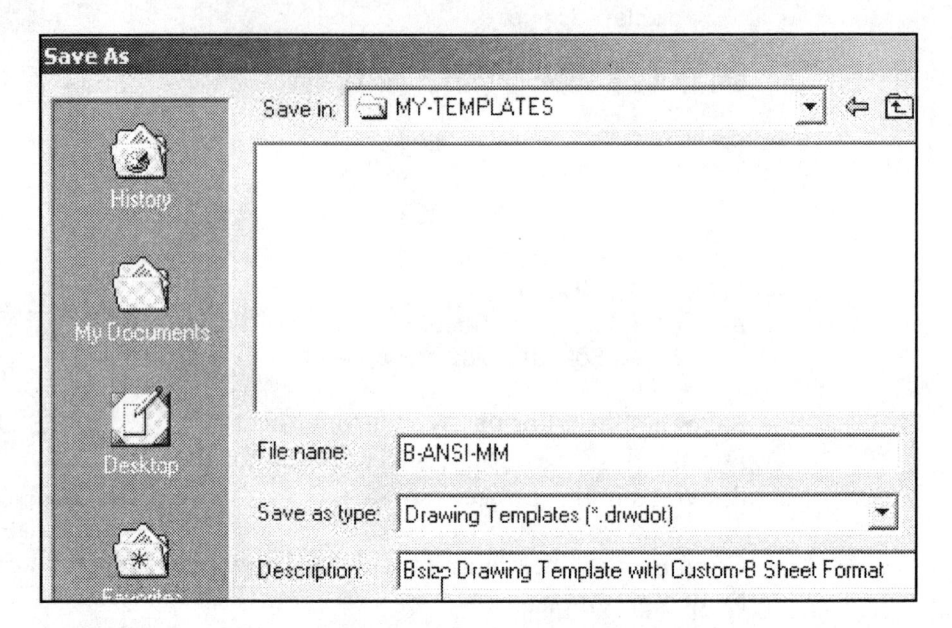

74) Select **ENGDESIGN-W-SOLIDWORKS\ MY-TEMPLATES** for Save In File Folder.

75) Enter **B-ANSI-MM** for File name.

76) Enter **B size DrawingTemplate with Custom B Sheet Format** for the Description.

77) Click **Save**.

78) Close all files. Click **Windows**, **Close All**.

79) Verify the Drawing Template. Click **File**, **New**.

80) Click **B-ANSI-MM** from the MY-TEMPLATES tab. The CUSTOM-B Sheet Format is contained with the Drawing Template. The current drawing name is Draw2. The current sheet name is Sheet1.

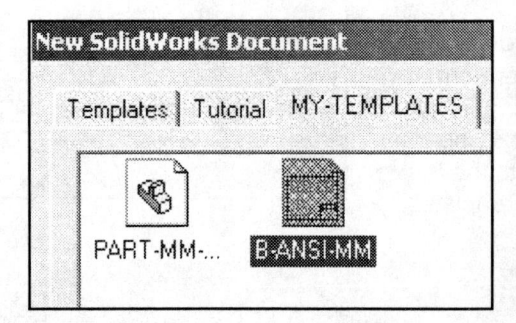

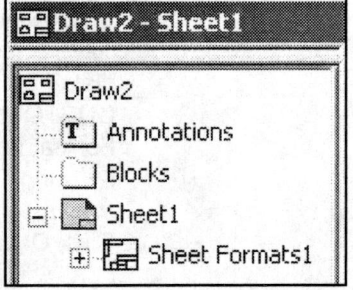

Create the GUIDE Drawing from the GUIDE Part

A drawing contains part views, geometric dimensioning and tolerances, notes and other related design information.

When a part is modified, the drawing automatically updates. When a dimension in the drawing is modified, the part is automatically updated.

Perform the following tasks before starting the GUIDE drawing:

- Verify the part. The drawing requires the associated part.

- View dimensions in each part. Step through each feature of the part and review all dimensions.

Change the overall width of the GUIDE from 80mm to 100mm.

81) Open the GUIDE part. Click **Open**.

82) Select **ENGDESING-W-SOLIDWORKS\PROJECTS** for Look in folder.

83) Select **Part** for Files of Type. Double-click **GUIDE**.

84) Double-click the **Base Extrude** feature in the GUIDE FeatureManager.

85) Double-click **80**.

86) Enter **100**.

87) Click **Rebuild** from the Standards toolbar.

88) Save the GUIDE part. Click **Save**.

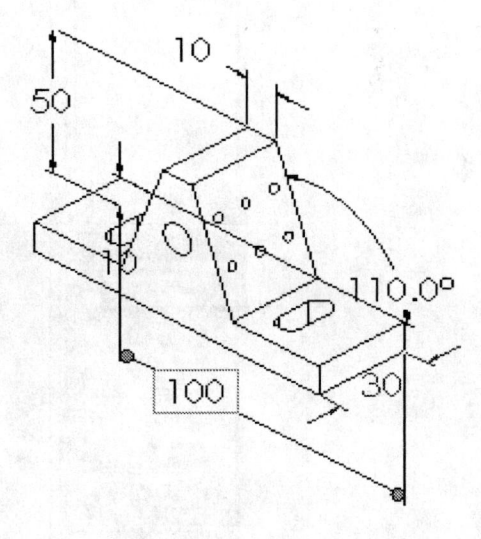

Fundamentals of Drawing **Engineering Design with SolidWorks**

Insert three standard views.

89) Display the drawing. Click **Window** from the Main menu. Click **Draw2-Sheet1**. The drawing must be in Edit Sheet mode to insert the drawing views. Click **Window**, **Tile Horizontally** from the Main menu.

90) Drag the **GUIDE Part** icon from the Feature Manager into the center of the drawing Graphics window.

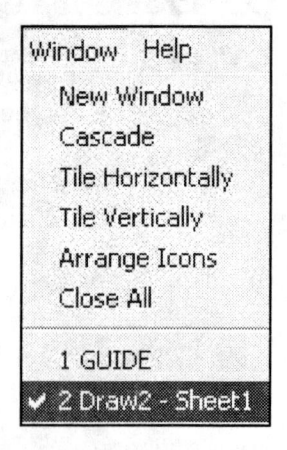

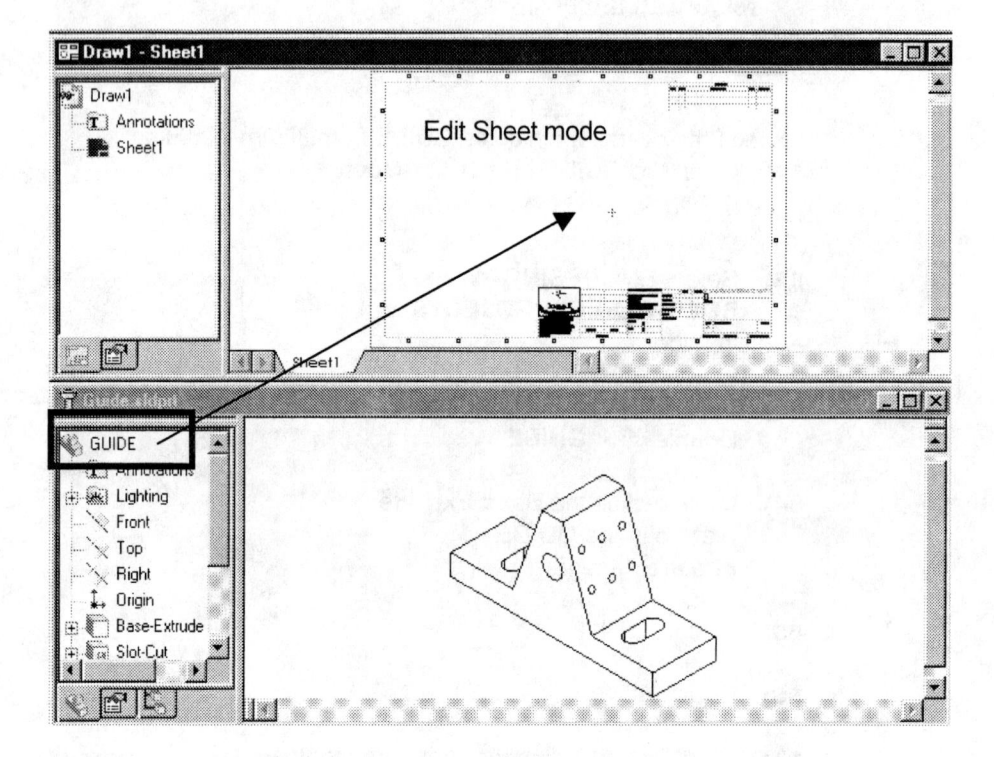

91) Click the **Maximize** icon for the GUIDE drawing window.

A part cannot be insert into a drawing when the Edit Sheet Format is selected. Edit Sheet Format displays all lines in blue.

The 3 standard drawing views flash once and then disappear. You are required to be in the Edit Sheet mode.

The three standard views are displayed in the Graphic window. The drawing views may be positioned too close to the title block.

Views can be added, deleted and moved. The default Display Mode is Hidden Lines Removed.

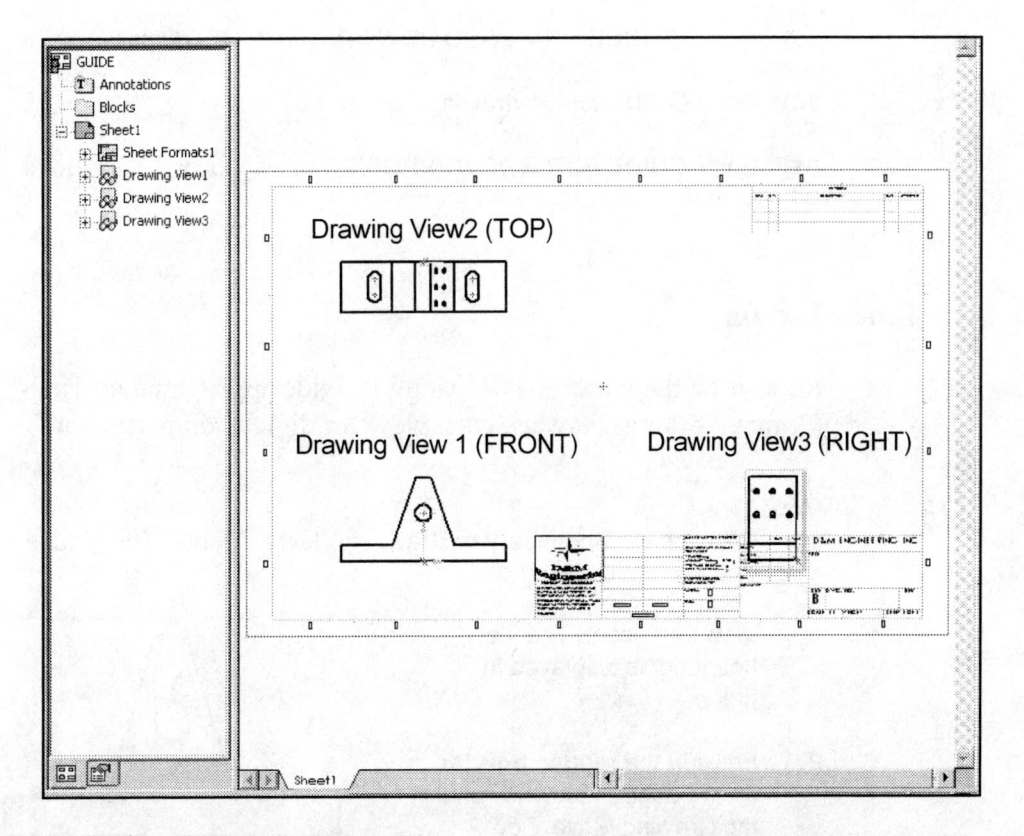

The mouse pointer provides feedback in both the Drawing Sheet and Drawing View modes. The mouse pointer displays the Drawing Sheet icon when the Sheet properties and commands are executed. The mouse pointer displays the Drawing View icon when the View properties and commands are executed.

Each drawing has a unique file name. In SolidWorks, drawing file names end with a .slddrw suffix. Part file names end with a .sldprt suffix.

A drawing or part file can have the same prefix. A drawing or part file cannot have the same suffix.

Example: Drawing file name: GUIDE.slddrw. Part file name: GUIDE.sldprt.

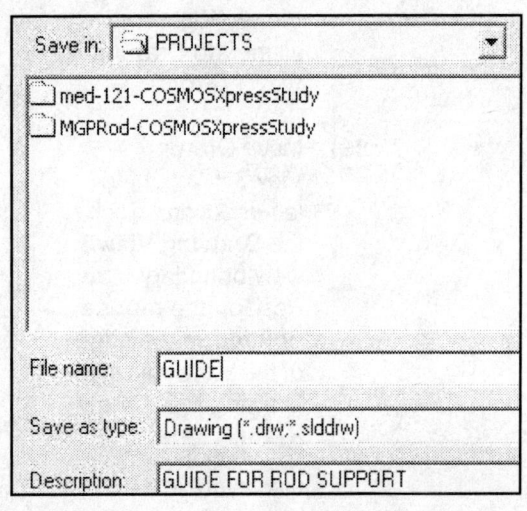

Save the Drawing.
92) Click **File**, **Save As** from the Main menu.

93) Select **ENGDESIGN-W-SOLIDWORKS\PROJECT** for the Save in File Folder.

94) Enter **GUIDE** for the drawing File name.

95) Enter **GUIDE FOR ROD SUPPORT** for Description. Click **Save**.

Move Views

Reposition the view on a drawing. Provide approximately 1in. - 2in., (25mm – 50mm) between each view for dimension placement.

Move Views.
96) Click the view boundary of **Drawing View1** (Front). The mouse pointer displays the Drawing View icon. The view boundary is displayed in green.

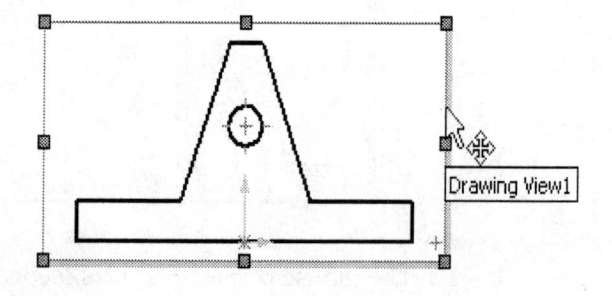

97) Position the **mouse pointer** on the edge of the view until the Drawing Move View icon is displayed.

98) Drag **Drawing View1** in an upward vertical direction. Drawing View2 (Top) and Drawing View3 (Right) move aligned to Drawing View1 (Front).

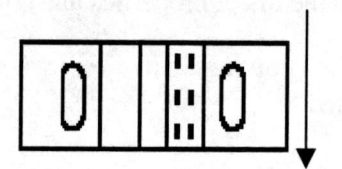

Click inside the view boundary. Drag Drawing View 1

Click inside the view boundary. Drag Drawing View3 left.

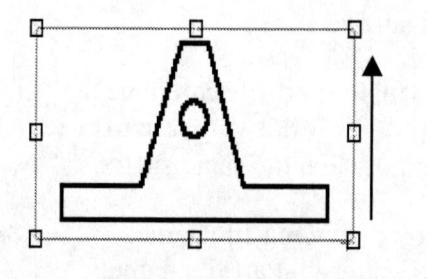

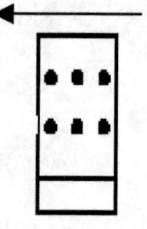

99) Move Drawing View3 in a right to left direction. Click the **Drawing View3** view boundary. Position the **mouse pointer** on the edge of the view until the Drawing Move View icon is displayed.

Engineering Design with SolidWorks   **Fundamentals of Drawing**

100) Drag **Drawing View3** in a right to left direction towards Drawing View1.

101) Move Drawing View2 in a downward vertical direction. Click the **view border**. Drag **Drawing View2** in a downward direction towards Drawing View1.

Named View

A Named view displays the part or assembly in various orientations. Add a Named view to the drawing at anytime.

Add an Isometric view to the drawing.
102) Click inside **Drawing View1** (Front). The view boarder is displayed in green.
Click **Named View** from the Drawing toolbar. The named views for the GUIDE are displayed. Click **Isometric** from the View Orientation text box. The Isometric view is placed on the mouse pointer.

103) Position the Isometric view. Click the **Graphics window** to the right of Drawing View2. Click **OK**.

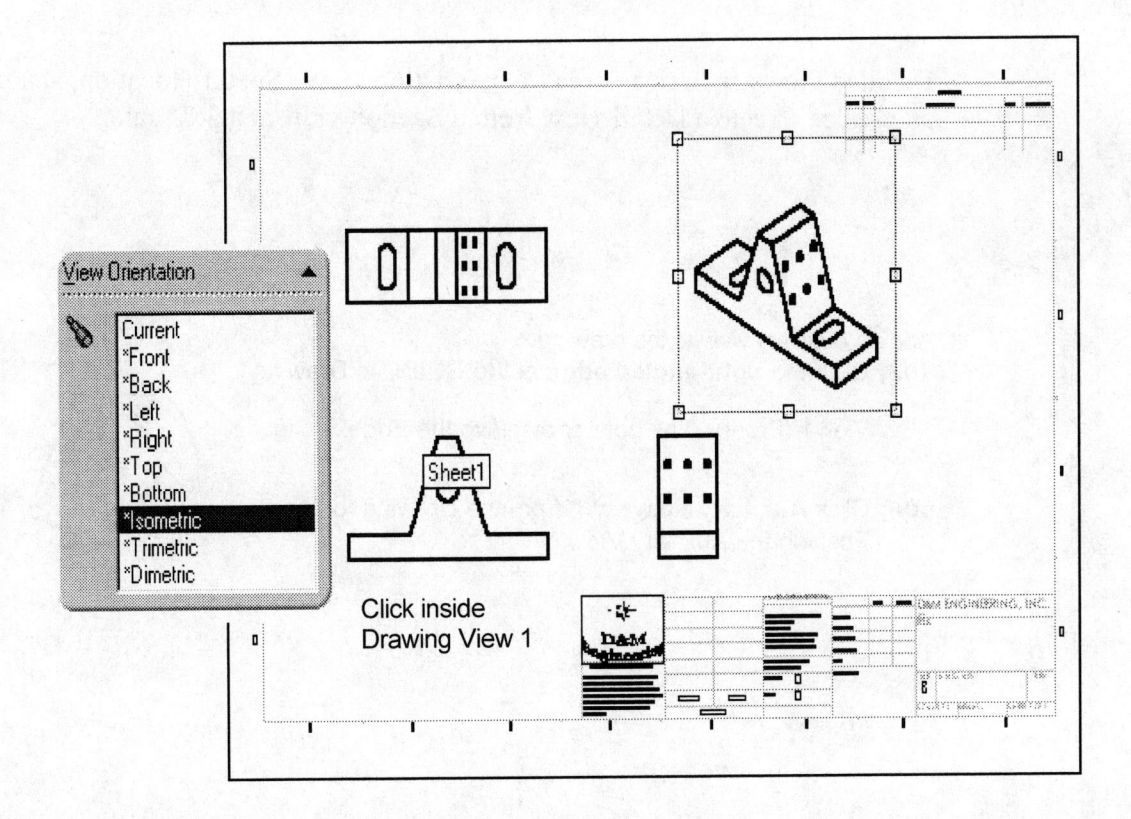

104) Save the GUIDE drawing. Click **Save**.

Additional Views

An Auxiliary view displays a plane parallel to an angled plane with true dimensions. A primary Auxiliary view is hinged to one of the six principle views. Create a primary Auxiliary view that references the Front view.

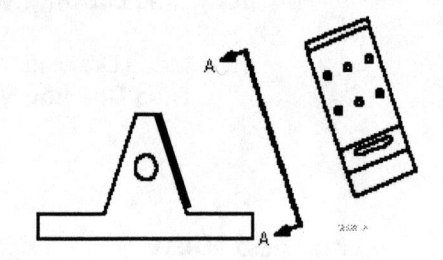

Section views display the interior features. Define a cutting plane with a sketched line in a view perpendicular to the Section view. Create a full Section view by sketching a section line in the Top view.

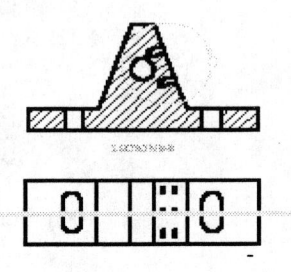

Detailed views enlarge an area of an existing view. Specify location, shape and scale. Create a Detail view from a Section view at a 3:2 scale.

Add an Auxiliary view to the drawing.
105) Click the **right-angled edge** of the GUIDE in Drawing View1 (Front). The pointer displays the Edge icon.

106) Click **Auxiliary View** from the Drawing toolbar. Position the Auxiliary View.

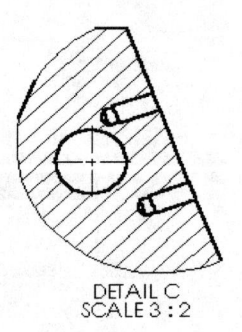

Engineering Design with SolidWorks **Fundamentals of Drawing**

107) Click a **location** to the right of Drawing View1 (Front). Click **OK** from the PropertyManager. **Click** and **drag** the section line A-A midpoint toward Drawing View1.

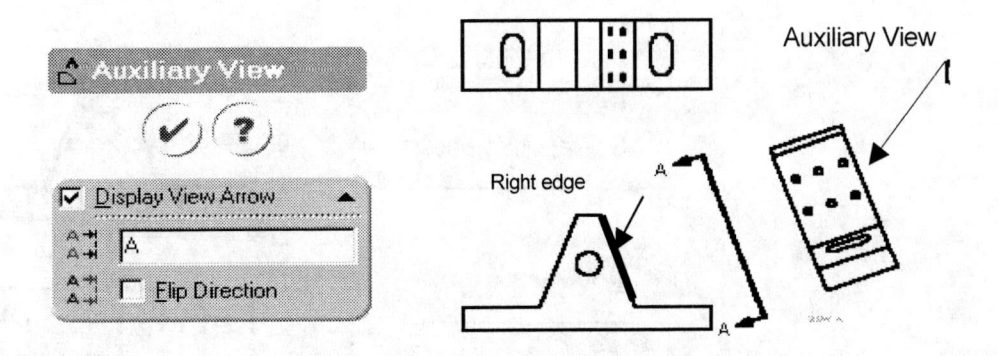

Add a Section view to the drawing.
108) Click **Drawing View2** (Top). Enlarge the view.

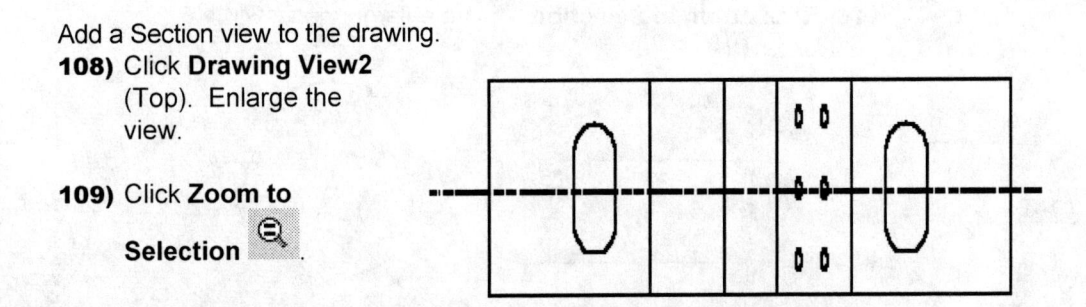

109) Click **Zoom to Selection**.

110) Click **Section View** from the Drawing toolbar.

111) Sketch a section Line through the **midpoints** of Drawing View2 (Top). The Section line extends beyond the left and right profile lines.

112) Position Section View B-B. Click a **location** above the Drawing View2 (Top) view. The section arrows point up. Click **Flip direction**, if required. Enter **B** for Section View Name in the Label text box.

113) Click **OK** from the PropertyManager.

114) Fit the drawing to the Graphics window. Press the **f** key.

Fundamentals of Drawing
Engineering Design with SolidWorks

Add a Detail view to the drawing.
115) Click the **Section view**.

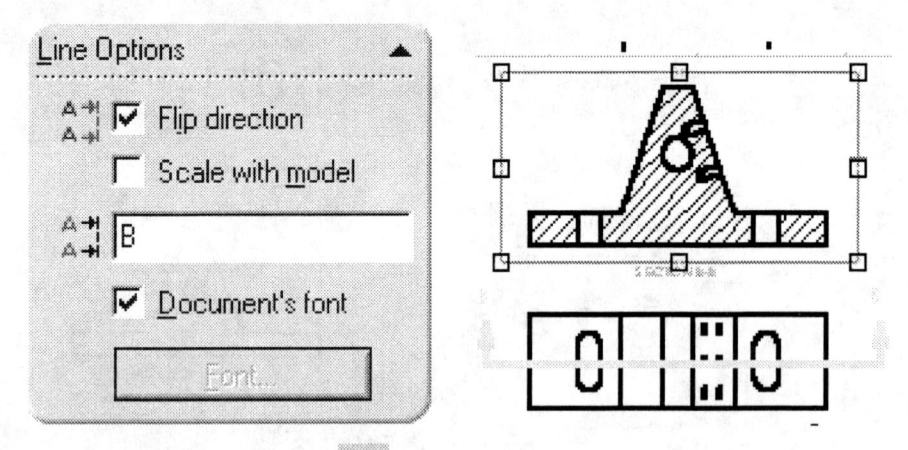

116) Click **Zoom to Selection** to enlarge the view.

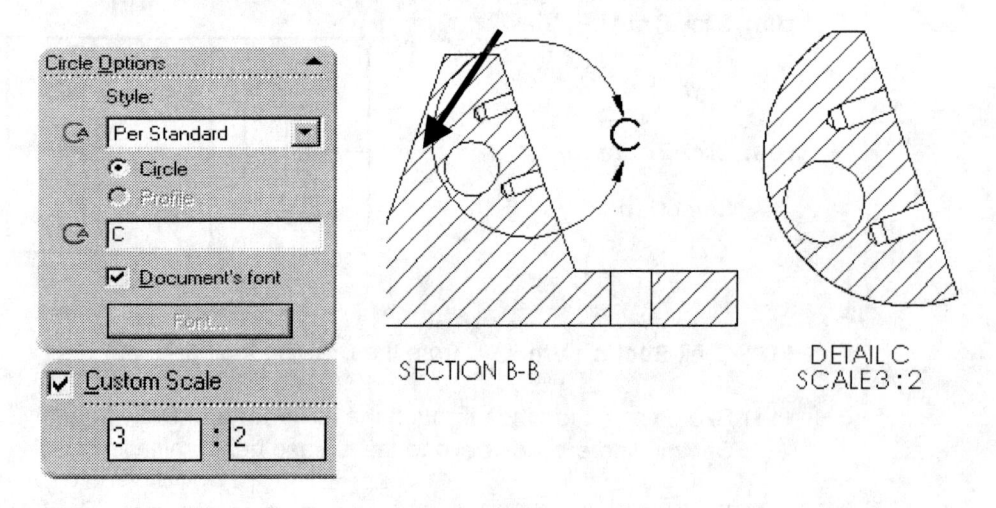

117) Click the **right-angled edge** between the two holes in Section View B-B. Click **Detail View** from the Drawing toolbar. The Circle Sketch tool is selected. Sketch the **Circle**. Include the 2 Tapped Holes in the sketch. Enter **C** for Detail View Name in the Label text box.

118) Position Detail View C. Click a **location** to the right of the Section view. Change the view Scale. Enter **3:2** in the Custom Scale text box. Click **OK** from the PropertyManager.

119) Fit the drawing to the Graphics window. Press the **f** key.

120) Save the GUIDE. Click **Save**.

The Drawing Views are complete. Move the views to allow for ample spacing for dimensions and notes.

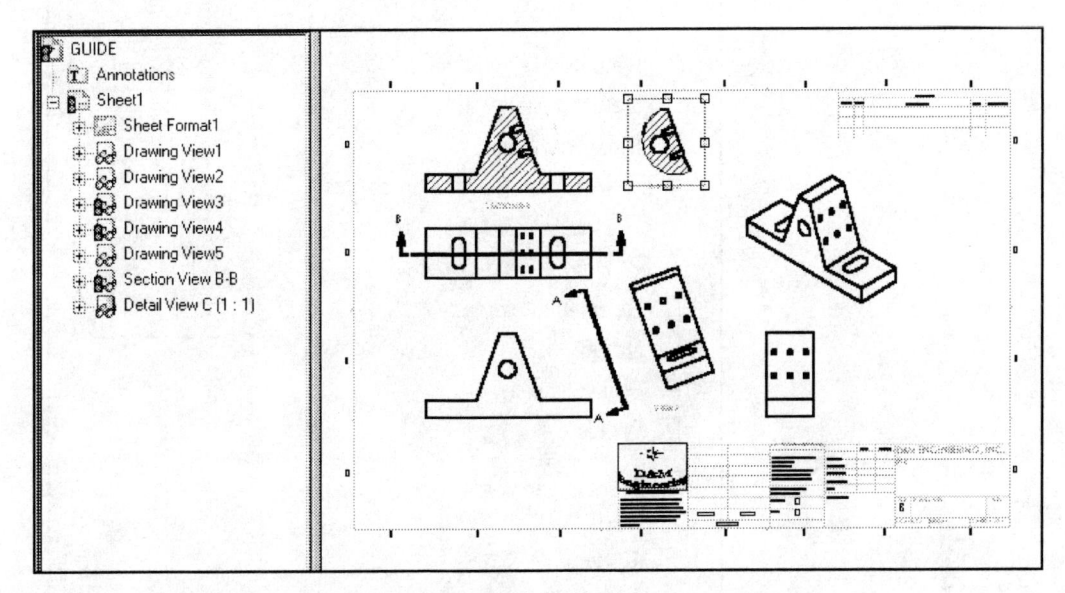

Insert part feature dimensions onto the Dims Layer.

Set the Dimension Layer.
121) Display the DIMS Layer. Click **DIMS** from the Layer list. Click the **Green Check mark** in the FeatureManager.

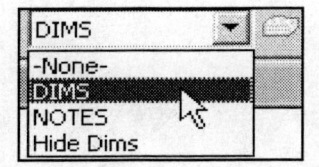

Insert Dimensions from the Part

Dimensions you created for each part feature are inserted into the drawing. Select the first dimensions to display for the Front view.

Note: Do not select the Import Items into all Views option. Dimension text is cluttered.

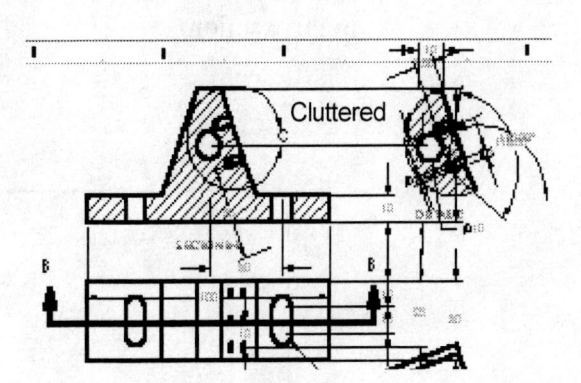

Insert dimensions.
122) Insert dimensions into the Drawing View1 (Front) view. Click **Drawing View1**. Click **Model Items**  from the Annotations toolbar. The Insert Model Items dialog box is displayed. Uncheck **Include items from hidden features**. Uncheck **Import items into all views**. Drawing View1 is listed in the Import into views list box. Click **OK**.

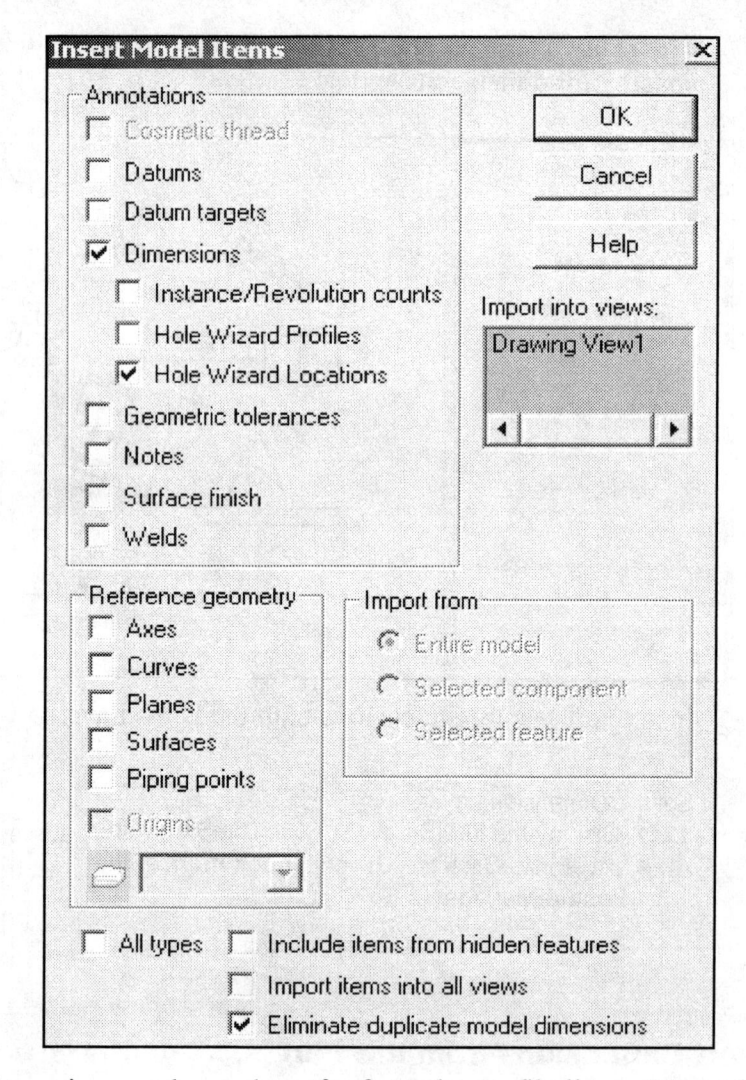

The dimensions are located too far from the profile lines. Move them later in this section.

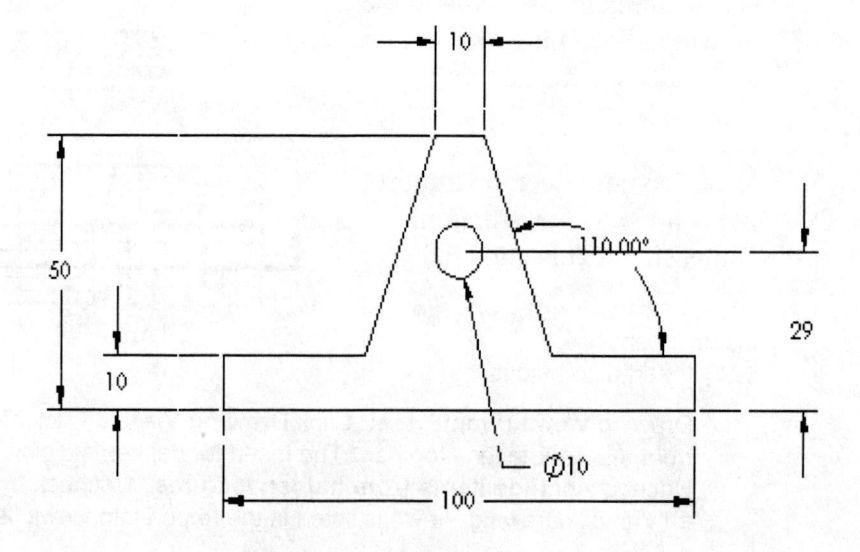

Drawing dimension location is dependent on:

- Feature dimension creation.
- Selected drawing views.

Move Dimensions in the Same View

Move dimensions within the same view. Use the mouse pointer to drag dimensions and leader lines to a new location.

Leader lines reference the size of the profile. A gap must exist between the profile lines and the leader lines. Shorten the leader lines to maintain a drawing standard. Use the green Arrow buttons to flip the dimension arrows.

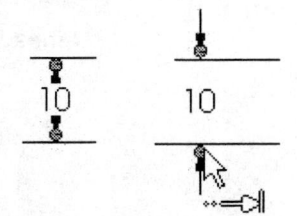

Move the linear dimensions in Drawing View1, (Front).

123) Zoom in on Drawing View1.

124) Click the vertical dimension text **29**. A green box is drawn around the text. The mouse pointer displays the Linear Dimension icon.

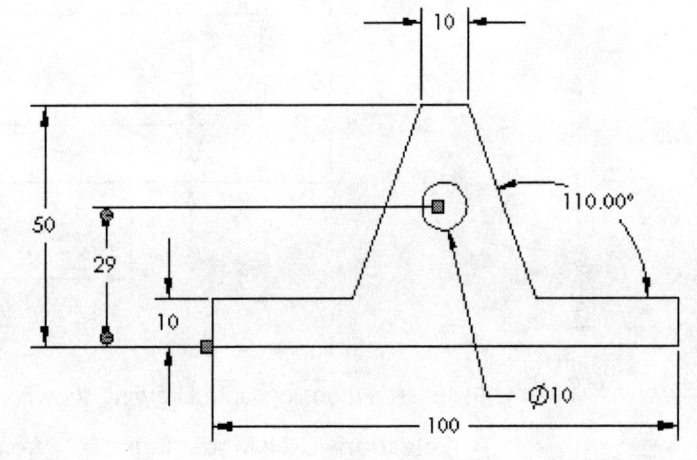

125) Drag the **dimension text** to the left. Click a **position** between the two arrowheads. Drag the **square green endpoint** to the vertex of the left corner. A gap is created between the extension line and the profile.

126) Click the vertical dimension **10**. Drag the **text** approximately 10mm's from the profile. The smallest linear dimensions are closest to the profile.

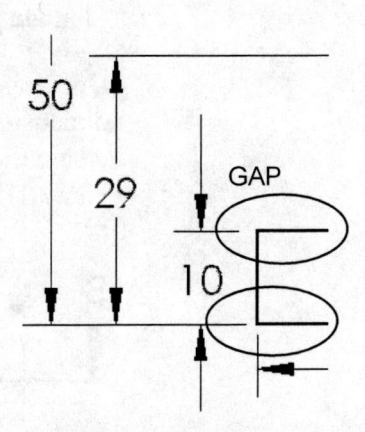

Note: A gap exists between the profile line and the leader lines. Drag the green endpoints to a vertex, to create a gap.

Display Hidden Lines in Drawing View1 and Drawing View2.
127) Position the **horizontal** and **angular** dimensions. **Flip** the arrows if required.

128) Hidden lines are displayed with thin dashed lines. Display the hidden lines. Click the **Drawing View1 boundary**. Click the **Drawing View1 boundary**. Click **Hidden Lines Visible** from the View toolbar. Fit the drawing to the Graphics window. Press the **f** key. Click inside **Drawing View2**. Click **Hidden Lines Visible**.

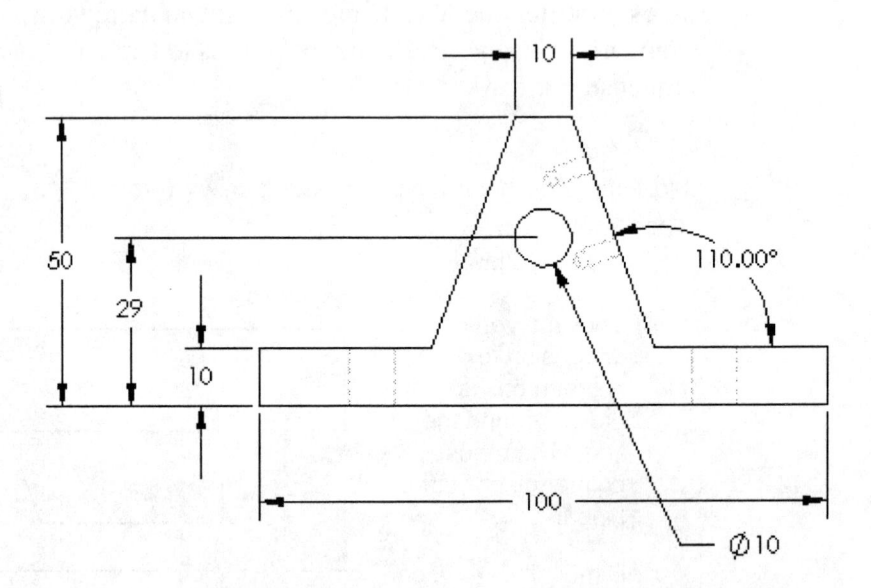

129) Insert dimensions into Drawing View2. Click **Drawing View2**. Click **Zoom to Selection**. Click **Model Items** from the Annotations toolbar. The Insert Model Items dialog box is displayed. Uncheck **Include items from hidden features**. Click **OK**.

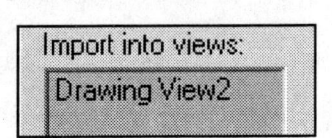

Drawing View2 displays crowded dimensions. Move the overall dimensions. Move the Slot Cut dimensions. Place dimensions in the view where they display the most detail. Move dimensions to the Auxiliary view. Hide the diameter dimensions and add Hole Callouts.

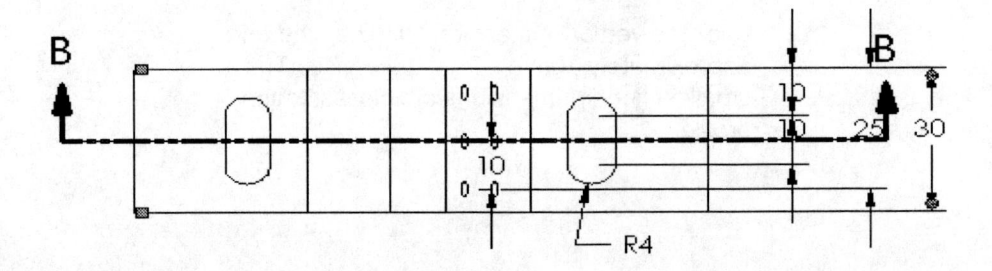

130) Press the **z** key until the vertical 30 dimension is displayed. Drag the overall depth dimension **30** from the right side to the left side of Drawing View2. Drag the **extension lines** off the profile. Click the **green dot** on the arrowhead to flip arrows to the inside.

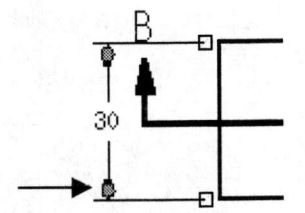

131) Drag the Slot Cut two vertical dimensions, **10** to the left of the Section arrow. Flip the **arrows** to the inside. Click the **dimension text** and drag the text outside the leader lines.

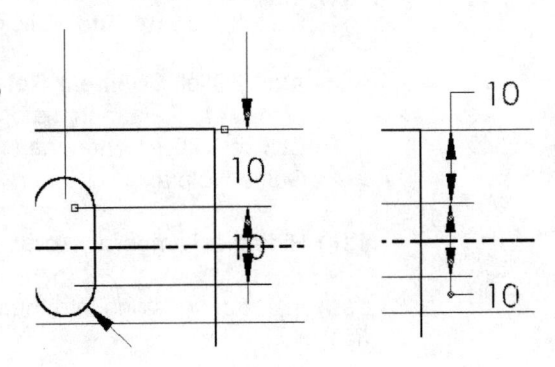

Partial Auxiliary View – Crop View

Create a Partial Auxiliary view from the Full Auxiliary view. Sketch a closed profile in an active Auxiliary view.

Create the Profile with a closed Spline. Create a Partial Auxiliary view. Crop the view. The 6mm dimension references the centerline from the Guide Hole. Sketch a centerline collinear with the Temporary Axis of the Guide Hole.

132) Sketch a closed profile in the active Auxiliary view. Click the **Auxiliary** view boundary. Click **Zoom to Selection**. An active view displays the green view boundary. Sketch a closed Spline. Click **Spline** from the Sketch tools toolbar. Click **six positions** clockwise to create the closed Spline. The first point is coincident with the last point.

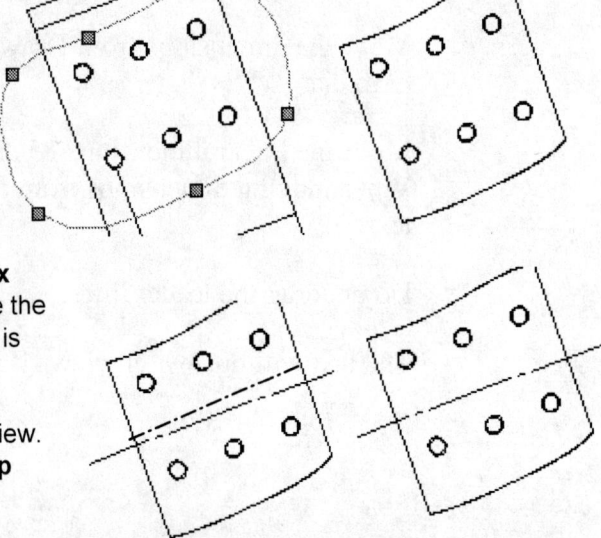

133) Display the Partial Auxiliary view. Click **Tools**, **Crop View**, **Crop** from the Main toolbar.

Fundamentals of Drawing   **Engineering Design with SolidWorks**

134) Click the **Centerline**.

135) Sketch a **centerline** between the two rows of tapped holes. Display the Guide-Hole axis.

136) Check **View**, **Temporary Axis**.

137) Right-click **Select**. Hold the **Ctrl** key down. Click the **centerline**. Click the **Temporary Axis**. Click **Collinear Relation** in the Add Relations dialog box. Release the **Ctrl** key. The centerline is collinear with the Guide Hole axis.

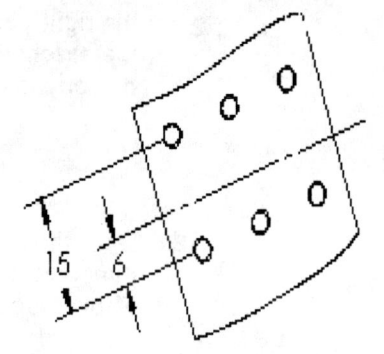

138) Hide the Temporary Axis. Click **View**, **Temporary Axis**.

139) Insert dimensions into the Auxiliary view. Click the **Auxiliary view boundary**.

140) Click **Model Items** from the Annotations toolbar. The Insert Model Items dialog box is displayed.

141) Uncheck **Include items from hidden features**.

142) Click **OK**.

143) Drag dimension **6** to the left of the Auxiliary view.

144) Drag dimension **15** to the left of dimension 6.

Move Dimensions to a Different View

Move the dimensions from Drawing View2 to Drawing View 5, (Auxiliary View).

Move the linear dimensions 25 and 10 that define the Linear Hole Pattern. When moving dimensions from one view to another, only drag the dimension text.

Do not drag the leader lines.

The text will not switch views if positioned outside the view boundary.

Move dimensions from the Top view to the Auxiliary view.

145) Press the **z key** approximately 4 times to view the dimension in Drawing View2.

146) Hold the **Shift** key down. Click the vertical dimension **25**. Drag the **dimension text** from Drawing View2 to Drawing View5. Release the **mouse button** and the **Shift** key when the mouse pointer is inside the Drawing View5 boundary.

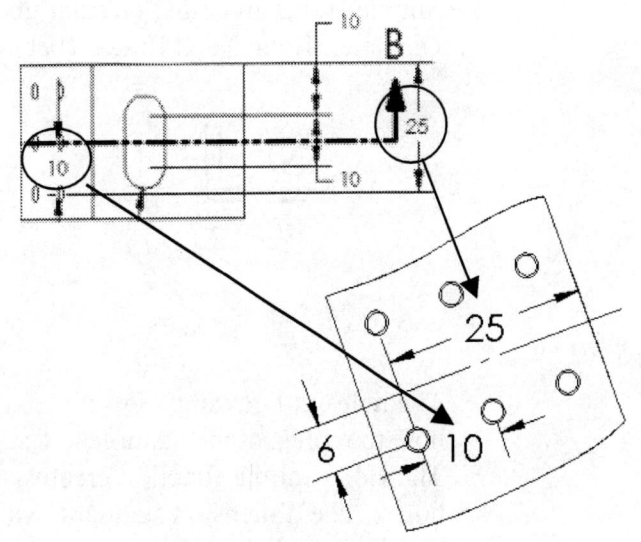

147) Repeat the above procedure for dimension **10**, between the two holes.

148) Drag the **dimensions** off the auxiliary profile. Position the **VIEW A** text centered below the view.

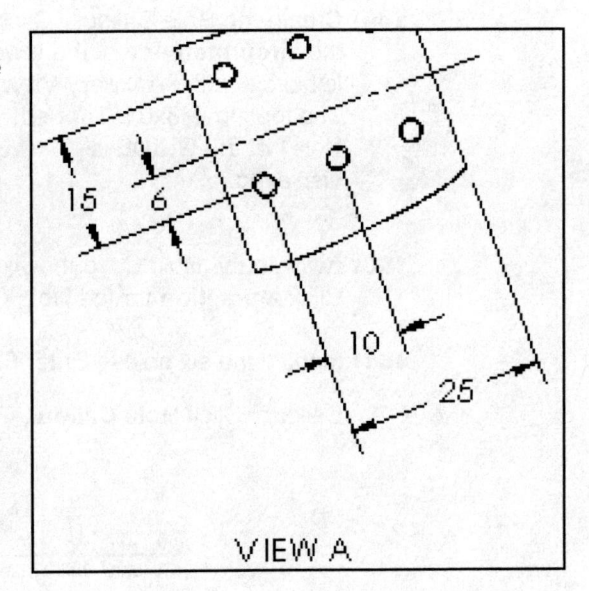

Dimension Holes

Simple Holes and other circular geometry are dimensioned in three ways: Diameter, Radius and Linear (between two straight lines).

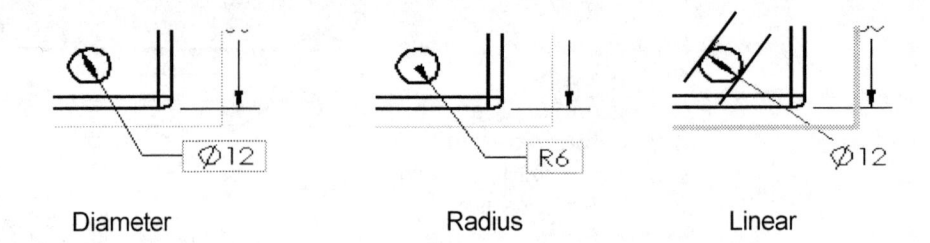

Diameter   Radius   Linear

The holes in Drawing View5 (Auxiliary) require a diameter dimension and a note to represent the six holes. Use the Hole Callout to dimension the holes. The Hole Callout function creates additional notes required to dimension the holes. The dimension standard symbols are displayed automatically when you use the Hole Wizard.

Dimension the Linear Pattern of Holes.

149) Create the Hole Callout. Click the **circumference** of the lower left circle in the Auxiliary View. The tool tip, M3x0.5 Tapped Hole1 of the GUIDE is displayed.

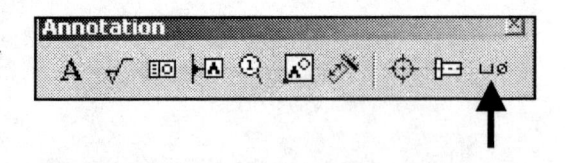

150) Activate the Hole Callout. Click **Hole Callout** in the Annotation Toolbar. Click a **position** for the Hole Callout text, left of the Auxiliary view.

151) Identify the six holes. Enter **6X** in the Dimension Text box. Deactivate the Hole Callout. Click **Hole Callout**.

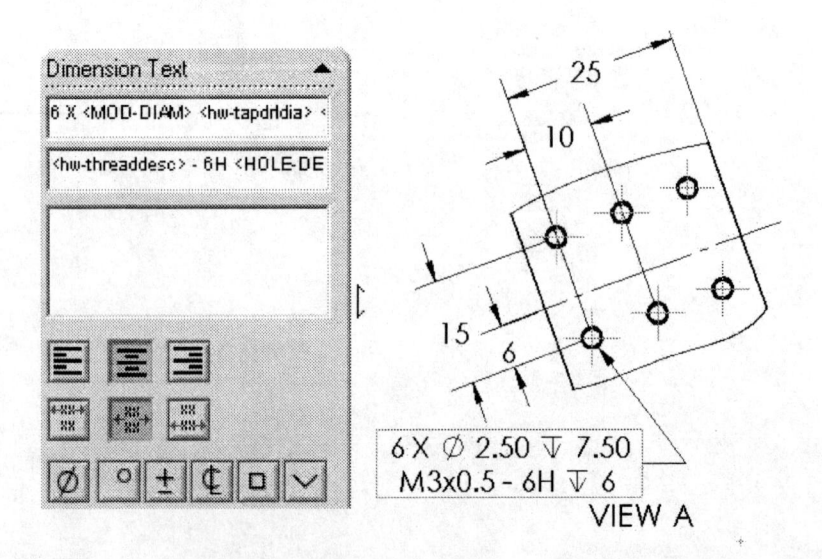

6 X ⌀ 2.50 ▽ 7.50
M3x0.5 - 6H ▽ 6

VIEW A

Remove the trailing zeros for millimeter display.

152) Click **Tools**, **Options**, **Document Properties**. Click **Remove**  Trailing zeroes: Remove from the Trailing zeroes list box.

153) Click **OK**.

154) Click **Rebuild**.

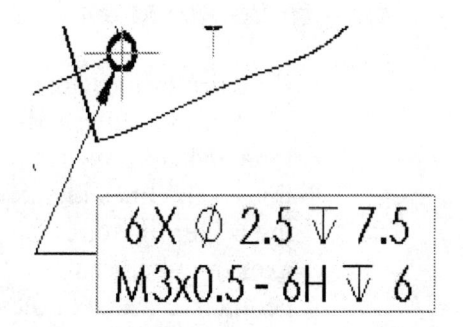

Symbols are located on the bottom of the Dimension Text dialog box. The current text is displayed in the text box. Example:

- <MOD-DIAM>: Diameter symbol ⌀.

- <HOLE-DEPTH>: Deep symbol.

- <HOLE-SPOT>: Counterbore symbol.

- <DIM>: Dimension value 3.

The mouse pointer displays the Hole Callout icon when the Hole Callout function is active.

155) Fit the drawing to the Graphics window. Press the **f** key.

156) Zoom in on Drawing View1.

157) Click the Guide Hole dimension ⌀**10** in Drawing View1. Enter text **THRU** in the Dimension Text box.

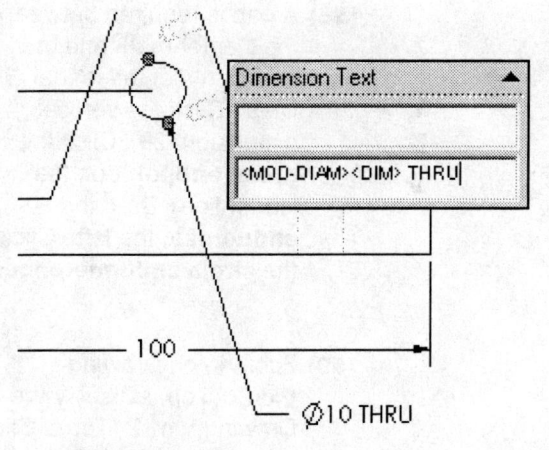

Access Center Marks, Hole Callouts and other Dimension Annotations through the Annotations toolbar menu. You can also access Annotations with the Right Mouse button, Dimension Annotations.

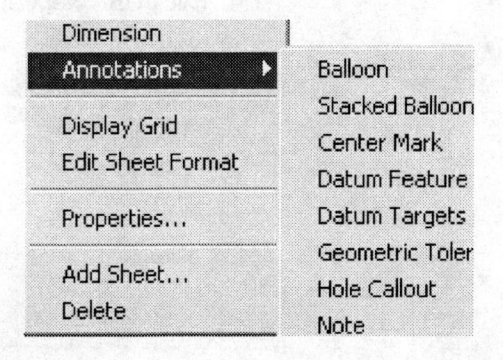

Fundamentals of Drawing   **Engineering Design with SolidWorks**

Create Center Marks

Hole centerlines are composed of alternating long and short dash lines. The lines identify the center of a circle, axes or cylindrical geometry. All six holes in Drawing View5, the Guide Hole in Drawing View1 and the Slot Cut arcs in Drawing View2 require Center marks.

Center marks represents two perpendicular intersecting centerlines.

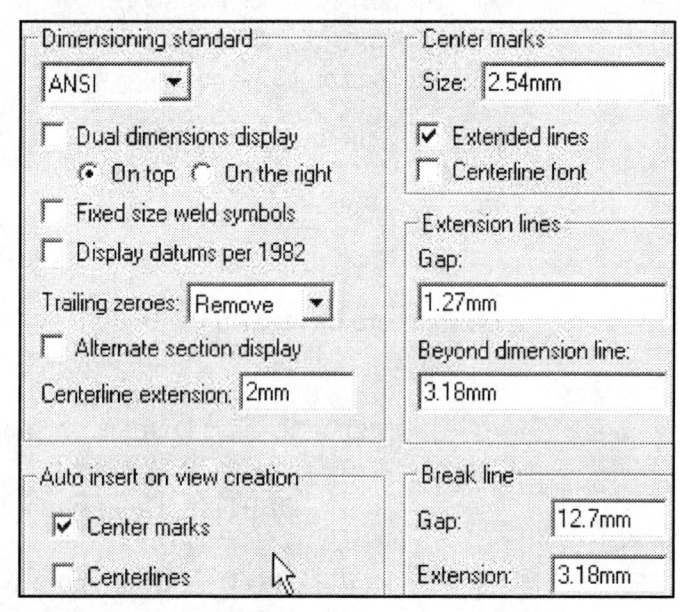

The default Size and Display of the Center mark is set in Tools, Options, Document Properties Center marks text box. The option Auto Insert on View Creation controls the display of Center marks within a new view.

Modify the Center marks.

158) A gap is required between the Center mark and the end points of the leader lines. Click the vertical dimension, **29**. Click the **green endpoint** on the leader line. Drag the **endpoint** to the left. Click the **circle circumference**.

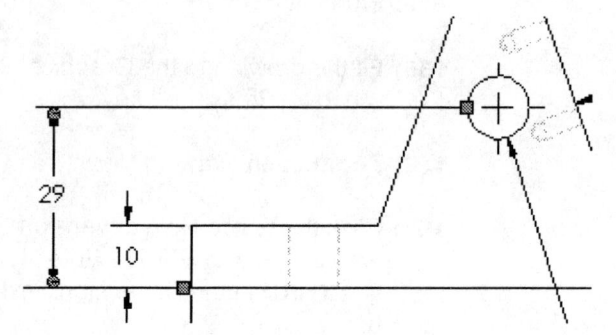

159) Zoom in on Drawing View2 (Top). Display the center cross of the Center mark for each arc in Drawing View2 (Top). Click the bottom left **Center mark**. Enter **1** for Mark size. Click **OK**. Repeat for the other **three Center marks**.

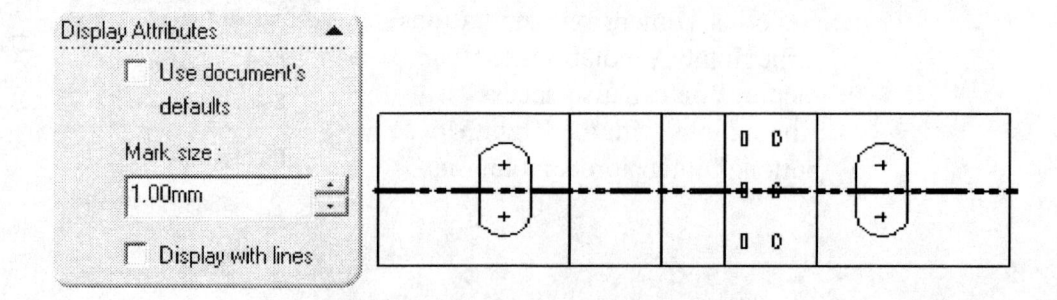

Engineering Design with SolidWorks — Fundamentals of Drawing

160) Delete the Center marks in the Auxiliary view. Click the **Auxiliary view boundary**. Click **Zoom to Selection**. Click the **first Center mark** in the lower left corner. Hold the **Ctrl** key down. Click the **five Center marks**. Release the **Ctrl** key. Press the **Delete** key.

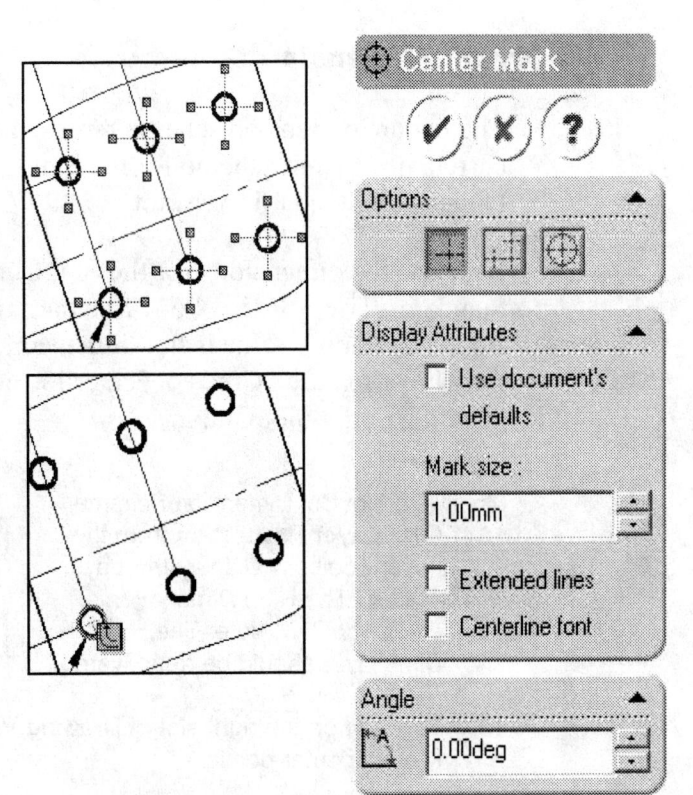

161) Add Center marks with centerlines. Click **Center mark** from the Annotations toolbar. Click **Single Center mark**. Uncheck **Use documented defaults**. Enter **1** for Mark size. Uncheck **Extended** lines. Click the **lower left Tapped Hole circumference**. The Center mark is displayed at an angle aligned to the Auxiliary view. Click the remaining **5 Tapped Holes**.

162) Display the Construction lines. Click the **Linear Center Mark** for Linear Patterns.

163) Click **OK**.

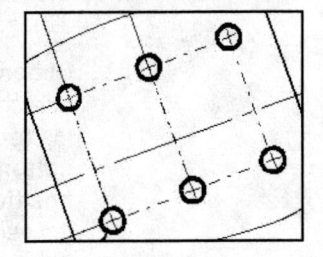

Note: The blue Propagate button appears on the first Center mark. Use the Propagate option for Center marks at 0 degrees.

PAGE 3-39

Fundamentals of Drawing   Engineering Design with SolidWorks

Modify the Dimension Scheme

The feature dimension scheme represents the design intent of the part. The current dimension scheme for the Slot Cut differs from the ASME 14.5M Dimension Standard for a slot.

Redefine the dimensions for the Slot Cut according to the ASME 14.5M Standard. The ASME 14.5M Standard requires the outside dimension of a slot. The Radius value is not dimensioned. The left Slot Cut was created with the Mirror feature. Create a centerline and dimension to complete the detailing of the Slot Cut.

Modify the Slot Cut Dimension Scheme.

164) Click **Layer Properties** from the Layer Toolbar. Display the Layer Status. The Hide Dims layer is off. Click **OK**. Note: The Hide Dims layer should be deactivated.

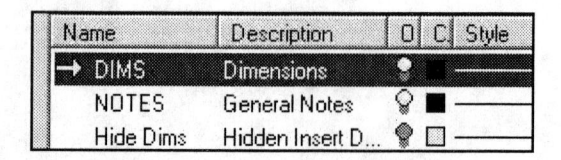

165) **Zoom in** on the right slot of Drawing View2. Hide the dimension between the two arc center points.

166) Right-click on **10**. Right-click **Properties**. Select **Hide Dims** from the Layer drop down list. The dimension text is placed on the Hidden Insert Dimensions layer. The Hidden Insert Dimensions layer is not activated. Click **OK**.

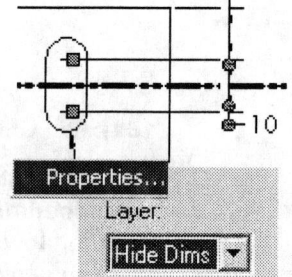

167) Create the vertical dimension. Click **Dimension** in the Sketch toolbar. Click the **top arc** and the **bottom arc**. Click a **position** to the right of the vertical profile line. The default arc conditions are measured from arc center point to arc center point.

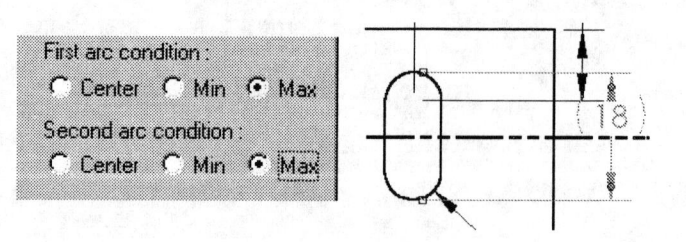

168) Modify the arc condition. Right-click in the **Graphics window**. Right-click **Properties**. Click **Max** for the First arc condition. Click **Max** for the Second arc condition. Click **OK**.

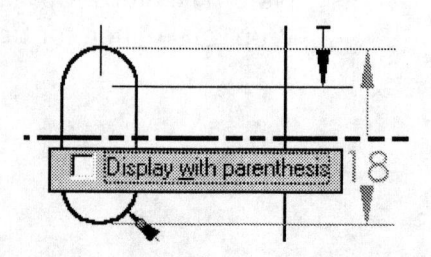

PAGE 3 - 40

169) The dimension is displayed with parenthesis as a referenced dimension. Remove the dimensions. Right-click on **(18)**. Click **Properties**. Uncheck the **Display with parenthesis** check box. Click **OK**.

170) Modify the Radius text. Left-click R4. Delete R<DIM> in the Dimension text box. Enter **2X R** for Dimension text. Do not enter the radius value.

171) Create the horizontal dimension. Click the **left vertical line** and **right vertical line** of the Slot Cut. Click a **position** below the horizontal profile line.

172) Click **Flip arrows**. 8 is located inside the leader lines. Right-click on **(8)**. Click **Properties**.

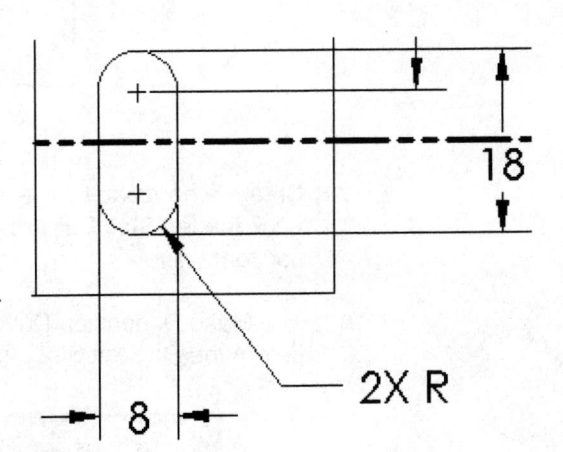

173) Uncheck the **Display with parenthesis** check box. Click **OK**.

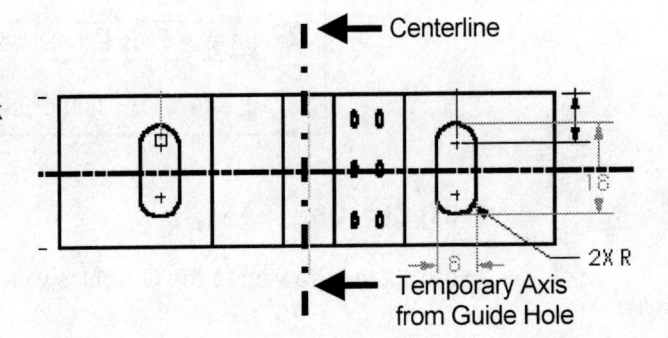

Create a centerline and reference dimension to locate the Slot Cut.
174) Right-click **Select**. Click the **Drawing View2** boundary.

175) Click **Zoom to Selection**.

176) Check **View**, **Temporary axis**.

177) Click **Centerline**.

178) Sketch a vertical **centerline** collinear with the Guide Hole axis in Drawing View2 (Top).

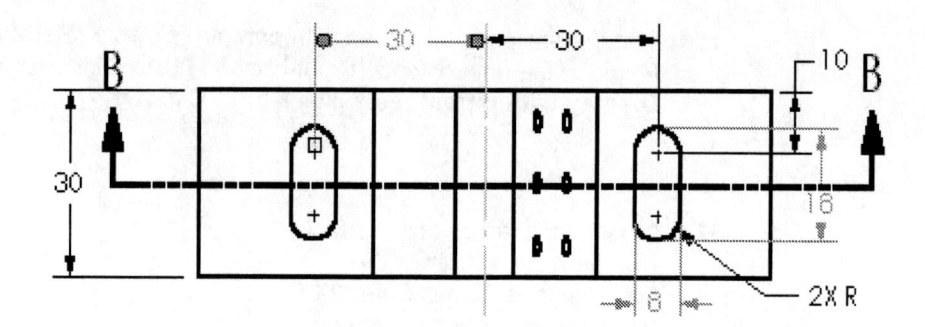

Fundamentals of Drawing   **Engineering Design with SolidWorks**

179) Create a horizontal dimension. Click **Dimension**. Click the **centerline**. Click the left **Slot Cut arc center point**. Click a **position** above the top horizontal line.

180) The Make Dimension Driven dialog box is displayed. The Mirror feature determines the left Slot location. Click **Make this Dimension Driven**.

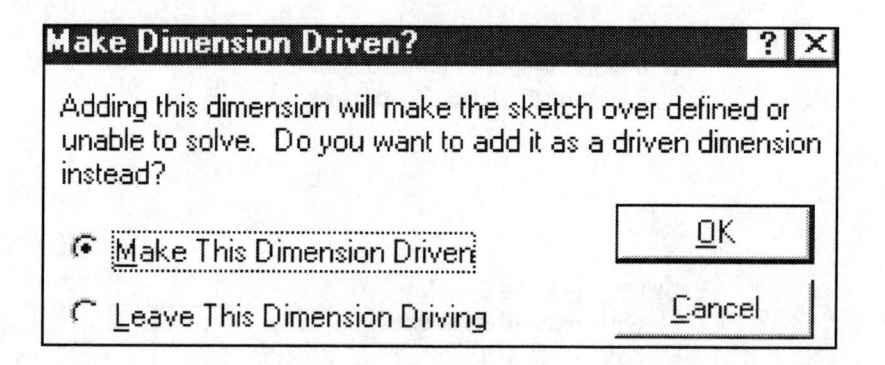

181) Click **OK**.

182) Fit the drawing to the Graphics window. Press the **f** key.

Create a reference dimension.

183) Activate the Section view. Click **Select**. Click **Section B view boundary**. Click **Centerline** from the Sketch Tools toolbar. Sketch **two vertical centerlines** through both Slot Cuts in the Section view.

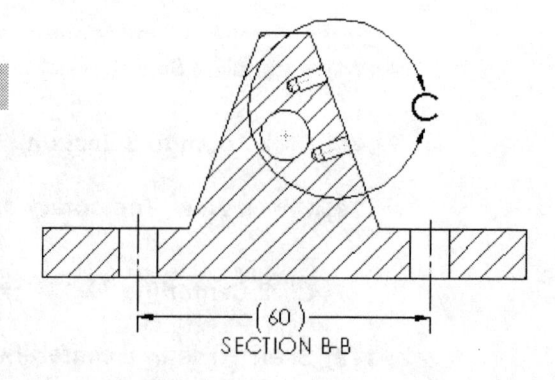

PAGE 3 - 42

184) Click **Dimension**. Click the **left centerline**. Click the **right centerline**. Click a **position** below the bottom horizontal line. Click **Make this Dimension Driven**. Click **OK**. The reference dimension is displayed with parenthesis.

185) Click **Select**. Move the SECTION B-B text downward. Click the **text** and **drag** downward below the profile line.

186) Hide the Temporary axis. Click **View**, **Temporary Axis**.

187) Save the GUIDE. Click **Save**. Click **Yes** to the question, "Save the document and the referenced models now?"

Clearance Fit for the ROD and GUIDE

In Project 2, you were introduced to the concept of tolerance and fit. The design requires the ROD to slide through the GUIDE. The shaft of the ROD is currently 10mm. The Hole of the GUIDE is 10mm. This is a problem! You need to create a Clearance fit for the ROD and GUIDE.

Do you remember what a Clearance fit is? A Clearance fit occurs when the shaft diameter of the ROD is smaller than the hole diameter of the GUIDE. There are many different classes of clearance fits. You can find various ANSI and ISO standards of Clearance fits in engineering textbooks or machine tool catalogs.

Dimension the GUIDE hole and ROD shaft for a Sliding Clearance fit. All below dimensions are in millimeters.

Use the following values:

Hole	Maximum	10.015mm.	
	Minimum	10.000mm.	
Shaft	Maximum	9.995mm.	
	Minimum	9.986mm.	
Fit	Maximum	10.015 – 9.986 = .029	Max. Hole – Min. Shaft.
	Minimum	10.000 – 9.995 = .005	Min. Hole – Max. Shaft.

Calculate the maximum variation:

Hole Max: 10.015 – Hole Min: 10.000 = .015 Hole Max. Variation.

Add the maximum and minimum hole dimensions.
188) Click the **Drawing View1 (Front) view boundary**. Click **Zoom to Selection**.

189) Click ⌀**10 THRU**.

190) Click the **drop down arrow** from the Tolerance list box in the Dimension PropertyManager.

191) Select **Limit**.

192) Enter **0.015** in the plus box.

193) Enter **0.000** in the minus box.

194) Click **OK**.

195) Improve view clarity. Drag the **10.015/10.000 dimension** to the right of the Guide Hole.

196) Save the GUIDE drawing. Click **Save**.

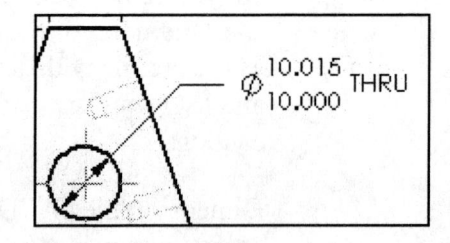

Add An Additional Feature

The design process is dynamic. We do not live in a static world. Create and add an edge Fillet feature to the GUIDE part. Insert new part dimensions into Drawing View1 (Front).

Create the Fillet.
197) Open the Guide part. Right-click in the **Drawing View1(Front)** view. Click **Open Guide.sldprt**.

198) Display Hidden Lines to select edges to fillet.

Click **Hidden Lines Visible**.

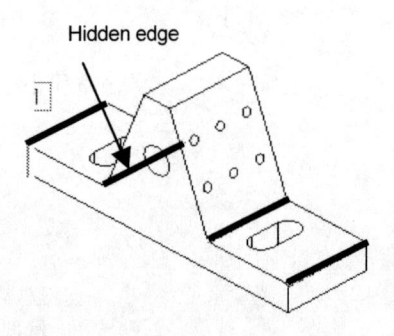

Engineering Design with SolidWorks Fundamentals of Drawing

199) Display the Isometric view. Click **Isometric**.

200) Create a Fillet/Round feature. Click the hidden edge. Click **Fillet**. The Fillet feature PropertyManager is displayed. Enter **1** in the Fillet Radius list box. Click the other **3 edges**. Each edge is added to the Items to Fillet list.

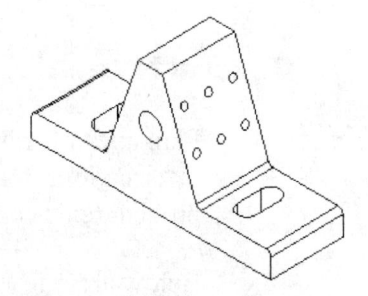

201) Display the Fillet. Click **OK**.

202) Save the GUIDE part. Click **Save**.

203) Open the Drawing. Right-click on the **GUIDE Part** icon in the FeatureManager. Click **Open Drawing**. Update the drawing. Click **OK**.

Display the GUIDE features.
204) Expand Drawing View1 in the FeatureManager. Click **Plus**.

205) Expand the GUIDE part. Click **Plus**.

206) Select the feature. Click **Fillet1**.

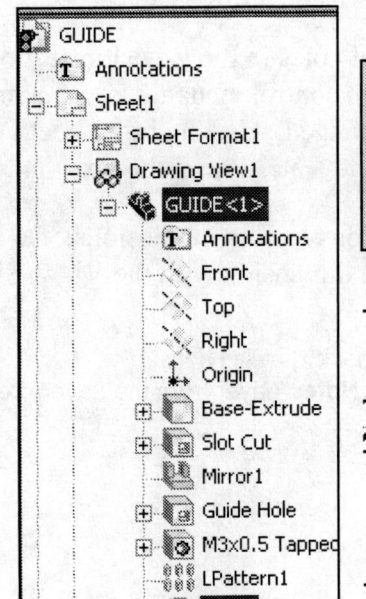

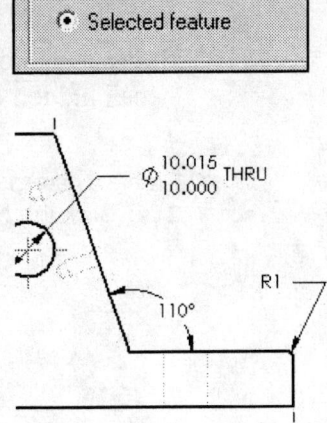

Insert dimensions.
207) Insert dimensions by Selected feature into Drawing View1 (Front). Click **Model Items** from the Annotations toolbar. Click **Selected feature** in the Import into views list box.

208) Click **OK**.

PAGE 3 - 45

209) The Fillet Radius is displayed in Drawing View1 (Front). Position the text. Click the **R1** text. Drag the **R1** text off the profile.

210) Save the GUIDE drawing. Click **Save**.

General Notes

Plan ahead for general drawing notes. Notes provide relative part or assembly information. Example: Material type, material finish, special manufacturing procedure or considerations, preferred supplier, etc.

Below are a few helpful guidelines to create general drawing notes:

- Use capital letters.
- Use left text justification.
- Font size should be the same as the dimension text.

Create parametric notes by selecting dimensions in the drawing. Example: Specify the Fillet Radius of the GUIDE as a note in the drawing. If the Radius is modified, the corresponding note is also modified.

Create general notes in the Revision section of the drawing. The documentation control manager issues a revision number when a drawing is initiated or changed. Use the Edit Sheet Format mode. Enter the Revision, Description and Date. Align the text.

Hide superfluous feature dimensions. Do not delete feature dimensions. Recall hidden dimension with the View, Show Hidden command.

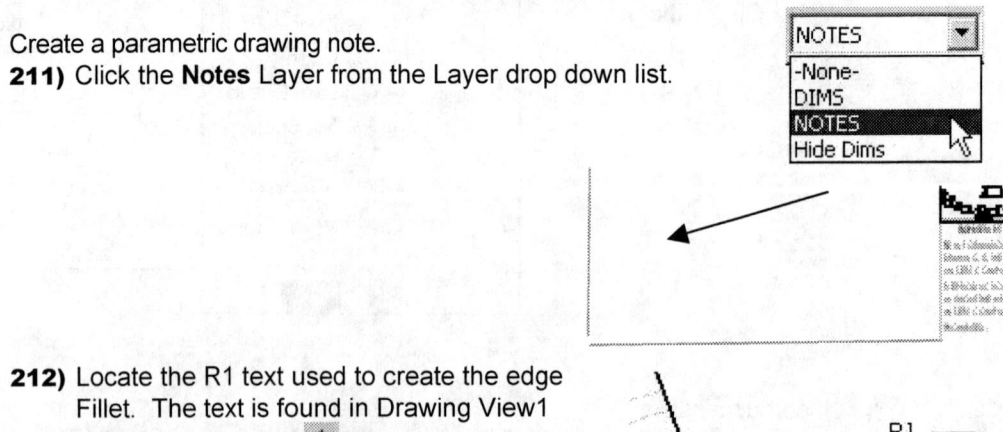

Create a parametric drawing note.
211) Click the **Notes** Layer from the Layer drop down list.

212) Locate the R1 text used to create the edge Fillet. The text is found in Drawing View1 (Front). Click **Note** A from the Annotations toolbar. Click a **start point** in the lower left corner of the Graphics window, away from the title block.

213) The Note PropertyManager is displayed. Type two lines of notes in the Note text box. For line 1: Enter **1. ALL ROUNDS**. Press the **Space** key. Select the **R1** text. The variable name "D1Fillet1@GUIDE@DrawingView1" is added to the text box. Press the **Space** key. Enter **MM**.

214) For line 2: Enter **2. REMOVE ALL BURRS.** on line 2. Display the Note. Click **OK**.

> 1. ALL ROUNDS "D1Fillet1@GUIDE@DrawingView1" MM.
> 2. REMOVE ALL BURRS.

215) The parametric note specifies that the radius of all Rounds is 1MM.

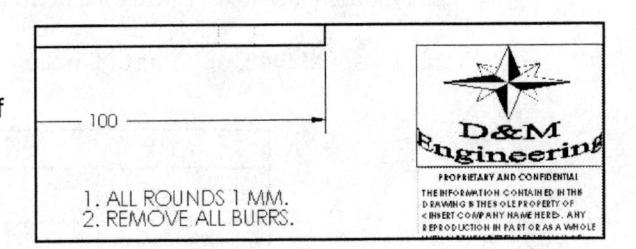

Do not double dimension a drawing with a note and the corresponding dimension. Hide the radial dimension.

216) Hide the R1 dimension on the Hidden Dims layer. Click the **R1** dimension in Drawing View1. Right-click **Properties**. Click **Hide Dims** from the Layer drop down list.

217) Verify the parametric note. Open the GUIDE part. Right-click **Open GUIDE.sldprt** in the Front view. Double-click **Fillet1** from the FeatureManager. Double-click the **R1** dimension.

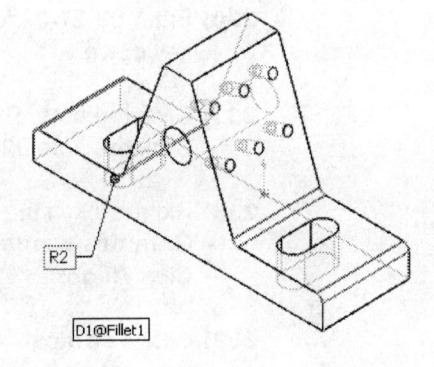

218) Modify the edge Fillet radius. Enter **2**. Click **Rebuild**. Click **Save**.

219) Right-click on the **GUIDE Part icon**. Click **Open drawing**. The parametric note displays 2MM. The updated views reflect the new Fillet radius.

220) Fit the drawing to the Graphics window. Press the **f** key.

221) Save the **GUIDE** part.

222) Save the **GUIDE** drawing.

223) Click **Save**.

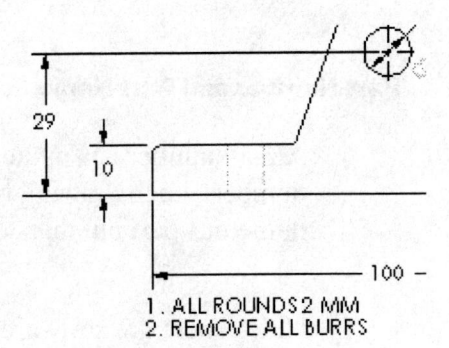

Fundamentals of Drawing   Engineering Design with SolidWorks

Revision Text and Align Text.

224) Right-click **Edit Sheet Format** [Edit Sheet Format].

225) Copy/Paste existing notes. **Zoom in** on the Revisions box in the upper right corner of the drawing. Click on the **REV** text. Press the **Ctrl C** key to Copy.

226) Locate the text. Click a **position** in the cell below. Press the **Ctrl V** key to Paste. The REV text is displayed in the second row.

227) Modify the text. Double-click on the **REV** text. Enter **1**.

228) Repeat the **Copy/Paste** function for **Description** and **Date**.

		REVISIONS
ZONE	REV.	DESCRIPTION
	1	ECO 32510 RELEASE TO MANUFACTURING

229) Enter **ECO 32510 RELEASE TO MANUFACTURING** for DESCRIPTION.

230) Enter **02-21-02** for date or the current date. Align the Date text. Hold the **Ctrl** key down.

231) Click the note, **DATE**. Click the note **02-21-02**.

232) Right-click in the **Graphics window**. Click **Align**.

233) Click **Leftmost**. The notes are leftmost aligned. Release the **Ctrl** key.

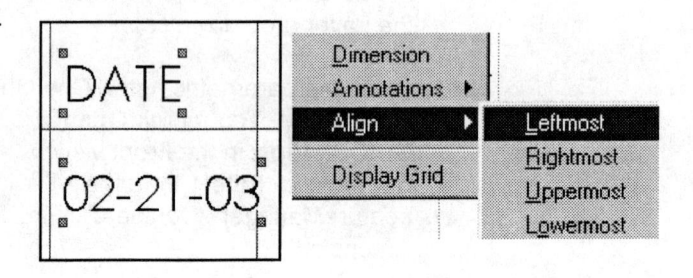

Part Number and Part Name

A part number is a numeric representation of the part. Each part has a unique number. Each drawing has a unique number. Drawings incorporate numerous part numbers or assembly numbers.

Note: There are software applications that incorporate unique part numbers to create and perform:

- Bill of Materials.
- Manufacturing procedures.

- Cost analysis.
- Inventory control / Just in Time, JIT.

As the designer, you are required to procure the part and drawing numbers from the documentation control manager.

In this project, use the following prefix codes to categories created parts and drawings. The part name, part number and drawing numbers are as follows:

Category:	Prefix:	Part Name:	Part Number:	Drawing Number:
Machined Parts	56-	GUIDE	56-A26	56-22222
		ROD	56-A27	56-22223
		PLATE	56-A28	56-22224
Purchased Parts	99-	FLANGE BOLT	99-FBM8x1.25	999-551-8
Assemblies	10-	GUIDE-ROD	10-A123	10-50123

Link notes in the title block to SolidWorks Properties. Create a note for the title of the drawing that is linked to the SolidWorks file name, GUIDE.

Create additional notes in the title block that complete the drawing.

Create a linked note.
234) Zoom in on the lower right corner of the drawing. Click **Note** from the Annotations toolbar. Click a **start point** in the TITLE section of the title block.

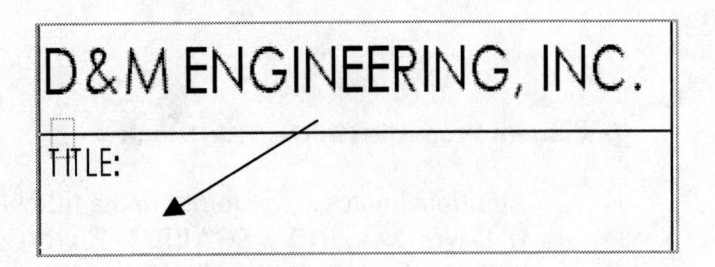

235) Click **Link to Property** from the Text Format box.

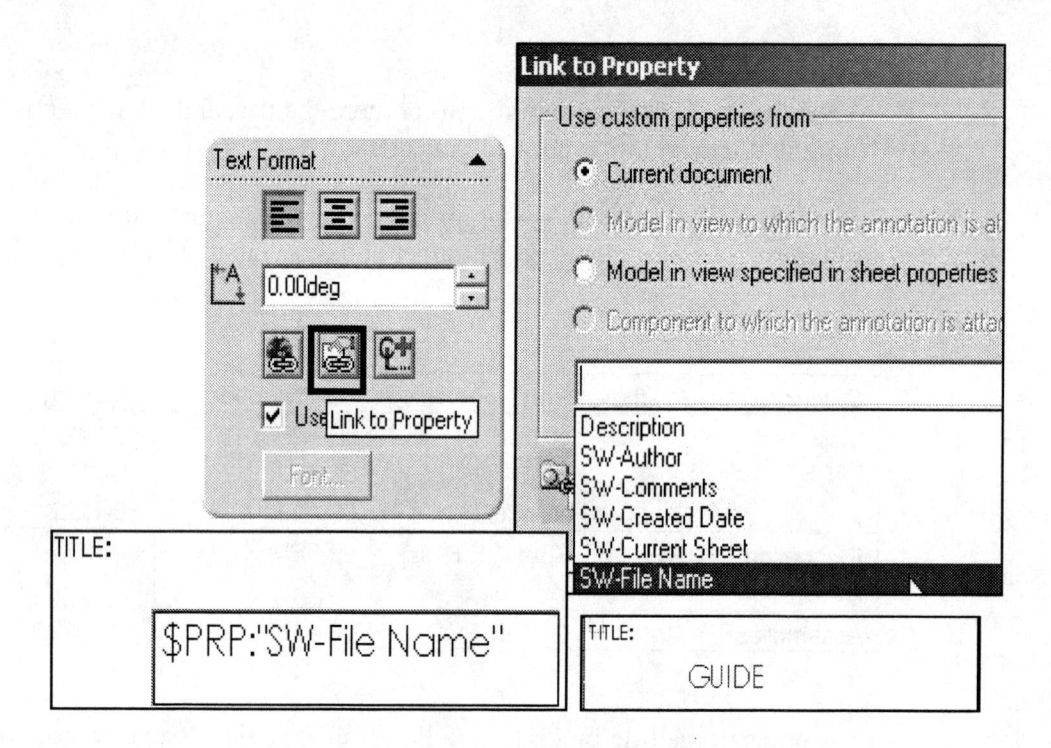

236) Select **SW-FileName** from the Link to Property drop down list.

237) Click **OK**. The variable $PRP"SW-File Name" is displayed in the Note text box.

238) Display the linked text in the Title. Click **OK**.

Document Properties in the Title Block

Additional notes are required in the title block. The text box headings: SIZE B, DWG. NO., REV., SCALE, WEIGHT and SHEET OF are entered in the SolidWorks default Sheet Format.

Properties are variables shared between documents and applications. Define the Document Properties in the GUIDE drawing.

Link the Document Properties to the notes in the title block.

Enter Summary information for the GUIDE drawing.

239) Click **File**, **Properties** from the Main menu. The Summary Information dialog box is displayed.

240) Enter you **initials** for Author. Example: GMP.

241) Enter **GUIDE, ROD** for Keywords.

242) Enter **GUIDE FOR CUSTOMER XYZ** for Comments.

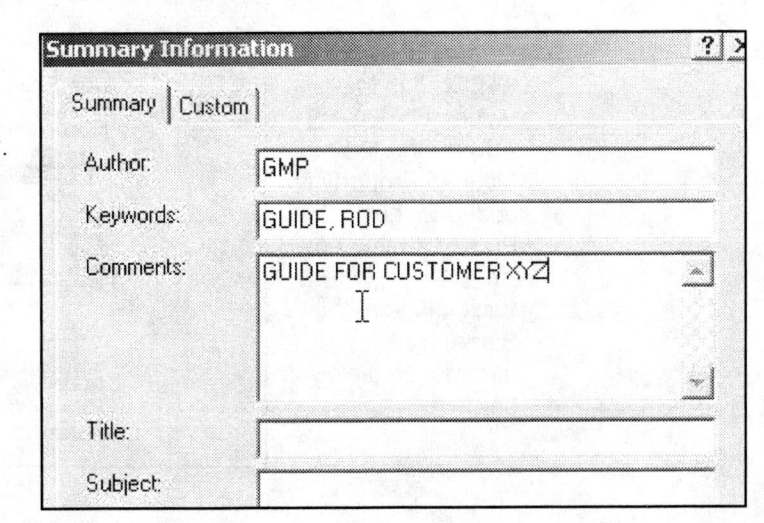

243) Define Custom Properties. Click the **Custom** Tab.

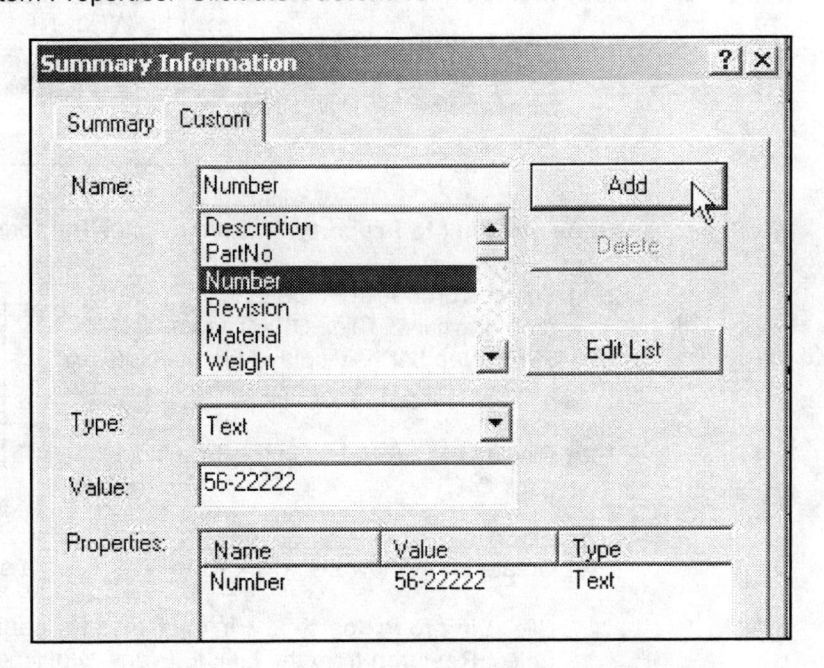

244) Add the predefined Properties. Select **Number** from the Name drop down list. Enter **56-22222** for Value.

245) Click the **Add** button. Number Property is added to the Properties text box.

246) Select **Revision** from the Name drop down list.

247) Enter **1** for Value.

248) Click the **Add** button. Revision Property is added to the Properties text box.

Fundamentals of Drawing   Engineering Design with SolidWorks

249) Select **Material** from the Name drop down list.

250) Enter **STAINLESS STEEL 303** for Value. Click the **Add** button. Material Property is added to the Properties text box.

251) Click **OK** from the Summary Information dialog box.

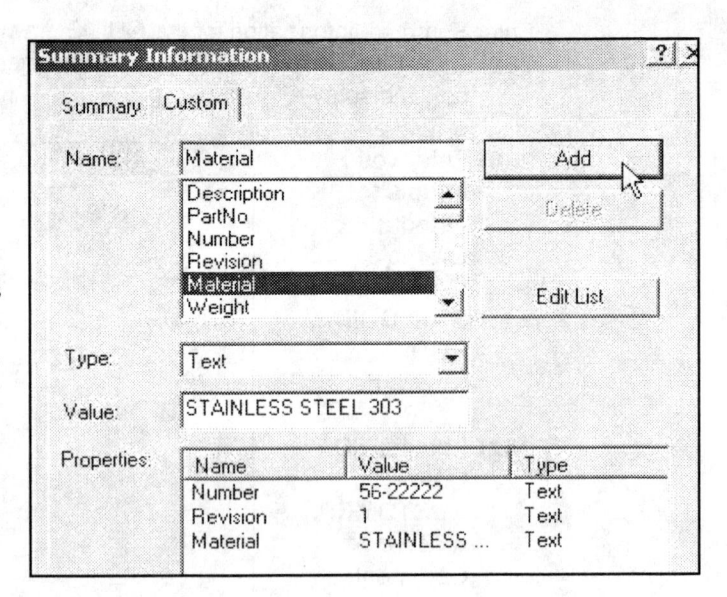

Link the Properties to the Drawing Notes.

252) Create the DWG. NO. Note.

Click **Note** A . Click a **start point** in the upper left hand corner below the DWG. NO. text.

253) Click **Link to Property** in the Notes PropertyManager Text Format box.

254) Select **Number** from the Link to Property drop down list. Click **OK**. The variable $PRP"Number" is displayed in the Note text box.

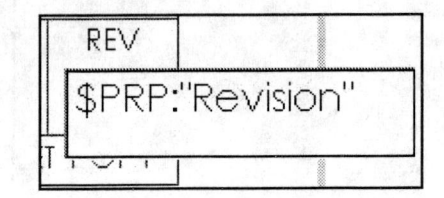

255) Display the linked text in the Title block. Click **OK**.

256) Create the REV Note. Click **Note** A . Click a **start point** below the REV text.

Click **Link to Property** in the Notes PropertyManager Text Format box. Select **Revision** from the Link to Property drop down list. Click **OK**. The variable $PRP"Revision" is displayed in the Note text box. Display the linked text in the Title block. Click **OK**.

257) Fit the drawing to the Graphics window. Click the **f** key.

258) **Zoom in** on the center of the Title block.

259) Double-click the **dash marks** in the MATERIAL text box. Click **Link to Property**. Select **Material** from the Link to Property drop down list. Click **OK**. The variable $PRP"Material" is displayed in the Note text box. Display the linked text in the Title block. Click **OK**.

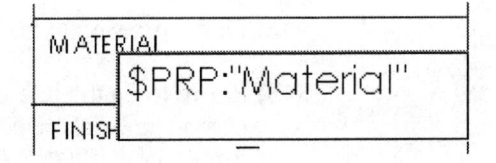

260) Create the DRAWN BY NAME Note. Click **Note**. Click a **start point** below the NAME text. Click **Link to Property**. Select **SW-Author** from the Link to Property drop down list. Click **OK**. The variable $PRP"SW-Author" is displayed in the Note text box.

261) Display the linked text in the Title block. Click **OK**.

262) Create the DATE Note. Click **Note**. Click a **start point** below the DATE text. Enter the current **date**.

263) Click **OK**. Adjust the Font size if required.

Complete the notes in the title block. The tolerance section in the title block displays information for an English drawing. Create a Metric drawing. Change the tolerance note text.

In Project 1, the design team decided that the ROD and GUIDE parts would be fabricated from 303 Stainless steel. There are numerous types of Stainless steels for various applications. Select the correct material for the application. This is critical!

When an assembly contains mating parts, document their relationship.

Add part numbers in the Used On section. Add your initials and the date.

Change the title block text using the Edit Sheet Format mode.

Fundamentals of Drawing Engineering Design with SolidWorks

Tolerance text.
264) Fit the drawing to the Graphics window. Press the **f** key.

265) Zoom in on the tolerance section of the title block. Double-click the text, "**DIMENSIONS ARE IN INCHES TOLERANCE**".

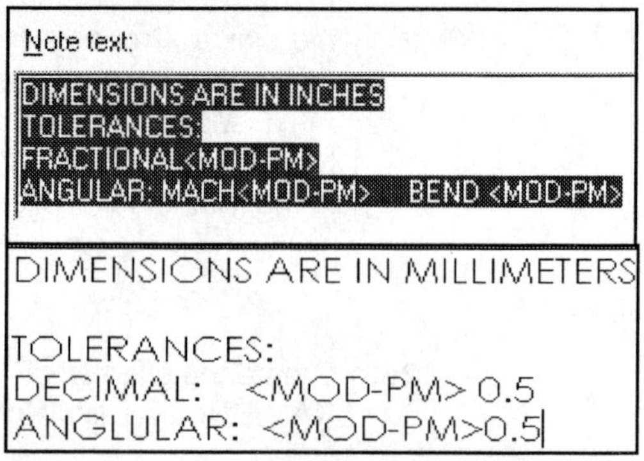

266) Edit the text in the note text box. **Delete** the word INCHES. Enter **MILLIMETERS**. Do not abbreviate units in the title block. **Delete** the line FRACTIONAL.

Enter the following lines:

DECIMALS ±0.5.

ANGULAR ±0.5.

Click **OK**.

Various symbols are available through the Symbol button in the Text dialog box. The ± symbol is found in the Modify Symbols list. The ± symbol is displayed as <MOD-PM>.

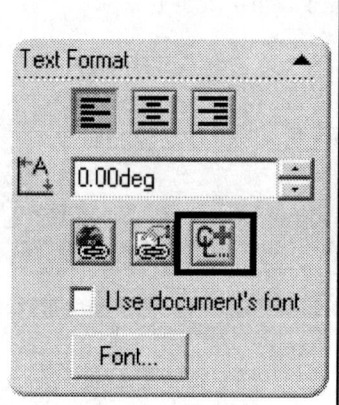

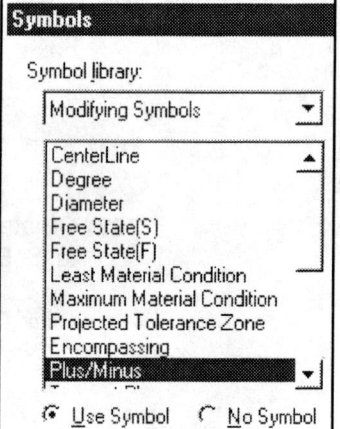

267) Create additional notes. Enter the drawing number for the GUIDE-ROD ASSY. Enter **10-50123** in the NEXT ASSY box.

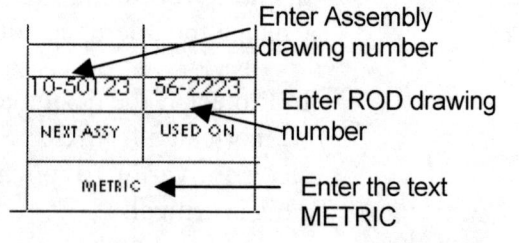

268) Enter the ROD drawing number. Enter **56-22223** in the USED ON box.

269) Enter the **METRIC** note below in the text APPLICATION.

270) Right-click in the **Graphics window**. Click **Edit Sheet**.

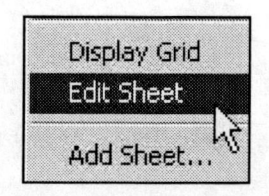

271) Fit the drawing to the Graphics window. Press the **f** key. Drag the **views** if required for spacing.

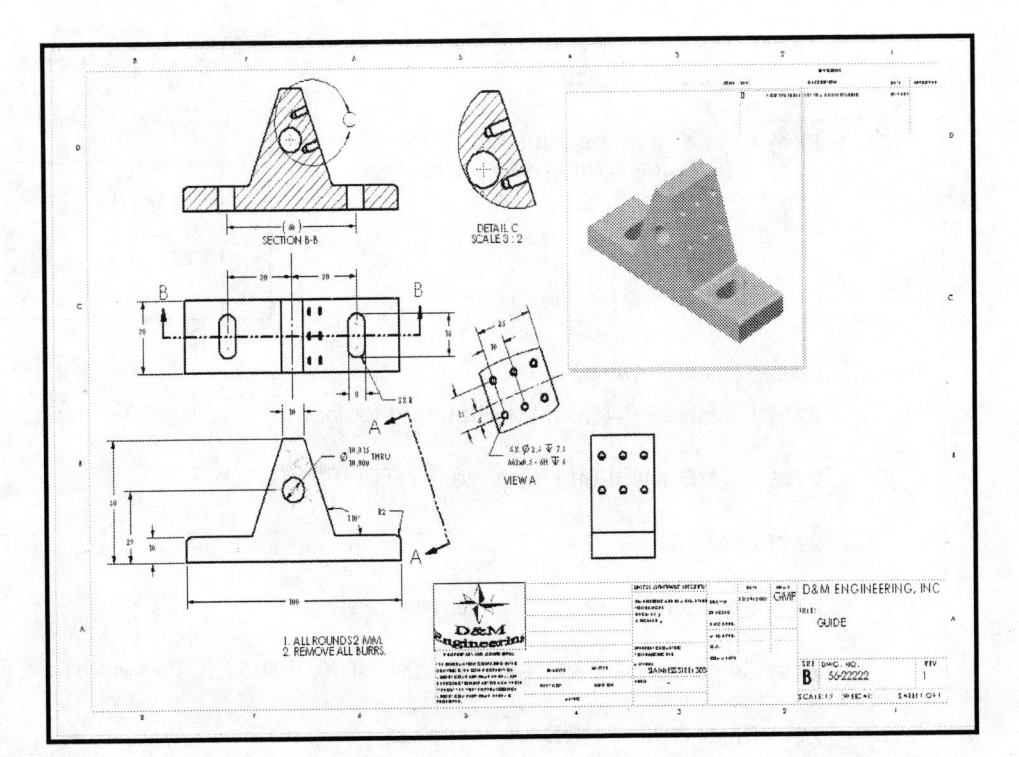

272) Save the GUIDE drawing. Click **Save**.

Exploded View and Bill of Materials

Add an Exploded view and Bill of Materials to the drawing.

Add the GUIDE-ROD assembly Exploded view.

The Bill of Materials reflects the components of the GUIDE-ROD assembly.

Create a drawing with a Bill of Materials. Perform the following steps:

- Create a new drawing from the B-ANSI-MM Drawing Template.
- Display the Exploded view of the assembly.
- Insert the Exploded view of the assembly into the drawing.
- Label each component with Balloon text.
- Create a Bill of Materials.

Fundamentals of Drawing   **Engineering Design with SolidWorks**

Let's start.

273) Close all parts and drawings. Click **Windows**, **Close All**.

274) Click **Yes** to the question, "Save changes to GUIDE-Sheet1?"

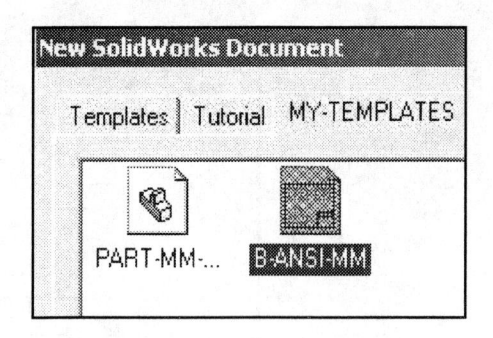

Create a new drawing.

275) Click **New** from the Standard toolbar.

276) Click **B-ANSI-MM** from the MY-TEMPLATE tab.

277) Click **OK**.

278) Save the drawing. Click **Save**.

279) Select **ENGDESIGN-W-SOLIDWORKS\PROJECTS** for Save in file folder.

280) Enter **GUIDE-ROD** for Filename.

281) Enter **GUIDE-ROD Assembly** for Description.

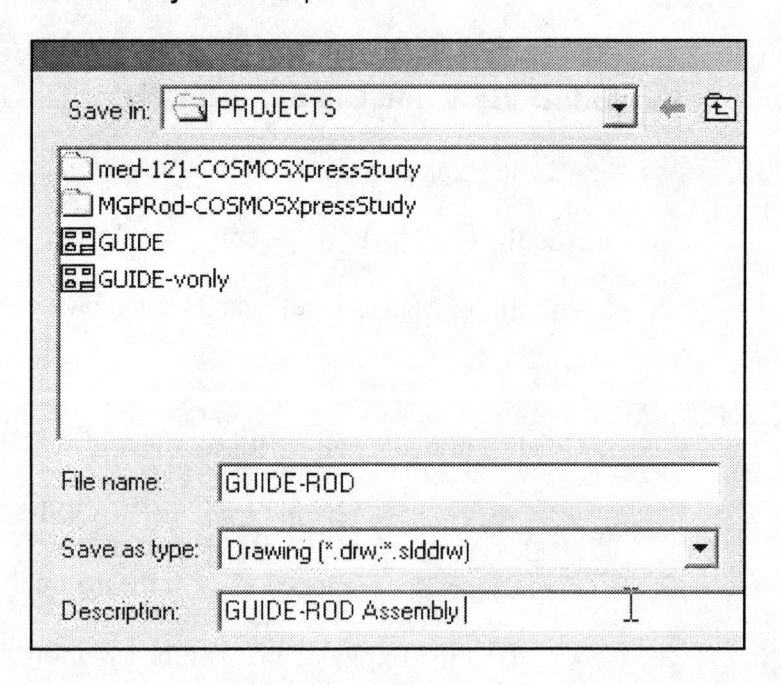

282) Click **Save**.

PAGE 3 - 56

283) Open the GUIDE-ROD assembly. Click **Open**. Select **Assembly** for Files of type. Click the **GUIDE-ROD**. The Lightweight and Preview boxes are checked. Click **Open**.

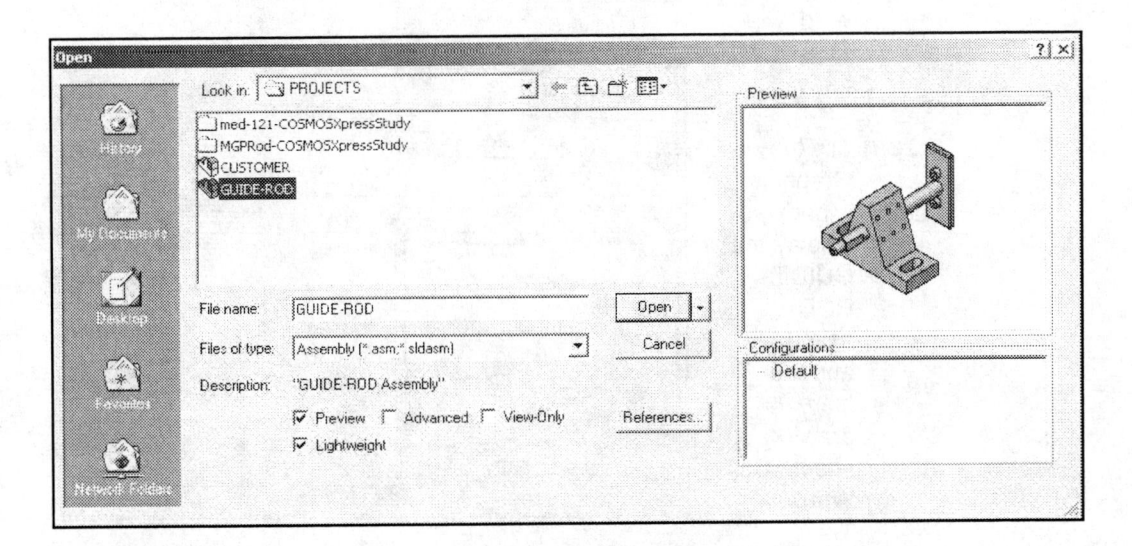

284) Resolve the Flange Bolts. Right-click **Flange Bolt<1>** Click **Set to Resolved**. Right-click **Flange Bolt<2>** Click **Set to Resolved**.

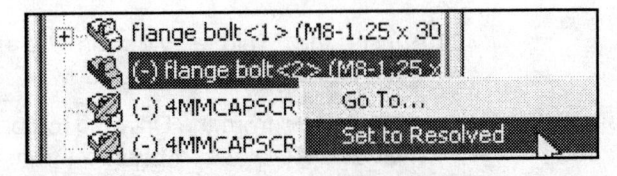

285) Split the **FeatureManager** if required.

286) Explode the view. Click **ConfigurationManager** at the bottom of the View Manager. Click the **Plus Sign** to the left of the Default entry.

287) Display the Exploded view in the Graphics widow. Double-click **Exp View1**.

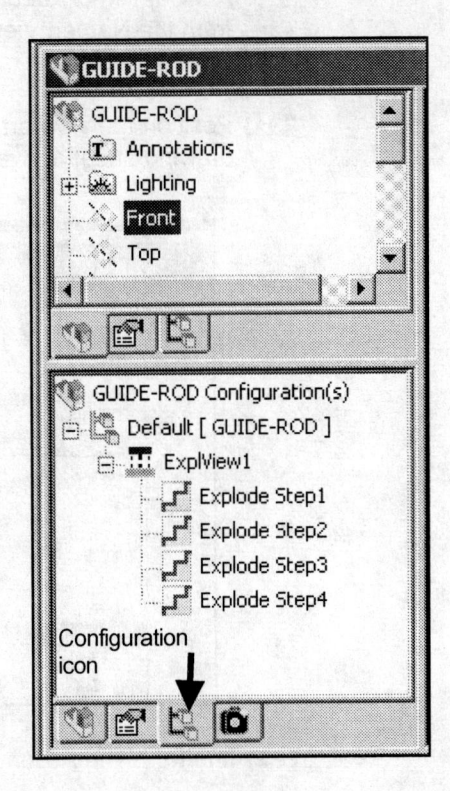

Fundamentals of Drawing   Engineering Design with SolidWorks

288) Fit the Exploded view to the Graphics window. Press the **f** key.

289) Tile the Graphics window to display the GUIDE-ROD assembly and the blank drawing sheet. Click **Window, Tile Horizontally**.

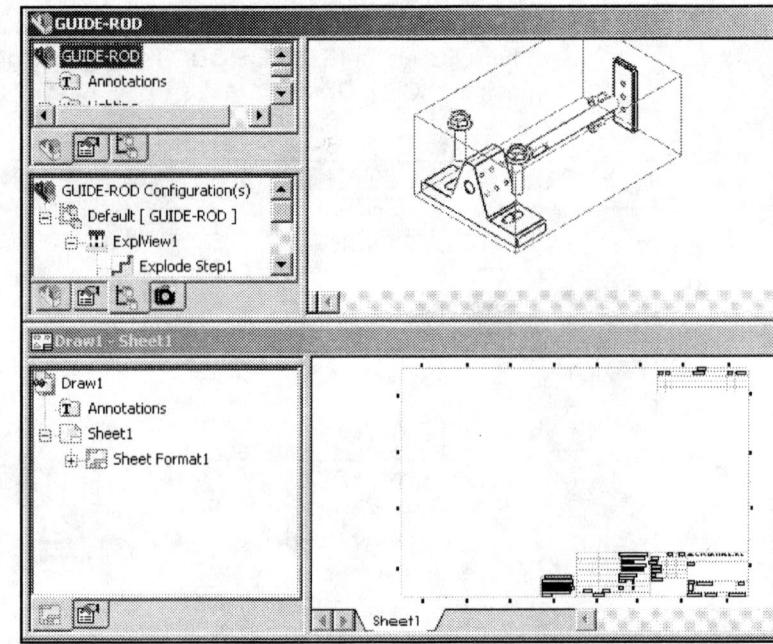

290) Insert the Exploded view into the drawing. Click the **Drawing** Graphics window. Click **Named Views** from the Drawing toolbar. Click the **GUIDE-ROD assembly icon** from the GUIDE-ROD assembly FeatureManager. Click **Isometric** from the Named View dialog box. The mouse pointer displays a + symbol.

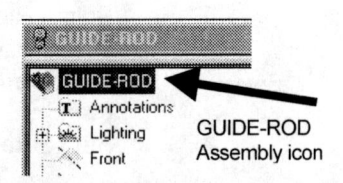

291) Position the view on the right side of the drawing. Click the **position** on the drawing sheet above the Title block.

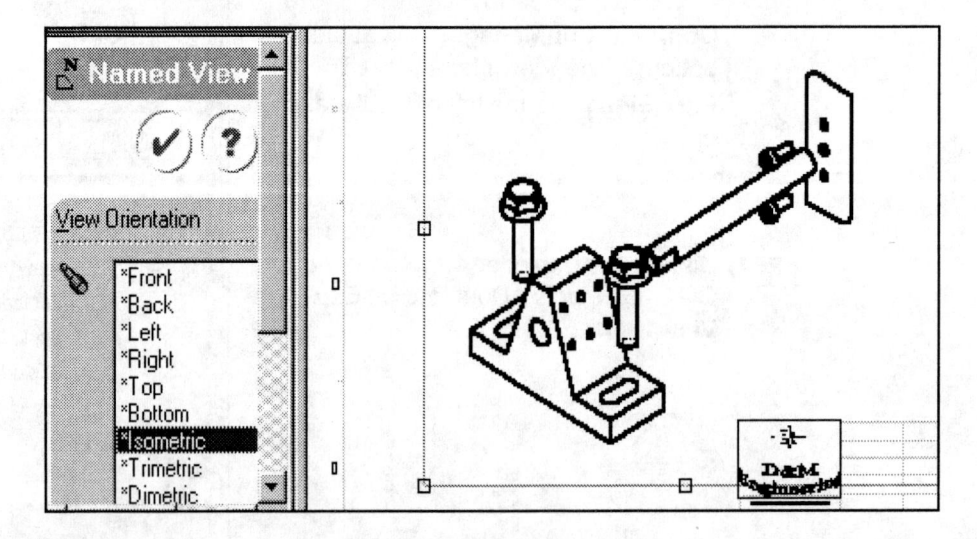

292) Minimize the Assembly window. **Maximize** the Drawing window.

PAGE 3 - 58

293) Position the view and resize the view border. **Move** the **view** to the left of the Revision block. The Revision block expands downward over the lifetime of the drawing.

294) Hide the Part Origins. Uncheck **View**, **Origins**.

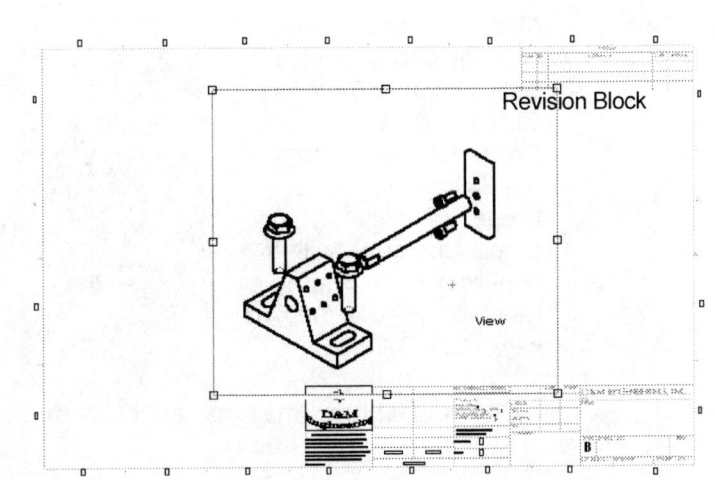

295) Save the GUIDE-ROD drawing. Click **Save**. Click **Yes** to the question, "Save the documents and referenced models now?"

Label each component.

296) Label each component with a unique item number. The item number is placed inside a circle. The circle is called Balloon text. Create the Balloon text. Click **Balloon** from the Annotations toolbar.

297) Position the first Balloon. Click the **angled face** of the GUIDE. A Balloon appears with a leader line and the number 1 inside a circle. Position the second Balloon. Click the **middle** of the ROD. Place the third Balloon. Click the **front face** of the PLATE. Position the forth Balloon. Click the **shaft** of the FLANGE BOLT. Position the fifth Balloon. Click the **shaft** of the CAPSCREW.

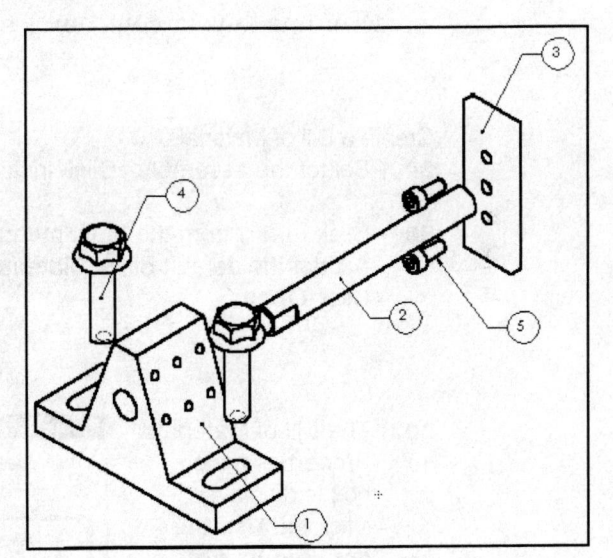

298) The Balloon text mode remains active until it is deactivated. Deactivate the Balloon text mode. Click **Balloon** from the Annotations toolbar.

Drag handles to move arrow or text

299) Reposition the Balloon text. Click the **Balloon** text. Drag the **text** by its handles off the profile.

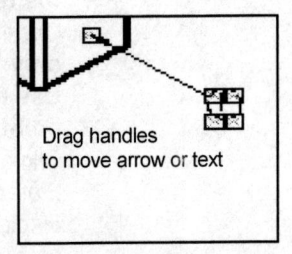

Fundamentals of Drawing	**Engineering Design with SolidWorks**

Modify the Arrow bend. Hold the **Ctrl** key down. Select the **five Balloons**. Click the **More Properties** button from the Balloon PropertyManager. Display the arrow with a bent line. Click **Bent Leader**. Release the **Ctrl** key. Click **OK**.

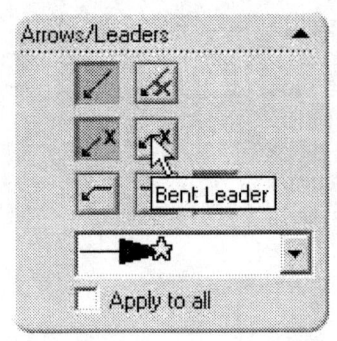

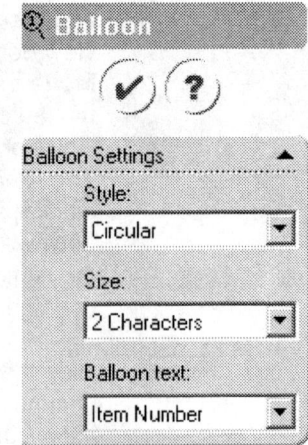

Note: Balloons may contain an item number and quantity.

Select the Balloon text. Click the More Properties button from the PropertyManager. The Properties dialog box for Balloon Notes is displayed.

Each Balloon references a part with a SolidWorks File Name. The SolidWorks File Name is linked to the Part Name in the Bill of Materials by default. Customize both the Balloon text and the Bill of Materials according to your company's requirements.

Create a Bill of Materials.
300) Select the assembly. Click inside the **Isometric** view.

301) Click **Insert** from the Main menu. Click **Bill of Materials**. Accept the default Bill of Material template, bomtemp.xls. Click **Open**.

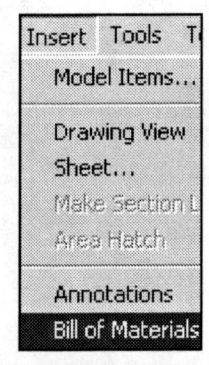

302) The Bill of Materials Properties dialog box is displayed. Click the **Use the documents note font when creating the table** checkbox. Click **Show Parts only**. Uncheck the **Use table anchor point** check box. Click **OK**. The Bill of Materials table is displayed in the current Isometric view.

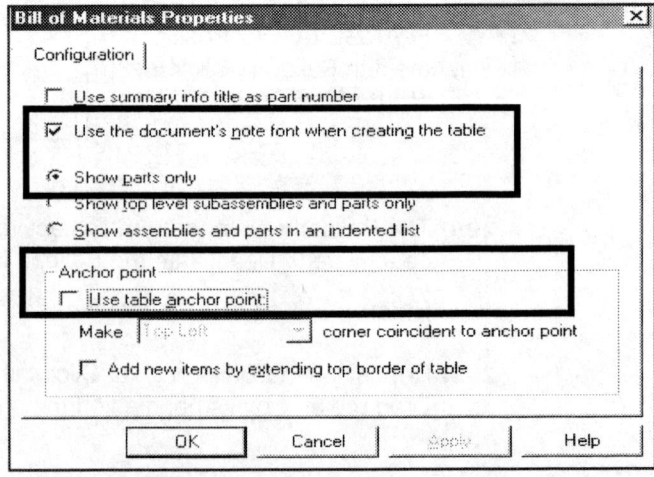

PAGE 3 - 60

Engineering Design with SolidWorks | Fundamentals of Drawing

303) Move the anchor point. Click the **Bill of Materials**. The grip points turn green. Fit the drawing to the Graphics window. Press the **f** key. Drag the **Bill of Materials** below the upper left hand corner of the Drawing.

ITEM NO.	QTY.	PART NO.	DESCRIPTION
1	1	GUIDE	GUIDE SUPPORT
2	1	ROD	ROD 10MM DIA x 100MM
3	1	PLATE	PLATE 56MM x 22MM
4	2	M8-1.25 x 30	
5	2	4MM CAP SCREW	CAP SCREW, 4MM

The Bill of Materials requires some editing. The current part file name determines the PART NO. values. The current part description determines the DESCRIPTION values. Redefine the PART NO. for the Bill of Materials.

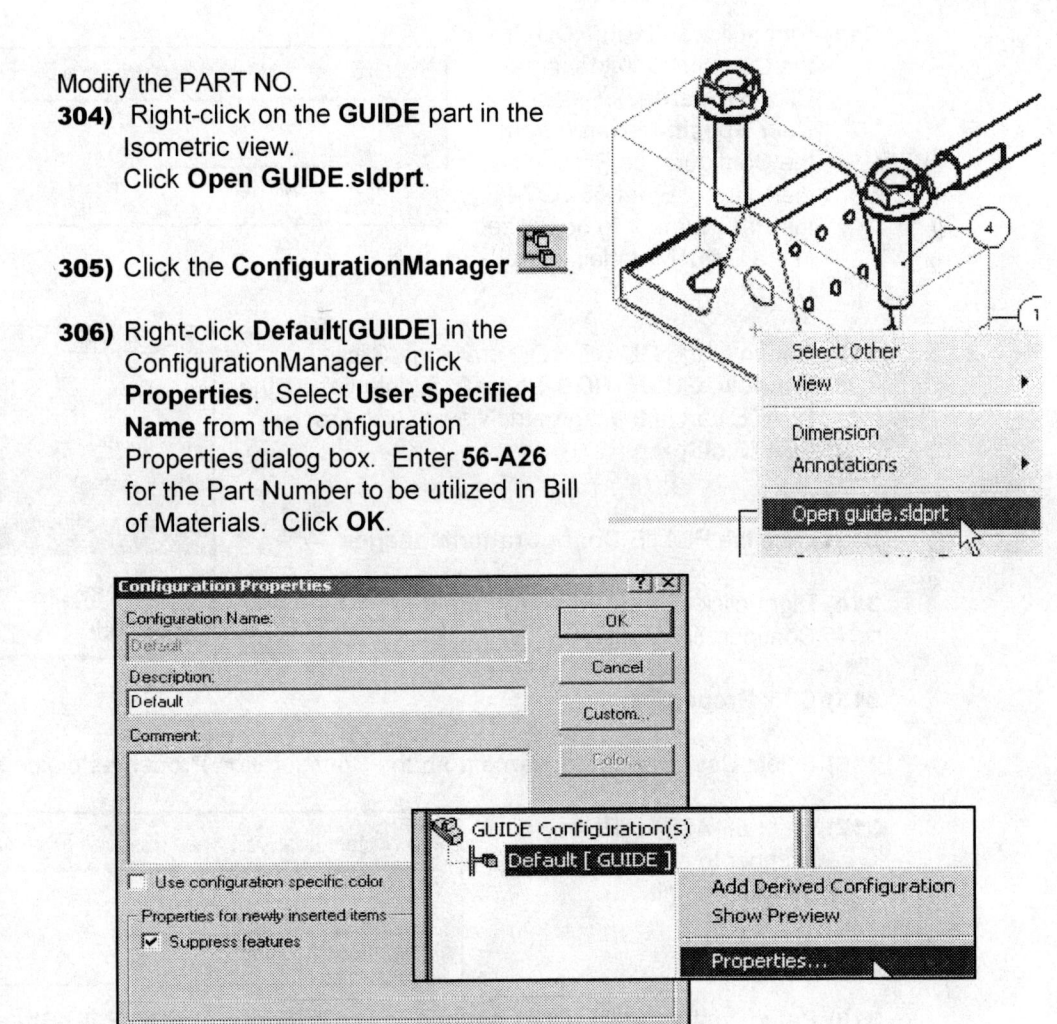

Modify the PART NO.
304) Right-click on the **GUIDE** part in the Isometric view. Click **Open GUIDE.sldprt**.

305) Click the **ConfigurationManager**.

306) Right-click **Default[GUIDE]** in the ConfigurationManager. Click **Properties**. Select **User Specified Name** from the Configuration Properties dialog box. Enter **56-A26** for the Part Number to be utilized in Bill of Materials. Click **OK**.

PAGE 3 - 61

Fundamentals of Drawing   **Engineering Design with SolidWorks**

307) Return to the GUIDE-ROD drawing. Click **Window**, **GUIDE-ROD-Sheet1**.

308) Right-click on the **ROD** part in the Isometric view.

309) Click **Open ROD.sldprt**.

310) Click the ROD **ConfigurationManager**.

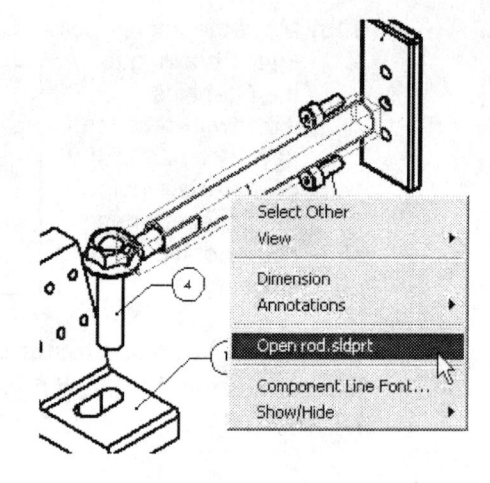

311) Right-click **Default[ROD]** from the Configuration Manager. Click **Properties**. Select the **User Specified Name** from the Configuration Properties dialog box. Enter **56-A27** for the Part Number to be utilized in the Bill of Materials. Click **OK**.

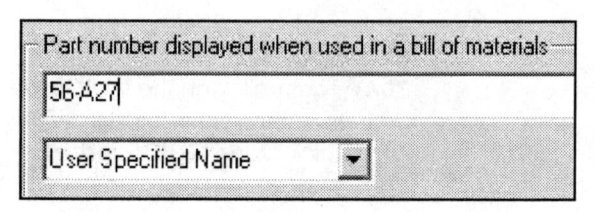

312) Return to the GUIDE-ROD drawing. Click **Window**, **GUIDE-ROD-Sheet1**. Right-click on the **PLATE** part in the Isometric view. Click **Open PLATE.sldprt**.

313) Click the PLATE **ConfigurationManager**.

314) Right-click **Default[PLATE]** from the ConfigurationManager.

315) Click **Properties**.

316) Select **User Specified Name** from the Configuration Properties dialog box.

317) Enter **56-A28** for Part Number to be utilized in the Bill of Materials.

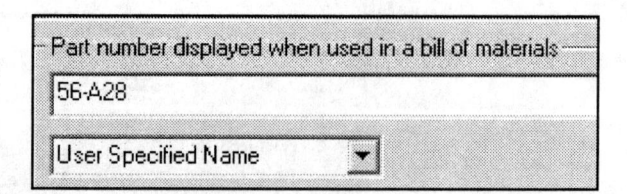

318) Click **OK**.

319) Return to the GUIDE-ROD drawing. Click **Window**, **GUIDE-ROD-Sheet1**. Right-click on the **FLANGE BOLT** part in the Isometric view.

320) Click **Open FLANGE BOLT.sldprt**.

PAGE 3 - 62

Engineering Design with SolidWorks   Fundamentals of Drawing

The FLANGE BOLT is a SolidWorks library part. Copy the part with a new name to the ENGDESIGN-W-SOLIDWORKS\PROJECTS file folder.

321) Save the FLANGE BOLT. Click **File**, **Save As**. Click **OK** to replace the FLANGE BOLT references in the GUIDE-ROD assembly with the new file name. Select the **ENGDESIGN-W-SOLIDWORKS\PROJECTS** file folder.

322) Enter **FLANGE-BOLT-8MM** for the File name.

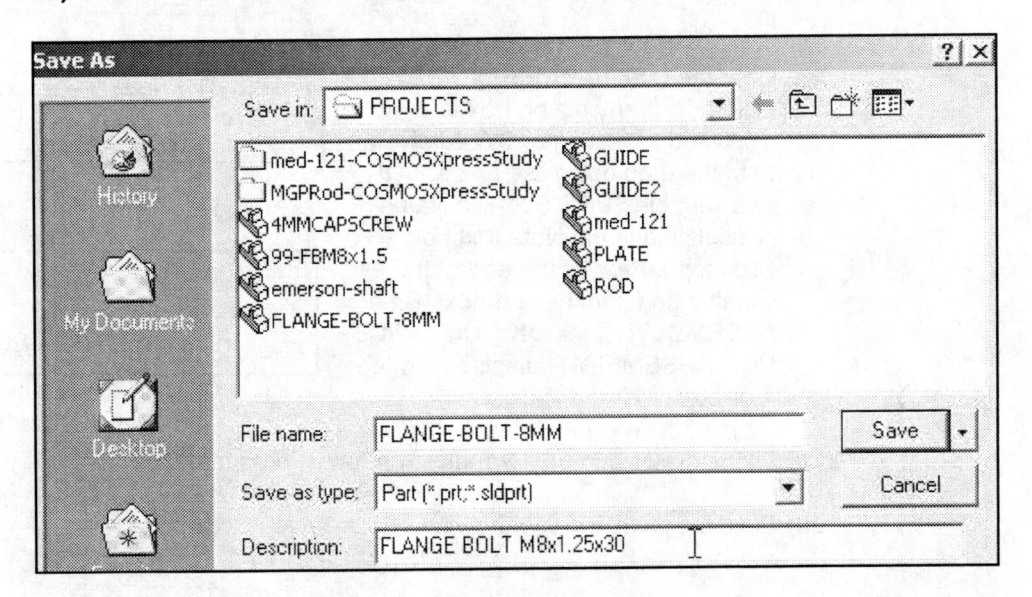

323) Enter **FLANGE BOLT M8x1.25x30** for Description. Click **Save**.

324) Return to the GUIDE-ROD drawing. Click **Window**, **GUIDE-ROD-Sheet1**. Update the Bill of Materials. Click **Rebuild**.

325) Fit the drawing to the Graphics window. Press the **f** key.

ITEM NO.	QTY.	PART NO.	DESCRIPTION
1	1	56-A26	GUIDE SUPPORT
2	1	56-A27	ROD 10MM DIA x 100MM
3	1	56-A28	PLATE 56MM x 22MM
4	2	99-FBM8x1.25	FLANGE BOLT M8x1.25x30
5	2	99-SHCS-25	CAP SCREW, 4MM

The FLANGE BOLT PART NO. and CAP SCREW PART NO. utilized in the Bill of Materials are explored in the project exercises.

Fundamentals of Drawing **Engineering Design with SolidWorks**

326) Zoom in on the Title block.

Create a linked note for the drawing Title.
327) Click a **start point** in the TITLE section of the title block. Click **Note** A from the Annotations toolbar. The Property dialog box is displayed.

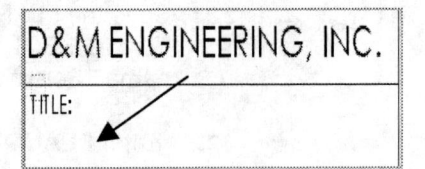

328) Click the **Link to Property**. The Link to Property dialog box is displayed. Select the **SW-FileName** from the drop down list. Click **OK**. The variable $PRP"SW-File Name" is displayed in the Note text box. Click a **position** at the end of the variable name in the text box. Enter **ASSEMBLY**. Click **OK**. The GUIDE-ROD ASSEMBLY is displayed in the Title box.

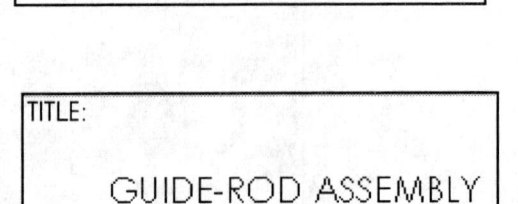

329) Fit the drawing to the Graphics window. Press the **f** key.

330) Save the GUIDE-ROD drawing. Click **Save**. Click **Yes**.

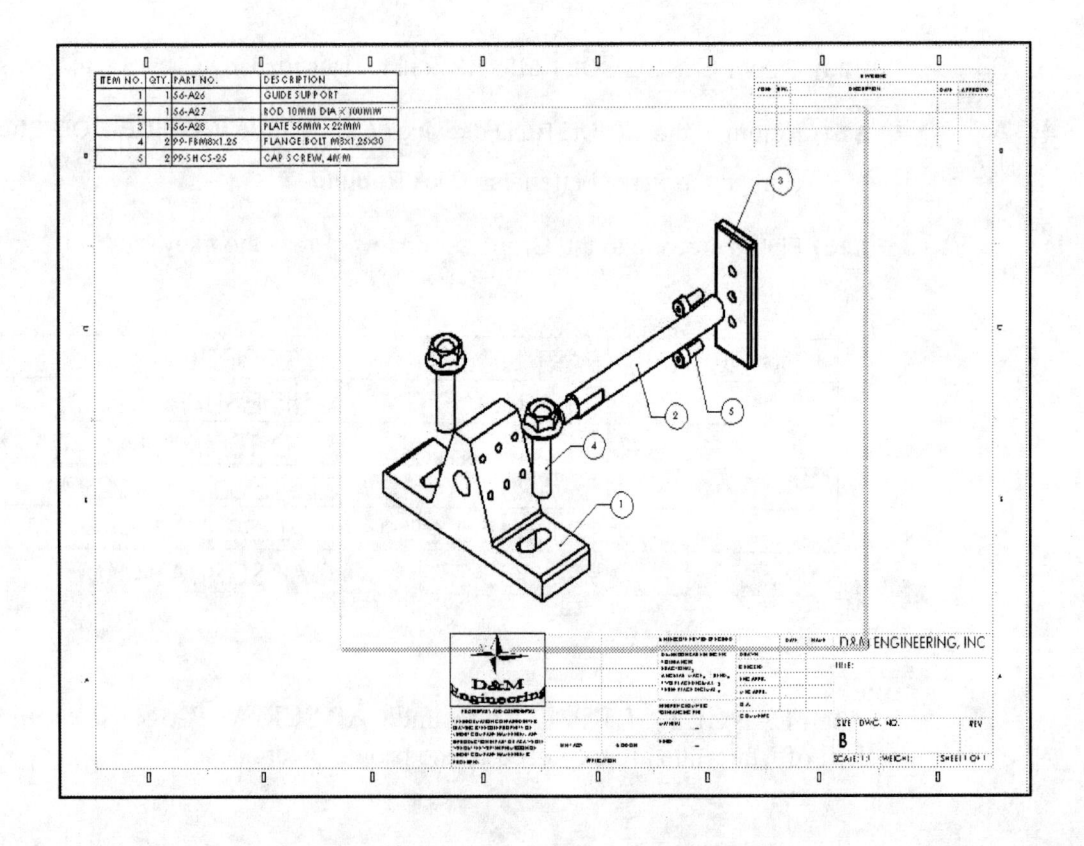

PAGE 3 - 64

Associative Part, Assembly and Drawing

The associative part, assembly and drawing share a common database. Verify the association between the part, assembly and drawing.

331) Open the GUIDE drawing. Click **Open**. Select **ENGDESIGN-W-SOLIDWORKS\PROJECTS** from the Look in folder list. Click **Drawing** from the Files of type drop down list. Double-click **GUIDE**.

332) Modify a dimension on the GUIDE drawing. Double-click on the vertical linear dimension **10** in Drawing View1 (Front). Enter **20** in the Modify dialog box.

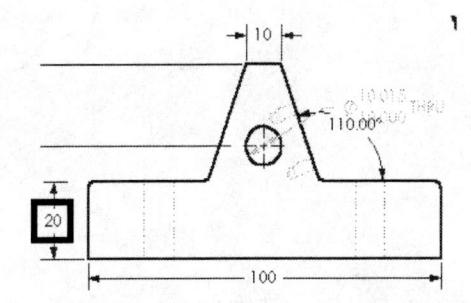

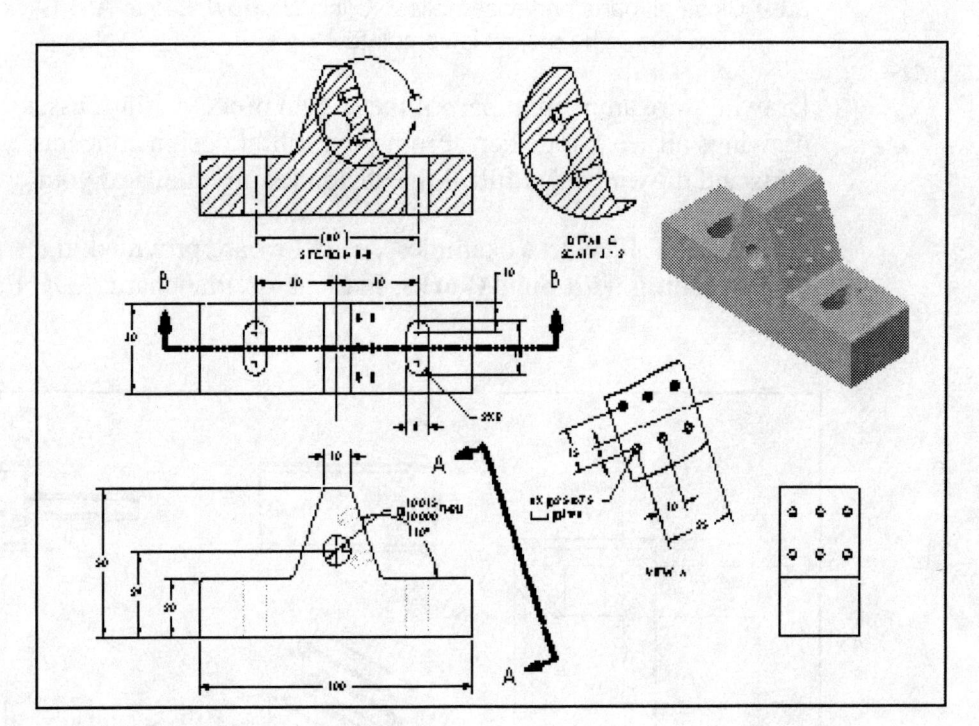

333) Click **Rebuild**. All views are updated.

334) View the updated part. Right-click **Drawing View1**. Click **Open GUIDE.sldprt**.

Fundamentals of Drawing   **Engineering Design with SolidWorks**

335) Open the GUIDE-ROD assembly. Click **Open**. Select **ENGDESIGN-W-SOLIDWORKS\PROJECTS** from the Look in folder list. Click **Assembly** from the Files of type drop down list. Double-click **GUIDE-ROD**. Click **Yes** to view the updated GUIDE.

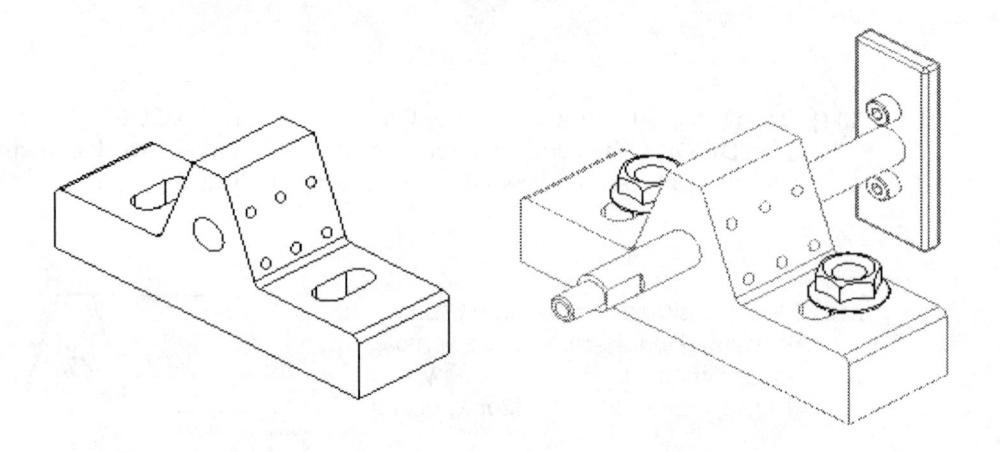

336) Close all parts and assemblies. Click **Window, Close All**. Click **Yes to All**. Click **Yes**. The project is complete.

Drawings are an integral part of the design process. Part, assemblies and drawings all work together. From your initial design concepts, you created parts and drawings that fulfill the design requirements of your customer.

Additional SolidWorks examples, see below are provided in the text **Drawing and Detailing with SolidWorks**, Planchard&Planchard, SDC Publications 2002.

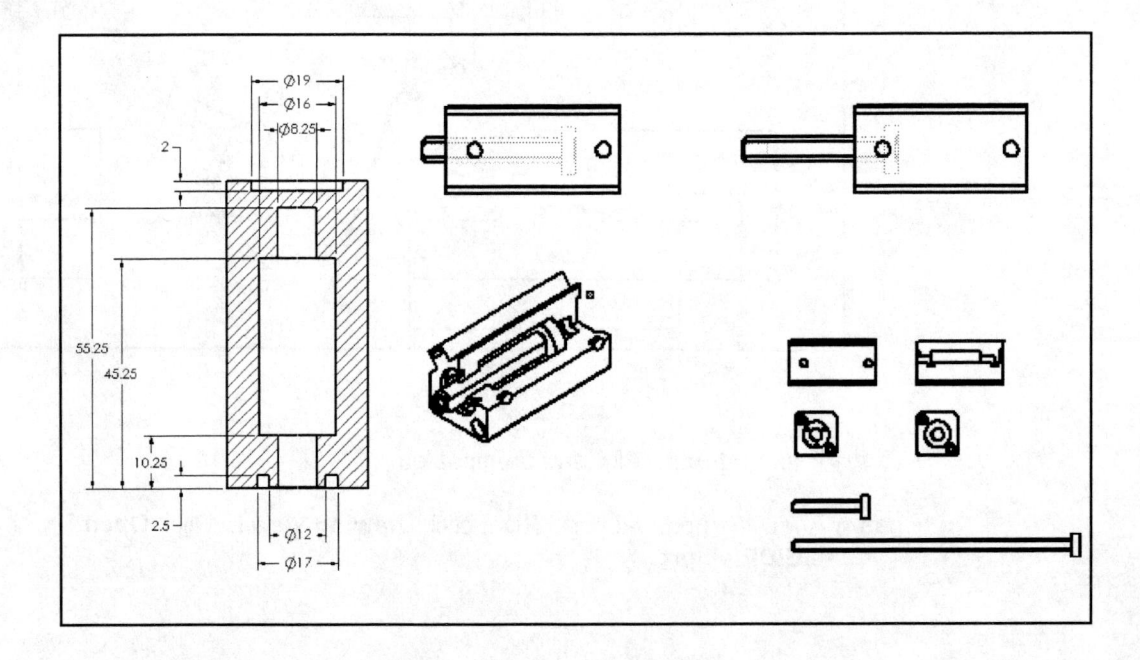

Project Summary

You produced two drawings: GUIDE drawing and the GUIDE-ROD assembly drawing. The drawings contained three standard views, (principle views) and an Isometric view. The drawings utilized a custom sheet format containing a company logo, title block and design intent sheet information. You incorporated the GUIDE part dimensions into the drawing.

You obtained an understanding of displaying the following views: Principle, Auxiliary, Detail and Section with the ability to insert, create and modify dimensions.

You used two major design modes in the drawings: Edit Sheet Format and Edit Sheet.

Project Terminology

Detailed view: Detailed views enlarge an area of an existing view. Specify location, shape and scale. You created a Detail view from a Section view at a 3:2 scale.

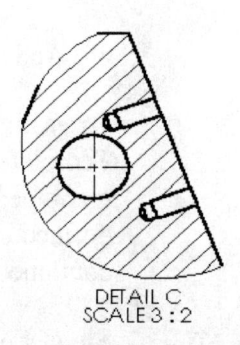

Section view: Section views display the interior features. Define a cutting plane with a sketched line in a view perpendicular to the Section view. You created a Section view by sketching a section line in the Top view.

Auxiliary view: Displays a plane parallel to an angled plane with true dimensions. A primary Auxiliary view is hinged to one of the six principle views. You created a primary Auxiliary view that references the Front view.

Crop View: Can be used to remove all but a bounded area from what is displayed.

Bill of Materials: In an assembly drawing, a bill of materials report can be automatically created and inserted onto the drawing sheet.

Notes: Notes can be used to add text with leaders or as a stand-alone text string. If an edge, face or vertex is selected prior to adding the note, a leader is created to that location.

Title block: Contains vital part or assembly information. Each company can have a unique version of a title block.

Edit Sheet Format Mode: Provides the ability to:

- Change the title block size and text headings.
- Incorporate a company logo.
- Add a drawing, design or company text.

Remember: A part cannot be insert into a drawing when the Edit Sheet Format mode is selected. Edit Sheet Format displays all lines in blue.

Edit Sheet Mode: Provides the ability to:

- Add or modify views.
- Add or modify dimensions.
- Add or modify text.

Drawing Template: The foundation of a SolidWorks drawing is the Drawing Template. Drawing size, drawing standards, company information, manufacturing and or assembly requirements, units and other properties are defined in the Drawing Template. In this project the Drawing Template contained the drawing Size, Document Properties and Layers.

Sheet Format: The Sheet Format is incorporated into the Drawing Template. The Sheet Format contains the border, title block information, revision block information, company name and or logo information, Custom Properties and SolidWorks Properties. In this project the Sheet Format contained the title block information.

Drawing file name: Drawing file names end with a .slddrw suffix.

Part file name: Part file names end with a .sldprt suffix. Note: A drawing or part file can have the same prefix. A drawing or part file cannot have the same suffix. Example: Drawing file name: GUIDE.slddrw. Part file name: GUIDE.sldprt

Drawing Layers: Contain dimensions, annotations and geometry.

View Appearance: There are two important factors which affect the appearance of views:

1. Whether the view is shown wireframe, hidden lines removed or hidden lines gray or dashed.
2. How tangent edges on entities such as fillets are displayed.

Leader lines: Reference the size of the profile. A gap must exist between the profile lines and the leader lines.

Hole centerlines: Are composed of alternating long and short dash lines. The lines identify the center of a circle, axes or cylindrical geometry.

Hole Callout: The Hole Callout function creates additional notes required to dimension the holes.

Center marks: Represents two perpendicular intersecting centerlines.

General Notes: Below are a few helpful guidelines to create general drawing notes:

- Use capital letters.
- Use left text justification.
- Font size should be the same as the dimension text.

Questions

1. Describe a Bill of Materials and its contents.

2. Name the two major design modes used to develop a drawing in SolidWorks.

3. What are the three sheet format options in SolidWorks?

4. Identify seven components that are commonly found in a title block.

5. How do you add an Isometric view to the drawing?

6. In SolidWorks, drawing file names end with a _____ suffix.

7. In SolidWorks, part file names end with a _____ suffix.

8. Can a part and drawing have the same name?

9. True or False. In SolidWorks, if a part is modified, the drawing is automatically updated.

10. True or False. In SolidWorks, when a dimension in the drawing is modified, the part is automatically updated.

11. Name three guidelines to create General Notes.

12. True or False. Most engineering drawings us the following font: Time New Roman – All small letters.

13. What are Leader lines? Provide an example.

14. Name the three ways that Holes and other circular geometry can be dimensioned.

15. Describe are Center Marks. Provide an example.

16. How do you calculate the maximum and minimum variation?

17. Describe the differences between a Drawing Template and a Sheet Template.

18. Describe the main differences between a Detailed view and a Section view.

Exercises

Exercise 3.1: L-Bracket.

Create a drawing for the L-BRACKET. You created the L-BRACKET Project 1, Exercise 1.1a – Exercise 1.1i.

Utilize Tools, Options, Document Properties to set Units and Font size.

Note: Dimensions are enlarged for clarity.

Insert a Front, Top and Right view.

Insert an Isometric view.

Insert Dimensions from the part.

Insert a Linked Note in the Title block.

Utilize the part filename.

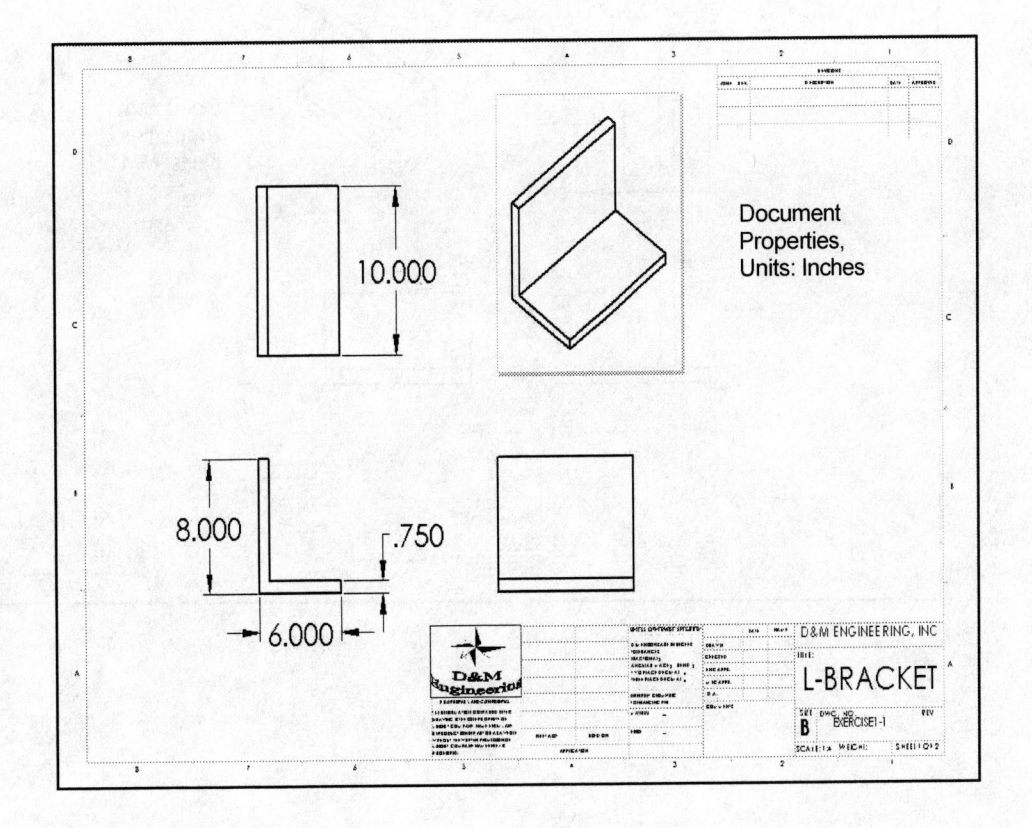

Exercise 3.2: T-Section.

Create a drawing for the T-SECTION. You created the T-SECTION in Project 1, Exercise 1.2a – Exercise 1.2f.

Insert a Front, Top and Right view.

Insert an Isometric view.

Insert Dimensions from the part.

Utilize Sketch Tools to add a centerline.

Insert a Linked Note in the Title block.

Utilize the part filename.

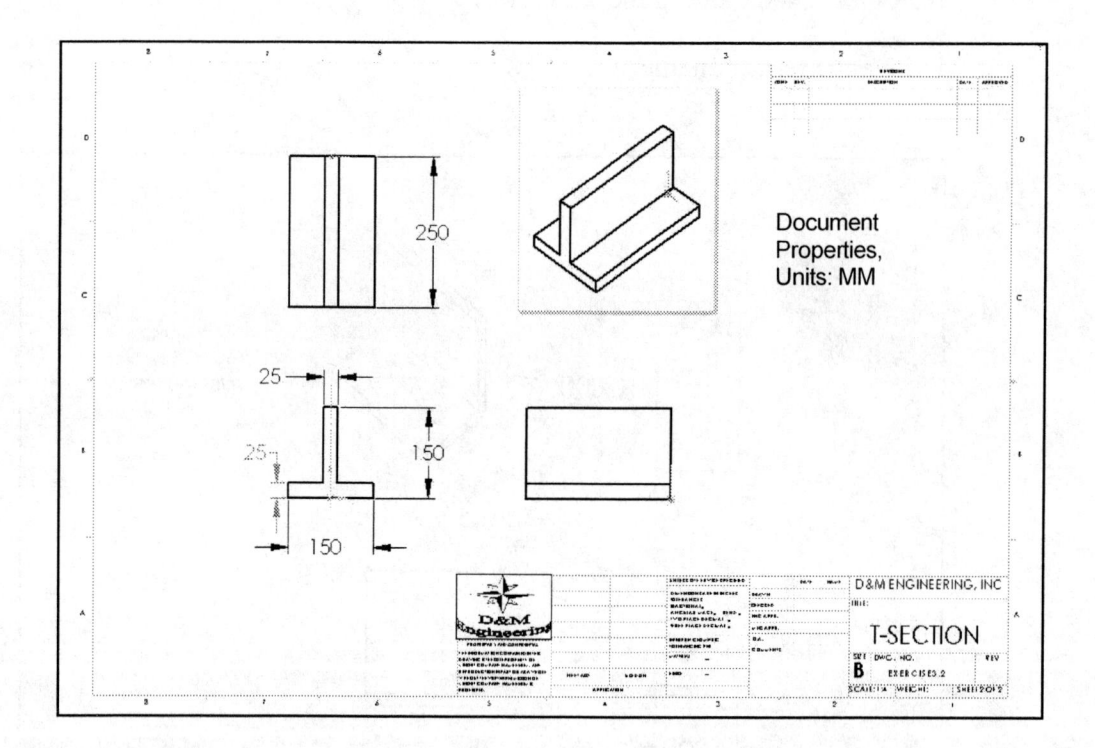

Exercise 3.3: Linkage Assembly.

Create four drawings for the mechanical parts required for the LINKAGE assembly, Project 1, Exercise 1.5a – Exercise 1.5d.

Note: Dimensions are enlarged for clarity. Utilize inch, millimeter or dual dimensioning. Insert a shaded Isometric view in the lower right corner of each drawing.

Exercise 3.3a FLAT BAR – 3HOLE.

Create the FLAT BAR – 3HOLE Drawing. Insert a Front view and a shaded Isometric view. Insert Dimensions from the part. Modify the hole dimension to include three holes.

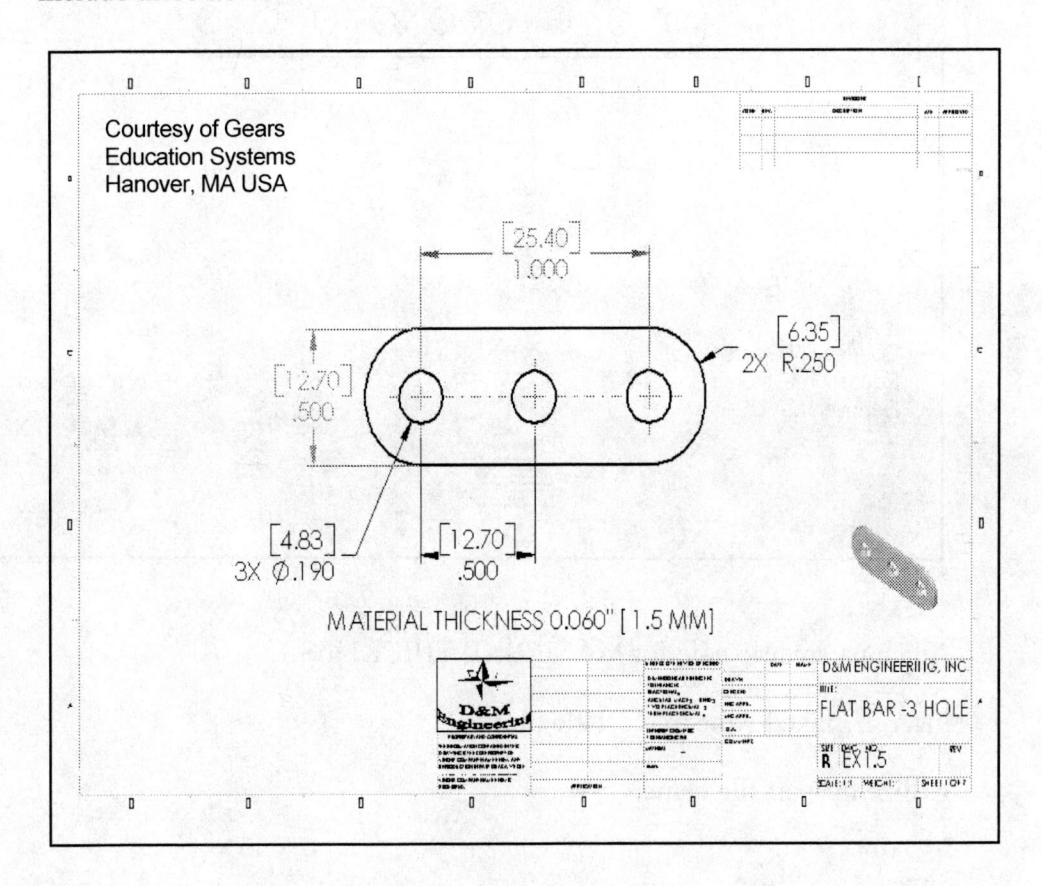

Add a parametric note for MATERIAL THICKNESS.

Insert a Linked Note in the Title block.

Utilize the part filename.

Exercise 3.3b FLAT BAR – 9 HOLE.

Create the FLAT BAR – 9HOLE Drawing. Insert a Front view and a shaded Isometric view. Insert Dimensions from the part. Modify the hole dimension to include nine holes.

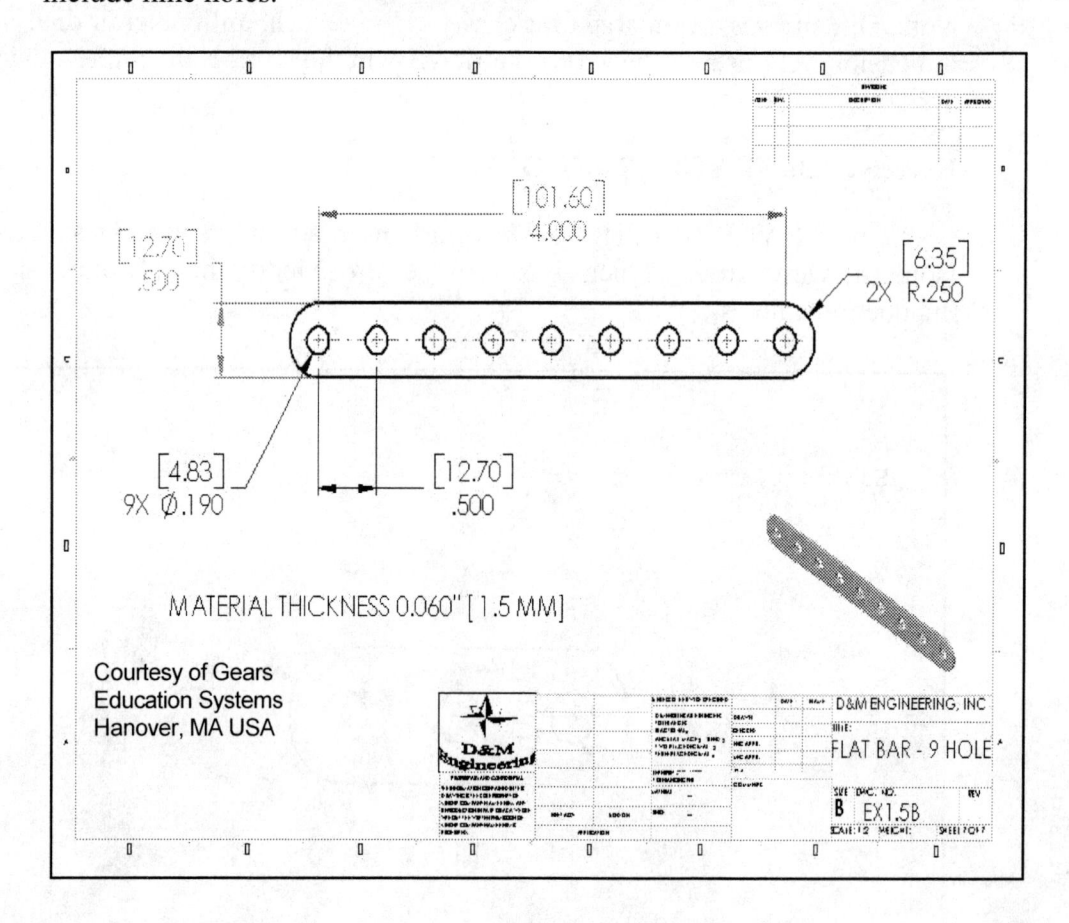

Add a parametric note for MATERIAL THICKNESS.

Insert a Linked Note in the Title block.

Utilize the part file name.

Exercise 3.3c: AXLE.

Create the AXLE Drawing. Insert a Front view, Right view and a shaded Isometric view. Insert Dimensions from the part. Add a centerline in the Right view.

Insert a Linked Note in the Title block.

Utilize the part file name.

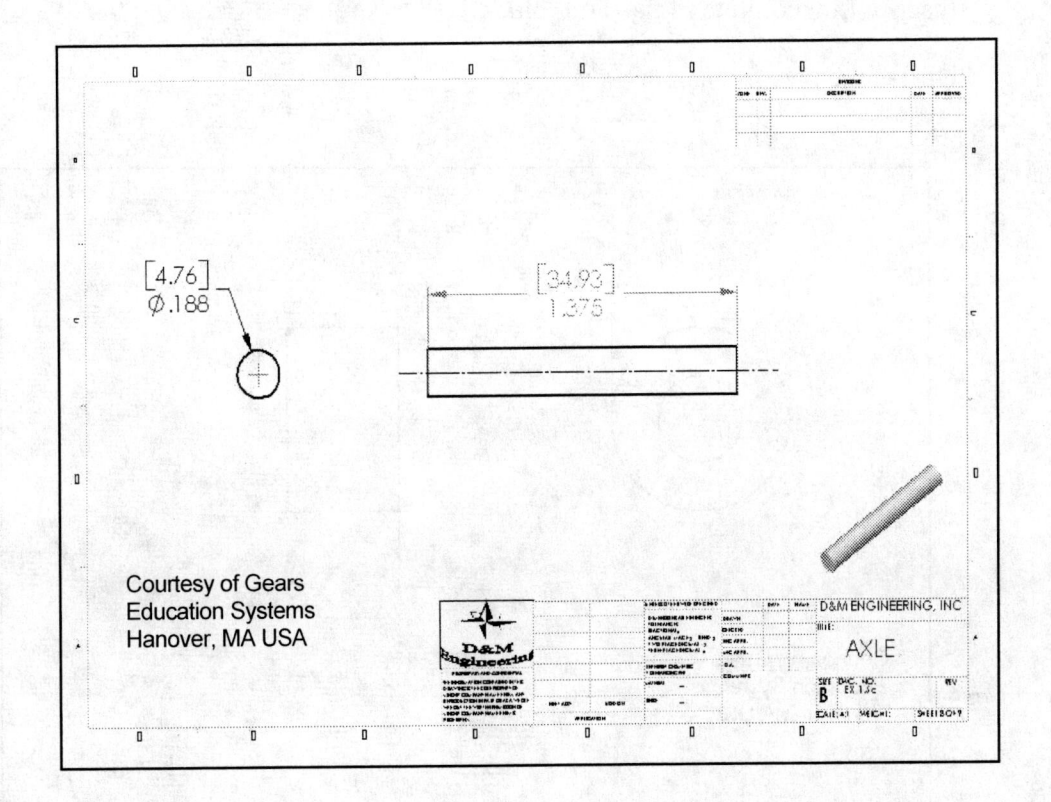

Exercise 3.3d: SHAFT COLLAR.

Create the SHAFT COLLAR Drawing.

Insert a Front view, Right view and a shaded Isometric view.

Insert Dimensions from the part.

Add a centerline in the Right view and a Center Mark in the Front view.

Insert a Linked Note in the Title block.

Utilize the part file name.

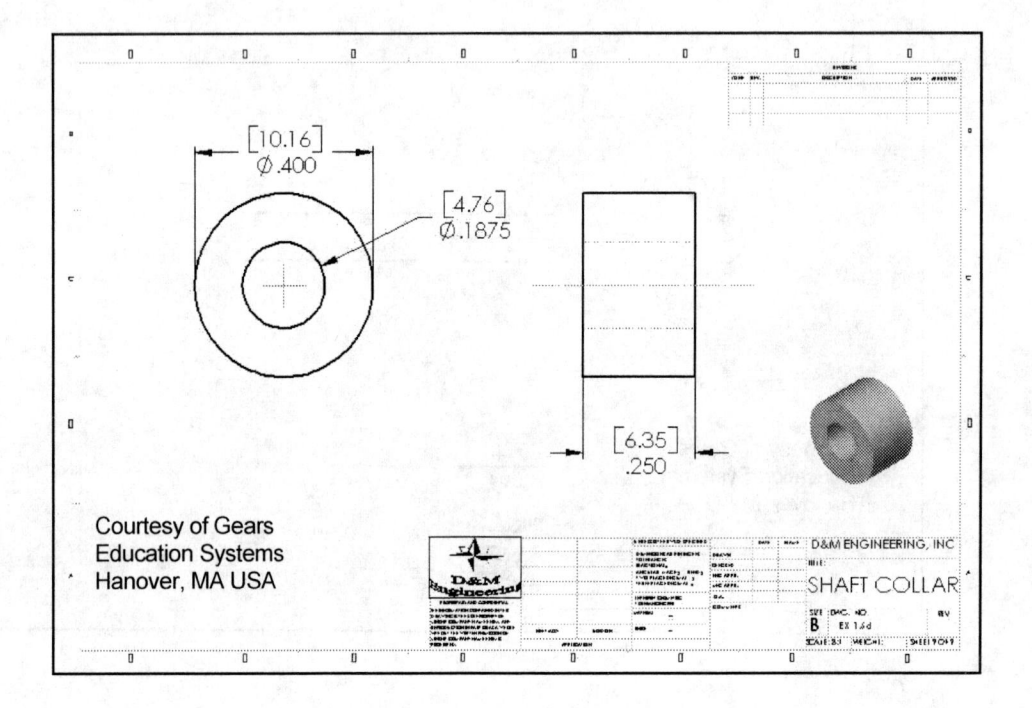

Exercise 3.4: Linkage Drawing.

a) Create a new drawing named, LINKAGE. Insert an Isometric view for the LINKAGE assembly created in Exercise 2.3a.

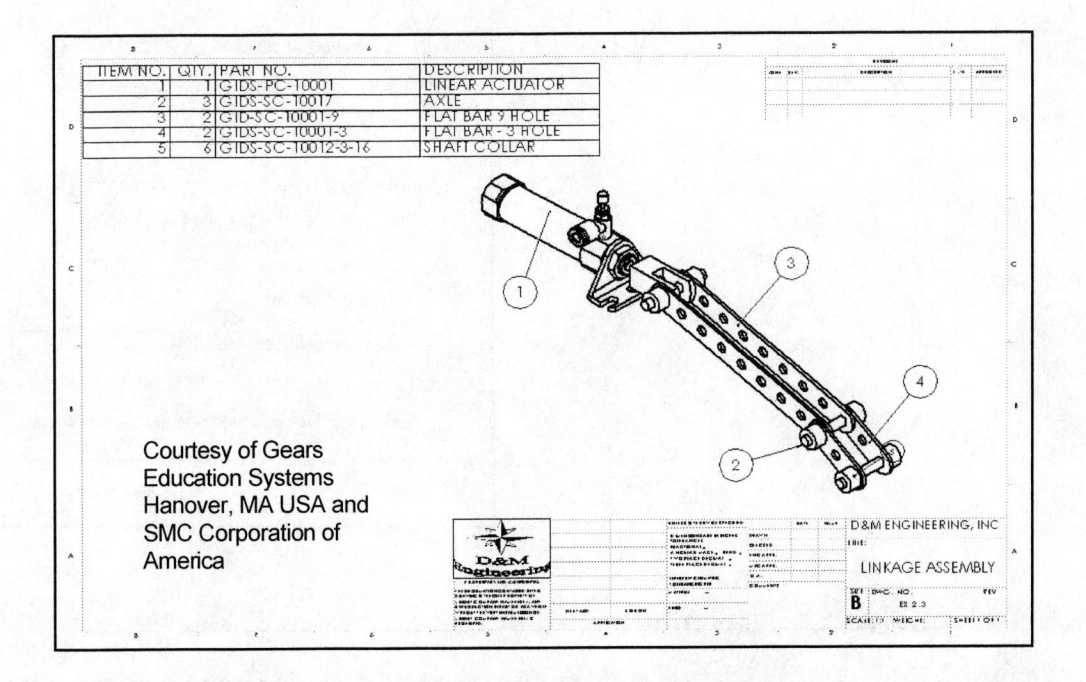

b) Open each mechanical part in the LINKAGE assembly. Define the PART NO. Property and the DESCRIPTION Property for the AXLE, FLAT BAR- 9HOLE, FLAT BAR – 3HOLE and SHAFT COLLAR. Save each part; save the LINKAGE Assembly to update the properties. Return to the LINKAGE Drawing.

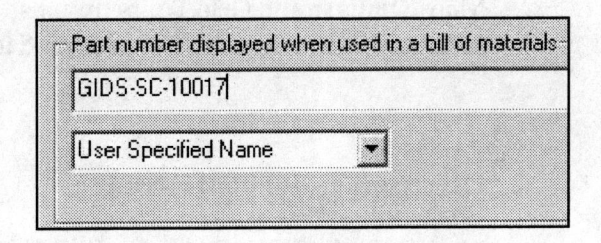

PART NO. Property	DESCRIPTION Property
GIDS-SC-10017	AXLE
GIDS-SC-10001-9	FLAT BAR – 9 HOLE
GIDS-SC-10001-3	FLAT BAR – 3 HOLE
GIDS-SC-10012-3-16	SHAFT COLLAR

c) Insert the Bill of Materials. Utilize the Show top-level subassemblies and parts only. Add Balloons. Select only the top-level components in the LINKAGE Assembly.

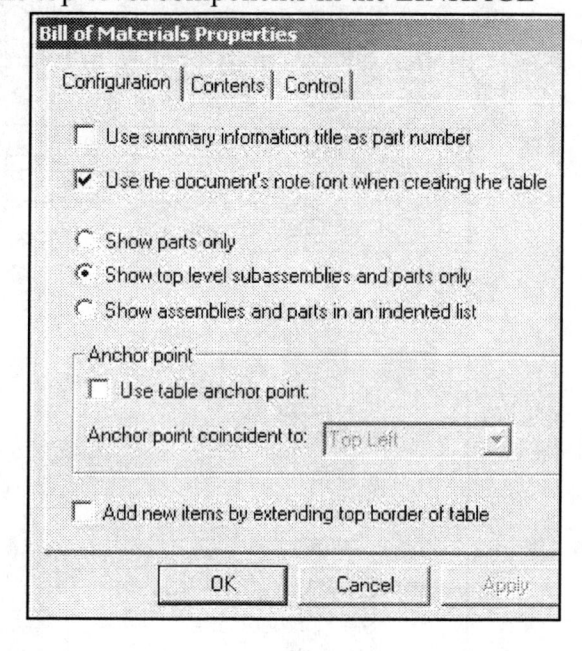

Exercise 3.5: Plate Drawing.

Create a Metric drawing from the PLATE part in the GUIDE-ROD assembly. Set Metric units in the Grid/Units options. Insert only the Top view into the drawing. Modify the dimensions. Add Notes, Hole Call Out and Center Marks.

Exercise 3.6: Bill of Materials.

Modify the PART NO. for the FLANGE BOLT and CAP SCREW in the GUIDE-ROD Assembly Bill of Materials.

| 4 | 2 | 99-FBM8x1.25 | FLANGE BOLT M8x1.25x30 |
| 5 | 2 | 99-SHCS-25 | CAP SCREW, 4MM |

Engineering Design with SolidWorks **Fundamentals of Drawing**

Exercise 3.7: VALVE PLATE.

Create the VALVE PLATE part as illustrated in the ASME 14.5M Geometric Dimensioning and Tolerancing Standard. Create the drawing according to the ASME 14.5M standard.

Geometric Tolerancing Symbols and Datum Planes annotations are found in the Annotations menu. The dimension text 12 and 24 are displayed as Basic dimensions under the Tolerance/Precision text box.

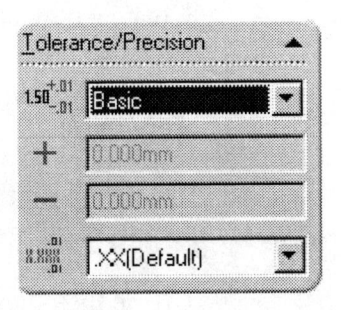

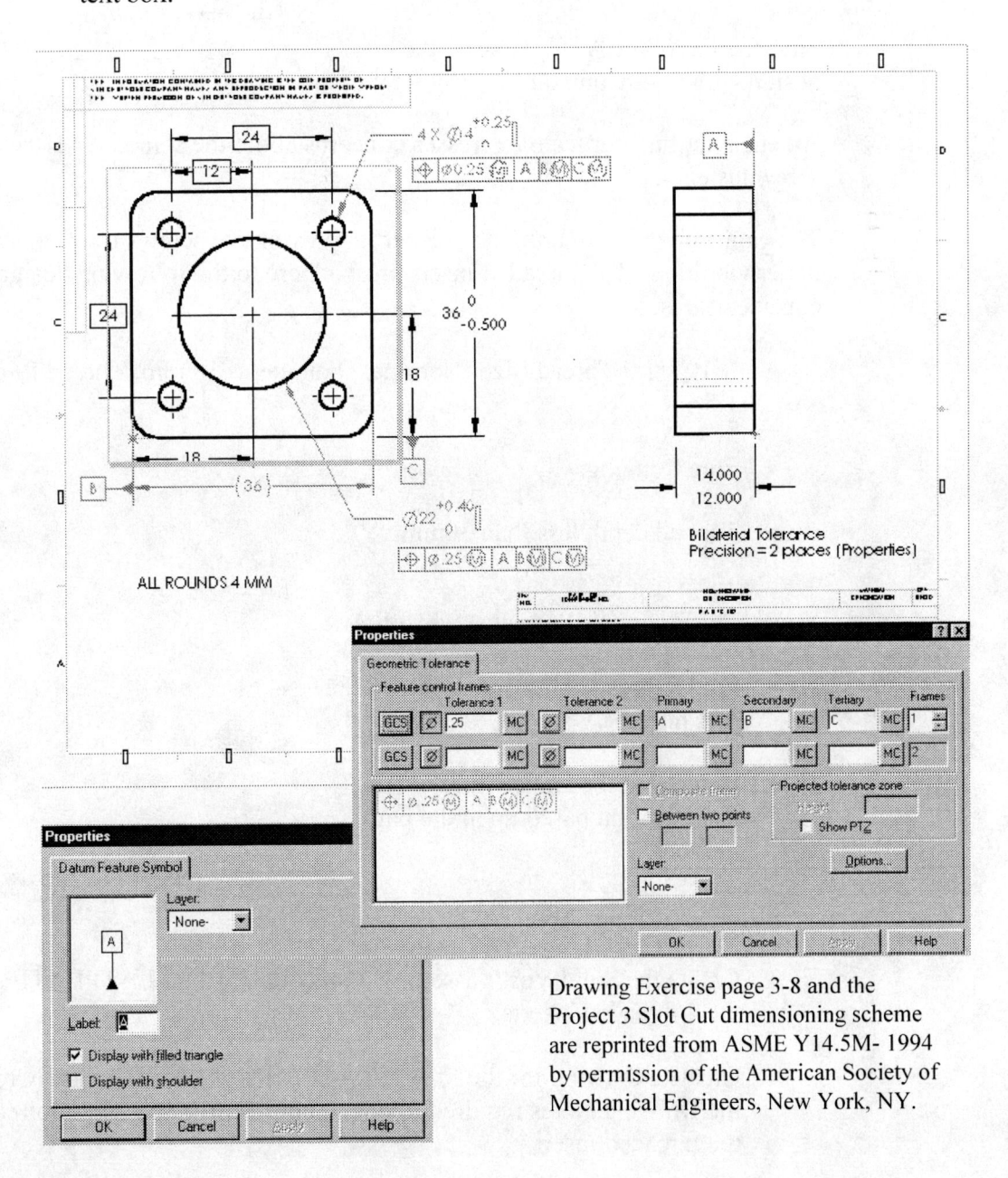

ALL ROUNDS 4 MM

Bilateral Tolerance
Precision = 2 places (Properties)

Drawing Exercise page 3-8 and the Project 3 Slot Cut dimensioning scheme are reprinted from ASME Y14.5M- 1994 by permission of the American Society of Mechanical Engineers, New York, NY.

Exercise 3.8: Industry Collaborative Exercise.

Create a new part called FLAT-PLATE. The FLAT-PLATE is 100mm x 100mm x 30mm. Add 4 - M10 x1.5 Thru holes.

Create the first hole with the HoleWizard.

Create the other three holes with a Linear Pattern.

Save the FLAT-PLATE. Review the design with your teammates.

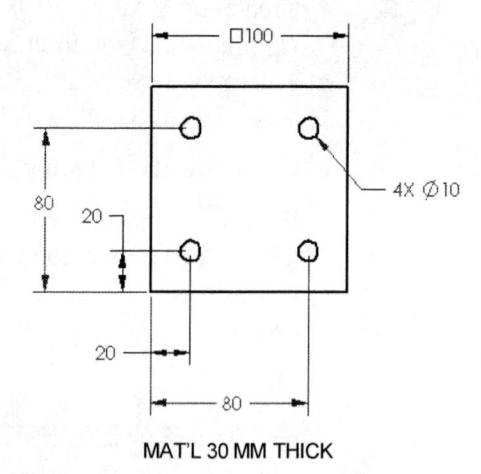

An engineer on your team expresses concerns about the standard M10x1.5 screw fastener.

The engineer specifies a threaded insert to prevent the screws from loosening under vibration. The threaded insert must adhere to the following design considerations:

- M10 x 1.5 Thread Size (Nominal Diameter = 10 mm, Thread Pitch = 1.5).

- Free Running.

- Tapped depth less than 30mm.

Your tasks are on this project are as follows:

- Find a threaded insert and record the part number.

- Determine the tapped hole specification based upon the part number.

- Download the SolidWorks part file.

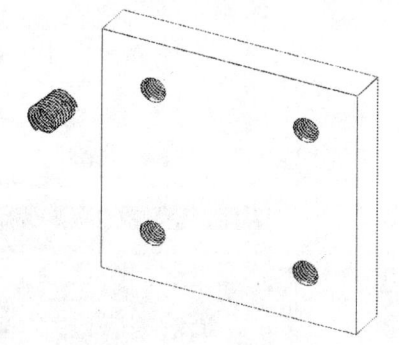

- Create the SolidWorks assembly that contains the FLATPLATE and the 4 threaded inserts.

- Create a drawing for the FLATPLATE with a Hole Call Out. Create the Drawing Notes required to document the drill and tap required for the threaded insert.

Engineering Design with SolidWorks **Fundamentals of Drawing**

Note: Emhart Fastening Teknologies, A Black & Decker Company, manufactures Heli-Coil® Screw Thread Inserts.

Obtain Heli-Coil® Screw Thread Insert information from the following URL: http:// www.emhart.com. Select the Heli-Coil® Selector button.

Select the Innovations button. Click the Emhart Mentor® SmartCAT button. Register and enter your email address.

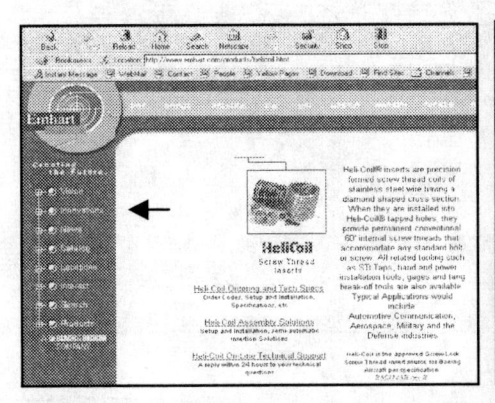

Click Heli-Coil®.

Click the Search Selector.

Enter the following product requirements:

- Thread size: M10x1.5
- Type: Free
- Length: 15mm
- Packaging: Bulk

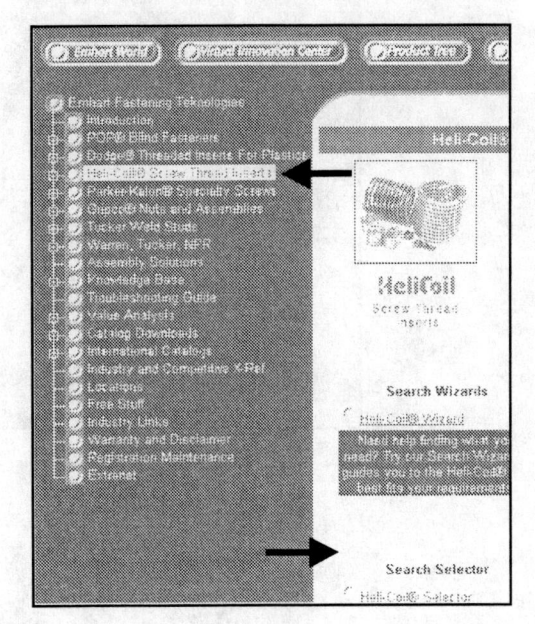

Fundamentals of Drawing   **Engineering Design with SolidWorks**

Obtain the part number, min/max outside diameter. Select the Spec button. Click the Product Tools button to obtain the drawing notes required for manufacturing. Click the CAD Drawings button to view and download the component.

Modify the FLATPLATE. Edit the Thru Hole. Change the Standard to HeliCoil® Metric Tapped Hole. Enter 21 for Tap Drill Type & Depth.

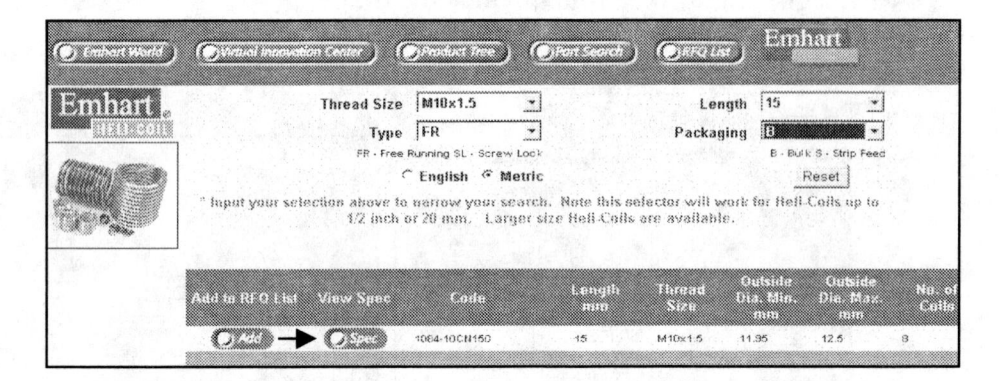

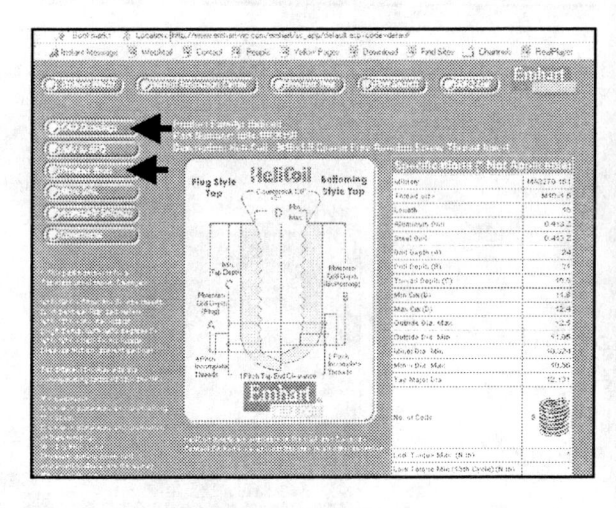

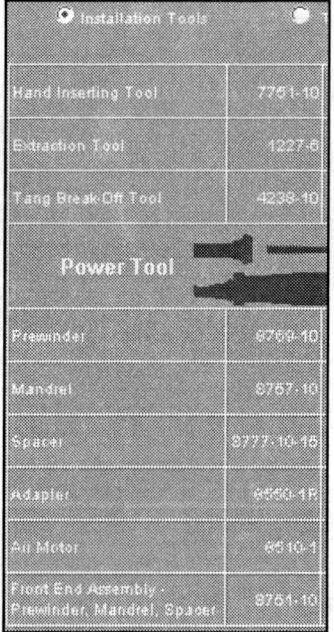

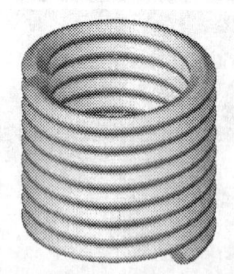

Components Courtesy of
Emhart Teknologies
New Haven, CT USA

Emhart® and Heli-Coil® are
registered trademarks of the
Black & Decker Company

Create a new assembly with the Heli-Coil® threaded insert and the FLATPLATE. P is the Pitch of the thread. P = 1.5.

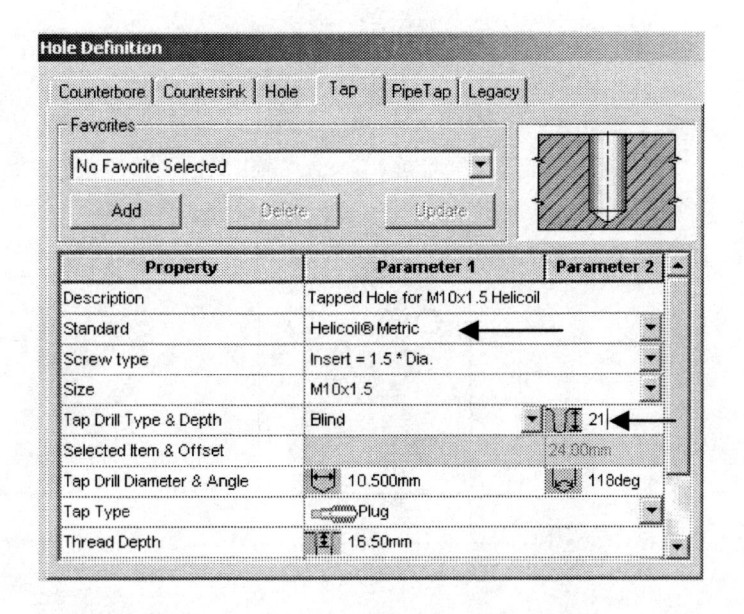

Create a Distance Mate between the two components.

Position the Heli-Coil® threaded insert between 0.75P to 1.5 P from the top face of the FLATPLATE.

Create a Coincident Mate between the Origin of the Heli-Coil® and Temporary Axis of the Tapped Hole.

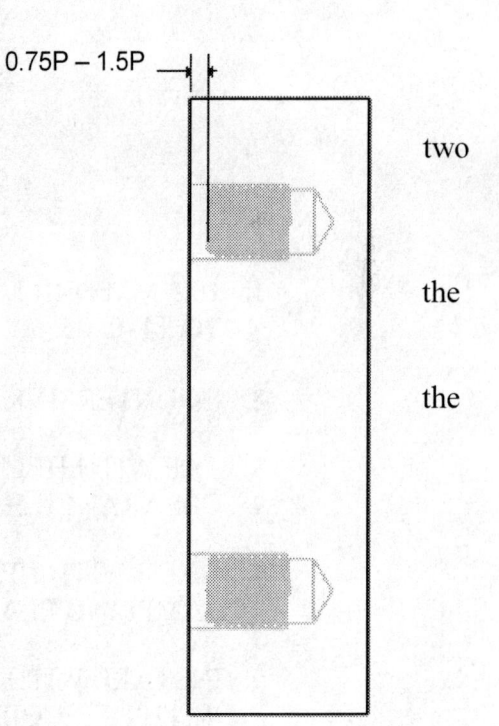

Create a drawing for the FLATPLATE. Add the following Notes to complete the drawing:

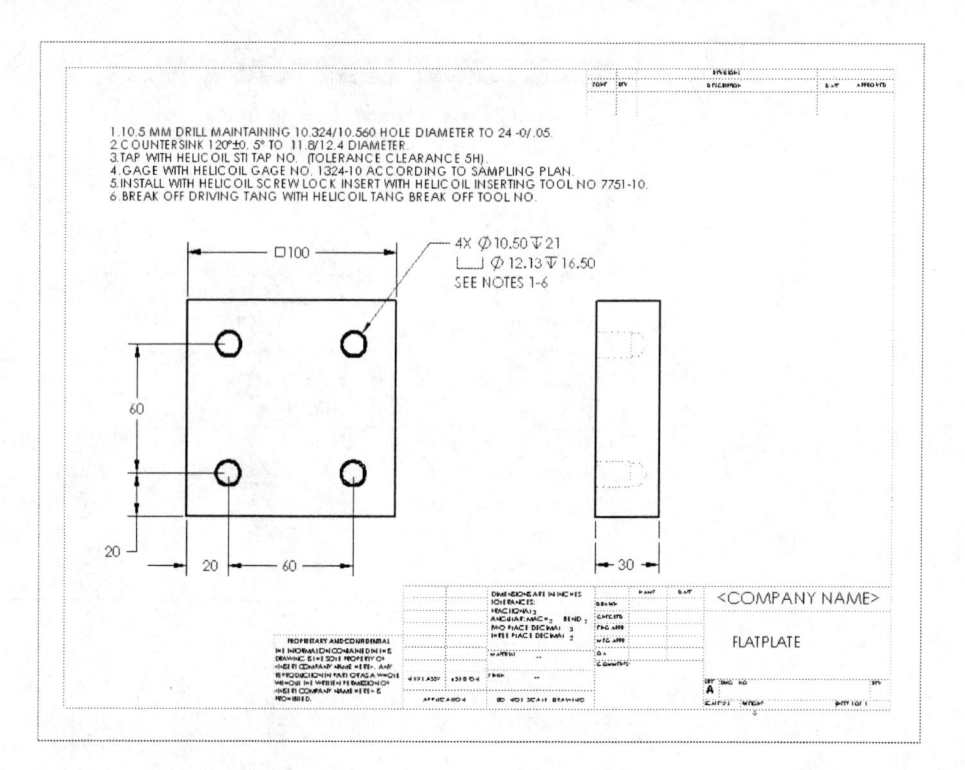

NOTES:

1. 10.5 MM DRILL MAINTAINING 10.324/10.560 HOLE DIAMETER TO 24 -0/.05.

2. COUNTERSINK 120°±0. 5° TO 11.8/12.4 DIAMETER.

3. TAP WITH HELICOIL STI TAP NO. 2087-10 (TOLERANCE CLEARANCE 5H).

4. GAGE WITH HELICOIL GAGE NO. 1324-10 ACCORDING TO SAMPLING PLAN.

5. INSTALL WITH HELICOIL SCREW LOCK INSERT 4184-10CN WITH HELICOIL INSERTING TOOL NO 7751-10.

6. BREAK OFF DRIVING TANG WITH HELICOIL TANG BREAK OFF TOOL NO. 4238-10.

Store Drawing notes for threaded inserts in a Notepad.txt document to conserve time.

Exercise 3.9: Industry Collaborative Exercise.

Create an eDrawing of the GUIDE drawing. A SolidWorks eDrawing is a compressed document that does not require the corresponding part or assembly. SolidWorks eDrawing is animated to display multiple views and dimensions. Review the eDrawing on-line Help for additional functionality.

Select Tools, Add-Ins, eDrawings. Click Publish eDrawing from the eDrawing toolbar. Click the Animate button. The Front view of the VALVEPLATE drawing is displayed. Click the Next button. Display the remaining views. Click the Play button. Save the GUIDE eDrawing. Email the drawing to a friend.

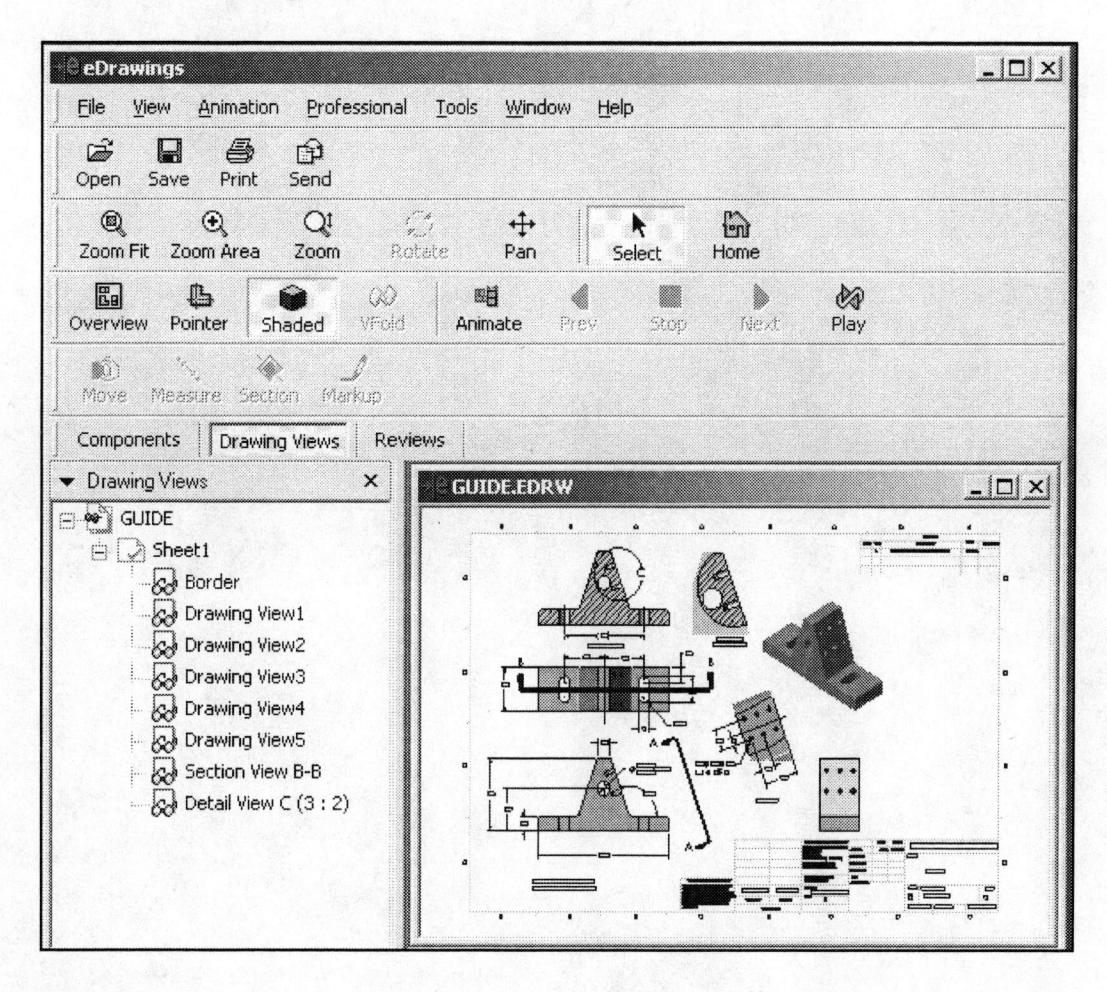

Notes:

Project 4

Extrude and Revolve Features

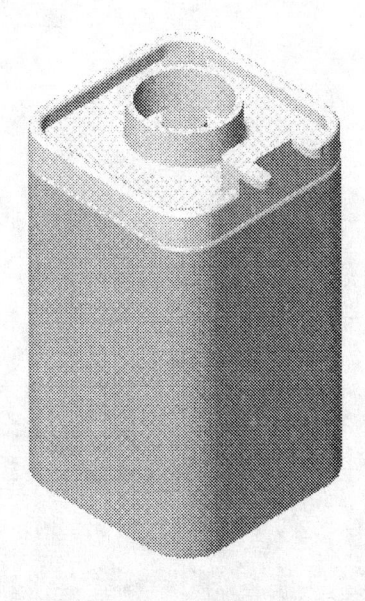

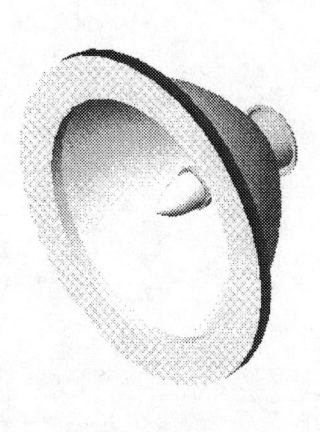

Below are the desired outcomes and usage competencies based on the completion of Project 4.

Project Desired Outcomes:	Usage Competencies:
A comprehensive understanding of the customer's design intent for a FLASHLIGHT assembly.	To comprehend the fundamental definitions and process of Feature-Based 3D Solid Modeling.
A design intent that is cost effective, serviceable and flexible for future manufacturing revisions.	Specific knowledge and understanding of the following Features: Extruded-Boss, Extruded Base, Extruded-Cut, Revolve Boss/Bass, Revolved Cut, Dome, Shell, Circular Pattern and Fillet.
Four key parts: • BATTERY. • BATTERYPLATE. • LENS. • BULB.	

Notes:

Project 4 – Extrude and Revolve Features

Project Objective

Design a FLASHLIGHT assembly that is cost effective, serviceable and flexible for future manufacturing revisions.

The FLASHLIGHT assembly consists of numerous parts. The team decides to purchase and model the following parts: One 6-volt BATTERY, LENS assembly, SWITCH and an O-RING. The LENS assembly consists of the LENS and the BULB.

In this project, create and model the following parts:

- BATTERY.
- BATTERYPLATE.
- LENS.
- BULB.

The other parts for the FLASHLIGHT assembly will be addressed in Project 5.

On completion of this project, you will be able to:

- Choose the best profile for sketching.
- Choose the proper sketch plane.
- Create a Template: English and Metric units.
- Set Document Properties.
- Set Units.
- Customize Toolbars.
- Add Dimensions.
- Add Geometric Relations.
- Create an Arc.

- Use the following SolidWorks features:
    - Extruded-Base.
    - Extruded-Cut.
    - Extruded-Boss.
    - Edge Fillets.
    - Face Fillets.
    - Revolved Boss.
    - Revolved Base.
    - Boss Revolve Thin.
    - Cut Revolve Thin.
    - Dome.
    - Shell.

Project Situation

You work for a company that specializes in providing promotional trade show products. The company is expecting a sales order for 100,000 flashlights with a potential for 500,000 units next year. Prototype drawings of the flashlight are required in three weeks.

You are the design engineer responsible for the project. You contact the customer to discuss design options and product specifications. The customer informs you that the flashlights will be used in an international marketing promotional campaign. Key customer requirements:

- Inexpensive reliable flashlight.

- Available advertising space of 10 square inches, 64.5 square centimeters.

- Light weight semi indestructible body.

- Self standing with a handle.

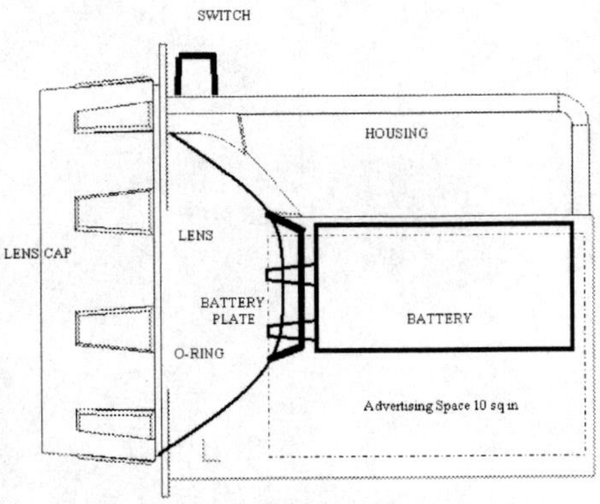

Figure 4.1

Your company's standard product line does not address the above key customer requirements. The customer made it clear that there is no room for negotiation on the key product requirements.

You contact the salesperson and obtain additional information on the customer and product. This is a very valuable customer with a long history of last minute product changes. The job has high visibility with great future potential.

In a design review meeting, you present a conceptional sketch. Your colleagues review the sketch. The team's consensus is to proceed with the conceptual design, Figure 4.1.

The first key design decision is the battery. The battery type directly affects the flashlight body size, bulb intensity, case structure integrity, weight, manufacturing complexity and cost.

Review two potential battery options:

- A single 6-volt lantern battery.

- Four 1.5-volt D cell batteries.

The two options affect the product design and specification. Think about it.

A single 6-volt lantern battery is approximately 25% higher in cost and 35% more in weight. The 6-volt lantern battery does provide higher current capabilities and longer battery life.

A special battery holder is required to incorporate the four 1.5 volt D cell configuration. This would directly add to the cost and design time of the FLASHLIGHT assembly, Figure 4.2.

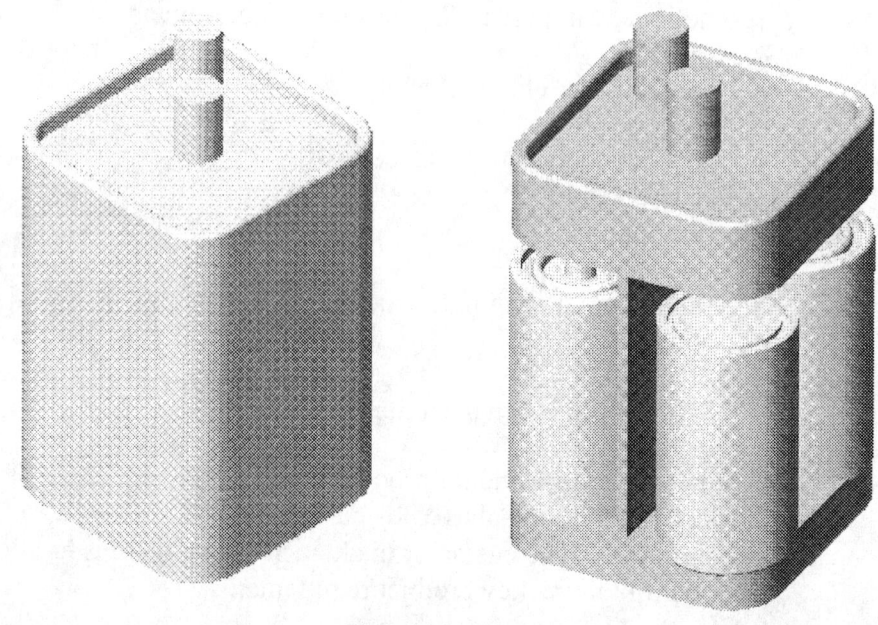

Figure 4.2

Time is critical.

For the prototype, you decide to use a standard 6-volt lantern battery. This eliminates the requirement to design and procure a special battery holder. However, you envision the 4-D cell battery model for the next product revision.

Design the FLASHLIGHT assembly to accommodate both battery design options.

Battery dimensional information is required for the design. Where do you go? Potential sources: product catalogs, company web sites, professional standards organizations, design handbooks and colleagues.

The team decides to purchase the following parts: 6-volt BATTERY, LENS ASSEMBLY, SWITCH and an O-RING. You will model the following purchased parts: BATTERY, LENS assembly, SWITCH and the O-RING. The LENS assembly consists of the LENS and the BULB.

Your company will design, model and manufacture the following parts: BATTERYPLATE, LENSCAP and HOUSING.

Purchased Parts:	Designed Parts:
BATTERY	BATTERYPLATE
LENS assembly	*LENSCAP
*SWITCH	*HOUSING
*O-RING	

*Parts addressed in Project 5.

Project Overview

Create four parts for the FLASHLIGHT assembly in this section, Figure 4.3a:

- BATTERY.
- BATTERYPLATE.
- LENS.
- BULB.

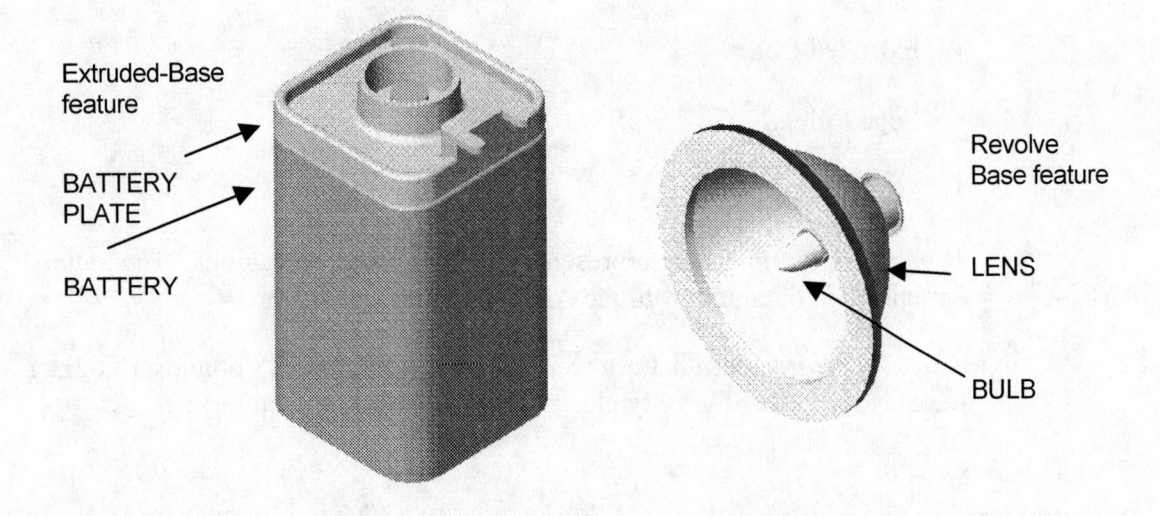

Figure 4.3a

Two major Base features are addressed in this project:

- Extrude – BATTERY and BATTERYPLATE.

- Revolve – LENS and BULB.

Note: Dimensions and features are used to illustrate the SolidWorks functionality in a design situation. Wall thickness and thread size have been increased for improved picture illustration. Parts have been simplified.

Four additional parts will be created in Project 5 for the final FLASHLIGHT assembly, Figure 4.3b.

- O-RING.

- LENSCAP.

- SWITCH.

- HOUSING.

Figure 4.3b

BATTERY

The BATTERY is a simplified representation of an OEM part. The BATTERY consists of the following features:

- Extruded Base.

- Extruded Cut.

- Edge Fillets.

- Face Fillets.

The battery terminals are represented as cylindrical extrusions. The battery dimension is obtained from the ANSI standard 908D.

Note: A 6-volt lantern battery weighs approximately 1.38 pounds, (0.62kg). Locate the center of gravity closest to the center of the battery.

BATTERY Feature Overview

Create the BATTERY, Figure 4.4a. Identify the required BATTERY features.

- Extruded Base: The Extruded Base feature is created from a symmetrical square sketch, Figure 4.4b.

- Fillet: The Fillet feature is created by selecting the vertical edges and the top face, Figure 4.4c and Figure 4.4e.

- Extruded Cut: The Extruded Cut feature is created from the top face offset, Figure 4.4d.

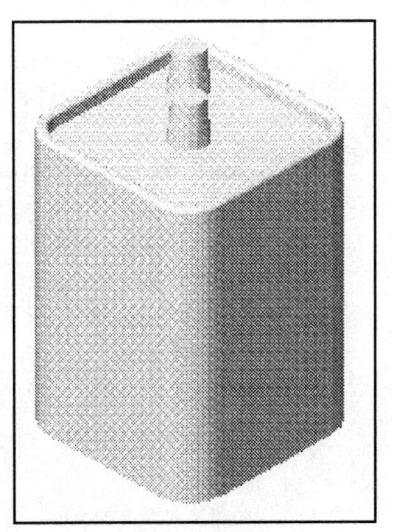

Figure 4.4a

- Extruded Boss: The Extruded Boss feature is created to represent the battery terminals, Figure 4.4f.

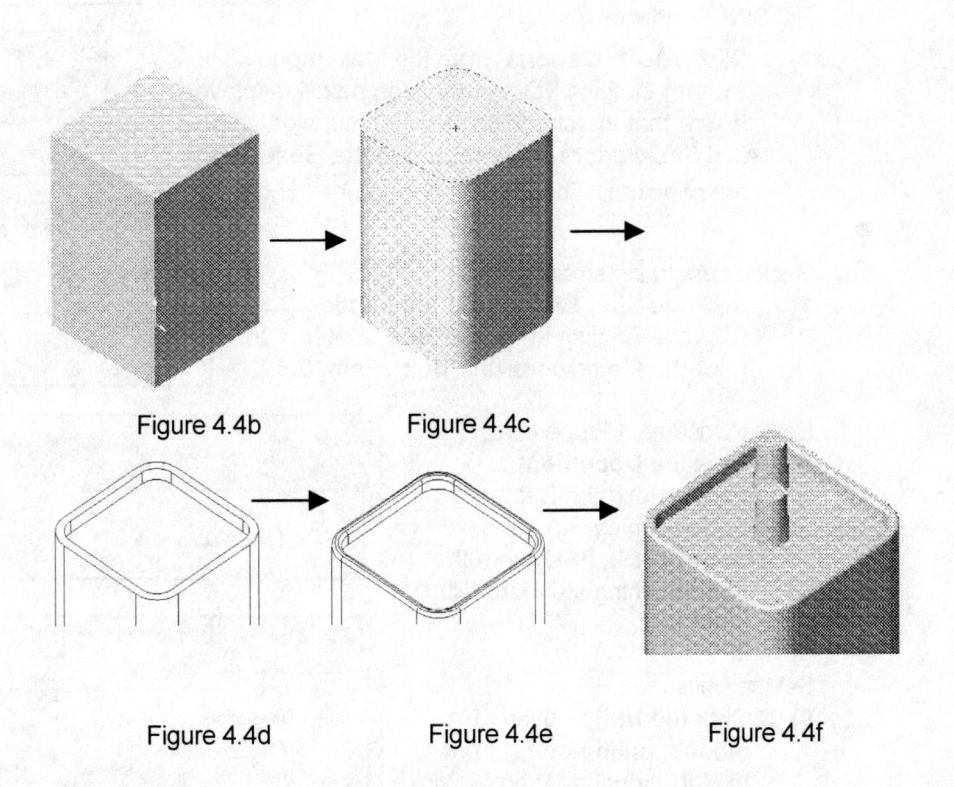

Figure 4.4b   Figure 4.4c

Figure 4.4d   Figure 4.4e   Figure 4.4f

Let's create the BATTERY.

Create the Template

Dimensions for the FLASHLIGHT assembly are provided both in English and Metric units. The Primary units are in inches. Three decimal places are displayed to the right of the decimal point. The Secondary units are in millimeters. Secondary units are displayed in brackets [x]. Two decimal places are displayed to the right of the decimal point. The PART-IN-ANSI Template contains System Options and Document Properties settings for the parts contained in the FLASHLIGHT ASSEMBLY. Substitute the PART-MM-ISO or PART-MM-ANSI Template to create the same parts in millimeters.

Create an English document template.

1) Click **New**. Click the **Part** icon from the Templates tab. Click **OK**. The Front, Top and Right reference planes are displayed in the Part1 Feature Manager.

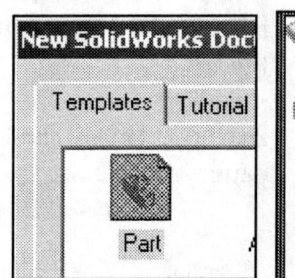

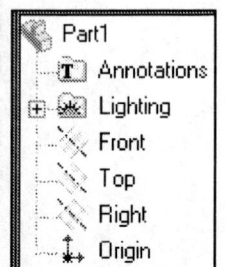

Set System options.

2) Click **Tools**, **Options**, from the Main menu. The System Options - General dialog box is displayed. Insure that the check box Input dimension value and Show errors every rebuild in the General box are checked. These are the default settings.

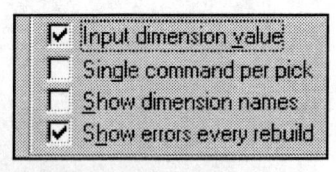

Set the Length increment.

3) Click the **Spin Box Increments** option. Click the English units **text box**. Enter **.100**. Click the Metric units **text box**. Enter **2.5**.

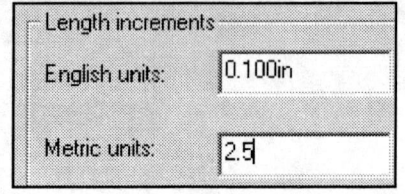

Set the Document Properties.

4) Click the **Document Properties** tab. Set the Dimensioning Standard. Select **ANSI**, **[ISO]** from the Dimensioning standard drop down list.

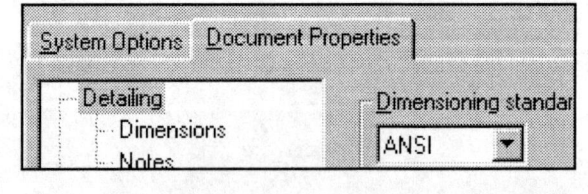

Set the Units.

5) Click the **Units** option. Enter **inches**, **[millimeters]** from the Linear units list box. Click the **Decimal** button. Enter **3**, **[2]** in the Decimal places spin box. Click **OK**.

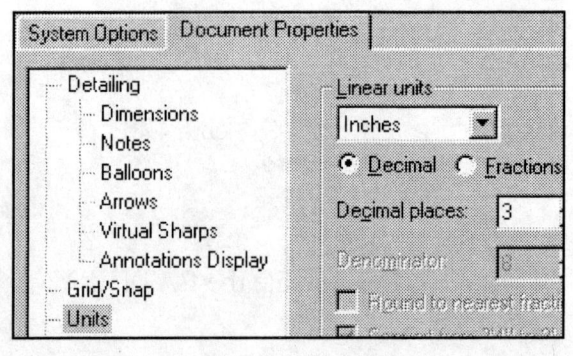

Save the Part Template.

6) Click **File** from the Main menu. Click **Save As**. Click **Part Templates (*.prtdot)** from the Save As type list box. Select **ENGDESIGN-W-SOLIDWORKS\MY-TEMPLATES** for Save in file folder. Enter **PART-IN-ANSI [PART-MM-ISO]** in the File name text box. Enter **English part template, units-inch, ANSI standard**, [**Metric part template, units-mm, ISO standard**] for Description. Click **Save**.

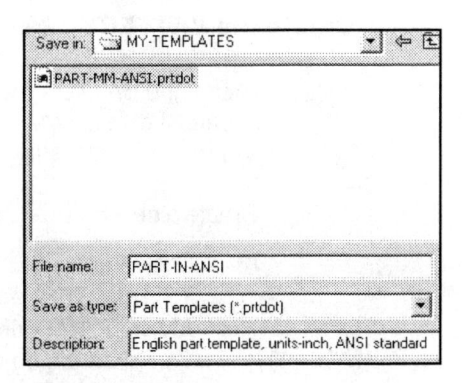

7) Close All documents. Click **Windows, Close All**. Click **NO** to Save documents.

ASMEY14.5M defines the types of decimal dimension display for inches and millimeters. The Primary units are in inches. Three decimal places are displayed to the right of the decimal point. The Secondary units are in millimeters. Secondary units are displayed in brackets [x]. Two decimal places are displayed to the right of the decimal point.

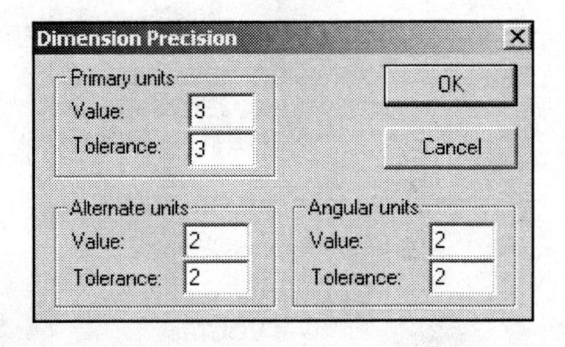

The precision is set to 3 decimal places for inches. Example: 2.700 is displayed. If you enter 2.7, the value 2.700 is displayed. The precision is set to 2 decimal places for millimeters. Example: [68.58] is displayed. For consistency, the inch part dimension values for the text include the number of decimal places required. The drawings utilizes the decimal dimension display as follows:

TYPES of DECIMAL DIMENSIONS (ASME Y14.5M):				
Description:	Example: MM	Description:	Example: INCH
Dimension is less than 1mm. Zero precedes the decimal point.	0.9 0.95	Dimension is less than 1 inch. Zero is not used before the decimal point.	.5 .56
Dimension is a whole number. No decimal point. Display no zero after decimal point.	19	Express dimension to the same number of decimal places as its tolerance. Add zeros to the right of the decimal point.	1.750
Dimension exceeds a whole number by a decimal fraction of a millimeter. Display no zero to the right of the decimal.	11.5 11.51	If the tolerance is expressed to 3 places, the dimension contains 3 places to the right of the decimal point.	

Create the BATTERY

Create the BATTERY with an Extruded Base feature. The Extruded Base feature uses a square sketch drawn centered about the Origin on the Top plane. Build parts with symmetric relationships. Use a line of symmetry in a sketch. Add geometric relationships.

Create a new part.

8) Click **New**. Click the **MY-TEMPLATES** tab. Click **PART-IN-ANSI**, [**PART-MM-ISO**] from the Template dialog box. Click **OK**.

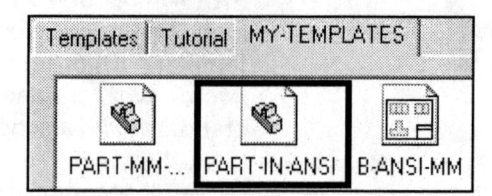

9) Save the empty part. Click **Save**. Select **ENGDESIGN-W-SOLIIDWORKS\PROJECTS** for Save in file folder.

10) Enter **BATTERY** for file name.

11) Enter **BATTERY, 6-VOLT** for Description.

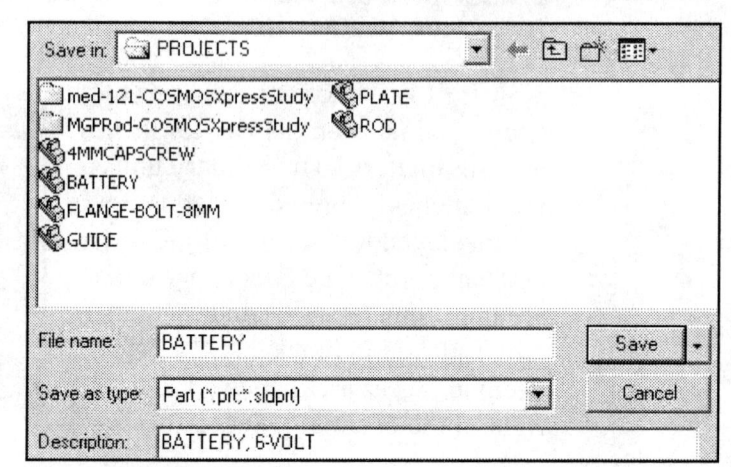

12) Click the **Save** button.

Create the Extruded Base feature.

13) Select the Sketch plane. Click the **Top** plane from the Feature Manager.

14) Create a new Sketch. Click **Sketch** from the Sketch toolbar.

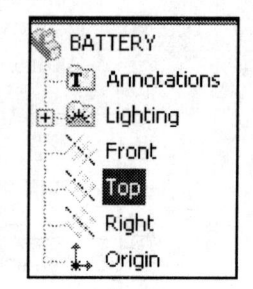

15) Display the Top view. Click **Top** from the Standards View toolbar.

16) Sketch the profile. Click **Rectangle**. Click the **first point** in the lower left quadrant. Click the **second point** in the upper right quadrant. The Origin is approximately in the middle of the Rectangle.

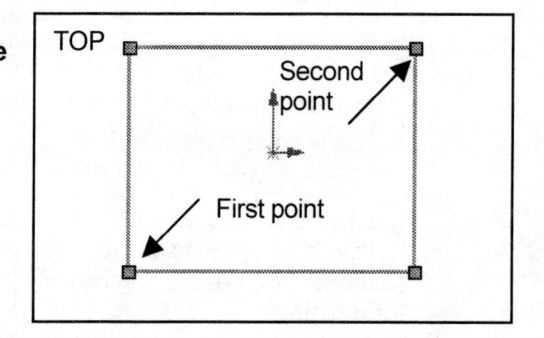

PAGE 4 - 12

Engineering Design with SolidWorks   Extrude and Revolve Features

17) Sketch the Centerline. Click **Centerline** from the Sketch Tools toolbar. Sketch a diagonal centerline from the **upper left corner** to the **lower right corner**.

The endpoints of the centerline are coincident with the corner points of the Rectangle.

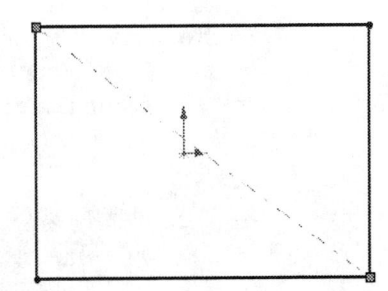

18) Add a dimension. Click **Dimension** from the Sketch toolbar. Select the **top horizontal line**. Click a **position** above the horizontal line. Enter **2.700, [68.58]** for width. Click the **Green Check Mark**.

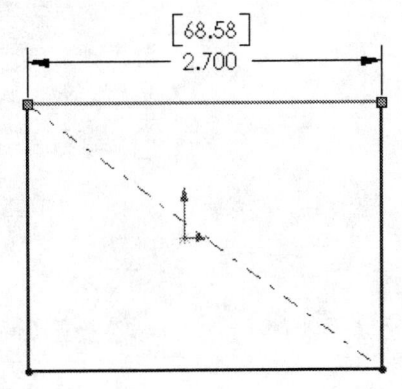

19) Add Geometric Relations. Click **Select**. Add a midpoint relation. Hold the **Ctrl** key down. Click the diagonal **centerline**. Click the **Origin**. Release the **Ctrl** key. Click the **Midpoint** button. Click **Close Dialog** from the Properties Manager.

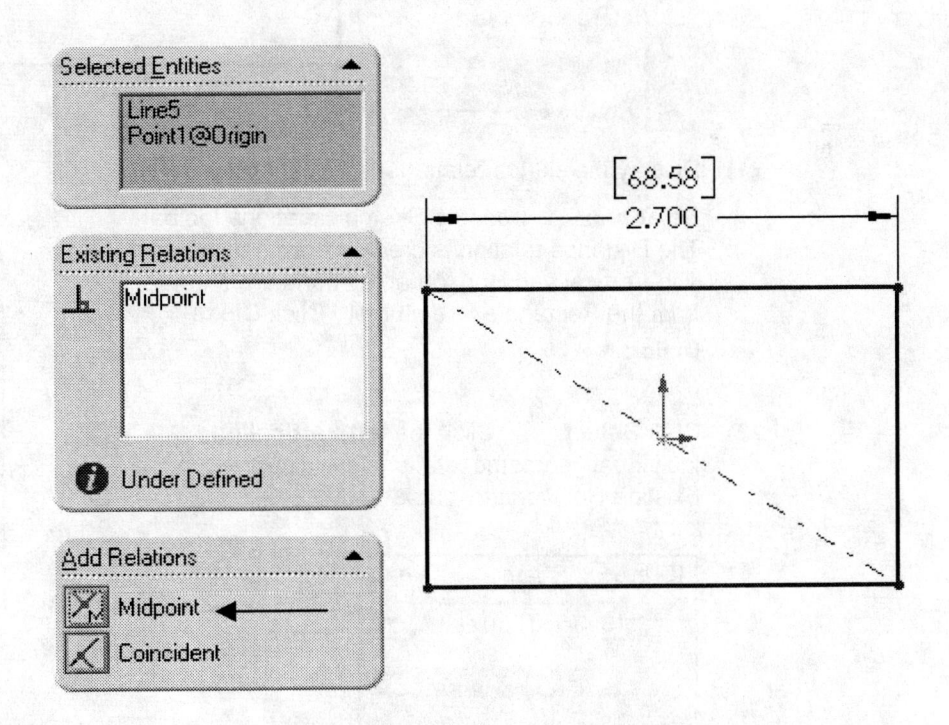

Note: The Line# may be different than the line numbers above. The Line# is dependent on the line number order creation.

PAGE 4 - 13

20) Add an equal relation. Click the **top horizontal line**. Hold the **Ctrl** key down. Click the **left vertical line**. Click the **Equal** button. Release the **Ctrl** key. Click **Close Dialog** ✔ from the Properties Manager. The black Sketch is fully defined.

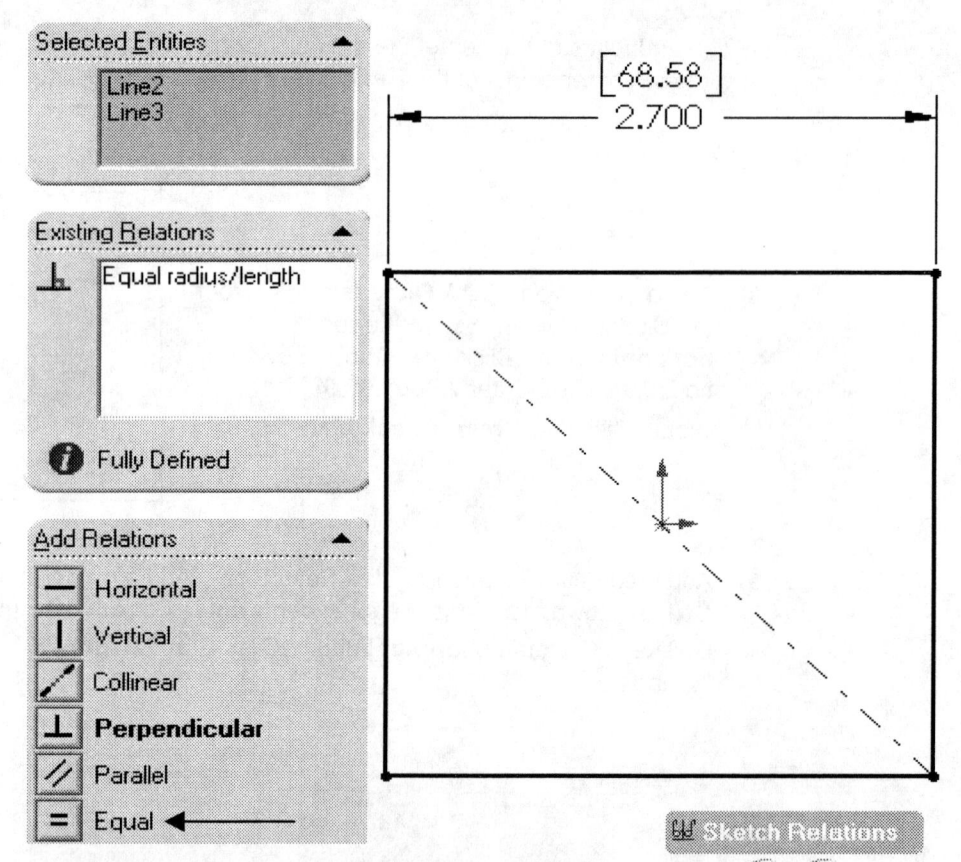

21) Display the sketch relations. Click **Display/Delete Relations** from the Sketch Relations toolbar. The Distance relation is created from a dimension. The Vertical and Horizontal relations are created from the Rectangle Sketch tool. Click **Close Dialog** ✔.

22) Click **Select**. Click the **left vertical line**. Individual geometric relations are displayed in the Existing Relations text box.

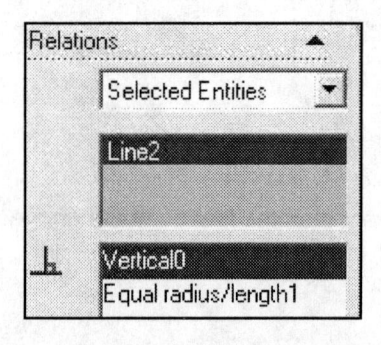

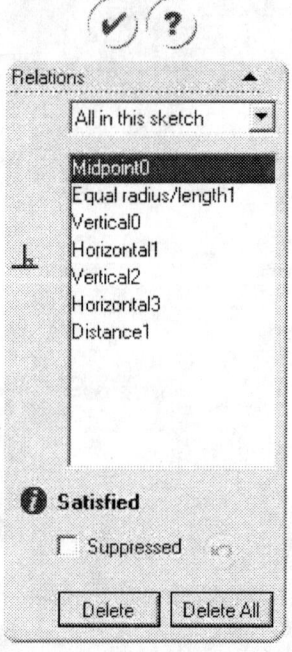

Engineering Design with SolidWorks — Extrude and Revolve Features

23) Extrude the Sketch. Click **Extruded Boss/Base**. Blind is the default Type option. Enter **4.100**, [**104.14**] for Depth. Display the Base-Extrude feature. Click **OK**.

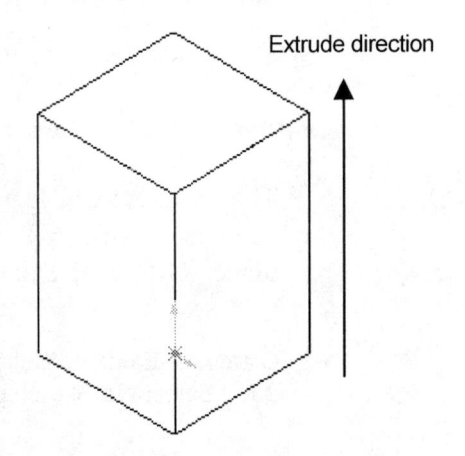

24) Fit the part to the Graphics window. Press the **f** key.

25) Rename **Extrude1** to **Base Extrude**.

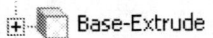

⊞ Base-Extrude

26) Save the BATTERY. Click **Save**.

Create the BATTERY – Use the Fillet Feature

The vertical sides on the BATTERY are rounded. Use the Fillet feature to round the 4 side edges.

Create a Fillet feature.

27) Display the part's hidden edges in gray. Click **Hidden Lines Visible** from the View toolbar.

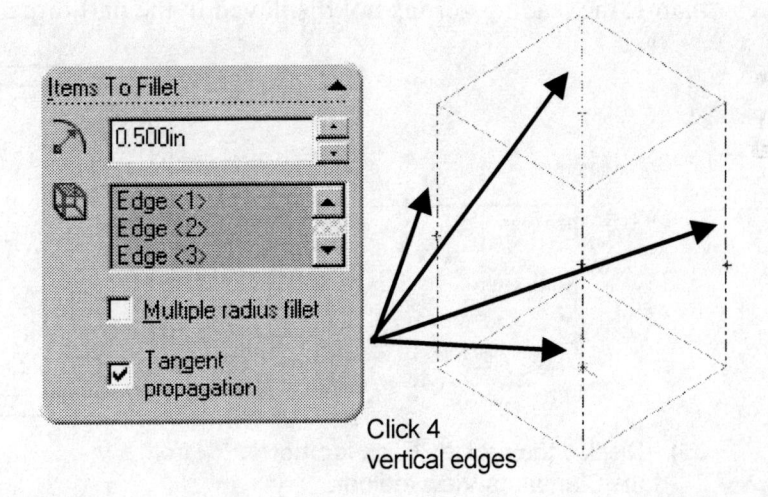

Click 4 vertical edges

28) Create a Fillet feature. Click the **left vertical edge**. Click **Fillet** from the Feature toolbar. Click the remaining **3 vertical edges**. Enter **.500**, [**12.7**] for Radius. Display the Fillet feature. Click **OK**.

29) Rename **Fillet1** to **Side-Fillets** in the Feature Manager.

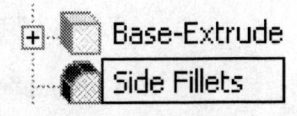

⊞ Base-Extrude
   Side Fillets

Extrude and Revolve Features Engineering Design with SolidWorks

30) Save the BATTERY. Click **Save**.

Create the BATTERY – Use the Extruded Cut Feature

The Extruded Cut feature removes material. An Offset Edge takes existing geometry, extracts it from an edge or face and locates it on the current sketch plane. Offset the existing Top face. Create a Cut feature.

Create the Extruded Cut feature. 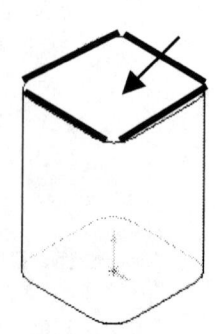
31) Select the Sketch plane. Click the **Top** face.

32) Create the Sketch. Click **Sketch**.

33) Display the face. Click **Top** from the Standards View toolbar.

34) Offset the existing geometry from the boundary of the Sketch plane. Click **Offset Entities** from the Sketch Tools toolbar. Enter **.150**, **[3.81]** for the Offset distance. Click the **Reverse** check box. The new Offset profile displays inside the original profile. Click **OK**.

Note: A leading zero is displayed in the spin box. For inch dimensions less than 1, the leading zero is not displayed in the part dimension.

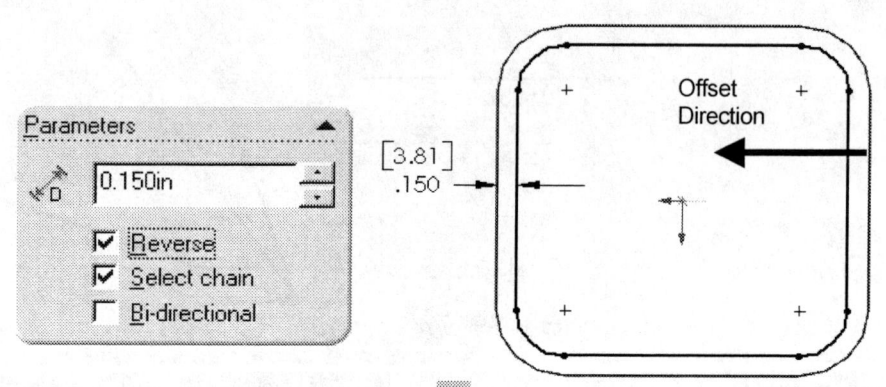

35) Display the profile. Click **Isometric** from the Standards View toolbar.

36) Extrude the Offset profile. Click **Extruded Cut** from the Feature toolbar. Enter **.200**, **[5.08]** for Depth of the Cut. Display Cut-Extrude1. Click **OK**.

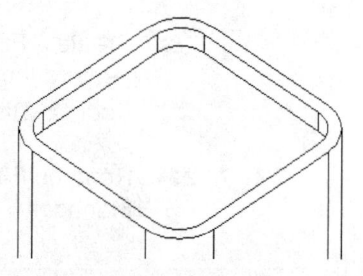

37) Rename **Cut-Extrude1** to **Top-Cut**.

PAGE 4 - 16

Engineering Design with SolidWorks   Extrude and Revolve Features

38) Save the BATTERY. Click **Save** 💾.

Create the Battery – Use the Fillet Feature on the Top Face

The Top outside edges of the Battery requires fillets. Use the top face of the Battery to create a constant radius Fillet feature. The top narrow face is small. Use the Face Selection Filter to select the faces of the Battery. Deactivate the filters to select the geometry.

Create the Fillet feature on the top face of the BATTERY.

39) Display the Selection Filter toolbar. Click **View** from the Main menu. Click **Tools, Selection Filter**.

40) Create the Fillet. Click **Filter Face** from the Selection Filter toolbar. Click the **top thin face**. Select **Fillet** from the Feature toolbar. Face<1> is displayed in the Edge fillet items box. Click **Constant Radius** for Fillet Type. Enter **.050**, [**1.27**] for Fillet Radius.

41) Display the Fillet on the inside and outside top edges of the BATTERY. Click **OK** ✔.

42) Deactivate the Face Filter. Click **Face Filter**.

43) Rename **Fillet2** to **Top Face Fillet**.

44) Save the BATTERY. Click **Save** 💾.

Note: Do not select a Fillet radius which is larger that the surrounding geometry.

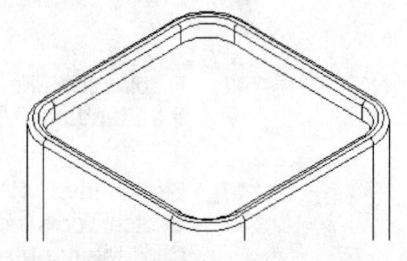

PAGE 4 - 17

Extrude and Revolve Features | **Engineering Design with SolidWorks**

Example: The top edge face width is .150, [3.81]. The Fillet is created on both sides of the face. A common error is to enter a Fillet too large for the existing geometry. A minimum face width of .200, [5.08] is required for a Fillet radius of .100, [2.54].

The following error occurs when the Fillet radius is too large for the existing geometry:

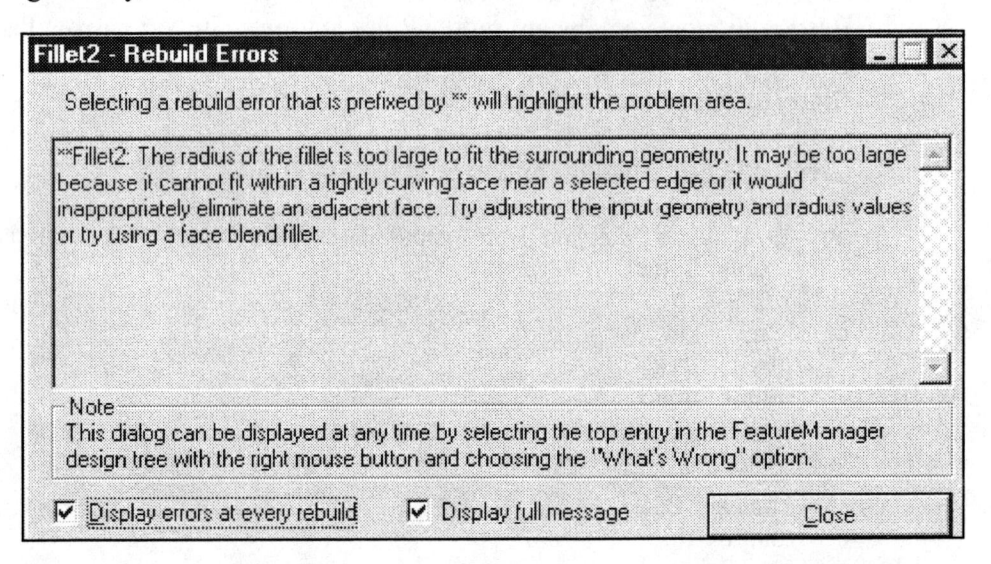

```
Fillet2 - Rebuild Errors

Selecting a rebuild error that is prefixed by ** will highlight the problem area.

**Fillet2: The radius of the fillet is too large to fit the surrounding geometry. It may be too large
because it cannot fit within a tightly curving face near a selected edge or it would
inappropriately eliminate an adjacent face. Try adjusting the input geometry and radius values
or try using a face blend fillet.

Note
This dialog can be displayed at any time by selecting the top entry in the FeatureManager
design tree with the right mouse button and choosing the "What's Wrong" option.

[x] Display errors at every rebuild    [x] Display full message         [ Close ]
```

Avoid the Fillet Rebuild error. Reduce the Fillet size or increase the face width.

Create the BATTERY – Use the Extruded Boss Feature

Two Battery Terminals are required. Conserve design time. Represent the terminals as a cylindrical Extruded Boss feature.

Create the Extruded Boss feature.
45) Select the Sketch plane. Click the **face** of the Top-Cut feature.

46) Create the Sketch. Click **Sketch**.

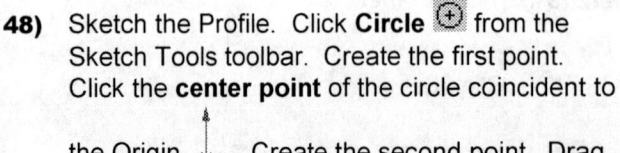

47) Display the Sketch plane. Click **Top** from the Standards View toolbar.

48) Sketch the Profile. Click **Circle** from the Sketch Tools toolbar. Create the first point. Click the **center point** of the circle coincident to the Origin. Create the second point. Drag the **mouse pointer** to the right. Click a **position** to the right of the Origin.

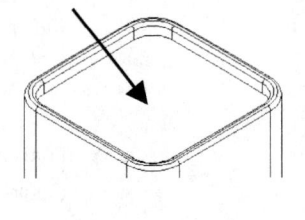

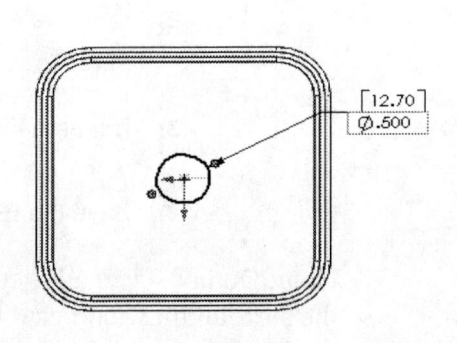

PAGE 4 - 18

Engineering Design with SolidWorks — Extrude and Revolve Features

49) Click **Dimension**. Select the **circumference** of the circle. Click a **position** below the bottom horizontal line. Enter **.500**, **[12.7]**. Click the **Green Check Mark**. The Sketch is fully defined. The sketch is displayed in black.

50) Copy the sketched circle. Click **Select**. Hold the **Ctrl** key down. Click the **circumference** of the circle. Drag the **circle** to the upper left quadrant. Create the second circle. Release the **mouse button**. Release the **Ctrl** key. The second circle is selected and is displayed in green.

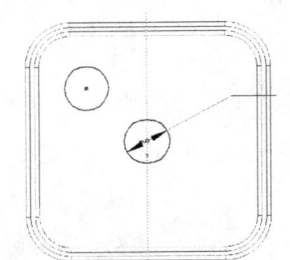

51) Add an equal relation. Hold the **Ctrl** key down. Click the **circumference of the first circle**. Both circles are selected. Click **Equal** from the Add Relations text box. Release the **Ctrl** key. Click **Close Dialog** from the Properties Manager.

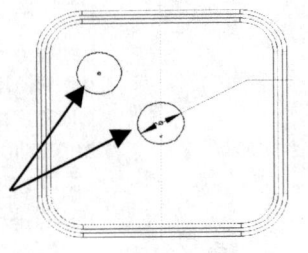

The dimension between the center points is critical. Dimension the distance between the two center points with an aligned dimension.

52) The Right plane is the dimension reference. Right-click the **Right** plane from the FeatureManager. View the plane. Click **Show**.

53) Add a dimension. Click **Dimension**. Click the **two center points** of the two circles. Click a **position** off the profile in the upper right corner. Enter **1.000**, **[25.4]** for the aligned dimension. Click the **Green Check Mark**.

The dimension text toggles between linear and aligned. An aligned dimension is created when the dimension is positioned between the two circles.

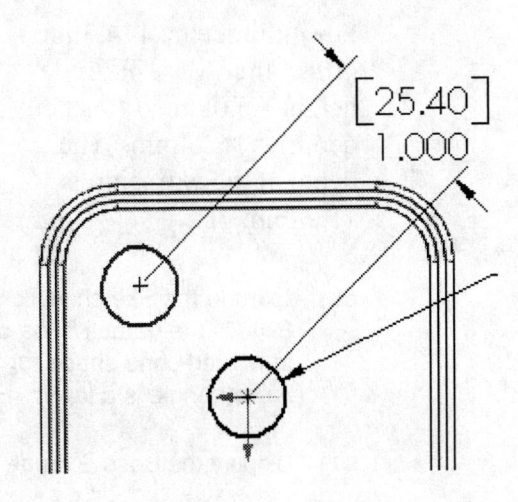

PAGE 4 - 19

Extrude and Revolve Features   Engineering Design with SolidWorks

54) Create an angular dimension. Click **Centerline**. Sketch a centerline between the **two circle center points**.

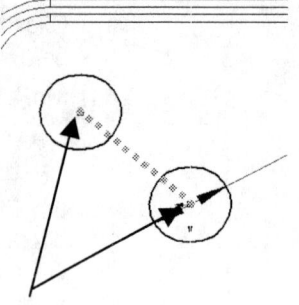

55) Create an acute angular dimension. Click **Dimension**. Click the **centerline** between the two circles. Click the **Right** plane(vertical line). Click a **position** between the centerline and the Right plane, off the profile. Enter **45**. Click the **Green Check Mark**.

Note: Acute angles are less than 90°. Acute angles are the preferred dimension standard.

The overall battery height is a critical dimension. The battery height is 4.500 inch, [114.30mm]. Calculate the depth of the extrusion:

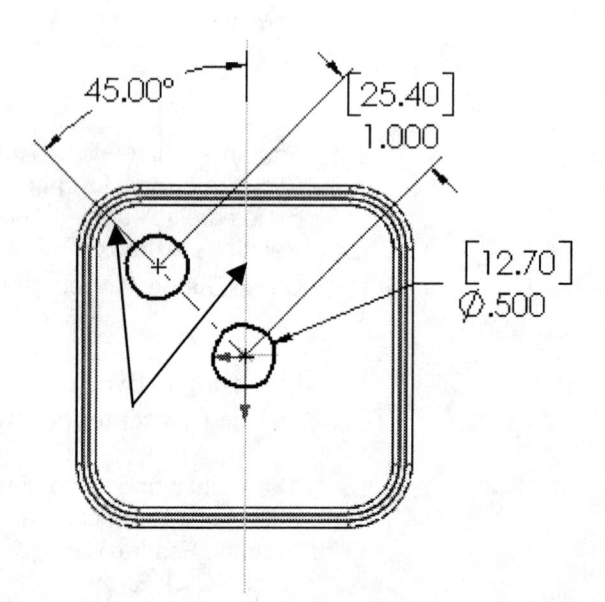

For inches: 4.500in. – (4.100in. Base-Extrude height – .200in. Offset cut depth) = .600in. The depth of the extrusion is .600in.

For millimeters: 114.3mm – (104.14mm Base-Extrude height – 5.08mm Offset cut depth) = 15.24mm. The depth of the extrusion is 15.24mm.

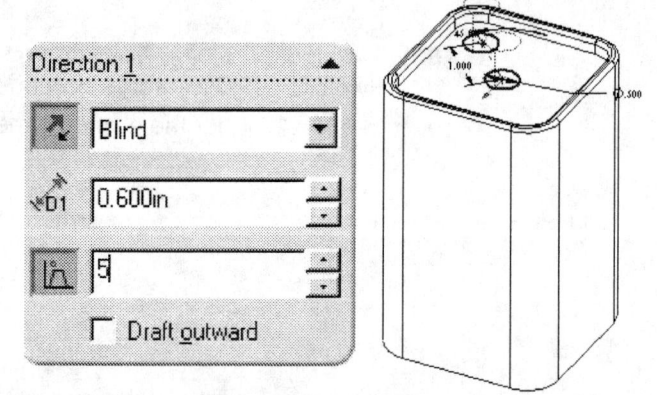

56) Extrude the Sketch. Click **Extruded Boss/Base** from the Feature toolbar. Blind is the default Type option. Enter **.600**, **[15.24]** for Depth. Create a truncated cone shape for the battery terminals. Click the **Draft ON/OFF** button. A draft angle is a taper. Enter **5** in the Draft Angle text box.

57) Display the Boss-Extrude feature. Click **OK**.

PAGE 4 - 20

Engineering Design with SolidWorks   **Extrude and Revolve Features**

58) Rename **Extrude2** to **Terminals**.

59) Rename **Sketch3** to **Sketch-TERMINALS**.

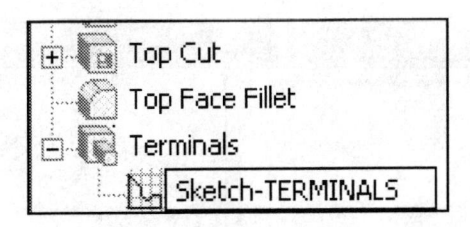

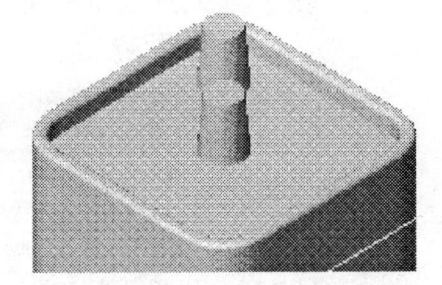

Note: Feature Names:

Every time you create a feature of the same feature type, the feature name is incremented by one. Example: Extrude1 is the first Extrude feature. Extrude2 is the second Extrude feature. If you delete a feature, rename a feature or exit a SolidWorks session, the feature numbers will vary from those illustrated in the text.

Measure the overall height.

60) Verify the overall height. Click **Tools**, **Measure** from the Main menu. Click **Right** from the Standard Views toolbar. Click the **top edge** of the battery terminal. Click the **bottom edge** of the battery. The overall height, Y is 4.500, [114.3]. Click **Close**.

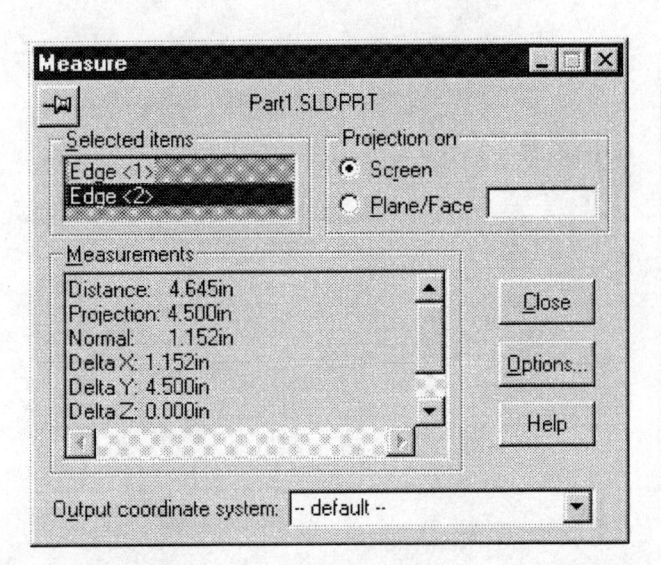

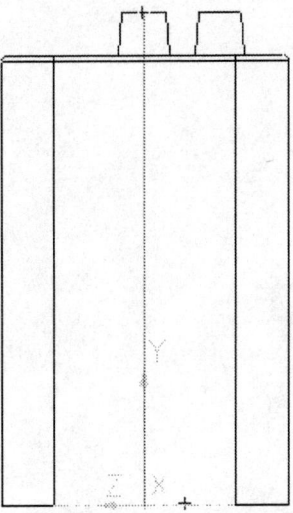

61) Hide all planes. Click **View** from the Main menu. Uncheck **Planes**.

62) Display the Trimetric view. Click **View Orientation** . Double-click **Trimetric**.

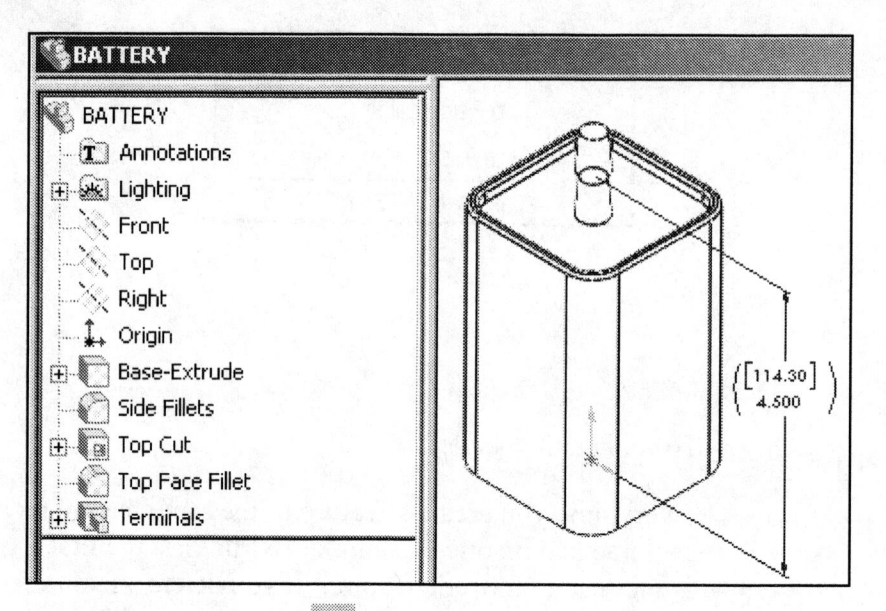

63) Save the BATTERY. Click **Save** .

BATTERYPLATE

The BATTERYPLATE is a critical part. The BATTERYPLATE:

- Aligns the LENS assembly.

- Creates an electrical connection between the SWITCH assembly, BATTERY and LENS.

Design the BATTERYPLATE, Figure 4.5.

Utilize features from the BATTERY to develop the BATTERYPLATE.

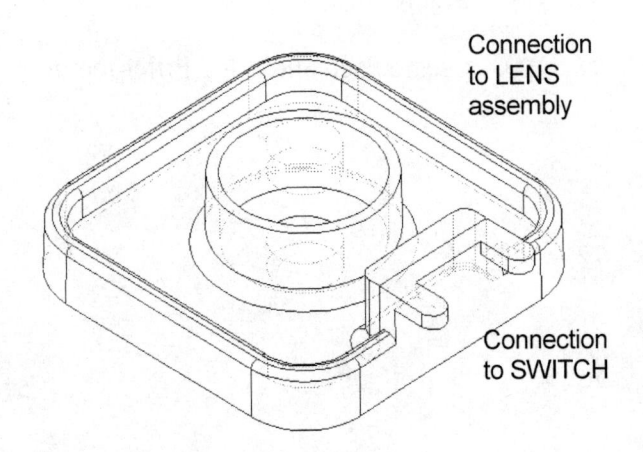

Figure 4.5 (Connection to LENS assembly; Connection to SWITCH)

BATTERYPLATE Feature Overview

Create the BATTERYPLATE. Modify the BATTERY features. Create two holes from the original sketched circles. Use the Extruded Cut feature, Figure 4.6.

Modify the dimensions of the Base feature. Add a 1-degree draft angle.

Note: A sand pail contains a draft angle. The draft angle assists the sand to leave the pail when the pail is flipped upside down.

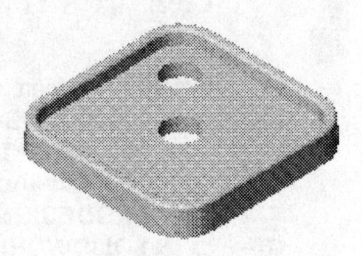

Figure 4.6

Create a new Extruded Boss Thin feature. Offset the center circular sketch, Figure 4.7.

The Extruded Boss Thin feature contains the LENS. Create an inside draft angle. The draft angle assists the LENS into the Holder.

Extruded Boss Thin Feature Holder for LENS

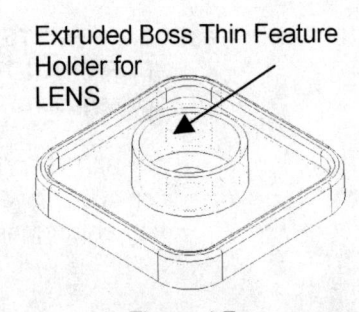

Figure 4.7

Extrude and Revolve Features   **Engineering Design with SolidWorks**

Create the first Extruded Boss feature using two depth directions, Figure 4.8.

Create the second Extruded Boss feature using sketched mirror geometry, Figure 4.9.

Create Face and Edge Fillet features to remove sharp edges, Figure 4.10.

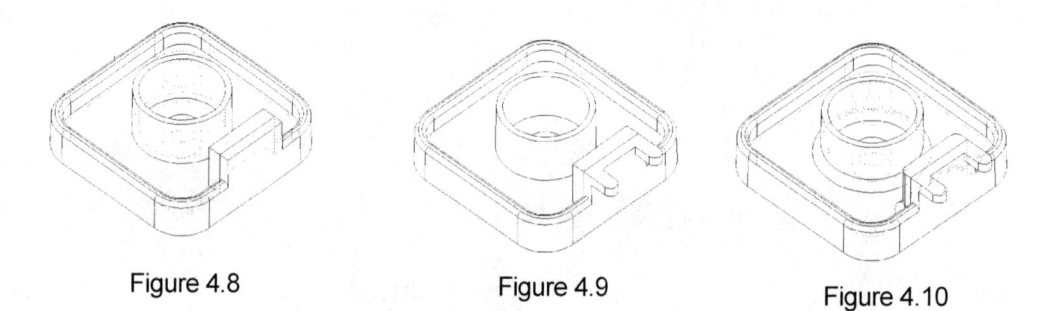

Figure 4.8   Figure 4.9   Figure 4.10

Let's create the BATTERYPLATE.

Create the BATTERYPLATE

Create the BATTERYPLATE from the BATTERY.

Create a new part.

64) Create the BATTERYPLATE from the BATTERY. Click **File**, **SaveAs**. Enter **ENGDESIGN-W-SOLIDWORKS\PROJECTS** for Save In File Folder.

65) Enter **BATTERYPLATE** for File name.

66) Enter **BATTERY PLATE FOR 6-VOLT** for Description. Click **Save**.

The BATTERYPLATE part icon is displayed at the top of the FeatureManager.

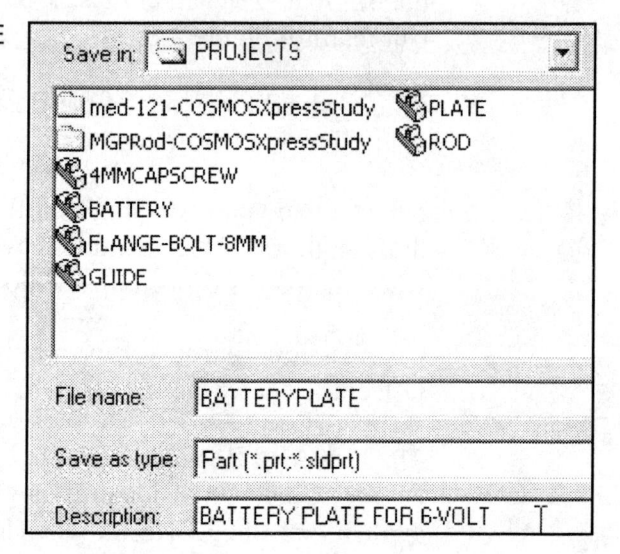

Create the BATTERYPLATE – Delete and Edit Features

Create two holes. Delete the Terminals feature and reuse the circle sketch.

Delete and Edit Features.

67) Remove the Terminals (Extruded Boss) feature. Right-click **Terminals** from the FeatureManager. Click **Delete Feature**. Click **Yes** from the Confirm Delete dialog box. Do not delete the two-circle sketch, Sketch-TERMINALS.

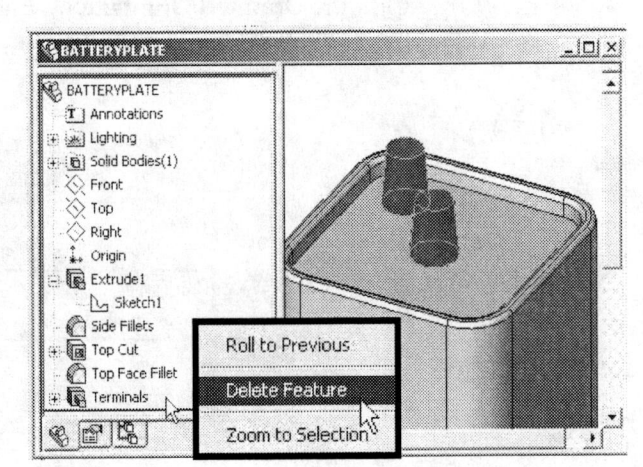

68) Create an Extruded Cut feature from the two circles. Click **Sketch-TERMINALS** from the FeatureManager. Click **Extruded-Cut**. Click **Through All** for the Depth.

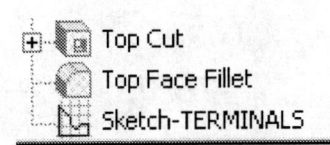

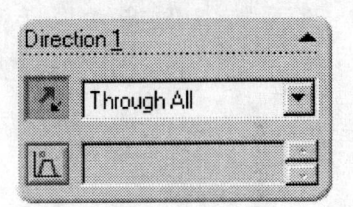

69) Create the cut holes. Click **OK**.

70) Rename **Cut-Extrude1** to **Holes**.

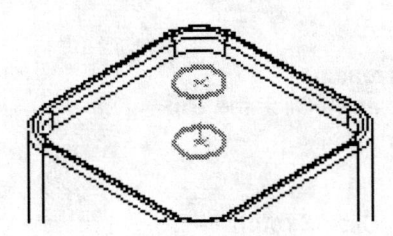

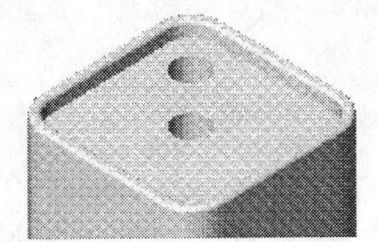

71) Save the BATTERYPLATE. Click **Save**.

Extrude and Revolve Features Engineering Design with SolidWorks

72) Edit the Base-Extrude feature. Right-click the **Base-Extrude** in the FeatureManager. Click **Edit Definition** from the Pop-up menu. Change the overall Depth to **.400**, **[10.16]**. Click the **Draft ON/OFF** button. Enter **1.00** in the Angle text box.

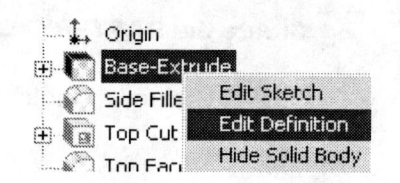

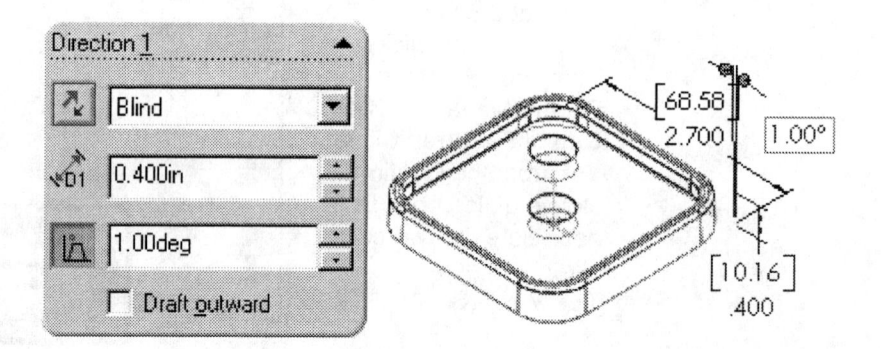

73) Display the modified Base feature. Click **OK** ✔.

74) Fit the model to the Graphics window. Press the **f** key.

75) Save the BATTERYPLATE. Click **Save** 🖫.

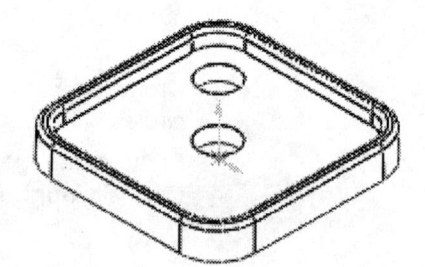

Create the BATTERYPLATE – Use the Extruded Boss Feature

The Holder is created with a circular Extruded Boss feature.

Create the Extruded Boss feature.
76) Select the Sketch plane. Click the **top face**.

77) Create the Sketch. Click **Sketch** ✏️. Offset the center circular edge. Click the **top circular edge** of the center Hole feature.

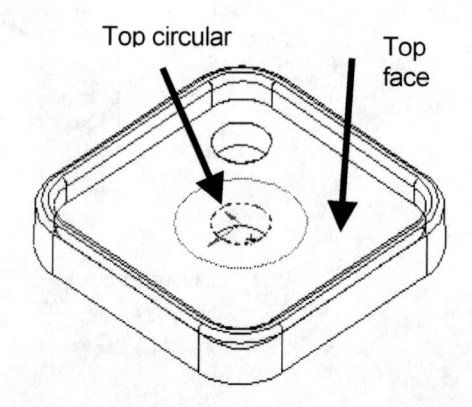

PAGE 4 - 26

Engineering Design with SolidWorks   **Extrude and Revolve Features**

78) Click **Offset Entities**. Enter **.300**, **[7.62]** for Distance. Click **OK**.

79) Create the second offset circle. Select the **offset circle**. Click **Offset Entities**. Enter **.100**, **[2.54]** for Distance. Click **OK** ✓.

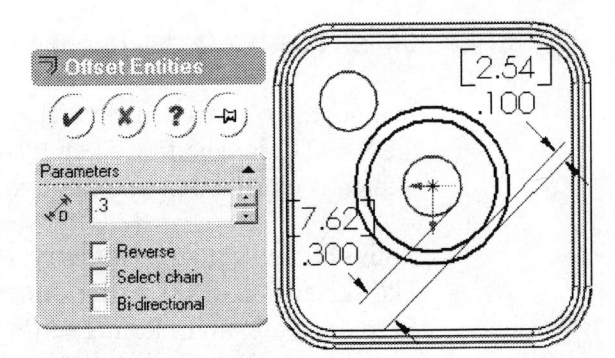

80) Extrude the Sketch. Click **Extruded Boss/Base**. Blind is the default Type option.

81) Enter **.400**, **[10.16]** for Depth. Click the **Draft ON/OFF** button. Enter **1** in the Angle text box.

82) Display the Extrude Boss feature. Click **OK** ✓.

83) Rename **Extrude1** to **Holder**.

84) Save the BATTERYPLATE. Click **Save**.

The outside face tapers inward and the inside face tapers outward when applying the Draft Angle to the two concentric circles.

Draft Angle displayed at 5 degrees

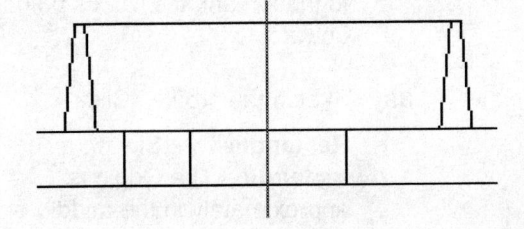

Create the BATTERYPLATE – Use the Extruded Boss Feature

The next two Extruded Boss features are used to connect the BATTERY to the SWITCH. The first Sketch is extruded in two directions. The second Sketch is extruded in one direction.

Both sketches utilize symmetry with the Origin and the Mirror Sketch Tool. The sketches utilize smaller dimensions than the current Grid Snap settings. Turn off the Snap to Points setting before you sketch the profiles.

Create the first Extruded Boss feature.
85) **Zoom** and **Rotate** the view to clearly display the inside right face. Click **Isometric**. Click **Shaded**.

Note: Press the arrow keys to rotate in 15-degree increments.

86) Select the Sketch plane. Click the **inside right face** of the Top Cut.

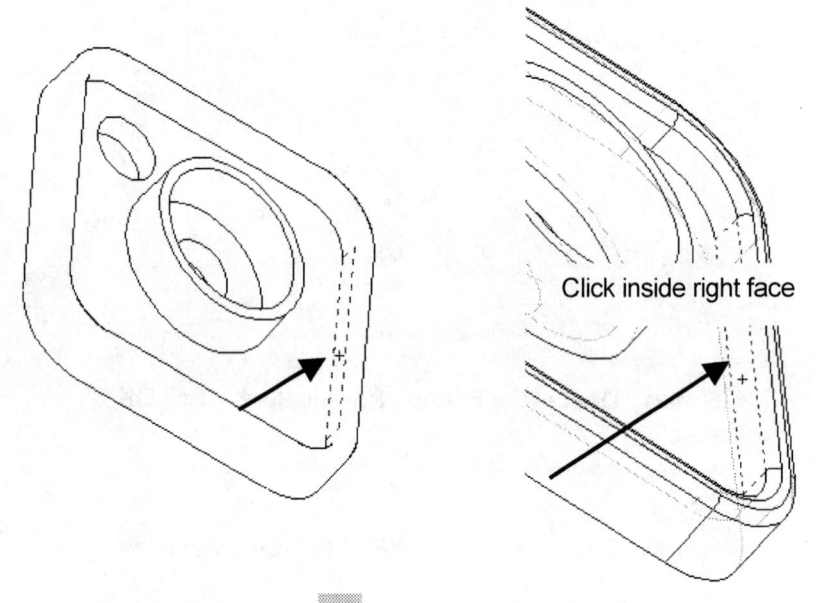

Click inside right face

87) Create the Sketch. Click **Sketch**.

88) Display the Right view. Click **Right**. The Z axis points to the left and the Y axis point upward.

Sketch a Rectangle

89) Sketch the profile. Click **Rectangle**. Sketch a **rectangle**. The Origin is approximately in the middle of the sketch.

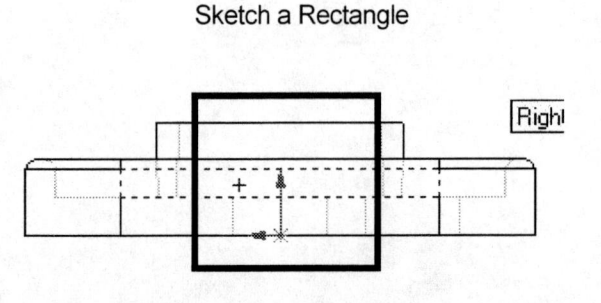

Engineering Design with SolidWorks   **Extrude and Revolve Features**

90) Add Geometric Relations. Click **Add Relations** from the Sketch toolbar. Click the **top edge** of the rectangle. Click the **top edge** of the BATTERYPLATE profile. Click the **Collinear** button from the Add Relations text box. Click **Close Dialog** from the Properties Manager.

91) Click the **bottom edge** of the rectangle. Click the **Origin** of the BATTERYPLATE profile. Click the **Midpoint** button from the Add Relations text box. Click **Close Dialog** from the Properties Manager. The sketch is symmetric about the Origin.

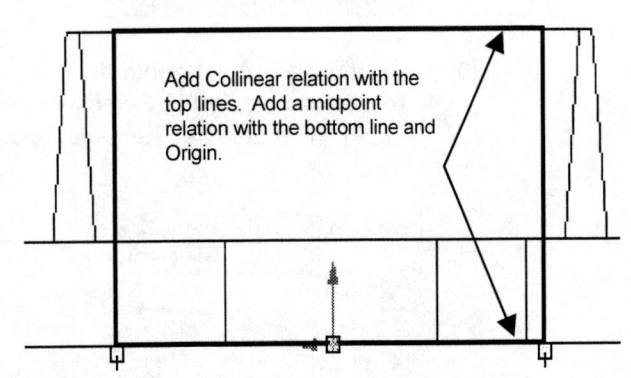

Add Collinear relation with the top lines. Add a midpoint relation with the bottom line and Origin.

Geometric relationships are captured as you sketch.

The mouse pointer icon displays the following relationships: Horizontal, vertical, coincident, midpoint, intersection, tangent and perpendicular.

Note: If Automatic Relations are not displayed, Click Tools from the Main menu. Click Options, General, Automatic Relations in the Sketch box. If Geometric Relations are not captured, utilize Add Relations from the Sketch toolbar. All BATTERYPLATE sketches are fully defined an displayed in black.

92) Dimension the Sketch. Click **Dimension**. Click the **bottom horizontal line**. Click a **position** below the Origin. Enter **1.000**, [**2.54**]. Click the **Green Check Mark**.

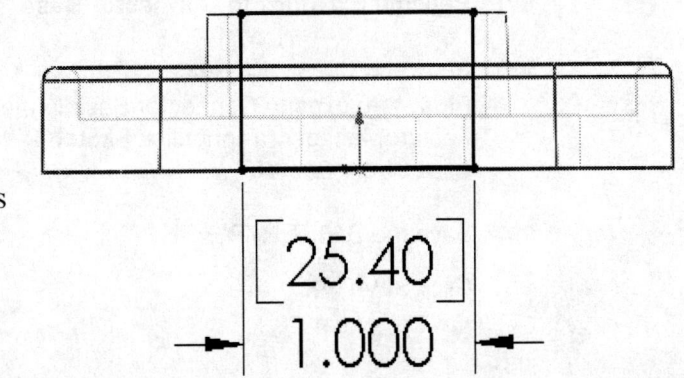

The black sketch is fully defined.

[25.40]
1.000

PAGE 4 - 29

Extrude and Revolve Features **Engineering Design with SolidWorks**

93) Display the Isometric view. Click **Isometric**.

94) Extrude the Sketch. Click **Extruded Boss/Base**. Create the first depth direction, Direction 1. Blind is the Type option. Enter **.400**, **[10.16]** for Depth. Click the **Draft ON/OFF** button. Enter **1.00** for Draft Angle. The sketch is extruded towards the Holes.

95) Create the second depth direction. Click the **Direction 2** check box. Select **Up to Surface** for Type. Select the outside **right face** for the second extruded depth. The Selected Items text box displays Face<1>.

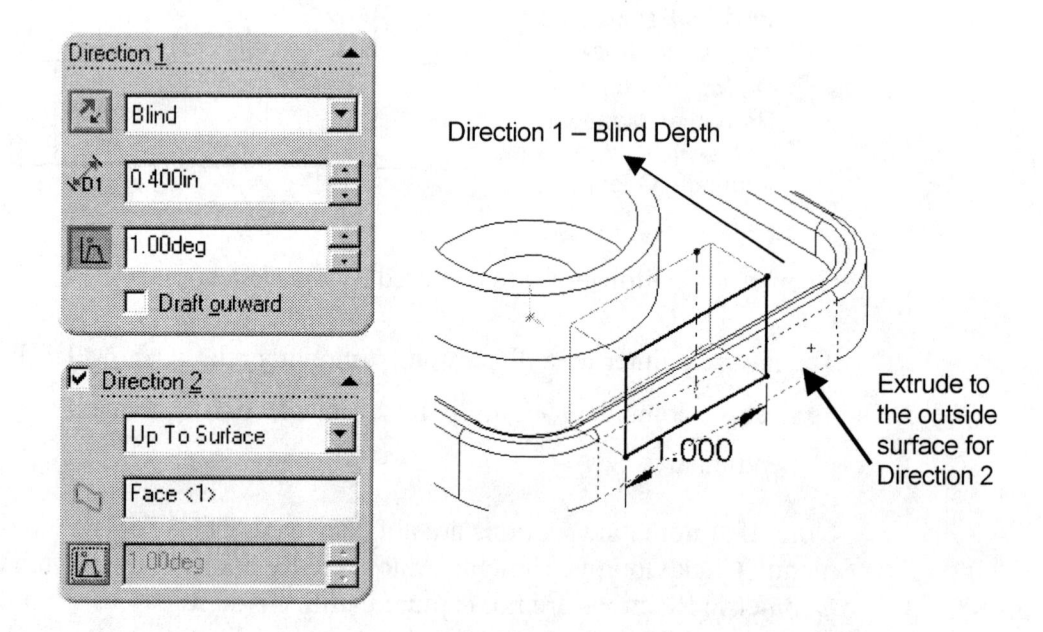

Direction 1 – Blind Depth

Extrude to the outside surface for Direction 2

96) Display the Extrude2 feature. Click **OK**.

97) Rename **Extrude2** to **Connector Base**.

98) Show the Connector Base sketch. Click **Plus** to expand Connector Base in the FeatureManager. Right-click **Sketch4**. Click **Show Sketch**.

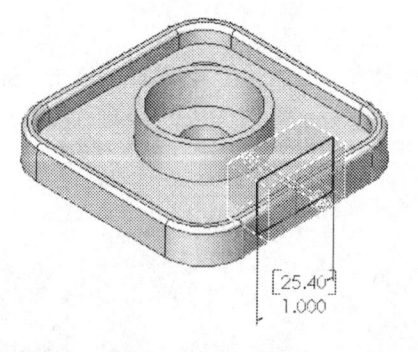

99) Save the BATTERYPLATE. Click **Save**.

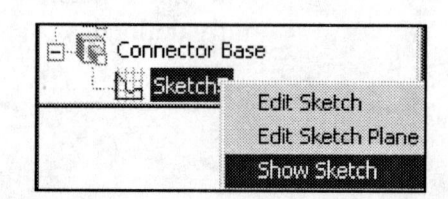

PAGE 4 - 30

Engineering Design with SolidWorks Extrude and Revolve Features

Create the second Extruded Boss feature.
100) Select the Sketch plane. Click the **top narrow face** of the Extruded Boss feature.

101) Create the Sketch. Click **Sketch**.
Display the Top view. Click **Top**.

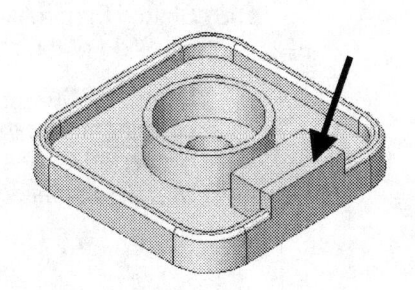

Convert a line segment from the Connector Base sketch to the current sketch plane.
102) Click the **vertical line** of the Connector Base Sketch. Click **Convert Entities** from the Sketch Tools toolbar.
Click **OK** to select single segment.

103) Sketch the centerline. Click **Centerline**. Sketch a horizontal centerline with the first point coincident to the **Origin**. The second point is coincident with the **midpoint** of the **rightmost vertical line**.

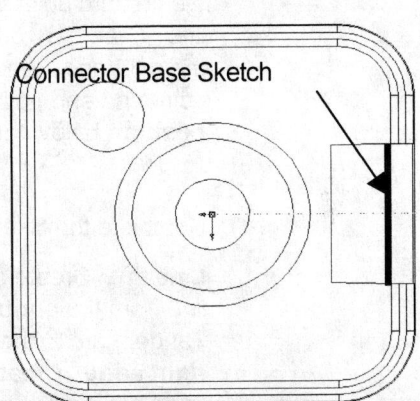

Connector Base Sketch

104) Create the mirrored centerline. Click **Sketch Mirror**. The centerline displays two parallel mirror marks.

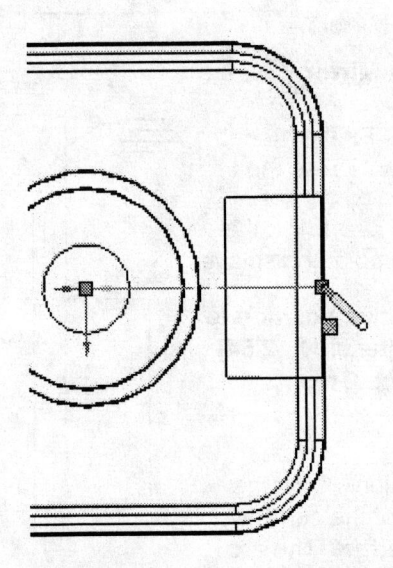

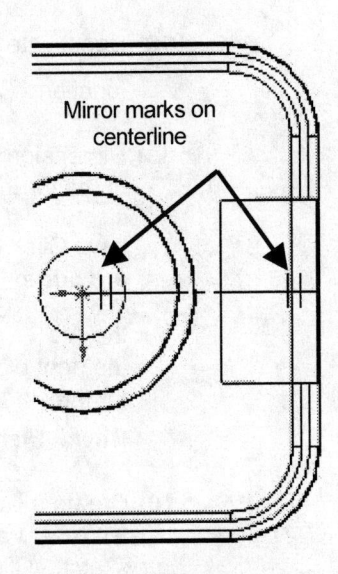

Mirror marks on centerline

PAGE 4 - 31

Extrude and Revolve Features   **Engineering Design with SolidWorks**

105) Sketch the profile. Create the Sketch on one side of the mirror centerline. Click **Line**. Create a **horizontal line** coincident with the endpoint of the converted Connector Base sketch. The line is automatically mirrored.

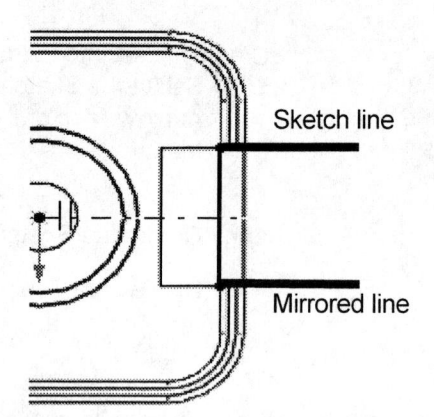

106) Create a Tangent Arc. Click **Tangent Arc**. Create the first arc point. Click the **endpoint** of the horizontal line. Create a 180° arc. Drag the **mouse pointer** to the right and then downward until the start point, center point and end point are vertically aligned. Click a **position** below the centerpoint.

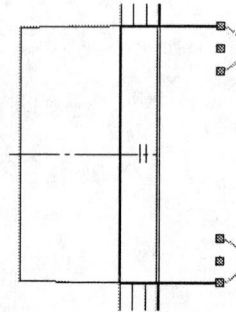

107) Complete the Sketch. Click **Line**. Create a horizontal line from the **endpoint** of the Tangent arc to the inside top **right edge**. Create a vertical line from the **endpoint** of the horizontal edge to the mirror **centerline**. The vertical line is collinear with the inside top right edge.

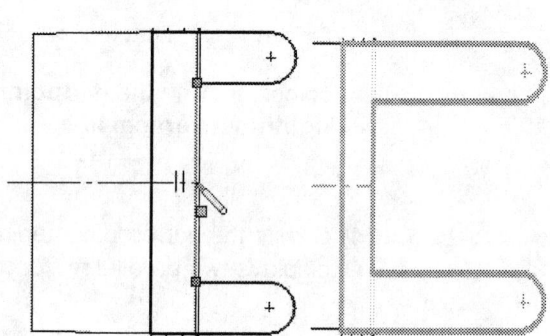

108) Deactivate the Sketch Mirror function. Click **Sketch Mirror**.

109) Dimension the Sketch. Fit the model to the Graphics window. Press the **f** key. Click **Dimension**. Click the **arc edge**. The mouse pointer displays the Arc icon. Click a **position** to the right of the arc. Enter **.100**, **[2.54]** for the Radius. Click the **Green Check Mark**.

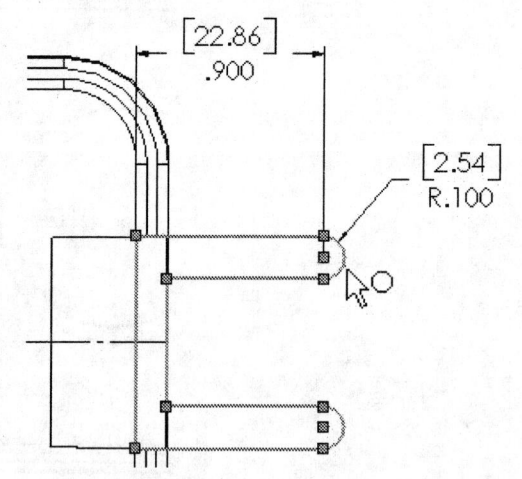

110) Create a linear dimension. Click the left most **vertical line** of the current sketch. Click the **arc edge**. The arc edge displays red. Click a position above the top horizontal line of the sketch.

PAGE 4 - 32

Note: Click the arc edge, not the arc center point to create a max. dimension. The linear dimension uses the arc center point as a reference. Modify the Properties of the dimension. The Maximize option references the outside tangent edge of the arc.

111) Right-click on the **linear dimension text**. Click **Properties** from the Pop-up menu. Click the **Max** button from the First arc condition option. Enter **1.000**, [**25.4**] in the Value list box. Display the dimension. Click **OK**. The black Sketch is fully defined.

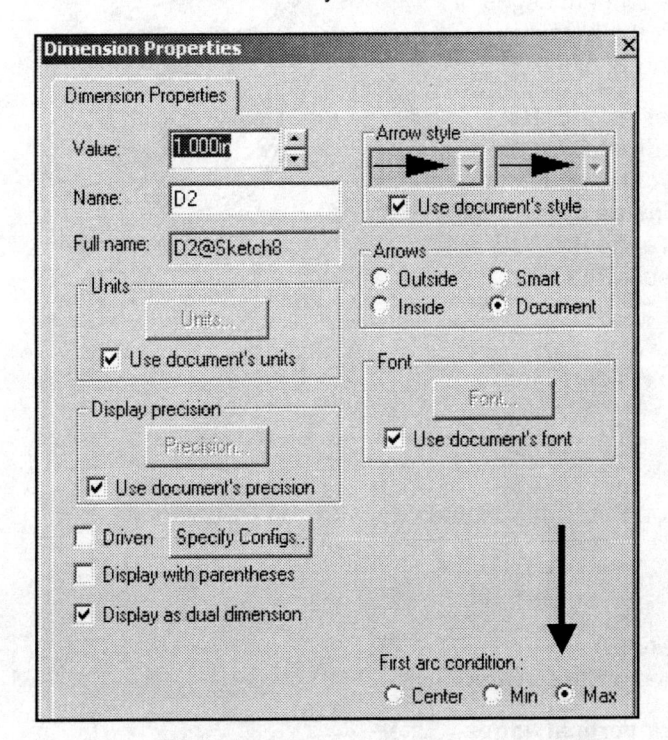

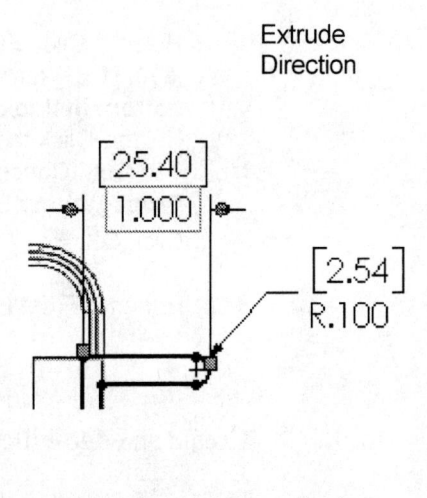

112) Display the Isometric view. Click **Isometric**.

113) Extrude the Sketch. Click **Extruded Boss/Base**. Blind is the default Type option. Enter **.100**, [**2.54**] for Depth. Click the **Reverse Direction** button. Display the Boss-Extrude feature. Click **OK**.

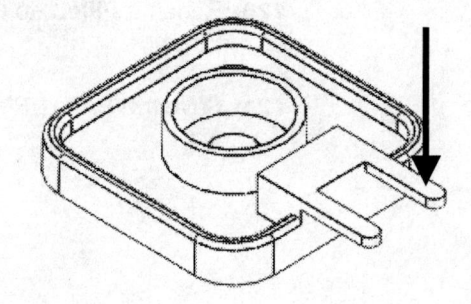

114) Rename **Boss-Extrude3** to **ConnectorSwitch**.

115) Save the BATTERYPLATE. Click **Save**.

116) Hide the ConnectorBase Sketch. **Expand** the ConnectorBase Feature in the FeatureManager. Right-click **Sketch4**. Click **Hide**.

Create the BATTERYPLATE - Edge Fillets

Fillet features are used to smooth rough edges. Two edge Fillets are required to produce the desired outcome.

Create a Fillet feature.
117) Create a fillet on the inside and outside edge of the Holder. Create a fillet on all inside tangent edges of the Top-Cut. Click the **bottom outside circular edge** of the Holder. Click **Fillet**. Enter .050, [1.27] for Radius. Click the **bottom inside circular edge** of the Holder. Click the **inside edge** of the Top Cut. Check **Tangent Propagation**. Display the Fillet. Click **OK**.

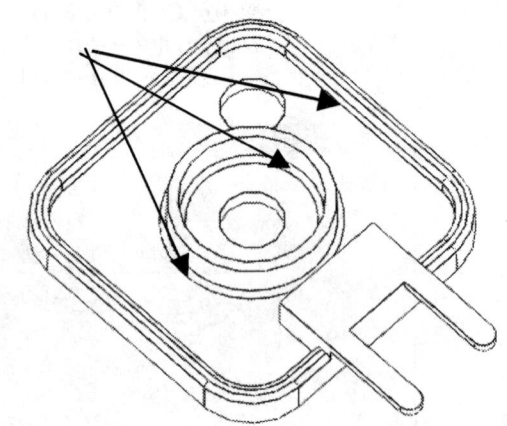

118) Rename **Fillet1** to **HolderFillet**.

Create an edge Fillet on the four vertical edges of the Connector.

Create the edge Fillet feature.
119) **Zoom in** on ConnectorBase. Click **Fillet**. Click the **four vertical edges**. Enter .050, [1.27] for Radius. Check **Tangent Propagation**. Click **OK**.

120) Rename **Fillet2** to **Connect Base Fillet Edge**.

121) Save the BATTERYPLATE. Click **Save**.

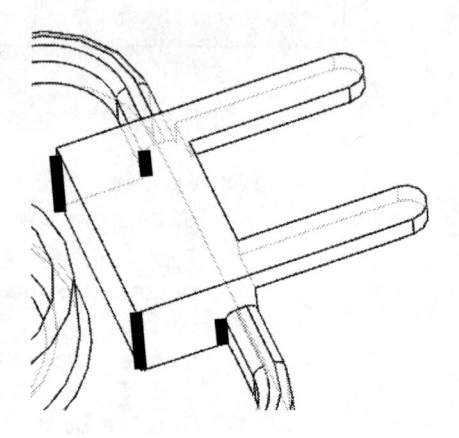

Engineering Design with SolidWorks — Extrude and Revolve Features

The FeatureManager displays all successful feature name icons. The BATTERYPLATE is completed.

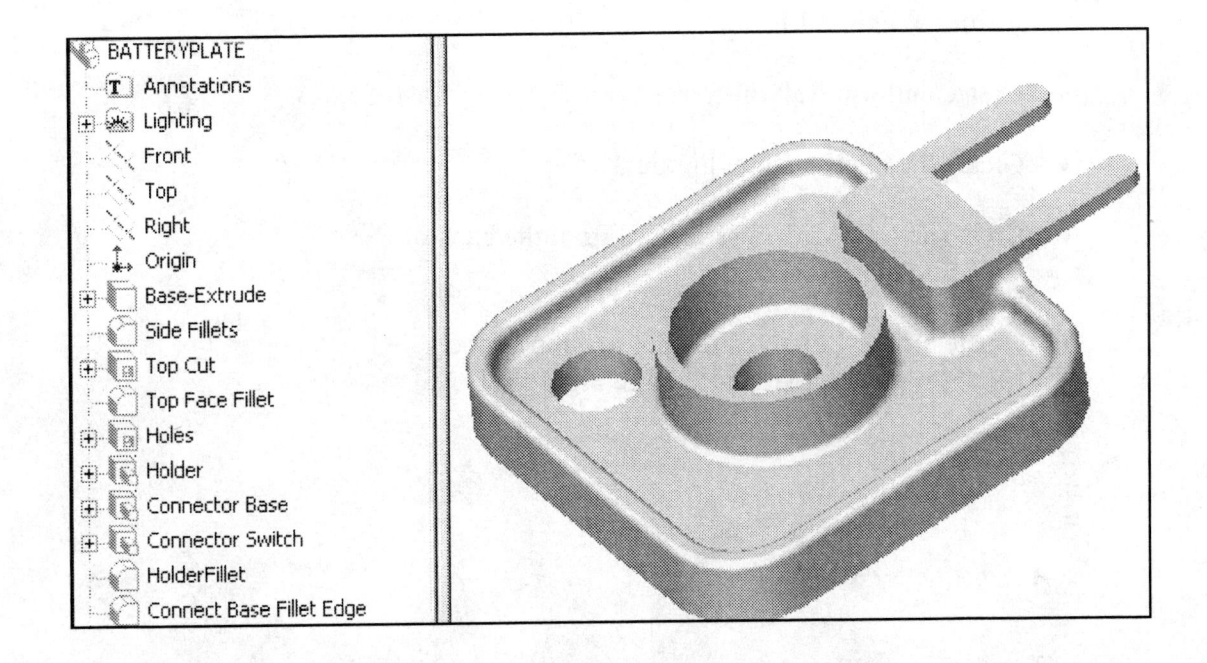

LENS

The LENS is a purchase part. Obtain dimensional information on the LENS assembly. Review the size, material and construction. Determine the key features of the LENS.

The Base feature for the LENS is a solid Revolved feature. A solid Revolved feature adds material.

The LENSANDBULB assembly is comprised of the LENS and BULB. The Revolved Base feature is the foundation for the LENS.

A Revolved feature is geometry created by rotating a sketched profile around a centerline. Close the Sketch profile for a solid Revolved feature. Do not cross the centerline.

Extrude and Revolve Features

Engineering Design with SolidWorks

LENS Feature Overview

- Create the LENS. Use the solid Revolved Base feature, Figure 4.14.

- Create uniform wall thickness.

- Create the Shell feature, Figure 4.15.

- Create an Extruded-Boss feature from the back of the LENS, Figure 4.16.

- Create a Thin-Revolved feature to connect the LENS to the BATTERYPLATE, Figure 4.17.

Figure 4.14

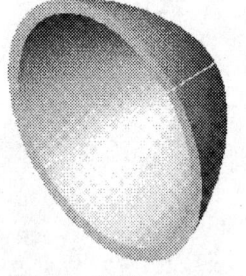

Figure 4.15

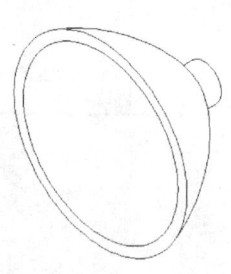

Figure 4.16

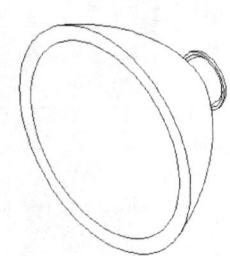

Figure 4-17

Create a Counterbore Hole feature with the HoleWizard, Figure 4.18.

The BULB is located inside the Counterbore Hole.

Create the front LensFlange feature.

Add a transparent LensShield feature, Figure 4.19.

Counter bore

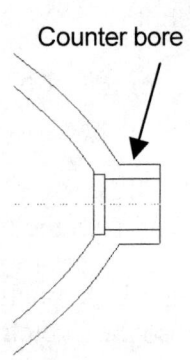

Figure 4.18

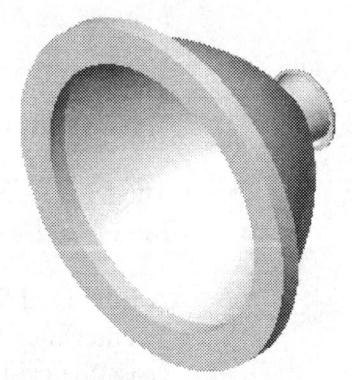

Figure 4.19

Create the LENS

Create the LENS with a Revolved Base feature. The solid Revolved Base feature requires a sketched profile and a centerline. The profile is located on the Right plane with the centerline collinear to the Top plane.

The profile lines reference the Top and Front planes. The curve of the LENS is created with a 3-point arc.

Create the LENS.

122) Click **New**. Click the **MY-TEMPLATES** tab. Click **PART-IN-ANSI**, [**PART-MM-ISO**] from the Template dialog box. Click **OK**.

123) Save the empty part. Click **Save**. Select **ENGDESIGN-W-SOLIIDWORKS\PROJECTS** for Save in file folder.

124) Enter **LENS** for file name.

125) Enter **LENS WITH SHIELD** for Description. Click the **Save** button.

126) View the planes. Right click on the **Front** plane in the FeatureManager. Click **Show**. Right click on the **Top** plane in the FeatureManager. Click **Show**.

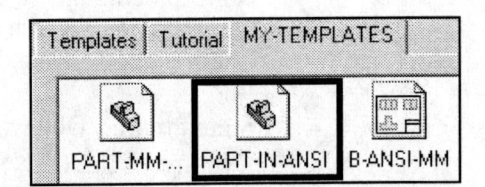

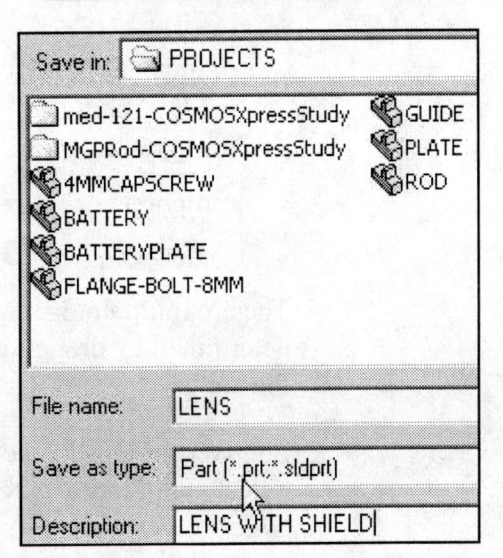

127) Select the Sketch plane. Click the **Right** plane from the FeatureManager.

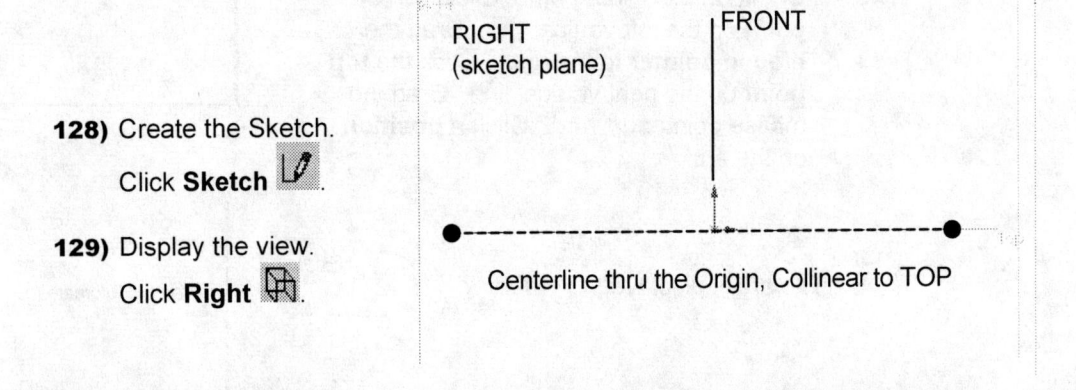

128) Create the Sketch. Click **Sketch**.

129) Display the view. Click **Right**.

RIGHT (sketch plane)
FRONT
Centerline thru the Origin, Collinear to TOP

Extrude and Revolve Features **Engineering Design with SolidWorks**

130) Sketch the centerline. Click **Centerline**. Sketch a horizontal **centerline** collinear to the Top plane, through the Origin.

131) Sketch the profile. Create three lines.

Click **Line**. Create the first line. Sketch a **vertical line** collinear to the Front plane coincident with the Origin. Create the second line. Sketch a **horizontal line** coincident with the Top plane. Create the third line. Sketch a **vertical line** approximately 1/3 the length of the first line.

Create an arc. Determine the curvature of the LENS.

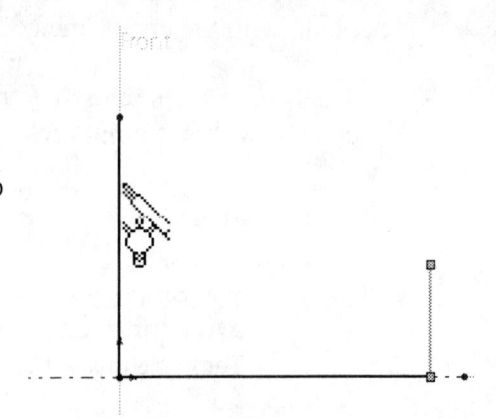

A 3 POINT Arc requires a:

- Start point.
- End point.
- Center point.

The arc midpoint is aligned with the center point. The arc position is determined by dragging the arc midpoint or center point above or below the arc.

SolidWorks On-line help contains an animation file to create a 3-point arc. Select Help, Index, Arc, 3Point. Run the animation. Select the AVI icon **Show Me**.

132) Create a 3 Point Arc. Click **3Pt Arc**. Create the arc start point. Click the **top point** on the left vertical line. Drag the **mouse pointer** to the right. Click the **top point** on the right vertical line. Drag the mouse pointer upward. Click a **position** on the arc.

The arc is displayed in green.

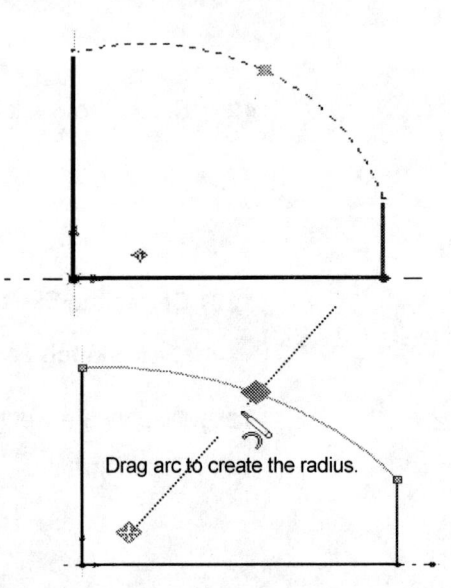

Drag arc to create the radius.

Engineering Design with SolidWorks **Extrude and Revolve Features**

133) Add geometric relationships. The arc is currently selected. Right-click **Select**. The arc is no longer selected. Create an Equal relationship. Hold the **Ctrl** key down. Click the **left vertical line**. Click the **horizontal line**. Click the **Equal** button. Release the **Ctrl** key.

134) Add dimensions. Click **Dimension**. Click the **vertical left line**. Click a **position** to the left of the profile. Enter **2.000**, [50.8]. Click the **Green Check Mark**.

135) Click the **vertical right line**. Click a **position** to the right of the profile. Enter **.400**, [10.16]. Click the **Green Check Mark**.

136) Click the **arc**. Click a **position** to the right of the profile. Enter **4.000**, [101.6]. Click the **Green Check Mark**. The sketch is fully defined. The sketch is displayed in black.

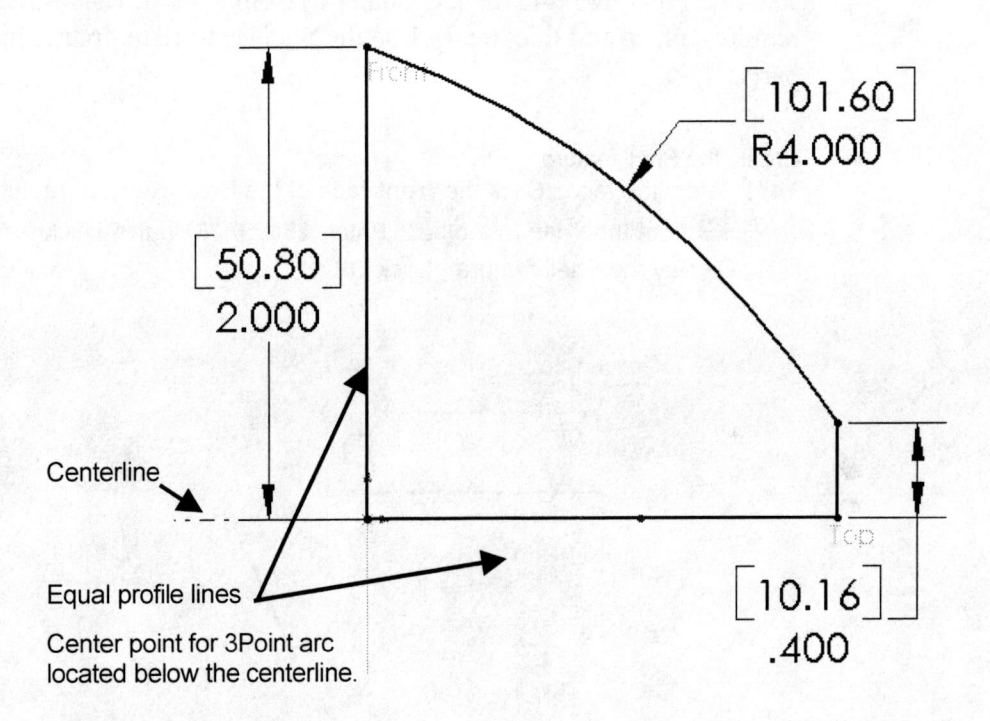

[101.60]
R4.000

[50.80]
2.000

Centerline

Equal profile lines

Center point for 3Point arc located below the centerline.

[10.16]
.400

137) Revolve the Sketch. Click **Revolve Boss/Base** from the Feature toolbar. The Revolve Feature dialog box is displayed. Accept the default option values. Click **OK**.

138) Rename **Revolve1** to **BaseRevolve**.

139) Save the LENS. Click **Save**.

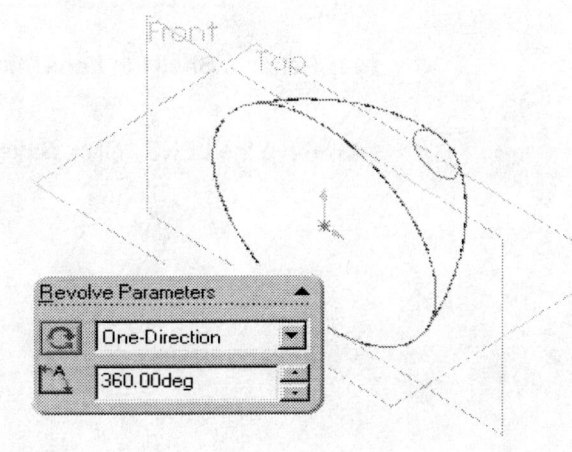

Revolve Parameters
One-Direction
360.00deg

PAGE 4 - 39

Revolve features contain an axis of revolution. The axis is critical to align other features.

140) Display the axis of revolution. Click **View** from the Main menu. Check **Temporary Axis**. A check mark is displayed next to the option. Hide the Temporary axis. Click **View**. Uncheck **Temporary Axis** to remove the check mark.

Solid Revolve features must contain a closed profile. Each revolved profile requires an individual sketched centerline.

Create the LENS – Use the Shell Feature

The Shell feature removes face material from a solid. The Shell feature requires a face and thickness. Use the Shell feature to create thin-walled parts.

Create the Shell feature.
141) Select the face. Click the **front face** of the Base-Revolve feature. Click **Shell** from the Feature toolbar. Enter **.250**, **[6.35]** in the Thickness text box. Display the Shell feature. Click **OK**.

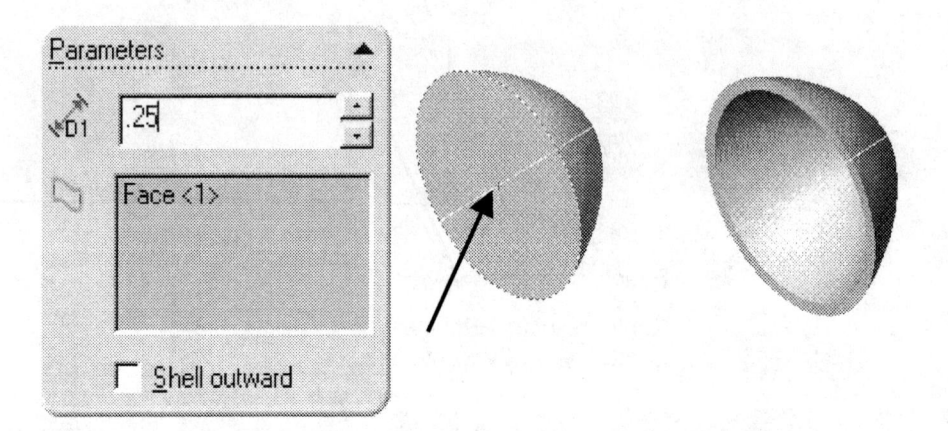

142) Rename **Shell1** to **LensShell**.

143) Save the LENS. Click **Save**.

Create the LENS – Use the Extruded Boss Feature

Create the LensNeck. Use the Extruded-Boss feature. The LensNeck houses the BULB base and is connected to the BATTERYPLATE. The feature extracts the back circular edge from the Base-Revolve feature.

Back edge

Create the Extruded Boss feature.
144) Rotate the Lens. Press the **left arrow** key approximately 5 times to display the back face. Select the Sketch plane. Click the **back face**.

145) Create the profile. Click **Sketch**. Extract the back face to the Sketch plane. Click **Convert Entities** from the Sketch Tools toolbar.

146) Extrude the Sketch. Click **Extruded Boss/Base**. Enter **.400**, **[10.16]** for Depth. Display the Boss-Extrude1 feature. Click **OK**.

147) Rename **Extrude1** to **LensNeck**.

148) Save the LENS. Click **Save**.

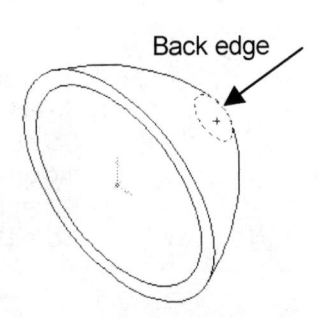

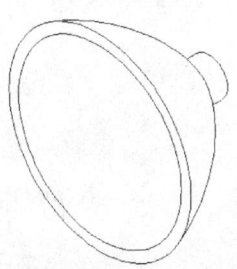

Create the LENS – Use the Hole Wizard Counterbore Hole Feature

The LENS requires a Counterbore Hole feature. Use the HoleWizard. The HoleWizard assists in creating complex and simple Hole features.

Specify the user parameters for the custom Counterbore Hole. Dimensions for the Counterbore Hole are provided both in inches and millimeters.

Create the Counterbore Hole.
149) Select the Sketch plane. Click **Front**. Click the small **inside back face** of the LensShell feature. Do not select the Origin.

150) Click **HoleWizard** from the Features toolbar. The Hole Definition dialog box is displayed. Click the **Counterbore** tab.

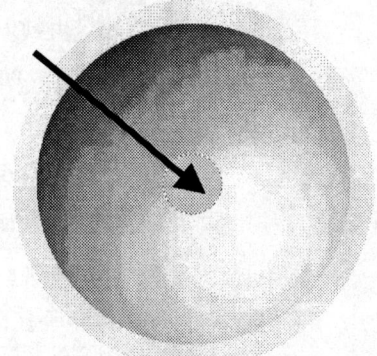

Note: For a metric hole, skip the next step.

For Inch Cbore Hole:

151) Select **Ansi Inch** for Standard. Enter **Hex Bolt** from the drop down list for Screw type. Select ½ from the drop down list for Size. Click **Through All** from the drop down list for End Condition & Depth. Accept the Hole Fit and Diameter value. Click the **C-Bore Diameter** value. Enter **.600**. Click the **C-Bore Depth** value. Enter **.200**.

Property	Parameter 1	Parameter 2
Description	CBORE for 1/2 Hex Head Bolt	
Standard	Ansi Inch	
Screw type	Hex Bolt	
Size	1/2	
End Condition & Depth	Through All	0.394in
Selected Item & Offset		0.000in
Hole Fit & Diameter	Normal	0.5312in
Angle at Bottom	118deg	
C'Bore Diameter & Depth	.6	.2

Note: For an inch hole, skip the next step.

For Millimeter Cbore Hole:

152) Select **Ansi Metric** for Standard. Enter **Hex Bolt** from the drop down list for Screw type. Select **M5** from the drop down list for Size. Click **Through All** from the drop down list for End Condition & Depth. Click the Hole Diameter value. Enter **13.5**. Click the **C-Bore Diameter** value. Enter **15.24**. Click the **C-Bore Depth** value. Enter **5**.

Property	Parameter 1	Parameter 2
Description	CBORE for M5 Hex Head Bolt	
Standard	Ansi Metric	
Screw type	Hex Bolt	
Size	M5	
End Condition & Depth	Through All	16.510mm
Selected Item & Offset		16.510mm
Hole Fit & Diameter	Normal	13.500mm
Angle at Bottom	118deg	
C'Bore Diameter & Depth	15.240mm	5.000mm

Engineering Design with SolidWorks — Extrude and Revolve Features

153) Add the new hole type to your favorites list. Click the **Add** button. Enter **CBORE FOR BULB**. Click **OK**.

154) Click **Next** from the Hole Definition dialog box. The Hole Placement dialog box is displayed. Do not select Finish at this time. Position the hole coincident with the Origin.

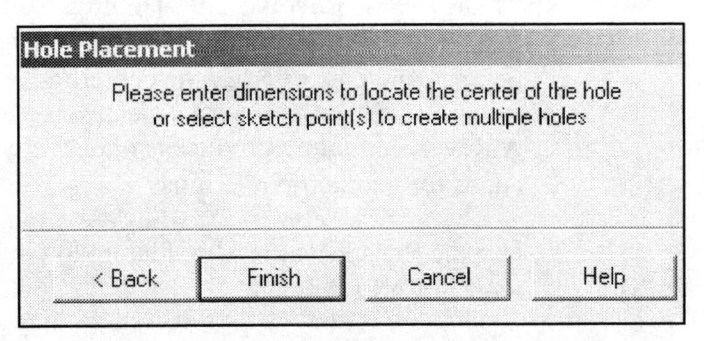

155) Click **Add Relations** from the Sketch toolbar. Click the **center point** of the Counterbore hole.

Click the **Origin**. Click **Coincident**. Complete the hole. Click **Finish** from the Hole Placement dialog box.

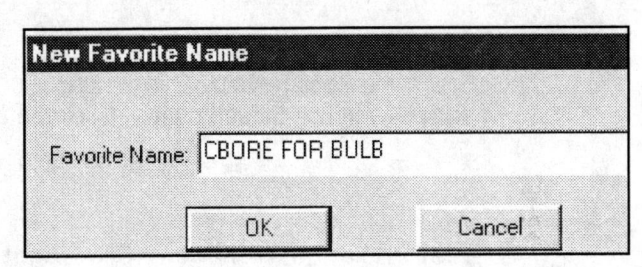

156) Expand the Hole. Click the **Plus Sign** to the left of the CBORE feature. Note: Sketch4 and Sketch5 created the CBORE feature.

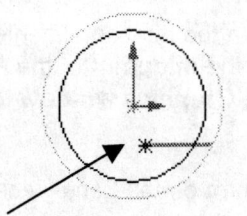

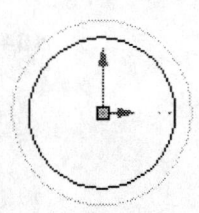

157) Display the Section view of the BulbHole through the Right plane. Click the **Right** plane from the FeatureManager. Click **View** from the Main menu. Click the **Display**, **SectionView**. Click the **Flip the Side to View** check box. Click **OK**.

158) Display the Full view. Click **View**, **Display**, **SectionView**.

159) Display the Temporary Axis. Click **View**. Check **Temporary Axis**. Click **Isometric**.

160) Rename **CBORE** to **BulbHole**.

161) Save the LENS. Click **Save**.

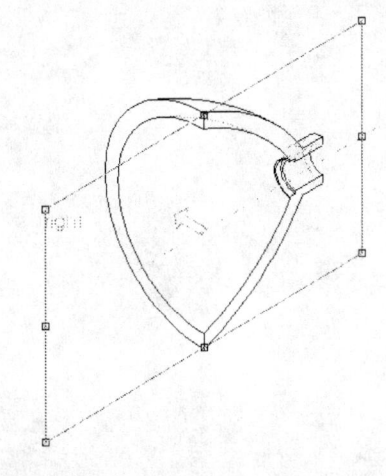

PAGE 4 - 43

Create the LENS – Use the Boss Revolve Thin Feature

Create a Boss Revolve Thin feature. Sketch a centerline. Sketch a profile. Rotate an open sketched profile around a centerline. The sketch profile must be open and cannot cross the centerline.

A Revolved feature produces silhouette edges in 2D views. Utilize silhouette edges for geometric relations.

Use the Boss Revolve Thin feature to connect the LENS to the BATTERYPLATE in the FLASHLIGHT.

Create the Boss Revolve Thin feature.
162) Select the Sketch plane. Click the **Right** plane in the FeatureManager. Create the Sketch. Click **Sketch**.

163) Display the Right view. Click **Right**.

164) Sketch the horizontal centerline. Click **Centerline**. Click the **Origin**. Click the **midpoint** of the right most vertical line of the Lens Neck. Double-click in the **Graphics window** to end the centerline.

165) Zoom in on the Lens Neck. Sketch the profile. Click **Centerpoint Arc**. Create the center point. Click the **top horizontal silhouette edge** of the Lens Neck. Create the start arc point. Click the **top right corner** of the LensNeck. Create the end arc point. Drag the **mouse pointer** counterclockwise to the left. Click a **position** above the centerpoint.

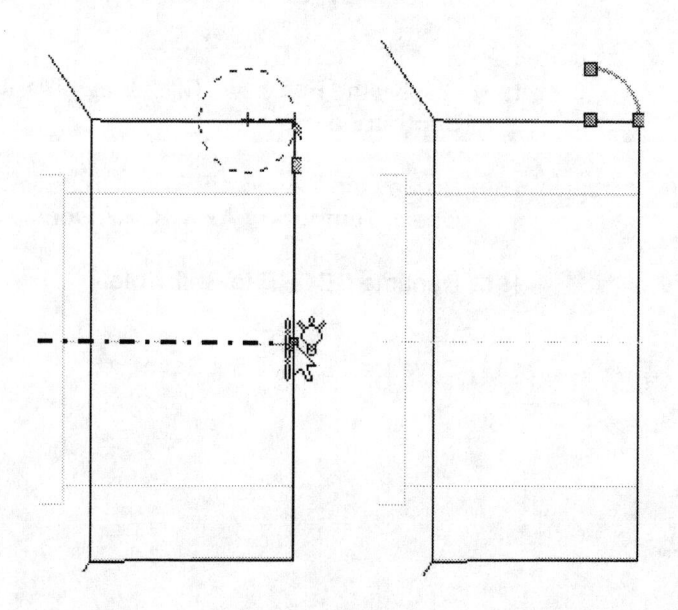

166) Add a dimension. Click **Dimension**. Click the arc. Click a **position** to the right of the profile. Enter **.100**, **[2.54]**. Click the **Green Check Mark**.

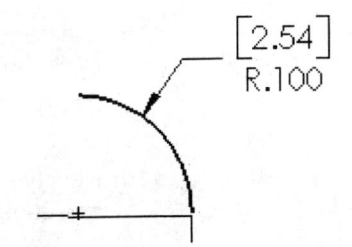

The sketched center point arc requires three geometric relationships: coincident, intersection and vertical. The three geometric relationships insure that the 90 degree center point of the arc is coincident with the horizontal silhouette edges of the Base-Revolve feature.

167) Add geometric relations. Create a coincident relation. Click **Select**. Hold down the **Ctrl** key. Click the **arc center point**. Click the **top horizontal line** (silhouette edge) of the LensNeck feature. Click the **Coincident** button. Release the **Ctrl** key. Click **Apply**.

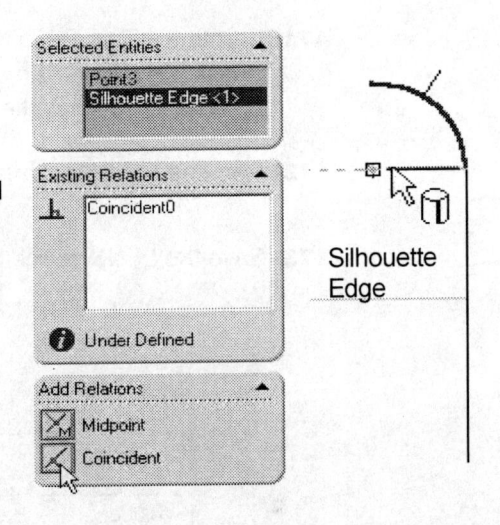

168) Create an intersection relation. Hold the **Ctrl** key down. Click the **arc start point**. Click the **rightmost vertical line** of the LensNeck feature. Click the **top horizontal line** (silhouette edge) of the LensNeck feature. Click the **Intersection** button. Release the **Ctrl** key. Click **Apply**.

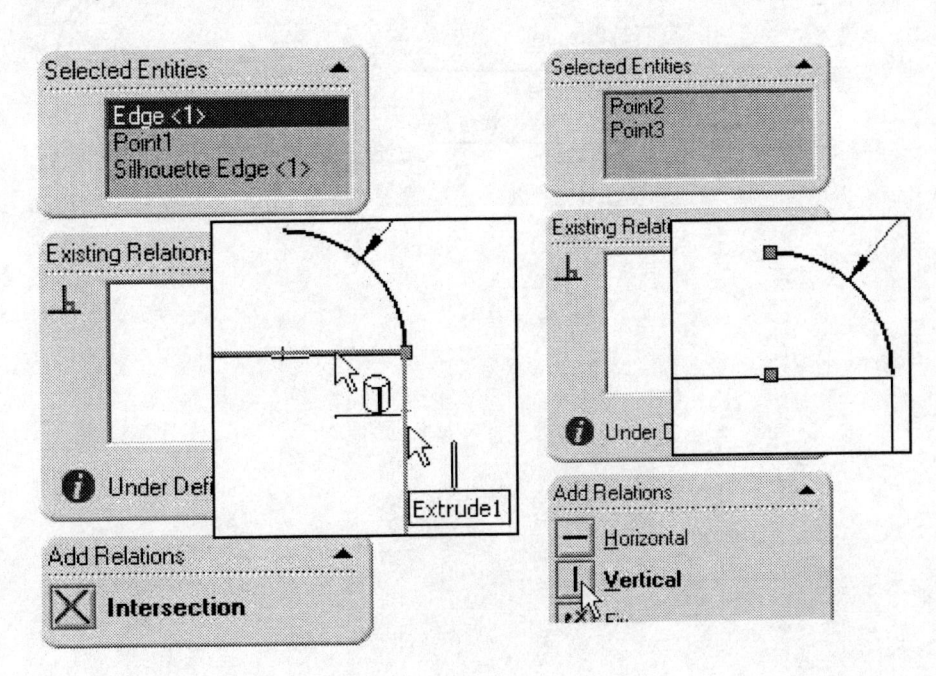

Extrude and Revolve Features   **Engineering Design with SolidWorks**

169) Create a vertical relation. Hold the **Ctrl** key down. Click the **arc center point**. Click the **arc end point**. Click the **Vertical** button. Release the **Ctrl** key. Click **Apply** ✔. The sketch is fully defined and is displayed in black.

170) Revolve the Sketch. Click **Revolve Boss/Base**. Click **No** to the warning message. The Thin Feature text box is displayed.

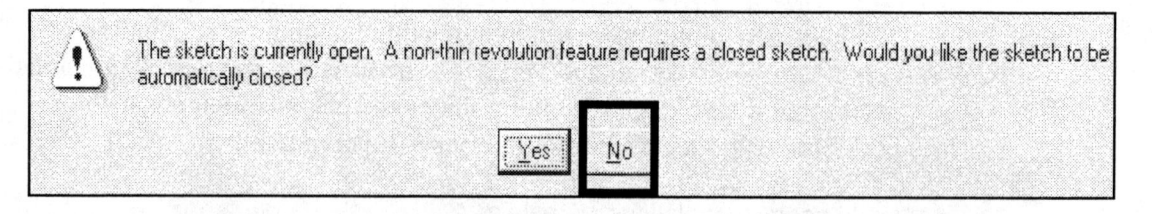

> ⚠ The sketch is currently open. A non-thin revolution feature requires a closed sketch. Would you like the sketch to be automatically closed?
>
> [Yes] [**No**]

171) Create the Thin-Revolved feature on both sides of the Sketch. Select **Mid-Plane** from the Thin Feature Type list box. Enter **.050**, **[1.27]** for Direction1 Thickness. Display the Revolve-Thin1 feature. Click **OK** ✔.

172) Rename **Revolve-Thin1** to **LensConnector**.

173) Save the LENS. Click **Save**.

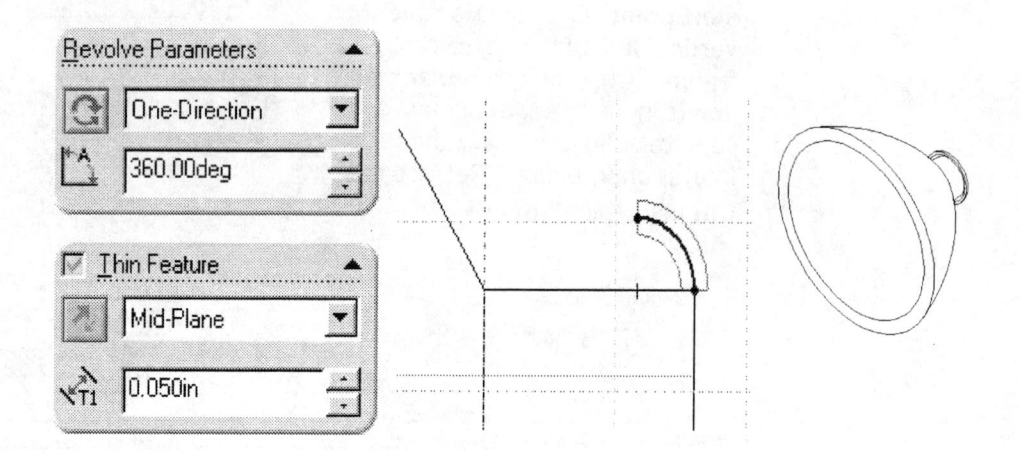

PAGE 4 - 46

Create the LENS – Use the Extruded Boss Feature

Use the Extruded-Boss feature to create the front LensCover. The feature extracts the front outside circular edge from the Base-Revolve feature. The front LensCover is a key feature for designing the mating part. The mating part is the LENSCAP.

Create the Extruded Boss feature.
174) Fit the model to the Graphics window. Press the **f** key. Click **Isometric**.

175) Select the Sketch plane. Click the **front circular face**.

176) Create the Sketch. Click **Sketch**.

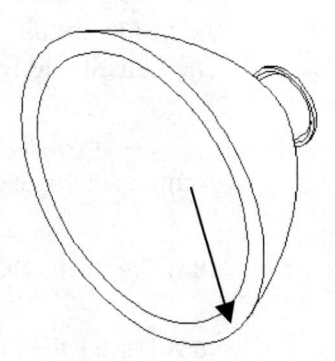

177) Display the Front view. Click **Front**.

178) Click the **outside circular edge**. Click **Offset Entities**. Click the **Bi-directional** check box. Enter **.250**, **[6.35]**. Click **OK**.

179) Extrude the Sketch. Click **Extruded Boss/Base**. Enter **.250**, **[6.35]** for Depth. Display the Boss-Extrude feature. Click **OK**.

180) Verify the position of the Boss Extrude. Click the **Top** view.

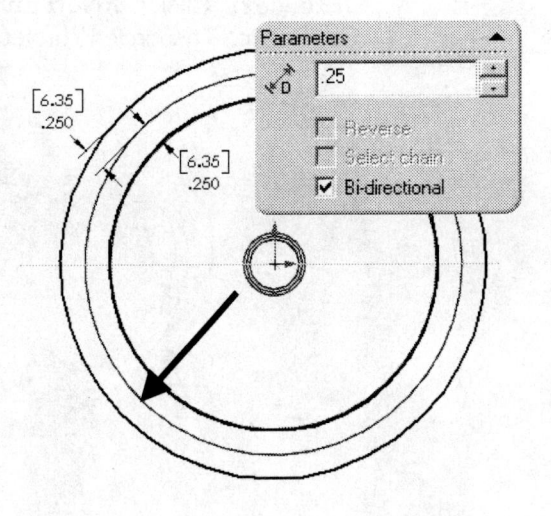

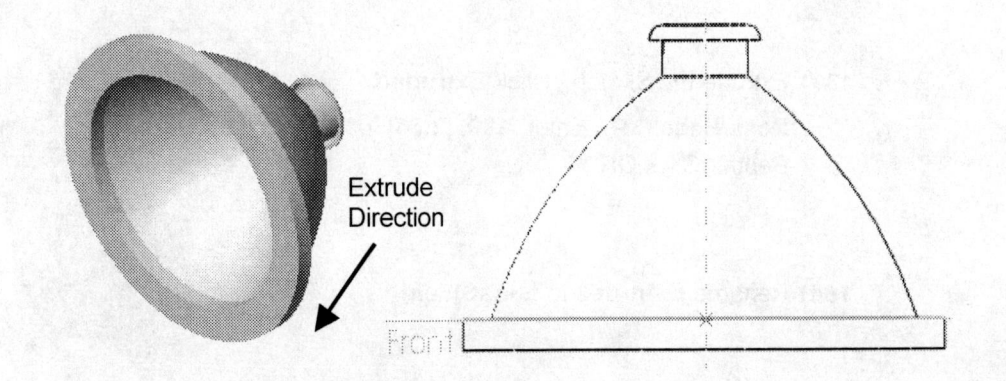

Extrude Direction

Extrude and Revolve Features   Engineering Design with SolidWorks

181) Rename **Extrude2** to **LensCover**.

182) Save the LENS. Click **Save**.

Create the LENS – Use the Extruded Boss Feature

Use an Extruded Boss feature to create the LensShield. The feature extracts the inside circular edge of the LensCover and places it on the Front plane. The LensShield feature is a transparent feature.

Create the Extruded Boss feature.
183) Select the Sketch plane. Click the **Front** plane from the FeatureManager.

184) Create the Sketch. Click **Sketch**.

185) Display the Front view. Click **Front**.

186) Sketch the profile. Click the **front inner circular edge** of the LensShield (Extrude2). Click **Convert Entities** from the Sketch Tool toolbar. Click **Isometric**. The circle is projected onto the Front Plane.

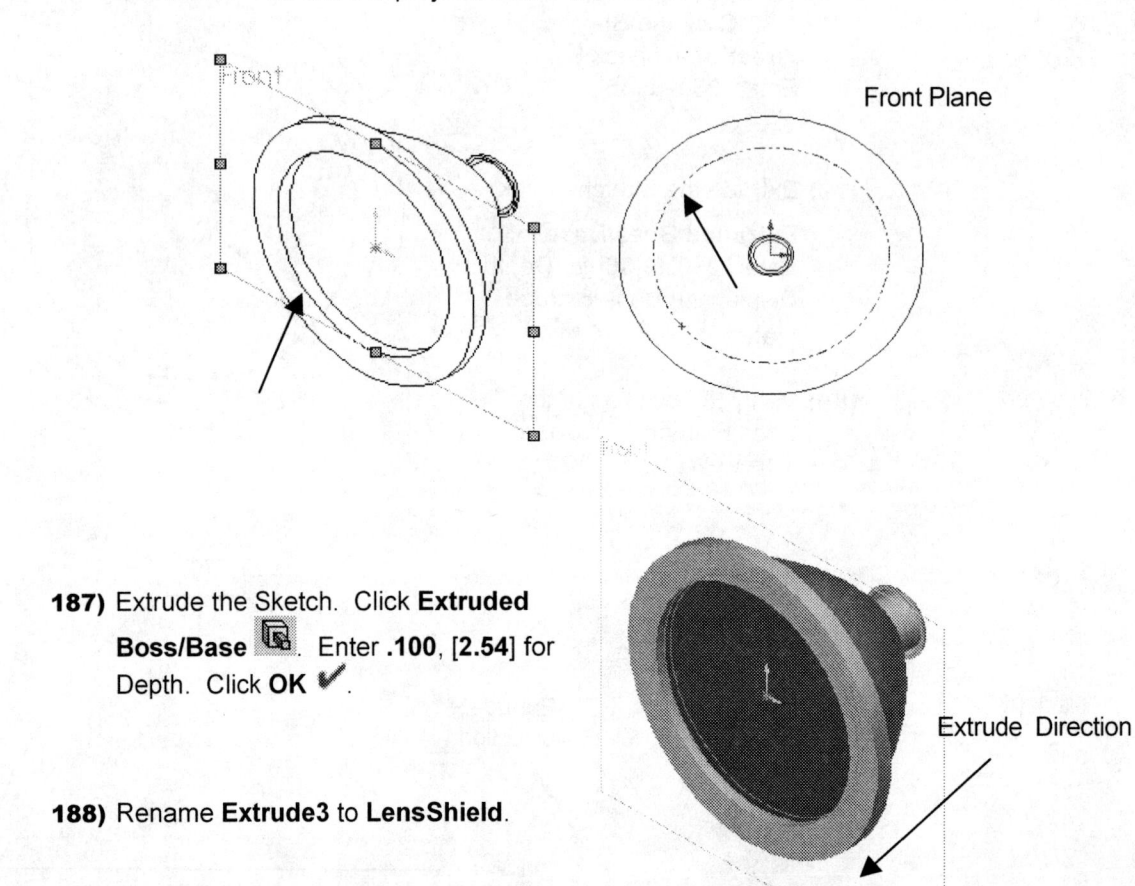

Front Plane

187) Extrude the Sketch. Click **Extruded Boss/Base**. Enter **.100**, [2.54] for Depth. Click **OK**.

Extrude Direction

188) Rename **Extrude3** to **LensShield**.

Engineering Design with SolidWorks — Extrude and Revolve Features

189) Add transparency to the LensShield. **Right-click** in the Graphics window. Click **Feature Properties**. The Feature Properties dialog box is displayed.

190) Click the **Color** button. The Entity Property dialog box is displayed. Click the **Advanced** button.

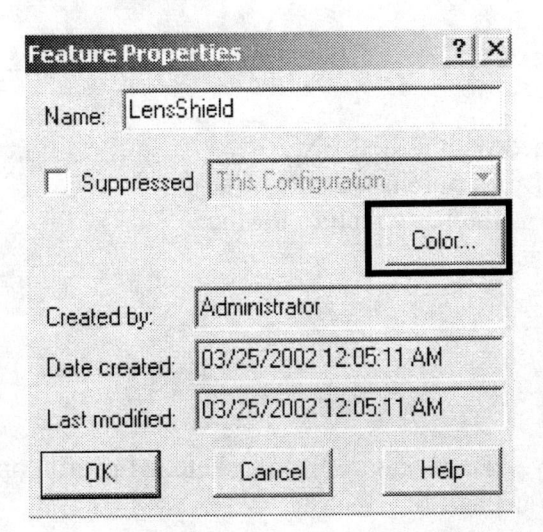

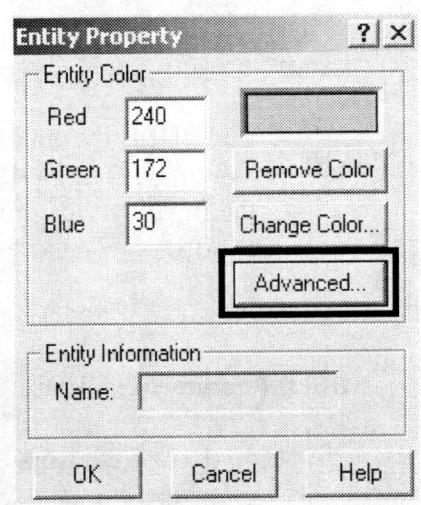

191) Set the transparency for the feature. Drag the **Transparency slider** to the far right side. Click **OK** from the Material Properties dialog box. Click **OK** from the Entity Property dialog box. Click **OK** from the Feature Properties box.

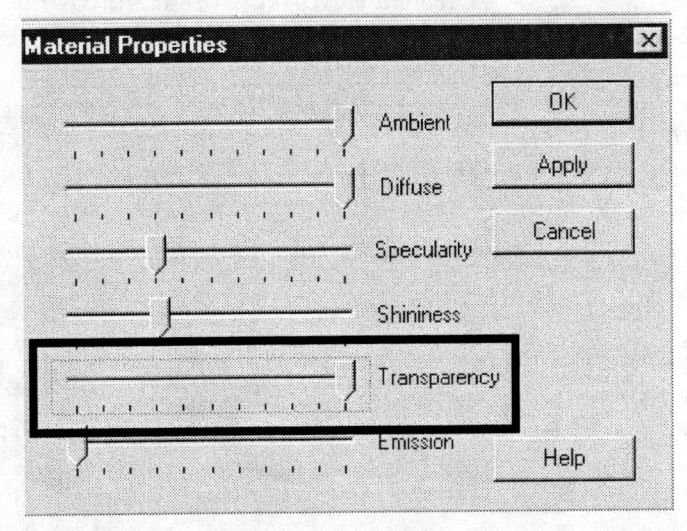

192) Display the transparent faces. Click **Shaded**. The faces are not transparent when the LensShield is selected. Click in the **Graphics window** to display the face transparency.

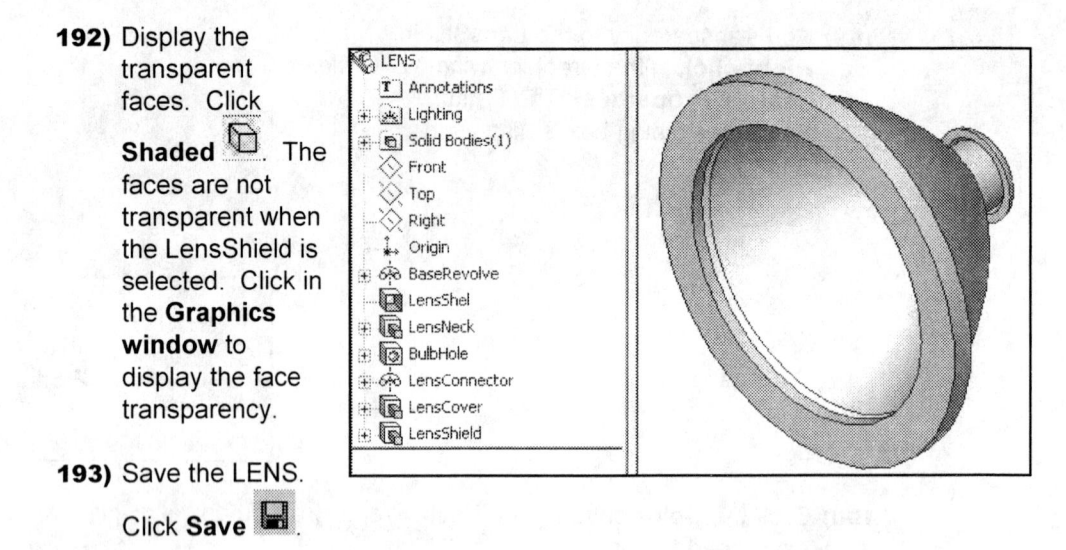

193) Save the LENS. Click **Save**.

BULB

The BULB is contained within the LENS assembly. The BULB is a purchased part. Use the Revolved feature as the Base feature for the BULB.

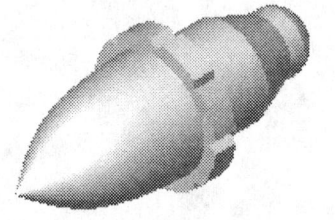

BULB Feature Overview

- Create the Revolved Base feature from a sketched profile on the Right plane, Figure 4.20a.

- Create a Revolved Boss feature using a B-Spline sketched profile. A B-Spline sketched profile is a complex curve, Figure 4.20b.

- Create a Revolved Cut Thin feature at the base of the BULB, Figure 4.20c.

- Create a Dome feature at the base of the BULB, Figure 4.20d.

- Create a Circular Pattern feature from an Extruded Cut, Figure 4.20e.

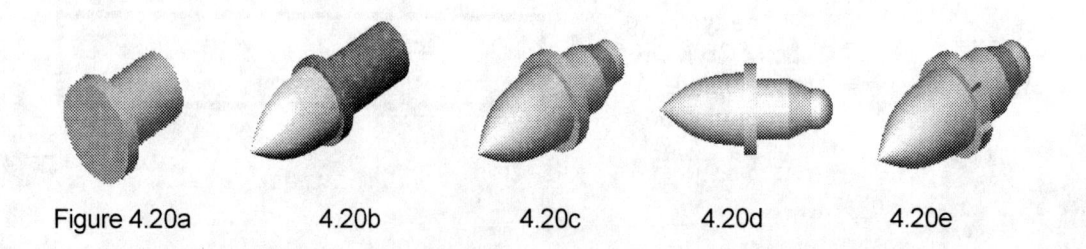

Figure 4.20a   4.20b   4.20c   4.20d   4.20e

Engineering Design with SolidWorks — Extrude and Revolve Features

Create the BULB – Use the Revolved Base Feature

The solid Revolved Base feature requires a centerline and a sketched profile. The flange of the BULB is located inside the Counterbore Hole of the LENS. Align the bottom of the flange with the Front plane. The Front plane mates against the Counterbore face.

Create a Revolved Base feature.

Create the BULB.

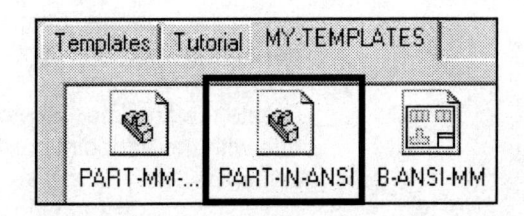

194) Click **New**. Click the **MY-TEMPLATES** tab. Click **PART-IN-ANSI**, [**PART-MM-ISO**] from the Template dialog box. Click **OK**.

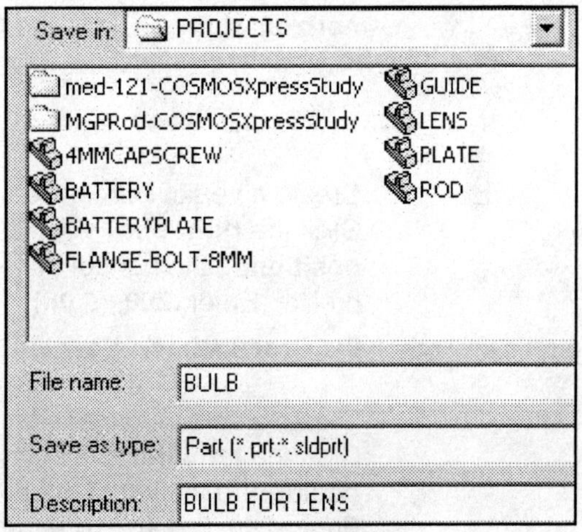

195) Save the empty part. Click **Save**. Select **ENGDESIGN-W-SOLIIDWORKS\PROJECTS** for Save in file folder.

196) Enter the file name. Enter **BULB**.

197) Enter **BULB FOR LENS** for Description. Click the **Save** button.

198) Select the Sketch plane. Click the **Right** plane from the FeatureManager. Create the Sketch. Click **Sketch**.

199) Display the three planes. Hold the **Ctrl** key down. Click **Front**, **Top** and **Right** from the FeatureManager. Right-click **Show**. Release the **Ctrl** key.

200) Display the Right view. Click **Right**.

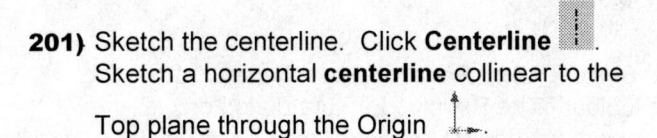

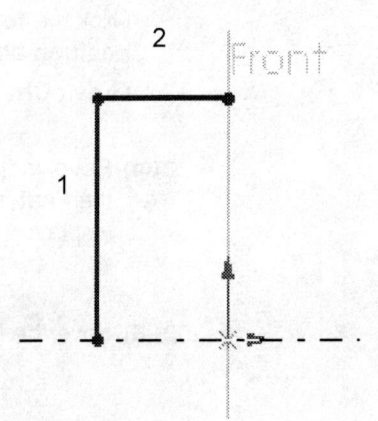

201) Sketch the centerline. Click **Centerline**. Sketch a horizontal **centerline** collinear to the Top plane through the Origin.

202) Sketch the profile. Create six lines. Click **Line**.

PAGE 4 - 51

Extrude and Revolve Features

Create the first line. Sketch a **vertical line** to the left of the Front plane.

Create the second line. Sketch a **horizontal line** with the endpoint coincident to the Front plane.

Create the third line. Sketch a short **vertical line** towards the centerline, collinear with the Front plane.

Create the forth line. Sketch a **horizontal line** to the right.

Create the fifth line. Sketch a **vertical line** with the endpoint collinear with the centerline.

Create the sixth line. Sketch a **horizontal line**. Close the sketch

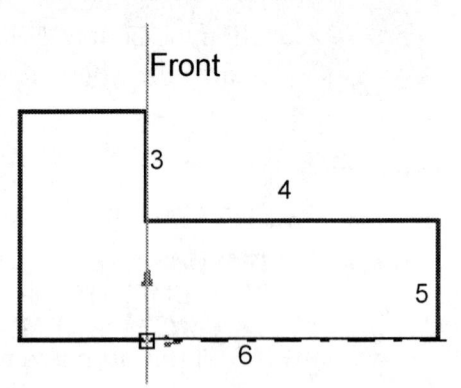

203) Add dimensions. Click **Dimension**.

Create a vertical dimension. Click the **right line**. Click a **position** to the right of the profile. Enter **.200**, **[5.08]**. Click the **Green Check Mark**.

Create a vertical dimension. Click the **left line**. Click a **position** to the left of the profile. Enter **.295**, **[7.49]**. Click the **Green Check Mark**.

Create a horizontal dimension. Click the **top left line**. Click a **position** above the profile. Enter **.100**, **[2.54]**. Click the **Green Check Mark**.

Create a horizontal dimension. Click the **top right line**. Click a **position** above the profile. Enter **.500**, **[12.7]**. Click the **Green Check Mark**.

204) Revolve the Sketch. Click **Revolve Boss/Base** from the Feature toolbar. The Revolve Feature dialog box is displayed. Accept the default option values. Click **OK**.

205) Save the BULB. Click **Save**.

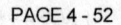

Create the BULB – Use the Revolved Boss Feature

The bulb requires a second solid Revolve feature. The profile utilizes a complex curve called a Spline (Non-Uniform Rational B-Spline or NURB). Draw Splines with control points. Adjust the shape of the curve by dragging the control points.

Create the Revolved Boss feature.
206) Select the Sketch plane. Click the **Right** plane. Create the Sketch. Click **Sketch**. Display the Right view. Click **Right**.

207) Sketch the centerline. Click **Centerline**. Sketch a **horizontal centerline** collinear to the Top plane, coincident to the Origin.

Sketch the profile. Click **Spline**. Sketch the start point. Click the **left vertical edge** of the Base feature.

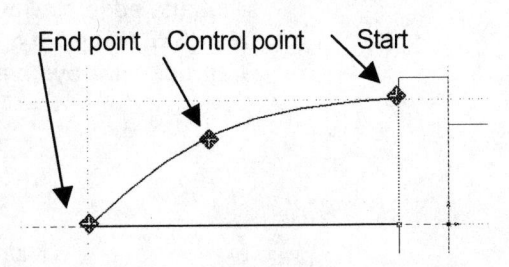

Sketch the control point. Drag the **mouse pointer** to the left of the Base feature. Click a **position** above the centerline. Sketch the end point. Drag the **mouse pointer** to the centerline. Double-click the **centerline** to end the Spline.

Note: Drag the mouse pointer upward and downward on the control points to modify the shape of the Spline.

208) Complete the profile. Sketch two lines.
Click **Line**. Create a horizontal line. Sketch a **horizontal line** from the Spline endpoint to the left edge of the Base-Revolved feature. Create a vertical line. Sketch a **vertical line** to the Spline start point, collinear with the left edge of the Base-Revolved feature.

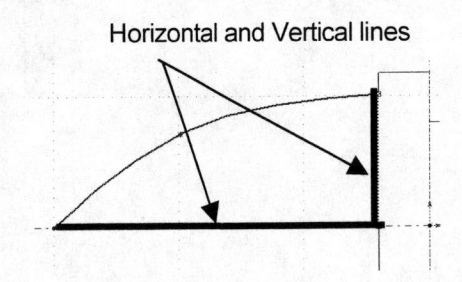

Note: You do not need dimensions to create a feature.

209) Revolve the Sketch. Click **Revolve Boss/Base** from the Feature toolbar. The Revolve Feature dialog box is displayed. Accept the default options. Display the Revolve feature. Click **OK**.

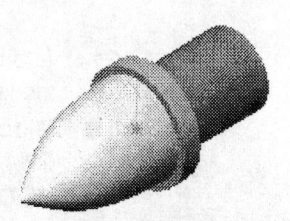

210) Save the BULB. Click **Save**.

Create the BULB – Use the Revolved Cut Thin Feature

A Revolved Cut Thin feature removes material by rotating an open sketch profile around a centerline.

Create a Revolved Cut Thin feature.
211) Select the Sketch plane. Click the **Right** plane from the FeatureManager. Create the profile. Click **Sketch**. Display the Right view. Click **Right**.

212) Sketch the centerline. Click **Centerline**. Sketch a **horizontal centerline** collinear to the Top plane, coincident to the Origin.

213) Sketch the profile. Click **Line**. Sketch a line from the **midpoint** of the top silhouette edge downward and to the right. Sketch a horizontal line with the .260, [6.6] end point coincident with the vertical **right edge**. Double-click in the **Graphics window** to end the line.

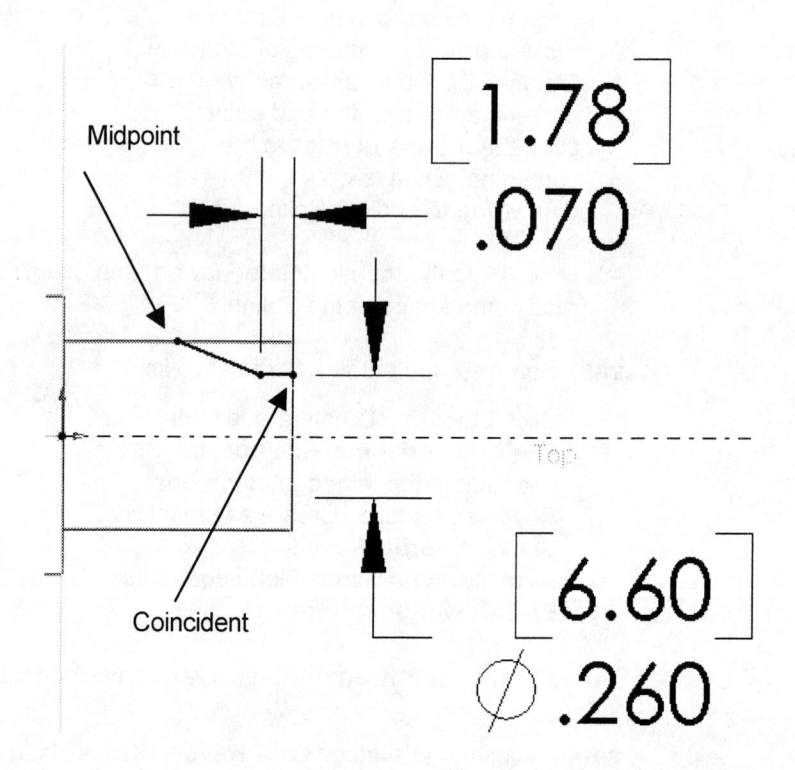

[1.78]
.070

[6.60]
⌀ .260

Midpoint
Coincident

214) Add relations. Click **Select**. Hold the **Ctrl** key down. Click the **start point** of the line. Click the top **Silhouette edge**. Release the **Ctrl** key. Click the **Midpoint** button. Click **Close Dialog**. Click **Select**. Hold the **Ctrl** key down. Click the **end point** of the line. Click the right **vertical edge**. Release the **Ctrl** key. Click the **Coincident** button. Click **Close Dialog**.

215) Add dimensions. Click **Dimension**. Create the diameter dimension. Click the **centerline**. Click the **short horizontal line**. Click a **position** below the centerline. Enter **.260**, **[6.6]**. Click the **Green Check Mark**.

216) Add a horizontal dimension. Click the **short horizontal line**. Click a **position** above the profile. Enter **.070**, **[1.78]**. Click the **Green Check Mark**. The Sketch is fully defined and is displayed in black.

Note: The ⌀.260 is displayed as a diameter dimension. Right-click Properties, uncheck the Display diameter check box to display a radius value.

217) Revolve the Sketch. Click **Revolved Cut** from the Feature toolbar. Click **No** to the Warning Message, "Would you like the sketch to be automatically closed?"

218) Expand the Thin Feature dialog box. The direction arrow points away from the centerline. Click the **Reverse Direction** button. Enter **.150**, **[3.81]** for Thickness. Display the Revolved Cut Thin feature. Click **OK**.

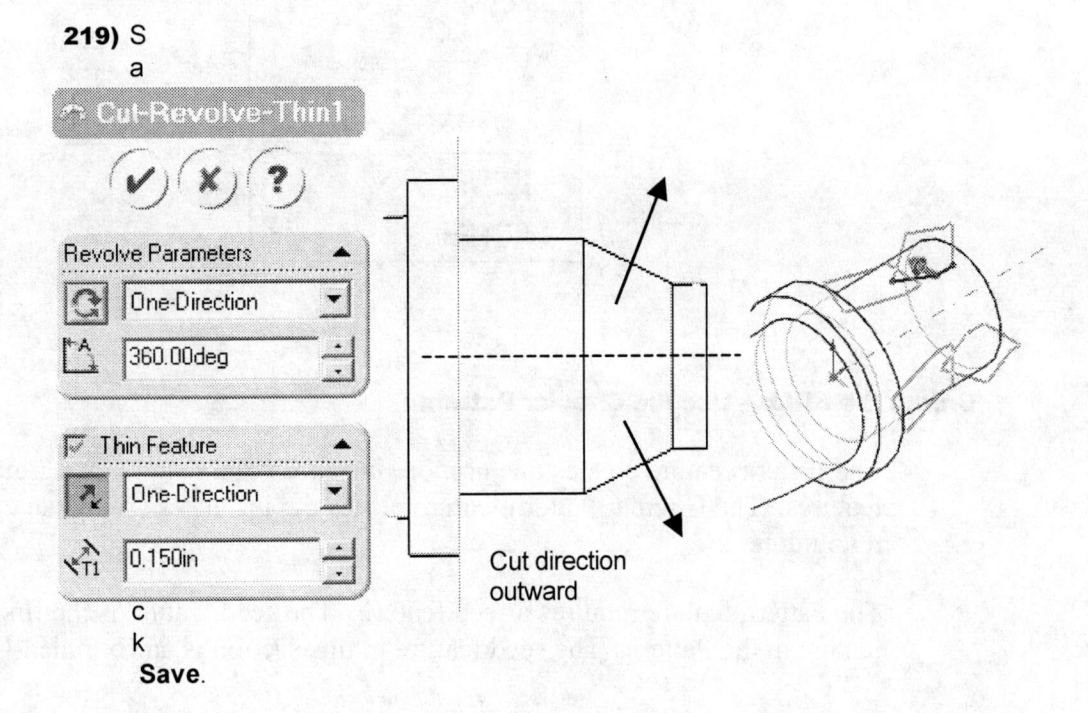

219) Save.

Cut direction outward

Create the BULB – Use the Dome Feature

A Dome feature creates spherical or elliptical shaped geometry. Use the Dome feature to create the Connector feature of the BULB.

Create the Dome feature.
220) Select the Sketch plane. Click the **back circular face** of the Revolve Cut Thin.

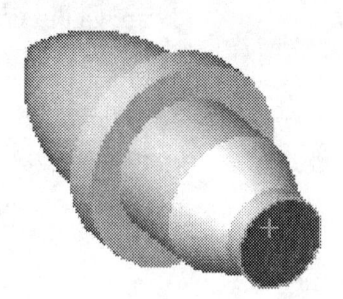

221) Click **Insert** from the Main menu. Click **Features**, **Dome**. The Dome dialog box is displayed. Enter **.100**, **[2.54]** for Height. Display the Dome. Click **OK** ✔.

222) Save the BULB. Click **Save**.

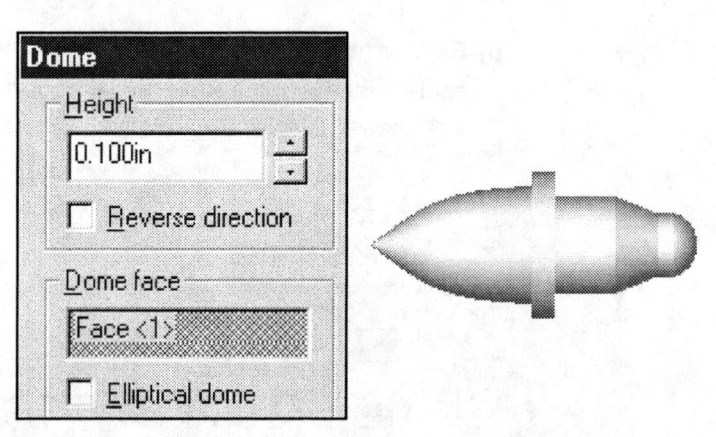

Create the BULB – Use the Circular Pattern

The Pattern feature creates one or more instances of a feature or a group of features. The Circular Pattern feature places the instances around an axis of revolution.

The Pattern feature requires a seed feature. The seed feature is the first feature in the Pattern. The seed feature in this section is an Extruded-Cut.

Create the Circular Pattern.
223) l**Zoom in** on the front flat face.

224) Select the Sketch plane. Click the **front face** of the Base feature.

225) Create the Sketch. Click **Sketch** ✎.

226) Extract the outside circular edge. Click **Select**. Click the **outside circular edge**. Click **Convert Entities** from the Sketch Tools toolbar.

227) Display the Front view. Click **Front**.

228) Display the Right plane. Click the **Right** plane in the FeatureManager. Right-click **Show**.

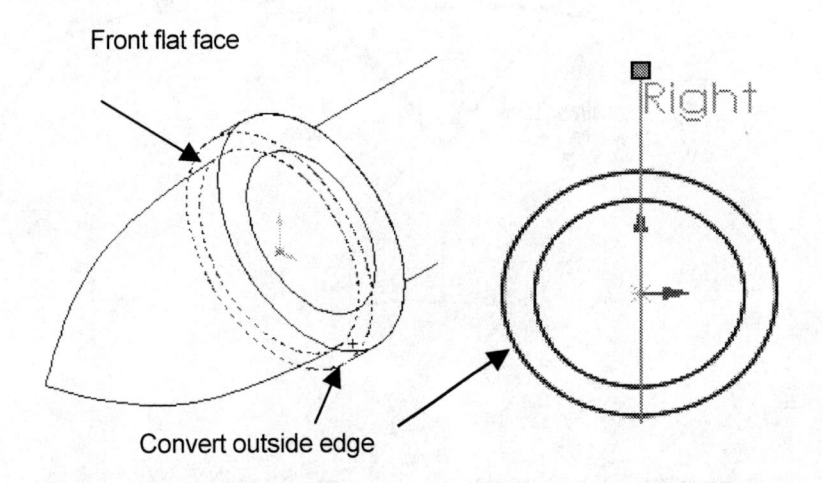

Front flat face

Convert outside edge

229) **Zoom in** on the top half of the BULB. Sketch the centerline. Click **Centerline**. Sketch a **vertical centerline** coincident with the top and bottom circular circles and coincident with the Right plane. The centerline is fully defined and is displayed in black.

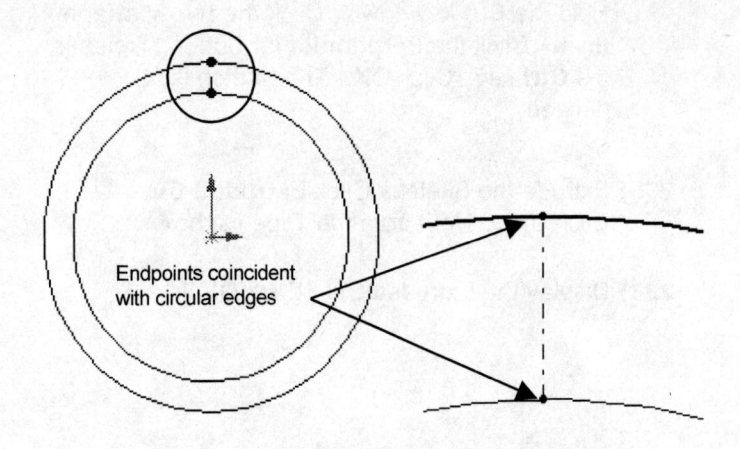

Endpoints coincident with circular edges

230) **Zoom** to display the centerline and the outside circular edge.

231) Sketch a V-shaped line. Click **Sketch Mirror**. Click **Line**. Create the first point. Click the **midpoint** of the centerline. Create the second point. Click the coincident **outside circle edge**.

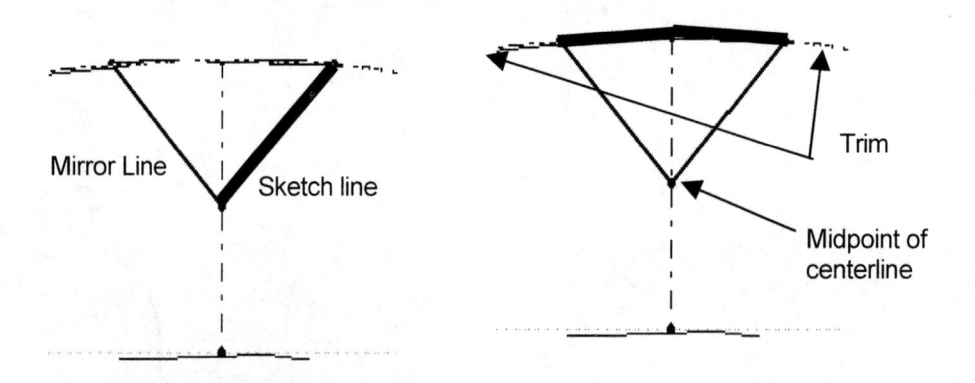

Mirror Line | Sketch line

Trim | Midpoint of centerline

232) Deactivate the Mirror. Click **Sketch Mirror**.

233) Trim the circle. Click **Trim**. Click the **outside circle** to the right.

234) Add the geometry relations. Right-click **Select**. Hold the **Ctrl** key down. Click the **two V shape lines**. Click the **Perpendicular** button. Release the **Ctrl** key. Click **OK**. The Sketch is fully defined.

235) Extrude the Sketch. Click **Extruded Cut**. Click **Up to Next** from the Type list box.

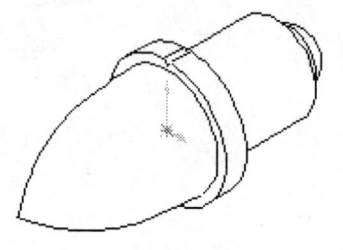

236) Display the Extruded Cut. Click **OK**.

Engineering Design with SolidWorks
Extrude and Revolve Features

237) Display the Temporary axis. Check **View**, **Temporary Axis** from the Main menu.

The Cut-Extrude is the seed feature for the Pattern.

238) Create the Pattern. Click **Isometric**. Click the **Cut-Extrude1** from the FeatureManager. Click **Circular Pattern**. The Circular Pattern dialog box is displayed. Click the pink **Pattern axis** text box. Click **Temporary Axis**. Create 4 copies of the Cut. Enter **4** in the Number of Instances spin box. Click the **Equal spacing** check box. Click the **Geometry pattern** check box.

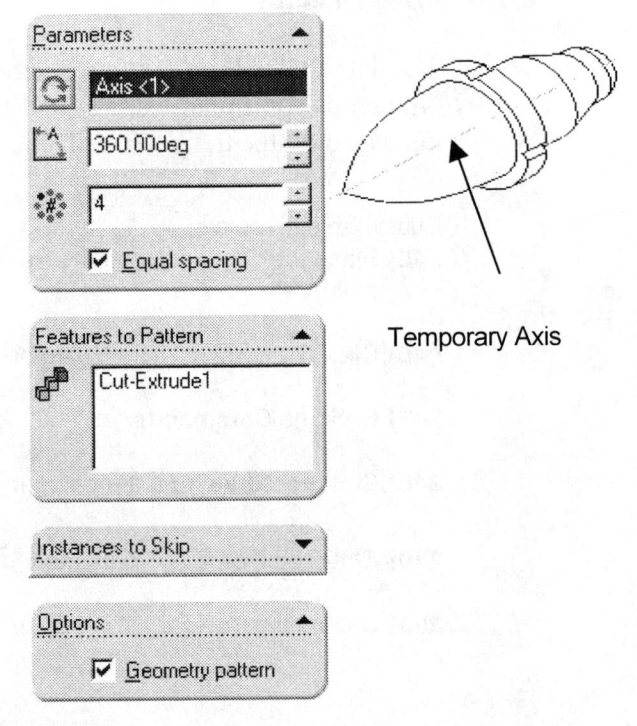

Temporary Axis

239) Display the Pattern feature. Click **OK**.

240) Edit the Pattern feature. Right-click on the **CirPattern1** from the Feature Manager. Click **Edit Definition**. Enter **8** in the Number of Instances spin box.

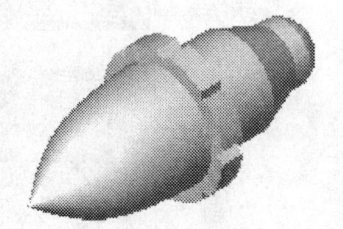

241) Display the updated Pattern. Click **OK**.

242) Hide the Temporary axis. Click **View** from the Main menu. Uncheck **Temporary Axis**. Hide the Planes. Uncheck **Planes** from the View menu.

243) Fit the model to the Graphics window. Press the **f** key.

244) Save the BULB. Click **Save**.

Customizing Toolbars

The default Toolbars contains numerous icons that represent basic functions. Numerous additional features and functions are available that are not displayed on the default Toolbars.

Customize the Toolbar.
245) Place the Dome icon on the Features Toolbar. Click **Tools** from the Main menu.

246) Click **Customize**. The Customize dialog box is displayed.

247) Click the **Commands** tab.

248) Click **Features** from the category text box.

249) Drag the **Dome** icon into the Features Toolbar.

250) Update the Toolbar. Click **OK** from the Customize dialog box.

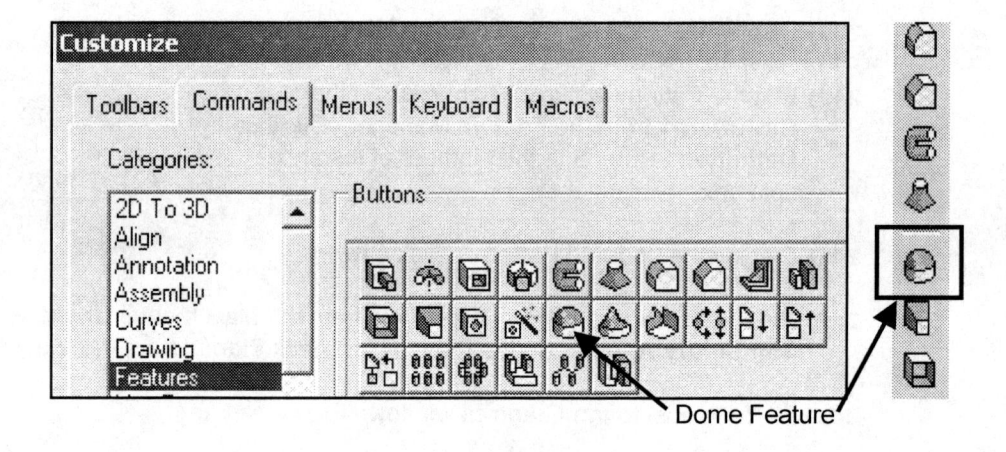

Dome Feature

Project Summary

You are designing a FLASHLIGHT assembly that is cost effective, serviceable and flexible for future design revisions. The FLASHLIGHT assembly is designed to accommodate two battery types: One 6-volt or 4-D cells.

The FLASHLIGHT assembly consists of numerous parts. The team decided to purchase and model the following parts: One 6-volt BATTERY, LENS assembly, SWITCH and an O-RING. The LENS assembly consists of the LENS and the BULB.

You created the following parts:

- BATTERY.

- BATTERYPLATE.

- LENS.

- BULB.

The other parts for the FLASHLIGHT assembly are addressed in Project 5.

You addressed two major Base features in this project:

- Extrude – BATTERY and BATTERYPLATE.

- Revolve – LENS and BULB.

Practice the enclosed exercises before you move on to the next Project.

Project Terminology

Draft angle: Draft is the degree of taper applied to a face. Draft angles are usually applied to molds or castings.

Geometric relationships: Are captured as you sketch. The mouse pointer icon displays the following relationships: Horizontal, vertical, coincident, midpoint, intersection, tangent and perpendicular.

Trim: Sketch tool that removes highlighted geometry.

Sketch Mirror: Sketch tool that mirrors sketch geometry to the opposite side of a sketched centerline.

Spline: A Spline is a complex curve.

Edit Sketch: Modify existing sketch geometry. Right-click the feature in the FeatureManager. Click Edit Sketch.

Edit Definition: Modify existing feature parameters. Right-click the feature in the FeatureManager. Click Edit Definition.

Project Features:

Extruded Base/Boss: Use to add material by extrusions. Requires a sketch on a sketch plane and a distance normal to the sketch plane. The draft angle option produces a taper in the normal direction.

Extruded Cut: Use to remove material from a solid. This is the opposite of the boss. Cuts begin as a 2D sketch and remove materials by extrusions.

Fillet: Removes sharp edges and faces of the BATTERY.

Revolved Base/Boss: Use to add material by revolutions. Requires a centerline and sketch on a sketch plane. Revolve features require an angle of revolution. The sketch is revolved around the centerline.

Revolved Cut: Use to remove material by revolutions. Requires a centerline and sketch on a sketch plane.

Thin option: The Thin option for the Revolved Boss and Revolved Cut utilizes an open sketch to add or remove material, respectfully.

HoleWizard: The HoleWizard is used to create specialized holes in a solid. It can create simple, tapped, counterbored and countersunk holes using a step-by-step procedure. Use the HoleWizard to create a countersink hole at the center of the PLATE. Use the HoleWizard to create tapped holes in the GUIDE.

Dome: Use to add a spherical or elliptical dome to a selected face.

Circular Pattern: A pattern repeats a feature in a circular array about an axis. The number of copies in a pattern is called instances.

Questions

1. Identify and describe the function of the following features:

    - Fillet.

    - Extruded Cut.

    - Extruded Boss.

    - Revolved Base.

    - Revolved Cut Thin.

2. Describe a symmetric relation.

3. Explain a Fillet Rebuild error message.

4. Describe an angular dimension.

5. What is a draft angle? Provide an example.

6. When is a draft angle used?

7. When do you use the Mirror command?

8. What is the function of the Shell feature?

9. An arc requires _____ points?

10. Name the required points of an arc?

11. When do you use the Hole Wizard feature?

12. What is a B-Spline?

13. Identify the required information for a Circular Pattern?

14. How do you add the Dome feature icon to the Feature Toolbar?

Exercises

Exercise 4.1: AIR RESERVOIR SUPPORT Assembly.

Your project team developed a concept sketch of the PNEUMATIC TEST MODULE Assembly. Develop the AIR RESERVOIR SUPPORT Assembly.

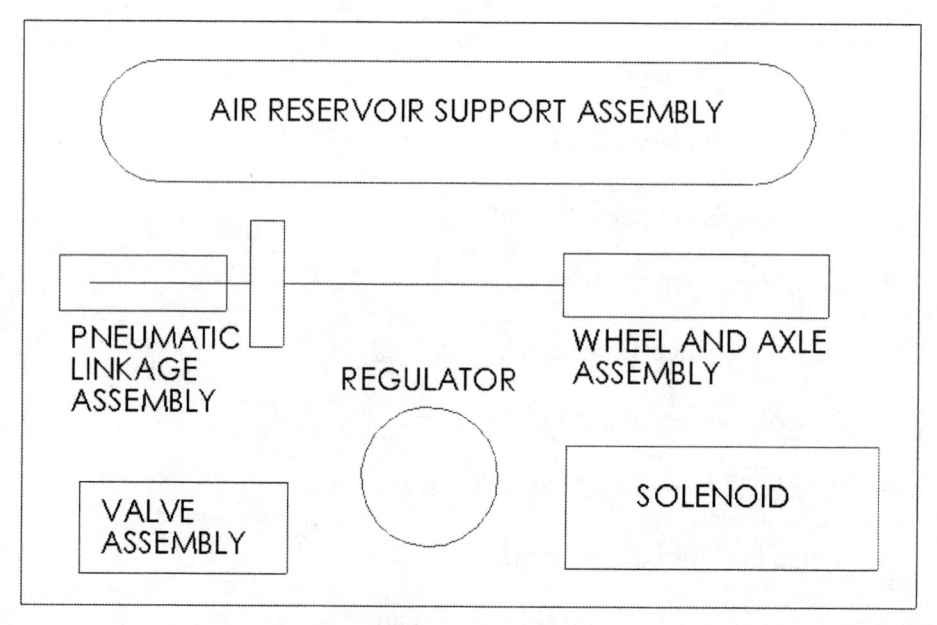

PNEUMATIC TEST MODULE Assembly Layout

Create three new parts:

- FLAT PLATE
- IM15-MOUNT
- ANGLE BRACKET

The SMC AIR RESERVOIR is a purchased part. If you downloaded the zipfile from the publisher's website, the assembly file is located in the **CD-ENGDESIGN-W-SW2003** folder, or can be downloaded from Gears Educational Systems (www.gearseds.com).

Obtain the MACHINE SCREW from SolidWorks Toolbox, your Hardware Vendor or create your own simplified screw part (See Exercise 4.1e).

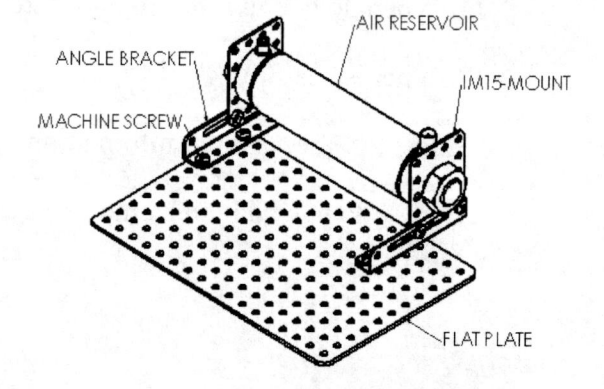

AIR RESERVOIR SUPPORT ASSEMBLY
Courtesy of Gears Educational Systems & SMC Corporation of America

Create a new assembly named, AIR RESERVOIR SUPPORT. Two IM15-MOUNT parts and two ANGLE BRACKETs hold the SMC AIR RESERVOIR. The ANGLE BRACKETs are fastened to the FLAT PLATE with MACHINE SCREWS.

Exercise 4.1a: FLAT PLATE Part.

Create the FLAT PLATE Part on the Top Plane. The FLAT PLATE is machined from 0.060 [1.5mm] Stainless Steel flat stock. The default units are inches.

The 8.688[220.68mm] x 5.688[144.48mm] FLAT PLATE contains a Linear Pattern of ∅.190[4.83mm] Thru Holes. The Holes are equally spaced, .500[12.70mm] apart. Utilize Geometric Pattern for the Linear Pattern Option.

Enter the FLAT PLATE Part Number in the Configuration Manager. Right-click Default. Click Properties. Enter GIDS-SC-10002 for Part number displayed when used in bill of materials. Select the User Specified Name from the list box.

Click the Custom button from the Configuration Properties dialog box. Enter FLAT PLATE for Description.

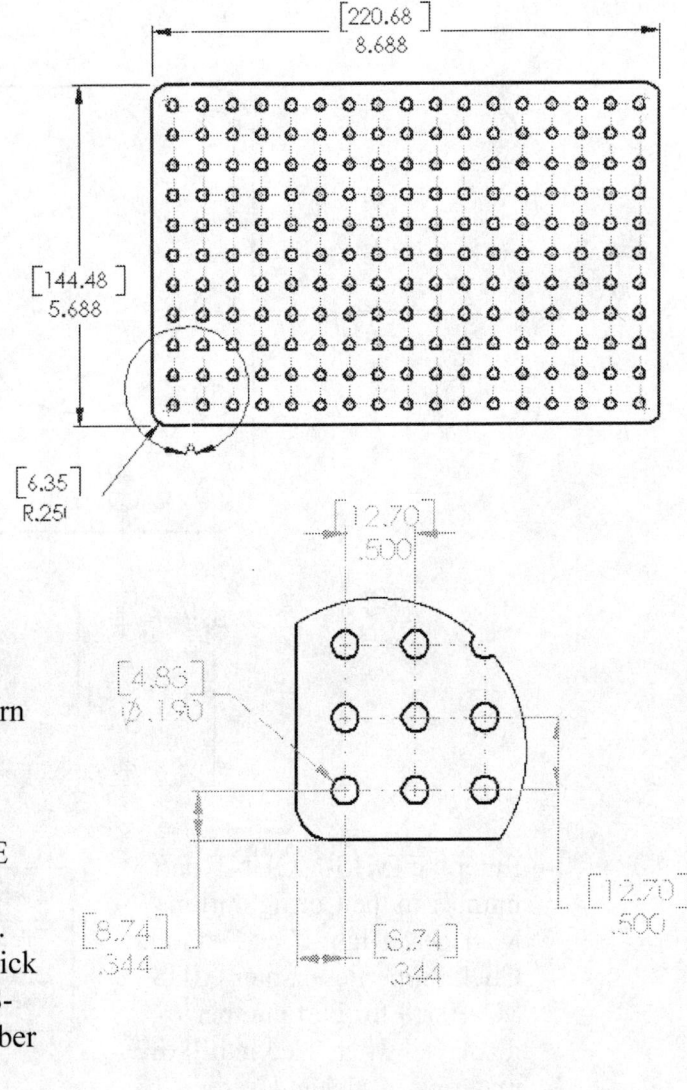

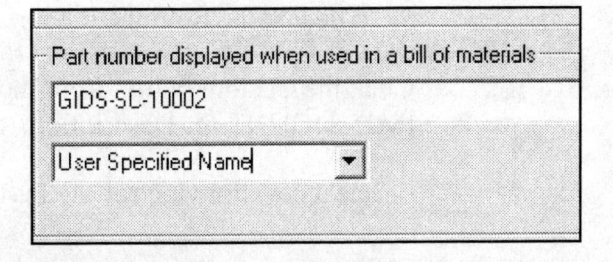

Create a new drawing for the FLAT PLATE.

Exercise 4.1b: IM15-MOUNT Part.

Create the IM15-MOUNT Part on the Right Plane. The IM15-MOUNT Part is machined from 0.060 [1.5mm] Stainless Steel flat stock. The default units are inches.

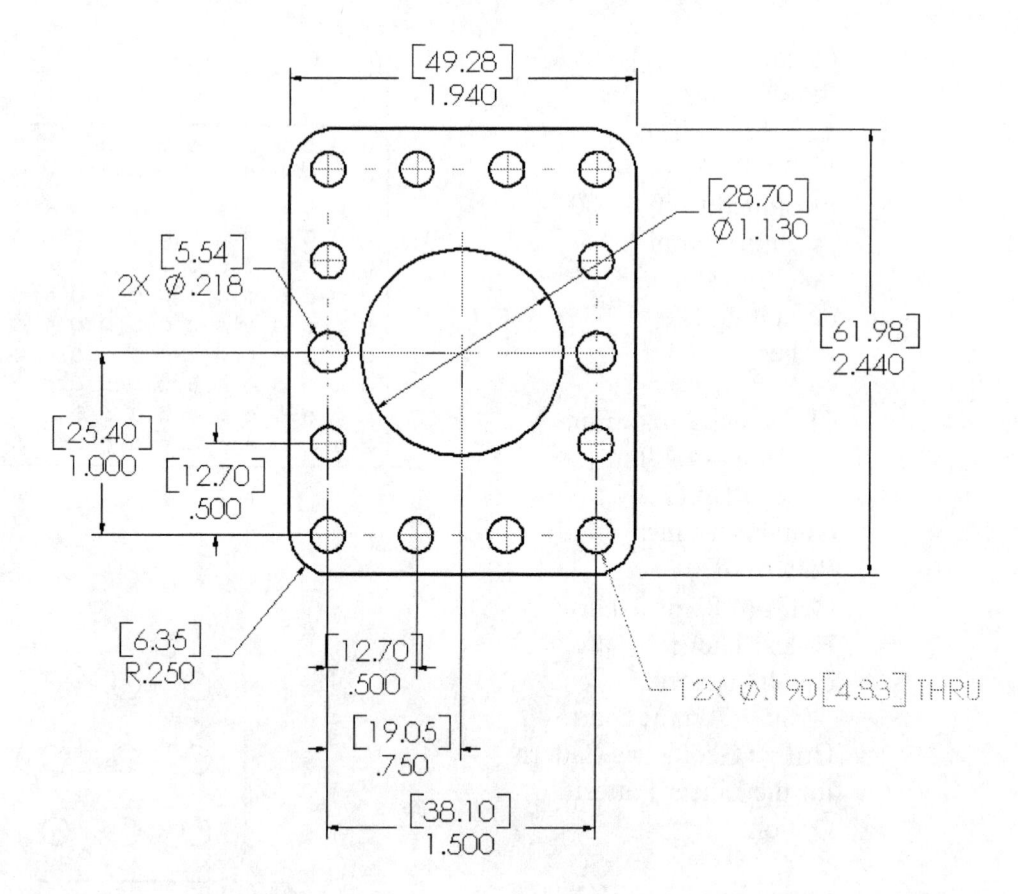

Enter the IM15-MOUNT part number in the Configuration Manager. Right-click Default. Click Properties. Enter GIDS-SC-10009 for Part number displayed when used in bill of materials. . Select User Specified Name from the list box.

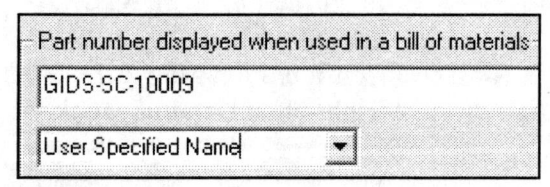

Click the Custom button from the Configuration Properties dialog box. Enter IM15 MOUNT for Description. Save the IM15-MOUNT Part.

Create a new drawing for the IM15 MOUNT.

Exercise 4.1c: ANGLE BRACKET Part.

Create the ANGLE BRACKET Part. The Base Extrude feature is sketched with an L-Shaped profile on the Right Plane. The ANGLE BRACKET Part is machined from 0.060 [1.5mm] Stainless Steel flat stock. The default units are inches.

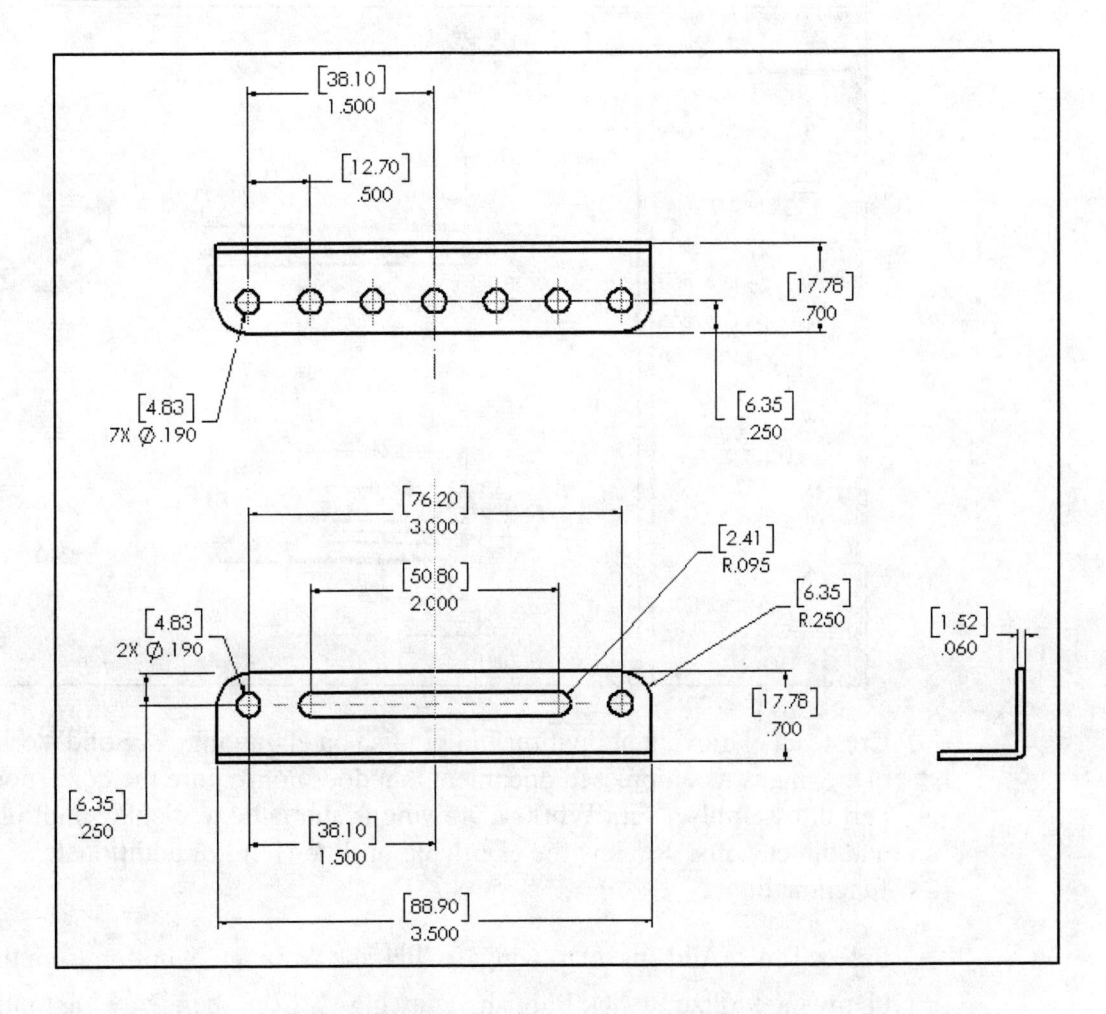

Enter the ANGLE BRACKET part number in the Configuration Manager. Right-click Default. Click Properties. Enter GIDS-SC-10007 for Part number. Select User Specified Name from the list box.

Click the Custom button from the Configuration Properties dialog box. Enter ANGLE BRACKET 7 HOLE for Description. Save the ANGLE BRACKET Part.

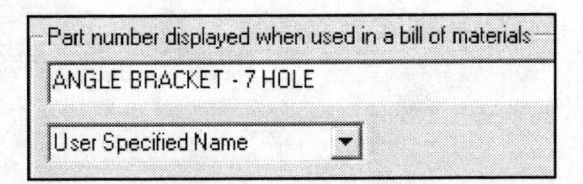

Exercise 4.1d: ANGLE BRACKET Drawing.

Create a new drawing named ANGLE BRACKET. Utilize the ANGLE BRACKET part created in Exercise 4.1c. Modify the Hole text to contain the total number of holes.

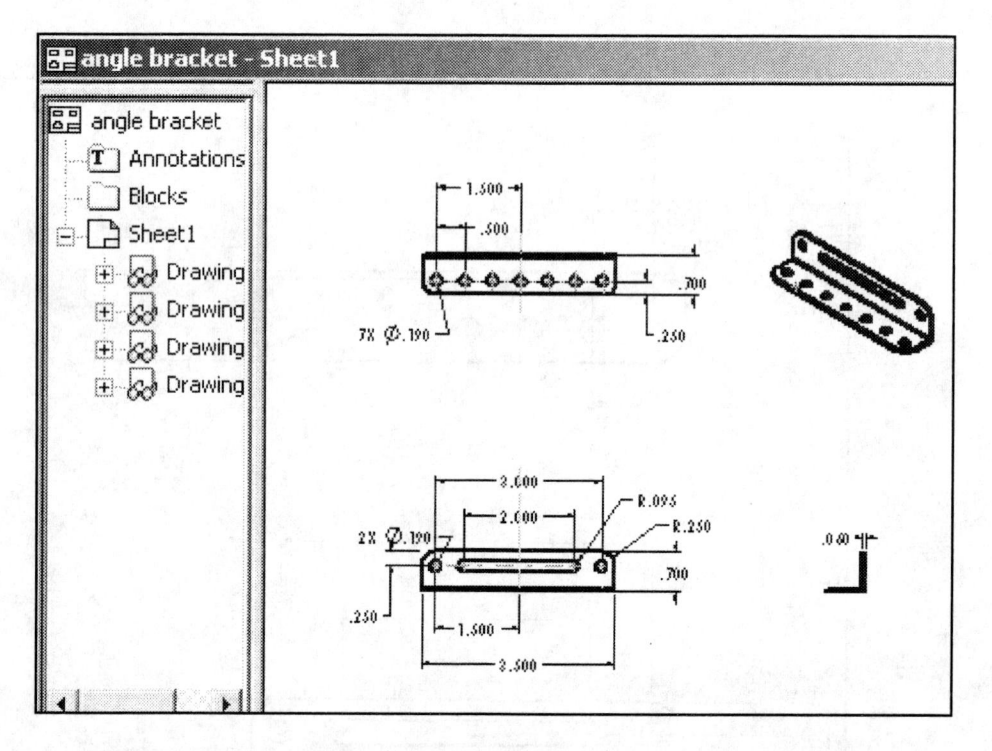

Create an eDrawing animation and publish an eDrawing. A SolidWorks eDrawing is a compressed document that does not require the corresponding part or assembly. SolidWorks eDrawing is animated to display multiple views and dimensions. Review the eDrawing on-line Help for additional functionality.

Select Tools, Add-Ins, eDrawings. Click the Animate button from the eDrawing toolbar. Click Publish eDrawing from the eDrawing toolbar. The Front view of the ANGLE BRACKET drawing is displayed. Click the Next button. Display the remaining views. Click the Play button. Save the ANGLE BRACKET eDrawing. Email the drawing to a friend.

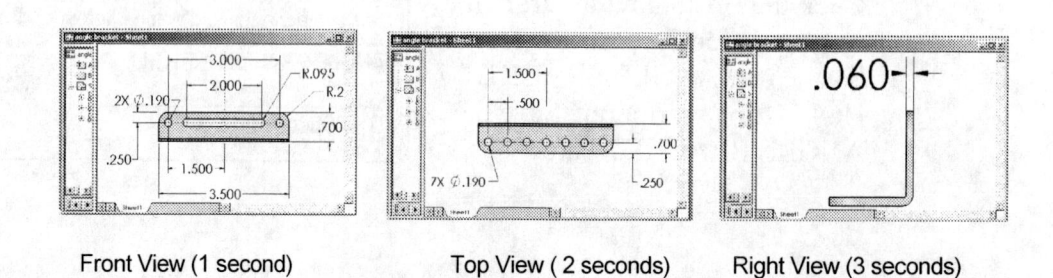

Front View (1 second)   Top View (2 seconds)   Right View (3 seconds)

Engineering Design with SolidWorks — Extrude and Revolve Features

Exercise 4.1e: MACHINE SCREW Part.

SolidWorks Toolbox is utilized in this project for machine screws, nuts and washers. Create the 10-24 x 3/8 MACHINE SCREW part if SolidWorks Toolbox is not available. Utilize a Revolved Feature. For metric size, utilize an M4x10 machine screw.

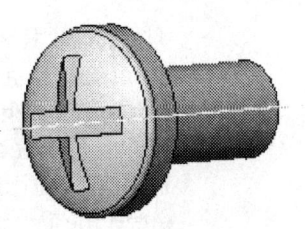

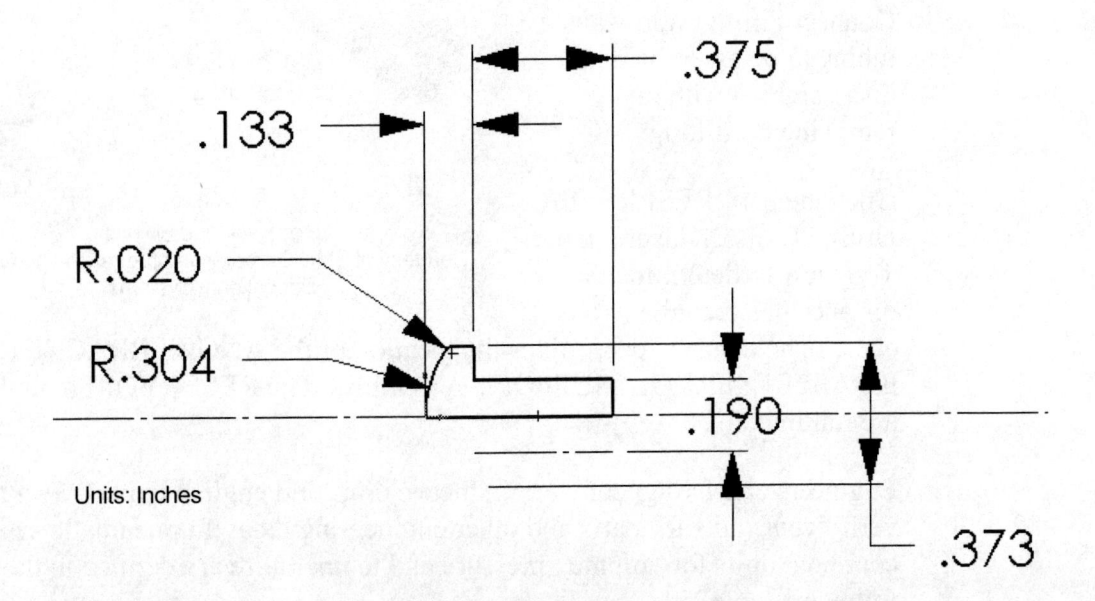

Units: Inches

Dimensions: .375, .133, R.020, R.304, .190, .373

Utilize an Extruded Cut feature with the Mid Plane option to create the Top Cut. The Top Cut is sketched on the Front Plane. Utilize the Convert Entities Sketch Tool to extract the left edge of the profile. Utilize Circular Pattern and the Temporary Axis to create 4 Top Cuts.

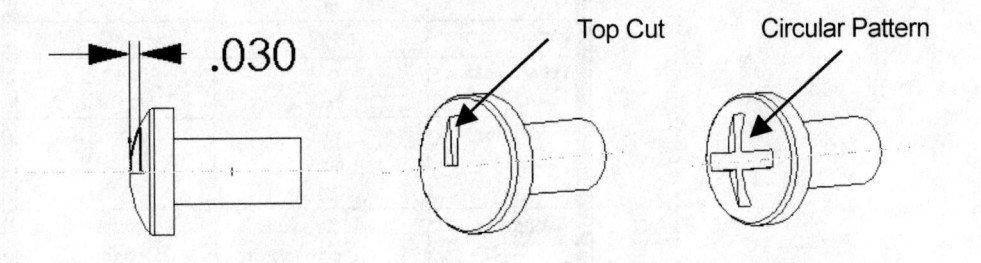

.030

Top Cut   Circular Pattern

Exercise 4.1f: RESERVOIR Assembly.

The RESERVOIR stores compressed air. Air is filled through a Schrader Valve. A Quick Connect Straight Fitting is utilized to supply air to the Pneumatic Test Module assembly. Quick Connect Fittings allow air tubing to be assembled and disassembled with out removing the fitting.

Open the part, RESERVOIR. Utilize Tools, Measure in the Top view to determine the distance between the two end caps. The distance determines the location of the ANGLE BRACKETS in the AIR RESERVOIR SUPPORT Assembly. The RESERVOIR default units are millimeters.

Engineers and Designers work in metric units and english units. Always verify your units for parts and other engineering data. In pneumatic systems, common units for volume, pressure and temperature are defined in the following table.

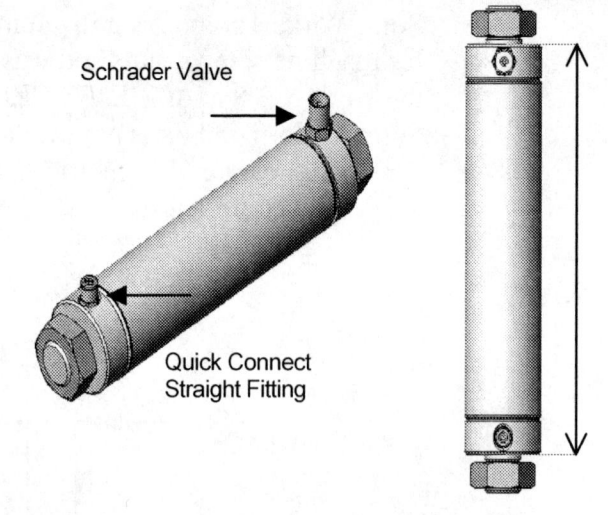

Reservoir and Fittings
Courtesy of SMC Corporation of America and Gears Educational Systems

Magnitude	Metric Unit (m)	English (e)
Mass	kg	pound
	g	ounce
Length	m	foot
	m	yard
	mm	inch
Temperature	°C	°F
Area, Section	m^2	sq.ft
	cm^2	sq.inch
Volume	m^3	cu.yard
	cm^3	cu.inch
	dm^3	cu.ft.
Volume Flow	$m^3 n / min$	scfm
	$dm^3 n /min$ (ℓ/min)	scfm
Force	N	pound force (ℓbf.)
Pressure	bar	ℓbf./sq.inch (psi)

Common Metric and English Units

The ISO unit of pressure is the Pa (Pascal). 1Pa = 1N/m.

Air is a gas. Boyle's Law states that with constant temperature, the pressure of a given mass of a gas is inversely proportional to its volume.

$P_1 / P_2 = V_2 / V_1$

$P_1 \times V_1 = P_2 \times V_2$

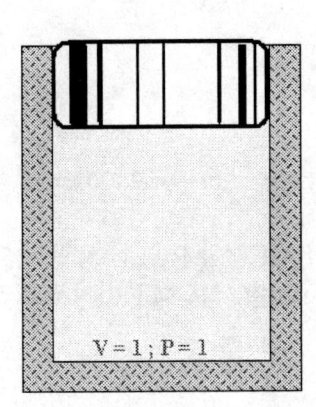

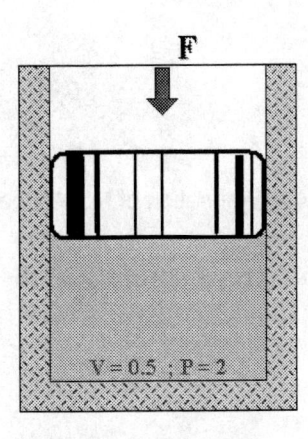

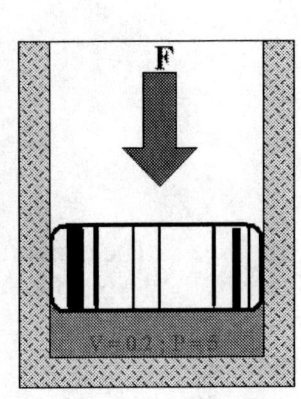

Illustration of Boyle's Law

Utilize COSMOSXpress to perform a simplified static analysis on the Reservoir.

Open the part, NCMTube. This is the first component in the Reservoir assembly.

Utilize Save As to copy NCMTube to NCMTube-Analysis. Suppress the Cut-Extrude1 and RodBore features.

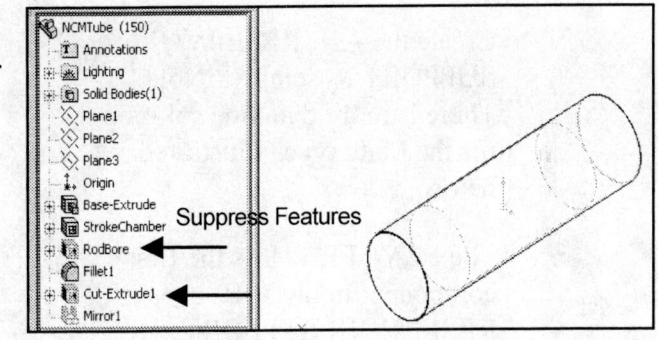

Run COSMOSXpress. Select Aluminum Alloy 1060 for material. Restrain both end circular faces.

Select Pressure for Type of Load. Select the outside cylindrical face. Enter 100 psi for Pressure Load. The direction arrow points outward.

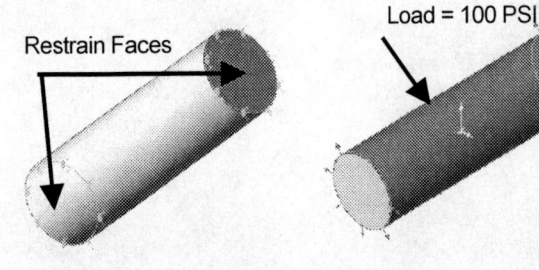

Run the analysis. The NCMTube is within the Factor of Safety.

Display the animation of the Deformation Plot and vonMises Stress Plot.

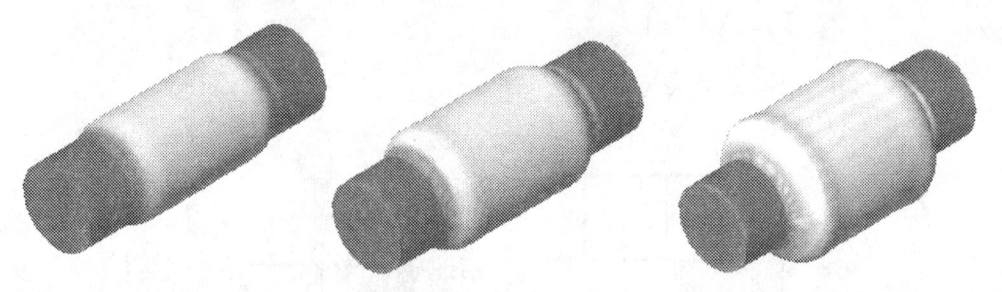

Animation of NCMTube part, magnified 2400 times

The maximum pressure rating on the SMC Air Reservoir is 250psi (1.70Mpa). Explain why the actual values vary from the NCMTube-Analysis part utilized with COSMOSXpress.

Exercise 4.1g: AIR RESERVOIR SUPPORT Assembly.

Create the AIR RESERVOIR SUPPORT Assembly. Note: There is more than one solution for the Mate types illustrated below.

The FLAT PLATE is the first component in the AIR RESERVOIR SUPPORT Assembly. Insert the FLAT PLATE. The FLAT PLATE is fixed to the Origin.

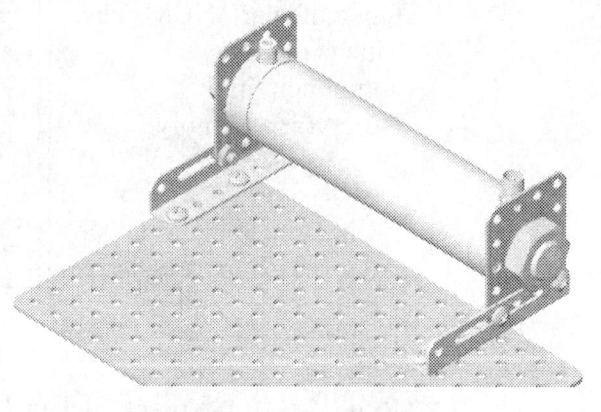

AIR RESERVOIR SUPPORT Assembly
Courtesy of Gears Educational Systems &
SMC Corporation of America

Insert the ANGLE BRACKET. Mate the ANGLE BRACKET to the FLAT PLATE.

The bottom flat face of the ANGLE BRACKET is coincident to the top face of the FLAT PLATE.

The center hole of the ANGLE BRACKET is concentric to the upper left hole of the FLAT PLATE.

The first hole of the ANGLE bracket is concentric with the hole in the 8th row, 1st column of the FLAT PLATE.

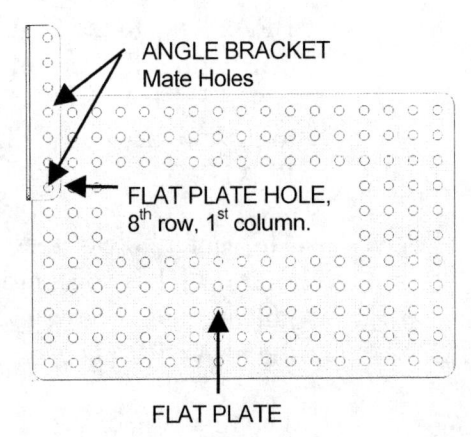

Insert the IM15-MOUNT. Mate the IM15-MOUNT. The IM15-MOUNT flat back face is coincident to the flat inside front face of the ANGLE BRACKET.

The bottom right hole of the IM15-MOUNT is concentric with the right hole of the ANGLE BRACKET.

The bottom edge of the IM15-MOUNT is parallel to bottom edge of the ANGLE BRACKET.

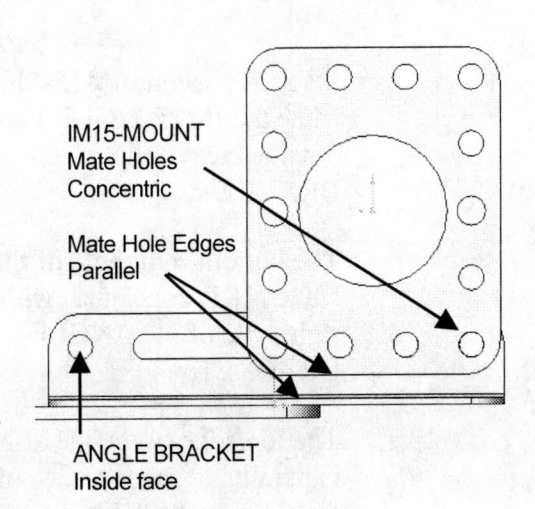

Insert the Reservoir Assembly. Mate the Reservoir Assembly.

The conical face of the Reservoir is concentric to the IM15-MOUNT center hole.

The left end cap of the Reservoir Assembly is coincident to the front face of the IM15-MOUNT.

The Hex Nut flat face is parallel to the top face of the FLAT PLATE.

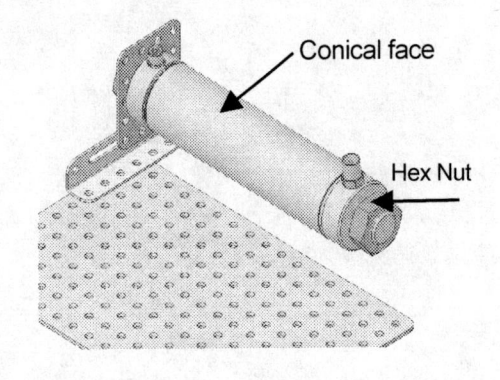

Extrude and Revolve Features **Engineering Design with SolidWorks**

Insert the second ANGLE BRACKET. Mate the ANGLE BRACKET to the FLAT PLATE. The bottom flat face of the ANGLE BRACKET is coincident to the top face of the FLAT PLATE.

The center hole of the ANGLE BRACKET is concentric with the hole in the 11th row, 13th column of the FLAT PLATE.

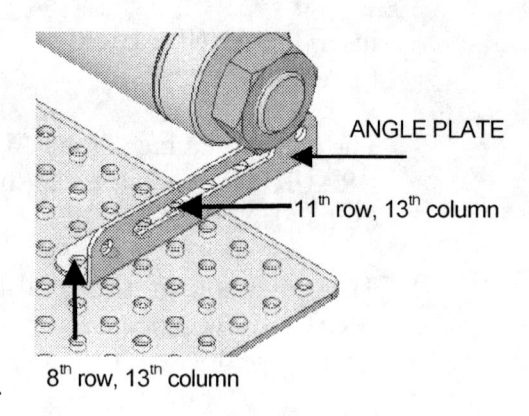

The first hole of the ANGLE bracket is concentric with the hole in the 8th row, 13th column of the FLAT PLATE.

Insert the second IM15-MOUNT. Mate the IM15-MOUNT to the outside face of the ANGLE BRACKET.

The bottom right hole of the IM15-MOUNT is concentric with the right hole of the ANGLE BRACKET.

The top edge of the IM15-MOUNT is parallel to the top edge of the ANGLE BRACKET.

The AIR RESERVOIR SUPPORT Assembly is complete. Save the assembly.

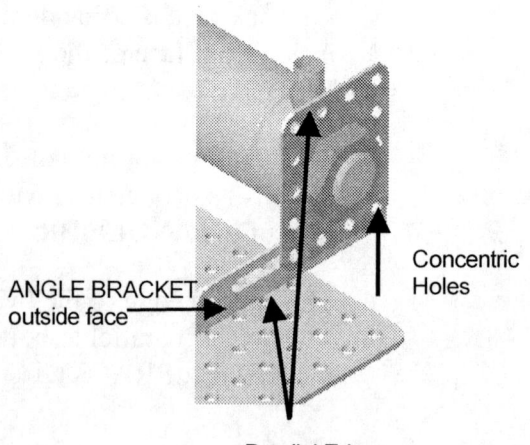

Exercise 4.2a: D-Size Battery Part.

Measure a D size battery to determine feature dimensions. Create the D-Size Battery.

Exercise 4.2b: BATTERY HOLDER Assembly.

Create a BATTERY HOLDER to contain 4-D size batteries. Design the BATTERY HOLDER to fit inside the HOUSING of the FLASHLIGHT.

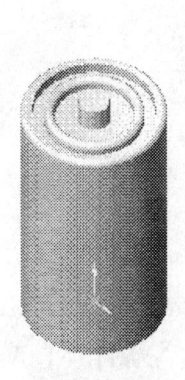

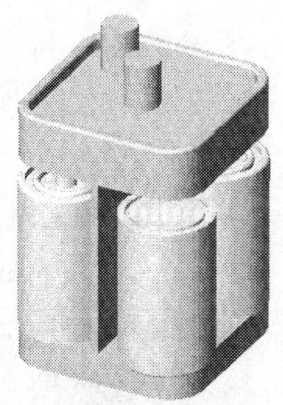

Exercise 4.2a
D-Size Battery

Exercise 4.2b
Battery Holder

Exercise 4.3: FLYWHEEL Assembly

Create a FLYWHEEL assembly. A SHAFT supports the WHEEL. The SHAFT connects two L-BRACKETS. The L-BRACKETS are mounted to a BASE PLATE. Use purchased parts to save time and cost. The only dimension provided is the WHEEL diameter.

Select a WHEEL diameter:

- 3in.
- 4in.
- 100mm

Find a material supplier using the WWW. See Exercise 4.7: www.globalspec.com.

Exercise 4.3 FLYWHEEL Assembly

FLYWHEEL Assembly Parts:

- BASE PLATE
- BUSHINGS
- L-BRACKET
- BOLTS
- SHAFT

Create manual sketches of the BASE PLATE and L-BRACKET.

Create the parts.

Create the FLYWHEEL Assembly.

Create the FLYWHEEL Drawing with Bill of Materials.

Exercise 4.4: TRAY and GLASS Parts.

Create a TRAY and GLASS. Use real objects to determine the overall size and shape of the Base feature. Below are a few examples.

Exercise 4.4 Tray and Glass

Exercise 4.5: JAR Assembly.

The JAR Assembly consists of two parts: JAR-BASE and JAR-COVER. Create a JAR-BASE. Save the JAR-BASE as a new part, JAR COVER. Use the dimensions from the JAR-BASE to determine the size of the JAR-COVER. Dimensions are not provide for this exercise.

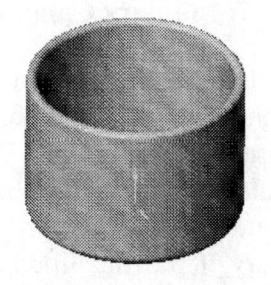

Exercise 4.5 Jar Assembly

Exercise 4.6: EMBOSSED STAMP Part.

Create an EMBOSSED-STAMP with your initials.

The initials are created with Extruded Sketched text. How do you create the text? Answer: Explore the command with SolidWorks on-line Help. Click Help. Click Index. Enter text. Click extruded text on model. Follow the instructions.

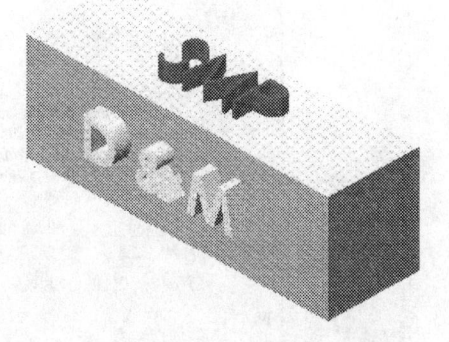

Exercise 4.6 Embossed Stamp

Exercise 4.7: Industry Collaborative Exercise.

Engineers and designers spend a great deal of time searching for product suppliers and part specifications. How do you obtain a supplier for the batteries used in this project? What are the overall dimensions and voltage of a D size battery compared to the current 6-volt battery design? Research suppliers and part information utilizing the URL: http:// www.globalspec.com. Enter Battery. Click the Find button. Select D for battery size.

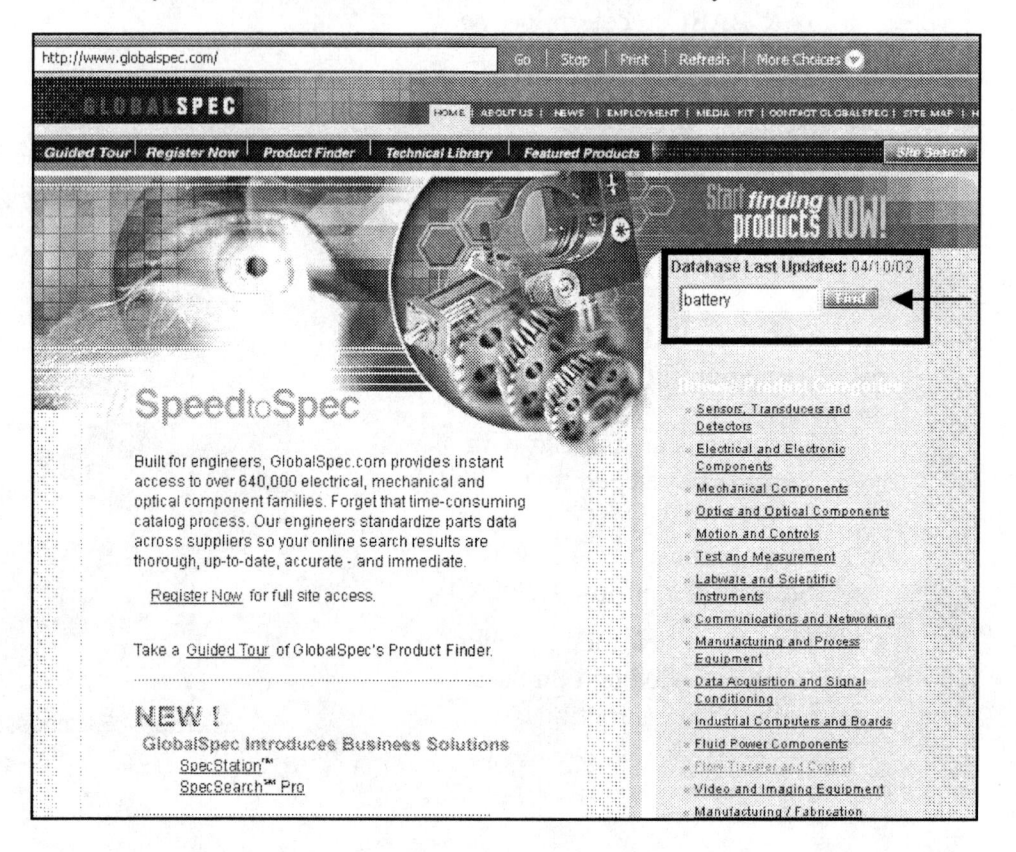

Gold Peak Industries of North America is the supplier. Select Search Results. Record the overall dimensions for the D size battery and voltage requirements.

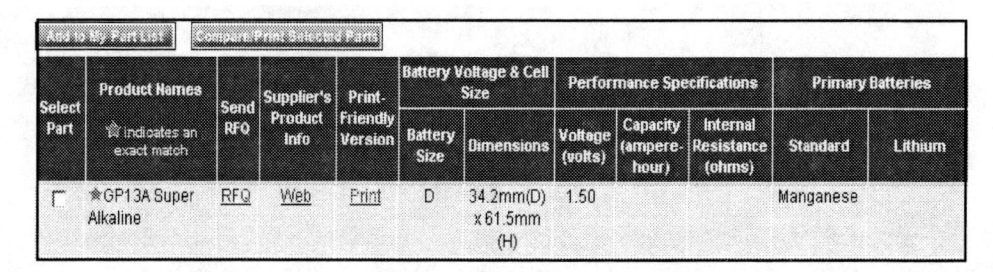

The second design option for the FLASHLIGHT assembly requires a battery holder and 4-D size batteries. Does a supplier for a 4-D battery holder exist? If so, list the name of the supplier, material and the overall size of the battery holder.

Engineering Design with SolidWorks   **Extrude and Revolve Features**

Exercise 4.8: Industry Collaborative Exercise.

PUMP Assembly

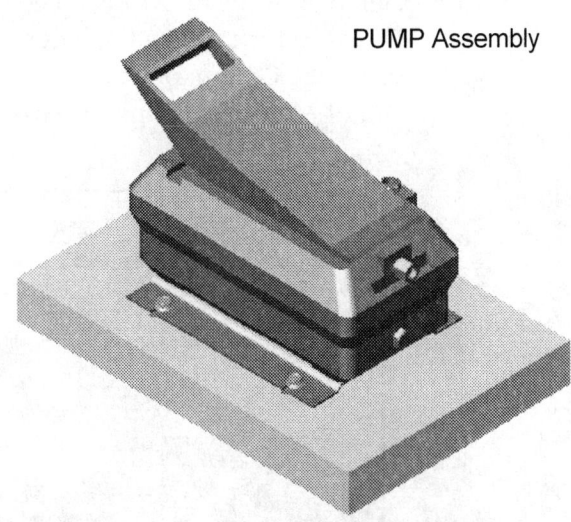

Enerpac (A Division of Actuant, Inc.) specializes in the manufacturing and distribution of high-pressure hydraulic tools and cylinders. Enerpac provides solutions for heavy lifting, pressing, pulling, and clamping for the construction, industrial maintenance and manufacturing industries.

a) Create the PUMP assembly. Your first task is to find a Turbo II® air hydraulic pump with a flow rate of 2.8 liters/min. Obtain the pump information and component from www.enerpac.com. The pump is mounted to a plate with four flange bolts. Manually sketch the top view of the mounting plate and the location of 4 slotted holes. Create the mounting plate part. Create a new assembly that contains the mounting plate, pump and flange bolts. What is the air pressure range required to operate this Turbo II® air-hydraulic pump?

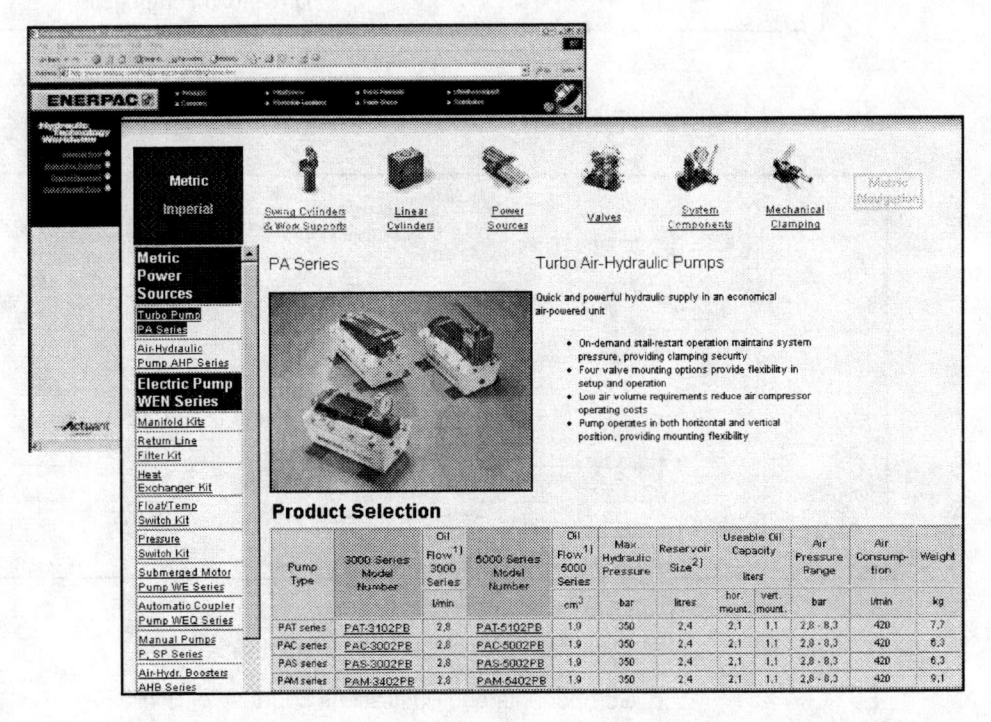

Components and illustrations courtesy of ENERPAC, Milwaukee, Wisconsin USA.

b) Your second task is to find a left turning swing cylinder with a maximum clamping force of 2.1 kN. The swing cylinder utilizes a standard clamp arm. Download the Swing Cylinder and the Clamp. Create the new assembly that mates the Clamp to the Swing Cylinder. Create an assembly drawing with a Bill of Materials listing the two components with part number and description.

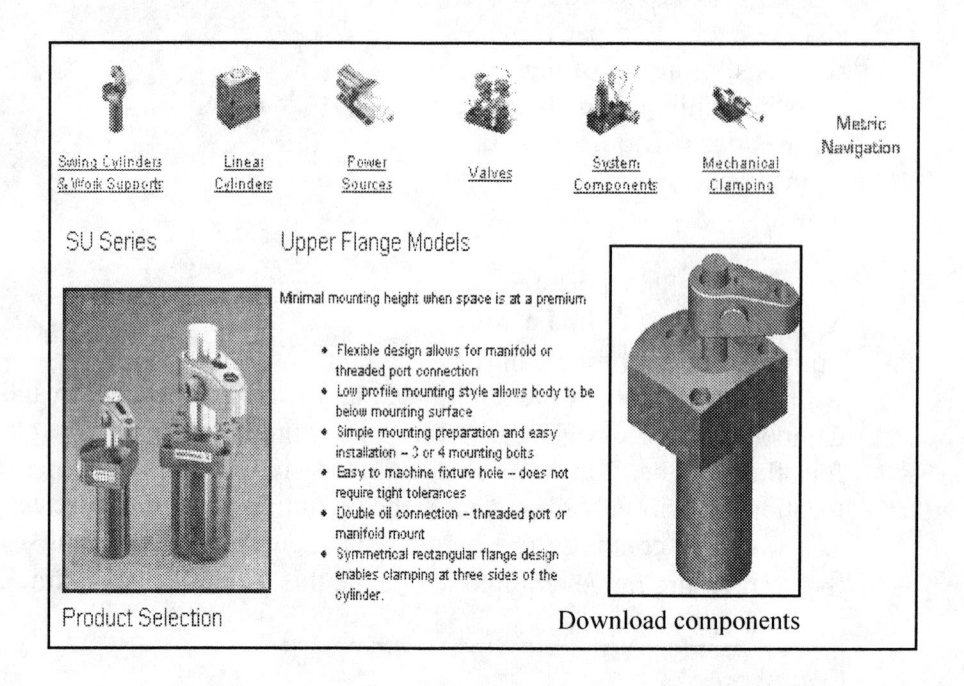

Clamping Force [1] (kN)	Stroke (mm) Clamp	Stroke (mm) Total	Left Turning	Right Turning	Cylinder Effective Area (cm²) Clamp	Cylinder Effective Area (cm²) Unclamp	Oil Capacity (cm³) Clamp	Oil Capacity (cm³) Unclamp	Max Oil Flow [1] (l/min)	Standard Clamp Arm	Long Clamp Arm
Single Acting			Model Number [2]							Sold Seperately	
2,1	8,1	16,5	SULS-22	SURS-22	0,77		1,31		0,2	CAS-22	CAL-22
4,9	9,9	22,6	SULS-52	SURS-52	1,81		4,10		0,4	CAS-52	CAL-52
8,0	11,9	22,1	SULS-92	SURS-92	3,16		6,88		1,0	CAS-92	CAL-92
10,7	12,7	28,4	SULS-121	SURS-121	4,06		11,47		1,6	CAS-121	CAL-122
17,4	14,0	27,9	SULS-202	SURS-202	7,10		19,99		2,3	CAS-202	CAL-202
33,1	16,0	30,0	SULS-352	SURS-352	12,39		37,20		3,9	CAS-352	CAL-352

Components and illustrations courtesy of ENERPAC, Milwaukee, Wisconsin USA.

Engineering Design with SolidWorks   Sweep and Loft Features

Project 5
Sweep and Loft Features

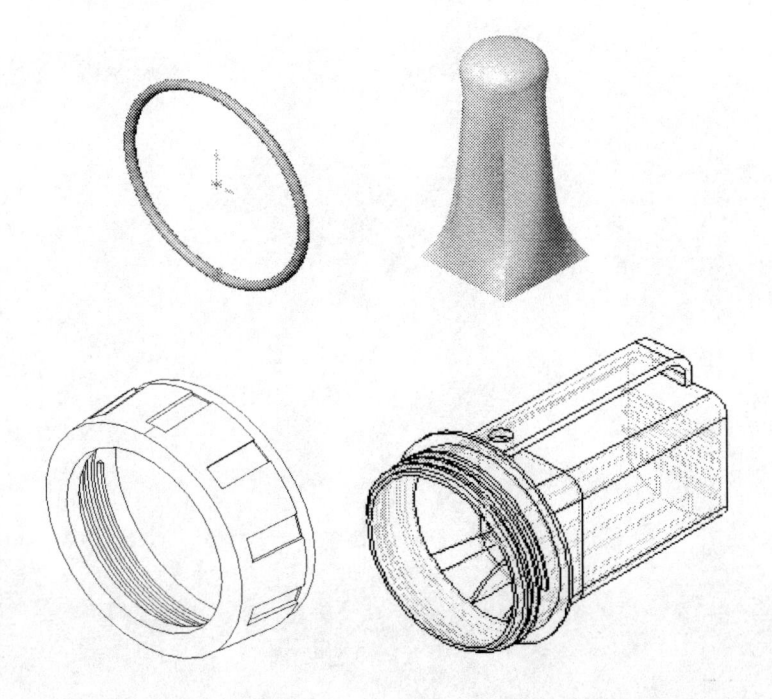

Below are the desired outcomes and usage competencies based on the completion of Project 5.

Project Desired Outcomes:	Usage Competencies:
Model four flashlight parts: O-RING, SWITCH, LENSCAP and HOUSING.	Comprehension of the fundamental definitions and process of Feature-Based 3D Solid Modeling using Sweeps and Lofts.
Four Assemblies: • LENSANDBULB assembly. • CAPANDLENS assembly. • BATTERYANDPLATE assembly. • FLASHLIGHT assembly.	Ability to combine multiple features to create components. Knowledge of additional assembly techniques.

Notes:

Project 5 – Sweep and Loft Features

Project Objective

Model four parts of the FLASHLIGHT assembly. Model the O-RING, SWITCH, LENSCAP and HOUSING.

The LENSCAP and HOUSING are designed plastic parts. The LENS and BATTERY is enclosed by the LENSCAP and HOUSING respectively. The HOUSING provides storage and support for the BATTERY.

The SWITCH and LENSCAP connect to the HOUSING. The LENSCAP thread fastens to the HOUSING thread.

Create four assemblies: Create the LENSANDBULB assembly, CAPANDLENS assembly, BATTERYANDPLATE assembly and the final FLASHLIGHT assembly. The LENSANDBULB, CAPANDLENS and BATTERYANDPLATE are subassemblies of the FLASHLIGHT assembly.

On completion of this project, you will be able to:

- Choose the best profile for sketching.

- Choose the proper sketch plane.

- Create a Template: English and Metric units.

- Set System Options.

- Set Document Properties.

- Set Units.

- Customize Toolbars.

- Export Files.

- Use the following SolidWorks features:

    o Sweep feature.

    o Loft feature.

    o Base-Extrude with a draft angle.

    o Revolve a Boss feature.

- Rib.
- Loft.
- Pattern.

Project Situation

Communications is a major component to a successful development program. Provide frequent status reports to the customer and to your team members.

Communicate with suppliers. Ask questions. Check on details. What is the delivery time for the BATTERY, LENS and SWITCH parts?

Talk to colleagues to obtain useful manufacturing suggestions and ideas.

Your team decided that plastic injection molding is the most cost effective method to produce large quantities of the desired part.

Investigate surface finishes that support the customers advertising requirement. You have two fundamental choices:

- Adhesive label.
- Silk-screen.

There are time, quantity and cost design constraints. For the prototype, use an adhesive label. In the final product, create a silk screen.

Investigate the options on O-Ring material. Common O-Ring materials for this application are Buna-N (Nitrile®) and Viton®. Be cognizant of compatibility issues between O-Ring materials and lubricants.

Project Overview

Model four parts in this project:

- O-RING.
- SWITCH.
- LENSCAP.
- HOUSING.

Two Base feature types are discussed in this project:

- Sweep feature – O-RING.
- Loft feature – SWITCH.

The LENSCAP and HOUSING are created with a combination of features.

Use the Sweep feature as the O-RING Base feature, Figure 5.1a.

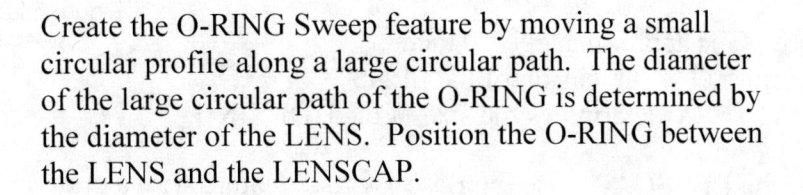

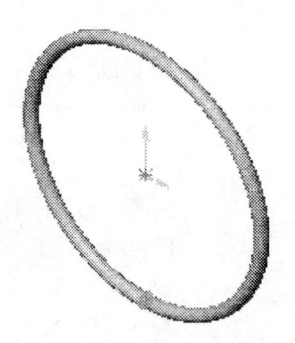

Figure 5.1a

Create the O-RING Sweep feature by moving a small circular profile along a large circular path. The diameter of the large circular path of the O-RING is determined by the diameter of the LENS. Position the O-RING between the LENS and the LENSCAP.

The SWITCH is a purchased part. The SWITCH is a complex assembly. Use the Loft feature to create the SWITCH, Figure 5.1b.

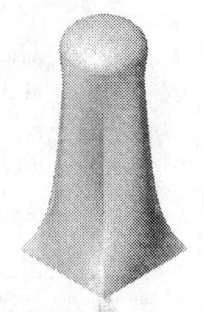

Figure 5.1b

The LENS and BATTERY is enclosed by the LENSCAP and HOUSING respectively. How do you design the LENSCAP and HOUSING to ease the transition from development to manufacturing? Answer: Review the fundamental design rules behind the plastic injection manufacturing process:

- Maintain a constant wall thickness. Inconsistent wall thickness creates stress.
- Create a radius on all corners. No sharp edges. Sharp edges create vacuum issues when removing the mold.
- Allow a minimum draft angle of 1 degree. Draft sides and internal ribs. Draft angles assist in removing the part from the mold.

Obtain additional information on material and manufacturing from material suppliers or manufacturers. Example: GE Plastics of Pittsfield, MA (www.geplastics.com & www.gepolymerland.com) provides design guidelines for selecting raw materials and creating plastic parts and components.

Review the three developed parts in Project 4:

- BATTERY.
- LENS.
- BATTERYPLATE.

The BATTERY and LENS are purchase parts. The BATTERYPLATE is a created part.

Position the BATTERY next to the LENS, Figure 5.2a.

The diameter of the LENSCAP must be slightly larger than the diameter of the LENS.

The overall height of the LENSCAP and HOUSING is approximately 7.0 inch, [177.8mm]. The dimension of the LENS and BATTERY determines the size of the HOUSING.

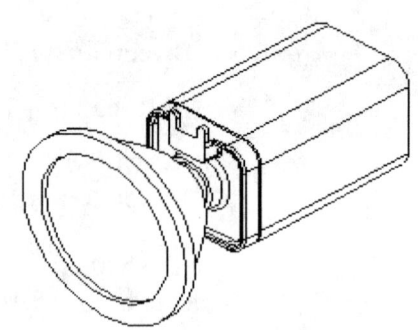

Figure 5.2a

Create the LENSCAP. Use the Base-Extrude feature with a draft angle, Figure 5.2b.

Create an internal thread in the LENSCAP. Use the Sweep feature and a sketched Helix/Spiral curve. Revolve a Boss feature around the outside of the LENSCAP. Use a Circular Pattern feature.

Create the HOUSING. The Base Extrude feature dimensions are determined from the LENS and LENSCAP. Use the Loft feature to transition a circular profile of the LENSCAP to a square profile of the BATTERY.

Figure 5.2b

Use the Extrude feature to enclose the BATTERY. Create a constant wall thickness with the Shell feature. Create the handle. Use the Sweep feature.

The Rib feature provides extra internal support and rigidity, Figure 5.2c.

Ribs maintain the proper alignment of the BATTERY and LENS.

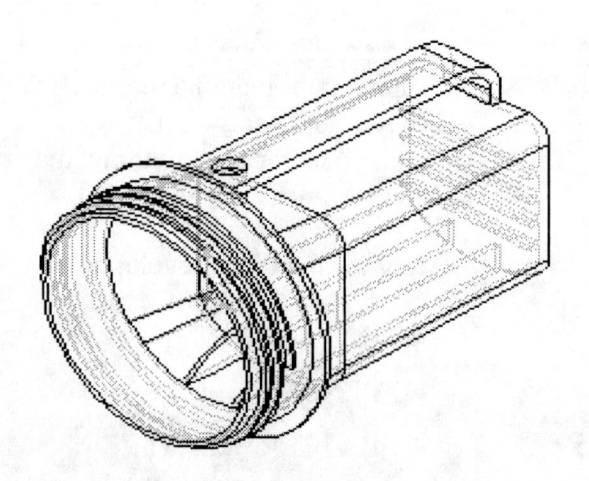

Figure 5.2c

Engineering Design with SolidWorks　　　　　　　　　　**Sweep and Loft Features**

Note: In this project, wall thickness, clearance fit and thread size are increased for improve illustration. Parts are simplified.

O-RING

The O-RING captures the LENS to the LENSCAP. Create the O-RING. Use the Sweep feature as the Base feature. A Sweep feature adds material by moving a profile along a path.

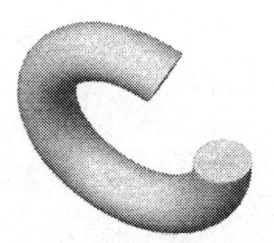

Figure 5.3

Note: The Sweep feature requires a cross section and a path. Example: Think of a doughnut with a bite, Figure 5.3.

The cross section is the small circle, "bite" of the doughnut.

The path is the large circle that constitutes the circumference of the donut.

O-RING Feature Overview

Create the O-RING. The default orientation of the O-RING is based on the orientation in the assembly.

Sketch a large circular path on the Front plane. Sketch the small cross section on the Right plane, Figure 5.4a. Combine the path cross section to create the Sweep feature, Figure 5.4b.

The diameter of the large circular path is 4.350 inch, [110.49mm]. The outside diameter of the LensCover feature is 4.466 inch, [113.44mm], Figure 5.4c.

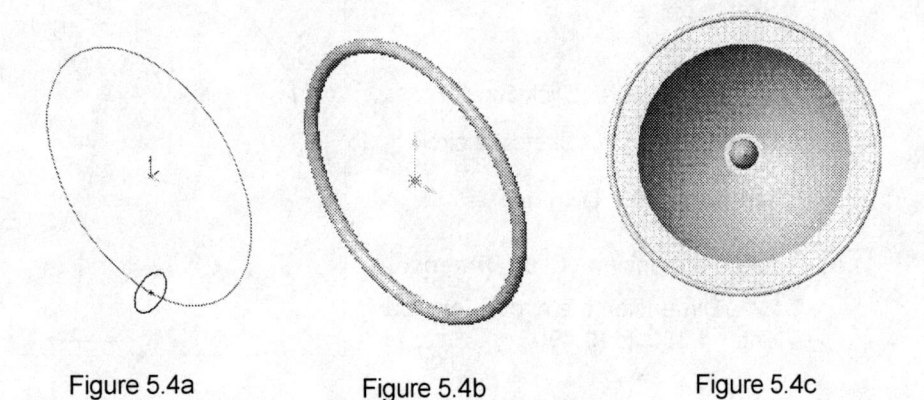

Figure 5.4a　　　Figure 5.4b　　　Figure 5.4c

Create the O-RING

Create the O-RING with a Sweep Base feature.

The Sweep-Base feature uses:

- A circular path sketched on the Front plane.

- A small cross section sketched on the Right plane.

Utilize the PART-IN-ANSI Template created in Project 4 for inch units. Utilize the PART-MM-ISO Template created in Project 4 for millimeter units. Millimeter dimensions are provided in brackets[x].

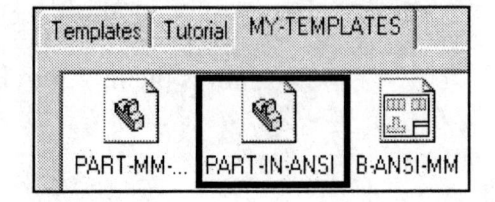

Create the O-RING.

1) Click **New**. Click the **MY-TEMPLATES** tab. Click **PART-IN-ANSI**, [**PART-MM-ISO**] from the Template dialog box. Click **OK**.

2) Save the empty part. Click **Save**. Select **ENGDESIGN-W-SOLIIDWORKS\PROJECTS** for the Save in file folder.

3) Enter **O-RING** for file name. Enter **O-RING FOR LENS** for Description. Click the **Save** button.

Create the Sweep path. 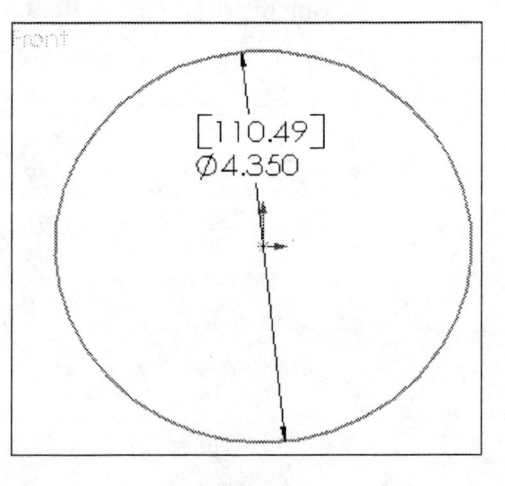
4) Select the Sketch plane. Click **Front** from the FeatureManager. The Front plane is the default Sketch plane.

5) Sketch the path. Click **Sketch**. Click **Circle**. Sketch a circle centered at the **Origin**.

6) Add a dimension. Click **Dimension**. Dimension the circumference. Enter **4.350**, [**110.49**].

7) Close the Sketch. Click **Sketch**. Sketch1 is displayed in the FeatureManager.

8) Rename Sketch1 to sketch-path.

Engineering Design with SolidWorks   Sweep and Loft Features

Create the Sweep profile.

9) Display the Isometric view. Click **Isometric**.

10) Select the Sketch plane. Click the **Right** plane from the FeatureManager.

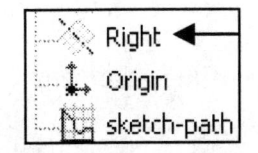

11) Sketch the profile (cross section). Click **Sketch**. Click **Circle**. Create a small **circle** left of the sketch-path.

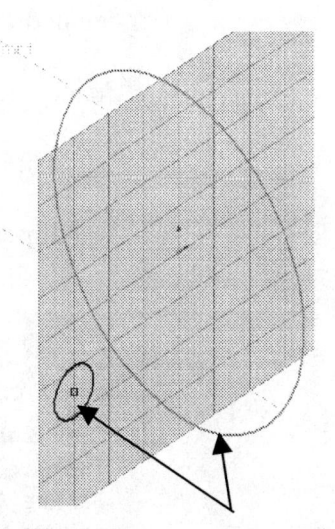

The Pierce geometric relation positions the center of the cross section on the sketched path.

12) Add geometric relations. Right-click **Select** in the Graphics window. Hold the **Ctrl** key down. Click the **small circle center point**. Click the **large circle circumference**. Click **Pierce**. Release the **Ctrl** key. The center point of the small circle pierces the sketch-path (large circle).

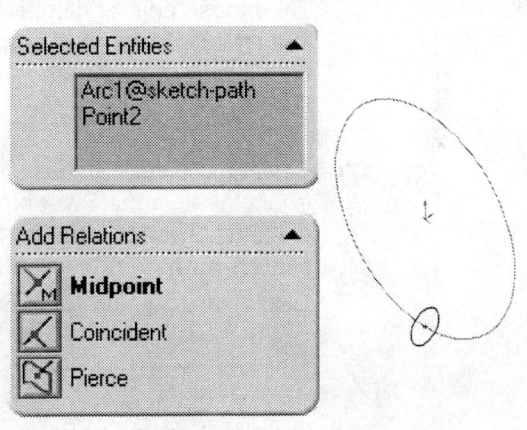

13) Add a dimension. Click **Dimension**. Dimension the small circle. Enter diameter. Enter **.125**, **[3.18]**.

14) Close the Sketch. Click **Sketch**.

15) Rename Sketch2 to sketch-profile.

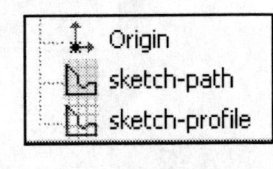

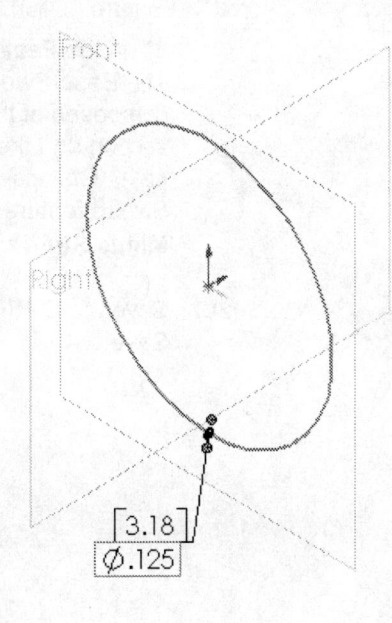

PAGE 5 - 9

Sweep and Loft Features   Engineering Design with SolidWorks

For detailed cross sections, perform the following steps to create the profile:

- Create a large cross section profile.

- Pierce the profile to the path.

- Add dimensions to create the true size.

Sweep the Profile.

16) Click **Sweep** from the Feature toolbar. The Sweep dialog box is displayed. Split the FeatureManager. Position the **mouse pointer** at the top of the FeatureManager.

   The mouse pointer displays the Split bar. Drag the **Split bar** half way down to display the FeatureManager and the Base-Sweep PropertyManager.

17) Select the Sweep profile. The sketch-profile is displayed in the Sweep profile text box. Click the **Sweep path** text box. Click the **sketch-path** from the FeatureManager. Display the Sweep. Click **OK**.

18) Rename Sweep1 to Base-Sweep.

19) Expand the Base-Sweep feature. Click the **Plus Sign** in the FeatureManager. The Base-Sweep is composed of the Sketch-section and the Sketch-path. Collapse the Base-Sweep feature. Click the **Minus Sign**.

20) Save the O-RING. Click **Save**.

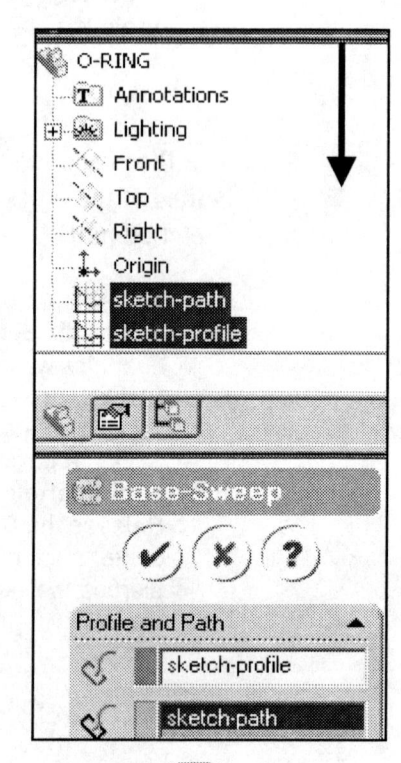

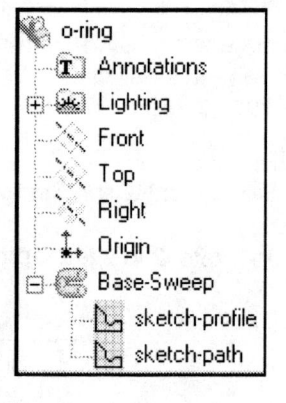

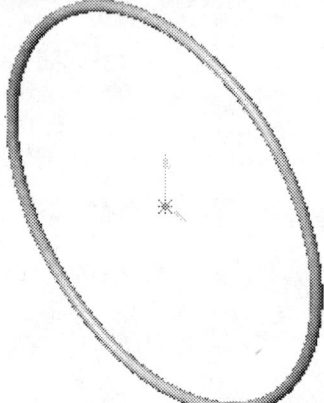

Engineering Design with SolidWorks   Sweep and Loft Features

SWITCH

The SWITCH is a purchased part. The SWITCH is a complex assembly. Create the outside casing. Create the SWITCH with the Base-Loft feature, Figure 5.5.

A Loft feature uses two or more cross sections to define geometry contours. Sketch each cross section on a plane. A path connects the cross sections to create the solid Loft feature.

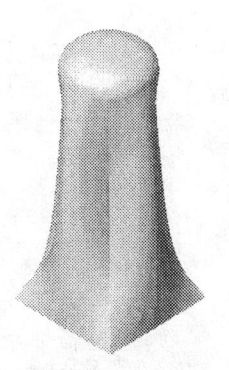

Figure 5.5

SWITCH Feature Overview

The orientation of the SWITCH is based on its position in the assembly. The SWITCH is comprised of three cross sections. Each cross section is sketched on a different plane, Figure 5.6a. Create the Loft feature, Figure 5.6b. Create a Dome feature to the top face of the Loft, Figure 5.6c.

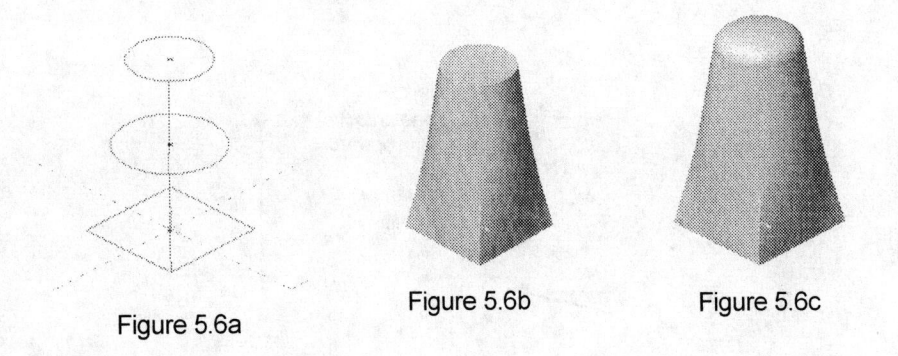

Figure 5.6a   Figure 5.6b   Figure 5.6c

Create the SWITCH – Use the Loft Feature

Create the SWITCH with a Loft feature. The Loft feature requires three cross sections. Sketch each cross section profile on a unique plane.

Create a new part.

21) Click **New**. Click the **MY-TEMPLATES** tab. Click **PART-IN-ANSI**, [**PART-MM-ISO**] from the Template dialog box. Click **OK**.

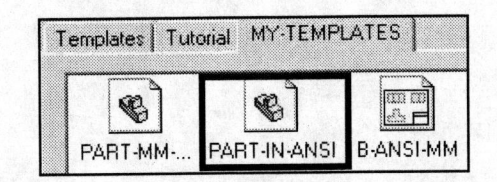

22) Save the empty part. Click **Save**. Select **ENGDESIGN-W-SOLIIDWORKS\PROJECTS** for the Save in file folder.

PAGE 5 - 11

Sweep and Loft Features **Engineering Design with SolidWorks**

23) Enter **SWITCH** for file name.

24) Enter **SWITCH, BUTTON STYLE** for Description. Click the **Save** button.

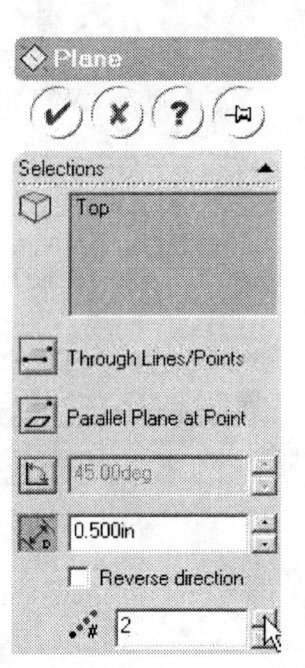

Create the Sketch planes.

25) Create the first Sketch plane. The first Sketch plane is the Top plane. Right-click the **Top** plane from the FeatureManager. Click **Show**.

26) Display the Isometric view. Click **Isometric**.

27) Create Plane1 and Plane2. Copy the Top plane. Hold the **Ctrl** key down. Click and Drag the **Top** plane upward. Create the offset Sketch plane. Release the **mouse button**. Release the **Ctrl** key. Enter **.5**, [**12.7**] for Distance. Enter **2** for # of Planes. Click **OK**.

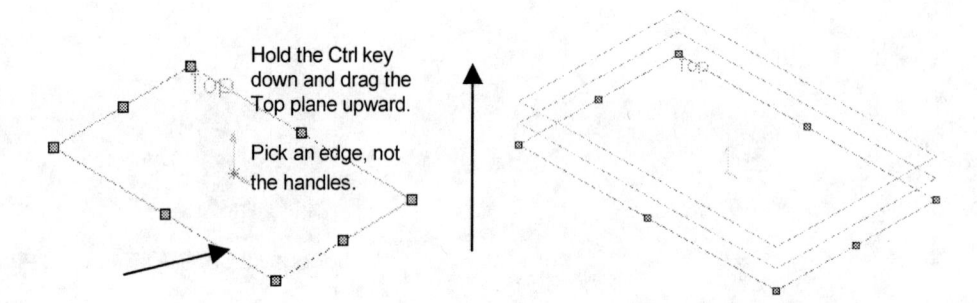

Hold the Ctrl key down and drag the Top plane upward.

Pick an edge, not the handles.

28) Display Sketch planes from the Front view. Click **Front**.

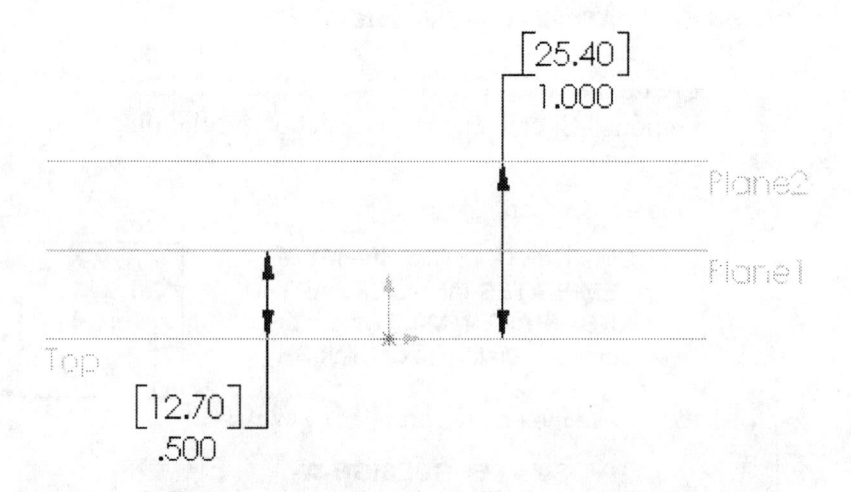

[25.40]
1.000

[12.70]
.500

PAGE 5-12

Engineering Design with SolidWorks — Sweep and Loft Features

29) Display the Sketch planes. Click **Isometric**.

Create Sketch1. Sketch1 is a square on the Top plane.

30) Select the Sketch plane. Click the **Top** plane.

31) Create the Sketch. Click **Sketch**.

[12.70]
.500

32) Display the Top view. Click **Top**.

33) Sketch the Profile. Click **Rectangle**. Sketch a **Rectangle** centered about the Origin.

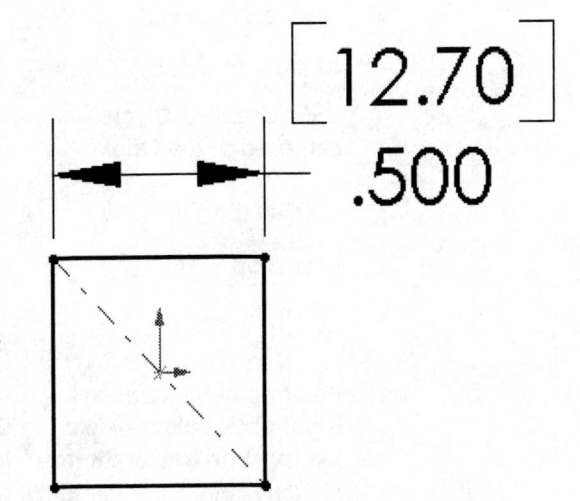

34) Click **Centerline**. Create a diagonal centerline between two opposite corner points of the rectangle. Click the two **corner points** of the rectangle. Double-click to **end** the centerline.

35) Add a Midpoint relation. Right-click **Select**. Hold the **Ctrl** key down. Click the **Origin**. Click the diagonal **centerline**. Click **Midpoint**. Release the **Ctrl** key. Click **Close Dialog**.

36) Add an Equal relation. Hold the **Ctrl** key down. Click the **left vertical line** and the **top horizontal line**. Click **Equal**. Release the **Ctrl** key. Click **Close Dialog**. All four sides are equal.

37) Add a dimension. Click **Dimension**. Dimension one line of the rectangle. Enter **.500**, **[12.7]**.

Rectangle on the Top plane.

Centered at Origin.

38) Close Sketch1. Click **Sketch**.

39) Rename Sketch1 to Sketch1-lower.

40) Display the Isometric View. Click **Isometric**.

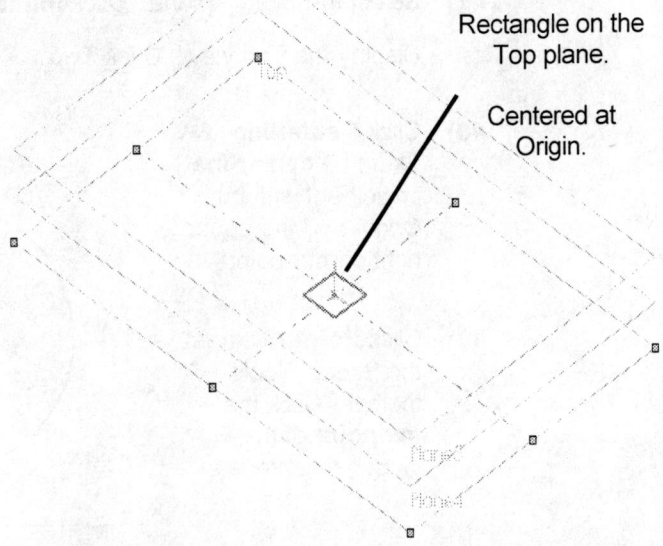

PAGE 5 - 13

Sweep and Loft Features   **Engineering Design with SolidWorks**

Create Sketch2. Sketch2 is a circle on Plane1.

41) Select the Sketch plane. Click **Plane1**.

42) Create the Sketch. Click **Sketch**. Click **Circle**. Create a **Circle** centered at the Origin.

Circle sketched Plane1

Centered at the Origin and Tangent to the Square

43) Display the Top view. Click **Top**.

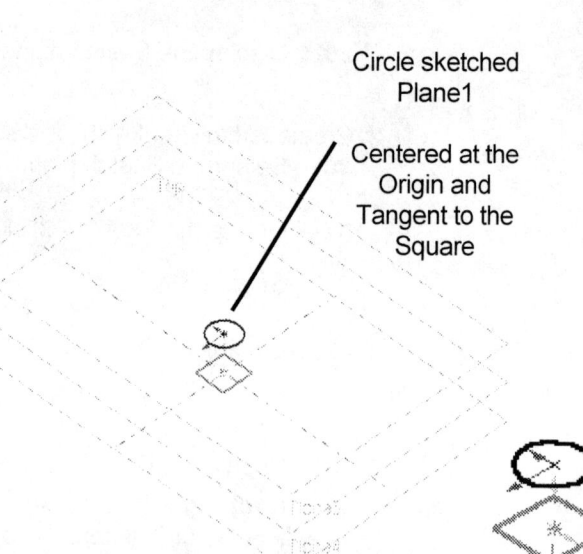

44) Add geometric relations.
Right-click **Select**. Hold the **Ctrl** key down. Click the top of the **circle**. Click the **top horizontal Sketch1-lower line**. Click the **Tangent** button. Release the **Ctrl** key. Click **Close Dialog** ✔.

45) Close Sketch2. Click **Sketch**.

46) Rename Sketch2 to Sketch2-middle.

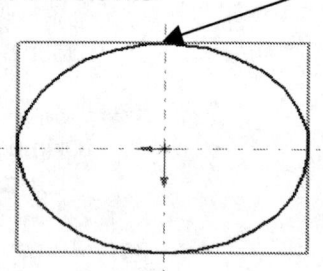

Create a Tangent relation between the circle and the line.

Create Sketch3. Sketch3 is a circle on Plane2.

47) Select the Sketch plane. Click **Plane2**. Create the Sketch. Click **Sketch**. Display the Top view. Click **Top**.

48) Click **Centerline**. Sketch a **centerline** coincident with the Origin and the upper right corner point.

49) Click **Point** from the Sketch Tools toolbar. Click the **midpoint** of the diagonal centerline.

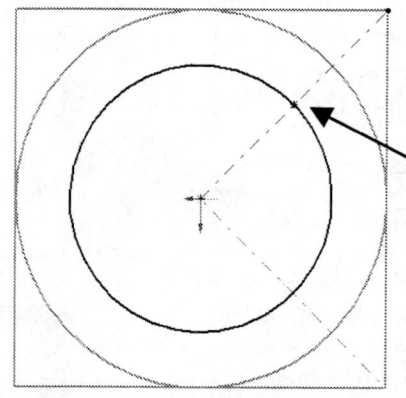

Centerline sketched from the Origin to the right corner point.

Add Point at the midpoint of the centerline.

Sketched Circle on Plane2 centered at Origin.

PAGE 5- 14

Engineering Design with SolidWorks **Sweep and Loft Features**

50) Click **Circle**. Create a **Circle** centered at the Origin to the midpoint of the centerline.

51) Close Sketch3. Click **Sketch**.

52) Rename Sketch3 to Sketch3-upper.

53) Hide the Planes. Click **View** from the Main menu. Uncheck **Planes**. Fit the model to the Graphics window. Press the **f** key.

Create the Loft.

54) Click **Loft** from the Feature toolbar. The Loft dialog box is displayed. Display the Isometric view. Click **Isometric**.

55) Split the FeatureManager. Drag the **Split bar** half way down to display the FeatureManager.

56) Select the Profiles inside the Graphics window. Click the **front corner** of Sketch1-lower. Click **Sketch2-middle**. Click **Sketch3-upper**.

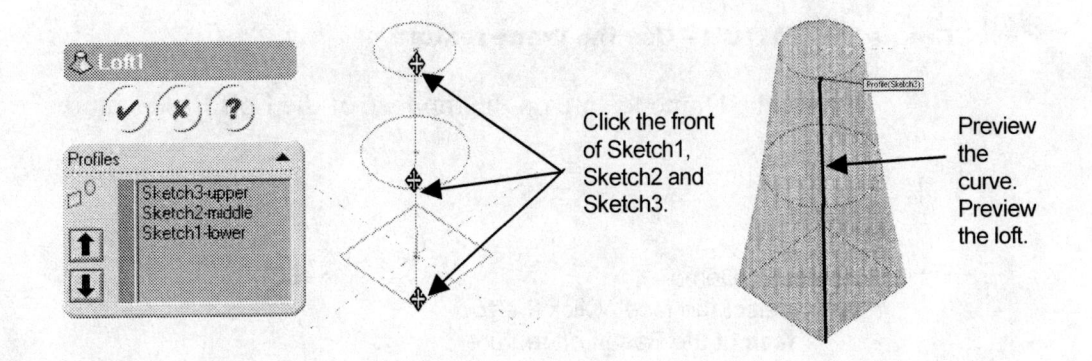

Use caution when selecting the profiles and location.

The system displays a preview curve and preview loft as the profiles are selected.

PAGE 5 - 15

Sweep and Loft Features   **Engineering Design with SolidWorks**

Use the Up button and Down button in the Loft dialog box to rearrange the order of the profiles. The correct order for the profiles is:

- Sketch1-lower.
- Sketch2-middle.
- Sketch3-upper.

Note: If you select an incorrect location on the Sketch, Right-click in the Graphics window. Click Clear Selections. Select the profiles again.

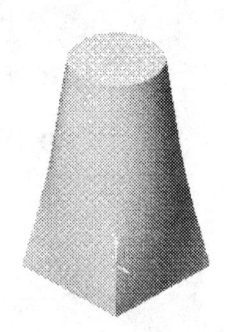

57) Display the Loft Base feature. Click **OK**.

58) Rename Loft1 to Base Loft.

59) Save the SWITCH. Click **Save**.

Create the SWITCH – Use the Dome feature

Create the Dome feature on the top face of the Loft Base feature.

Create the Dome.
60) Select the face. Click the **top face** of the Base-Loft feature.

61) Click **Dome** from the Feature toolbar. The Dome dialog box is displayed. Enter **.100**, **[2.54]**. Display the Dome.

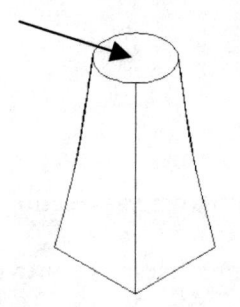

62) Click **OK**.

63) Expand the Base-Loft feature. Click the **Plus Sign** in the FeatureManager. The Base-Loft is composed of: Sketch1-upper, Sketch2-middle and Sketch3-lower.

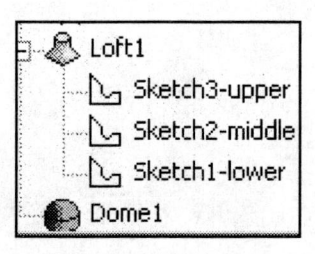

Engineering Design with SolidWorks
Sweep and Loft Features

64) Modify the Base-Loft feature. Double-click on the **Base-Loft**. Double-click on the Plane1 offset dimension, **.500**, [**12.70**]. Enter **.125**, [**3.18**].

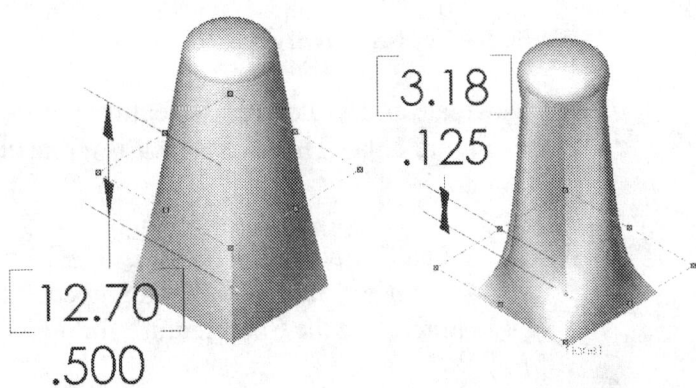

65) Click **Rebuild**.

66) Save the SWITCH. Click **Save**.

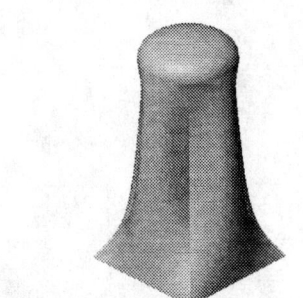

LENSCAP

The LENSCAP is a plastic part used to position the LENS to the HOUSING.

An O-RING is positioned between the LENS and LENSCAP, Figure 5.7.

How do you sketch the LENSCAP? Sketch the LENSCAP on the Front plane.

Identify the key dimensions from the LENS, diameter and depth to create the LENSCAP.

Figure 5.7

How do you determine the LENSCAP depth? Measure the depth of the LENS. The LENS is approximately 3inchs, [76.2mm] deep.

Position half of the LENS inside the LENSCAP. Position the other half within the HOUSING, Figure 5.8.

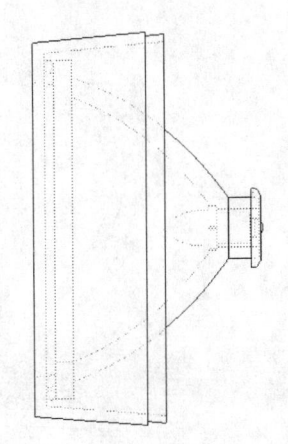

Figure 5.8

LENSCAP Feature Overview

- Create an Extruded Base feature with a circular profile on the Front plane, Figure 5.9a. The LENSCAP is a plastic part. Use a Draft angle for plastic parts.

- Add an Extruded Cut feature. The Extruded Cut feature should be equal to the diameter of the LENS Revolved-Base feature, Figure 5.9b. Create a Shell feature. Use the Shell feature for a constant wall thickness, Figure 5.9c.

- Add a Revolved-Cut feature on the back face, Figure 5.9d.

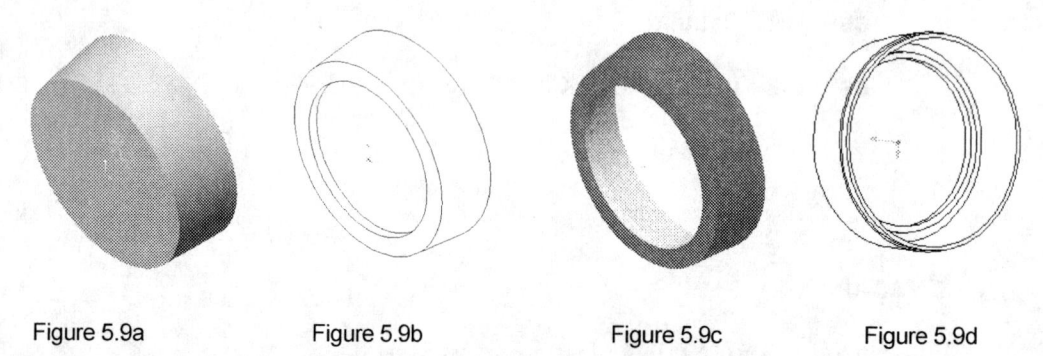

Figure 5.9a   Figure 5.9b   Figure 5.9c   Figure 5.9d

- Create an Extruded Thin Cut feature. Use the Extruded Thin Cut feature to grip the outside of the LENSCAP, Figure 5.9e.

- Create a Pattern feature. Use the Pattern feature is to create multiple instances of the Extruded Cut feature, Figure 5.9f.

- Create a thread using a sketched Helical Curve feature and a Sweep feature, Figure 5.9g.

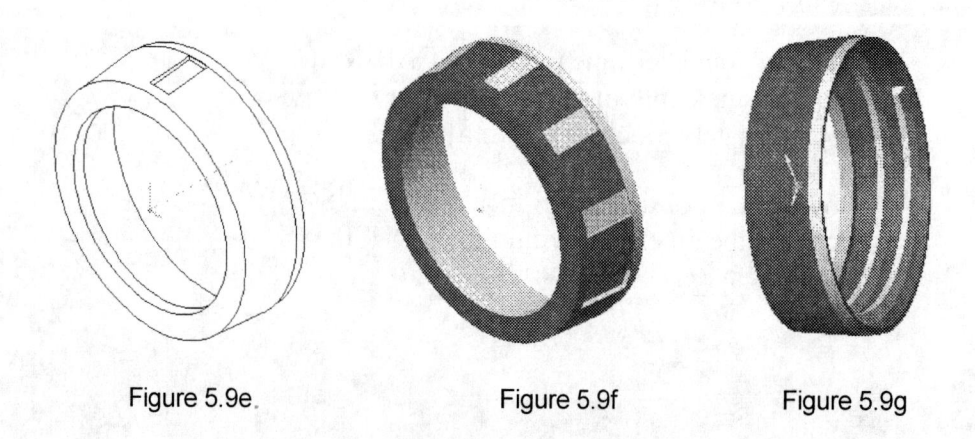

Figure 5.9e   Figure 5.9f   Figure 5.9g

Engineering Design with SolidWorks — Sweep and Loft Features

Create the LENSCAP

Create a new part.

67) Click **New**. Click the **MY-TEMPLATES** tab.

68) Click **PART-IN-ANSI**, [**PART-MM-ISO**] from the Template dialog box.

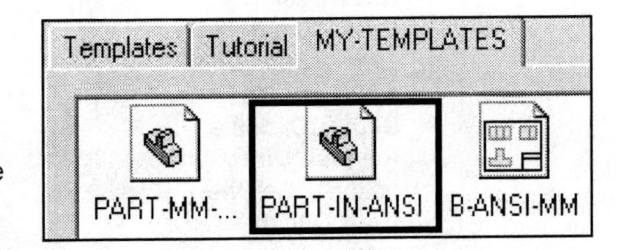

69) Click **OK**.

70) Save the empty part. Click **Save**.

71) Select **ENGDESIGN-W-SOLIIDWORKS\ PROJECTS** for the Save in file folder.

72) Enter **LENSCAP** for file name.

73) Enter **LENSCAP for 6V FLASHLIGHT** for Description.

74) Click the **Save** button.

Create the Extruded Base feature.

75) Select the Sketch plane. Click the **Front** plane.

76) Create the Sketch. Click **Sketch**. Click **Circle**. Create a circle centered at the **Origin**.

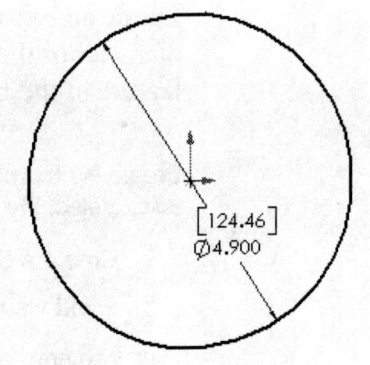

77) Add the dimension. Click **Dimension**.

78) Click the **circumference** of the circle.

79) Enter **4.900**, [**124.46**].

PAGE 5 - 19

Sweep and Loft Features   Engineering Design with SolidWorks

80) Extrude the Sketch. Click **Extruded Boss/Base**. Blind is the default Type option. Enter **1.725**, **[43.82]** for Depth. Click the **Reverse Direction** button. Click the **Draft On/Off** button. Enter **5 deg** for Angle. Click the **Draft outward** check box.

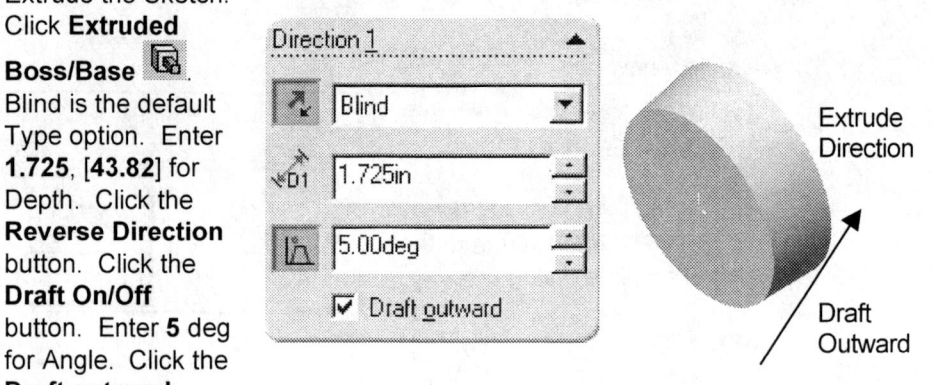

81) Display the Extruded-Base feature. Click **OK**.

82) Rename Extrude1 to Base Extrude.

83) Save the LENSCAP. Click **Save**.

Create the LENSCAP – Use the Extruded Cut Feature

Create an Extruded Cut feature on the front face of the Base feature. The diameter of the Extruded Cut equals the diameter of the Base-Revolved feature of the LENS.

Create the Extruded Cut Feature.
84) Select the Sketch plane. Click the **front face**. Create a Sketch. Click **Sketch**. Click the **Front** view. Click **Circle**. Create a circle centered at the Origin.

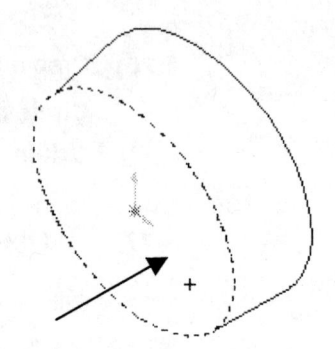

85) Add a dimension. Click **Dimension**.

Click the **circumference** of the circle. Enter **3.875**, **[98.43]**.

86) Extrude the Sketch. Click **Extruded Cut**. Blind is the default Type option. Enter **.275**, **[6.99]** for Depth. Click the **Draft On\Off** button.

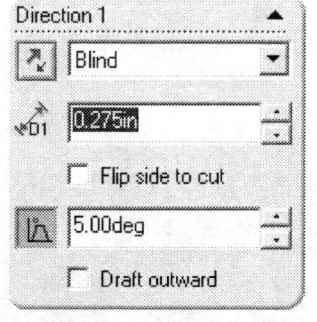

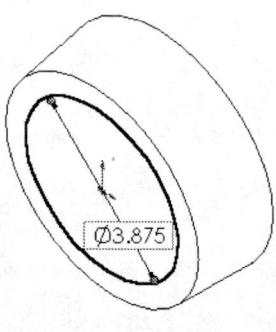

87) Enter **5** deg for Angle. Display the Cut-Extrude feature. Click **OK** ✔.

88) Rename Cut-Extrude1 to Front-Cut.

89) Save the LENSCAP. Click **Save**.

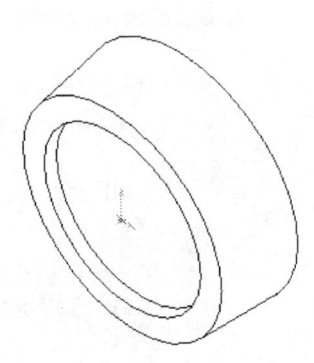

Create the LENSCAP – Use the Shell Feature

The Shell feature removes face material from the solid LENSCAP.

90) Click **Rotate**. Rotate the LENSCAP to view the part. Deactivate Rotate. Click **Rotate**. Click **Isometric**.

Create the Shell.

91) Click **Shell** from the Feature toolbar. The Shell Feature dialog box is displayed. Click the **front face** of the Front-Cut. Press the **left arrow** 5 times. Click the **back face** of the Base Extrude. Enter **.150**, [3.81] for Thickness.

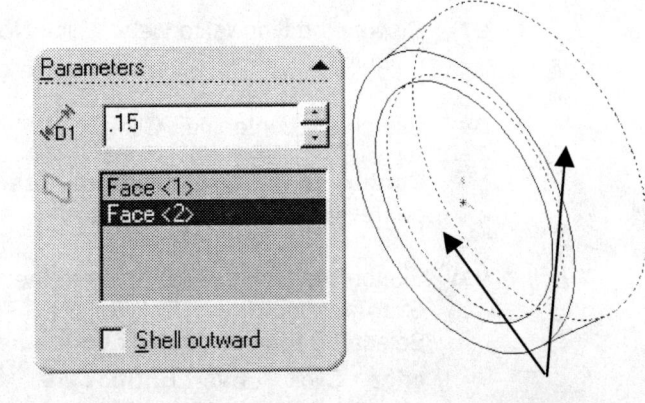

92) Display the Shell. Click **OK** ✔.

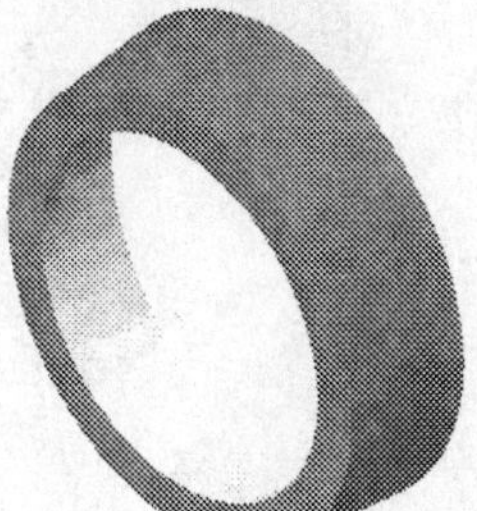

93) Display the inside of the Shell. Click **Right**. Click **Hidden Lines Visible**. Note: Use the inside gap created by the Shell feature to seat the O-RING in the assembly.

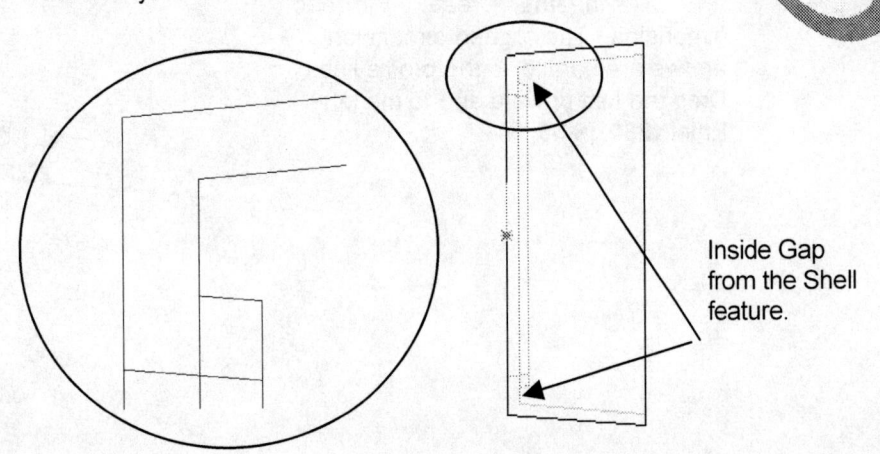

Inside Gap from the Shell feature.

Sweep and Loft Features   **Engineering Design with SolidWorks**

94) Save the LENSCAP. Click **Save**.

Create the LENSCAP – Use the Revolved Cut Thin Feature

The Revolved Cut Thin feature removes material by rotating a sketched profile around a centerline.

Utilize the Convert Entities Sketch tool to create the profile.

Create the Revolved Cut Thin feature.
95) Select the Sketch plane. Click the **Right** plane.

96) Create the Sketch. Click **Sketch**.

97) Display the Normal to view. Click **NormalTo** from the Standard View toolbar.

98) Sketch the centerline. Click **Centerline**. Sketch a **horizontal centerline** through the Origin.

99) Sketch the profile. Right-click in the **Graphics window**. Click **Select**. Select the **top silhouette outside edge**. Click **Convert Entities** from the Sketch toolbar.

Drag the left end point 2/3 towards the right endpoint.

100) Create a short line. Drag the **left endpoint** 2/3 towards the right endpoint. Release the **mouse button**.

101) Add a dimension. Click **Dimension**. Click the **line**. Create an aligned dimension. The aligned dimension arrows are parallel to the profile line. Drag the text upward and to the left. Enter **.250**, [**6.35**].

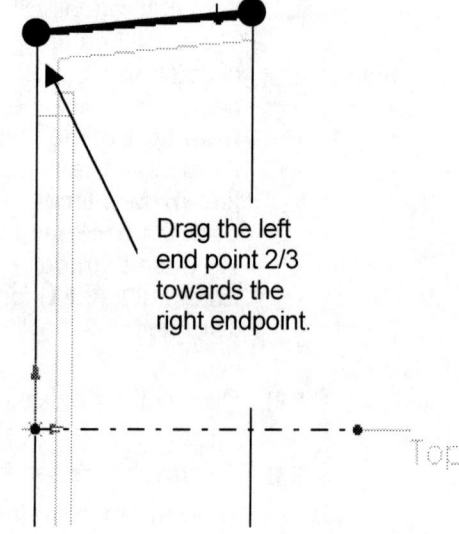

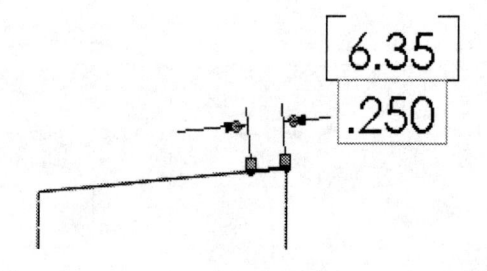

PAGE 5-22

Engineering Design with SolidWorks — Sweep and Loft Features

102) Revolve the Sketch. Click **Revolve Cut** from the Feature toolbar. Do not close the Sketch. The warning message states; "The sketch is currently open." Click **No**. Click the **Reverse** checkbox in the Thin Feature box. Enter **.050**, **[1.27]** for Direction 1 Thickness.

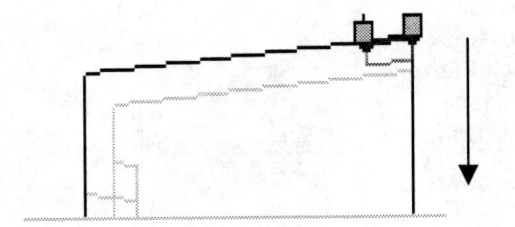

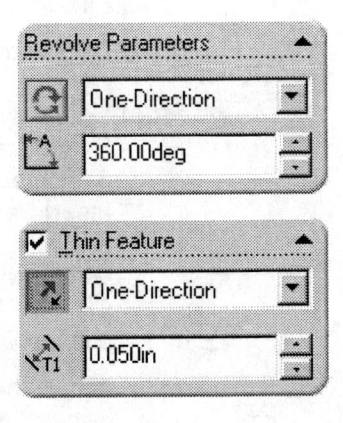

103) Display the Cut-Revolve-Thin feature. Click **OK** ✔.

104) Display the backside of the Revolve-Thin feature. **Rotate** the part. Click **Isometric**.

105) Rename Cut-Revolve-Thin1 to BackCut.

106) Save the LENSCAP. Click **Save**.

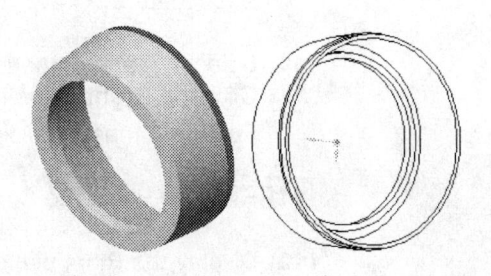

Create the LENSCAP – Use the Circular Pattern

A Pattern creates one or more instances of a feature or group of features. A Circular Pattern requires a seed feature and an axis of revolution.

The seed feature in this example is an Extruded Cut Thin feature created on a new Surface Reference plane. Create the following:

- A Surface Reference plane for the Sketch plane.
- The seed feature as an Extrude-Thin feature.
- The Circular Pattern.

Create the Surface Reference Plane.

107) Display the planes in the Isometric view. Click **Isometric**.

108) Create a Surface Reference plane for the Sketch plane. Click **Insert** from the Main menu. Click **Reference Geometry**. Click **Plane**. Click the **On Surface** button. Drag the **Split bar** ⸺ half way down to display the FeatureManager and the Plane PropertyManager.

109) Select the Surface. Click the **top outside conical face** of the Extruded-Base feature.

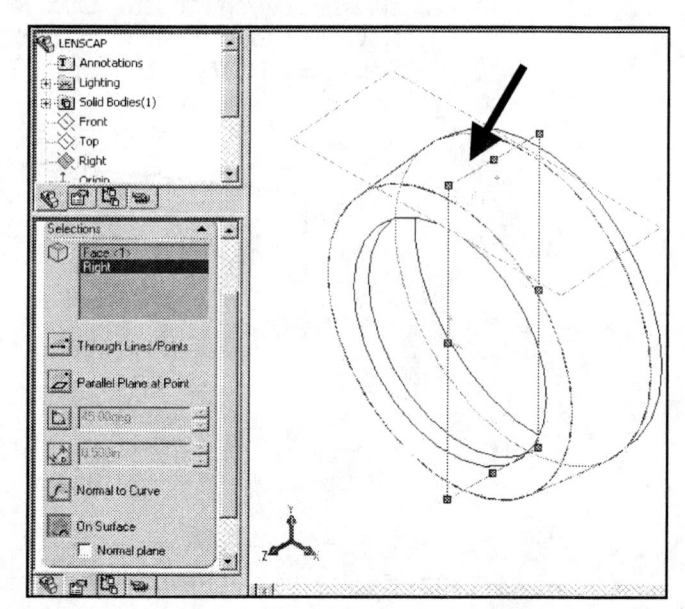

110) Select a plane perpendicular to the surface. Click the **Right** plane from the FeatureManager. Click **OK**.

111) Rename Plane1 to SurfacePlane.

112) Display the Right plane. Click the **Right** plane. Right-click **Show**.

113) Save the LENSCAP. Click **Save**.

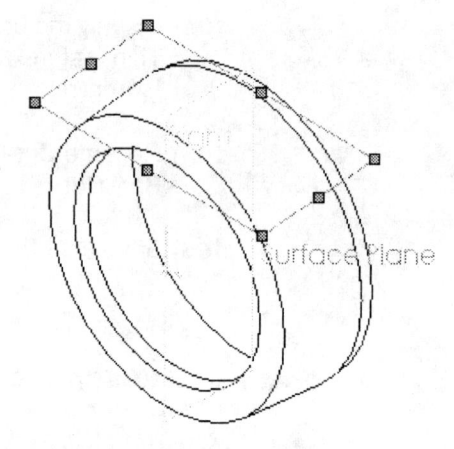

Create the seed feature.

114) Select the Sketch plane. Click the **SurfacePlane**. Click **NormalTo**.

115) Create the Sketch. Click **Sketch**. Click **Line**. Sketch a **vertical line** coincident with the Right plane. The first point is coincident with the Back Cut. The second point is coincident with the circular edge of the FrontCut.

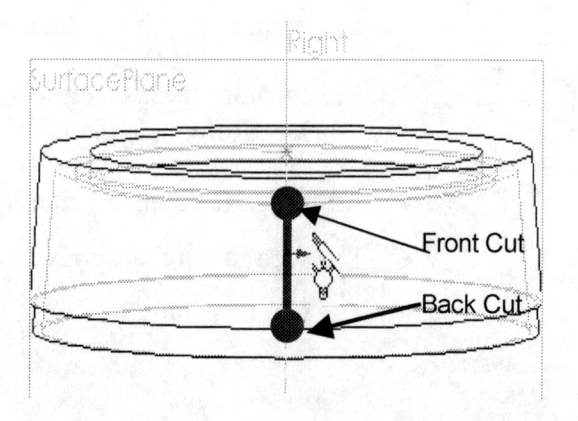

PAGE 5-24

Engineering Design with SolidWorks — Sweep and Loft Features

116) Add Relations. Right-click **Select**. Hold the **Ctrl** key down. Click the **bottom end point** of the line. Click the **Back Cut edge**. Click **Coincident**. Release the **Ctrl** key. Click **Close Dialog** ✔.

117) Hold the **Ctrl** key down Click the **top end point** of the line. Click the **Front Cut edge**. Click **Coincident**. Release the **Ctrl** key. Click **Close Dialog** ✔.

118) Extrude the Sketch. Click **Extruded Cut**. Click the **Thin Feature** check box. Select **Mid-Plane** for Thin Feature Type. Enter **.500**, **[12.7]** for Thickness. Select **Offset From Surface** from the Direction 1 Type drop down list. Select the Base Extrude outside **face** for surface. Enter **.100**, **[.254]** for Offset Distance. Display the Extruded Cut Thin feature.

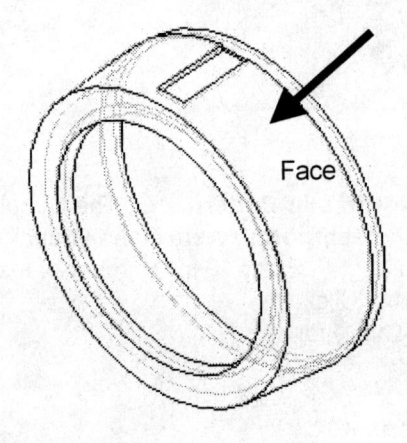

119) Click **OK** ✔.

120) Click Isometric.

121) Rename Cut-Extrude-Thin to Seed-Cut.

122) Save the LENSCAP. Click **Save**.

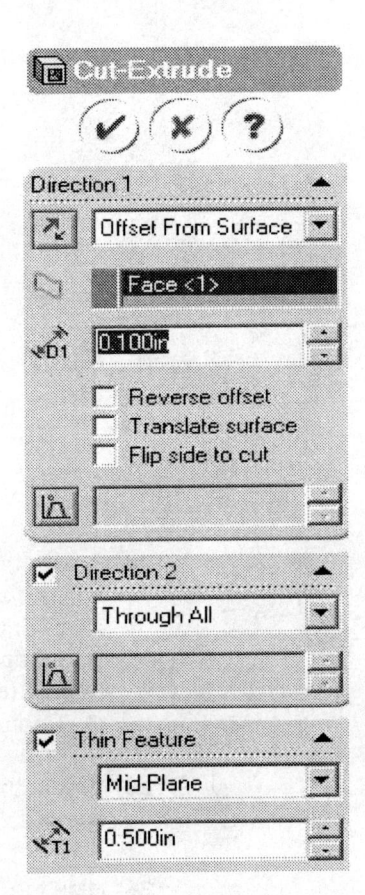

Create the Circular Pattern.

123) Display the Temporary Axis. Click **View** from the Main menu. Check **Temporary Axis** from the View menu. Hide the Planes. Click **View** from the Main menu. Uncheck **Planes**.

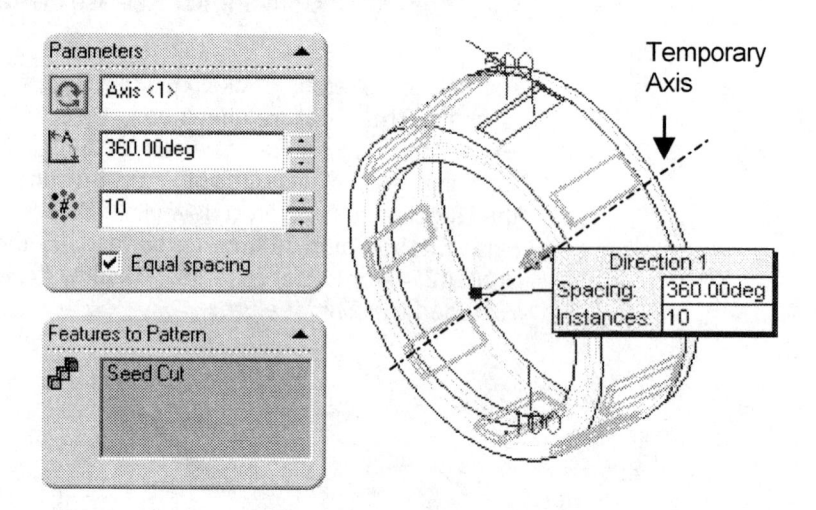

124) Create the Pattern. Click **Circular Pattern**. The Circular Pattern dialog box is displayed. Click the **Temporary Axis** in the Graphics window for Pattern axis. Enter **360** for Total angle. Enter **10** for Number of Instances. Click the **Equal spacing** check box. Click **Seed Cut** (Cut Extrude-Thin1) for the Features to Pattern.

125) Create the Circular Pattern. Click **OK**.

126) Hide the Temporary axis. Click **View** from the Main menu. Uncheck **Temporary axis**.

127) Save the LENSCAP. Click **Save**.

Suppress Feature

A suppressed feature is a feature that is not displayed.

Hide features to improve clarity. Suppressed features to improve model Rebuild time.

128) Suppress the SeedCut feature. Right-click on the **SeedCut** feature in the FeatureManager. Click **Suppress**. The SeedCut and the CirPattern features are displayed in gray in the FeatureManager.

The Circular Pattern feature is suppressed. The CirPattern feature is a child of the SeedCut feature.

Create the LENSCAP – Use the Sweep Feature

The LENSCAP requires threads. Use the Sweep feature to create the required threads. The O-RING Base-Sweep feature utilized a circular path and a sketched cross section.

The thread requires a spiral path. This path is called the Threadpath. The thread requires a sketched cross section. This cross section is called the Threadsection.

Note: Coils and springs use helical curves.

There are numerous steps required to create a thread. The plastic thread on the LENSCAP requires a smooth lead in.

The thread is not flush with the back face. Use an offset plane to start the thread.

Create a new offset Sketch plane, ThreadPlane. Thread cross sections are normally small and detailed. Create a thread. Construction lines control the dimensions of the thread. Below are the steps to create the thread:

- Create a new plane for the start of the thread.

- Create the thread path.

- Create a large thread cross section for improve visibility.

- Create the Sweep feature.

- Reduce the size of the thread cross section.

Create the Offset Reference Plane.

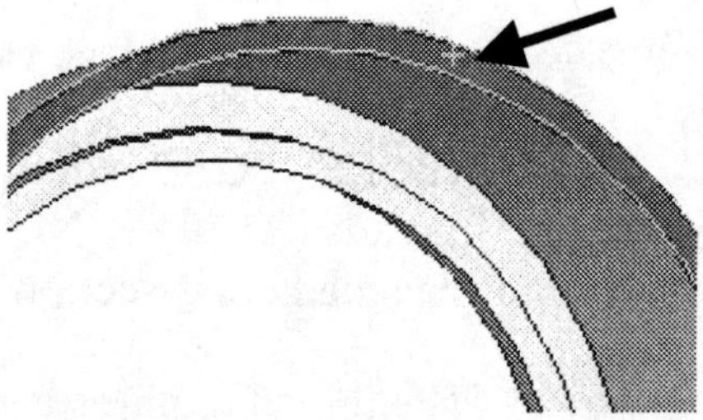

129) Display the back face of the LENSCAP. **Rotate** and **Zoom** the LENSCAP. Click the **narrow back face** of the Base Extrude feature.

130) Display the Right view. Click **Right**

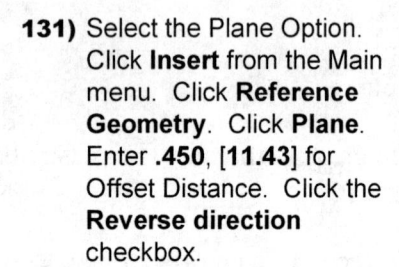

131) Select the Plane Option. Click **Insert** from the Main menu. Click **Reference Geometry**. Click **Plane**. Enter **.450**, **[11.43]** for Offset Distance. Click the **Reverse direction** checkbox.

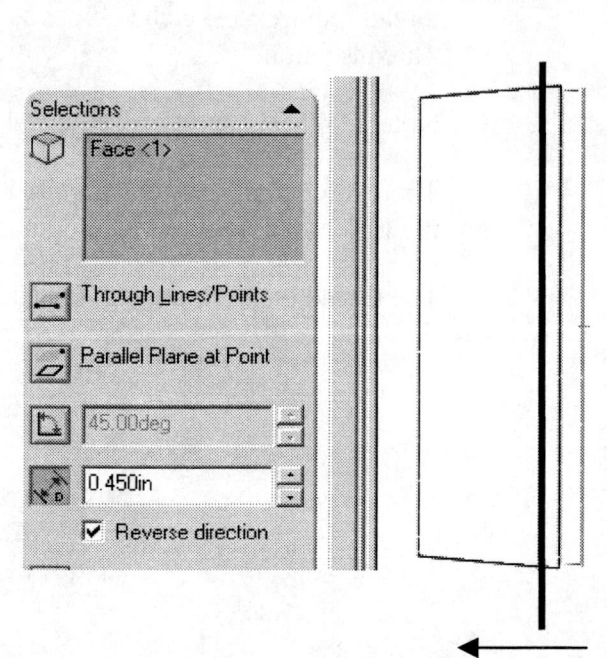

132) Create the plane. Click **OK**.

133) Rename Plane2 to Threadplane.

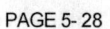

Engineering Design with SolidWorks

Sweep and Loft Features

134) Display the Isometric view with the hidden lines removed. Click **Isometric**.

135) Click **HiddenLinesRemoved**. The current Threadplane is displayed in green.

Create the Thread path.

136) Create the Sketch. Click **Sketch**. Extract the edge to the Threadplane. Click the **back inside circular edge** of the Shell. Click **Convert Entities**.

Back inside circular edge

137) Click the **Top** view. The circular edge is displayed on the ThreadPlane.

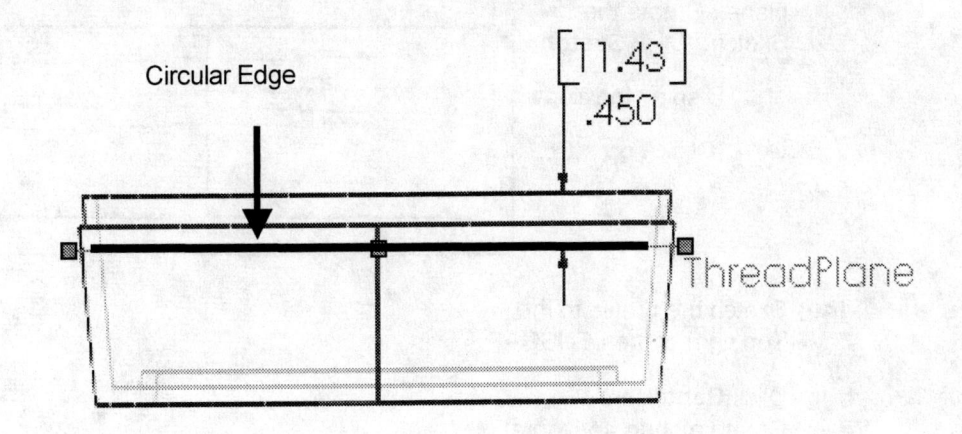

Circular Edge

[11.43]
.450

ThreadPlane

PAGE 5 - 29

Sweep and Loft Features   **Engineering Design with SolidWorks**

Create the Helix/Spiral curve path. Click **Insert** from the Main menu. Click **Curve**. Click **Helix/Spiral**. The Helix Curve dialog box is displayed. Enter **.250**, **[6.35]** for Pitch. Enter **2.5** for Revolution. Click the **Taper Helix** check box. Enter **5** for Angle. Uncheck the **Taper outward** check box. Enter **0** for Starting angle. Click the **Reverse direction** checkbox. Display the Helical/Spiral path. Click **OK**.

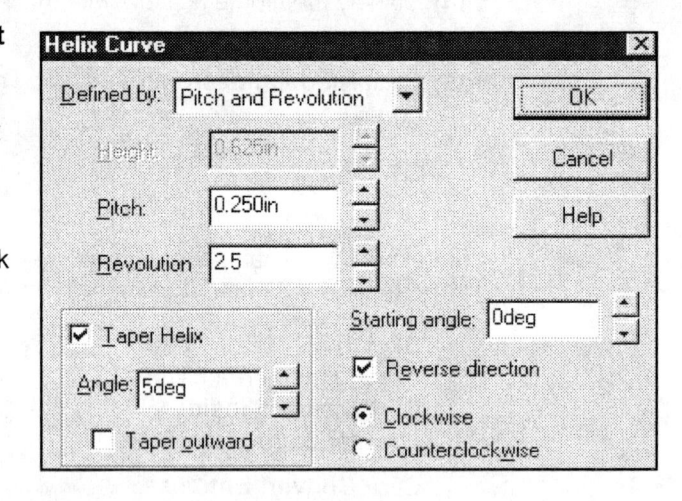

138) Rename Helix1 to ThreadPath.

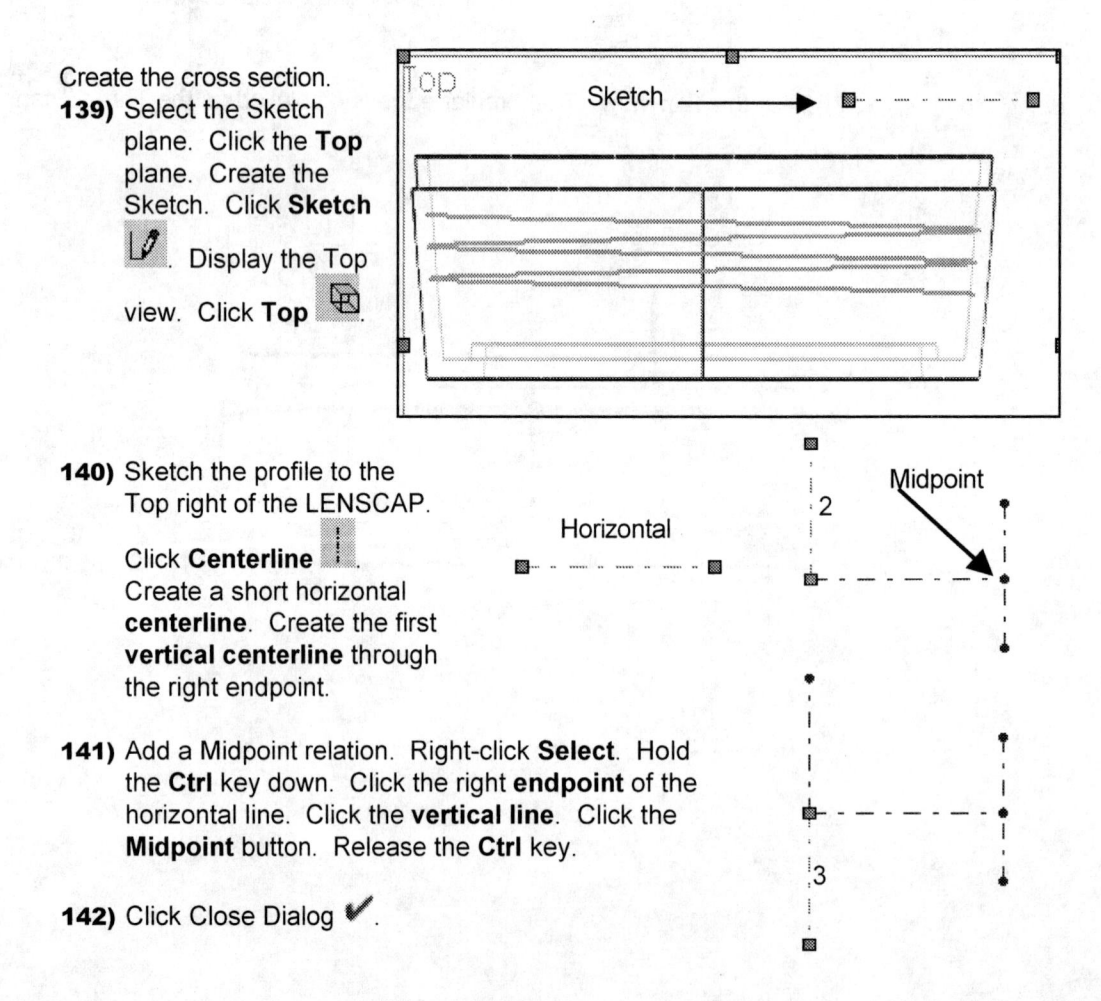

Create the cross section.
139) Select the Sketch plane. Click the **Top** plane. Create the Sketch. Click **Sketch**. Display the Top view. Click **Top**.

140) Sketch the profile to the Top right of the LENSCAP.

Click **Centerline**. Create a short horizontal **centerline**. Create the first **vertical centerline** through the right endpoint.

141) Add a Midpoint relation. Right-click **Select**. Hold the **Ctrl** key down. Click the right **endpoint** of the horizontal line. Click the **vertical line**. Click the **Midpoint** button. Release the **Ctrl** key.

142) Click Close Dialog.

PAGE 5-30

Engineering Design with SolidWorks   **Sweep and Loft Features**

143) Create the second **vertical centerline** with the start point coincident with the left horizontal endpoint. Drag the **centerline** upward until it is approximately the same size as the left vertical line.

144) Create the third **vertical centerline** with the start point coincident with the left horizontal endpoint. Drag the **centerline** downward until it is approximately the same size as the left vertical line.

145) Add an Equal relation. Right-click **Select**. Hold the **Ctrl** key down. Click the **three vertical lines**. Click the **Equal** button. Release the **Ctrl** key.

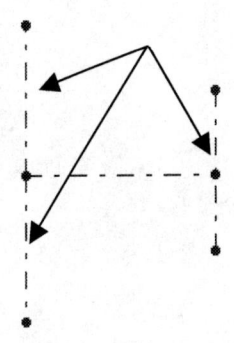

146) Click Close Dialog.

147) Add dimensions. Click **Dimension**. Click the **left vertical endpoints**. Enter **.500**, [12.7].

148) Sketch the profile. Click **Line**. The profile is a trapezoid. Click the 4 **endpoints** of the vertical centerlines to create the trapezoid.

149) Double-click the **first point** to close and end the line. Maintain the centerlines as part of the Sketch with geometric relationships.

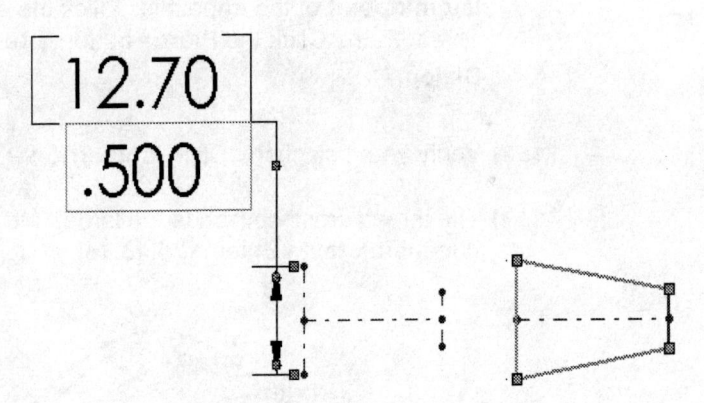

PAGE 5 - 31

150) Add an Equal relation. Right-click **Select**. Hold the **Ctrl** key down. Click the **left**, **top** and **bottom lines** of the trapezoid. Click the **Equal** button. Release the **Ctrl** key. Click **Close Dialog** ✔.

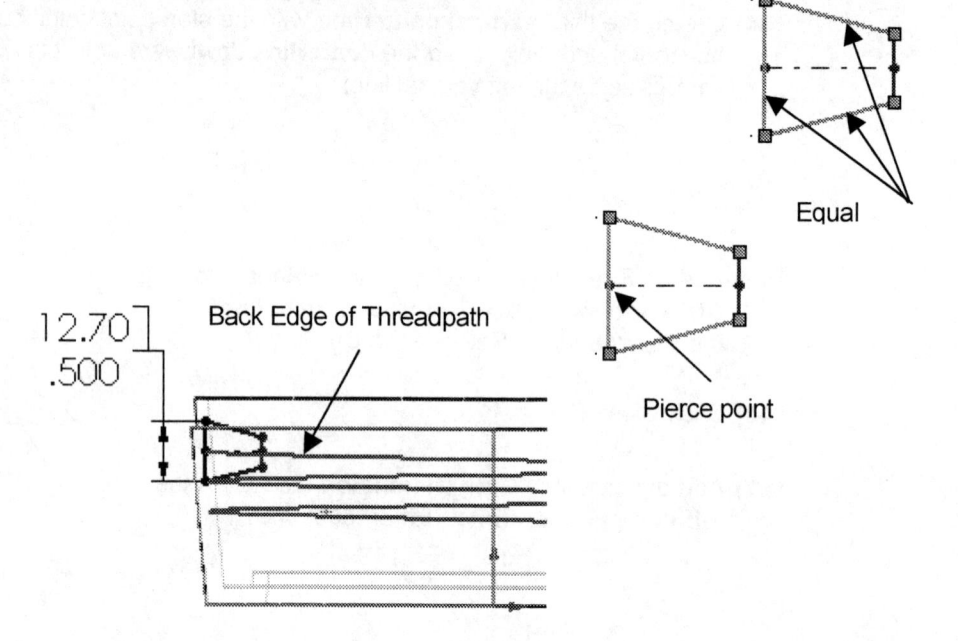

151) Add a Pierce relation. Right-click **Select**. Hold the **Ctrl** key down. Click the **left midpoint** of the trapezoid. Click the **starting left back edge** of the ThreadPath. Click the **Pierce** button. Release the **Ctrl** key. Click **Close Dialog** ✔.

152) Verify your selection. Click **Isometric**. Display the Sketch. Click **Apply**.

153) The thread cross section is too large. Modify the left line. Double click the **dimension text**. Enter **.125**, [3.18].

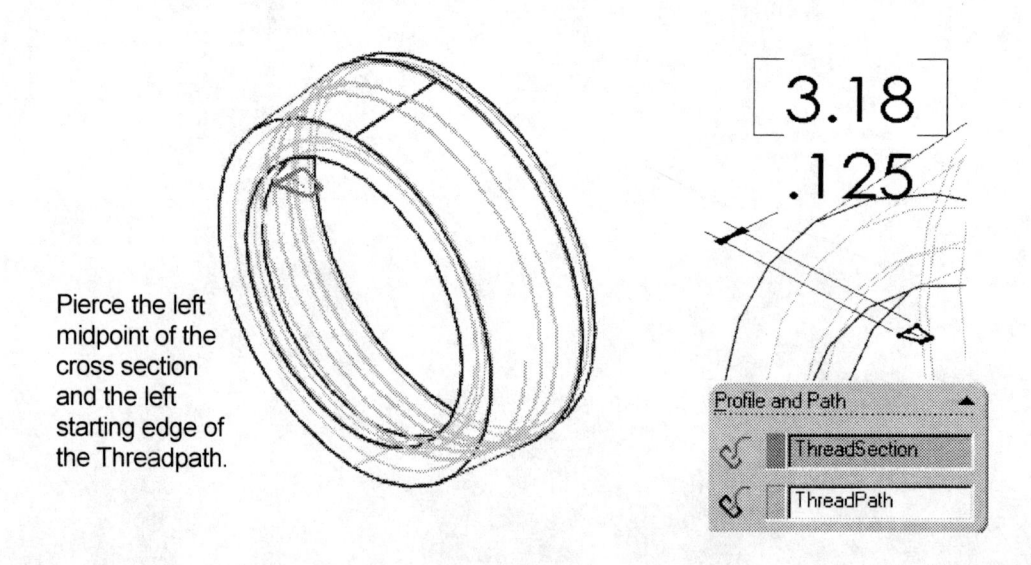

Pierce the left midpoint of the cross section and the left starting edge of the Threadpath.

Engineering Design with SolidWorks — Sweep and Loft Features

154) Close the Sketch. Click **Sketch**.

155) Rename Sketch6 to Threadsection.

156) Create the Sweep. Click **Sweep** from the Feature toolbar. The Sweep dialog box is displayed.

157) Split the FeatureManager. Drag the **Split bar** half way down to display the FeatureManager.

158) The Threadsection is the displayed in the Sweep profile text box. Click the **Sweep path** text box.

159) Click the helical **Threadpath** from the FeatureManager.

160) Display the Sweep. Click **OK**.

161) Rename **Sweep1** to **Thread**.

162) Expand the Thread feature. Click the **Plus Sign** in the FeatureManager. The Thread feature is composed of the following: Threadsection and Threadpath.

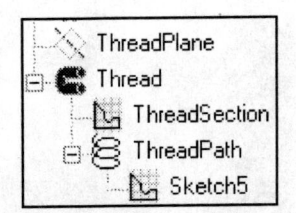

163) Expand the Threadpath. Click the **Plus Sign**. The Threadpath contains the circular Sketch and the definition of the Helical curve.

PAGE 5 - 33

Note: If the Threadsection geometry intersects itself, the cross section is too large. Reduce the cross section size and create the Sweep feature again.

164) Restore the CirPattern feature. Right-click on the **CirPattern 1** feature. Click **Unsuppress**.

165) The LENSCAP is complete. Save the LENSCAP. Click **Save**.

HOUSING

The HOUSING provides storage and support for the BATTERY. The SWITCH and LENSCAP connect to the HOUSING.

Do you remember the original customer requirements?

- An inexpensive reliable flashlight.
- Available advertising space of 10 square inches, [64.5 square centimeters].
- A light weight semi indestructible body.
- Self standing with handle.

The HOUSING must meet the above customer requirements. In a design situation, you may not have all of the required dimensions. Where do you start?

Engineering Design with SolidWorks Sweep and Loft Features

The LENS, BATTERYPLATE and BATTERY are internal components to the HOUSING, Figure 5.11.

The LENSCAP thread fastens to the HOUSING thread.

Create a circular Extruded Base feature that fits inside the LENSCAP and contains the LENS.

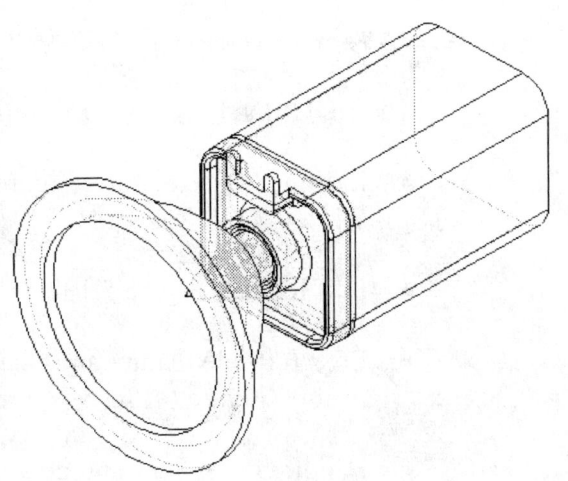

Figure 5.11

In this exercise, create a loose fit between the BATTERY and the HOUSING.

Review the mating components and determine the dimensions that require modification.

HOUSING is circular at LENS. Fits inside threaded LENSCAP.

SWITCH access thru HOUSING

HOUSING is rectangular at the base and contains the BATERY.

Loft Feature

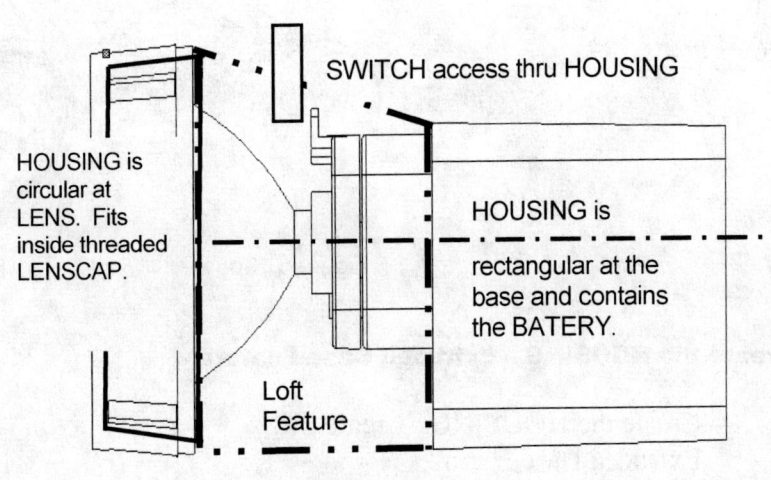

Figure 5.12

The LENS and the BATTERY are purchased parts. You cannot modify their size or shape!

How do you transition the HOUSING from a rectangular shape to a circular shape? Answer: Create a Loft feature, Figure 5.12.

The HOUSING supports the threaded LENSCAP and LENS.

The SWITCH fits through an opening in the HOUSING.

Internal Ribs are added to the HOUSING for structural integrity.

HOUSING Feature Overview

The HOUSING is composed of the following features:

- Extruded Base feature, Figure 5.13a.

- Loft feature, Figure 5.13b.

- Boss Extrude feature and Shell feature, Figure 5.13c.

- Handle Sweep and Thread-Sweep features, Figure 5.13d.

- Rib feature and Linear Pattern, Figure 5.13e.

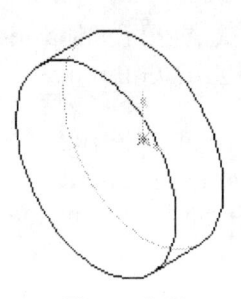

Figure 5.13a 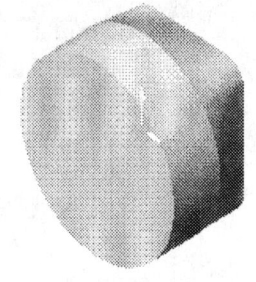   Figure 5.13b

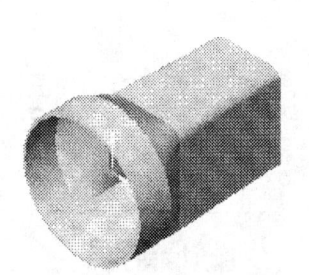

Figure 5.13c 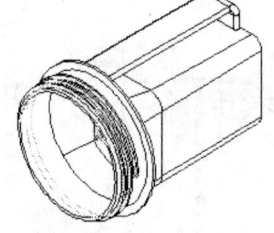   Figure 5.13d 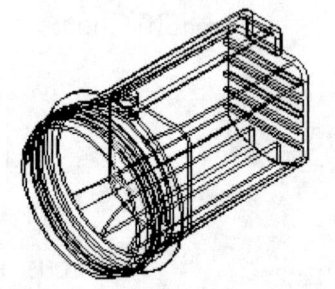   Figure 5.13e

Create the HOUSING – Extruded Base Feature

Create the HOUSING. Create the Extruded Base feature.

Create the new part.

166) Click **New**. Click the **MY-TEMPLATES** tab. Click **PART-IN-ANSI**, [**PART-MM-ISO**] from the Template dialog box. Click **OK**.

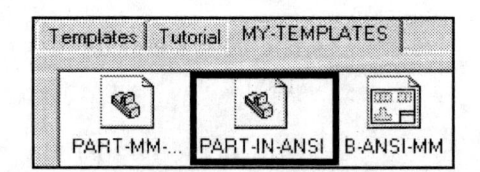

167) Save the empty part. Click **Save**. Select **ENGDESIGN-W-SOLIIDWORKS\ PROJECTS** for the Save in file folder.

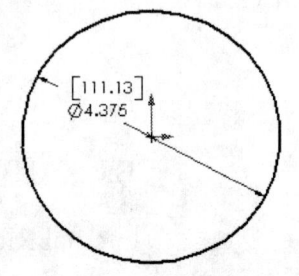

[111.13]
⌀4.375

Engineering Design with SolidWorks   Sweep and Loft Features

168) Enter **HOUSING** for file name.

169) Enter **HOUSING FOR 6VOLT FLASHLIGHT** for Description. Click the **Save** button.

170) Select the Sketch plane. Click **Front**. The Front plane is the default Sketch plane.

171) Create the Sketch. Click **Sketch**. Click **Circle**. Create a circle centered at the **Origin**.

172) Add a dimension. Click **Dimension**. Dimension the diameter of the circle. Enter **4.375**, **[111.13]**.

173) Extrude the Sketch. Click **Extruded Boss/Base**. The Extrude Feature dialog box is displayed. Blind is the default Type option. Enter **1.300**, **[33.02]** for Depth.

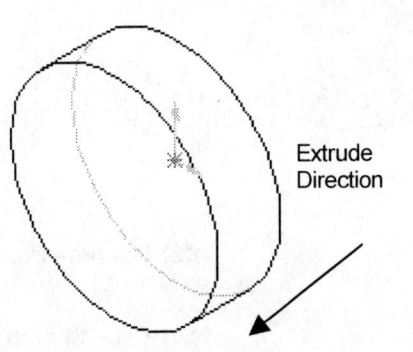

174) Display the Base-Extrude feature. Click **OK**.

175) Rename Extrude1 to Base Extrude.

176) Save the HOUSING. Click **Save**.

Extrude Direction

Create the HOUSING – Use the Loft Feature

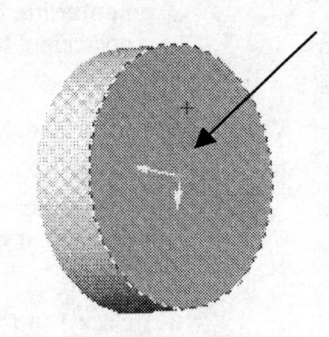

The Loft feature is composed of two profiles. Create the first profile from the back face of the Base-Extrude feature.

Create the second profile on an Offset plane. Close the Sketch.

Create the first profile.
177) Select the Sketch plane. Click the **back face** of the Base-Extrude feature.

178) Click **Sketch**. Extract the entire face. Click **Convert Entities** from the Sketch Tools toolbar.

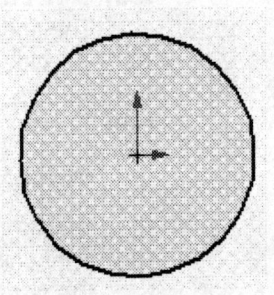

179) Close the Sketch. Click **Sketch**.

180) Rename Sketch2 to Sketch Circle.

PAGE 5-37

Create the second profile.

181) Create the Sketch plane. Click the **back face**. Click **Insert** from the Main menu. Click **Reference Geometry**, **Plane**. Enter **1.300**, **[33.02]** for Offset Distance.

182) Verify the plane position. Click **Top**. Click **OK**.

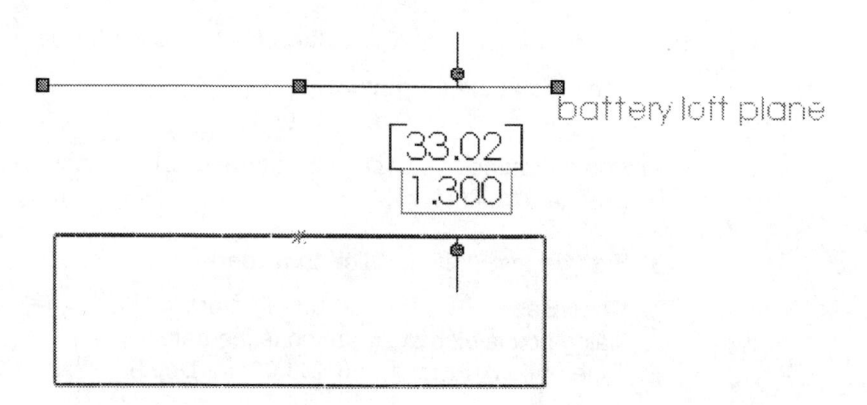

183) Rename Plane1 to Battery Loft Plane.

184) Click **Sketch**. Click **Back**. Extract the circular edge. Click the **circumference of the circle**. Click **Convert Entities**.

185) Click **Centerline**. Sketch a **vertical centerline** collinear with the Right plane coincident to the Origin.

Convert the outside edge.

186) Click **Sketch Mirror**. Click **Line**. Sketch a **horizontal line** above the Origin. The left end point is coincident with the vertical centerline.

187) Click Tangent Arc. Sketch a 90 degree arc.

188) Click **Line**. Sketch a **vertical** line. The endpoint of the vertical line is coincident with the edge of the circle.

Trim the left and right edges.

189) Deactivate the Mirror. Click **Sketch Mirror**.

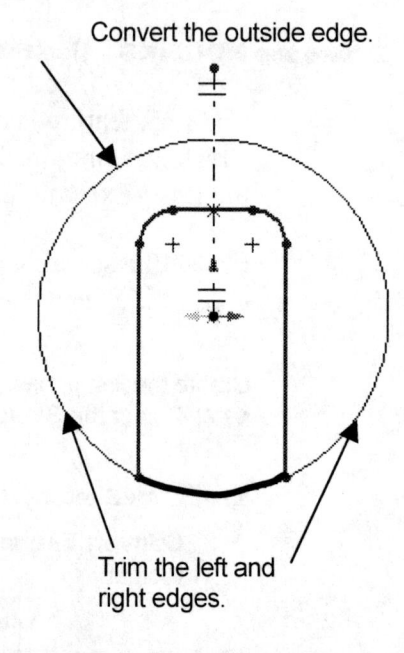

190) Trim unwanted geometry. Click **Sketch Trim**. Click the far **right edge of the circle**. Click the far **left edge of the circle**.

Add dimensions. Click **Dimension**. Create the **horizontal** dimension. Click the **left vertical line**. Click the **right vertical line**. Enter **3.100**, **[78.74]**. Create the **vertical** dimension from the Origin to the top horizontal line. Enter **1.600**, **[40.64]**. Create the **radial** dimension. Enter **.500**, **[12.7]**.

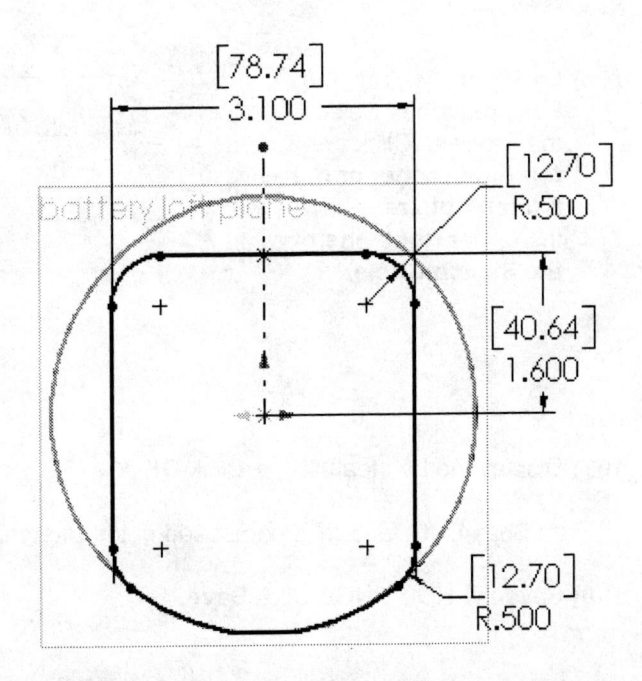

The FLASHLIGHT components must remain aligned to a common centerline.

191) Remove all sharp edges. Add fillets to the lower corners of the Sketch. Click **Sketch Fillet**. Enter **.500**, **[12.7]**. Click the **lower left point**. Click the **lower right point**.

192) Click **OK**.

193) Close Sketch3. Click **Sketch**.

194) Rename Sketch3 to Sketch Square.

Sweep and Loft Features   **Engineering Design with SolidWorks**

Create the Loft.

195) Display the Isometric view. Click **Isometric**.

196) Click **Loft** from the Feature menu. Select the Profiles. Click the **top right corner of the Sketch Square**. Click the **upper right side of the Sketch Circle**.

197) Display the Loft feature. Click **OK**.

The Boss-Loft1 feature is composed of the Sketch Square and the Sketch Circle.

198) Save the HOUSING. Click **Save**.

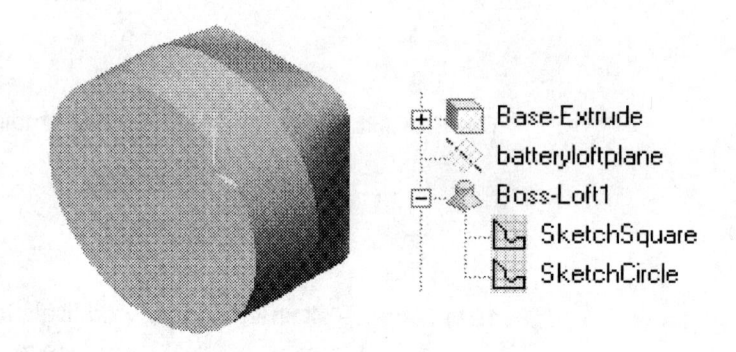

Create the HOUSING – First Extruded Boss Feature

Create the first Extruded Boss feature from the square face of the Loft. How do you estimate the depth of the Extruded Boss feature?

Answer: The Extruded Base feature of the BATTERY is 4.1inches, [104.14mm]. Ribs are required to support the BATTERY.

Design for Rib construction. Use a 4.4inch, [111.76mm] depth as the first estimate. Adjust the estimated depth dimension later if required in the FLASHLIGHT assembly.

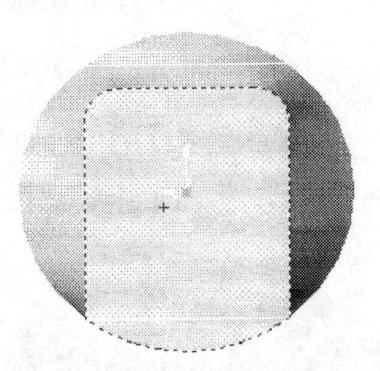

Create the first Extruded Boss feature.
199) Select the Sketch plane. Click the **Back** view. Click the **back face** of the Loft.

200) Create the Sketch. Click **Sketch**. Extract the entire face. Click **Convert Entities** from the Sketch Tools toolbar.

201) Extrude the Sketch. Click **Extruded-Boss/Base**. Enter **4.400**, **[111.76]** for Depth. Check the **Draft On/Off** button. Enter **1** for Draft Angle.

202) Display the Boss feature. Click **OK**. Display the **Right** view.

203) Rename Extrude2 to Boss-Battery.

204) Save the HOUSING. Click **Save**.

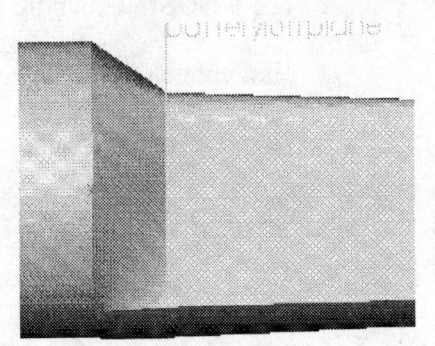

Sweep and Loft Features   Engineering Design with SolidWorks

Create the HOUSING – Use the Shell Feature

The Shell feature removes material. Use the Shell feature to remove the front face of the HOUSING.

205) Create the Shell feature. Click **Isometric**. Click **Shell** from the Feature toolbar. Click the **front face** of the Base Extrude feature. The Shell Feature dialog box is displayed.

206) Enter **.100**, **[2.54]** for Thickness.

207) Display the Shell. Click **OK**.

208) Save the HOUSING. Click **Save**.

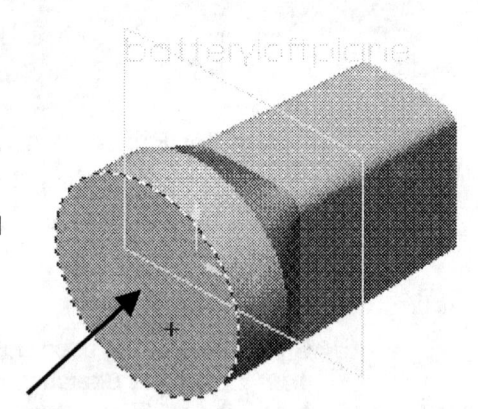

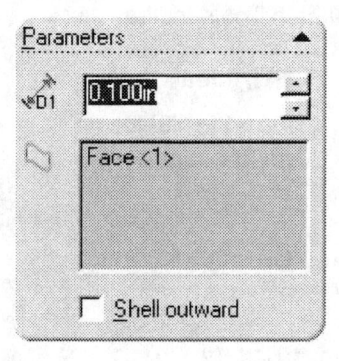

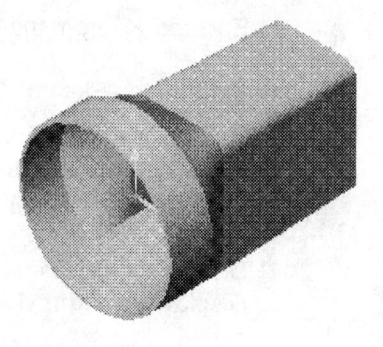

The HOUSING handle is a solid.

First create the Shell feature.

Then create the Sweep feature.

Create the HOUSING - Second Extruded Boss Feature

The second Extruded Boss feature creates a solid circular ring on the back circular face of the Extruded Base feature.

The solid ring is a cosmetic stop for the LENSCAP and provides rigidity at the transition of the HOUSING.

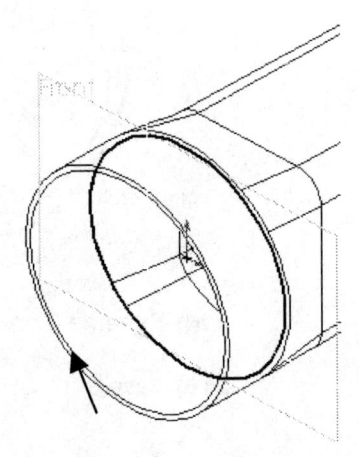

Create the second Extruded Boss.
209) Select the Sketch plane. Click the **Front** plane.

210) Create the Sketch. Click **Sketch**. Create the inside circle. Click the **front inside circular edge** of Shell1. Click **Convert Entities**.

211) Create the outside circle. Click **Circle**. Create a **circle** centered at the Origin.

212) Add a dimension. Click **Dimension**. Click the **circle**. Enter **5.125**, **[130.18]**.

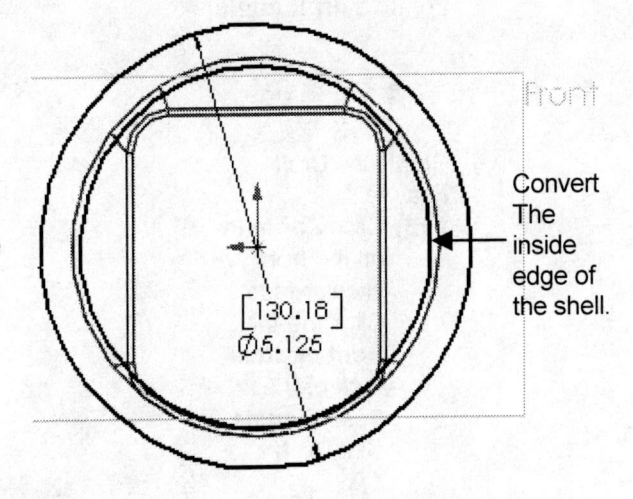

213) Extrude the Sketch. Click **Extruded Boss/Base**. Enter **.100**, **[2.54]** for Depth. Display the Boss feature. Click **OK**.

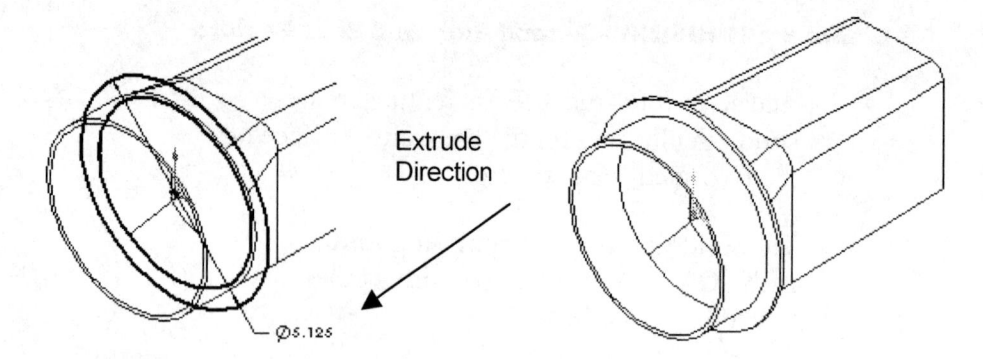

Sweep and Loft Features   Engineering Design with SolidWorks

Extrude Direction

⌀5.125

214) Rename Extrude3 to Boss-Stop.

215) Save the HOUSING. Click **Save**.

Create the HOUSING – Use the Draft Feature

A 5 degree draft is required to insure proper thread mating between the LENSCAP and the HOUSING. The LENSCAP Extruded Base feature has a 5 degree draft angle.

The outside face of the Extruded Base feature HOUSING requires a 5 degree draft angle. The inside HOUSING wall does not require a draft angle. The Extruded Base feature has a 5 degree draft angle. Use the Draft feature to create a draft angle.

Type of Draft
Neutral Plane

Draft Angle
5.00deg

Neutral Plane
Face<1>

Faces to Draft
Face<2>

Face propagation:
None

Create the Draft Feature.
216) Click **Zoom In** on the front circular face. Click the thin **front circular face** of the Base Extrude. Clrk **Draft**. The Draft Feature dialog box is displayed.

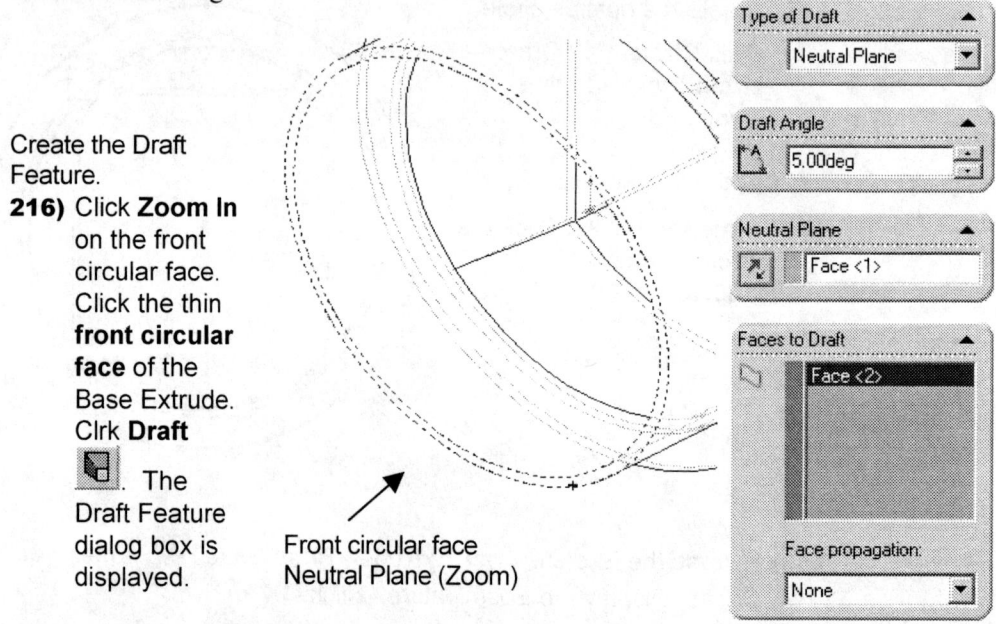

Front circular face Neutral Plane (Zoom)

217) The front circular face is displayed in the Neutral plane text box. Click the **Faces to draft** text box.

218) Click the **outside face**. Enter **5** for Draft Angle. Display the Draft. Click **OK** ✔.

PAGE 5-44

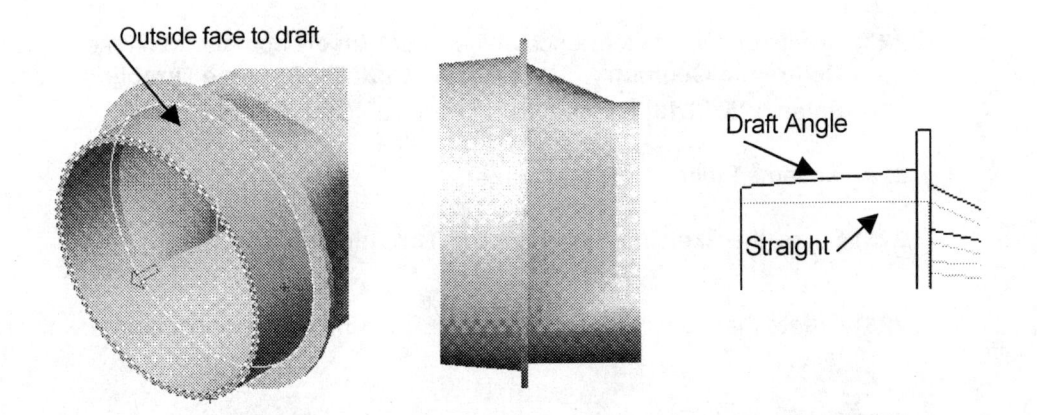

219) Display the Draft Angle and the straight interior. Click the **Right** view. Click **Hidden Lines Visible**.

220) Save the HOUSING. Click **Save**.

Create the HOUSING – Use the Thread with Sweep Feature

The HOUSING requires a thread. Create the threads for the HOUSING on the outside Draft feature of the Housing. Create the thread with the Sweep feature. The thread requires two sketches:

- Threadpath.

- Threadsection.

Create the Threadpath sketch.
221) Click **Isometric**.

222) Click the **front flat circular face** of the HOUSING.

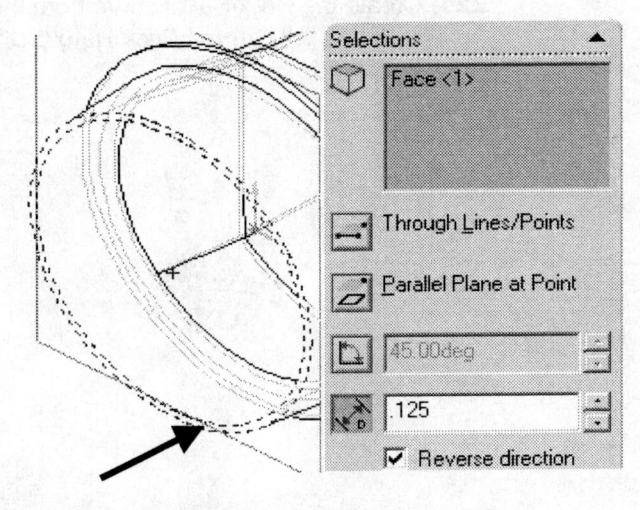

Sweep and Loft Features   **Engineering Design with SolidWorks**

223) Create an Offset Reference plane. Click **Insert** from the Main menu. Click **Reference Geometry**. Click **Plane**. Click the **Reverse Direction** check box. Enter **.125**, **[3.18]** for the Offset Distance. Click **OK** ✔.

224) Rename Plane2 to Threadplane.

225) Select the Sketch plane. Click the **Threadplane**.

226) Create the Sketch. Click **Sketch**. **Zoom in** on the front narrow flat circular face.

227) Select the **front outside circular edge** of the Base Extrude.

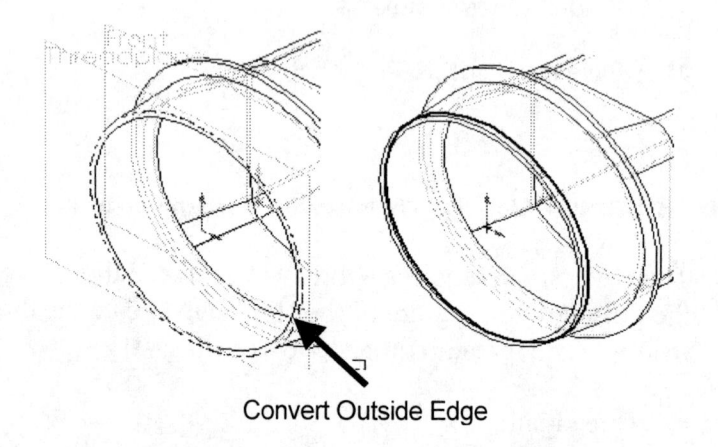

Convert Outside Edge

228) Click **Convert Entities**. The circular edge is displayed on the Threadplane.

229) Create the Helix/Spiral curve from the circular edge. Click **Insert** from the Main menu. Click **Curve**. Click **Helix/Spiral**. The **Helix Curve** dialog box is displayed.

Engineering Design with SolidWorks   Sweep and Loft Features

230) Enter **.250**, **[6.35]** for Pitch. Enter **2.5** for Revolution. Click the **Taper Helix** check box. Enter **5** for Angle. Click the **Taper Outward** check box. Enter **180** in the Starting angle spin box. Click the **Reverse direction** checkbox. Display the Helix/Spiral curve. Click **OK**.

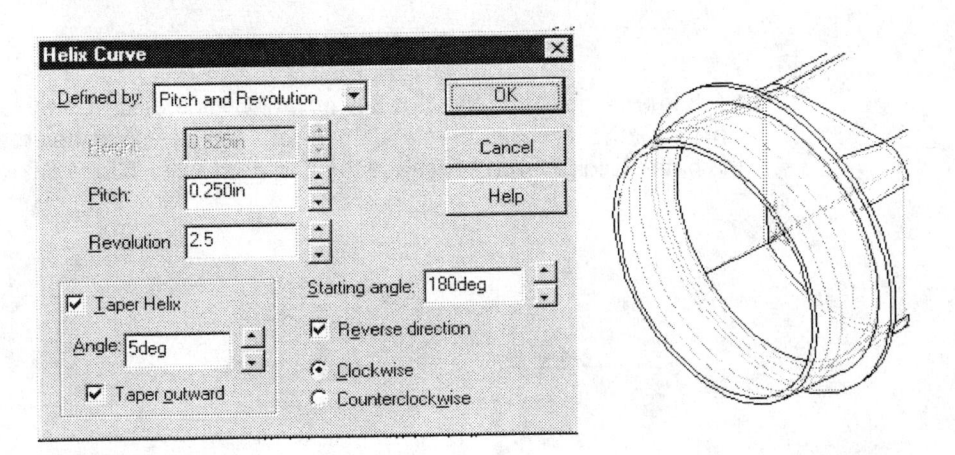

231) Rename Helix1 to Threadpath.

232) Save the HOUSING. Click **Save**.

Create the Threadsection sketch for the HOUSING. Copy the created LENSCAP sketched cross section.

Copy the Threadsection from the LENSCAP.
233) **Open** the LENSCAP. **Expand** the Thread Sweep feature from the FeatureManager. Click the **Threadsection** sketch. Click **Edit** from the Main menu. Click **Copy**. **Close** the LENSCAP.

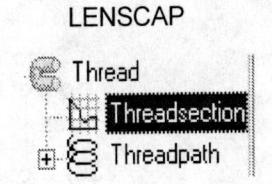

234) Click the **Top** plane in the HOUSING FeatureManager. Click **Edit** from the Main menu. Click **Paste**. The Threadsection is displayed on the Top plane. The new Sketch7 name is added to the FeatureManager.

Paste the Threadsection on the Top plane.

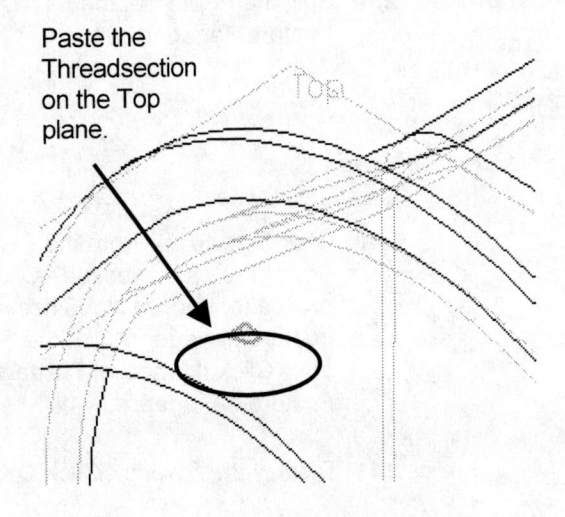

235) Rename Sketch7 to Threadsection.

PAGE 5 - 47

Sweep and Loft Features Engineering Design with SolidWorks

236) Pierce Threadsection to Threadpath. Right-click on **Threadsection**. Click **Edit Sketch**. Click the **ThreadSection**. Click **Zoom to Selection**. Click the **Midpoint** of the Threadsection.

237) Click **Isometric**. Hold the **Ctrl** key down. Click the **right back edge of the Threadpath**. The Pierce relation positions the center of the cross section on the path. Click **Pierce**. Release the **Ctrl** key. Click **Close Dialog**.

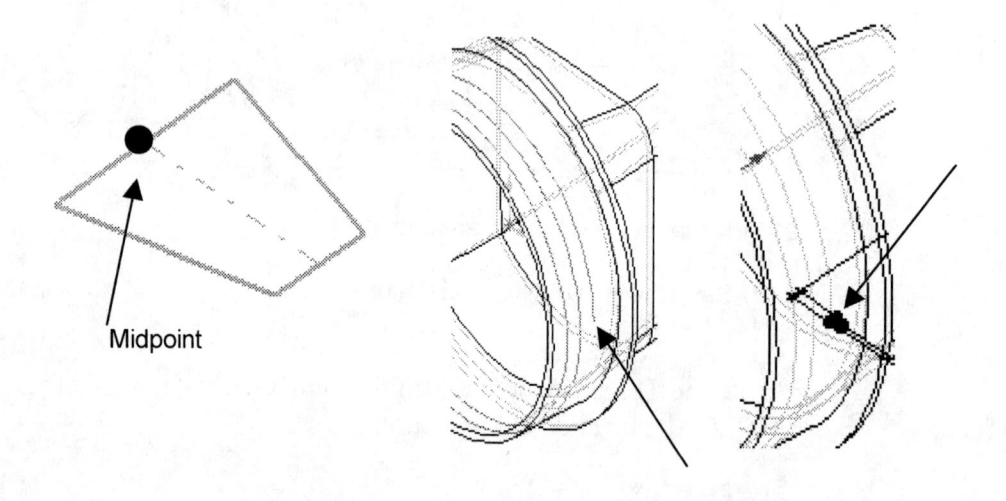

Midpoint

Caution: Do not click the front edge of the Thread path. The Thread is created out of the HOUSING.

238) Close the Sketch. Click **Sketch**.

239) Split the FeatureManager. Drag the **Split bar** half way down to display the FeatureManager.

Create the Sweep.

240) Click **Sweep** from the Feature toolbar. Select the cross section. Click the **Sweep profile** text box. Click **Threadsection** from the FeatureManager. Click the **Sweep path** text box. Click the helical **Threadpath** from the FeatureManager.

241) Display the Sweep. Click **OK**.

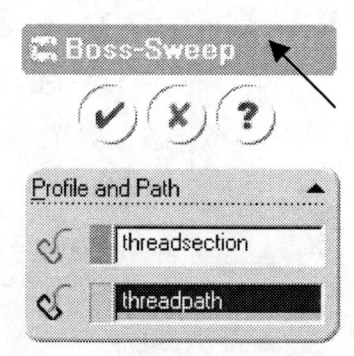

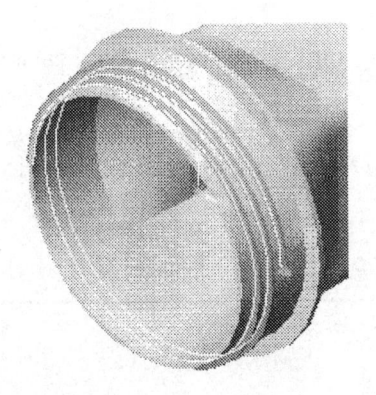

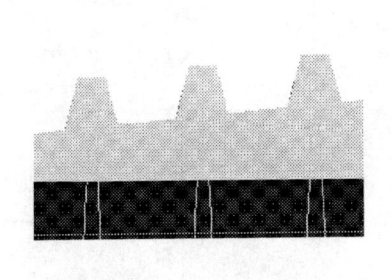

Engineering Design with SolidWorks　　　　　　　　　Sweep and Loft Features

An external thread is created by piercing the midpoint of the Threadsection to the right side of the Threadpath.

The Threadplane allows for a smooth lead. The Threadplane offset dimension is modified in the assembly by adjusting the mating threads of the LENSCAP and HOUSING.

Creating a Threadplane provides flexibility to the design.

242) Rename **Sweep1** to **Thread**.

243) Save the HOUSING. Click **Save**.

Note: Conserve design time. Use the Right-mouse button to invoke the Zoom and Pan commands. View the mouse pointer feedback for entity confirmation.

Create the HOUSING – Use the Sweep Feature

Create the handle with the Sweep feature. The Sweep feature consists of a sketched path and cross section profile.

Sketch the path on the Right plane. The sketch uses edges from existing features.

Sketch the cross section on the back circular face of the Boss-Stop feature.

Create the Sweep path.
244) Select the Sketch plane. Click **Right** from the FeatureManager.

245) Create the Sketch. Click **Sketch**.

Sweep and Loft Features Engineering Design with SolidWorks

246) Drag the **bottom point** upward on the vertical line.

247) Click **Line**. Sketch a **horizontal line** below the top of the Boss Stop. Sketch a **vertical line** to the right edge of the Housing.

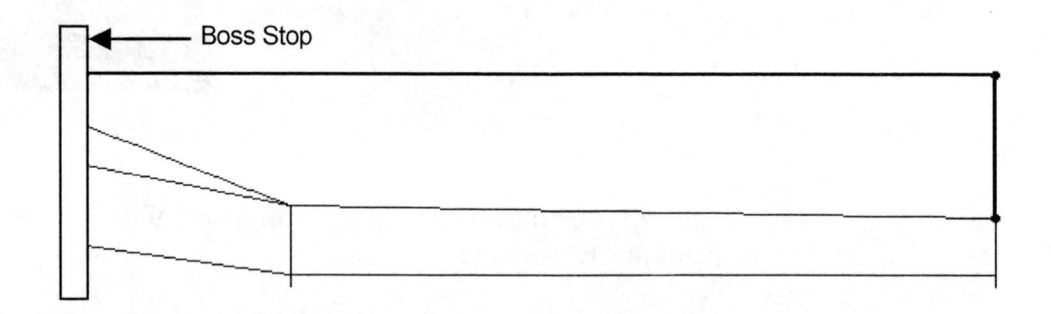

248) Create a 2D Fillet. Click **2D Fillet**. Click the **upper right corner** of the sketch lines. Enter **.500**, **[12.7]** for Radius. Click **Close**.

249) Add a dimension. Click **Dimension**. Add a vertical dimension from the **Origin** to the horizontal line. Enter **2.500**, **[63.5]**.

250) Add relations. Right-click **Select**. Hold the **Ctrl** key down. Click the **left end point** of the horizontal line. Click the **right vertical edge** of the Boss Stop. Click **Coincident**. Release the **Ctrl** key. Click **Close Dialog**.

251) Hold the **Ctrl** key down. Click the **bottom end point** of the vertical line. Click the **right vertical edge** of the Housing. Click the **horizontal edge** of the Housing. Click **Intersection**. Release the **Ctrl** key. Click **Close Dialog**.

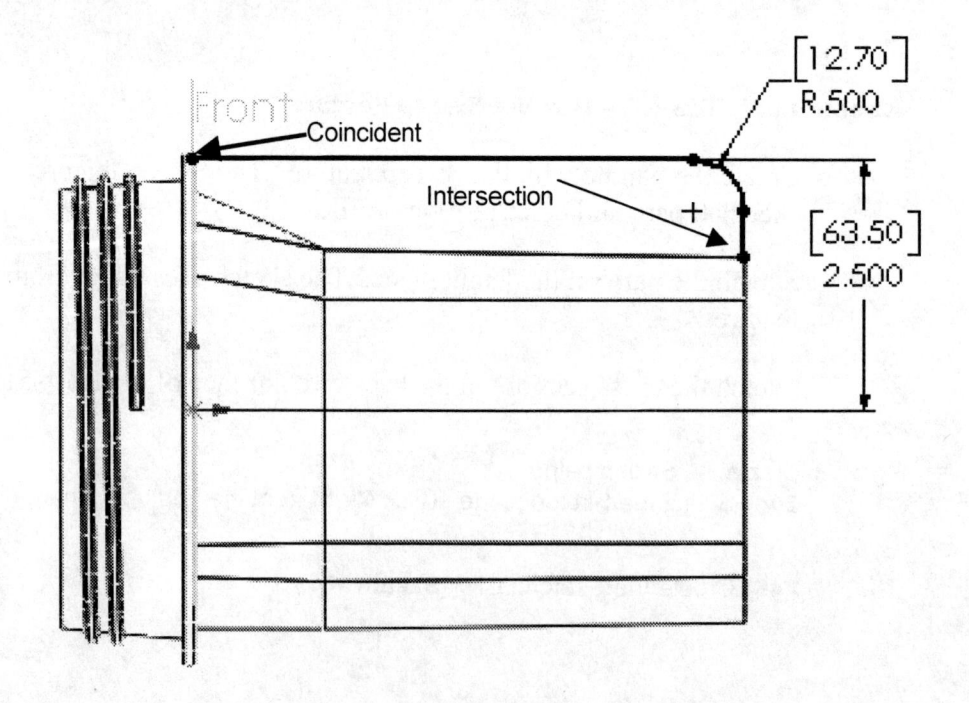

PAGE 5- 50

Engineering Design with SolidWorks **Sweep and Loft Features**

252) Close the Sketch. Click **Sketch**.

253) Rename Sketch8 to HandlePath.

Create the Sweep Profile.

254) Select the Sketch plane. Click **Back**. Select the **back circular face** of the Boss-Stop feature.

255) Sketch the second profile. Click **Sketch**. Click **Centerline**. Sketch a **vertical centerline** collinear with the Right plane coincident to the Origin.

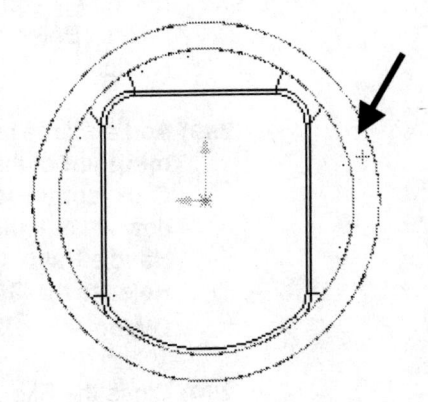

256) **Zoom in** on the top of the Boss Stop. Click **Sketch Mirror**. Click **Line**. Sketch a **horizontal** line. Click **Tangent Arc**. Sketch a **180 degree** arc. Click **Line**. Sketch a **horizontal** line to close the sketch.

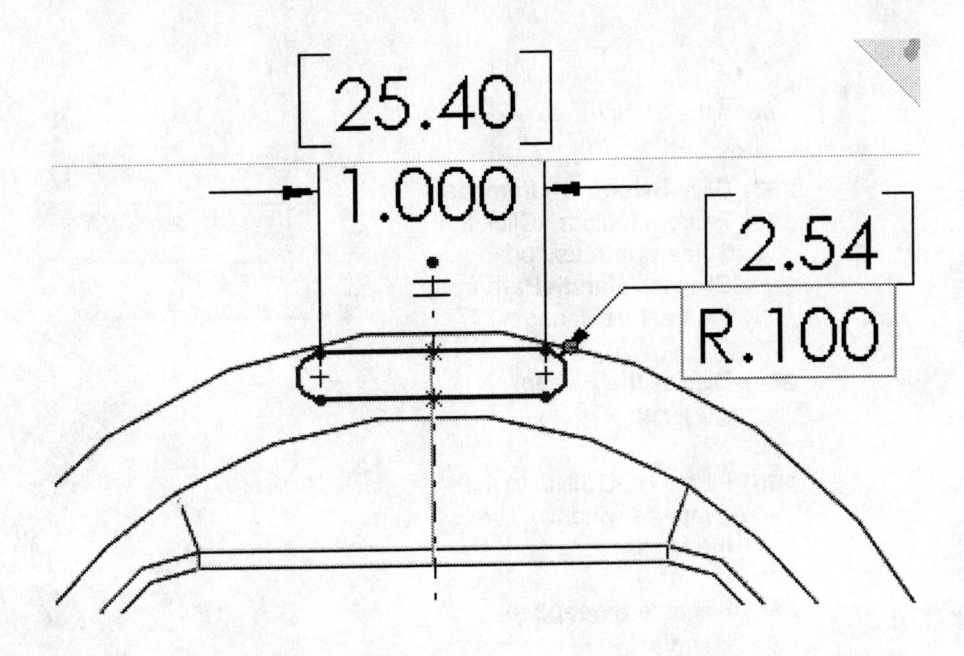

PAGE 5 - 51

Sweep and Loft Features Engineering Design with SolidWorks

257) Deactivate the Mirror. Click **Sketch Mirror**.

258) Add dimensions. Click **Dimension**. Create the **horizontal** and **radial dimensions.** Enter **1.000**, **[25.4]** between the arc center points. Enter **.100**, **[2.54]** for Radius.

259) Add a Pierce relation. Click the **top midpoint** of the Sketch profile. Click **Isometric**. Hold the **Ctrl** key down. Click the **line** from the Handle Path. Click **Coincident**. Release the **Ctrl** key. Click **Close Dialog**.

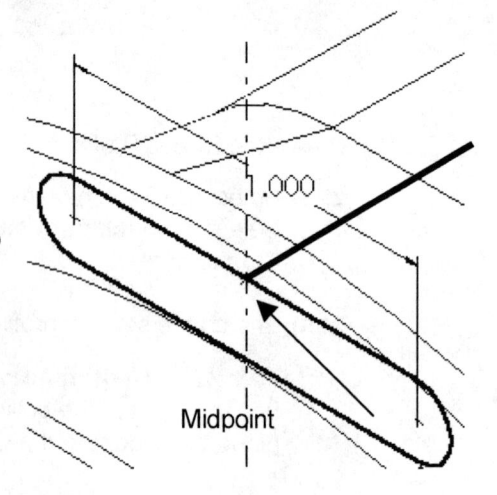

260) Close the Sketch. Click **Sketch**.

261) Rename Sketch9 to HandleProfile.

262) Split the FeatureManager. Drag the **Split bar** half way down to display the FeatureManager.

Sweep the Sketch.

263) Click **Sweep** from the Feature toolbar. Click the **Sweep path** text box. Click the **HandlePath** from the FeatureManager.

264) Display the Sweep. Click **OK**.

265) Fit the HOUSING to the Graphics window. Press the **f** key.

266) Rename **Sweep2** to **Handle**.

267) Save the HOUSING. Click **Save**.

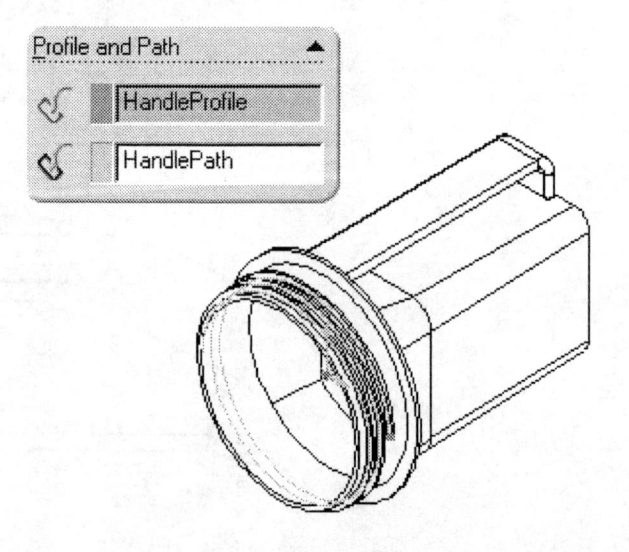

Engineering Design with SolidWorks   Sweep and Loft Features

Create the HOUSING – Use the Extruded Cut Feature for the SWITCH

Insert the SWITCH into the HANDLE of the HOUSING.

Create an Extruded Cut in the HANDLE for the SWITCH.

Create the Extruded Cut feature.
268) Select the Sketch plane. Click the **top face** of the Handle.

269) Create the Sketch. Click **Sketch**. Click **Normal To**. Right click the **Right plane**. Click **Show**. Right-click the **Front plane**. Click **Show**.

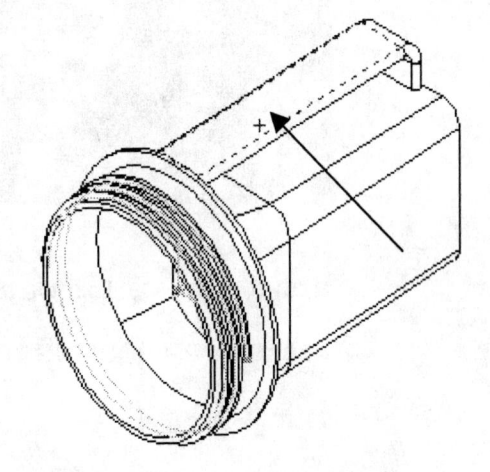

270) Create a circle. Click **Circle**. Center the **circle** collinear to the Right plane.

271) Add dimensions. Click **Dimension**. Enter **.510**, **[12.95]** for diameter. Enter **.450**, **[11.43]** for the distance from the Front plane.

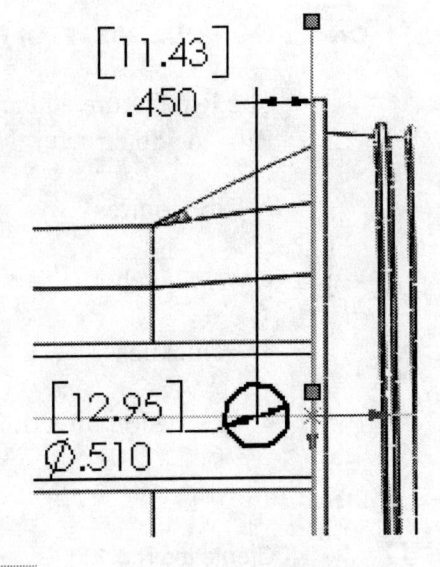

272) Extrude the Sketch. Click **Extruded Cut**. Click the **UpTo Surface** option from the Type list box.

273) Rotate the model to view the inside Shell1. Click the **top inside face** of the Shell1, below the Sketch.

274) Click **OK**.

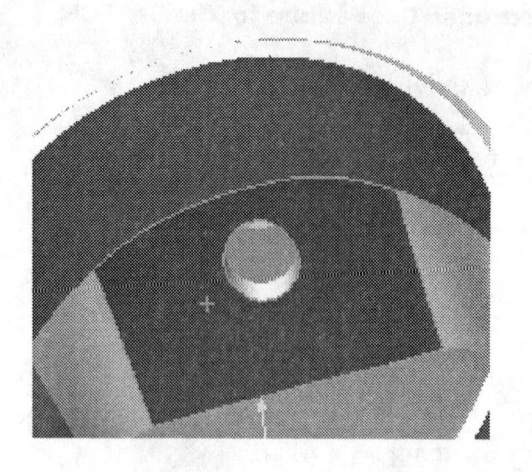

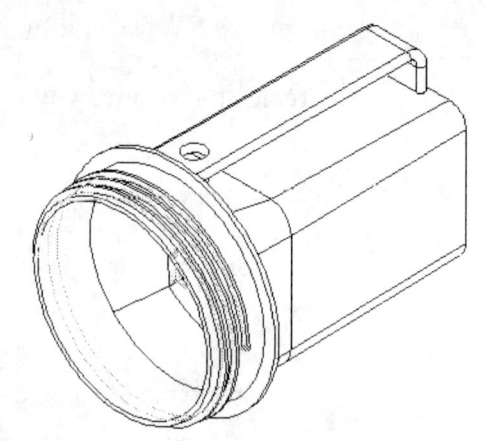

275) Rename Cut Extrude to SwitchHole.

276) Save the HOUSING. Click **Save**.

Create the HOUSING - First Rib Feature

The Rib feature adds material between contours of existing geometry. Use Ribs to add structural integrity to a part.

A Rib requires:

- A Sketch.

- Thickness.

- Extrusion direction.

Create the Rib.
277) Select the Sketch plane. Click the **Top** plane.

278) Display all hidden lines to avoid unwanted relationships. Click **Hidden Lines Visible**.

Engineering Design with SolidWorks — Sweep and Loft Features

279) Create the Sketch. Click **Sketch**. Display the Top view. Click **Top**. Click **Line**. Sketch a horizontal **line**. The endpoints of the sketch are collinear with the inside wall of the Shell feature.

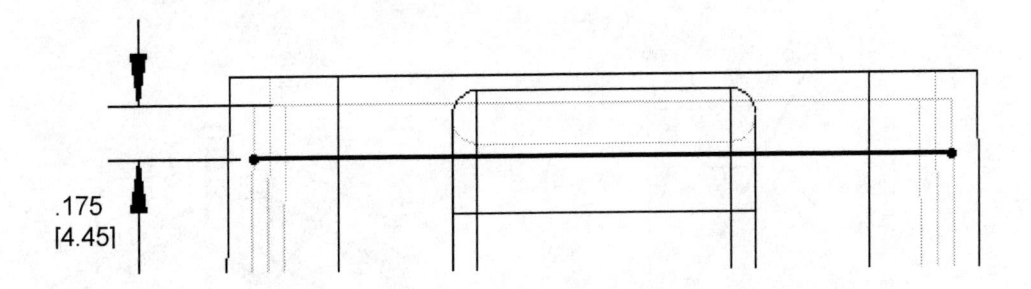

.175
[4.45]

280) Add a linear dimension. Click **Dimension**. Click the **line**. Click the **inner back edge**. Enter **.175**, **[4.45]**.

281) Create the Rib. Click **Rib** from the Feature toolbar. Create the Rib on both sides of the Sketch plane. Click the **Both Sides** button. Enter **.100**, **[2.54]** for Rib Thickness. Click the **Parallel to Sketch** button. The Rib direction arrow points to the back. Flip the material side if required. Click the **Enable draft** check box. Click the **Draft On/Off** button. Enter **1.00** for Draft Angle.

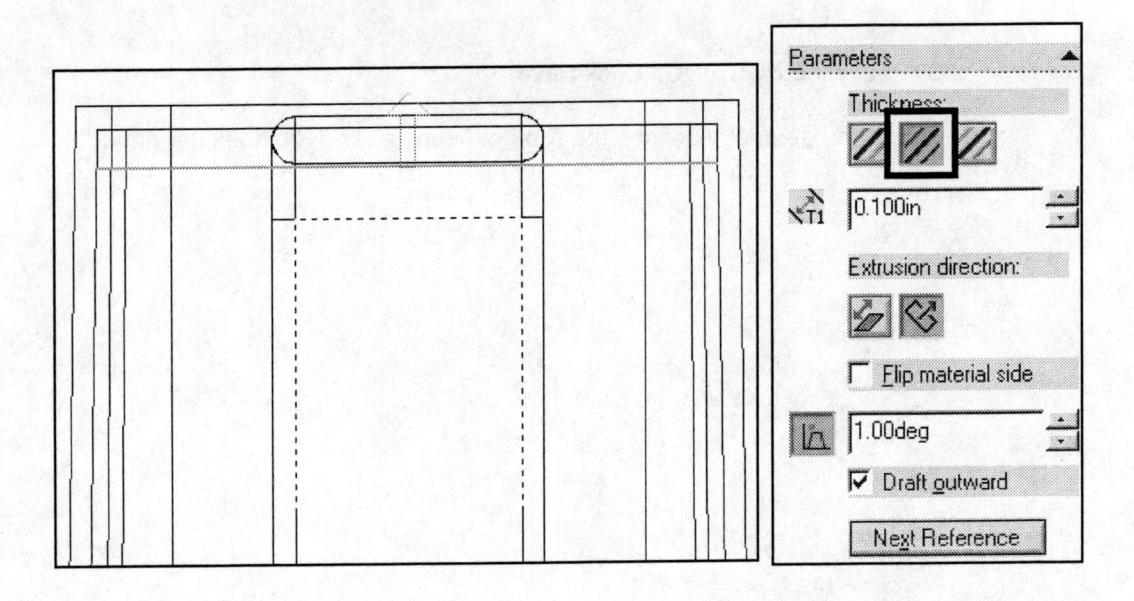

Note: Click the Flip material side check box if the direction arrow does not point towards the back.

282) Display the Rib. Click **OK** ✔.

283) View the Rib. Click **Right**. Click **Isometric**.

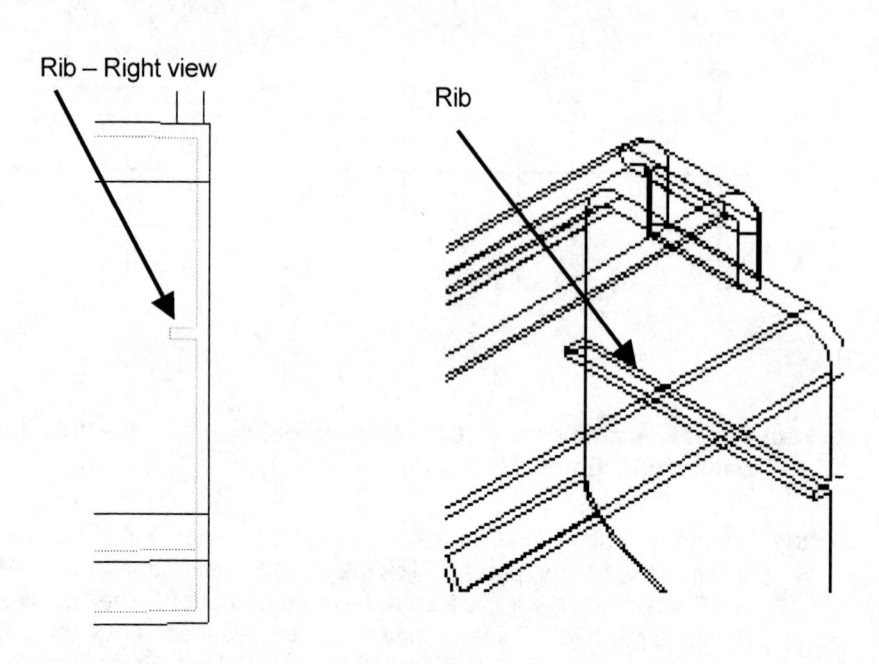

Rib – Right view

Rib

284) Save the HOUSING. Click **Save**.

Existing geometry defines the Rib boundaries. The Rib does not penetrate through the wall.

Create the HOUSING – Use Linear Pattern for the Ribs

The HOUSING requires multiple Ribs to support the BATTERY.

A Linear Pattern creates multiple Instances of a feature along a straight line.

Create the Linear Pattern in two directions along the same vertical edge of the HOUSING.

Create a Linear Pattern of the Rib feature.

285) Click **Linear Pattern**. The Linear Pattern PropertyManager is displayed.

286) Click **Rib1** in the Features to Pattern box. Click the **Direction 1 Pattern Direction** text box. Click the **hidden upper back vertical edge** of Shell1. The direction arrow points upward. Click the Reverse direction button if required.

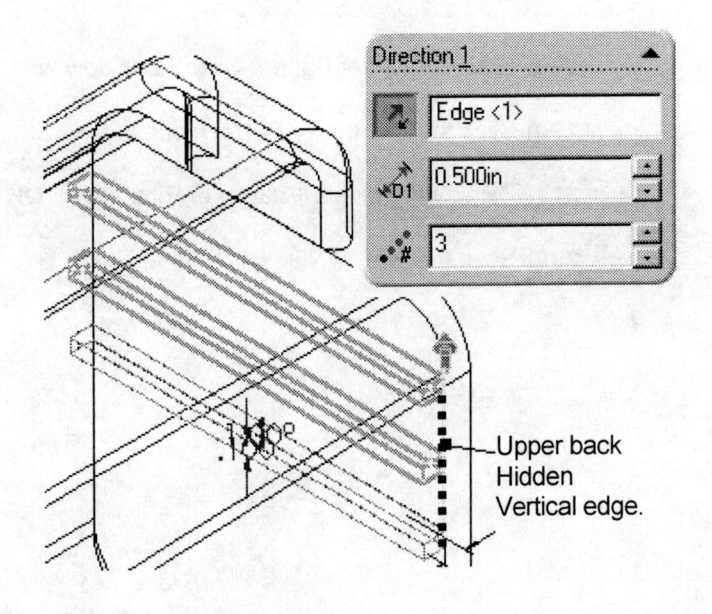

287) Enter **.500**, **[12.7]** for Spacing. Enter **3** for Number of Instances.

288) Expand the Direction 2 properties. Click the hidden **lower back vertical edge** of Shell1. The direction arrow points downward. Enter **.500**, **[12.7]** for Spacing. Enter **3** for Number of Instances. Select the **Pattern seed only** check box.

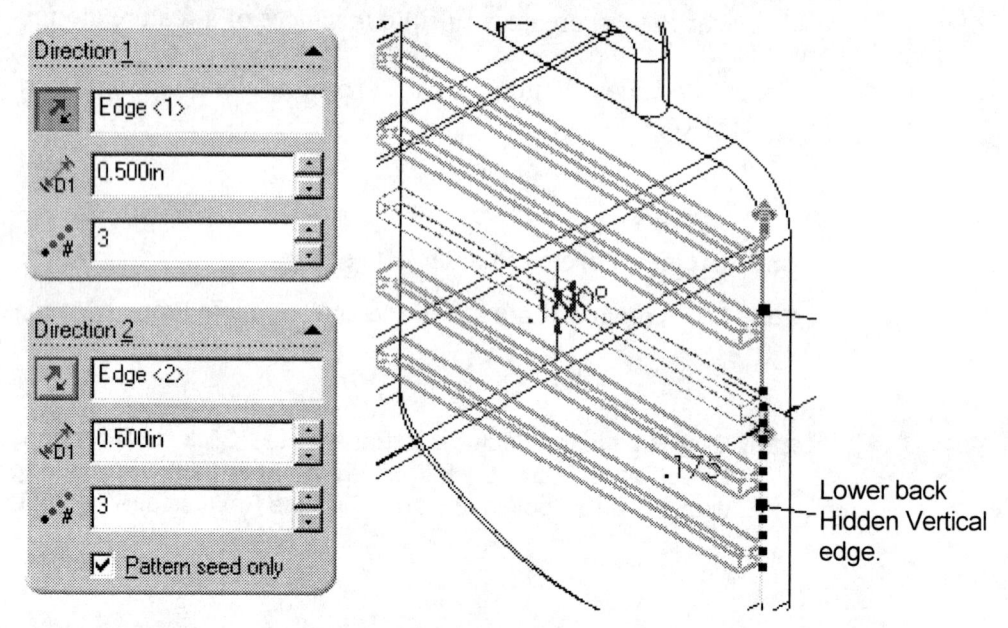

289) Drag the Linear Pattern **Scroll bar** downward to display the Options box.

290) Click the **Geometry Pattern** check box.

291) Display the Linear Pattern of Ribs. Click **OK**.

292) Save the HOUSING. Click **Save**.

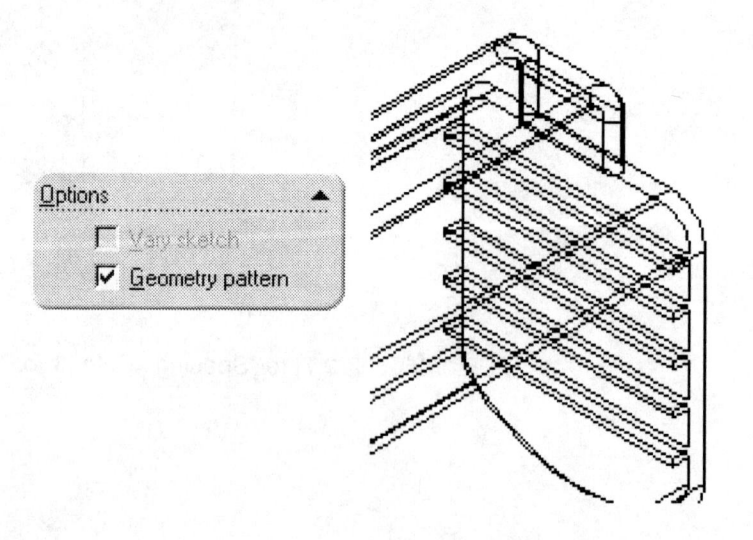

Create the HOUSING - Second Rib Feature

The Second Rib feature supports and centers the battery.

Create the Plane.
293) Click **Shaded**.

294) **Zoom in** on the back of the Handle.

295) Create the Sketch plane. Click **Insert** from the Main menu. Click **Reference Geometry**, **Plane**. Click **Parallel Plane at Point**. The Point-Plane option requires a point and a plane. Click the **Right** plane from the FeatureManager. Click the **vertex** (point) at the back right of the handle.

296) Display the plane. Click **OK** ✔.

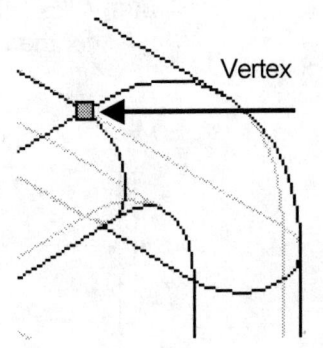

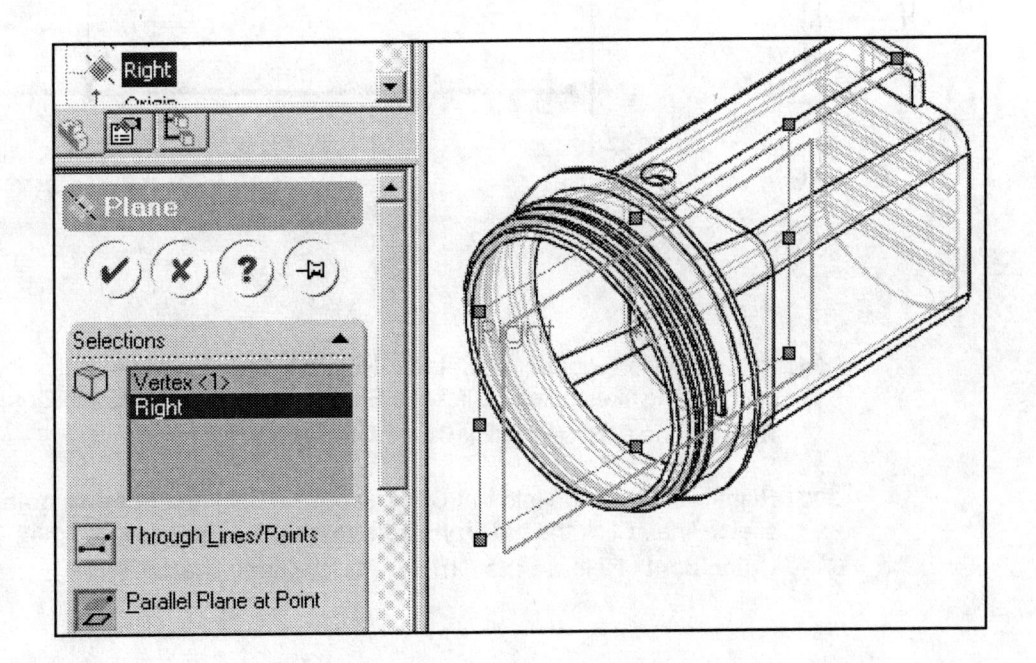

297) Rename Plane1 to LongRibPlane.

Create the second Rib.
298) Select the Sketch plane. Click the **LongRibPlane**. Display the Right view. Click **Right**.

299) Create the Sketch. Click **Sketch**. Click **Line**. Sketch a **horizontal line**. Do not select the edges of the Shell1 feature.

300) Click **Dimension**. Click the **horizontal line**. Click **Origin**. Click a location for the vertical linear dimension **text**. Enter **1.300**, [33.02].

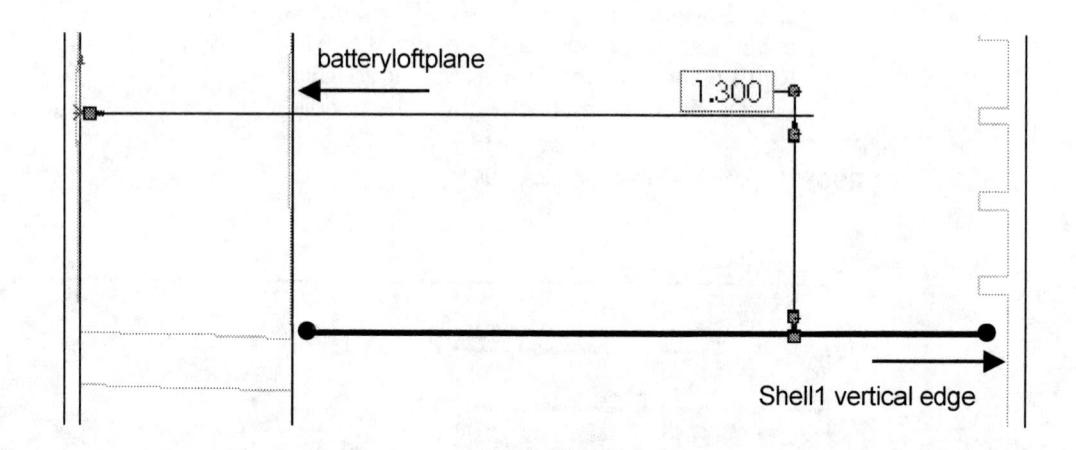

301) Add relations. Right-click **Select**. Hold the **Ctrl** key down. Click the **right end point** of the sketch line. Click the Shell1 **vertical edge**. Click **Coincident**. Release the **Ctrl** key. Click **Close Dialog**.

302) Right-click **Select**. Hold the **Ctrl** key down. Click the **left end point** of the sketch line. Click the **batteryloftplane** from the FeatureManager. Click **Coincident**. Release the **Ctrl** key. Click **Close Dialog**.

Note: When the sketch and reference geometry become complex, create dimensions by selecting reference planes in the FeatureManager.

For design flexibility, dimension the Rib from the Origin. Not from the bottom HOUSING surface.

Engineering Design with SolidWorks — Sweep and Loft Features

303) Click **Tangent Arc**. Click the **left end** of the horizontal line. Click the **intersection** of the Shell1 and Boss Stop features. **Double-click** to end the arc.

304) Add relations. Right-click **Select**. Hold the **Ctrl** key down. Click the **end point** of the arc. Click the Shell1 **horizontal inside wall**. Click the left **vertical Boss-Stop edge**. Click **Intersection**. Release the **Ctrl** key. Click **Close Dialog**.

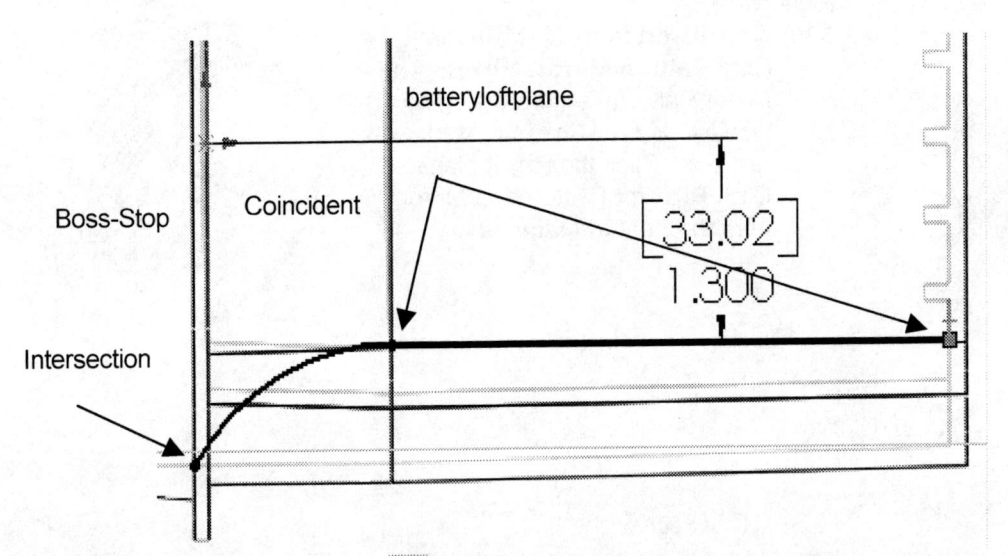

305) Create the Rib. Click **Rib** from the Feature toolbar. The Rib Property Manager is displayed. Create the Rib on both sides of the Sketch plane. Click the **Both sides** button. Enter **.075**, **[1.91]** for Rib Thickness.

306) Create the draft. Click the **Draft On/Off** button. Enter **1** for Angle. Click the **Draft outward** check box. The direction arrow points towards the bottom. Click the Flip material side check box if required.

307) Display the Rib. Click **OK**.

308) Display the Isometric. Click **Isometric**.

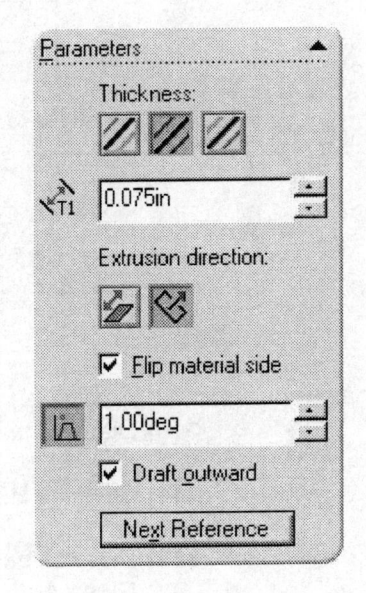

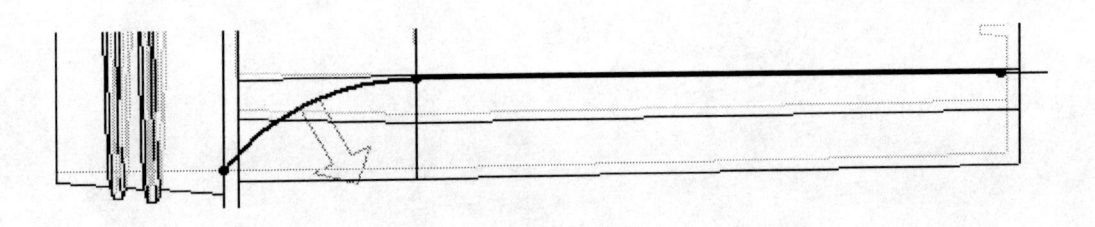

PAGE 5 - 61

Sweep and Loft Features

Engineering Design with SolidWorks

Mirror the Second Rib

An additional Rib is required to support the BATTERY.

Mirror the second Rib feature about the Right plane.

309) Click **Insert** from the Main menu. Click **Pattern/Mirror**, **Mirror**. The Mirror Pattern Feature dialog box is displayed. Click the **Mirror plane** text box. Click the **Right** plane. Click **Rib2** for Features to Mirror from the FeatureManager.

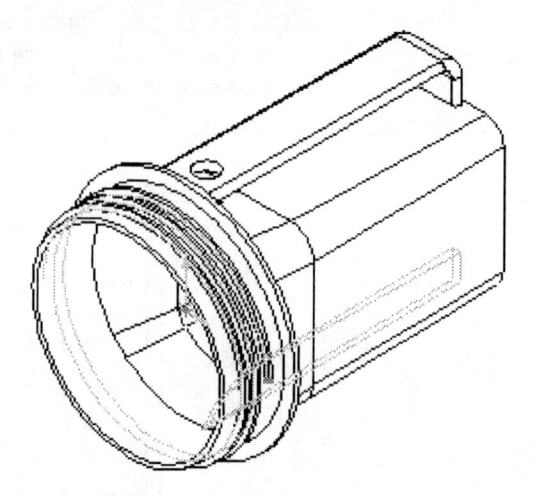

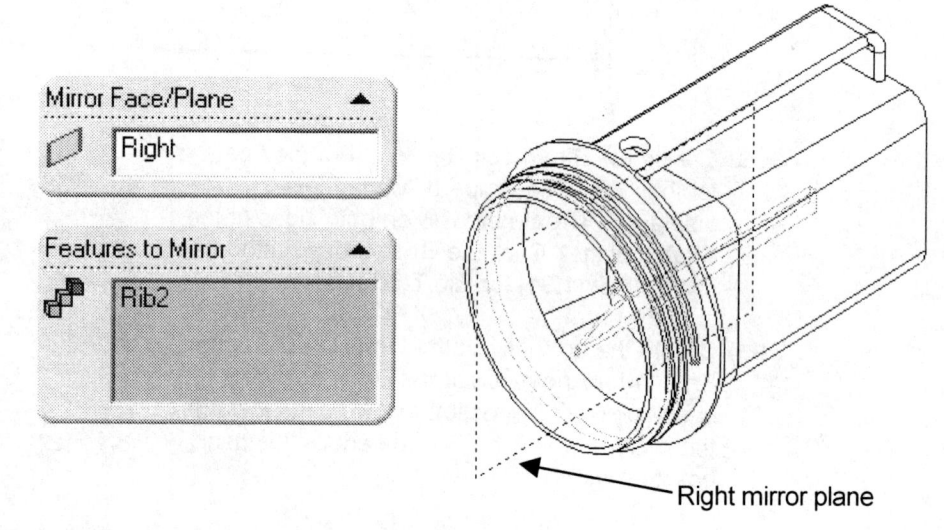

Right mirror plane

310) Display the Mirror Pattern feature. Click **OK**.

311) Save the HOUSING. Click **Save**.

312) Close all parts before you begin the next assembly exercise. Click **Window**, **Close All**.

PAGE 5- 62

Engineering Design with SolidWorks — Sweep and Loft Features

The parts for the FLASHLIGHT are complete!

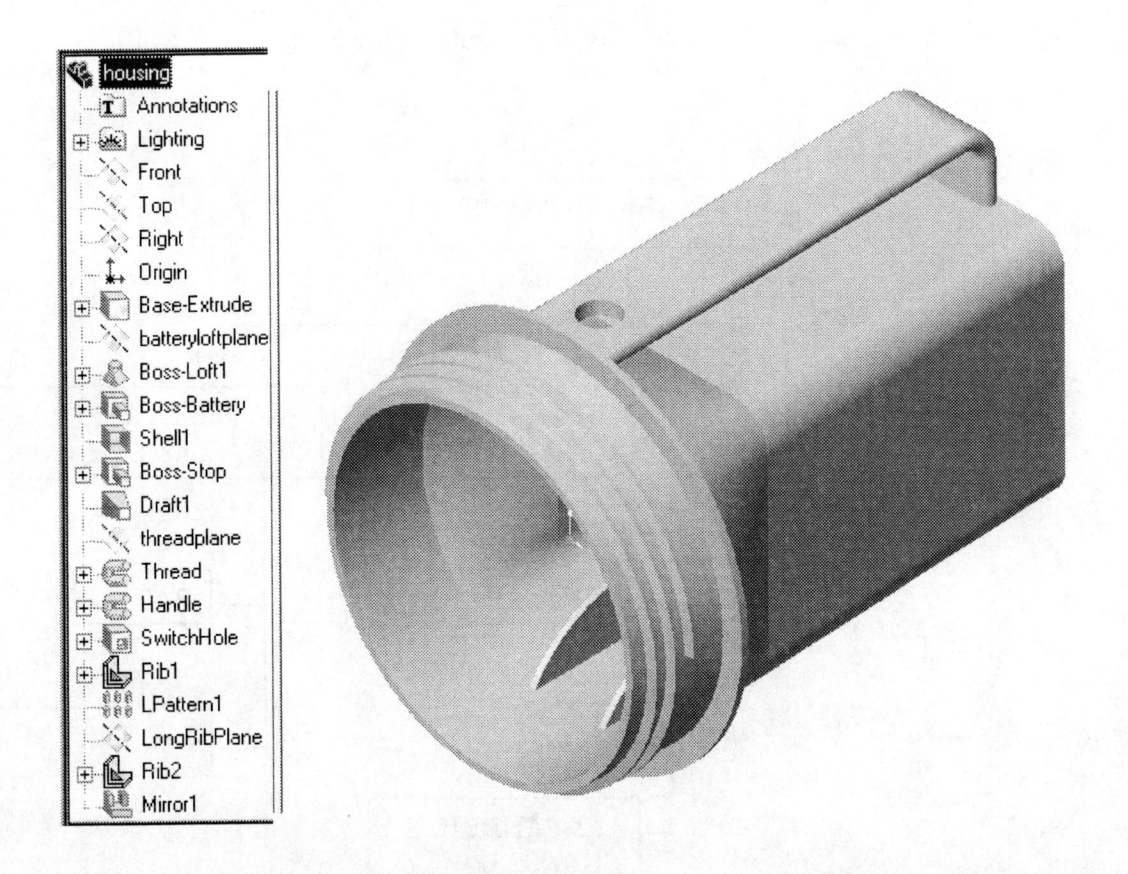

Sweep and Loft Features Engineering Design with SolidWorks

FLASHLIGHT Assembly

Plan the sub-assembly component layout, Figure 5.14a.

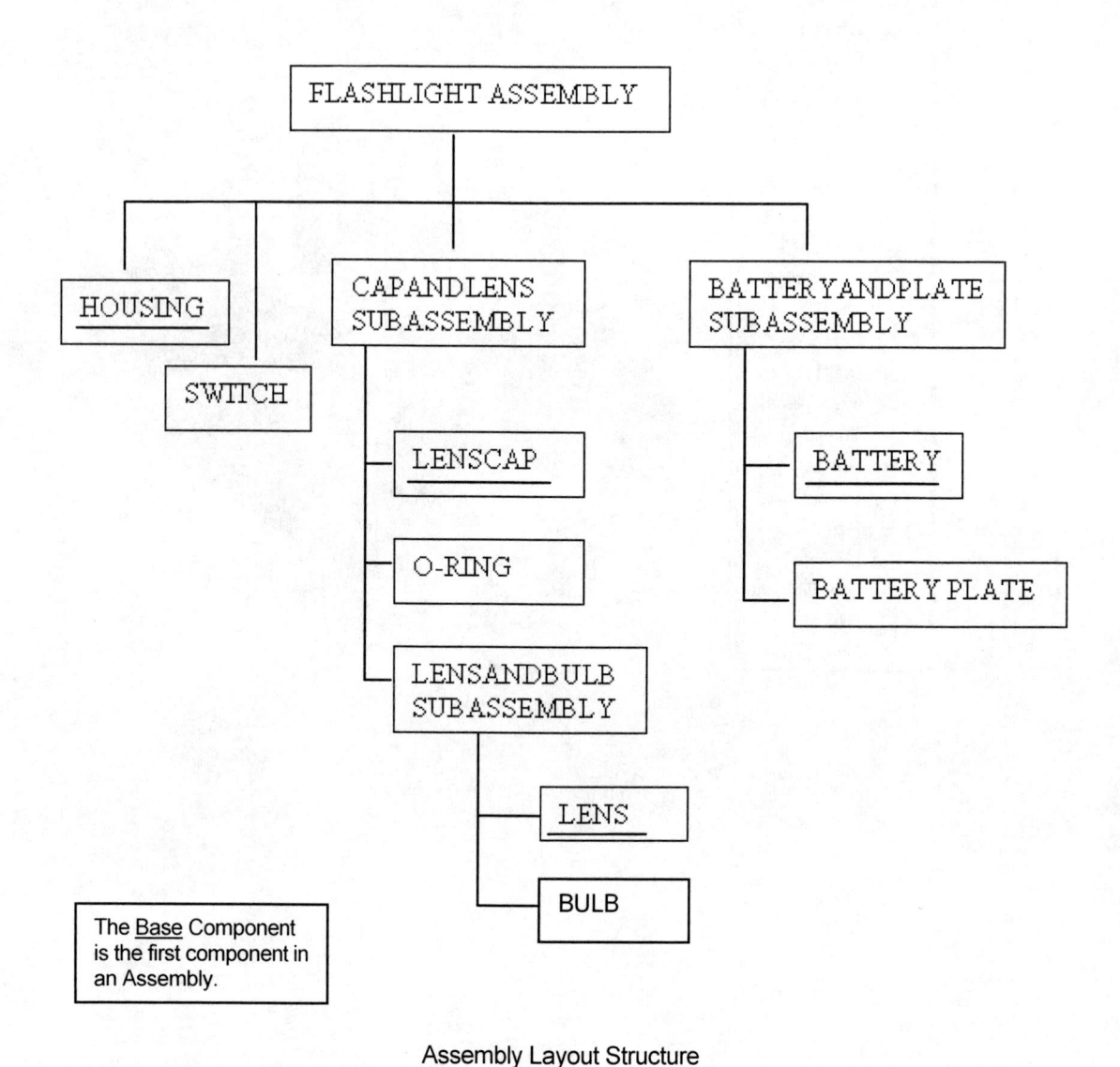

```
                    ┌─────────────────────┐
                    │ FLASHLIGHT ASSEMBLY │
                    └─────────────────────┘
                              │
         ┌────────────────────┼────────────────────┐
    ┌─────────┐      ┌──────────────────┐   ┌────────────────────┐
    │ HOUSING │      │    CAPANDLENS    │   │  BATTERYANDPLATE   │
    └─────────┘      │   SUBASSEMBLY    │   │    SUBASSEMBLY     │
         │           └──────────────────┘   └────────────────────┘
    ┌─────────┐              │                       │
    │ SWITCH  │              ├──┌─────────┐          ├──┌──────────┐
    └─────────┘              │  │ LENSCAP │          │  │ BATTERY  │
                             │  └─────────┘          │  └──────────┘
                             │                       │
                             ├──┌─────────┐          └──┌───────────────┐
                             │  │ O-RING  │             │ BATTERY PLATE │
                             │  └─────────┘             └───────────────┘
                             │
                             └──┌──────────────────┐
                                │   LENSANDBULB    │
                                │   SUBASSEMBLY    │
                                └──────────────────┘
                                         │
                                         ├──┌──────┐
                                         │  │ LENS │
                                         │  └──────┘
                                         │
                                         └──┌──────┐
                                            │ BULB │
                                            └──────┘
```

> The <u>Base</u> Component is the first component in an Assembly.

Assembly Layout Structure
Figure 5.14a.

PAGE 5- 64

Engineering Design with SolidWorks　　　　　　　　　　　Sweep and Loft Features

FLASHLIGHT Assembly Overview

The FLASHLIGHT assembly steps are as follows:

- Create the LENSANDBULB sub-assembly from the LENS and BULB, Figure 5.14a. The LENS is the Base component.

- Create the BATTERYANDPLATE sub-assembly from the BATTERY and BATTERYPLATE, Figure 5.14b. The BATTERY is the Base component.

- Create the CAPANDLENS sub-assembly from the LENSCAP, O-RING and LENSANDBULB sub-assembly, Figure 5.14c. The LENSCAP is the Base component.

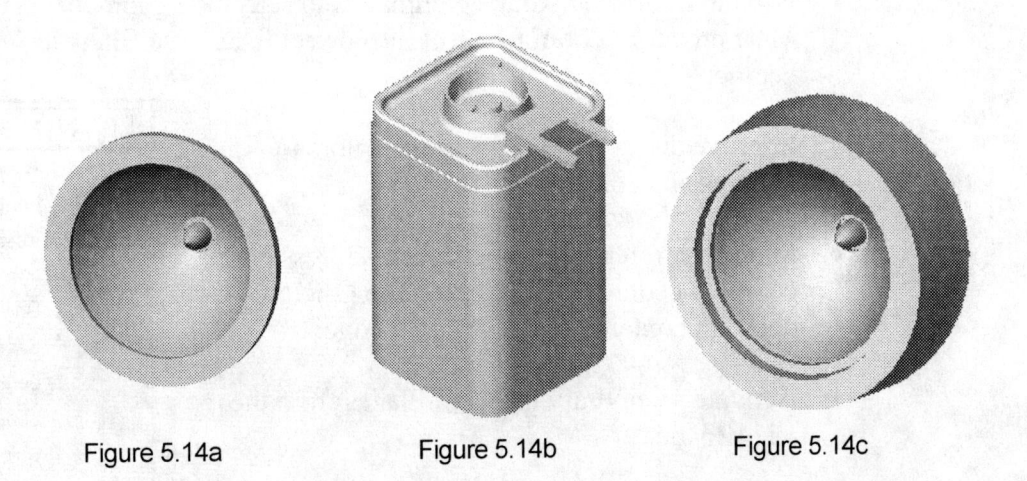

Figure 5.14a　　　　Figure 5.14b　　　　Figure 5.14c

- Create the FLASHLIGHT assembly. The HOUSING is the Base component, Figure 5.14d. Add the SWITCH, CAPANDLENS and BATTERYANDPLATE, Figure 5.14e.

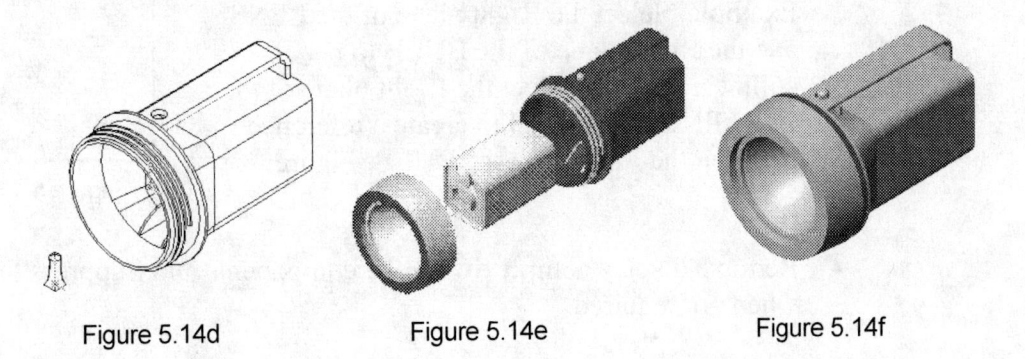

Figure 5.14d　　　　Figure 5.14e　　　　Figure 5.14f

- Modify the dimensions to complete the FLASHLIGHT assembly, Figure 5.14f.

PAGE 5 - 65

Assembly Techniques

Assembly modeling requires practice and time. Below are a few helpful hints and techniques to address the Bottom up design modeling approach.

- Create an assembly layout structure. The layout structure will organize the sub-assemblies and components and save time.

- Insert sub-assemblies and components as lightweight components. Lightweight components save on file size, rebuild time and complexity.

    Set Lightweight components in the Tools, Options command.

- Use the Zoom and Rotate commands to select the geometry in the Mate process. Zoom to select the correct face. Use filters to select the geometry.

- Improve display. Apply various colors to features and components.

- Mate with reference planes when addressing complex geometry. Example: The O-RING does not contain a flat surface or edge.

- Activate Temporary axis and Planes from the View menu.

- Select reference planes from the FeatureManager. Expand the component to view the planes.

    Example: Select the Right plane of the LENS and the Right plane of the BULB to be collinear. Do not select the Right plane of the HOUSING if you want to create a reference between the LENS and the BULB, Figure 5.15.

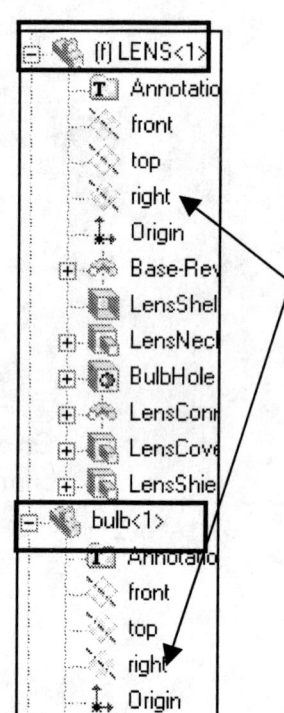

Figure 5.15

- Remove display complexity. Hide components and Suppress features when not required.

- Use the Move Component and Rotate Component commands before mating. Position the component in the correct orientation.

- Conserve time. Use Preview during the Mate operation. Suppress unwanted features. Create additional flexibility into a mate. Use a Distance Mate with a zero value. Flip the direction if required.

- Remove unwanted entries. Use the Delete key from the Assembly Mating Items Selected text box.

- Verify the position of the mated components. Use Top, Front, Right and Section views.

- Use caution when you view the color red in an assembly. Red indicates that a part is being edited in the context of the assembly.

- Avoid unwanted references. Verify your selections with the PropertyManager.

Create an Assembly Template

An Assembly Document Template is required to create the Flashlight Assembly and its sub-assemblies.

For English Components: Create an Assembly Document Template using Inch units.

For Metric Components: Create an Assembly Document Template using millimeter units.

Create an assembly template.

313) Click **New**. Click **Assembly** from the Template dialog box. Click **OK**.

314) Set the Assembly Document Template options. Click **Tools**, **Options**, **Document Properties**.

For millimeter assembly template, go to step 321.

315) For English components: Select **ANSI** from the Detailing option.

316) Click **Units**. Enter **Inches** from the Linear units list box. Enter **3** in the Decimal places spin box.

317) Click **OK**.

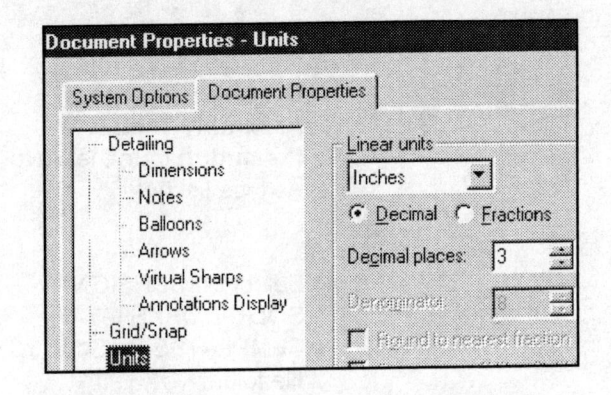

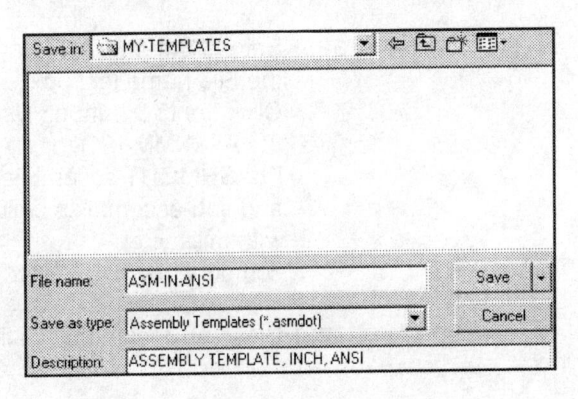

318) For English components: Save the assembly template. Click **File**, **SaveAs**. Click the **Assembly Template (*.asmdot)** from the Save As type list box. Select **ENGDESIGN-W-SOLIDWORKS\MY-TEMPLATES** for Save in file folder.

319) Enter **ASM-IN-ANSI** in the File name text box.

320) Enter **ASSEMBLY TEMPLATE, INCH, ANSI** for Description. Click the **Save** button. Use the ASM-IN-ANSI for all FLASHLIGHT assemblies and sub-assemblies built with inch components.

321) For Millimeter components: Select **ISO** from the Detailing option.

322) Click **Units**. Enter **Millimeters** from the Linear units list box. Enter **2** in the Decimal places spin box.

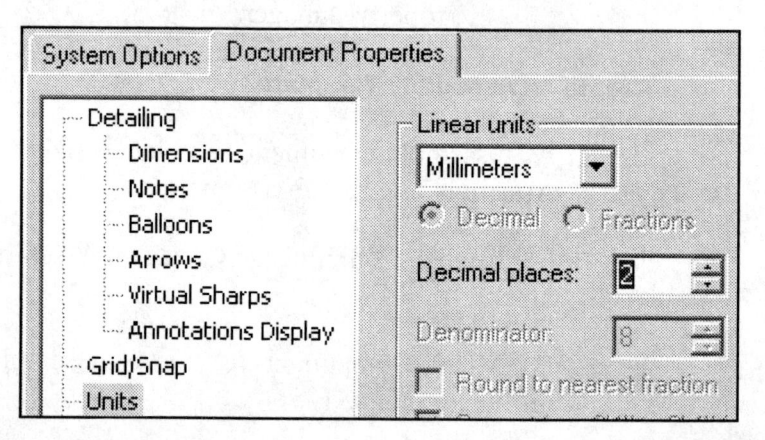

323) Save the assembly template. Click **File**, **SaveAs**. Click the **Assembly Template (*.asmdot)** from the Save As type list box.

324) Select ENGDESIGN-W-SOLIDWORKS\MY-TEMPLATES for Save in file folder.

325) Enter **ASM-MM-ANSI** in the File name text box. Click the **ISO** button. Use the ASM-MM-ISO for all FLASHLIGHT assemblies and sub-assemblies built with millimeter components.

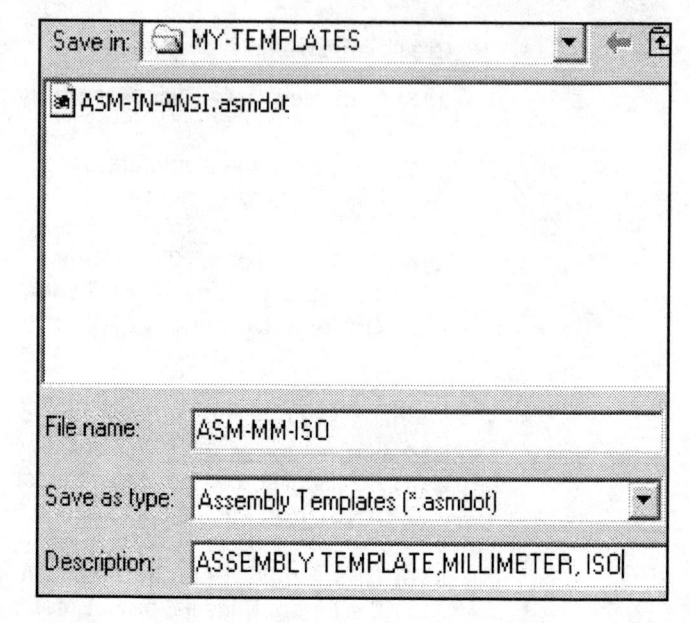

326) Close all documents. Click **Windows**, **Close All**.

Engineering Design with SolidWorks — Sweep and Loft Features

LENSANDBULB Sub-assembly

Create the LENSANDBULB sub-assembly. The LENS is the Base component. LENSANDBULB sub-assembly mates the BULB component to the LENS component.

Create the LENSANDBULB sub-assembly.

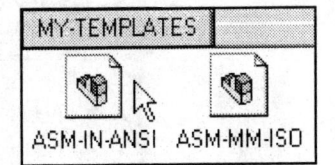

327) Click **New** from the Standard toolbar. The New SolidWorks Documents dialog box is displayed.

328) Click the **MY-TEMPLATES** tab.

329) Click the **ASM-IN-ANSI**, [**ASM-MM-ISO**] template. Click **OK**.

330) Click **Save**. Select the **ENGDESIGN-W-SOLIDWORKS\PROJECTS** file folder.

331) Enter **LENSANDBULB** for File name.

332) Enter LENS AND BULB ASSEMLBY FOR 6VOLT FLASHLIGHT for DESCRIPTION.

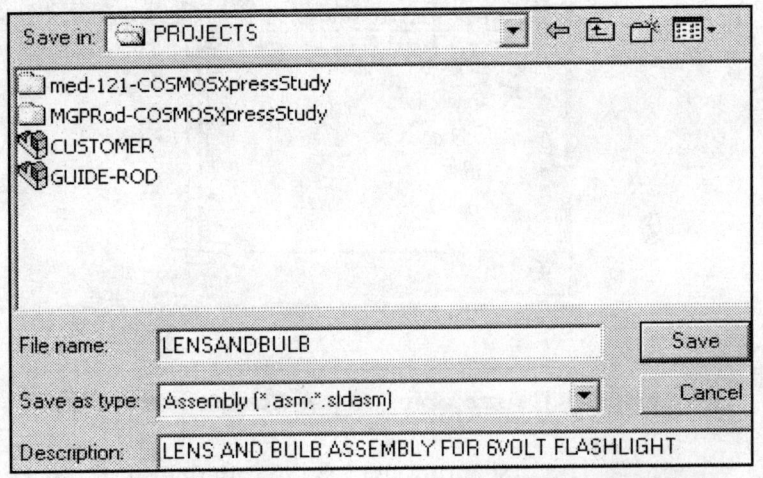

333) Save the file. Click the **Save** button.

334) Open the LENS. Click **Open**. Select **LENS** from the ENGDESIGN-W-SOLIDWORKS\PROJECTS file folder.

335) Open the BULB. Click **Open**. Select **BULB**.

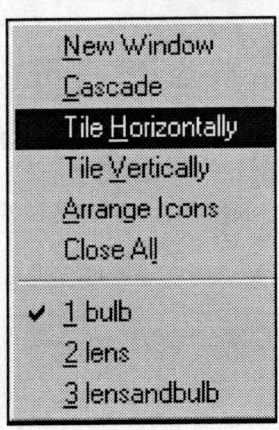

PAGE 5 - 69

336) Click **Window** from the Main menu. The following documents are displayed: BULB, LENS and LENSANDBULB. Display the assembly and components. Click **Tile Horizontally** from the Window menu. Review the windows. The Origin feature of the LENSANDBULB assembly is displayed in the Graphics window. The FeatureManager is displayed on the left side of the three graphics windows.

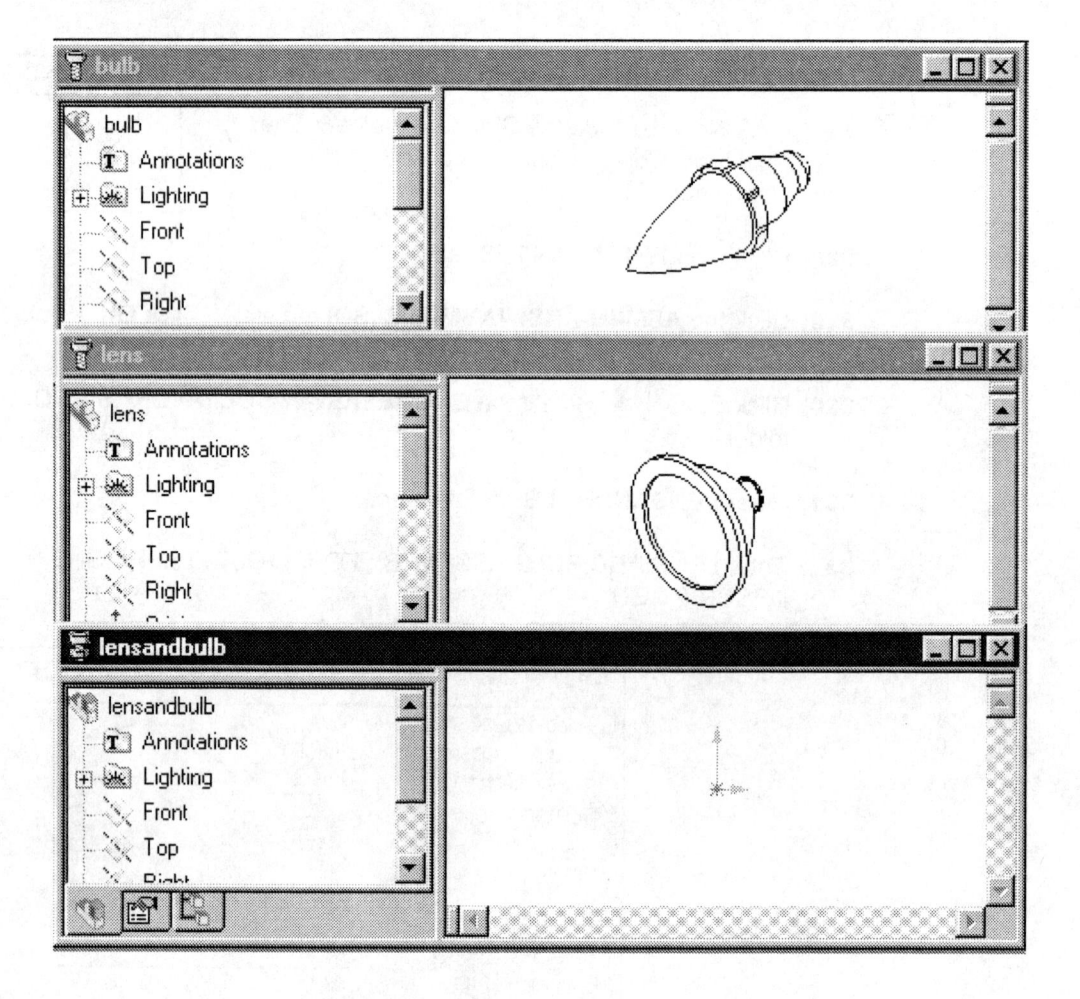

LENSANDBULB Assembly – Insert Components

The first component is the foundation of the assembly. The LENS is the first component. Use the following techniques to add components o the assemblies:

- Click Insert from the Main menu. Click Component. Click From File.

- Drag components from Windows Explorer.

- Drag components from the Feature Palette window.

- Drag components from the Open part files.

Engineering Design with SolidWorks — Sweep and Loft Features

Add the first component.
337) Click the **LENSANDBULB** in the Graphics window. Click **View**. Check **Origins**.

338) Position the LENS inside the assembly Graphics window. Click the **LENS** lens icon from the top of the FeatureManager.

339) Drag the **LENS** icon to the Origin of the **LENSANDBULB** assembly window. The mouse pointer displays when position on the Origin. Display the LENS component in the assembly Graphics window. Release the **mouse button**.

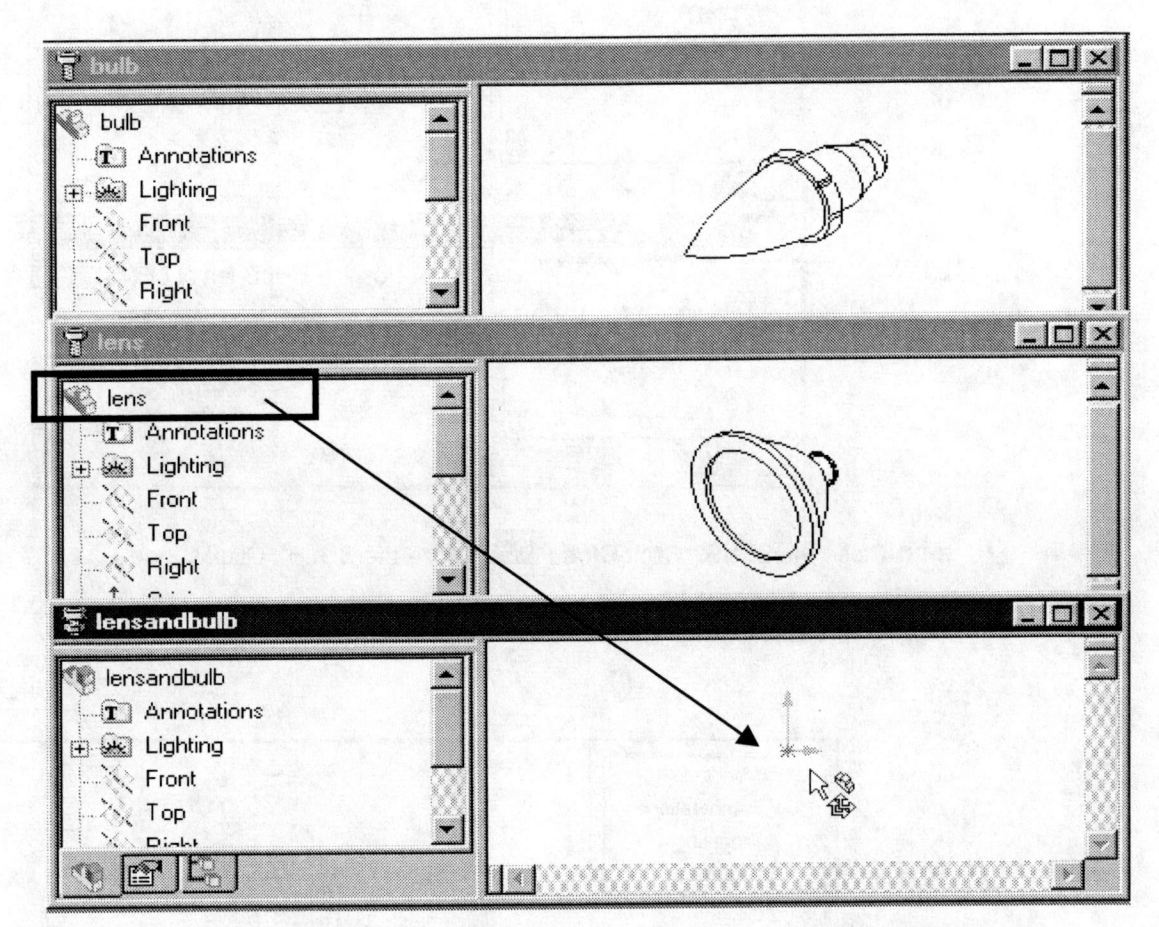

Note: The front view of the LENS is displayed in the LENSANDBULB assembly. The LENS name is added to the LENSANDBULB assembly FeatureManager with the symbol (f).

The symbol (f) represents a fixed component.

A fixed component cannot move and is locked to the assembly Origin.

To remove the fixed state, Right-click a component name in the FeatureManager. Click Float.

Sweep and Loft Features Engineering Design with SolidWorks

Add the second component.

340) Click the **BULB** 🔧 bulb icon from the top of the FeatureManager.

341) Drag the **BULB icon** to the LENSANDBULB Graphics window. The mouse pointer displays 🔧 when positioned inside the LENSANDBULB Graphics window.

342) Display the BULB component. Release the **mouse button**.

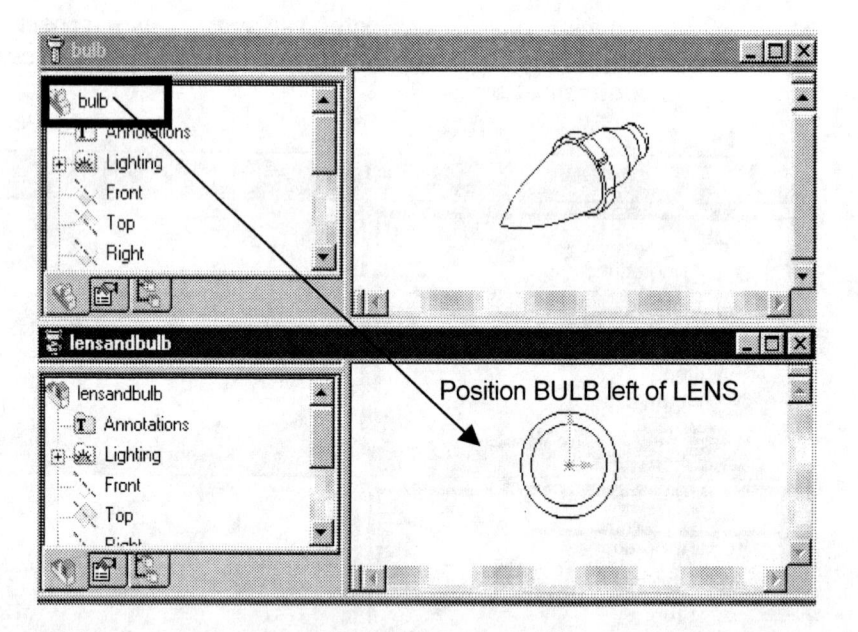

343) Close the LENS. Click **Close** ⊠. Close the BULB. Click **Close** ⊠.

344) Display the Isometric view. Click **Isometric** 🔲 in the LENSANDBULB window.

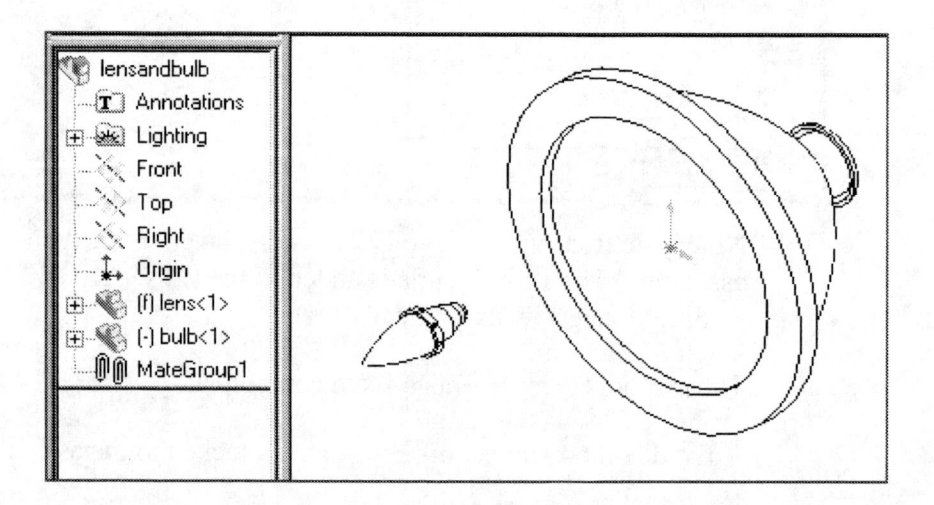

PAGE 5-72

Engineering Design with SolidWorks   **Sweep and Loft Features**

345) Enlarge the assembly window. Click **Maximize** in the upper right hand corner of the LENSANDBULB window.

346) Fit all components in the Graphics window. Press the **f** key.

347) Hide the planes. Click **View**. Uncheck **Planes**.

348) Save LENSANDBULB. Click **Save**.

Review the FeatureManager Syntax.
349) Double click the **LENS** component inside the FeatureManager. The LENS component lists features. Example: Base-Revolve, Shell, Lens neck. Sketches are contained within the LENS features that display the Plus sign icon.

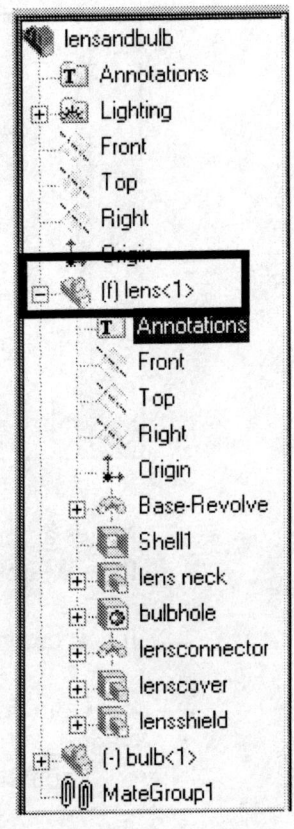

Review the BULB component syntax in the FeatureManager.

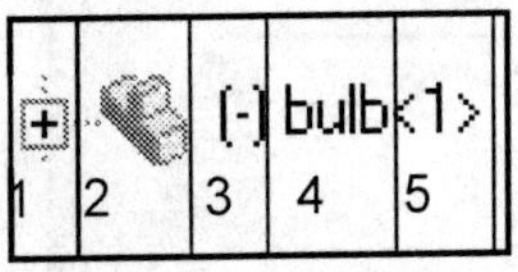

1. A Plus sign icon indicates that additional feature information is available. A minus sign indicates that the feature list is fully expanded.

2. A component icon indicates that the BULB is a part. The assembly icon indicates that the LENSANDBULB is an assembly. Sub-assemblies display the same icon as an assembly.

3. Column 3 identifies the Component State:

    - A minus sign (−) indicates that the component is under defined and requires additional information.

    - A plus sign (+) indicates that the component is over defined.

    - A fixed symbol (f) indicates that the component does not move.

    - A question mark (?) indicates that additional information is required.

4. BULB - Name of the component.

5. The symbol <#> indicates the number of copies in the assembly. The symbol <1> indicates the original component, "BULB" in the assembly.

Sweep and Loft Features   **Engineering Design with SolidWorks**

Move the BULB Component.

350) Components located in an assembly are free to move. Click the **BULB** in the Graphics window. Click **Move Component** from the Assembly toolbar. The PropertyManager is displayed on the left side of the Graphics window. The mouse pointer displays.

351) Position the **BULB** in front of the LENS. Click **OK**.

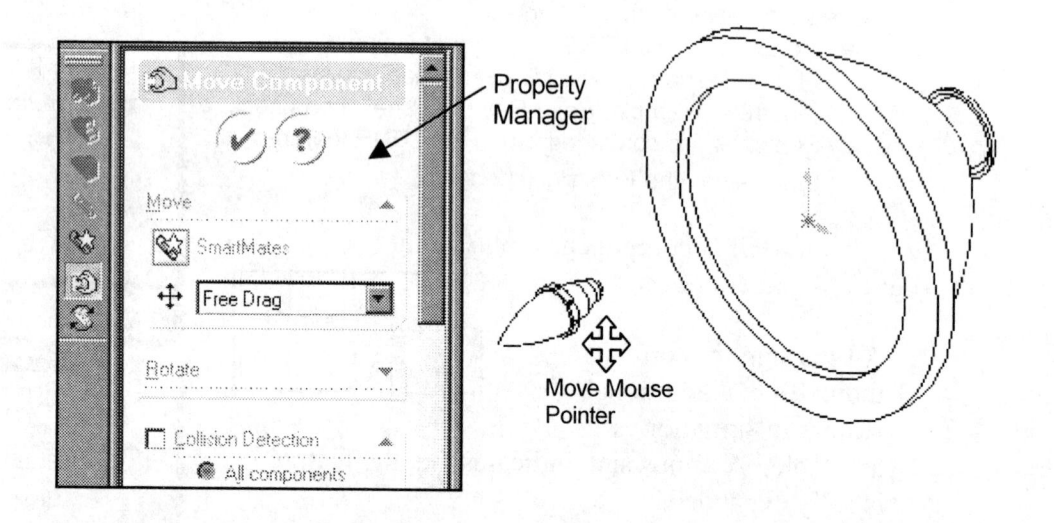

Property Manager

Move Mouse Pointer

Mates are relationships that align and fit components in an assembly. Create three Mates in this section.

Insert components into an assembly with various intuitive options:

- o Coincident.
- o Parallel.
- o Tangent.
- o Concentric.
- o Distance.
- o Angle.
- o Perpendicular.

Establishing the correct component relationship in an assembly requires forethought on component interaction.

Engineering Design with SolidWorks **Sweep and Loft Features**

Suppress the Lens shield feature to view all surfaces during the mate process.

Suppress the Lens shield feature.
352) Right-click **Lens Shield** in the LENS FeatureManager. Click **Feature Properties**. Check the **Suppressed** box. Click **OK**.

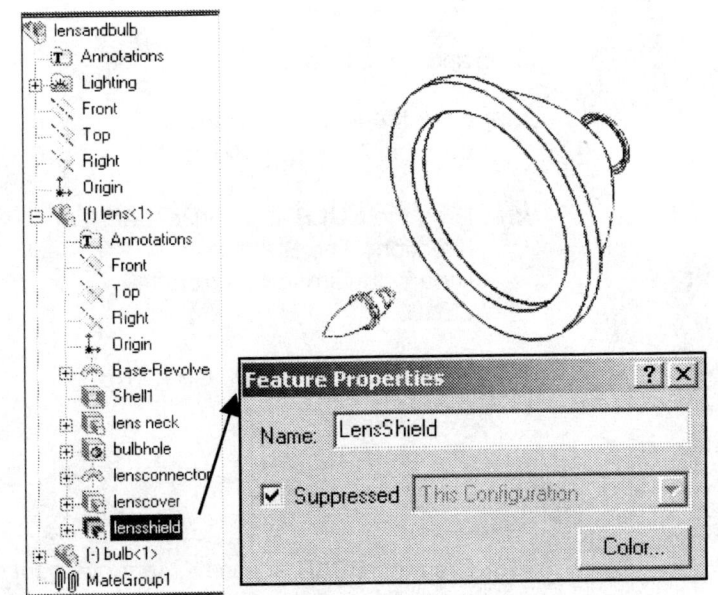

Create the first Mate.
353) Click **Mate** from the Assembly toolbar. The Assembly Mating PropertyManager is displayed. Click the **Mate PropertyManager** **Mate** icon to display the FeatureManager.

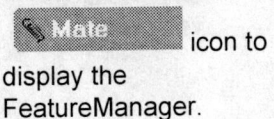

354) Create a Coincident Mate. Click the **Right** plane of the **LENS** from the FeatureManager. Expand the **BULB** FeatureManager. Click the **Right** plane of the **BULB** from the FeatureManager. Click **Preview**. Click **OK**.

The Right plane of the LENS and the Right plane of the BULB are coincident.

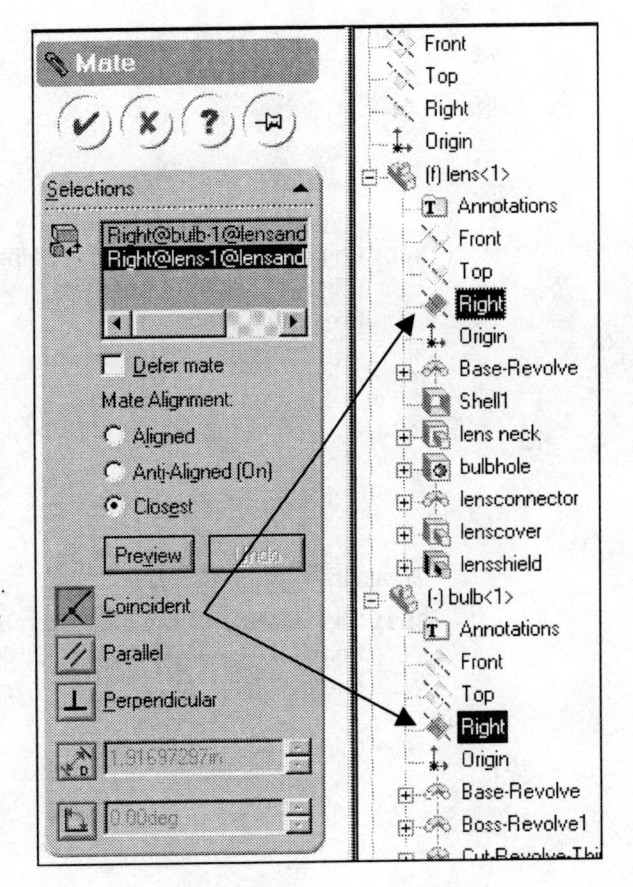

Mates remove degrees of freedom.

Sweep and Loft Features

Move and Rotate the BULB.

355) Click **Move Component** from the Assembly toolbar.

356) Drag the **BULB** in a horizontal direction. The BULB travels linearly towards/away from the LENS.

357) Display the Top view. Click **Top**.

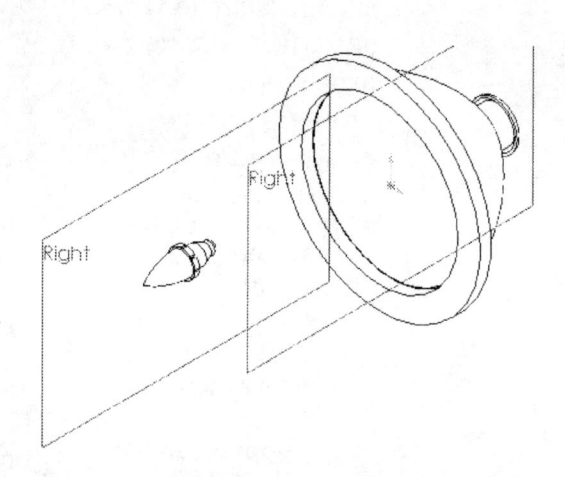

358) Drag the **BULB** and view the movement constrained to the Right plane. Click **Isometric**.

359) Click the **Rotate** arrow from the Move PropertyManager. Drag the **BULB** in a vertical direction. The BULB rotates about its Origin.

360) Position the **BULB** in front of the LENS. Click **OK** from the PropertyManager.

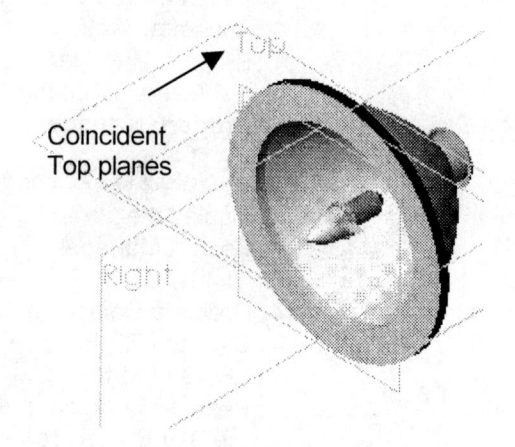

Create the second Coincident Mate.

361) Click **Mate** from the Assembly toolbar. Click the **Top** plane of the LENS from the FeatureManager. Click the **Top** plane of the BULB from the FeatureManager.

362) Click **Preview**. Click **OK**.

Engineering Design with SolidWorks

Sweep and Loft Features

When selecting faces, position the mouse pointer in the middle of the face. Do not position the pointer near the edge of the face. If the wrong face or edge is selected, click the face or edge again to remove it from the Items Selected text box.

Right-click in the Graphics window. Click Clear Selections to remove all geometry from the Items Selected text box. The next step is an example of the Select Other option.

Select hidden geometry.
363) Create the third Mate. **Zoom in** and **Rotate** on the counter bore hole. Click **Mate**. Select the **Counterbore face** of the LENS.

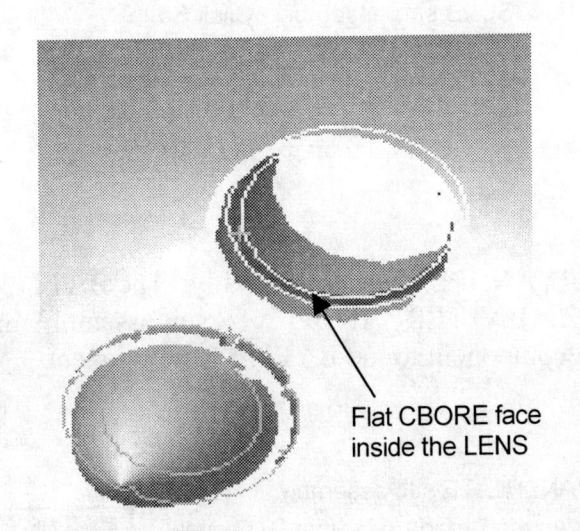

Flat CBORE face inside the LENS

364) Create a Distance Mate. Click the **bottom flat face** of the BULB. Click the **Distance** button. Enter **0** for Distance. Click **Preview**. Click **OK**.

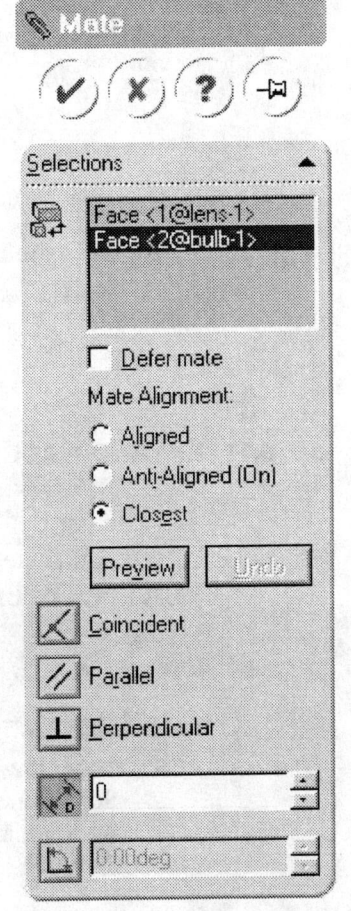

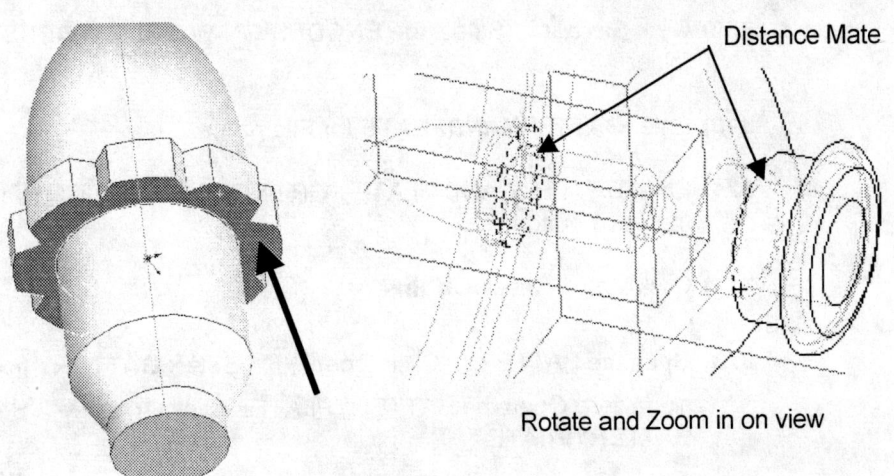

Distance Mate

Rotate and Zoom in on view

PAGE 5 - 77

365) The LENSANDBULB sub-assembly is fully defined. Display the mate types. Double-click on **Mates** in the FeatureManager.

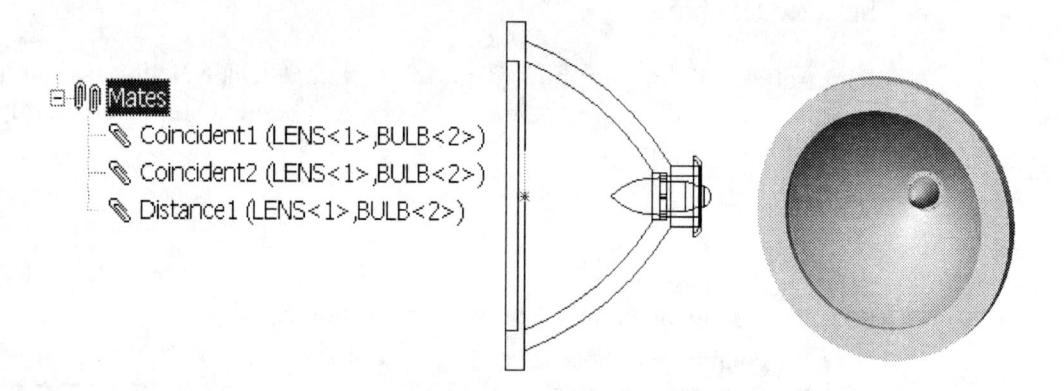

366) Save the LENSANDBULB sub-assembly. Click **Save**.

367) Close all parts and assemblies. Click **Windows, Close All**.

BATTERYANDPLATE

Create the BATTERYANDPLATE sub-assembly. The BATTERY is the Base component. The BATTERYANDPLATE sub-assembly mates the BATTERYPLATE component to the BATTERY component.

Create the BATTERYANDPLATE sub-assembly.

368) Click **New** from the Standard toolbar. The New SolidWorks Documents dialog box is displayed. Click the **MY-TEMPLATES** tab. Click the **ASM-IN-ANSI**, [ASM-MM-ISO] template. Click **OK**.

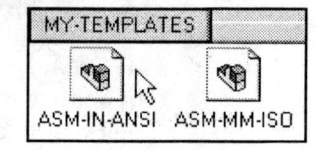

369) Click **Save**. Select the **ENGDESIGN-W-SOLIDWORKS\PROJECTS** file folder.

370) Enter **BATTERYANDPLATE** for File name.

371) Enter BATTERY AND PLATE FOR 6VOLT FLASHLIGHT for Description. Click the Save button.

372) Click View. Check **Origins**.

373) Open the BATTERY. Click **Open**. Select **BATTERY** from the PROJECTS file folder. Open the BATTERYPLATE. Click **Open**. Select **BATTERYPLATE**.

Engineering Design with SolidWorks | Sweep and Loft Features

374) Click **Window** from the Main menu. The following documents are displayed:

- o BATTERY.
- o BATTERYPLATE.
- o BATTERYANDPLATE.

Display the assembly and components. Click **Tile Horizontally** from the Window menu.

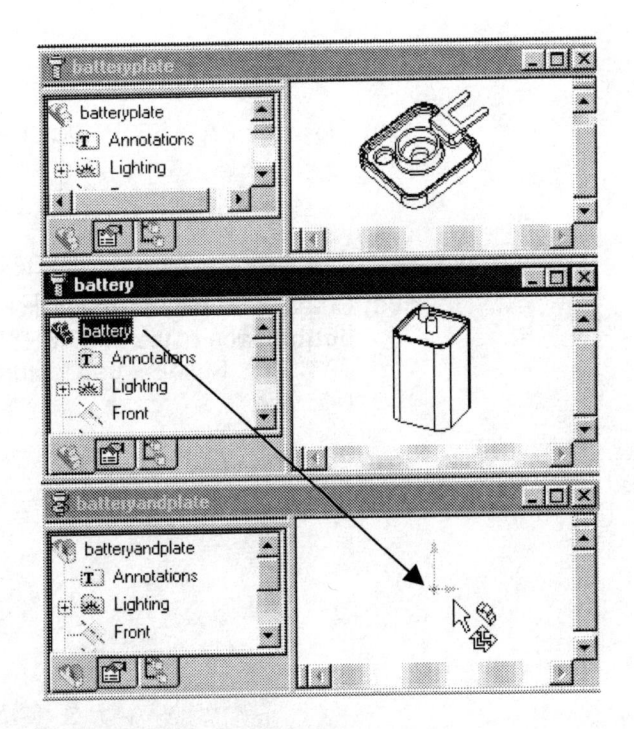

Add the first component.

375) Position the BATTERY inside the assembly Graphics window. Click the BATTERY icon from the top of the FeatureManager. Drag the **BATTERY** icon to the **Origin** of the **BATTERYANDPLATE** assembly window. The mouse pointer displays when position on the Origin.

Add the second component.

376) Click the **BATTERYPLATE** icon from the top of the FeatureManager. Drag the **BATTERYPLATE** component to the BATTERYANDPLATE Graphics window, to the left of the BATTERY. The mouse pointer displays when positioned inside the BATTERYANDPLATE Graphics window.

377) Display the Isometric view. Click **Isometric** in the BATTERYANDPLATE window.

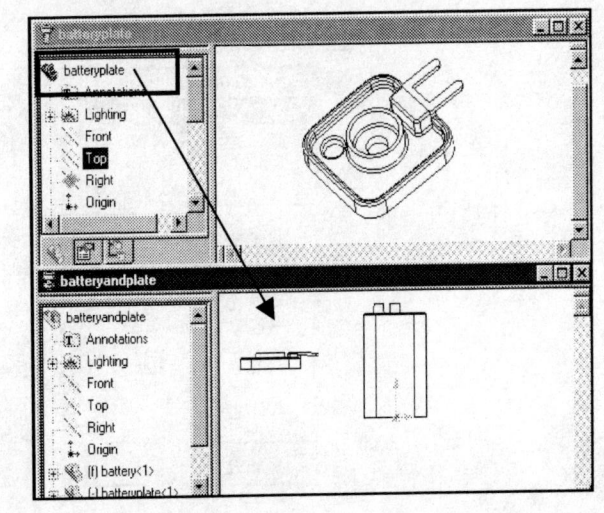

PAGE 5 - 79

Sweep and Loft Features Engineering Design with SolidWorks

378) Close the BATTERYPLATE. Click **Close** ☒.

379) Close the BATTERY. Click **Close** ☒.

Create the first Mate between the BATTERYPLATE and the BATTERY components.

380) Create a Distance Mate. Click **Mate** from the Assembly toolbar. Click the **bottom face** of the BATTERYPLATE. Click the **top narrow flat face** of the BATTERY BaseExtrude feature. Click the **Distance** button. Enter **0** for Distance.

381) Click **Preview**. Click **OK**.

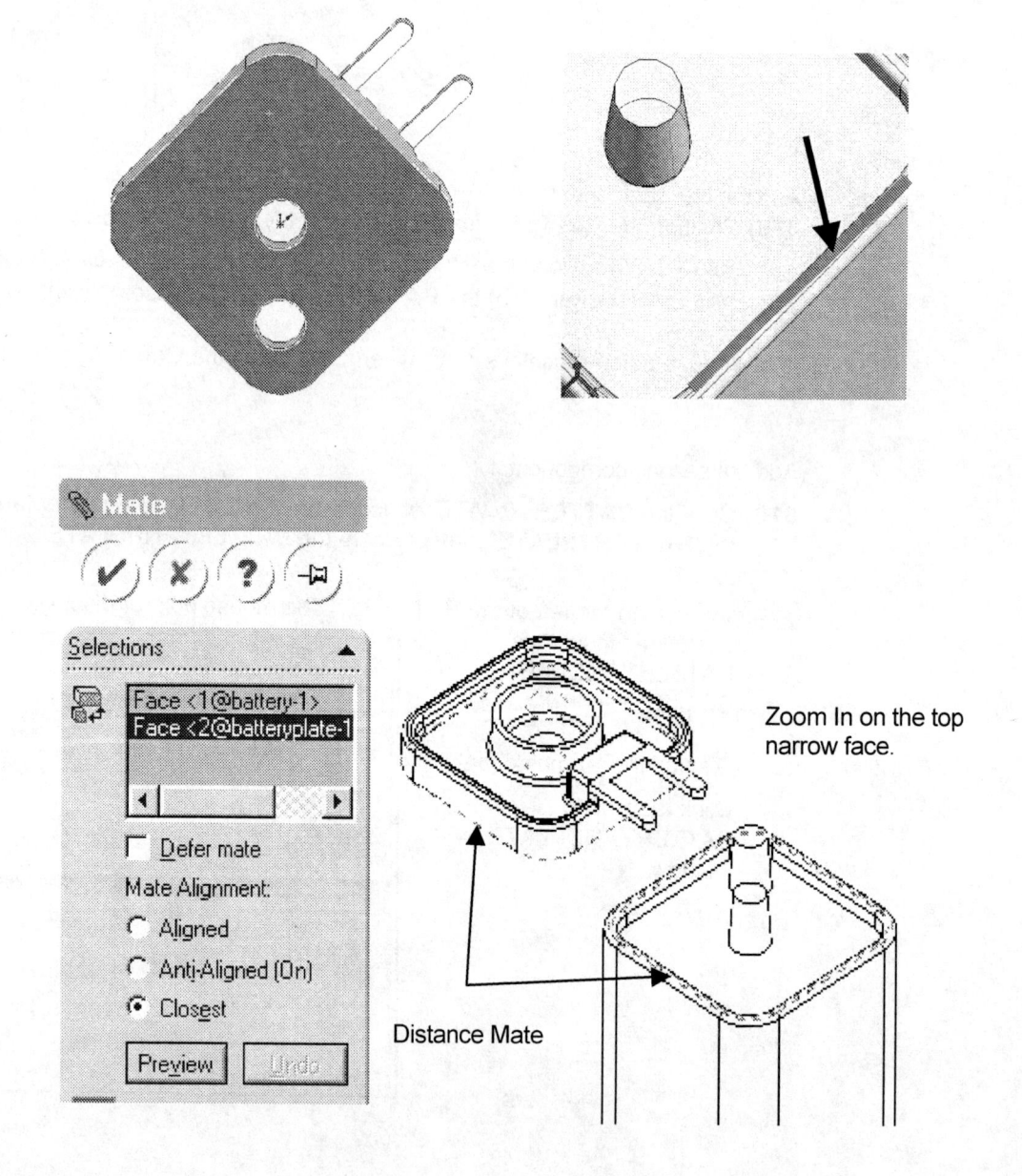

Zoom In on the top narrow face.

Distance Mate

PAGE 5-80

Create the second Mate.

382) Click **Mate** from the Assembly toolbar. Click the **Mate PropertyManager**

Mate icon to display the FeatureManager.

383) Create a Coincident Mate. Click the **Right** plane of the BATTERY. Click the **Right** plane of the BATTERYPLATE. Click **Preview**. Click **OK**.

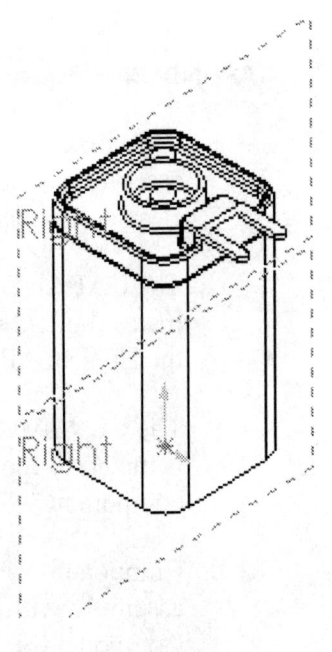

Create the third Mate.

384) Click **Mate** from the Assembly toolbar. Create a Concentric Mate. Click the **cylindrical face** Terminal feature of the BATTERY. Click the **cylindrical face** Holder feature of the BATTERYPLATE. Click **Preview**. Click **OK**. The BATTERYANDPLATE sub-assembly is fully defined.

385) Save the BATTERYANDPLATE. Click **Save**.

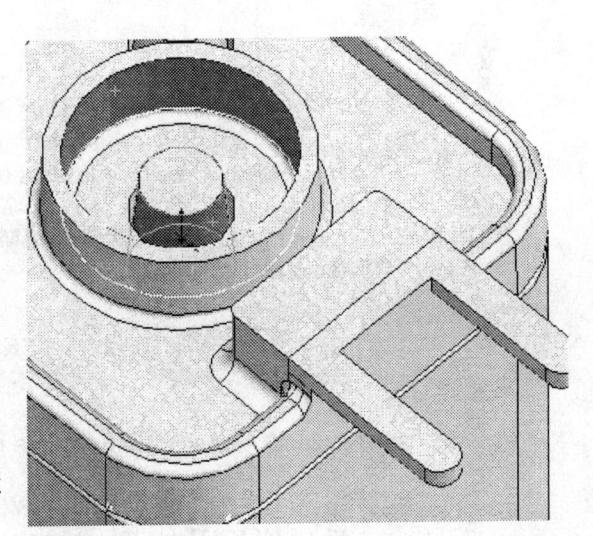

386) Close all documents. Click **Windows, Close All**.

CAPANDLENS Sub-assembly

Create the CAPANDLENS sub-assembly. The LENSCAP is the Base component.

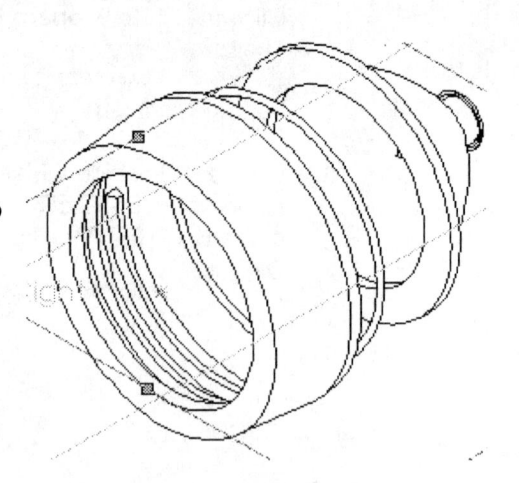

The CAPANDLENS sub-assembly mates the O-RING component to the LENSCAP component.

The LENSANDBULB sub-assembly is mated to the LENSCAP component.

Components are inserted into the assembly with the Insert, Component, From File command.

Create the CAPANDLENS sub-assembly.

387) Click **New** from the Standard toolbar. The New SolidWorks Documents dialog box is displayed. Click the **MY-TEMPLATES** tab. Click the **ASM-IN-ANSI**, [**ASM-MM-ISO**] template. Click **OK**.

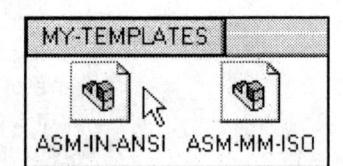

388) Click **Save**. Select the **ENGDESIGN-W-SOLIDWORKS\PROJECTS** file folder.

389) Enter **CAPANDLENS** for File name.

390) Enter **LENSCAP AND LENS FOR 6VOLT FLASHLIGHT** for Description. Click the **Save** button.

391) Click **View**. Check **Origins**.

392) Insert the Base component. Click **Insert** from the Main menu. Click **Component**, **From File**. Click **Isometric**.

393) Enter **LENSCAP**. Click **Open**. Click the **Origin** in the CAPANDLENS Graphics window.

Engineering Design with SolidWorks   **Sweep and Loft Features**

394) Insert the second component. Click **Insert** from the Main menu.

395) Click Component, From File.

396) Enter **O-RING**.

397) Click **Open**. Click a **position** behind the LENSCAP.

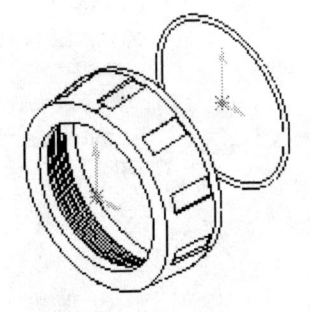

398) Insert the third component. Click **Insert** from the Main menu. Click **Component**, **From File**.

399) Enter LENSANDBULB.

400) Click **Open**. Click a **position** behind the O-RING.

Note: LENSANDBULB is an assembly. Select File type ".sldasm" from the Open dialog box.

Caution: Select the correct reference.

Expand the LENSCAP and O-RING. Click the Right plane within the LENSCAP. Click the Right plane within the O-RING.

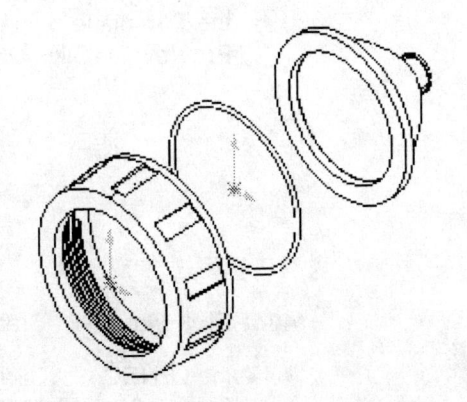

Do not select the Right plane from the top level FLASHLIGHT assembly. You will create unwanted references.

401) Right-click on the **LENSANDBULB** sub-assembly in the FeatureManager. Click **Hide Components**.

Create the Mates between the LENSCAP and O-RING component.

402) Click **Mate** from the Assembly toolbar. Create a Coincident Mate. Click the **Right** plane of the LENSCAP. Click the **Mate**

Mate icon to display the FeatureManager. Click the **Right** plane of the O-RING. Click **Preview**. Click **OK**.

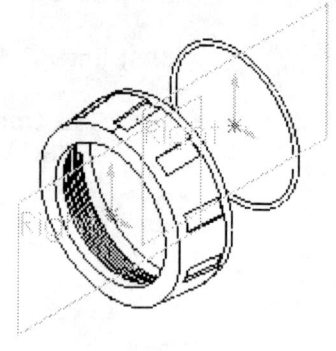

403) Click **Mate**. Create a Coincident Mate. Click the **Top** plane of the LENSCAP. Click the **Top** plane of the O-RING. Click **Preview**. Click **OK**.

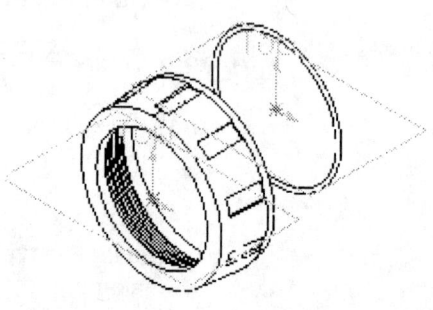

404) Click **Mate**. Create a Distance Mate. Click the Shell1 **back inside face** of the LENSCAP. Click the **Front** plane of the O-RING. Click **Distance**. Enter **.125/2**, **[3.18/2]** for Distance.

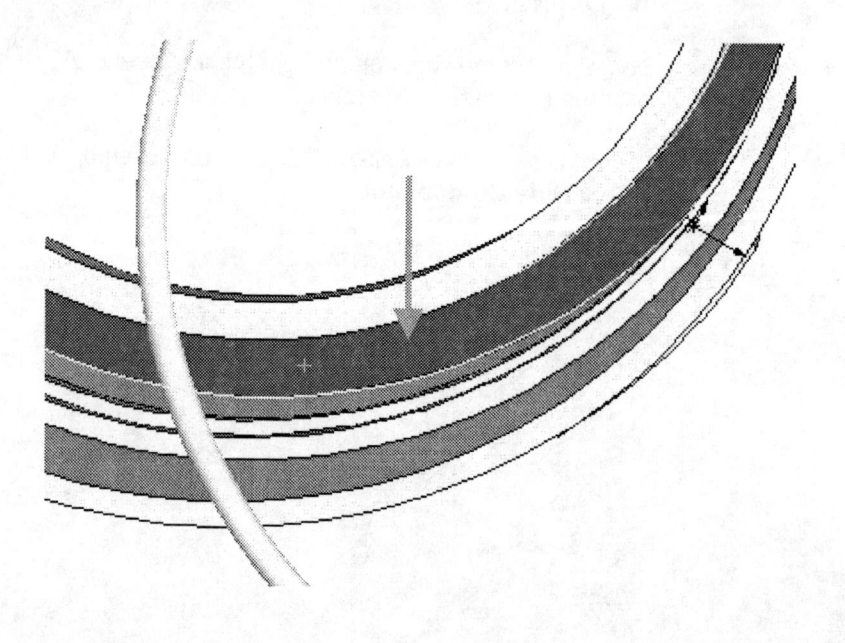

Engineering Design with SolidWorks   **Sweep and Loft Features**

405) Click the **Right** view. Click **Preview**.

406) If required, click the **Flip the dimension to the other side** check box. Click **OK**.

How is the Distance Mate, .0625, [1.588] calculated?

Answer:

O-RING Radius (.1250in/2) = .0625in.

O-RING Radius [3.18mm/2] = [1.589].

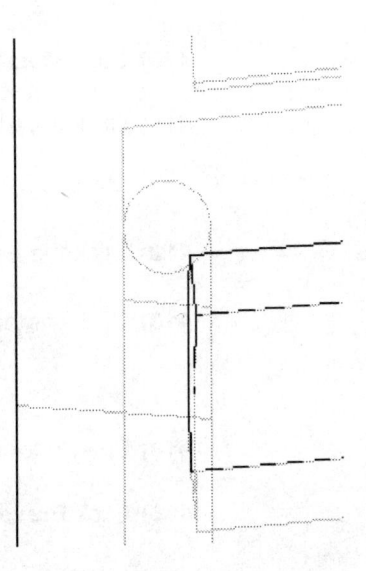

Note: The Distance Mate option offers additional flexibility over the Coincident Mate option. The Distance Mate value can be modified.

407) Display the LENSANDBULB. Right-click **LENSANDBULB** in the FeatureManager. Click **Show Components**.

Create the Mates between the LENSCAP and the LENSANDBULB sub-assembly.

408) Click **Mate**. Create a Coincident Mate. Click the **Right** plane of the LENSCAP. Click the **Right** plane of the LENSANDBULB. Click **Preview**. Click **OK**.

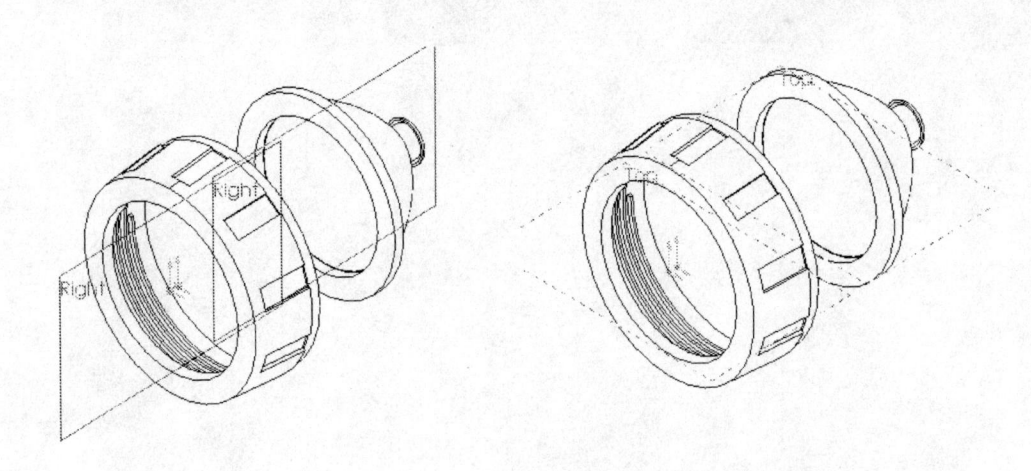

409) Click **Mate**. Create a Coincident Mate. Click the **Top** plane of the LENSCAP. Click the **Top** plane of the LENSANDBULB. Click **Preview**. Click **OK**.

410) Click **Mate**. Create a Distance Mate.

411) Click the flat **narrow back face** of the LENSCAP.

412) Click the **front flat face** of the LENSANDBULB.

413) Click **Distance**.

414) Enter **0** for Distance.

415) Click **Preview**. Click **OK**.

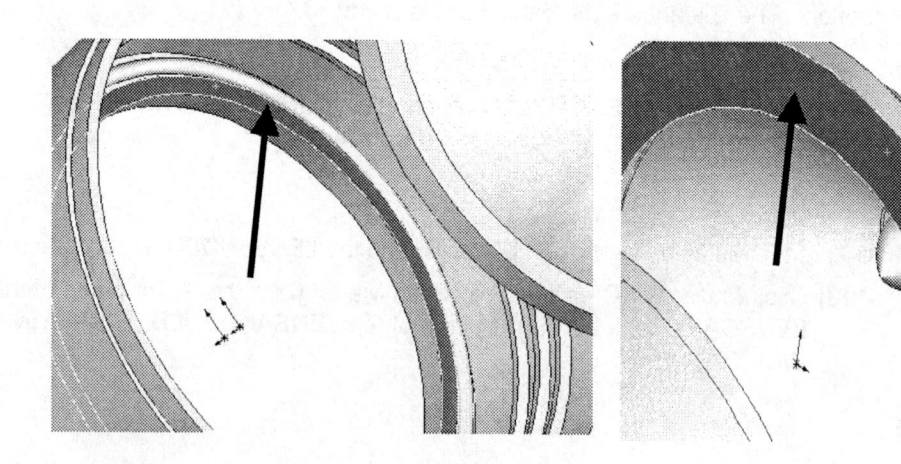

Engineering Design with SolidWorks — Sweep and Loft Features

416) The CAPANDLENS sub-assembly is fully defined. Confirm the location of the O-RING. Click the CAPANDLENS **Right** plane.

417) Click **View** from the Main menu.

418) Click **Display**.

419) Check Section View. Click Flip the Side to view.

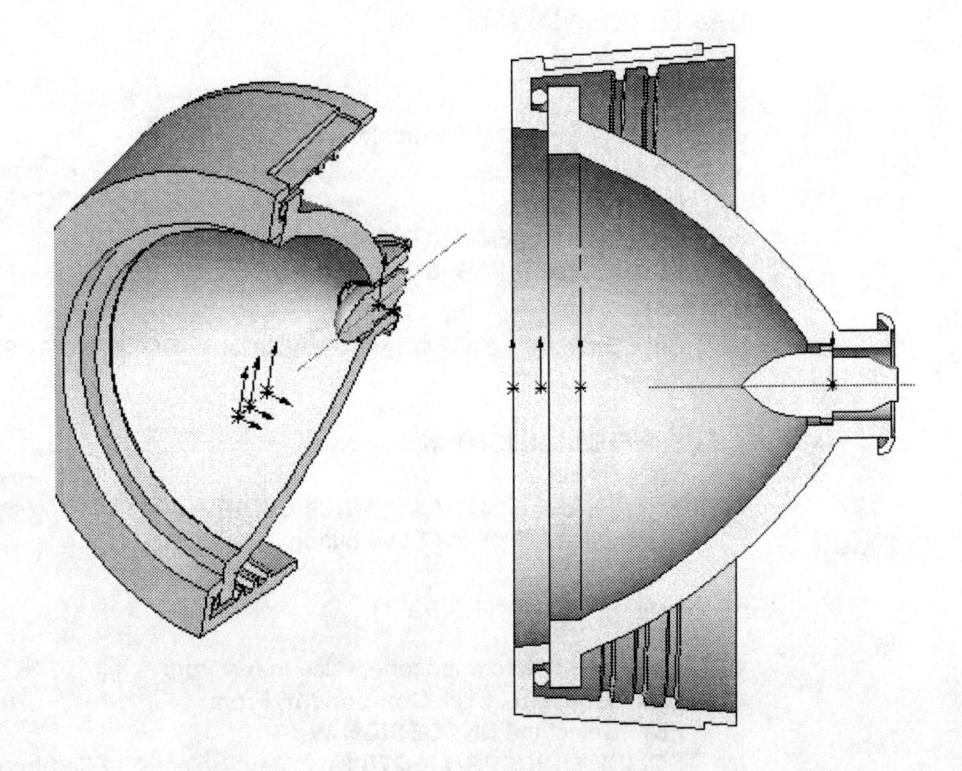

420) Remove the Section view. Click **View**, **Display**. Uncheck **Section View**.

421) Close all documents. Click **Windows**, **Close All**.

422) Save the CAPANDLENS sub-assembly. Click **Save**.

Sweep and Loft Features   **Engineering Design with SolidWorks**

Complete the FLASHLIGHT Assembly

Create the FLASHLIGHT assembly. The HOUSING is the Base component. The FLASHLIGHT assembly mates the HOUSING to the SWITCH component.

The FLASHLIGHT assembly mates the CAPANDLENS and BATTERYANDPLATE.

Create the FLASHLIGHT assembly.

423) Click **New** from the Standard toolbar. The New SolidWorks Documents dialog box is displayed. Click the **MY-TEMPLATES** tab. Click the **ASM-IN-ANSI**, [**ASM-MM-ISO**] template. Click **OK**.

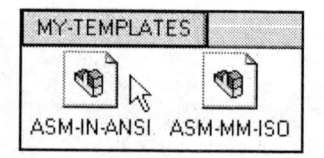

424) Click **Save**. Select the **ENGDESIGN-W-SOLIDWORKS\PROJECTS** file folder.

425) Enter **FLASHLIGHT** for File name.

426) Enter **FLASHLIGHT ASSEMBLY 6VOLT** for Description. Click the **Save** button.

427) Click **View**. Check **Origins**.

428) Insert the Base component. Click **Insert** from the Main menu. Click **Component**, **From File**. Select the **ENGDESIGN-W-SOLIDWORKS\PROJECTS** file folder. Select **All Files** from Files of Type list box. Enter **HOUSING**. Click **Open**. Click the **Origin** of the FLASHLIGHT. Click **Isometric**.

429) Insert the second component. Click **Insert** from the Main menu. Click **Component, From File**. Enter **SWITCH**. Click **Open**. Click a **position** in front of the HOUSING.

Create the Mates between the HOUSING and the SWITCH component.

430) Click **Mate**. Click the **Keep Visible** icon. Create a Coincident Mate. Click the **Right** plane of the HOUSING. Click the **Right** plane of the SWITCH. Click **Coincident**. Click **Preview**. Click **OK**.

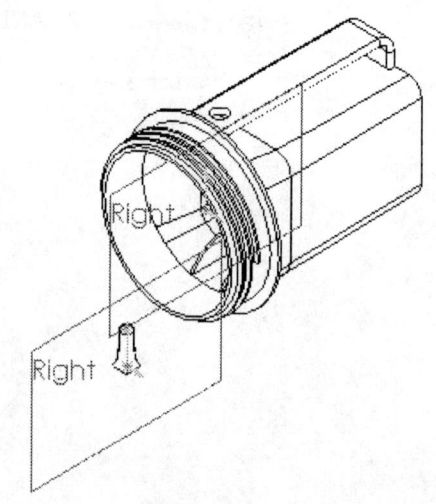

Engineering Design with SolidWorks Sweep and Loft Features

431) Click **View** from the Main menu. Check **Temporary axis**. Click **Mate**. Create a Coincident Mate.

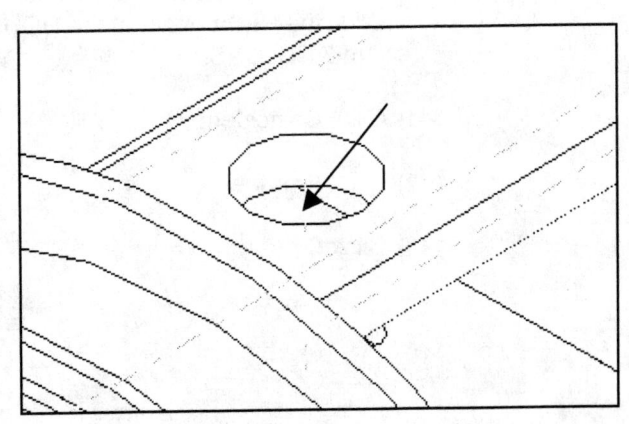

432) Click the **temporary axis** inside the Switch Hole of the HOUSING.

433) Click the **Front** plane of the SWITCH.

434) Click Coincident.

435) Click **Preview**. Click **OK**.

436) Create a Distance Mate. Click the **top face** of the Handle. Click the **Vertex** on the Dome of the SWITCH. Click **Distance**. Enter **.200**, **[5.08]** for Distance. Click **Preview**. Check Flip Direction if required. Click **OK**.

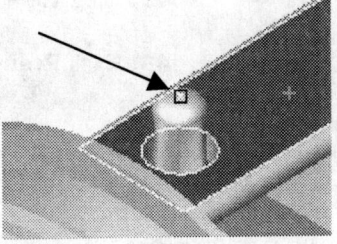

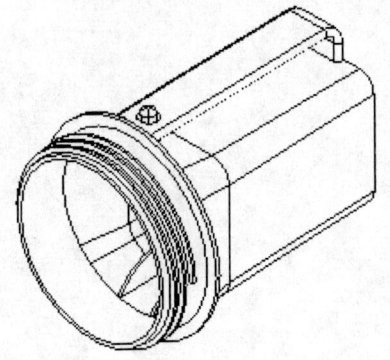

437) Save the FLASHLIGHT assembly. Click **Save**.

438) Insert the CAPANDLENS sub-assembly. Click **Insert** from the Main menu. Click **Component**, **From File**.

439) Enter **CAPANDLENS**. Click **Open**. Click a **position** in front of the HOUSING.

Sweep and Loft Features **Engineering Design with SolidWorks**

Create the Mates between the HOUSING component and the CAPANDLENS sub-assembly.

440) Create a Coincident Mate. Click **Mate**. Click the **Keep Visible** icon. Click the **Right** plane of the HOUSING. Click the **Right** plane of the CAPANDLENS.

441) Click Coincident.

442) Click **Preview**.

443) Click **OK**.

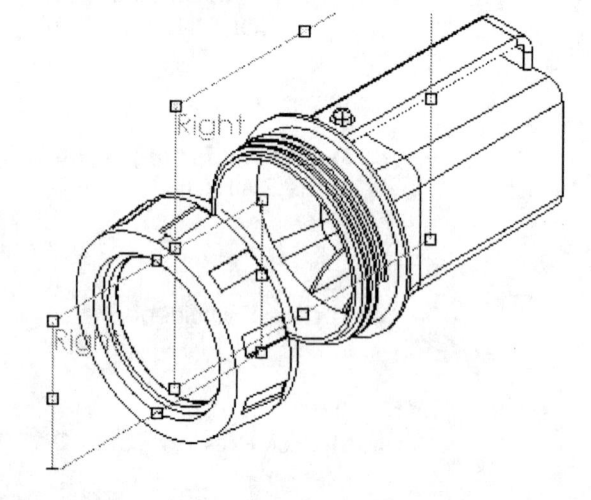

444) Create a Coincident Mate. Click the **Top** plane of the HOUSING. Click the **Top** plane of the CAPANDLENS.

445) Click Coincident.

446) Click **Preview**. Click **OK**.

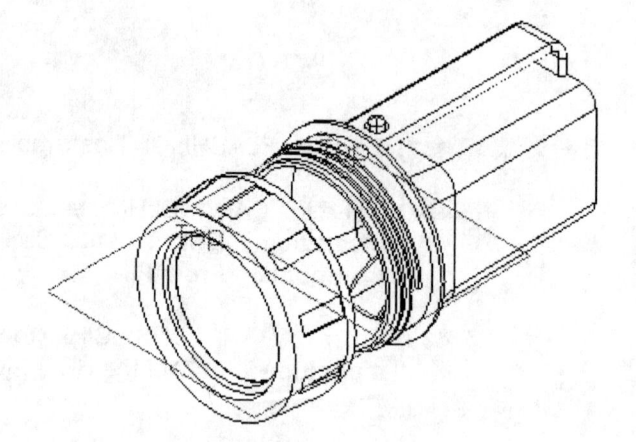

PAGE 5-90

Engineering Design with SolidWorks **Sweep and Loft Features**

447) Create a Distance Mate. Click the **front face of the Boss-Stop** on the HOUSING. Rotate the view. Press the **Left Arrow Key**.

Click the **back face** of the CAPANDLENS. Click **Distance**. Enter **0** for Distance.

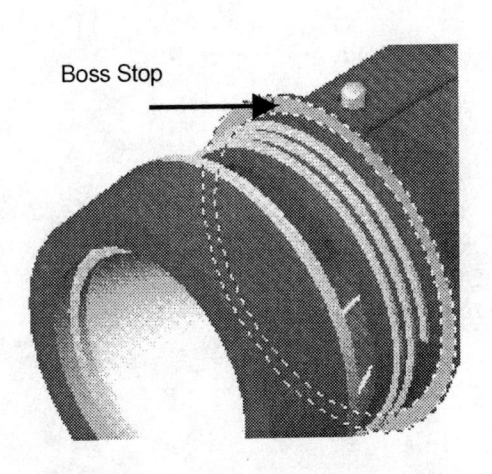

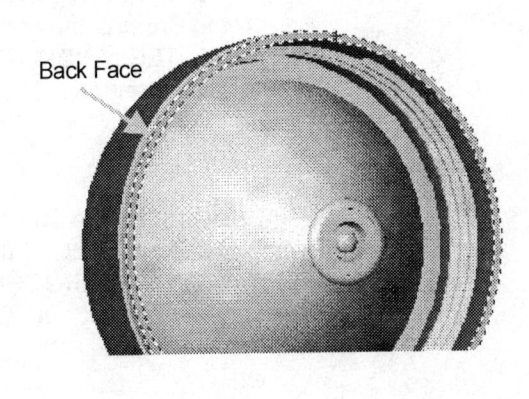

Boss Stop

Back Face

448) Save the FLASHLIGHT assembly. Click **Save**.

449) Insert the BATTERYANDPLATE sub-assembly. Click **Insert** from the Main menu. Click **Component, From File**. Enter **BATTERYANDPLATE**. Click **Open**. Click a **position** to the left of the HOUSING.

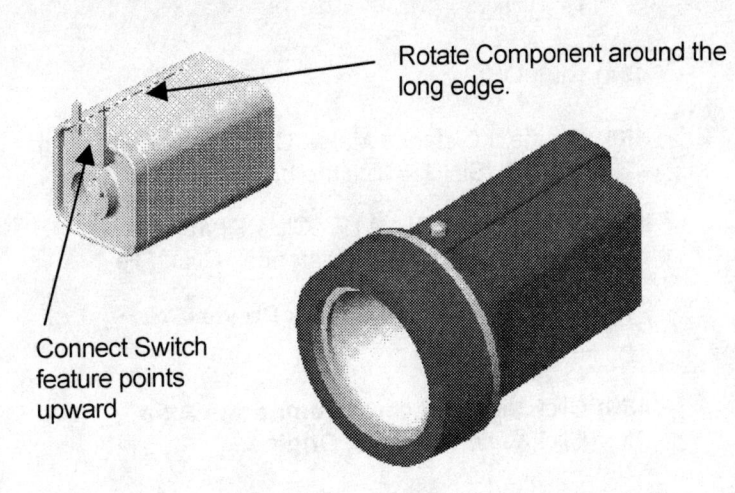

Rotate Component around the long edge.

Connect Switch feature points upward

450) Rotate the part. Click **Rotate Component** from the Assembly toolbar. Rotate the **BATTERYANDBATTERYPLATE** until the ConnectorSwitch feature is vertical. Click **OK**.

Sweep and Loft Features **Engineering Design with SolidWorks**

Create the Mates between the HOUSING component and the BATTERYANDPLATE.

451) Create a Coincident Mate. Click **Mate**.

Click the **Keep Visible** icon. Click the **Right** plane of the HOUSING. Click the **Front** plane of the BATTERYANDPLATE. Click **Coincident**. Click **Preview**. Click **OK**.

452) Create a Coincident Mate. Click the **Top** plane of the HOUSING. Click the **Right** plane of the BATTERYANDPLATE. Click **Coincident**. Click **Preview**. Click **OK**.

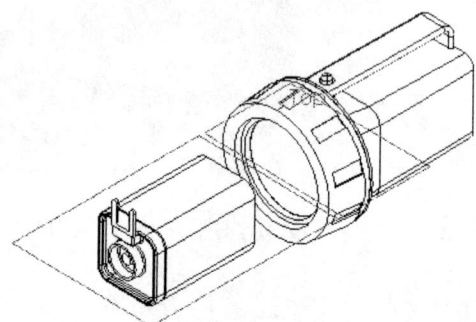

453) Position the BATTERYANDPLATE in front of the HOUSING. Click **Move Component** from the Assembly toolbar. Drag the **BATTERYANDPLATE** out of the HOUSING.

454) Click **OK**.

455) Create a Distance Mate. Click the **back face** of the HOUSING. Click the **bottom face** of the BATTERYANDPLATE. Click **Distance**. Enter **.275**, **[6.99]** for Distance. Click Flip Direction if required. Click **Preview**. Click **OK**.

456) Click **View**. Uncheck **Temporary Axis**. Click **View**. Uncheck **Origins**.

457) **Rebuild** the FLASHLIGHT.

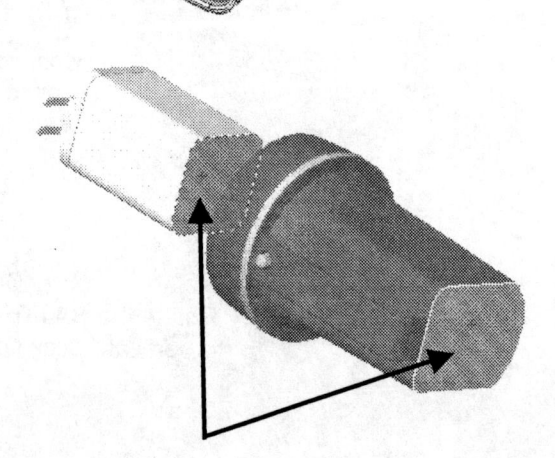

458) Save the FLASHLIGHT. Click **Save**. Click **Yes** to update all components.

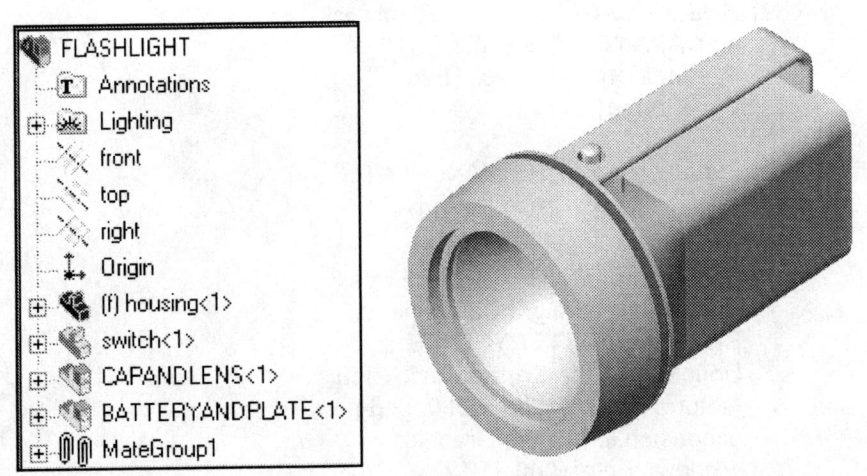

FLASHLIGHT displayed in PhotoWorks and saved in TIF format.

Addressing Design Issues

There are interference issues between the FLASHLIGHT components. Address these design issues:

- Reduce the SwitchConnector feature size on the BATTERYPLATE.

- Adjust Rib2 on the HOUSING. Test with the Interference Check command.

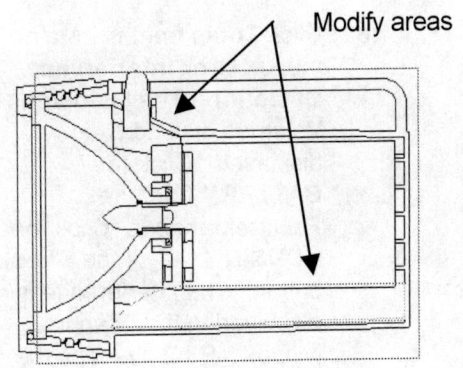

The ConnectorSwitch feature of the BATTERYPLATE is too long. Contain the ConnectorSwitch within the HOUSING.

Sweep and Loft Features **Engineering Design with SolidWorks**

Check for Interference.

459) Hide the CAPANDLENS. Right click on **CAPANDLENS** in the FeatureManager. Click **Hide Component**.

460) Expand the BATTERYANDPLATE sub-assembly. Click **Plus** ⊞.

461) Expand the BATTERYPLATE. Resolve Lightweight Components. Right-click **BATTERYPLATE**. Double-click the **ConnectorSwtich** feature. Double click **1.000**, **[2.54]** dimension text in the Graphics window. Enter **.500**, **[12.7]**.

462) Click **Rebuild**.

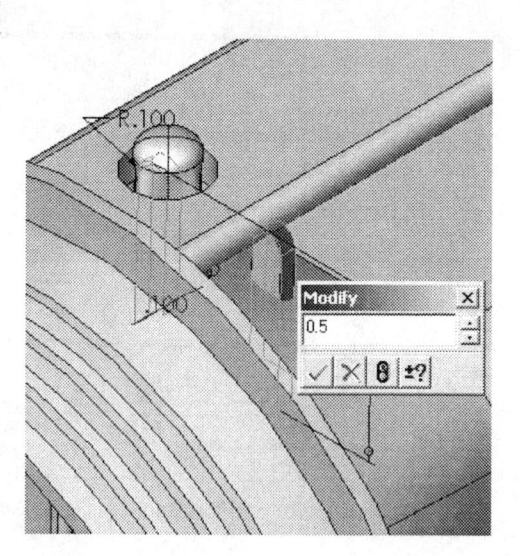

There is an interference fit issue between the HOUSING and the BATTERY.

463) Click **Tools** from the Main menu. Click **Interference Detection**. The Interference Volumes dialog box is displayed. Click the **BATTERY** from the FeatureManager. Click the **HOUSING**. Click the **Check** button. The interference is displayed in the Graphics window. Rib2 overlaps the BATTERY by .050, [1.27].

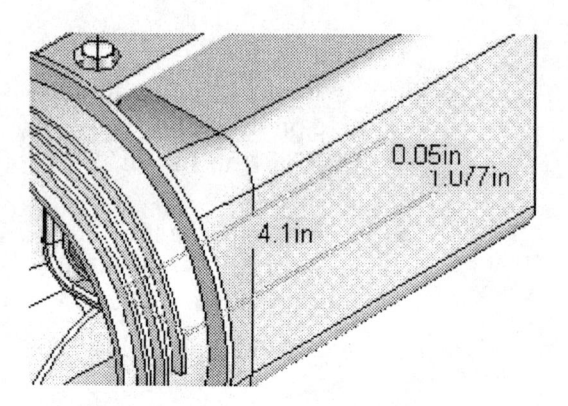

464) Click **Close**.

Engineering Design with SolidWorks | Sweep and Loft Features

465) Expand the **HOUSING** in the FeatureManager. Double-click on the **Rib2** feature. Double click **1.300**, [**33.02**]. Enter **1.350**, [**34.29**]. Click **Rebuild**.

Note: Interference must exist between the BULB and the BATTERY to create an electrical connection.

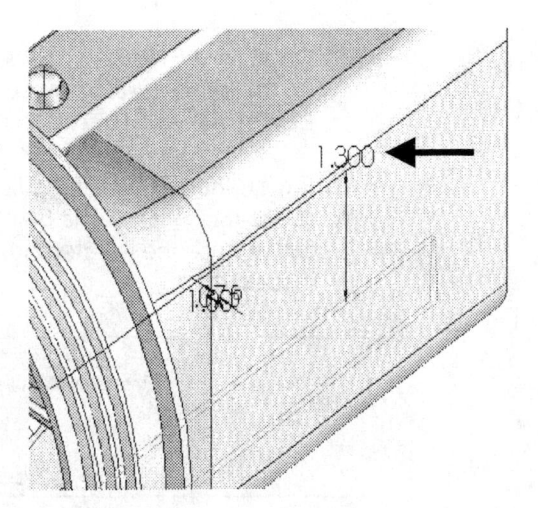

466) Display the CAPANDLENS. Right-click on **CAPANDLENS** in the FeatureManager. Click **Show Components**. The FLASHLIGHT design is complete.

467) Save the FLASHLIGHT. Click **Save**.

468) Click **YES** to the question, "Rebuild the assembly and update the components".

Export Files

You receive a call from the sales department. They inform you that the customer increased the initial order to 200,000 units. However, the customer requires a prototype to verify the design in six days.

What do you do? Answer: Contact a Rapid Prototype supplier.

You export three SolidWorks files:

- HOUSING.
- LENSCAP.
- BATTERYPLATE.

Use the Stereo Lithography (STL) format. Email the three files to a Rapid Prototype supplier.

Example: Paperless Parts Inc. (www.paperlessparts.com).

A Stereolithography (SLA) supplier provides physical models from 3D drawings. 2D drawings are not required. Export the LENSCAP.

Sweep and Loft Features Engineering Design with SolidWorks

469) Open the LENSCAP. Expand the **CAPANDLENS** sub-assembly. Right-click on the **LENSCAP** from the FLASHLIGHT FeatureManager. Click **Open LENSCAP.SLDPRT**.

470) Export the LENSCAP. Click **File**, **SaveAs**. A warning message states that, "Lens-cap.SLDPRT is being referenced by other open documents. "Save AS" will replace these references with the new name. Create a copy of LENSCAP. Click **OK**.

> C:\Program Files\SolidWorks\lenscap.SLDPRT is being referenced by other open documents. "Save As" will replace these references with the new name. Check "Save As Copy" in the "Save As" dialog if you wish to maintain existing references.
>
> [OK]   [Cancel]

471) Click **STL Files (*.stl)** from the Save as Type drop down list. The Save dialog box displays new options.

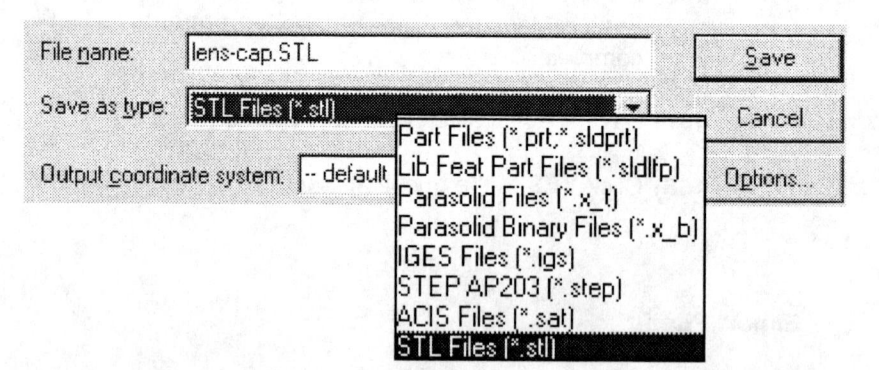

File name: lens-cap.STL
Save as type: STL Files (*.stl)
- Part Files (*.prt;*.sldprt)
- Lib Feat Part Files (*.sldlfp)
- Parasolid Files (*.x_t)
- Parasolid Binary Files (*.x_b)
- IGES Files (*.igs)
- STEP AP203 (*.step)
- ACIS Files (*.sat)
- STL Files (*.stl)

Output coordinate system: -- default

[Save] [Cancel] [Options...]

472) Click the **Options** button in the lower right corner of the Save dialog box. The STL Export Options dialog is displayed.

473) Click **Binary** from the Output format box. Click **Course** from the Quality box.

474) Display the STL triangular faceted model. Click **Preview**.

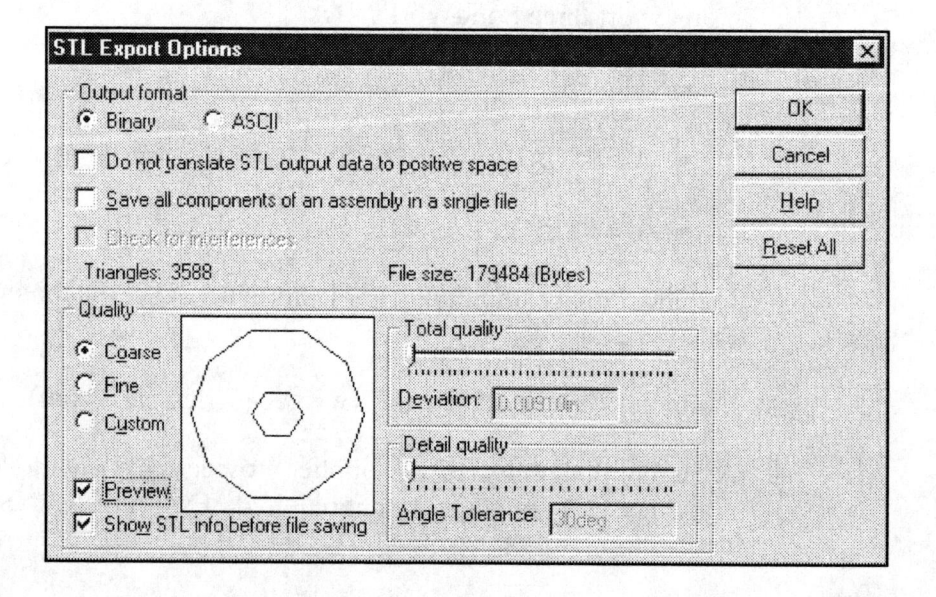

STL Export Options

Output format:
- (•) Binary   ( ) ASCII
- [ ] Do not translate STL output data to positive space
- [ ] Save all components of an assembly in a single file
- [ ] Check for interferences

Triangles: 3588   File size: 179484 (Bytes)

Quality:
- (•) Coarse
- ( ) Fine
- ( ) Custom

Total quality: ....
Deviation: 0.00910in

Detail quality: ....
Angle Tolerance: 30deg

[x] Preview
[x] Show STL info before file saving

[OK] [Cancel] [Help] [Reset All]

PAGE 5-96

Engineering Design with SolidWorks
Sweep and Loft Features

475) Drag the **dialog box** to the left to view the Graphics window.

476) Create the binary STL file. Click **OK** from the STL dialog box. Click **Save** from the Save dialog box. A status report is provided. Save the file LENS-CAP.STL. Click **Yes**.

You receive the three SLA physical models for the supplier.

Assemble the rapid prototype models with purchased components.

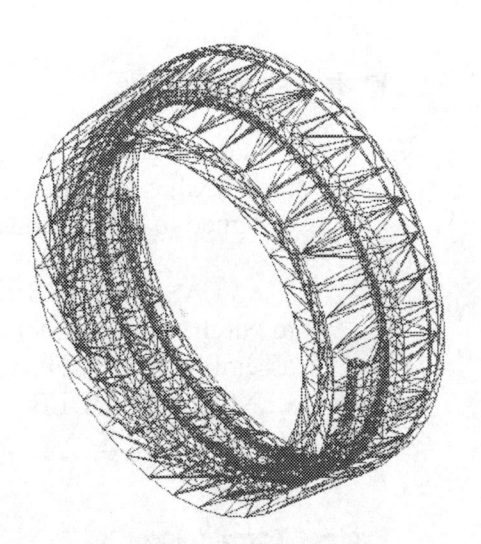

Results: A flashlight assembly delivered to the customer in six days. Success!

SolidWorks eDrawings provides a facility for you to animate, view and create compressed documents to send to colleagues, customers and vendors.

477) Click File, Publish eDrawing2003.

478) Click the **Play**. View the animation.

479) Click **Stop**. Save the eDrawing. Click **Save**.

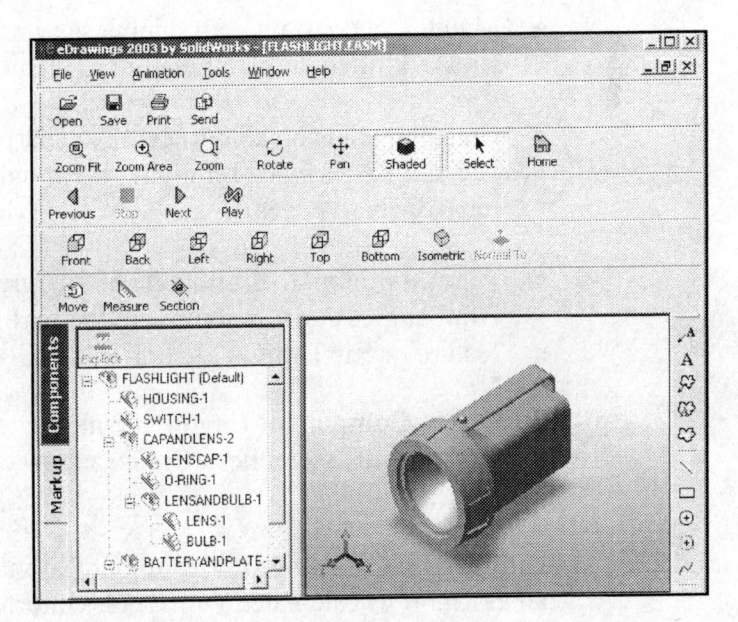

It is time to go home. The telephone rings. Guess who? The customer is ready to place the order. Tomorrow you will receive the purchase order. You think about the concerns of manufacturing, purchasing and shipping.

In the rush to create the prototype, you forgot about the packaging, the part numbers and the silk screen vendor.

Congratulations, the project is just beginning! Let's try some more examples.

Project Summary

You designed a FLASHLIGHT assembly that is cost effective, serviceable and flexible for future design revisions. The FLASHLIGHT assembly is designed to accommodate two battery types: One 6-volt or 4-D cells.

The FLASHLIGHT assembly consists of numerous parts. The team decided to purchase and model the following parts: One 6-volt BATTERY, LENS assembly, SWITCH and an O-RING. The LENS assembly consists of the LENS and the BULB.

Project Terminology

Plastic injection manufacturing: Review the fundamental design rules behind the plastic injection manufacturing process:

- Maintain a constant wall thickness. Inconsistent wall thickness creates stress. Utilize the Shell feature to create constant wall thickness.

- Create a radius on all corners. No sharp edges. Sharp edges create vacuum issues when removing the mold. Utilize the Fillet feature to remove sharp edges.

- Allow a minimum draft angle of 1 degree. Draft sides and internal ribs. Draft angles assist in removing the part from the mold. Utilize the Draft feature or Rib Draft angle option.

Assembly Component Layout: Plan the top-level assembly. Organize parts into smaller subassemblies. Create a flow chart or manual sketch to classify components.

Interference Detection: The amount of interference between components in an assembly is calculated with Tools, Interference Detection.

Stereo Lithography (STL) format: STL format is the type of file format requested by Rapid Prototype manufacturers.

Component: When a part or assembly is inserted into a new assembly, it is called a component.

eDrawings: SolidWorks eDrawings application provides a facility to animate, view and create compressed documents.

Assembly Techniques: Methods utilized to create efficient and accurate assemblies.

Project Features

Sweep: A Sweep Boss/Base feature adds material. A Sweep Cut removes material. A Sweep requires a profile sketch and a path sketch. A Sweep feature moves a profile along a path.

The O-RING part utilized a Sweep Base feature. The LENSCAP and HOUSING utilized a Sweep Boss feature with a Helix Curve to create the Thread.

Loft feature: A Loft Boss/Base feature adds material. A Loft Cut removes material. A Loft transitions two or more profiles on separate planes to create the feature.

The HOUSING part utilized a Loft Boss feature to transition a circular profile of the LENSCAP to a square profile of the BATTERY.

Rib: Adds material between contours of existing geometry. Use Ribs to add structural integrity to a part.

Pattern: A Pattern creates one or more instances of a feature or group of features. A Circular Pattern creates instances of a seed feature about an axis of revolution. A Temporary Axis was utilized to create the Circular Pattern of the LENSCAP Seed Cut.

A Linear Pattern creates instances of a seed feature in a rectangular array, along one or two edges. The Linear Pattern was utilized to create multiple instances of the HOUSING Rib1 feature.

Suppressed: A feature or component that is not loaded into memory. A feature is suppressed or unsuppressed. A component is suppressed or resolved. Suppressed features and components to improve model Rebuild time.

Base-Extrude: Create the LENSCAP using the Base-Extrude feature with a draft angle. The Base Extrude feature dimensions are determined from the LENS and LENSCAP.

Revolved Cut Thin: Removes material by rotating a sketched profile around a centerline.

Extruded Thin Cut Used to grip the outside of the LENSCAP.

Helix/Spiral Curve: A Helix is a curve with pitch. The Helix is created about an axis. Used to create a thread for the LENSCAP.

Questions

1. Identify the function of the following features:

   - Sweep.
   - Revolved Cut Thin.
   - Loft.
   - Rib.
   - Pattern.

2. Describe a suppressed feature.

3. Why would you suppress a feature?

4. The Rib features require a Sketch, thickness and a _____ direction.

5. What is a Pierce relation?

6. What is the advantage of the Distance Mate option over the Coincident Mate option?

7. How do you create a thread using the Sweep feature? Provide an example.

8. How do you create a Linear Pattern? Provide an example.

9. Describe five proven assembly modeling techniques. Can you add a few more?

10. How do you determine an interference between components in an assembly?

Exercises

Exercise 5.1: QUATTRO-SEAL-O-RING.

Create the QUATTRO-SEAL-O-RING part as one Sweep feature. Create a 100mm diameter Circle on the Front plane for the path, Sketch1. Create the symmetric cross section on the Top plane for the profile, Sketch2.

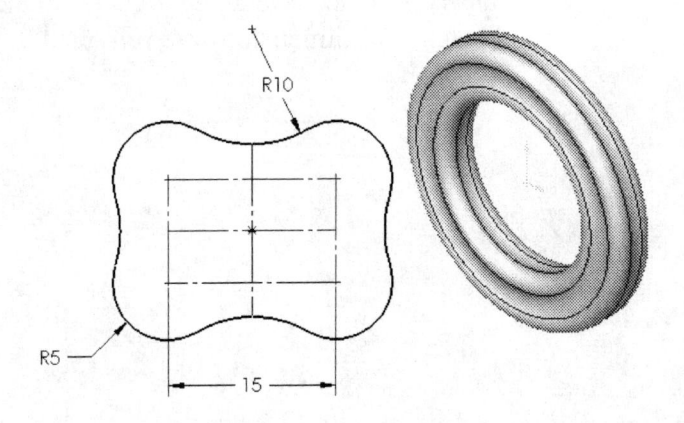

Exercise 5.2: HOOK.

Create the HOOK part. Utilize a Sweep Base feature. Create the profile, Sketch1 on the Front plane. Create the path, Sketch2 on the Top plane.

Create the bottom Thread as a Sweep Cut feature.

Create Plane1, offset .020[.51mm] from the Top Plane.

Create the Helix Curve for the path. Pitch 0.050 in, Revolution 4.0, Starting Angle 0.0 deg. Sketch a circle ∅.020[.51mm] on the Right plane for the Thread profile.

Chamfer the bottom face.

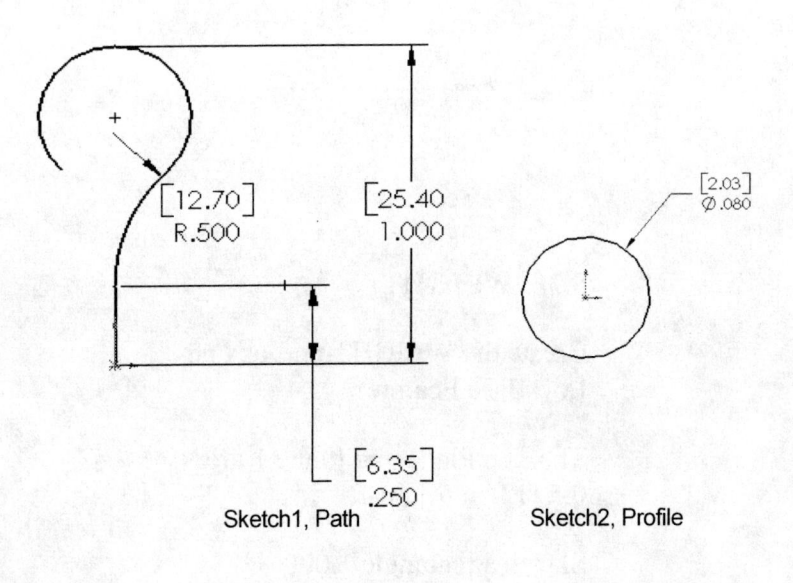

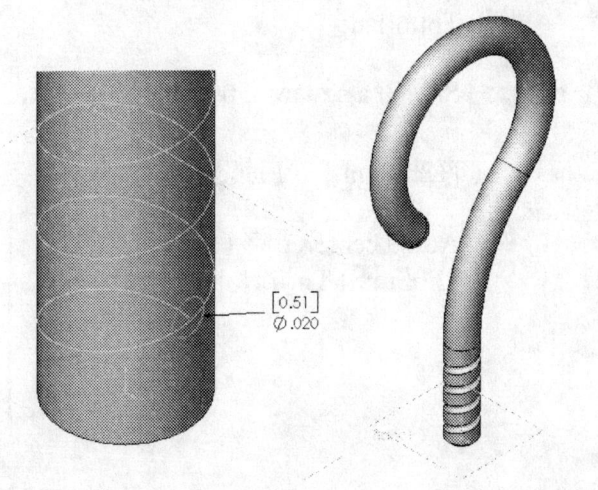

Sweep and Loft Features Engineering Design with SolidWorks

Exercise 5.3a: OFFSET-LOFT.

Create the OFFSET-LOFT part as one Loft Base Feature. Sketch a square on the Top Plane. Sketch a Circle on Plane1. Utilize geometric relations in the sketch. No dimensions are provided.

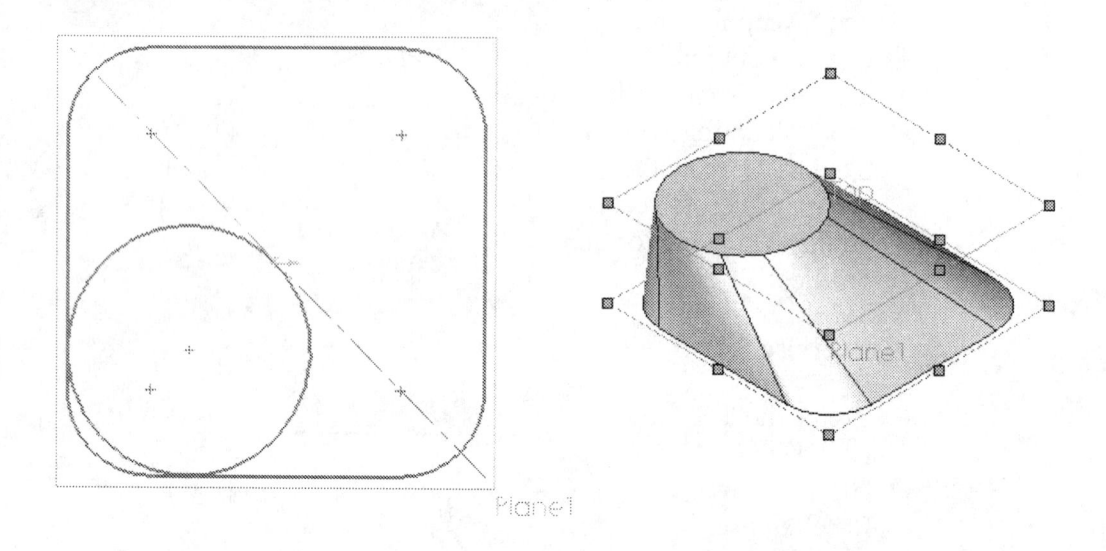

5.3b: WEIGHT

Create the WEIGHT part as one Loft Base Feature.

The Top Plane and Plane 1 are 0.5 [12.7mm] apart.

Sketch a rectangle 1.000 [25.40mm] x .750 [19.05] on the Top Plane.

Sketch a square .500

[12.70mm]on Plane 1.

Add a centered ⌀.080 [2.03mm] Thru Hole.

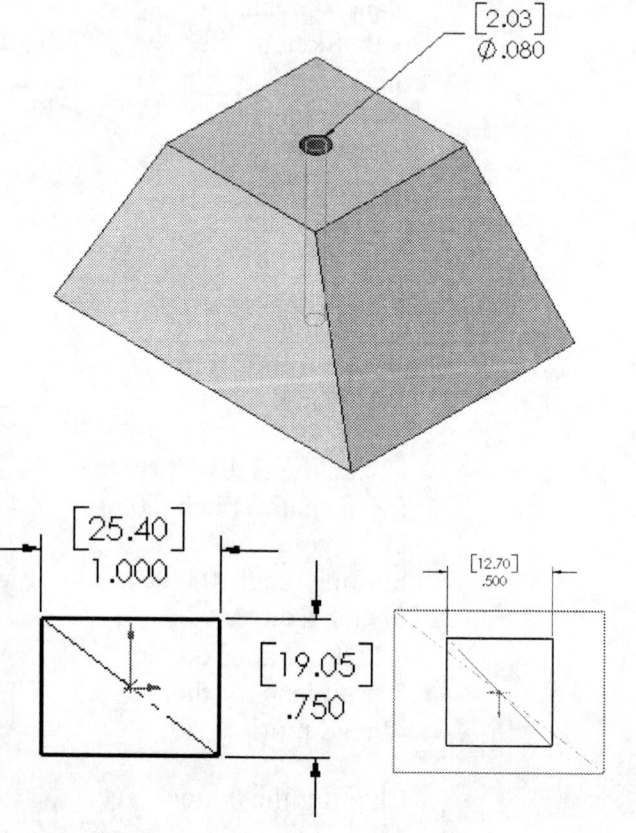

PAGE 5- 102

Exercise 5.4: FLASHLIGHT DESIGN CHANGES.

Review the HOUSING.

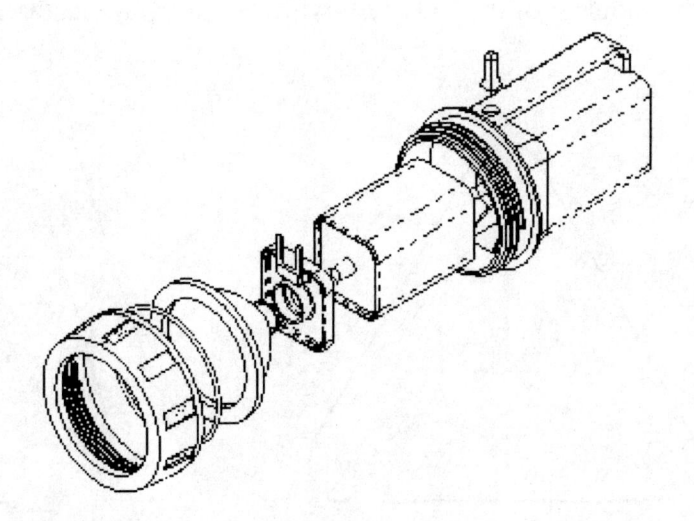

The mold for the HOUSING requires that all sharp edges are removed.

- Add Fillet features throughout the FeatureManager to remove the sharp edges from the inside and outside of the HOUSING.

- The HOUSING LongRibs are too thick. The current Ribs cause problems in the mold. Divide the current Rib thickness by 2.

- The HOUSING Handle is too short. A large human hand cannot comfortably hold the Handle. Redesign the Sweep path of the Handle.

- Create an Exploded view of the FLASHLIGHT assembly.

- Create a Drawing for the FLASHLIGHT assembly.

- Utilize Custom Properties to add Part No. and Description for each FLASHLIGHT component.

- Add a Bill of Materials to the Drawing.

Exercise 5.5: Triangular Shaped Bottle.

A plastic BOTTLE is created from a variety of SolidWorks features. Create the shoulder of the BOTTLE with the Loft Base feature.

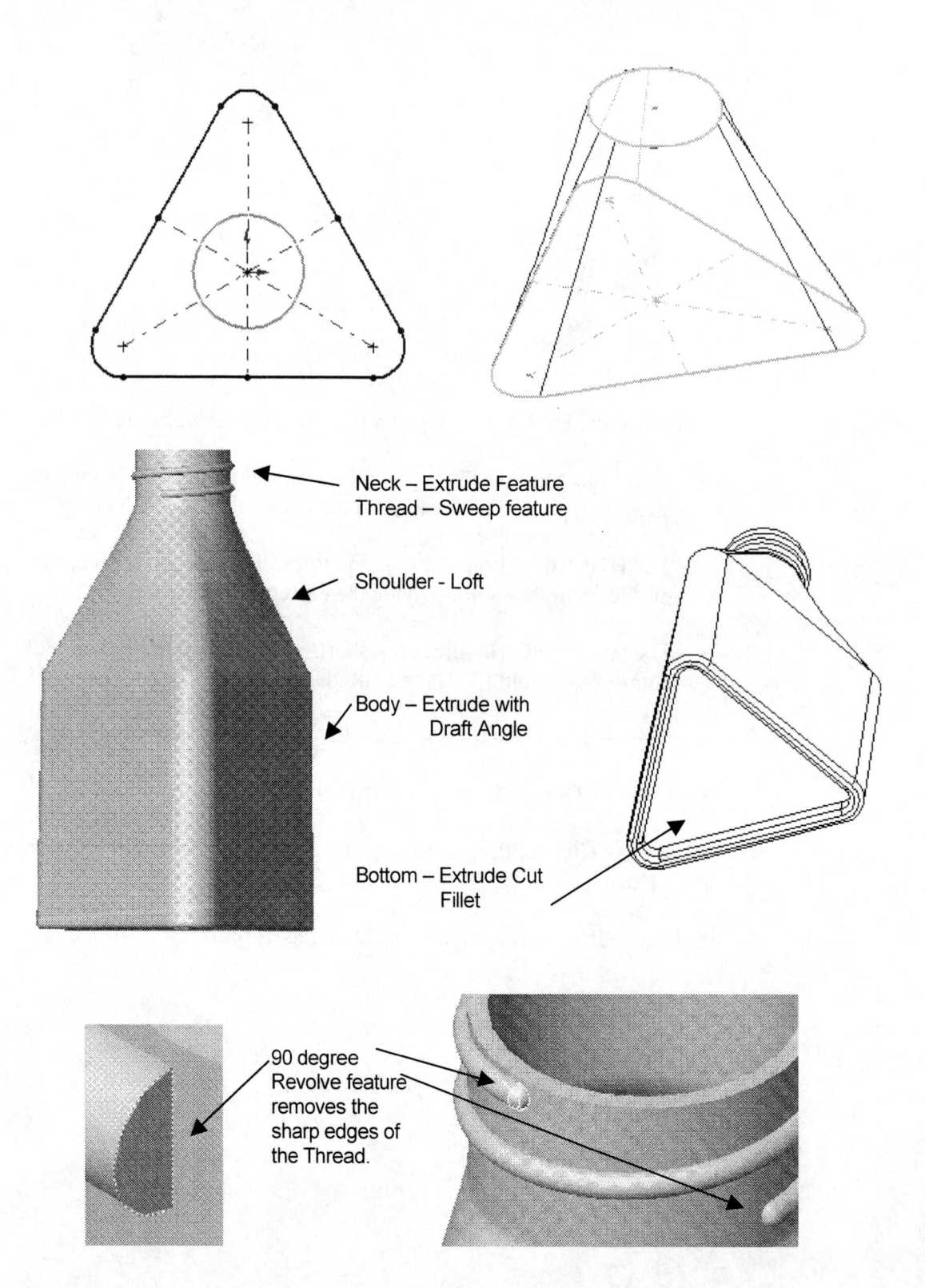

Neck – Extrude Feature
Thread – Sweep feature

Shoulder - Loft

Body – Extrude with Draft Angle

Bottom – Extrude Cut
Fillet

90 degree Revolve feature removes the sharp edges of the Thread.

Exercise 5.6: WHEEL-AND-AXLE

The WHEEL-AND-AXLE assembly is utilized to transform linear motion from the Pneumatic Linkage assembly (Exercise 2) into rotational motion of the WHEEL. The counter-clock wise rotational motion of the WHEEL is utilized to lift the applied WEIGHT.

The WHEEL-AND-AXLE assembly consists of the eleven unique parts:

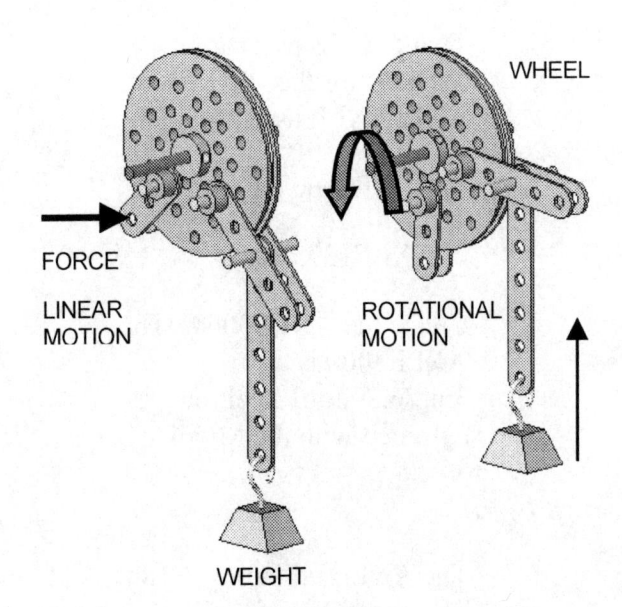

WHEEL-AND-AXLE Assembly
Courtesy of GEARS EDUCATIONAL SYSTEMS

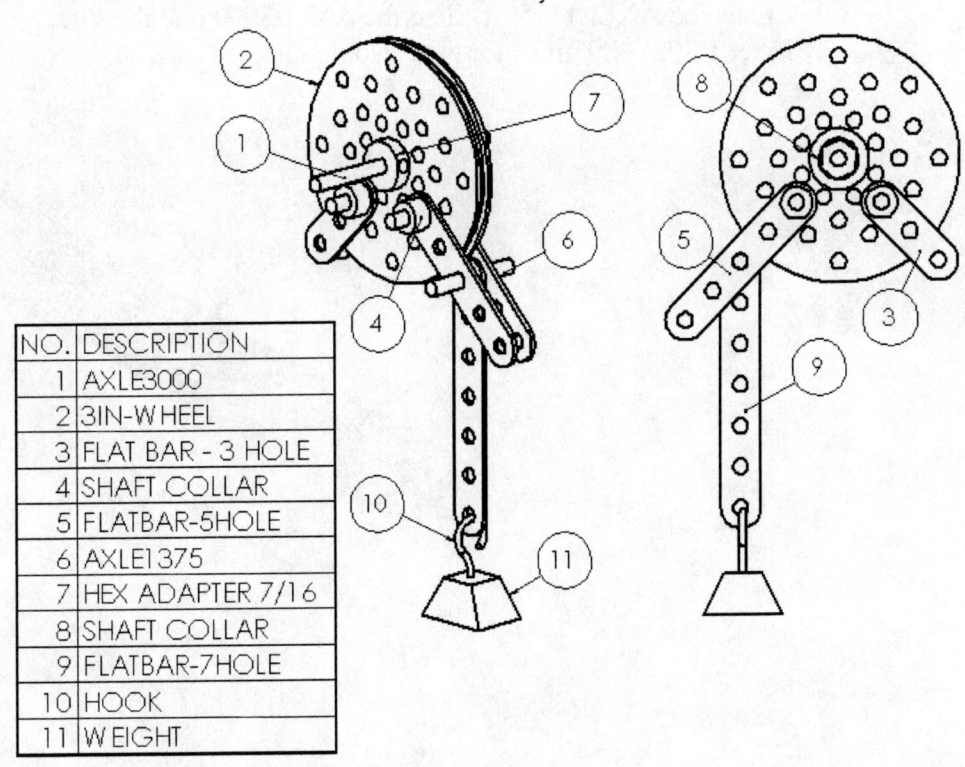

The WEIGHT (Exercise

NO.	DESCRIPTION
1	AXLE3000
2	3IN-WHEEL
3	FLAT BAR - 3 HOLE
4	SHAFT COLLAR
5	FLATBAR-5HOLE
6	AXLE1375
7	HEX ADAPTER 7/16
8	SHAFT COLLAR
9	FLATBAR-7HOLE
10	HOOK
11	WEIGHT

5.3b) and HOOK (Exercise 5.2) are components in the LINK-AND-HOOK subassembly. The other parts dimensions are provided. Read the entire exercise before you begin.

Exercise 5.6a: AXLE Part

Two AXLEs parts are required for the WHEEL-AND-AXLE assembly:

1) AXLE3000 Part.

2) AXLE1375 Part.

Create the AXLE3000. The AXLE3000 is a 3in[76.20mm] steel rod extruded with the Midplane.

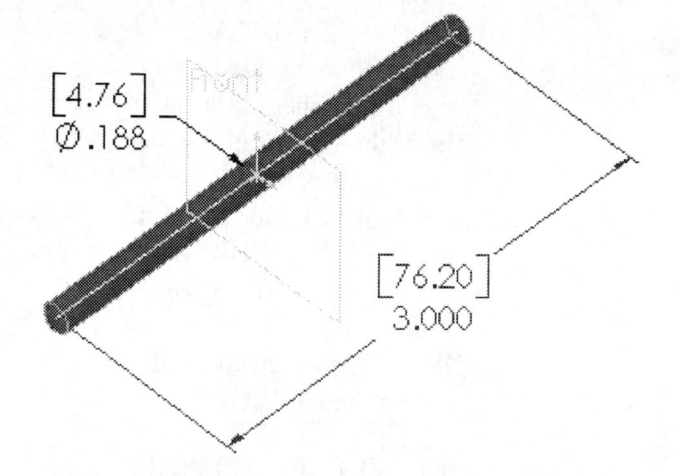

The AXLE utilizes a Ø.188 [4.76mm] circular sketch on the Front Plane. The Front Plane is utilized between the AXLE3000 and other mating components in the WHEEL-AND-AXLE assembly.

Create the AXLE1375. Utilize the AXLE3000 and the File, Save As Option. Enter 1.375[34.93mm] for extruded depth.

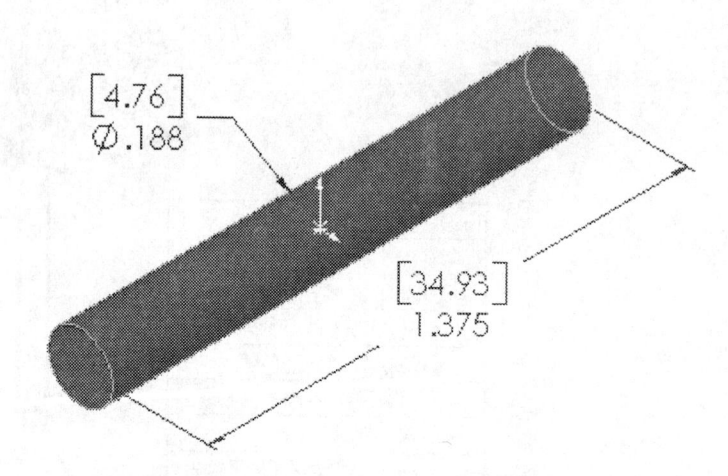

Exercise 5.6b: FLATBAR Part.

Three FLATBAR parts are required for the WHEEL-AND-AXLE assembly:

1) FLATBAR–3HOLE Part.

2) FLATBAR–5HOLE Part.

3) FLATBAR–7HOLE Part.

The FLATBAR–3HOLE Part was created in Exercise 1.5a.

Utilize the FLATBAR-3HOLE Part with the File, Save As option to create the FLATBAR-5HOLE and FLATBAR-7HOLE.

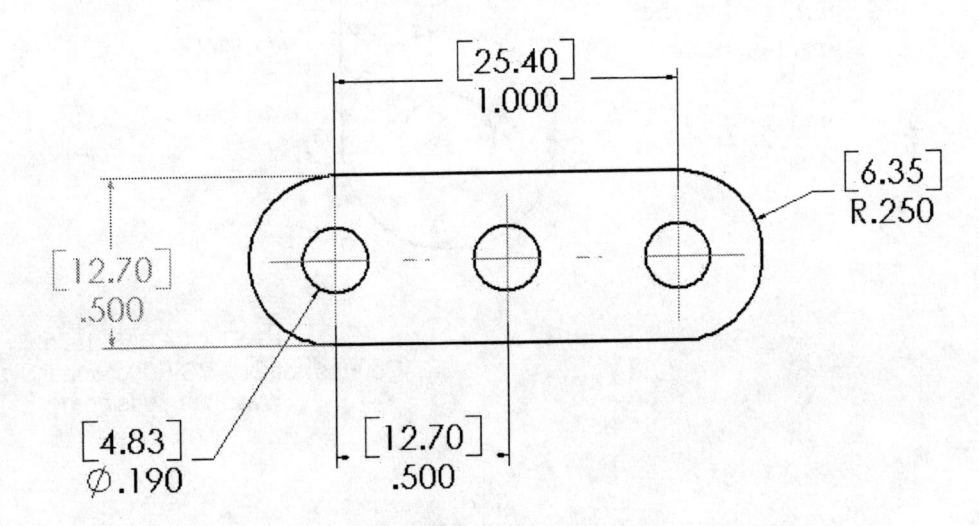

FLAT BAR, 3 HOLE
Courtesy of GEARS Educational Systems, LLC
Hanover, MA USA
www.gearseds.com

Utilize the Front Plane for the Sketch Plane.

Utilize a Linear Pattern for the holes. The holes are space .500[12.70mm] apart.

The three parts are manufactured from 0.060in [1.5mm] Stainless Steel.

Exercise 5.6c: SHAFT COLLAR Part.

Two SHAFT-COLLAR parts are required for the WHEEL-AND-AXLE assembly:

1) SHAFT-COLLAR-1875 Part.

2) SHAFT-COLLAR-5000 Part.

The SHAFT-COLLAR Part was created in Exercise 1.5d. Utilize File, Save to Save the part with the new name, SHAFT-COLLAR-1875. Utilize File, Save As. Enter SHAFT-COLLAR-5000 for file name.

Create the SHAFT COLLAR-1875 part. Utilize the Front plane for the Sketch plane.

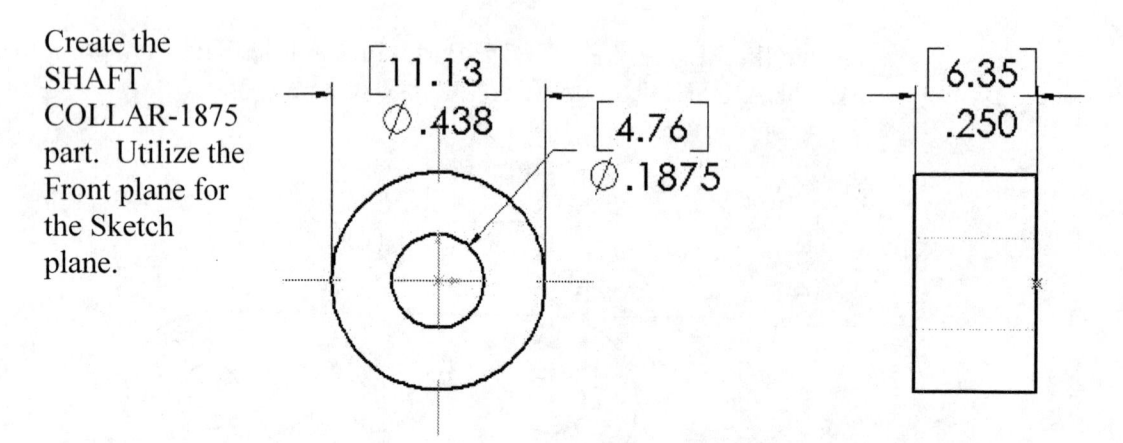

SHAFT-COLLAR-1875
Courtesy of GEARS Educational Systems
www.gearseds.com

Create the SHAFT-COLLAR-5000. Modify feature dimensions.

The ID is .5000[12.70mm].

The OD is .750[19.05mm].

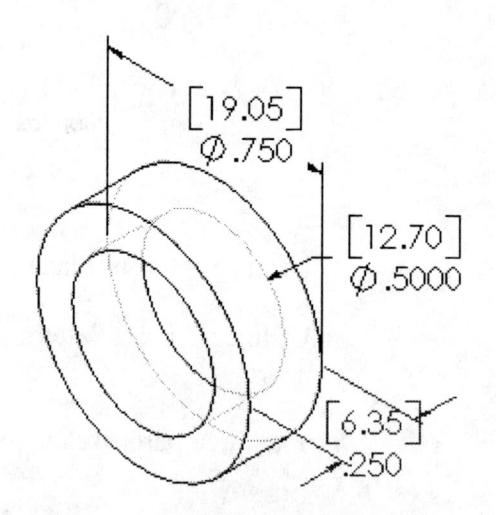

SHAFT-COLLAR-5000
Courtesy of GEARS Educational Systems
www.gearseds.com

Exercise 5.6d: HEX ADAPTER Part.

Create the HEX ADAPTER Part. Utilize the Hole Wizard to create an 8-32 UNC Tapped Hole on the Right Plane with the Thru All option. Mirror the Tapped Hole about the Right Plane.

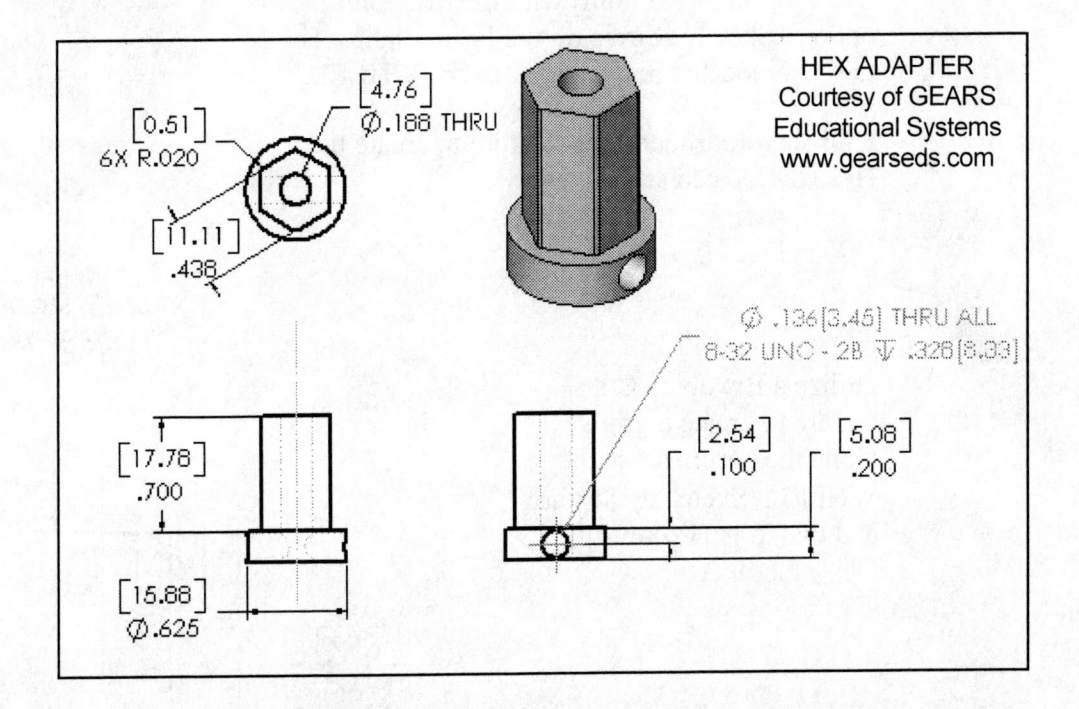

Sweep and Loft Features   **Engineering Design with SolidWorks**

Exercise 5.6e: WHEEL Part.

Create the WHEEL Part.

Utilize an Extruded Base feature with the Depth of 0.25 [6.35mm] with the MidPlane option. Sketch a circle on the Front Plane. The Origin is located at the center of the WHEEL.

Utilize an Extruded Boss feature to create the Hex Cut located at the center.

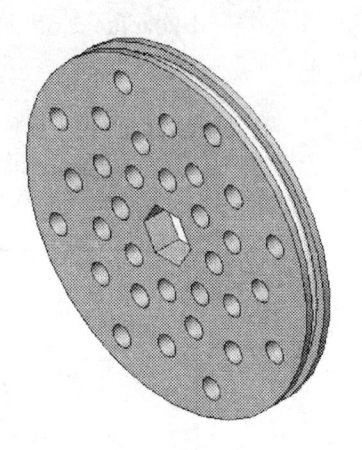

WHEEL
Courtesy of GEARS
Educational Systems

Utilize a Revolved Cut feature to create a groove along the perimeter of the WHEEL. Sketch a 180° arc and two .078 [1.98mm]. Close the Sketch.

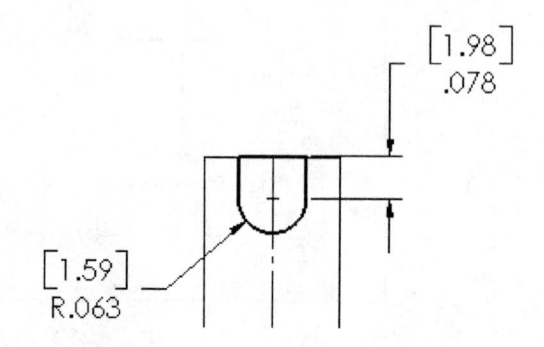

Groove Profile

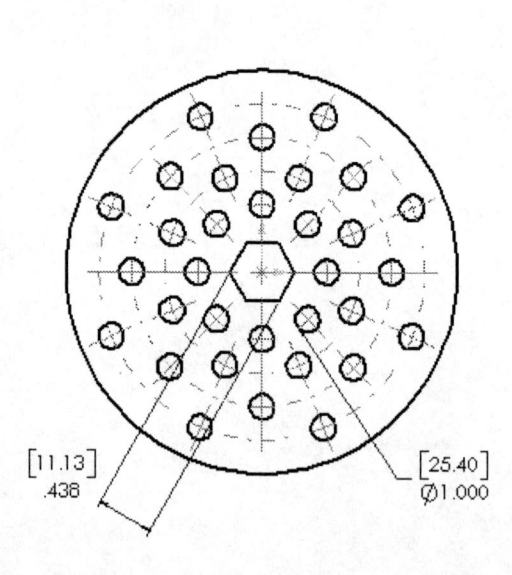

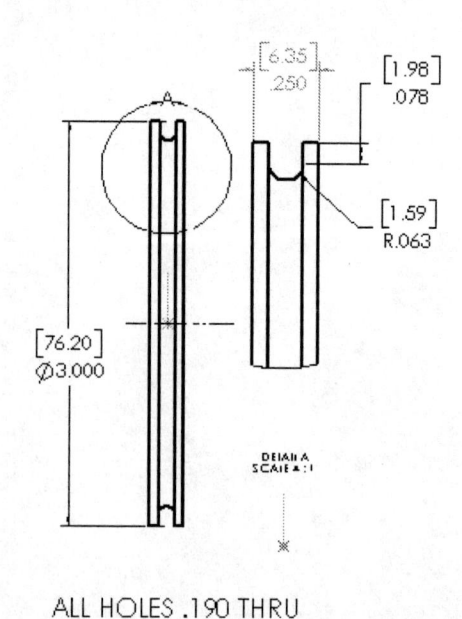

ALL HOLES .190 THRU

PAGE 5-110

Engineering Design with SolidWorks — Sweep and Loft Features

Four bolt circles spaced 0.5[12.7mm] apart locate 8 - ⌀.190[4.83mm] holes. Utilize two Extruded Cut features. Locate the first Extrude Cut seed hole on the first bolt circle and third bolt circle.

Sketch two construction line bolt circles, 1.000[25.4mm] and 2.000[50.80mm]. Sketch a 45° centerline from the Origin to the second circle. Sketch two ⌀.190[4.83mm] circles at the intersection of the bolt circle and the 45° centerline. Extrude the profile to create the holes.

Locate the second Extrude Cut seed hole on the second bolt circle and forth bolt circle. Sketch two construction line bolt circles, 1.500[38.10mm] and 2.500[63.50]. Sketch a 22.5° centerline from the Origin to the second circle. Sketch two ⌀.190[4.83mm] circles at the intersection of the bolt circle and the 22.5° centerline. Extrude the profile to create the holes.

Utilize a Circular Pattern. Select both Extrude Cut features and the Temporary axis to pattern all the holes.

Insert a Reference Axis. Click the Top Plane and the Right Plane. Axis1 is positioned through the Hex Cut centered at the Origin. Drag the Axis1 handles outward to extend the length. Utilize the Reference Axis1 during the assembly mate process.

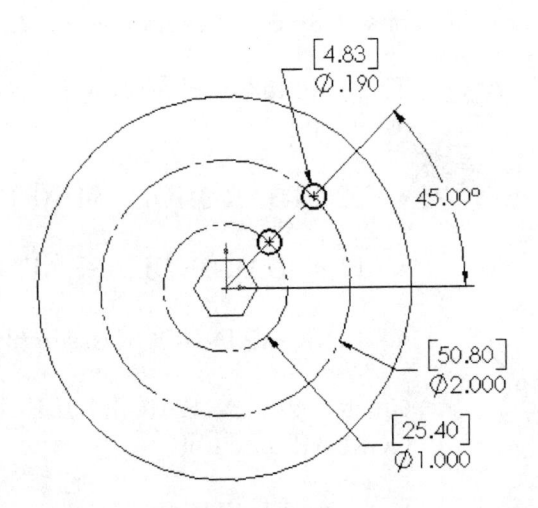

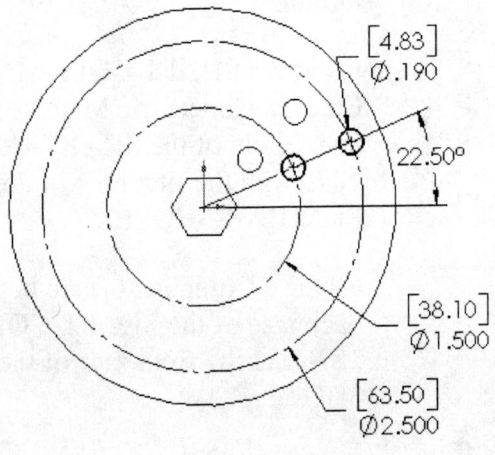

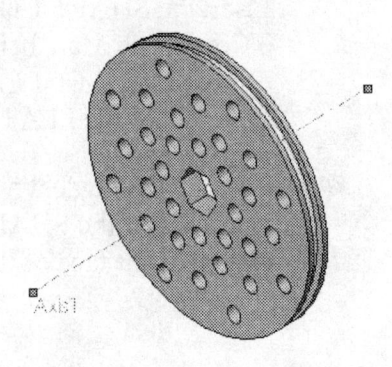

Exercise 5.6f: Sub Assemblies for WHEEL-AND-AXLE.

There are three subassemblies contained in the WHEEL-AND-AXLE assembly:

- FLATBAR-3HOLE-SHAFT COLLAR assembly.

- FLATBAR-5HOLE-SHAFT COLLAR assembly.

- LINK-AND-HOOK assembly.

Create the FLATBAR-3HOLE-SHAFT-COLLAR assembly.

Insert the FLATBAR-3HOLE part. The FLATBAR-3HOLE part is fixed to the Origin.

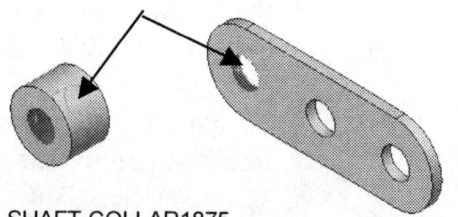

Insert the SHAFT-COLLAR-1875. Create a Concentric Mate between the inside hole of the SHAFT-COLLAR-1875 with the first hole of the FLATBAR-3HOLE.

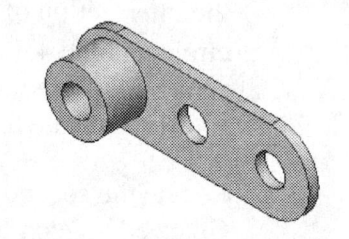

Create a Coincident Mate between the back face of the SHAFT-COLLAR-1875 and the front face of the FLATBAR-3HOLE.

Create the FLATBAR-5HOLE-SHAFT-COLLAR assembly. Insert the FLATBAR-5HOLE. The FLATBAR-5HOLE part is fixed to the Origin.

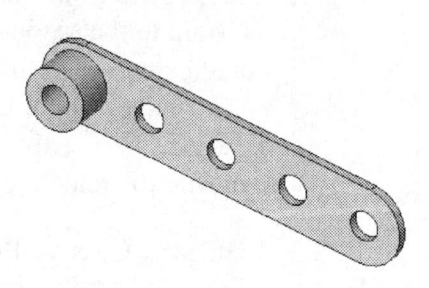

Insert the SHAFT-COLLAR-1875. Create a Concentric Mate between the inside hole of the SHAFT-COLLAR-1875 with the first hole of the FLATBAR-5HOLE.

Create a Coincident Mate between the back face of the SHAFT-COLLAR-1875 and the front face of the FLATBAR-5HOLE.

Create the LINK-AND-HOOK assembly.

Insert the AXLE-1375 part. The AXLE-1375 part is fixed to the Origin.

Insert the FLATBAR-7HOLE part. Create a Concentric Mate between the first hole of the FLATBAR-7HOLE and the shaft of the AXLE-1375.

Create a Distance Mate between the FLATBAR-7HOLE Front Plane and the AXLE-1375 Front Plane. The FLATBAR-7HOLE is symmetric about the AXLE-1375 Front Plane.

Insert the HOOK. Position the HOOK in the last hole of the FLATBAR-7HOLE.

Insert the WEIGHT. Position the WEIGHT .25 [6.35mm] above the bottom face of the HOOK.

Note: There is more than one mating technique for this exercise. Incorporate symmetry into the assembly. Divide large assemblies into smaller subassemblies.

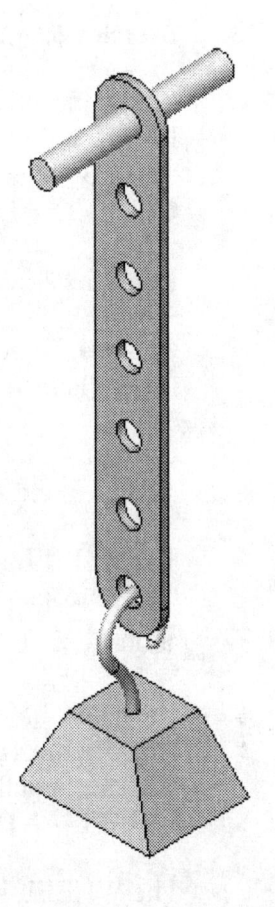

Exercise 5.6g WHEEL-AND-AXLE Assembly.

Create the WHEEL-AND-AXLE Assembly. Insert the first component, AXLE3000. The AXLE3000 is fixed to the WHEEL-AND-AXLE Origin. Display the Temporary Axis.

Insert the second component, WHEEL.

Create a Coincident Mate between the WHEEL Front Plane to the AXLE Front Plane.

Create a Coincident Mate between the WHEEL AXIS1 to the AXLE3000 Temporary Axis.

The WHEEL is free to rotate about the AXLE3000. Utilize Rotate Component to rotate the WHEEL about AXIS1.

Create a 22.5° Angle Mate between the Top Plane of the WHEEL and the Top Plane of the WHEEL-AND-AXLE assembly.

The WHEEL is static and does not rotate. Rename Angle Mate to Wheel-Static. The Mate process is easier when parts do not rotate. Add parts with the Wheel-Static Angle Mate in the Resolved state. Suppress the Wheel-Static Angle Mate to simulate dynamic motion in the final step.

Locate the mating holes on the WHEEL for the FLATBAR-3HOLE-SHAFT-COLLAR assembly and the FLATBAR-5HOLE-SHAFT-COLLAR assembly.

Insert the FLATBAR-3HOLE-SHAFT-COLLAR assembly. Create a Concentric Mates with the first hole of the FLATBAR-3HOLE and the WHEEL hole located at 135° from the X-axis.

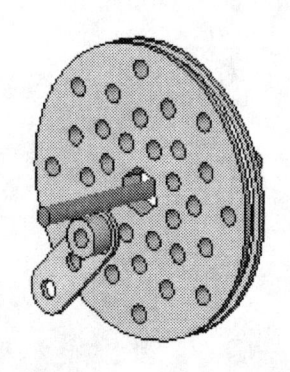

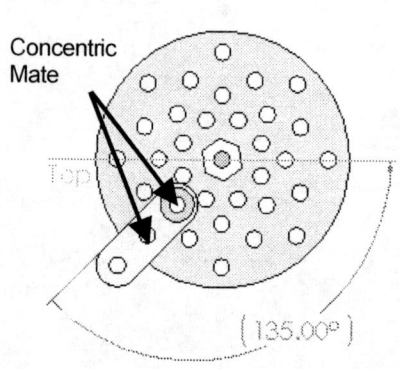

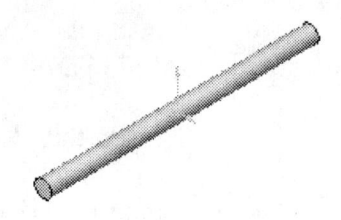

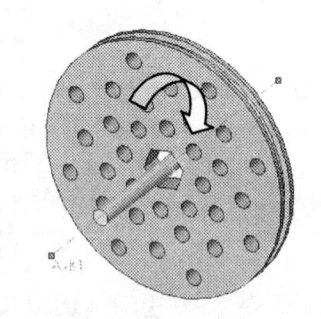

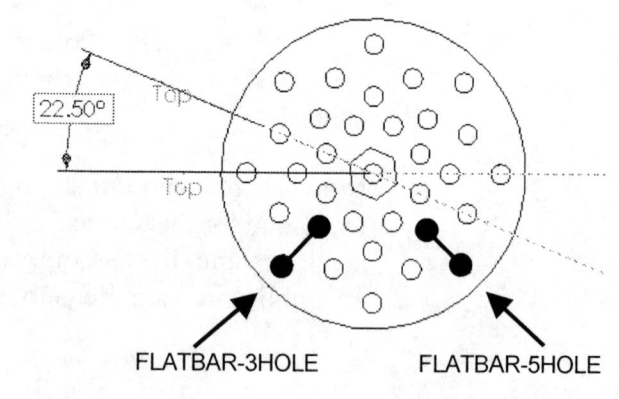

Create a Concentric Mates with the second hole of the FLATBAR-3HOLE and the WHEEL hole located at 135° from the X-axis.

Create a Coincident Mate between the back face of the FLAT-BAR-3HOLE-SHAFT-COLLAR and the front face of the WHEEL.

Insert the FLATBAR-5HOLE-SHAFT-COLLAR assembly. Create two Concentric Mates and one Coincident Mate. The two FLATBARs are 90° apart.

Insert the second FLATBAR-3HOLE-SHAFT-COLLAR assembly coincident with the back face of the WHEEL. Add two Concentric Mates and one Coincident Mate.

Insert the second FLATBAR-5OLE-SHAFT-COLLAR assembly coincident with the back face of the WHEEL. Add two Concentric Mates and one Coincident Mate.

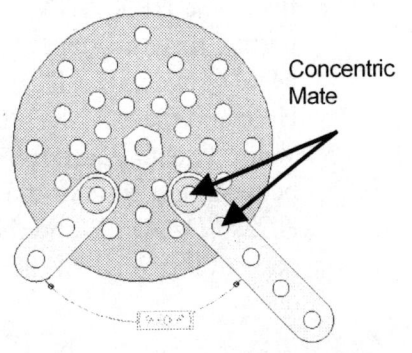

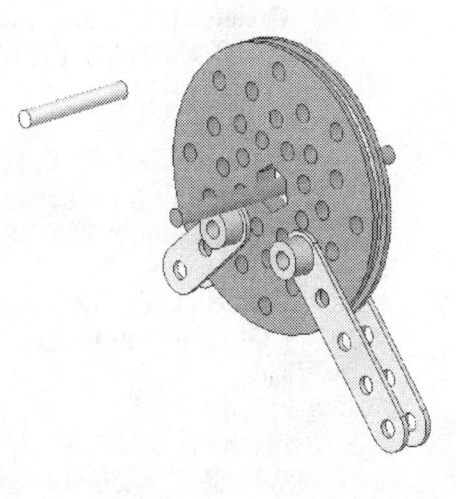

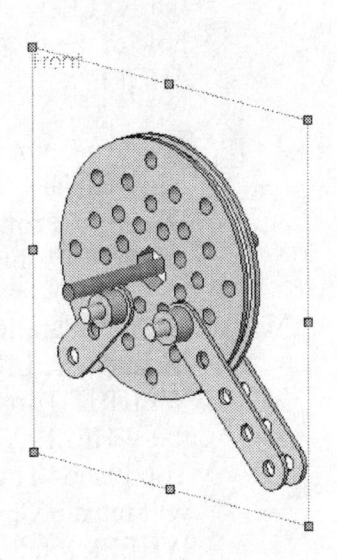

Insert the AXLE-1375 part. Create a Concentric Mate between the AXLE-1375 conical face and the left SHAFT-COLLAR hole. Create a Coincident Mate between the AXLE-1375 Front plane and the WHEEL Front plane. The AXLE-1375 is symmetric about the WHEEL Front Plane.

Insert the AXLE-1375 part. Create a Concentric Mate between the AXLE-1375 conical face and the right SHAFT-COLLAR hole. Create a Coincident Mate between the AXLE-1375 Front plane and the WHEEL Front plane.

Insert the HEX ADAPTER part. Create a Concentric Mate between the Thru Hole and the AXLE-3000 shaft. Create Parallel Mate between the edge of the HEX ADAPTER and the edge of the WHEEL hex cut.

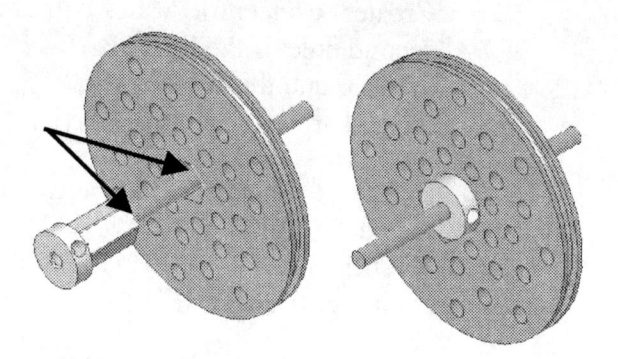

Create a Coincident mate between the bottom face of the HEX ADAPTER Base Extrude and the front face of the WHEEL

Insert the SHAFT-COLLAR-5000 part. Create a Concentric Mate between the SHAFT-COLLAR-5000 and the AXLE3000.

Create a Coincident Mate between the SHAFT-COLLAR-5000 back face and the WHEEL back face.

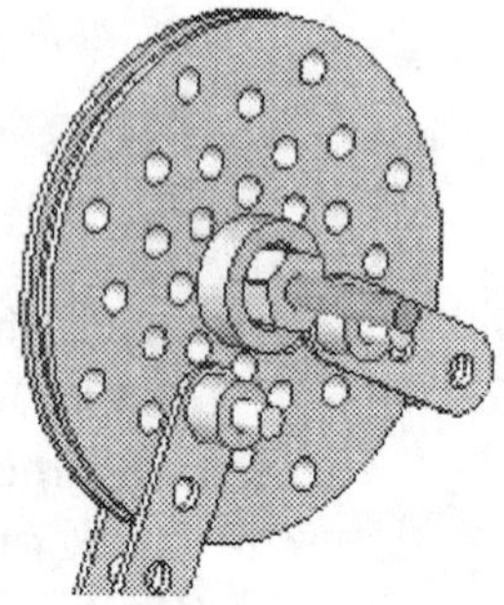

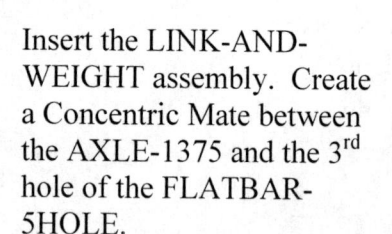

Insert the LINK-AND-WEIGHT assembly. Create a Concentric Mate between the AXLE-1375 and the 3rd hole of the FLATBAR-5HOLE.

Create a Coincident Mate between the LINK-AND-WEIGHT Front Plane and the WHEEL Front Plane.

Create a Parallel Mate between the LINK-AND-WEIGHT Top Plane and the WHEEL-AND-AXLE Top Plane. The WEIGHT will remain vertical during WHEEL rotate.

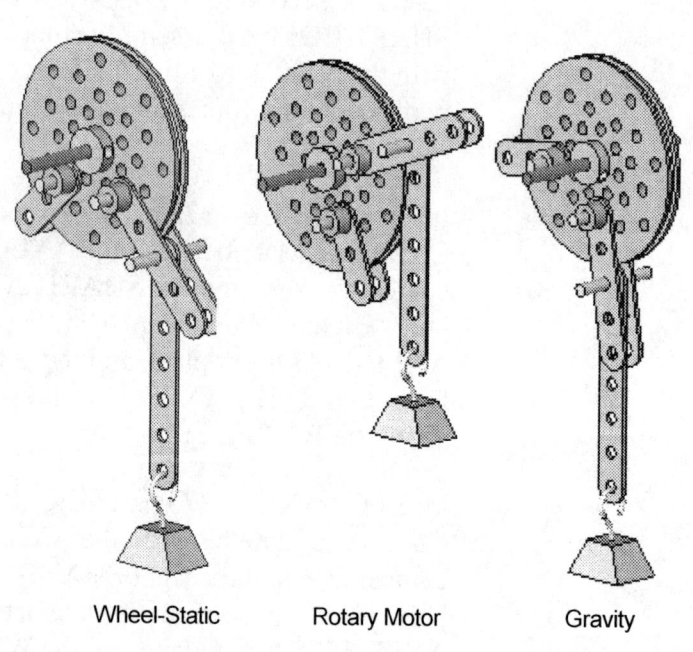

Wheel-Static   Rotary Motor   Gravity

Resolve the Wheel-Static Angle Mate. Apply a counter-clockwise Rotary Motor Simulation to the front face of the WHEEL. Play the simulation. The WHEEL rotates counter-clockwise and the WEIGHT translates in an upward vertical direction. Suppress the Rotary Motor.

Apply Gravity Simulation to the FLAT-BAR-7HOLE front face. Play the simulation. The WEIGHT translates in a downward vertical direction and the WHEEL rotates clockwise.

Suppress the Wheel-Static Angle Mate. The WHEEL returns to the default rest position. The WHEEL-AND-AXLE assembly is complete.

Create the WHEEL-AND-AXLE drawing. Add PART NO. and DESCRIPTION Property Values to all parts.

Create a Bill of Materials. Utilize the Show Parts Only option in the Bill of Materials dialog box.

ITEM NO.	QTY.	PART NO.	DESCRIPTION
1	1	GIDS-SC-10017	AXLE
2	1	GIDS-SC-10014	3IN-WHEEL
3	2	GIDS-SC-10001-3	FLAT BAR - 3 HOLE
4	5	GIDS-SC-10012-3-16	SHAFT COLLAR
5	2	GIDS-SC-10001-5	FLATBAR-5HOLE
6	3	GIDS-SC-10017	AXLE1375
7	1	GIDS-SC-10013	HEX ADAPTER 7/16
8	1	GIDS-SC-10001-7	FLATBAR-7HOLE
9	1	DM-10000-02	HOOK
10	1	DM-10000-01	WEIGHT

Exercise 5.7: Industry Collaborative Exercise.

You now work with a team of engineers on a new industrial application. The senior engineer specifies a ½HP AC Motor. The Parts Department stocks the motor. Create the SPEED REDUCER Assembly. Safely reduce the input speed of the motor at a ratio of 30:1. An enclosed gear drive, called a Speed Reducer, safely reduces the speed of the motor.

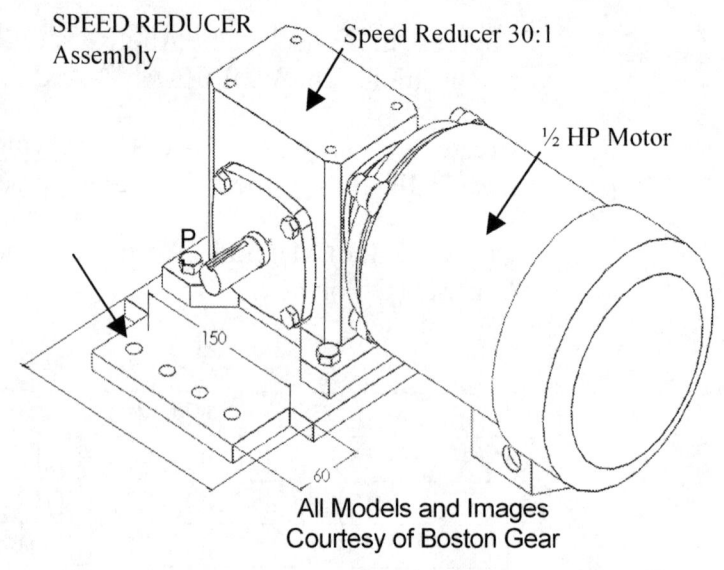

SPEED REDUCER Assembly
Speed Reducer 30:1
½ HP Motor

All Models and Images Courtesy of Boston Gear

The enclosed gear drive is a purchased part. The Motor is fastened to the Speed Reducer.

Determine the parts to design. PLATE-SR contains a 150mm x 60mm notched section with 4 equally spaced holes. The Speed Reducer is mounted to the PLATE-SR with 4 fasteners. All holes on PLATE-SR use the same fastener.

You receive vendor information from the Purchasing Department. The ½ HP Motor is manufactured by Boston Gear, Quincy, MA. The Motor part number is FUTF. No drawings of the Motor exist in the Engineering Department.

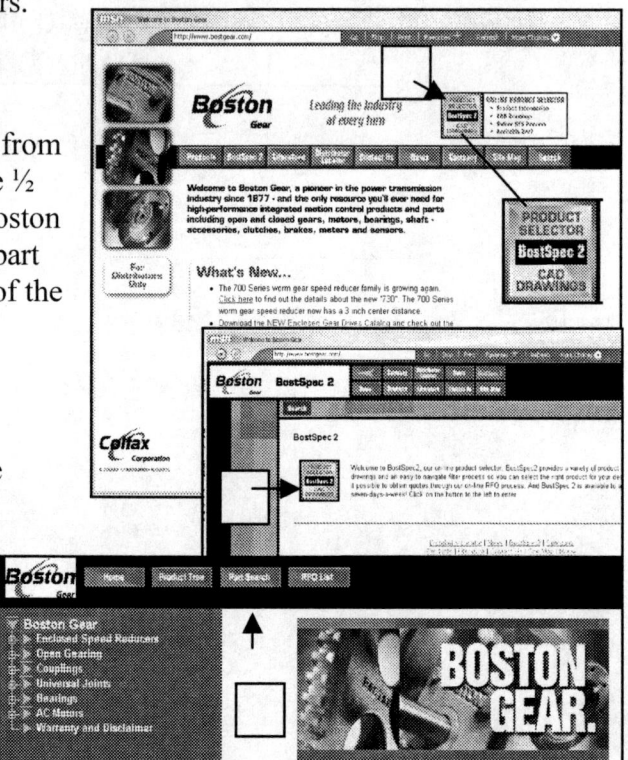

Visit the Boston Gear website: www.bostongear.com. Click the BostSpec2 button. Click the BostSpec2 button again. The Boston Gear Product Tree is displayed to the left. Click the Part Search Button. Enter FUTF for the part number.

Engineering Design with SolidWorks **Sweep and Loft Features**

Click the View Spec Sheet button to display the motor specification. Click the 3D CAD button to view the part.

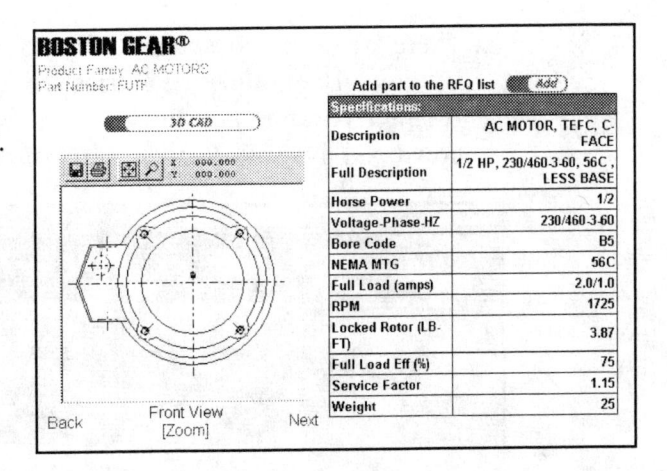

Click the 3D Formats button. Select the SolidWorks 2001Plus Assembly to download.

Find the Enclosed Speed Reducer. Click the Product Tree button. Click the Worm Gear Speed Reducers, Single Reduction option. Click the Type B – G icon in the second row.

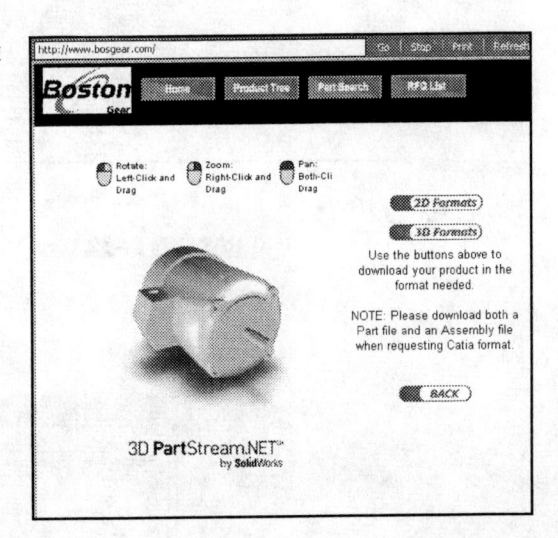

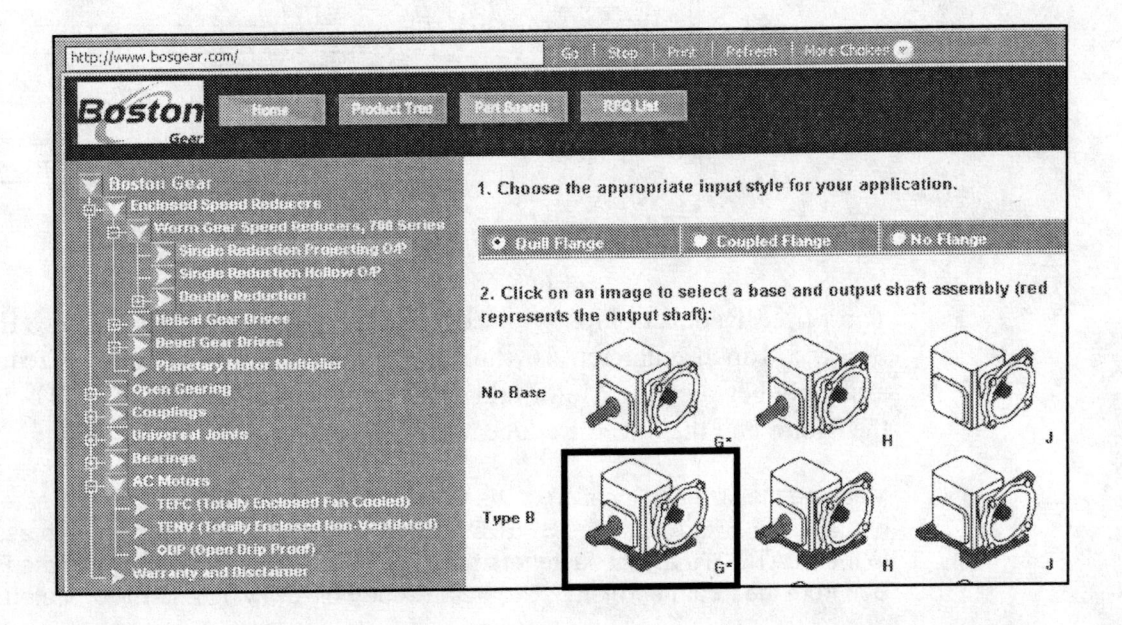

There are over 100 Speed Reducers listed. Select ½ in the Motor HP (Class I) list box. The number of entries reduces to 10. Select 30 in the Ratio list box. A single part number remains. View the Speed Reducer specification and download the SolidWorks 2001Plus Assembly.

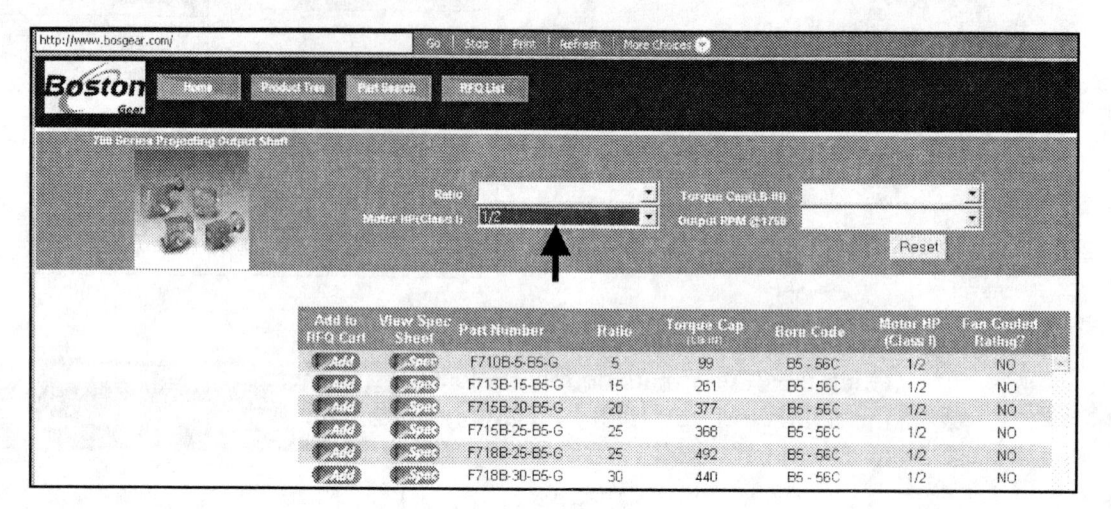

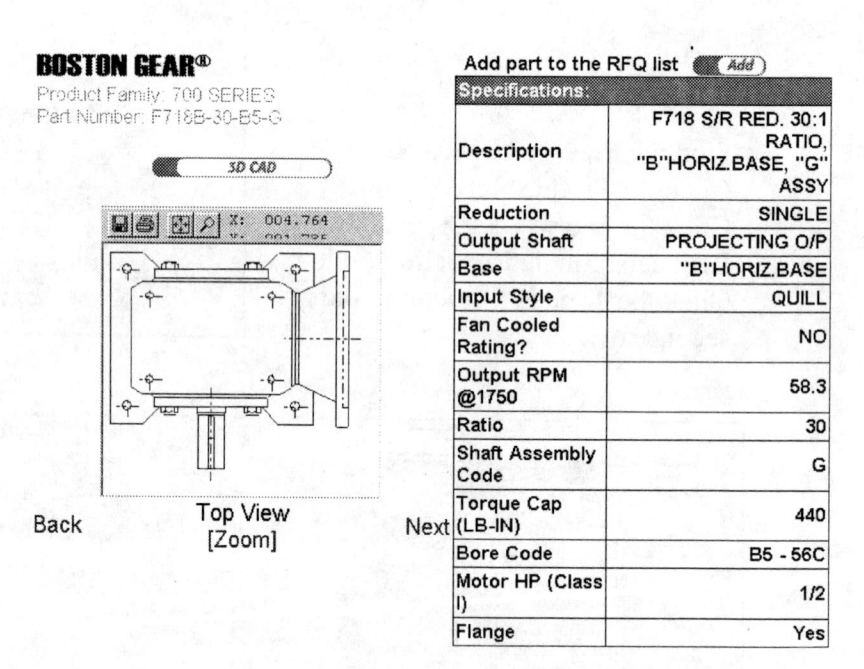

The Speed Reducer ships with HHCS (4) 3/8-16 x 1 inch, (Heavy Hex Cap Screw 3/8 inch diameter, 16 threads per inch, 1 inch length). Create a sub assembly between the Motor and Speed Reducer. Mate the 4 HHCS between the Motor and the Speed Reducer.

Measure the size and location of the Speed Reducer mounting holes. Create the part, PLATE-SR. Create the SPEED REDUCER assembly. Mate the sub assembly to the PLATE-SR. Utilize fasteners from the SolidWorks/Toolbox or use the Flange Bolt from the Feature Palette. Create an assembly drawing and Bill of Materials.

Project 6

Top Down Assembly Modeling

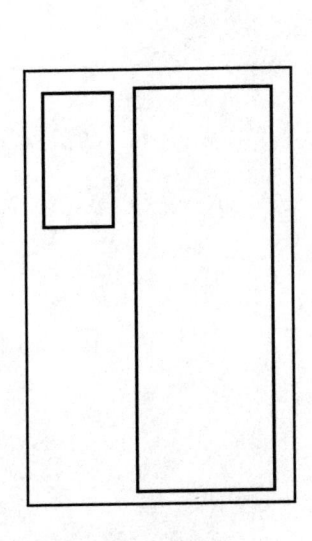

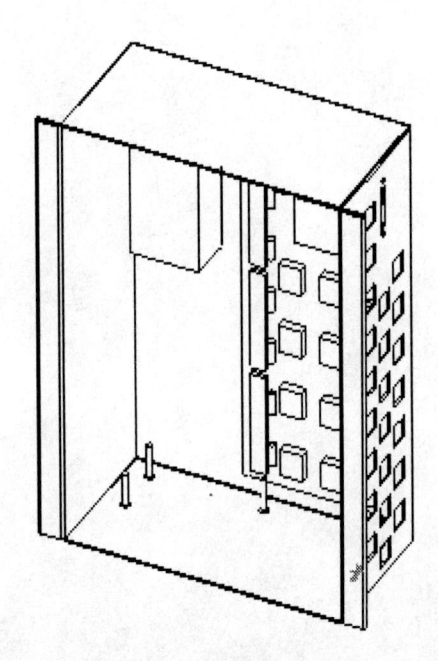

Below are the desired outcomes and usage competencies based on the completion of Project 6.

Project Desired Outcomes:	Usage Competencies:
Create three different BOX sizes: • Small. • Medium. • Large.	Ability to create a new Top Down assembly and layout sketch.
	Knowledge to create and modify components developed In-Context of an assembly.
	An understanding of sheet metal features.
	Ability to incorporating non-SolidWorks models into an assembly.

Notes:

Project 6 – Top Down Assembly Modeling

Project Objective

Create three different BOX sizes utilizing the Top Down assembly modeling approach. The Top Down approach is a conceptual approach used to develop products from within the assembly. Major design requirements are translated into sub-assemblies or individual components and key relationships.

You will model sheet metal components in their 3D formed state. You will manufacture the sheet metal components in their 2D flatten state.

On completion of this project, you will be able to:

- Create a new Top Down assembly and layout sketch

- Choose the best profile for the Layout sketch.

- Export a file using CircuitWorks.

- Choose the proper sketch plane.

- Insert sheet metal bends to transform a solid part into a sheet metal part.

- Create a Hole.

- Create a Linear Pattern of Holes.

- Insert IGES format PEM® self-clinching fasteners obtained from the Internet.

- Use a Component Pattern to maintain a relationship between the Holes and the fasteners.

- Set units.

- Set System Options.

- Create Link Values

- Use and produce Equations.

- Generate a Die Cutout Palette feature.

- Use the Palette forming tool directory.

- Address manufacturing considerations.
- Create an Equal Relation.
- Select the correct PEM® Fastener for the assembly.
- Use the following SolidWorks features:
    - Base Extrude feature.
    - Extrude Boss.
    - Shell the Extrude
    - Rib.
    - HoleWizard.

Project Situation

You now work for a different company. Life is filled with opportunities. You are part of a global project design team that is required to create a family of electrical boxes for general industrial use. You are the project manager.

You receive a customer request from the Sales department for three different size electrical boxes.

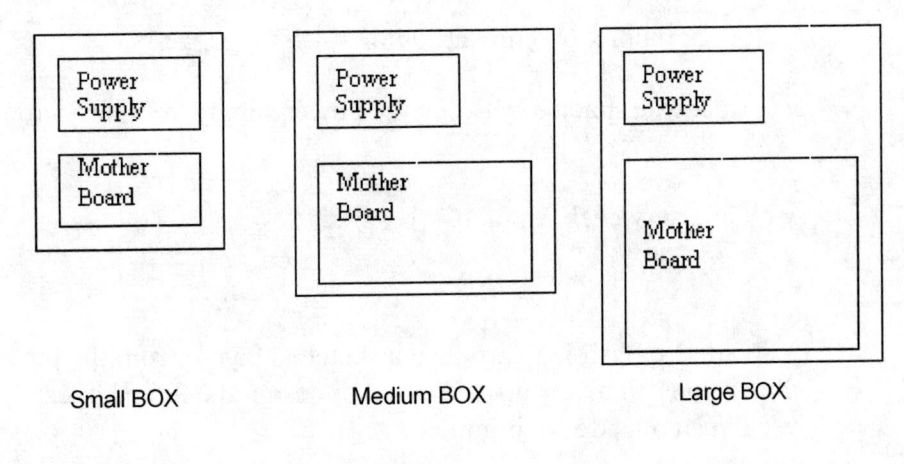

Small BOX   Medium BOX   Large BOX

Figure 6.1

Delivery time to the customer is a concern. You work in a concurrent engineering environment.

Your company is expecting a sales order for 5,000 units in each requested BOX configuration.

The BOX contains the following key components:

- Power supply.

- Motherboard.

The size of the power supply is the same for all boxes. There are three different BOX sizes, Figure 6.1.

In other words, the size of the power supply is the constant. The size of the BOX is a variable.

The available space for the motherboard is dependent upon the size of the BOX. The depth of the three boxes is 100mm.

You contact the customer to discuss and obtain design options and product specifications. Key customer requirements:

- Three different BOX sizes:
    - 300mm x 400mm x 100mm   Small.
    - 400mm x 500mm x 100mm   Medium.
    - 550mm x 600mm x 100mm   Large.

- Adequate spacing between the power supply, motherboard and internal walls.

- Field serviceable.

You are responsible to produce a sketched layout from the provided critical dimensions. You are also required to design the three boxes. The BOX is used in an outside environment.

Top Down Design Approach

Top down design approach is a conceptual approach used to develop products from within the assembly. Major design requirements are translated into sub-assemblies or individual components and key relationships, Figure 6.2.

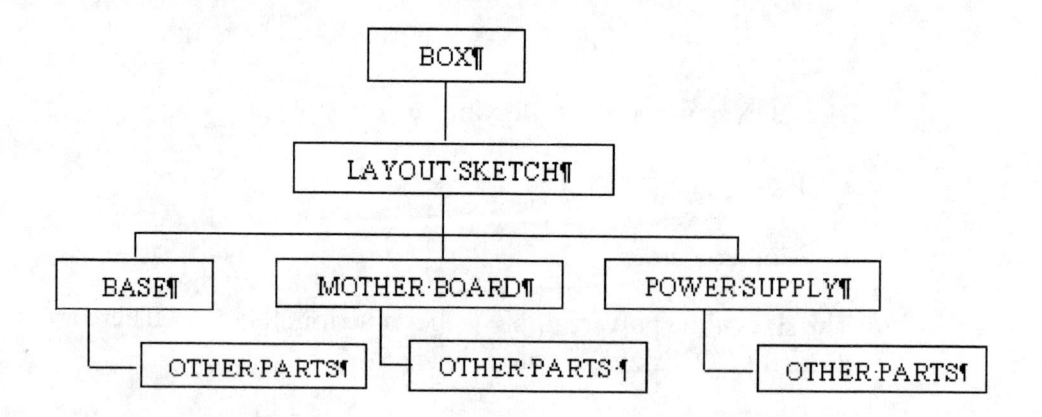

Figure 6.2

Start with a Layout sketch.

The Layout sketch specifies the location of the key components.

Create or add additional components to complete the BOX.

Use a combination of Top Down and Bottom Up approaches in the BOX assembly.

Consider the following in a preliminary design product specification:

- What are the major components in the design? The motherboard and power supply are the major components.

- What are the key design constraints? Three different BOX sizes specified by the customer.

- How does each part relate to the other? From past experience and discussions with the electrical engineering department, a 25mm physical gap is required between the power supply and the motherboard. A 20mm physical gap is required between the internal components and the side of the BOX.

- How will the customer use the product? The customer does not disclose the specific usage of the BOX. The customer is in a very competitive market.

- What is the most cost-effective material for the product? Aluminum is cost-effective, strong, relatively easy to fabricate, corrosion resistant and non-magnetic.

Incorporate the design specifications into the BOX assembly.

Use a Layout sketch, solid parts and sheet metal parts.

Obtain additional parts from your vendors.

BOX Assembly Overview

Create the Layout sketch in the BOX assembly. Insert a new part, MOTHERBOARD. Convert edges from the Layout sketch to develop the outline of the MOTHERBOARD. Use the outline sketch as the Base Extrude feature for the MOTHERBOARD component.

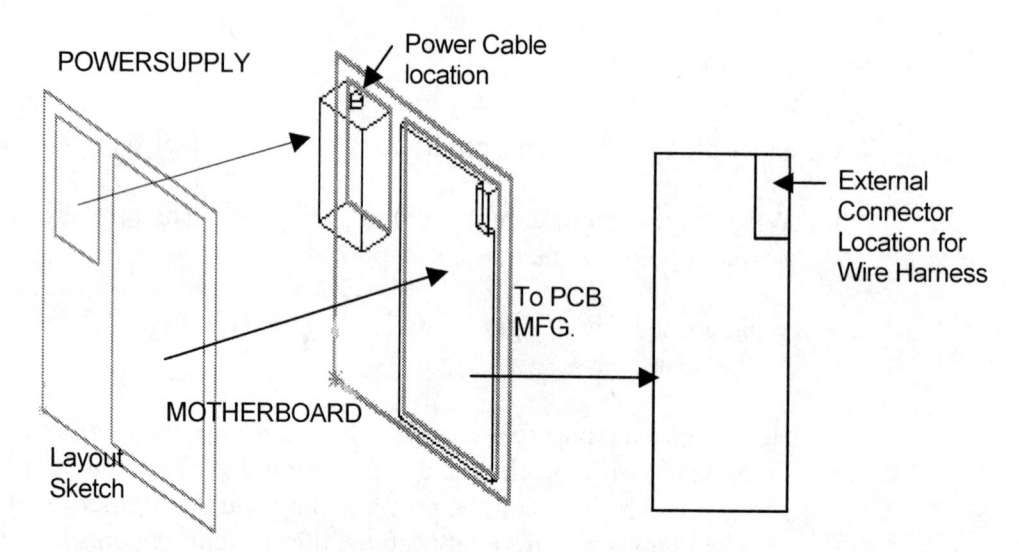

An Extrude Boss locates a key electrical connector for a wire harness. Design the location of major electrical connections early in the assembly process.

Export the MOTHERBOARD outline to CircuitWorks. Inform the Printed Circuit Board manufacturer the maximum size of the MOTHERBOARD and the location of key electrical components.

Insert a new part, POWERSUPPLY. Convert edges from the Layout sketch to develop the outline of the POWERSUPPLY.

Use the outline sketch as the Base Extrude feature for the POWERSUPPLY component.

An Extruded Boss feature represents the location of the power cable.

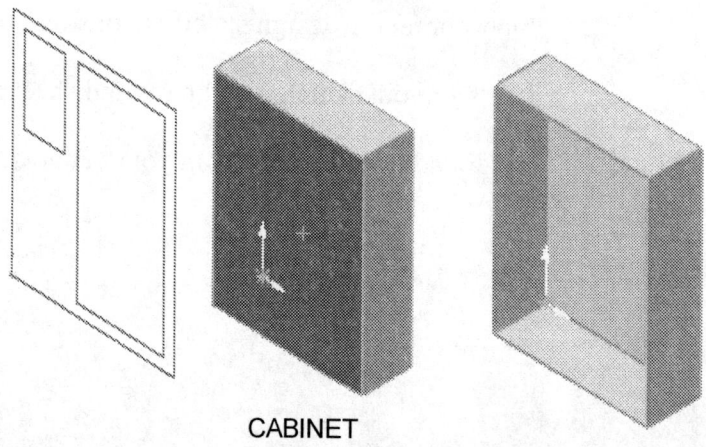

CABINET

Add additional features to complete the POWERSUPPLY.

Insert a new part, CABINET. Convert edges from the Layout.

Use the outline sketch as the Base Extrude feature for the CABINET component.

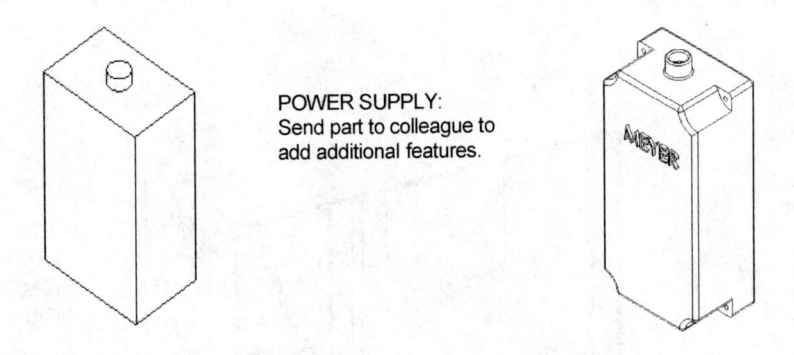

POWER SUPPLY:
Send part to colleague to add additional features.

Shell the Extrude feature. Add Rib features to the 4 corners. Insert Sheet Metal Bends to transform a solid part into a sheet metal part.

Sheet metal components utilize features to create flanges, cuts and forms. Model sheet metal components in their 3D formed state. Manufacture sheet metal components in their 2D flatten state.

Work with your sheet metal manufacturer. Discuss cost effective options. Create a sketched pattern to add square cuts instead of formed louvers.

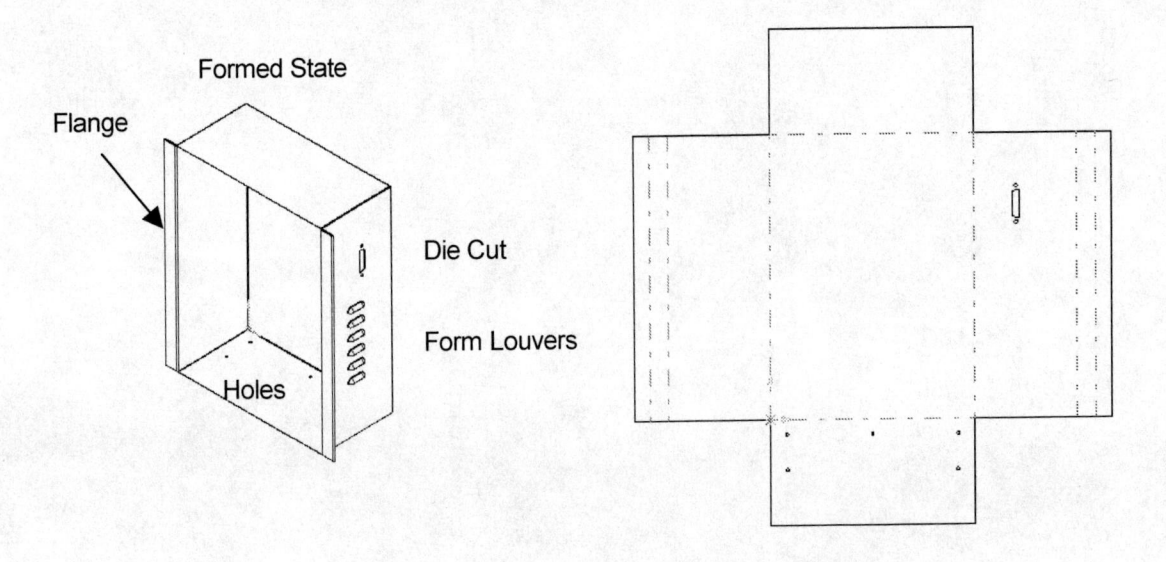

Formed State

Flange

Die Cut

Form Louvers

Holes

Create a Linear Pattern of Holes for the CABINET.

Insert IGES format PEM® self-clinching fasteners obtained from the Internet.

Use a Component Pattern to maintain a relationship between the Holes and the fasteners.

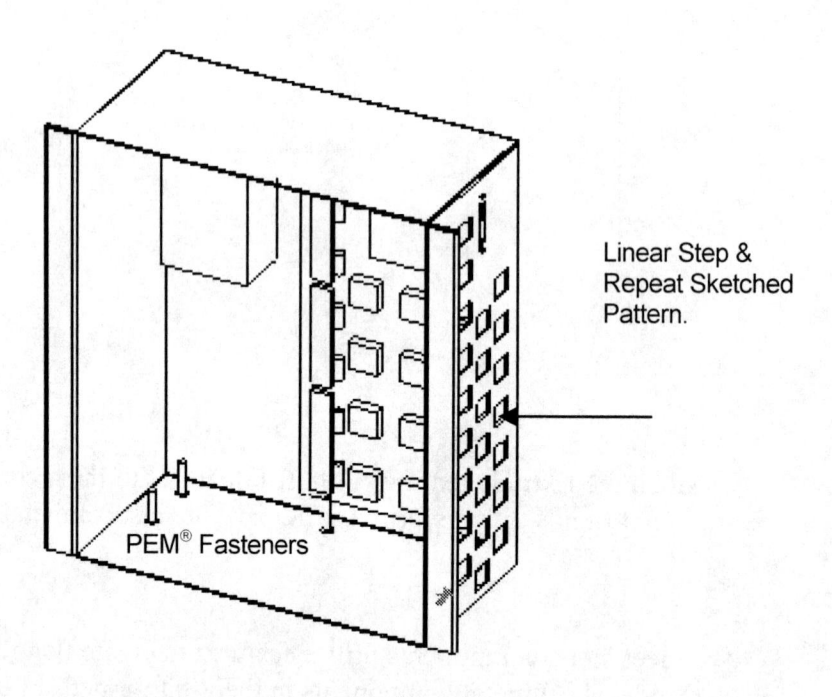

Linear Step & Repeat Sketched Pattern.

PEM® Fasteners

Layout Sketch

Utilize the Layout Sketch to develop component space allocations and relations. Components and assemblies reference the Layout Sketch.

The BOX contains the following key components:

- Power supply.

- Motherboard.

The minimum physical spatial gap between the motherboard and the power supply is 25mm.

The minimum physical spatial gap between the motherboard, power supply and the internal sheet metal BOX wall is 20mm.

After numerous discussions with the electrical engineer, you standardize on a Power Supply size: 150mm x 75mm x 50mm.

Create the Assembly Template for the BOX.

1) Click **New**.

2) Click **Assembly** from the New Document dialog box.

3) Click **OK**.

4) Set the Assembly Document Template options. Click **Tools**, **Options**, **Document Properties**.

5) For English components: Select **ANSI** from the Dimensioning Standard option.

6) Click **Units**. Enter **Millimeters** from the Linear units list box.

7) Enter **2** in the Decimal places spin box.

8) Click **OK**.

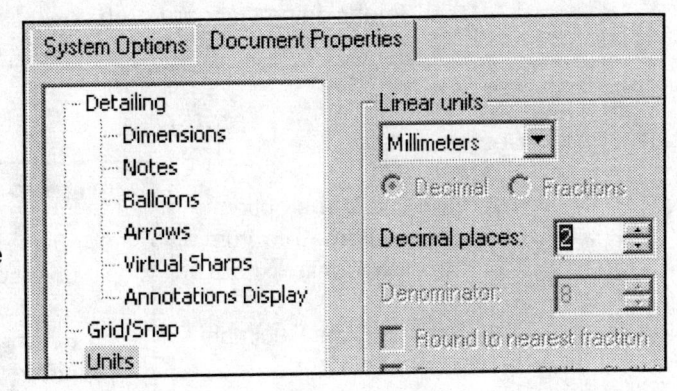

9) Save the assembly template. Click **File**, **SaveAs**. Click the **Assembly Template (*.asmdot)** from the Save As type list box.

10) Select **ENGDESIGN-W-SOLIDWORKS\MY-TEMPLATES** for the Save in file folder.

11) Enter **ASM-MM-ANSI** in the File name text box.

12) Enter ASSEMBLY TEMPLATE MILLIMETER, ANSI for Description.

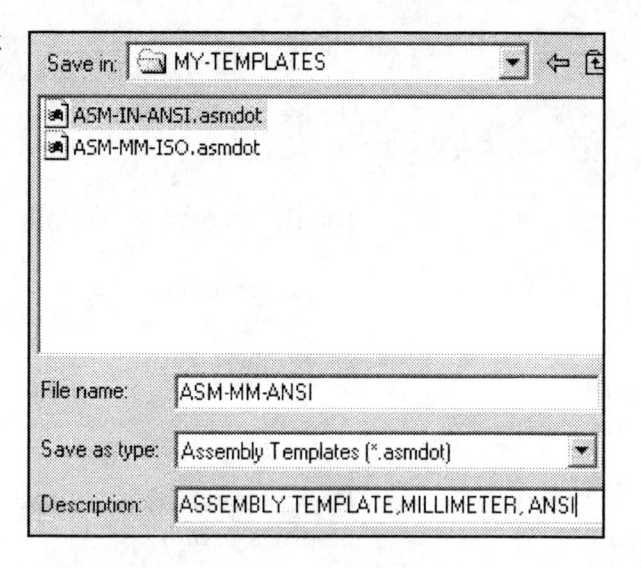

13) Click the **Save** button. Use the ASM-MM-ANSI for all BOX assemblies and sub-assemblies.

If your MY-TEMPLATES folder contains the PART-MM-ANSI Template, go to Step 21.

Create the Part Template for the BOX.

14) Click **New**. Click the **Part** icon from the Templates tab. Click **OK**. The Front, Top and Right reference planes are displayed in the Part1 Feature Manager.

15) Click **Tools**, **Options**. Click the **Document Properties** tab. Set the Dimensioning Standard. Select **ANSI** from the Dimensioning standard drop down list.

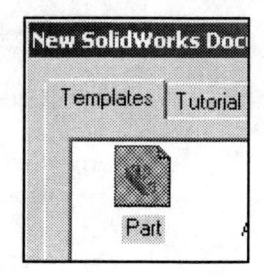

Set the Units.

16) Click the **Units** option. Enter **millimeters** from the Linear units list box.

17) Enter **2** in the Decimal places spin box.

18) Click **OK**.

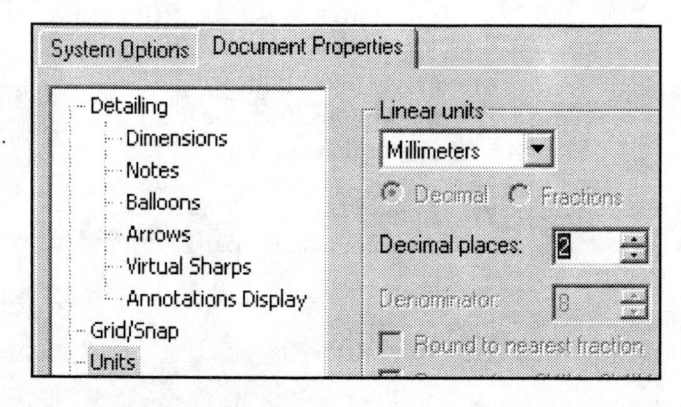

Engineering Design with SolidWorks — Top Down Assembly Modeling

Save the Part Template.
19) Click **File** from the Main menu. Click **Save As**.

20) Click **Part Templates (*.prtdot)** from the Save As type list box.

21) Select ENGDESIGN-W-SOLIDWORKS\MY-TEMPLATES for Save in file folder.

22) Enter **PART-MM-ANSI** in the File name text box.

23) Enter Metric part template, units-mm, ANSI standard for Description. Click Save.

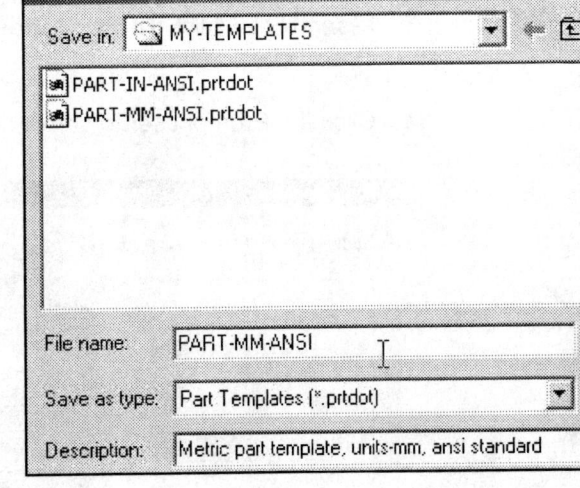

24) Close All documents. Click **Windows**, **Close All**.

Create the BOX Assembly.
25) Click **New**.

26) Click the **MY-TEMPLATES** tab.

27) Click **ASM-MM-ANSI**. Assembly template.

28) Click **OK**.

29) Save the BOX assembly. Click **Save**.

30) Select ENGDESIGN-W-SOLIDWORKS\PROJECTS for the file folder.

31) Enter **BOX** in the file name box.

32) Enter **BOX ASSEMBLY** for Description. Click the **SAVE** button.

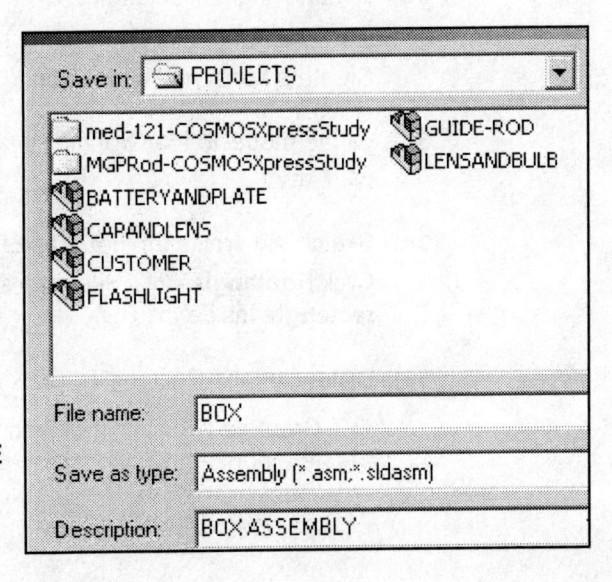

PAGE 6 - 13

Set the System Options.
33) Use the PART-MM-ANSI and ASM-MM ANSI Templates for all parts and assemblies. Click **Tools**, **Options**. Click **Default Templates** from the System Options box.

34) Click the Prompt user to select document template button. Click OK.

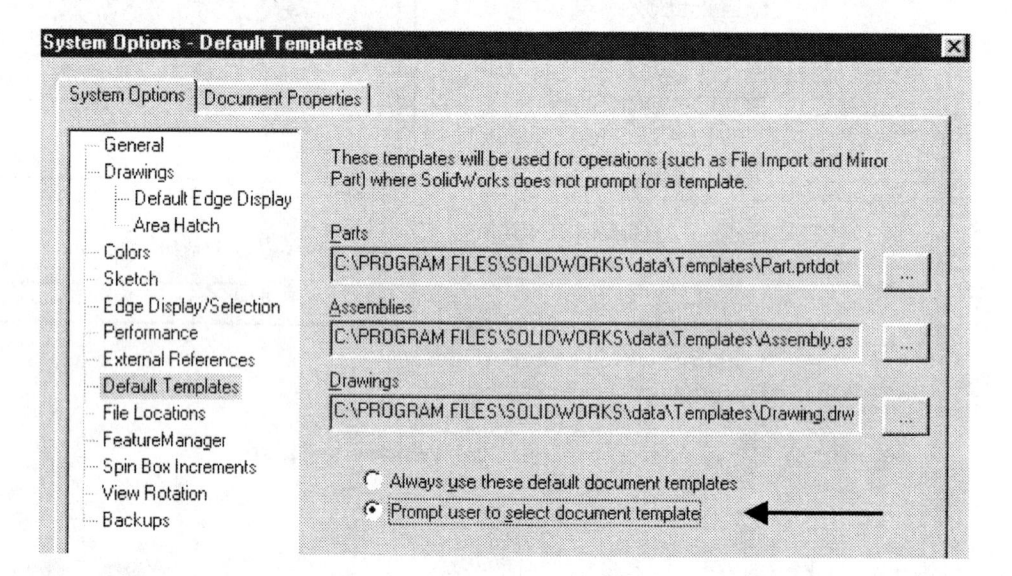

Create the Layout Sketch.
35) Select the Sketch plane. The Front plane is the default Sketch plane. Click **Sketch**. Sketch the profile of the BOX. Click **Rectangle**. Position the first point of the rectangle at the **Origin**.

36) Dimension the Rectangle. Click **Dimension**. Create a vertical dimension. Enter **400**. Create a horizontal dimension. Enter **300**.

37) Fit the model to the Graphics window. Press the **f** key.

38) Sketch the profile for the POWER SUPPLY. Click **Rectangle**. Sketch a small **rectangle** inside the BOX.

39) Dimension the Rectangle. Click **Dimension**. Create a vertical dimension. Enter **150**. Create a horizontal dimension. Enter **75**.

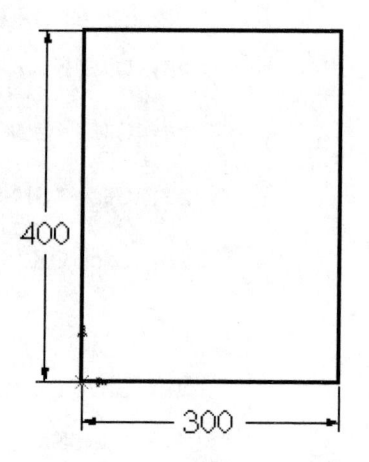

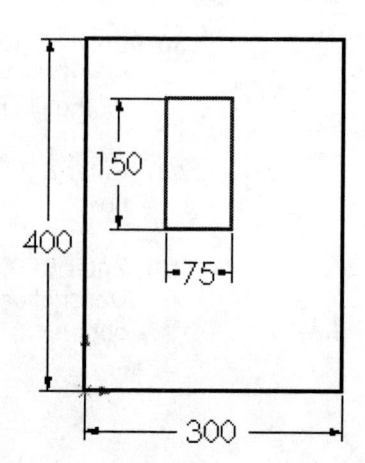

PAGE 6 - 14

Engineering Design with SolidWorks **Top Down Assembly Modeling**

40) Dimension the POWER SUPPLY 20mm from the edge of the BOX. Create a horizontal dimension. Click the BOX **vertical edge**. Click the POWERSUPPLY **vertical edge**. Enter **20**. Create a horizontal dimension. Enter **20**.

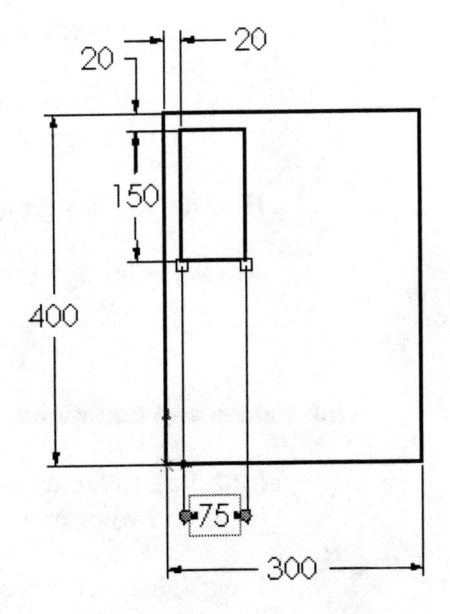

41) Sketch the profile for the MOTHERBOARD. Click **Rectangle**. Sketch a **rectangle** to the right of the POWERSUPPLY.

42) Click **Dimension**. Create a horizontal dimension between the **right edge** of the MOTHERBOARD and the **left edge** of the POWERSUPPLY. Enter **25**.

43) Create a vertical dimension between the **top edge** of the MOTHERBOARD and the **top edge** of the BOX. Enter **20**.

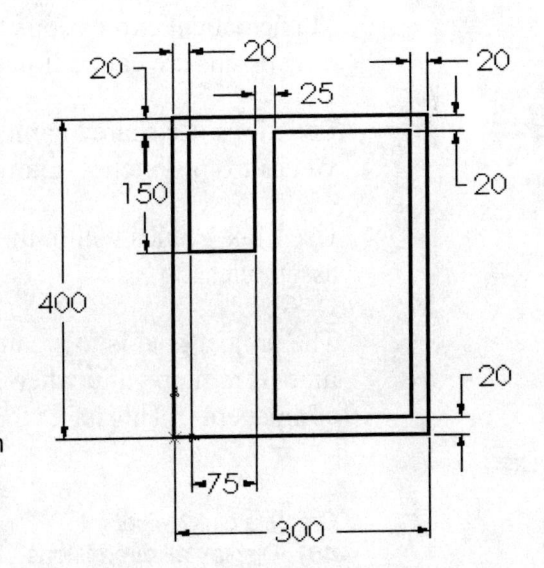

44) Create a horizontal dimension from the **right edge** of the MOTHERBOARD and the **right edge** of the BOX. Enter **20**.

45) Create a vertical dimension between the **bottom edge** of the MOTHERBOARD and the **bottom edge** of the BOX. Enter **20**. Sketch1 is fully defined and is displayed in black.

46) Close the Sketch. Click **Sketch**.

47) Save the BOX assembly. Click **Save**.

Top Down Assembly Modeling   Engineering Design with SolidWorks

What happens when the size of the motherboard changes?

How do you insure that the BOX maintains the required 20mm spatial gap between the internal components and the BOX boundary?

How do you design for future revisions?

Answer: Through Link Values.

Link Values and Equations

Link Values are used to define equal relations. Create an equal relation between two or more sketched dimensions and or features with a Link Value.

Link Values require a shared name.

Mathematical expressions that define relationships between parameters and or dimensions are called Equations.

Equations use shared names to control dimensions. Use Equations to connect values from sketches, features, patterns and various parts in an assembly.

Use Link Values within the same part. Use Equations in different parts and assemblies.

The project goal is to create three boxes of different sizes. Insure that the models remain valid when dimensions change for various internal components. This is key!

Create a Link Value.
48) Display all dimensions. Right-click the **Annotations** folder. This is the first entry in the FeatureManager.

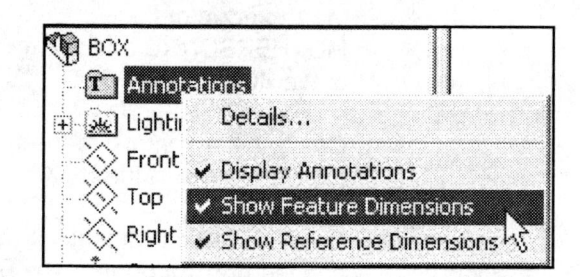

49) Click Show Feature Dimensions.

50) Right click on the lower right vertical dimension **20**.

51) Click **Link Values**.

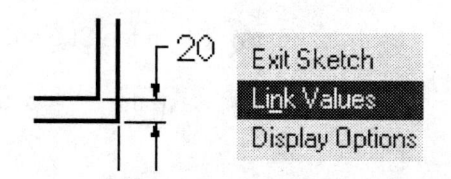

Engineering Design with SolidWorks | **Top Down Assembly Modeling**

52) The Shared Values dialog box is displayed. Enter **gap** in the Name text box. Gap is the Link Value name.

53) Click **OK**.

54) Link the remaining 20mm dimensions. Right click on the upper vertical dimension **20**.

55) Click **Link Values**. The Shared Values dialog box is displayed.

56) Click the **drop down arrow** from the Name text box. Select **gap**.

57) Repeat for the dimensions that are **20**.

58) Test the Link Values. Double-click the lower right **20** dimension. Enter **10**. Click **Checkmark** ✓ in the Modify dialog box. The five Link Values change.

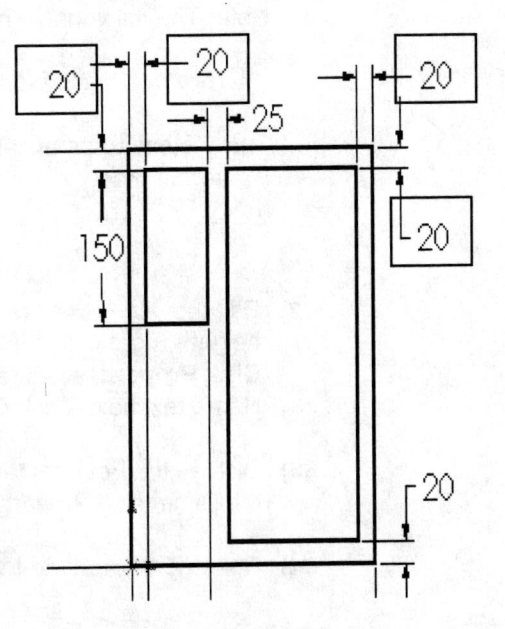

59) Return to the original value. Click the lower right **10** dimension. Enter **20**. Click **Checkmark** ✓. All Link Values are equal to 20.

Each dimension has a unique variable name. The names are used as Equation variables.

The default names are based on the Sketch, Feature or Part. Feature names do not have to be changed. Rename variables for clarity when creating numerous equations.

Top Down Assembly Modeling Engineering Design with SolidWorks

Edit the dimension Name for overall box width.
60) Right-click the horizontal dimension, **300**. Click **Properties**.

61) Enter **box-width** in the Name text box. Click **OK**.

Note: The full variable name is:

"box-width@Sketch1"

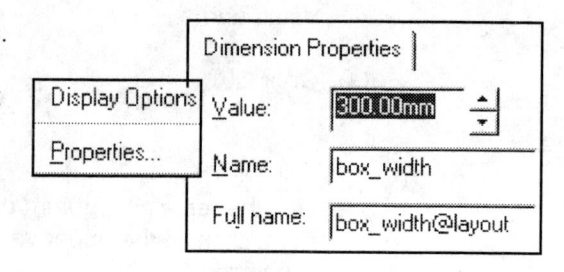

The system automatically appends Sketch1. If features are created or deleted in a different order, your variable names will be different.

62) Edit the dimension Name for the overall box height. Right-click the vertical dimension, **400**. Click **Properties**. Enter **box-height** in the Name text box. Click **OK**.

63) Display the FeatureManager. Click **OK** from the Dimension PropertyManager.

64) Rename **Sketch1** to **Layout**.

65) Display the Isometric view. Click **Isometric**.

66) Hide all dimensions. Right-click the **Annotations** folder. Uncheck **Show Feature Dimensions**.

67) Save the BOX assembly. Click **Save**. Click **Yes** to the Rebuild Now message. The Layout Sketch is complete.

Save both the assembly and the referenced components when you exit.

Open the assembly before opening individual components referenced by the assembly when you start a new session of SolidWorks.

Solid components, sheet metal components or a combination of solid and sheet metal components utilize a Layout Sketch.

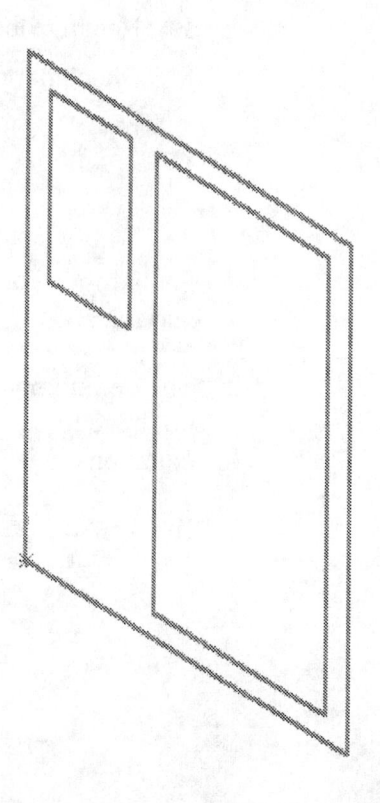

PAGE 6 - 18

Insert Component – MOTHERBOARD

The MOTHERBOARD requires the greatest amount of lead time to design and manufacture. The outline of the MOTHERBOARD is created. Create the MOTHERBOARD component from the Layout sketch.

The MOTHERBOARD represents the special constraints of a blank Printed Circuit Board (PCB).

A rough design of a critical connector is located on the MOTHERBOARD.

CircuitWorks from Zeal Solutions (www.circuitworks.co.uk) is a fully integrated data interface between SolidWorks and PCB Design systems.

As the project manager, your job is to create the board outline with the corresponding dimensions from the Layout sketch.

Export the MOTHERBOARD data in an industry-standard Intermediate Data Format (IDF) from SolidWorks.

The IDF file is sent to the PCB designer to populate the board with the correct 2D electronic components.

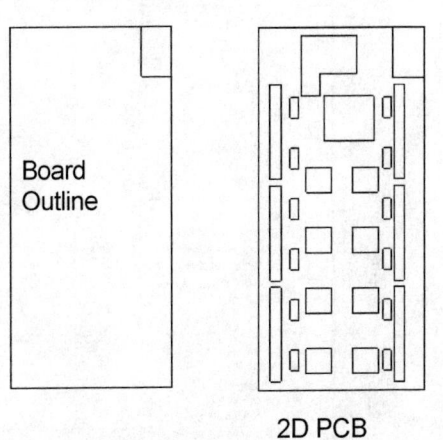

CircuitWorks works with either industry-standard IDF file or PADS file, and produces a 3D SolidWorks Assembly of the MOTHERBOARD. IDF and PADS are common file formats utilized in the PCB industry.

The MOTHERBOARD is fully populated with the components at the correct height. Your colleagues use the MOTHERBOARD assembly to develop other areas of the BOX assembly.

An engineer develops the wire harness from the MOTHERBOARD to electrical components in the BOX.

A second engineer uses the 3D geometry from each electrical component on the MOTHERBOARD to create a heat sink.

As the project manager, you move components that interfere with other mechanical parts, cables and or wire harness.

You distribute the updated information to your colleagues and manufacturing partners.

A new IDF file containing the component positions is sent back to the PCB manufacturer for fabrication.

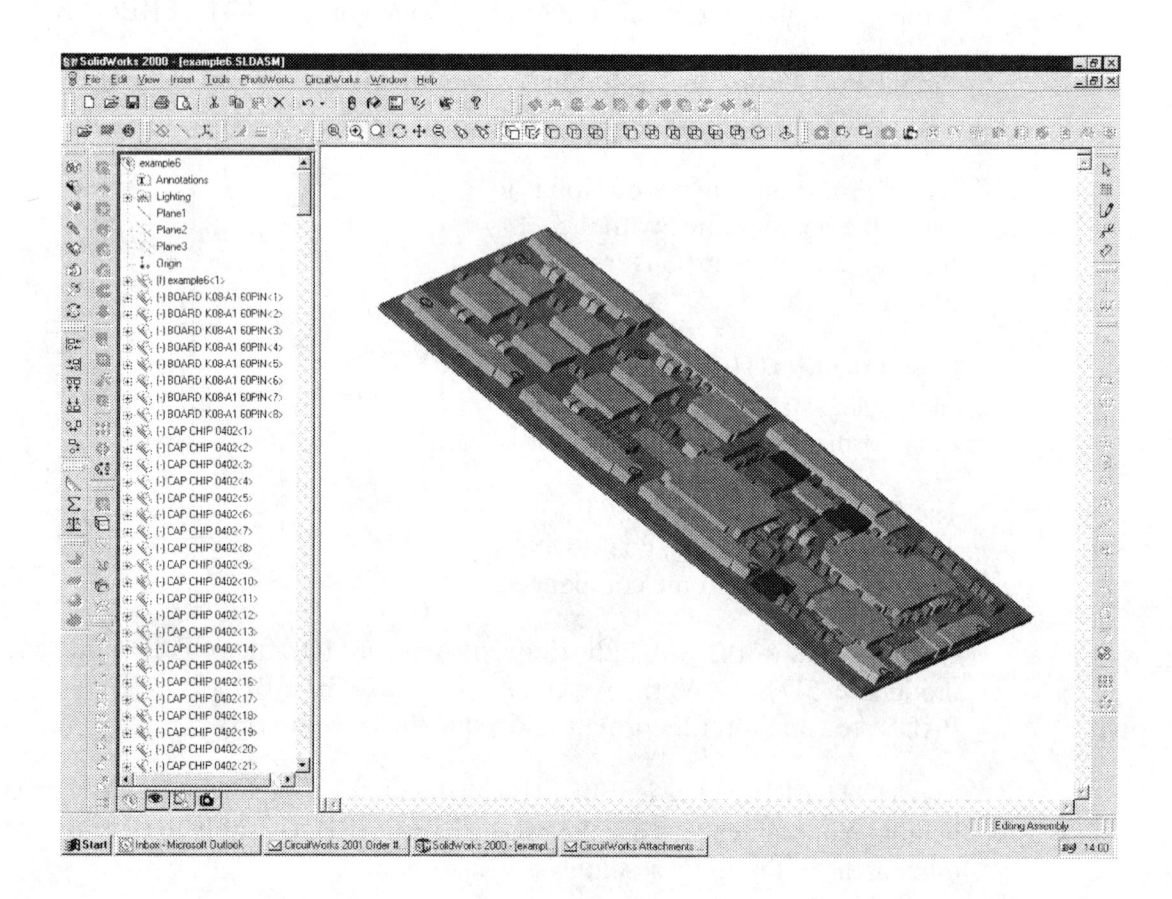

Create the MOTHERBOARD component.
68) Click **Insert** from the Main menu.

69) Click **Component**.

70) Click **New Part**.

71) Double-click the **PART-MM-ANSI** template icon.

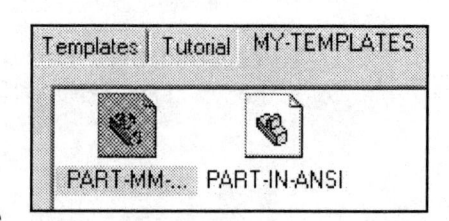

Engineering Design with SolidWorks — Top Down Assembly Modeling

72) The Save As dialog box is displayed. Enter **MOTHERBOARD** in the Filename text box.

73) Enter MOTHERBOARD FOR BOX ASSEMBLY for Description.

74) Click **Save**.

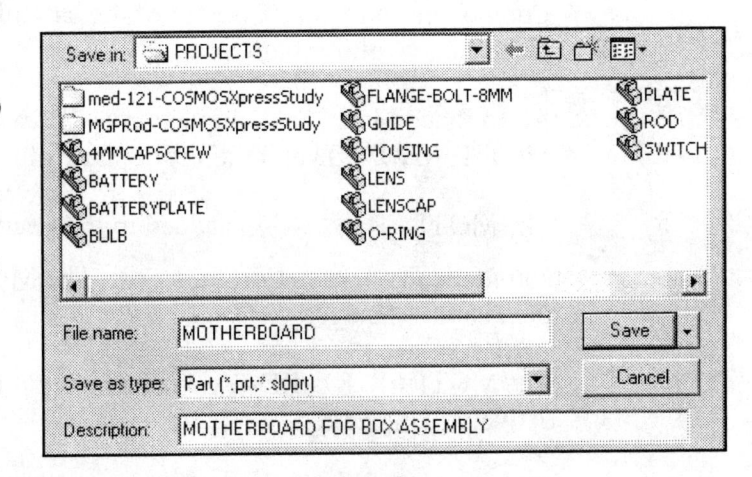

The Component Pointer is displayed on the mouse pointer. The MOTHERBOARD component is empty and requires a sketch plane.

75) Click the **Front** plane of the BOX in the FeatureManager. The Front plane of the MOTHERBOARD component is mated with the Front plane of the BOX. Sketch1 is the active sketch.

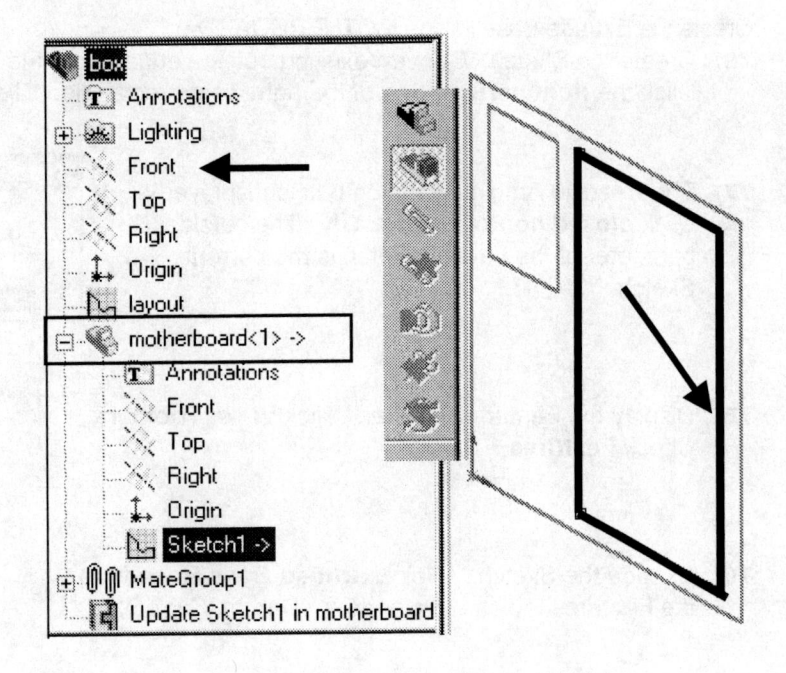

Components added In-Context of the assembly automatically receive an In Place Mate within MateGroup1.

An In Place Mate is a coincident mate between the front plane of the MOTHERBOARD and the Front plane of the BOX.

The MOTHERBOARD is added to the FeatureManager. The system automatically selects Edit Part . The MOTHERBOARD text appears in the FeatureManager.

The MOTHERBOARD text is displayed in blue to indicate that the part is actively being edited.

The current Sketch plane is the Front plane. The current sketch name is Sketch1. The name is indicated on the current graphics window title,

"Sketch1 of MOTHERBOARD - in- BOX."

MOTHERBOARD is the name of the component created in the context of the assembly BOX. The system automatically selects Sketch .

Create the Extrude Base for the MOTHERBOARD.
76) Create the Sketch. Convert existing outside edges from the Layout Sketch. Click the **right vertical line** of the right inside rectangle. Click **Convert Entities** .

77) The Resolve Ambiguity dialog box is displayed. Click **closed contour**. Click **OK**. The outside perimeter of the Layout Sketch is the current Sketch.

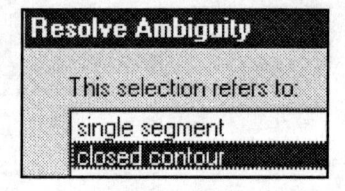

78) Display the Features Toolbar. Click **View**, **Toolbars**. Check **Features**.

79) Extrude the Sketch. Click **Extruded Boss/Base** from the Features toolbar. Enter **10** for Depth. Click **OK**.

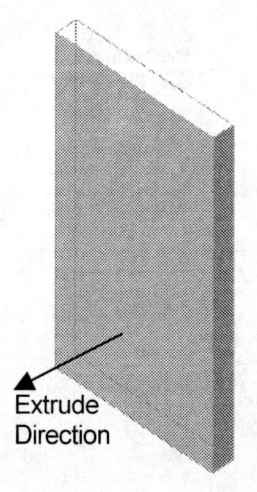

Extrude Direction

Engineering Design with SolidWorks   **Top Down Assembly Modeling**

80) Rename Extrude1 to Base Extrude.

The MOTHERBOARD name is displayed in blue. The part is being edited In Context of the BOX assembly.

81) Display the Extruded-Base feature. **Expand** the MOTHERBOARD part icon in the FeatureManager. The features are displayed in blue.

82) Return to the BOX assembly. Click **Edit Part** from the Assembly toolbar. The MOTHERBOARD is displayed in black.

83) Save the BOX Assembly. Click **Save**.

84) Click **Yes** to save the referenced models.

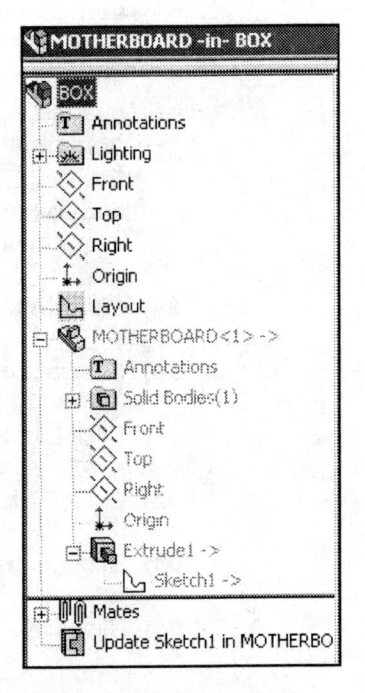

Additional features are required that do not reference the Layout Sketch or other components in the BOX assembly.

An Extruded Boss feature indicates the approximate position of an electrical connector. The actual measurement of the connector or type of connector has not been determined.

Perform the following steps to avoid unwanted assembly references:

- Open the part. Add features. Save the part.

- Open the Assembly. Save the Assembly.

Create the Extruded Boss for the MOTHERBOARD.
85) Right-click **MOTHERBOARD** in the FeatureManager.

86) Click Open MOTHERBOARD.sldprt.

87) Fit the model to the Graphics window. Press the **f** key.

88) Create the Sketch. Click the **front face** of the Extruded-Base. Click **Sketch**. Click the outside **right vertical edge**. Click **Convert Entities**. Drag the **bottom endpoint** of the convert line three quarters upward. Click the **top horizontal edge**. Click **Convert Entities**. Drag the **left endpoint** of the converted line three quarters of the way to the right.

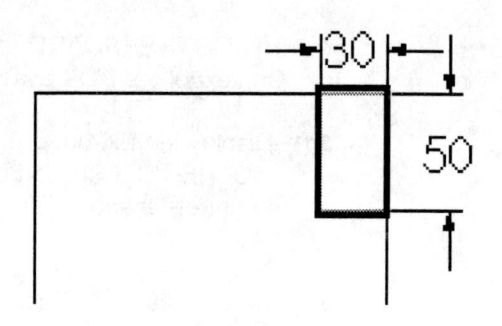

89) Click **Line**. Sketch a **vertical line**.

90) Sketch a **horizontal line** to complete the rectangle.

91) Click **Dimension**. Enter **30** for the horizontal dimension.

92) Enter **50** for the vertical dimension.

93) Extrude the Sketch. Click **Extruded Boss/Base**. Enter **10** for Depth.

94) Click **OK**.

95) Display an Isometric view. Click **Isometric**.

Use color to indicate electrical connectors. Color the face of the Extruded Boss feature.

96) Color the front face of the connector. Click the **front face**.

97) Click **Color**.

98) Select **yellow**.

99) View the color. Click **Apply**.

100) Click **OK**.

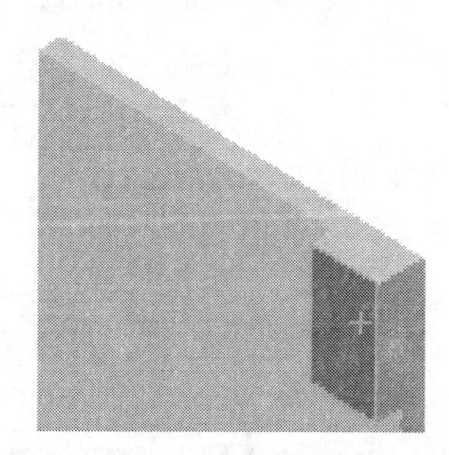

Engineering Design with SolidWorks   **Top Down Assembly Modeling**

101) Rename Extrude2 to Connector1.

102) Save the MOTHERBOARD. Click **Save**.

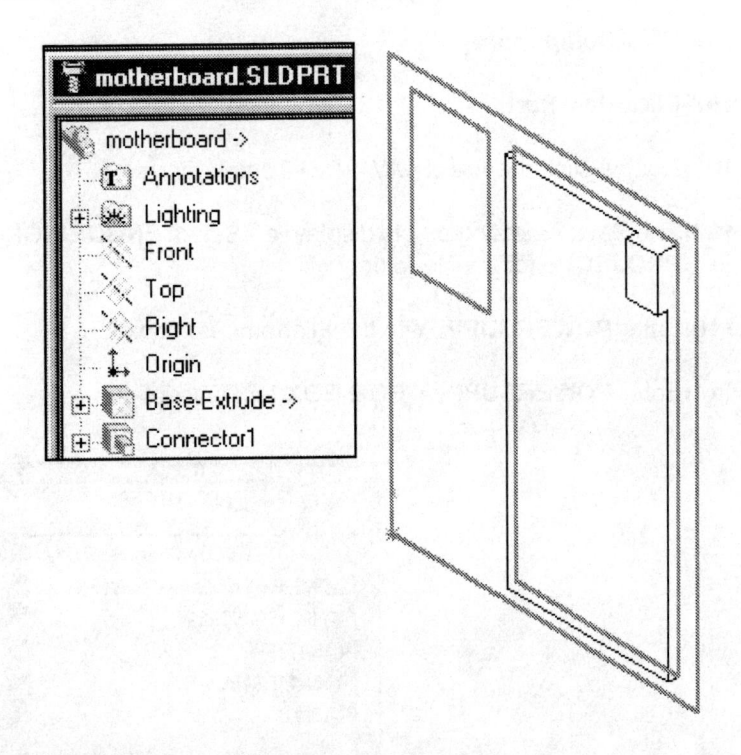

103) Return to the BOX assembly. Click **Window**, **BOX**.

104) Click **Yes** to rebuild model. The MOTHERBOARD contains the new Extrude Boss feature.

INSERT COMPONENT – POWERSUPPLY

Create the POWERSUPPLY Component.
105) Click **Insert** from the Main menu.

106) Click **Component**.

107) Click **New Part**.

108) Double-click the PART-MM-ANSI Template.

109) The Save As dialog box is displayed. Select **ENGDESIGN-W-SOLIDWORKS\PROJECTS** for the file folder.

110) Enter **POWERSUPPLY** in the Filename text box.

111) Enter **POWERSUPPLY FOR BOX** for Description.

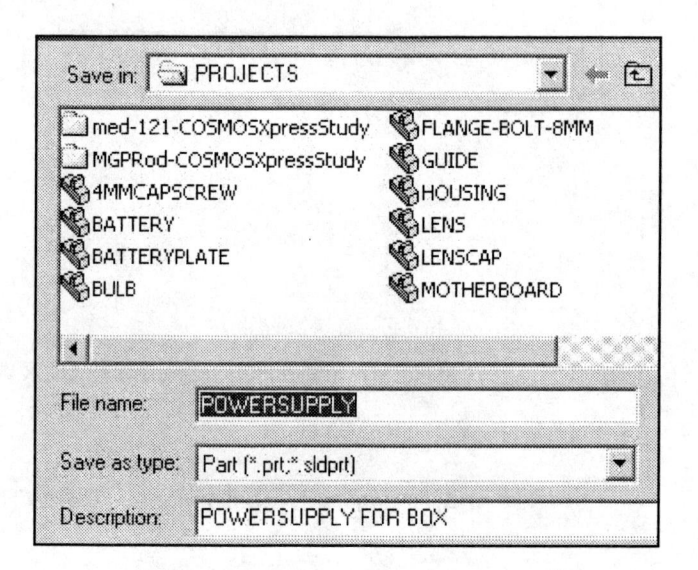

112) Click **Save**.

The Component Pointer is displayed on the mouse pointer. The POWERSUPPLY component is empty and requires a sketch plane.

113) Click the **Front** plane of the BOX in the FeatureManager for the In Place Mate plane. The Front plane of the component is mated with the Front plane of the BOX. Sketch1 is the active sketch.

The system automatically selects Edit Part.

The POWERSUPPLY text appears in the FeatureManager.

The POWERSUPPLY text is displayed in blue. Blue indicates that the part is being edited.

Engineering Design with SolidWorks   **Top Down Assembly Modeling**

The current Sketch plane is the Front plane.

The current sketch name is Sketch1. The name is indicated on the current graphics window title,

"Sketch1 of POWERSUPPLY - in- BOX."

POWERSUPPLY is the name of the component created In Context of the assembly BOX.

Create the Extrude Base for the POWERSUPPLY.
114) Create the Sketch. Convert existing outside edges from the Layout Sketch. Click the **right vertical line** off the PowerSupply sketch.

115) Click Convert Entities.

116) The Resolve Ambiguity dialog box is displayed. Click **closed contour**.

117) Click **OK**. The outside perimeter of the small left box from the Layout is the current Sketch.

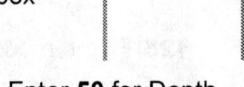

118) Extrude the Sketch. Click **Extruded Boss/Base**. Enter **50** for Depth.

119) Click **OK**.

POWERSUPPLY is displayed in blue. The part is being edited In Context of the BOX assembly.

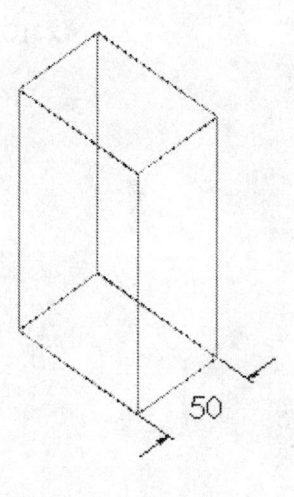

120) Return to the BOX assembly. Click **Edit Part** from the Assembly toolbar. The POWERSUPPLY is displayed in black.

PAGE 6 - 27

121) Save the BOX Assembly. Click **Save**.

122) Click **Yes** to save the referenced models.

The Extruded Boss feature represents the location of the cable that connects to the POWERSUPPLY.

Think about where the cables and wire harness connects to key components. You do not have all of the required details for the cables?

In a concurrent engineering environment, create a simplified version early in the design process. No other information is required from the BOX assembly to create additional features for the POWERSUPPLY.

Open the POWERSUPPLY in Part mode.

Create the Extrude Boss for the POWERSUPPLY.
123) Right-click **POWERSUPPLY** in the FeatureManager.

124) Click Open POWERSUPPLY.sldprt.

125) Fit the model to the screen. Press the **f** key.

126) Display the Top view. Click **Top**.

127) Create the Sketch. Click the **top face** of the Extruded Base feature. Sketch a diagonal **Center line**. Align the **endpoints** with the corners of the POWERSUPPLY.

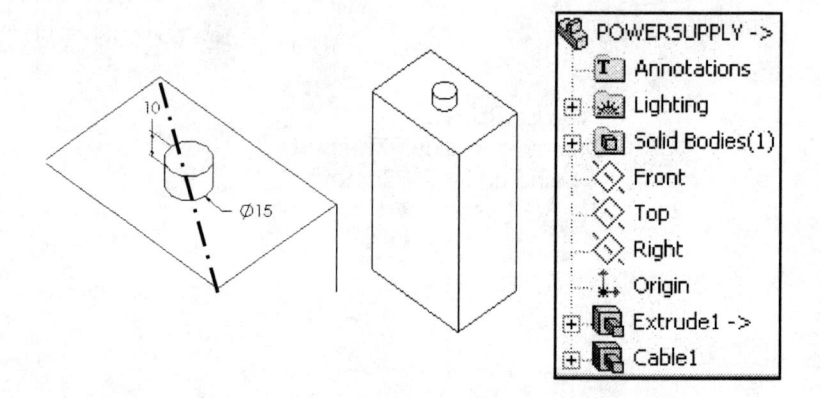

128) Sketch a **Circle**. Click **Select**. Click **Add Relation**. Add a Midpoint relation. Click the **center point of the circle** and the **centerline**. Click **Midpoint**.

129) Enter **15** for diameter.

130) Click **OK**.

Engineering Design with SolidWorks — Top Down Assembly Modeling

131) Extrude the Sketch. Click **Extruded Boss/Base**. Enter **10** for Depth. Click **OK**.

132) Rename Extrude 2 to Cable1.

133) Save the POWERSUPPLY. Click **Save**.

134) Return to the BOX assembly. Click **Window**, **BOX**. Click **Yes** to rebuild the assembly. The POWERSUPPLY contains the new Extrude Boss feature, Cable1.

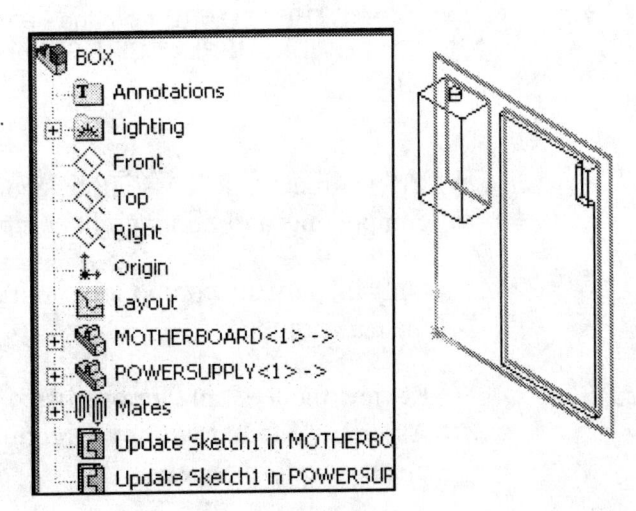

Test the Layout dimensions for the BOX.

135) Display all dimensions. Right-click the **Annotations folder**.

136) Click Show Feature Dimensions.

137) Double-click on the overall width dimension, **300**.

138) Enter **550** for the horizontal dimension.

139) Double-click on the overall vertical dimension, **400**.

140) Enter **600** for the vertical dimension.

141) Click **Rebuild**.

142) Return to **300** x **400** for the width and height.

143) Click **Rebuild**.

144) Hide all dimensions. Right-click the **Annotations folder**.

145) Uncheck Show Feature Dimensions.

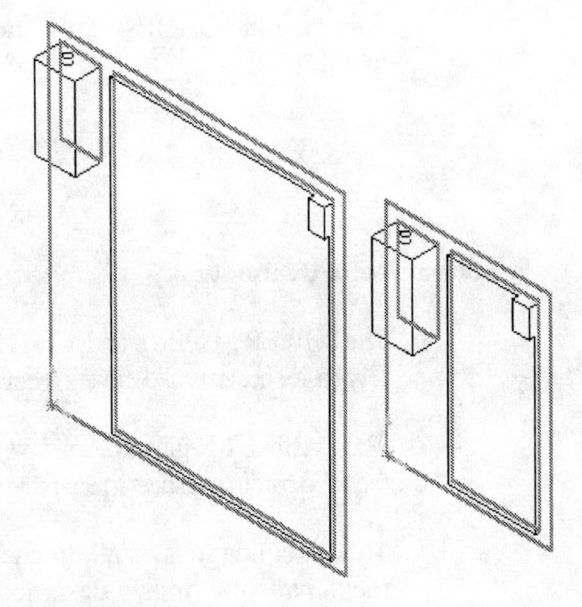

146) Suppress the MOTHERBOARD and POWERSUPPLY. Right-click **MOTHERBOARD** from the FeatureManager. Click **Suppress**. Right-click **POWERSUPPLY**. Click **Suppress**.

Electro-mechanical assemblies contain hundreds of parts. Suppress components and lightweight components save rebuild time.

Only display the components and geometry required to create a new part or to mate a component.

Review the System Options listed in Large Assembly Mode to improve rebuild time and display performance.

Electro-Mechanical assemblies contain parts fabricated from sheet metal.

Fabricate the CABINET for the BOX from sheet metal.

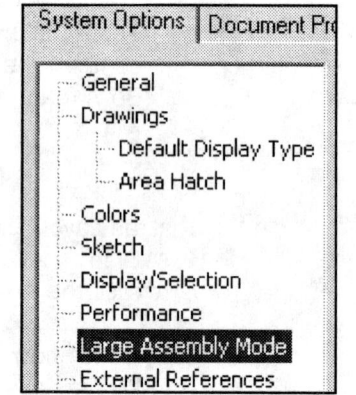

Sheet Metal Overview

The POWER SUPPLY and MOTHERBOARD are solid parts. You need to understand a few basic sheet metal definitions before starting the next part, CABINET.

The CABINET begins as a solid part. The Rip feature and Insert Bends feature transforms the solid part into a sheet metal part.

Talk to colleagues. Talk to sheet metal manufacturers. Review other sheet metal parts previously designed by your company.

Material Thickness

Sheet metal parts are fabricated from a flat piece of raw material. The material thickness does not change.

The material is cut, formed and folded to produce the final part.

Design State

There are two design states for sheet metal parts:

- Formed.

- Flat.

Work in the formed 3D state and then flatten the part to display the manufactured flat state.

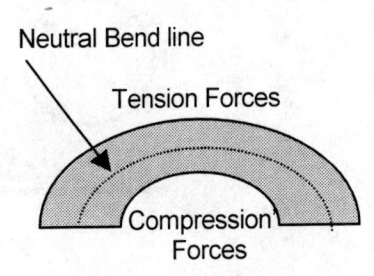

Figure 6.4

Neutral Bend Line

Example: Use a flexible eraser, 50mm or longer. Bend the eraser in a U shape. The eraser displays tension and compression forces.

The area where there is no compression or tension is called the neutral axis or neutral bend line, Figure 6.4.

Developed Length

Assume the material has no thickness. The length of the material formed into a 360° circle is the same as its circumference.

The length of a 90° bend would be ¼ of the circumference of a circle, Figure 6.5.

Circumference of a circle = $2\pi R$

$L = 2\pi R$

Length of a ¼ circle = $2\pi R*(90/360) = \pi R/2$

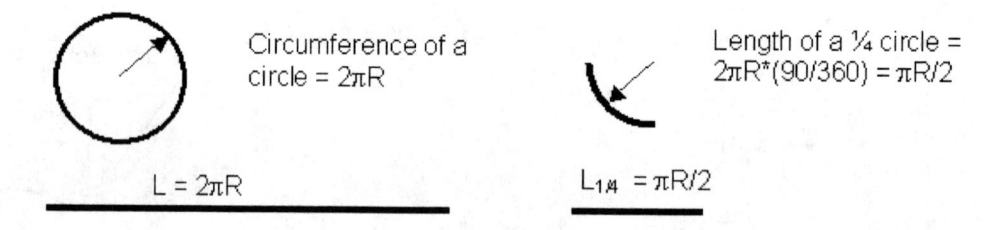

$L_{1/4} = \pi R/2$

Figure 6.5

In the real world, materials do have thickness. Materials develop different lengths when formed in a bend depending on their thickness.

There are three major properties which determines the length of a bend:

- Bend radius.
- Material thickness.
- Bend angle.

The distance from the inside radius of the bend to the neutral bend line is labeled δ, Figure 6.6.

90° Bent Material

Flat Developed Length

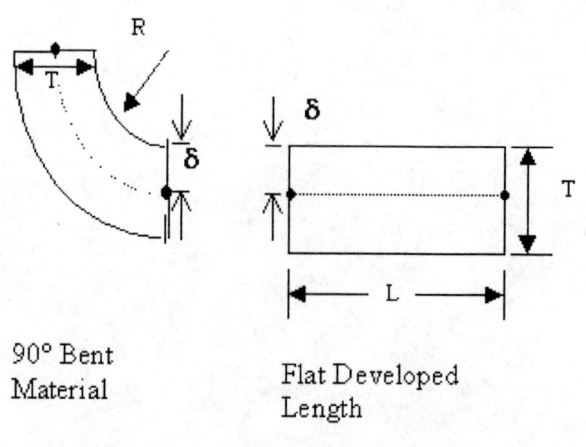

Figure 6.6

The symbol 'δ' is the Greek letter delta. The amount of flat material required to create a bend is greater than the inside radius and depends upon the neutral bend line.

The true developed flat length L, is measured from the endpoints of the neutral bend line.

Example:

T is the Material thickness.

Create a 90° bend with an inside radius of R:

$L = \frac{1}{2} \pi R + \delta T$

The ratio between δ and T is called the K-factor. Let K = 0.41 for Aluminum:

$K = \delta/T$

$\delta = KT = 0.41T$

$L = \frac{1}{2} \pi R + 0.41T$

U.S. Sheet metal shops use their own numbers.

Example:

One shop may use K = 0.41 for Aluminum, versus another shop uses K = 0.45.

Use tables, manufactures specifications or experience.

Relief

Sheet metal corners are subject to stress.

Excess stress will tear material. Remove material to relieve stress.

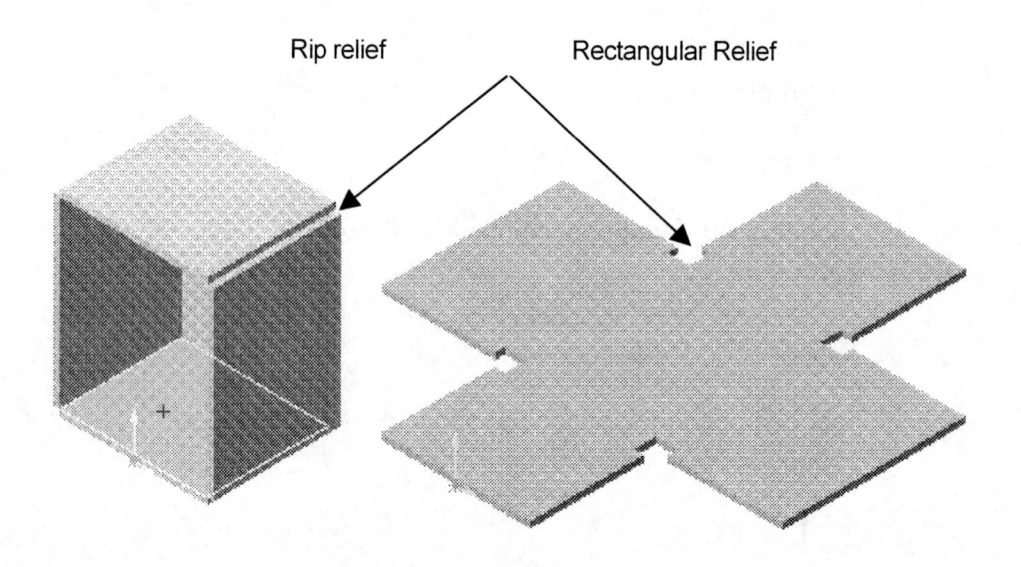

Rip relief   Rectangular Relief

Insert Component – CABINET

Create the CABINET component inside the assembly and attach it to the Front plane. The CABINET component references the Layout sketch.

Create the CABINET component as a sheet metal Base Flange feature in the context of the BOX assembly.

Add additional sheet metal Edge Flange features.

Add dies, louvers and cuts to complete the CABINET.

Insert the Base Component.
147) Create the CABINET component. Click **Insert** from the Main menu.

148) Click **Component**.

149) Click **New Part**.

150) Double-click **PART-MM-ANSI**.

151) The Save As dialog box is displayed. Enter **CABINET** in the Filename text box.

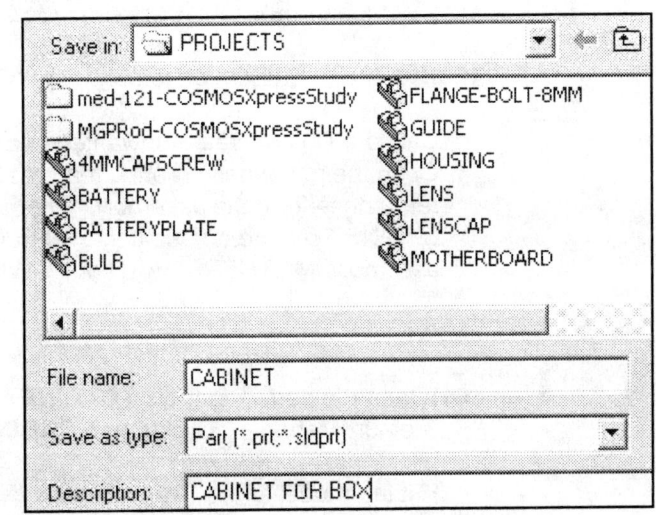

152) Enter **CABINET FOR BOX** for Description.

153) Click **Save**. The Component Pointer is displayed on the mouse pointer. The CABINET component is empty and requires a sketch plane.

154) Click the **Front** plane of the BOX in the FeatureManager for the In Place Mate plane. The Front plane of the component is mated with the Front plane of the BOX. CABINET is the name of the component created in the context of the BOX assembly. The system automatically selects Sketch.

Create the Base feature for the CABINET.
155) Create the Sketch. Convert existing outside edges from the Layout Sketch. Click the **right vertical edge** of the BOX.

156) Click Convert Entities.

157) The Resolve Ambiguity dialog box is displayed. Click **closed contour**.

158) Click **OK**. The outside perimeter of the Layout Sketch is the current Sketch.

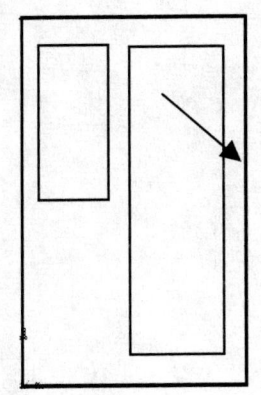

Top Down Assembly Modeling   **Engineering Design with SolidWorks**

159) Extrude the Sketch. Click **Extruded Boss/Base**. Enter **100**. Click **OK**.

160) Return to the BOX assembly. Click **Edit Part**.

> Caution: Do not create unwanted geometry references. Open the part when creating features that require no references from the assembly. The features of the CABINET require no additional references from the BOX assembly, MOTHERBOARD or POWERSUPPLY.

161) Open the part. Right-click **CABINET** from the FeatureManager. Click **Open CABINET.sldprt**.

162) Fit the model to the Graphics window. Press the **f** key.

163) Click the **Isometric** view.

164) Create the Shell. Click the **front face** of the Extruded Base feature. Click **Shell**.

165) Check the **Shell OutWard** box.

166) Enter **1** mm for Thickness.

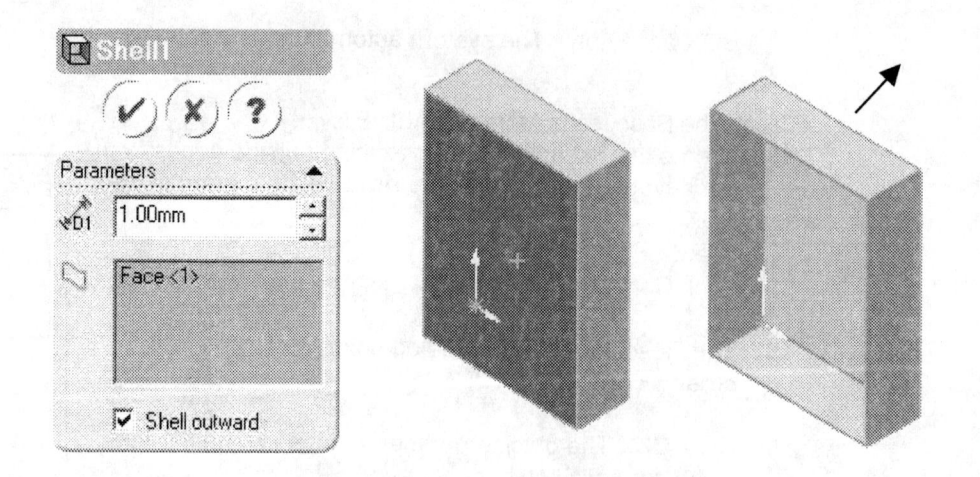

167) Click **OK**.

168) Display the Sheet metal toolbar. Click **View**, **Toolbars**, **Sheet metal**.

Create the Rip Feature and Insert Sheet Metal Bends

The Rip feature creates a cut along the edges of the Extruded Base feature.

Rip the Extruded Base feature along the four edges. The Bend feature creates sheet metal bends.

Specify bend parameters: bend radius, bend allowance and relief.

Select the bottom face to remain fixed during bending and unbending.

Create the Rip.

169) Click **Rip** from the Sheet Metal toolbar.

170) Click the inside lower left edge.

171) **Rotate** the part to view the inside edges.

172) Click the inside upper left edge.

173) Click the inside upper right edge.

174) Click the inside lower right edge.

175) Click **OK**.

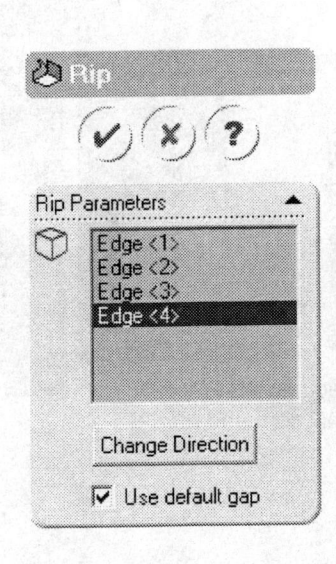

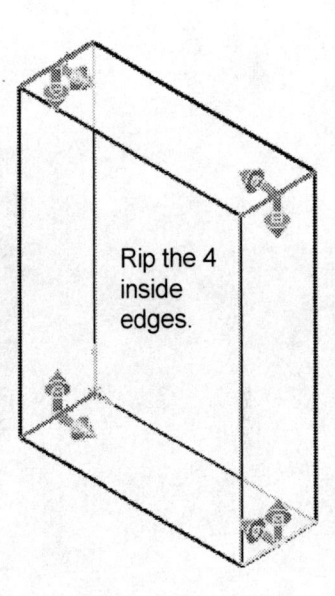

Rip the 4 inside edges.

Create the Sheet Metal Bends.
176) Click the inside **bottom face** to remain fixed.

177) Click **Insert Bends** from the Sheet metal toolbar. Accept the default valued for Bend Parameters, Bend Allowance and Auto Relief.

178) Click **OK**.

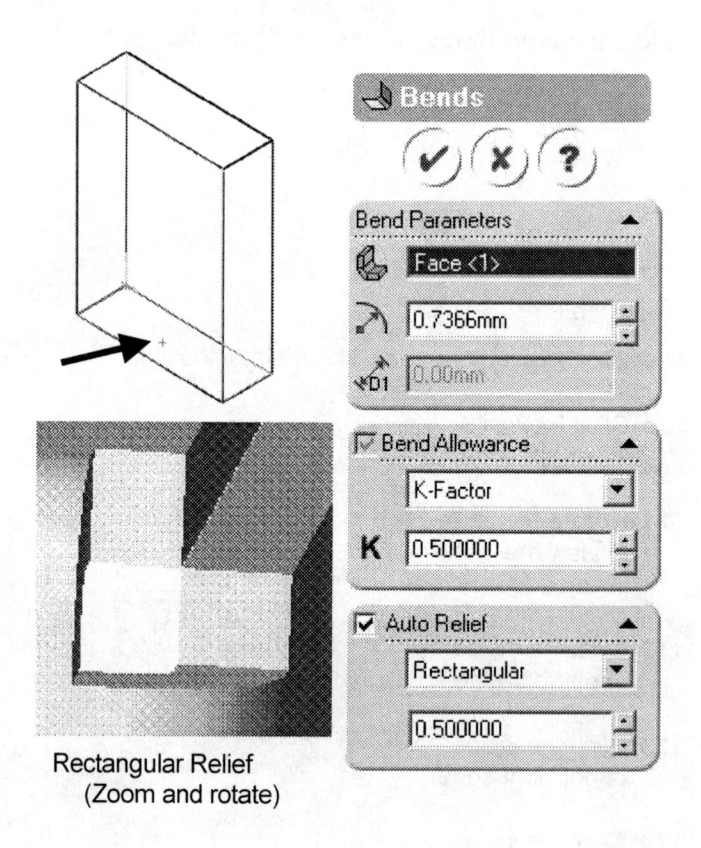

Rectangular Relief
(Zoom and rotate)

179) Display the bends. Click **OK** to the system message, "Auto Relief cuts were made to one or more bends.

180) **Zoom in** on the Rectangular relief in the upper back corner.

181) Fit the model to the screen. Press the **f** key.

Display the Flat State.

182) Display the part in its flat manufactured state. Click **Flattened**. The Rollback bar moves upward between Flatten-Bends1 and Process-Bends1.

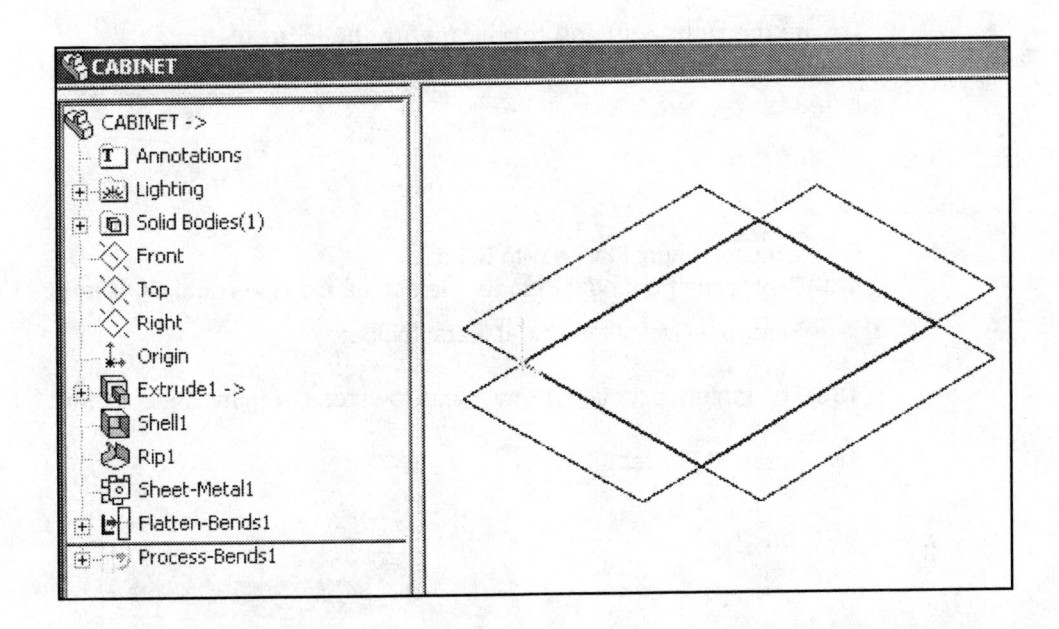

183) Display the part in its fully formed state. Click **Flatten**.

184) Save the CABINET. Click **Save**.

Create the Flange Walls

Create the right flange wall and left flange wall with the Edge Flange feature.

Select the inside edges when creating bends.

Create the right hem and left hem with the Hem feature.

Create the front right Flange with Bend.
185) Create the front right Flange. Select the **front vertical right edge**. Click **Edge Flange** from the Sheet Metal toolbar.

186) The feature direction arrow points towards the right. Select **Blind** for Type.

187) Enter **30** for Depth.

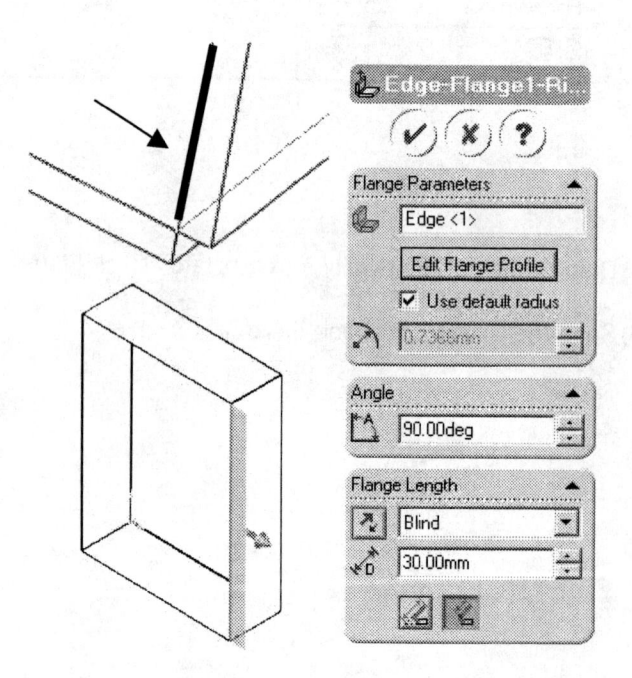

188) Accept all other defaults. Click **OK**.

Create the Hem

189) Select the **right edge** of the **right flange**.

190) Click **Hem** from the Sheet metal toolbar.

191) Click the **Reverse** direction button.

192) Enter **10** for Hem length.

193) Click **OK**.

194) Create the **left Edge Flange** wall. Select the **front vertical left edge**. Click **Edge Flange**. The feature direction arrow points towards the left. Select **Blind** for Type.

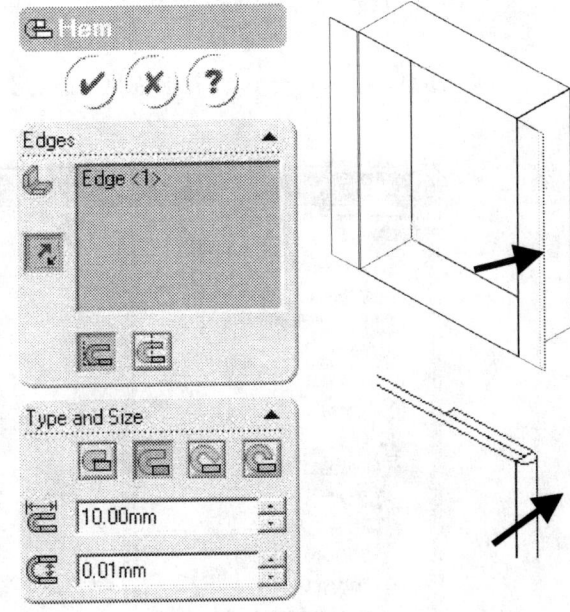

195) Enter **30** for Depth.

196) Accept all other defaults. Click **OK**.

197) Select the **left edge** of the left flange.

198) Click **Hem** from the Sheet metal toolbar.

199) Click the **Reverse** direction button.

200) Enter **10** for Hem length.

201) Click **OK**.

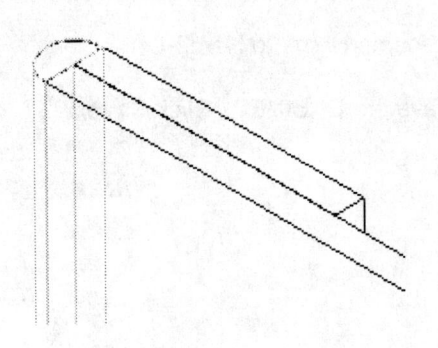

202) Display the part in its flat manufactured state. Click **Flat State**. Click the **Top** view.

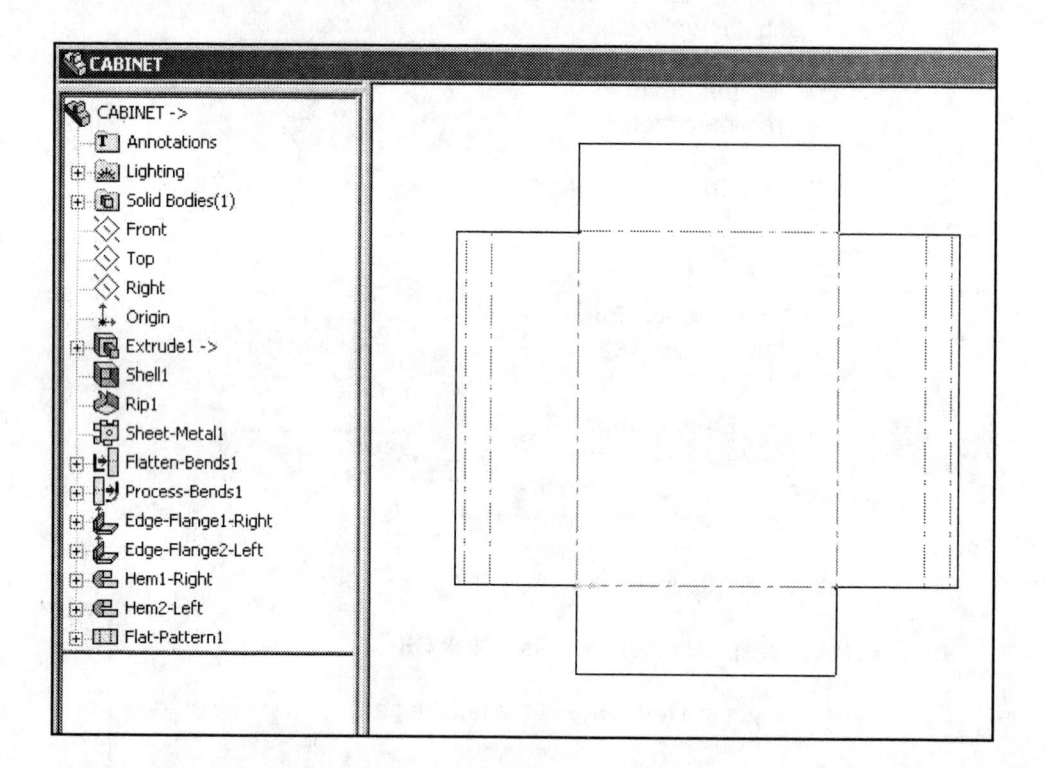

203) Display the part in its fully formed state. Click **Flat State**.

204) Click the **Isometric** view.

205) Rename Edge-Flange1 to Edge-Flange-Right.

206) Rename Edge-Flange2 to Edge-Flange-Left.

207) Rename **Hem1** to **Hem1-Right**.

208) Rename Hem2 to Hem2-Left.

209) Save the CABINET. Click **Save**.

Verify the sheet metal component.

210) Return to the BOX assembly. Click **Window**, **BOX**.

211) Click **Yes** to update the assembly.

212) Double-click **Layout** in the FeatureManager.

213) Double-click the **300** horizontal dimension. Enter **500**. Click **Rebuild**.

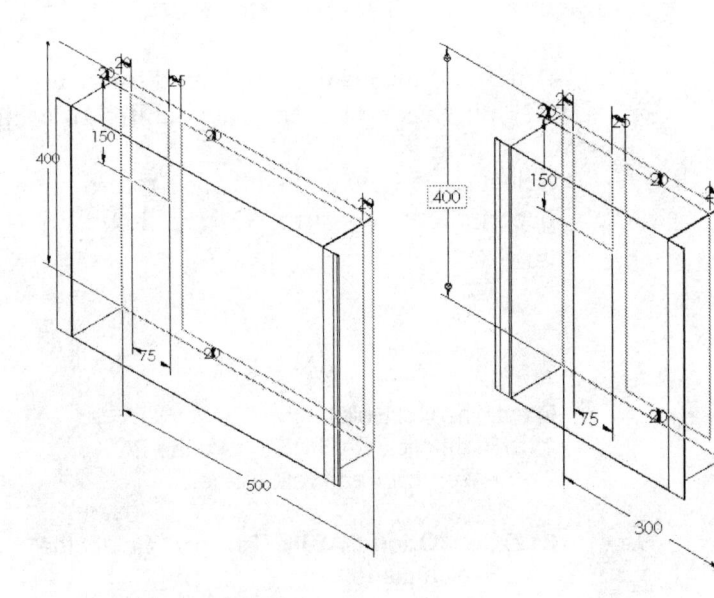

214) Click **OK**.

215) Return to the original Layout dimensions, 300. Click **Undo**.

Create a Hole and a Pattern of Holes

Sheet metal holes are created through a punch or drill process. Each process has advantages and disadvantages:

- Cost.

- Time.

- Accuracy.

Investigate a Linear Pattern of holes. Select a self-clinching threaded fastener that is inserted into the sheet metal during the manufacturing process.

The fastener requires a thru hole in the sheet metal before insertion.

Holes should be of equal size and utilize common fasteners. Why? You need to insure a cost effect design that is price competitive.

Your company must be profitable with their designs to insure financial stability and future growth.

Another important reason for fastener commonality and simplicity is the customer. The customer or service engineer does not want to supply a variety of tools for different fasteners.

A designer needs to be prepared for changes. You proposed two fasteners. Ask Purchasing to verify availability of each fastener.

Select a M5 hole and wait for a return phone call from the Purchasing department to confirm. Design flexibility is key!

Create the first Hole.
216) Right click **CABINET** from the BOX Assembly FeatureManager.

217) Click **Open CABINETsldprt**. Select the Sketch plane.

218) Click the **inside bottom face**. Create the Hole with the Hole Wizard.

219) Click **HoleWizard** from the Features toolbar.

220) Click **Hole tab**.

221) Select **ANSI Metric** for the Standard Property.

222) Select **M5.0** from the Size text box.

223) Select **Thru All** for Hole Depth. Hole1 is positioned on the inside bottom face based upon the selection point.

224) Click **Next**. DO NOT CLICK THE FINISH BUTTON FROM THE HOLE PLACEMENT DIALOG BOX AT THIS TIME.

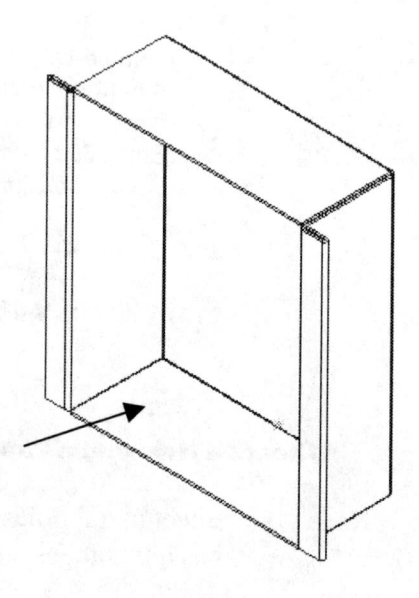

225) Dimension Hole1. Click **Dimension** . Click **Top** . Click the **Origin** of the Extruded-Base feature.

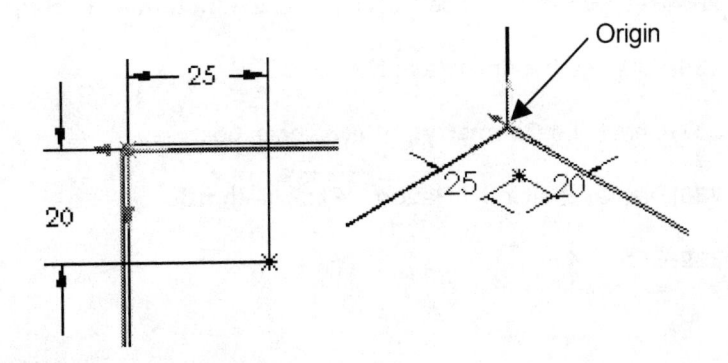

226) Click the Hole **center point**. Enter **25** for the horizontal dimension.

227) Click the **Origin**.

228) Click the Hole **center point**. Enter **20** for the vertical dimension.

229) Click **Finish** from the Hole Placement dialog box.

Create a Linear Pattern of Holes.
230) Create a Linear Pattern of Hole1. Click **Linear Pattern** . Create the pattern in the first direction.

231) Click the **back inside edge** for the Direction 1 selected text box. The first direction arrow points to the right.

232) Enter **125** for Spacing.

233) Enter **3** for Number of Instances.

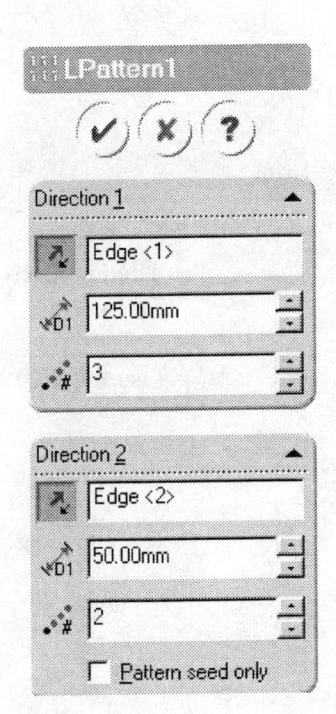

234) Create the Linear Pattern in Direction 2. Click the **left edge** for the Direction 2 selected text box. Enter **50** for Spacing. Enter **2** for Number of Instances.

235) Remove an Instance. Click inside the **Instances to Skip** text box.

236) Click the front middle hole.

237) Check the **Geometry pattern** check box.

238) Display the Linear Pattern. Click **Isometric**.

239) Click **OK**.

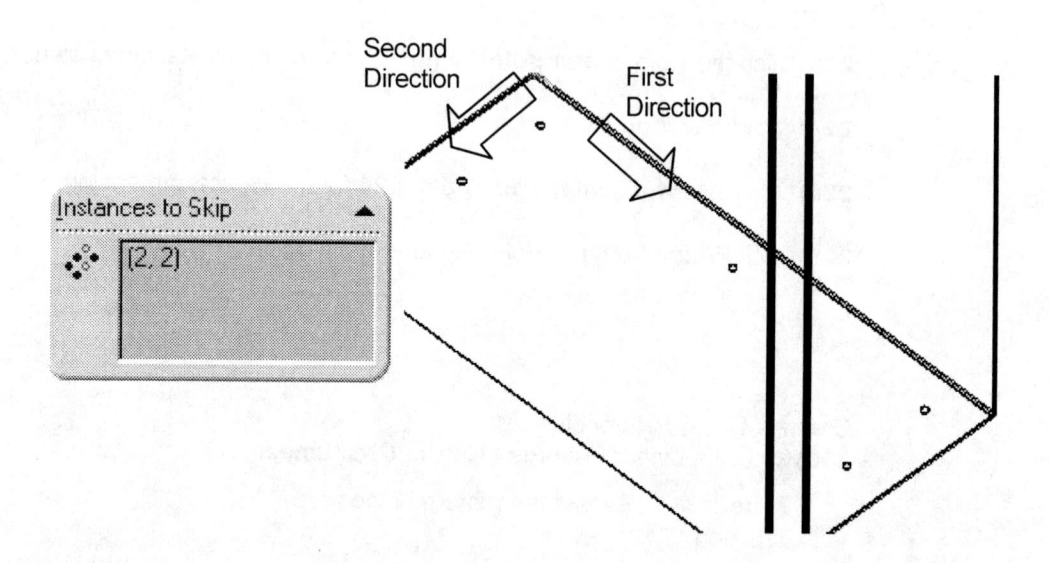

The Geometry pattern option copies faces and edges of the seed feature. Type options such as Up to Surface are not copied.

Use the Geometry pattern to improve system performance.

Engineering Design with SolidWorks — Top Down Assembly Modeling

240) Display the Flat State. Click **Flatten**.

241) Display the Formed State. Click **Flatten**.

242) Save the CABINET. Click **Save**.

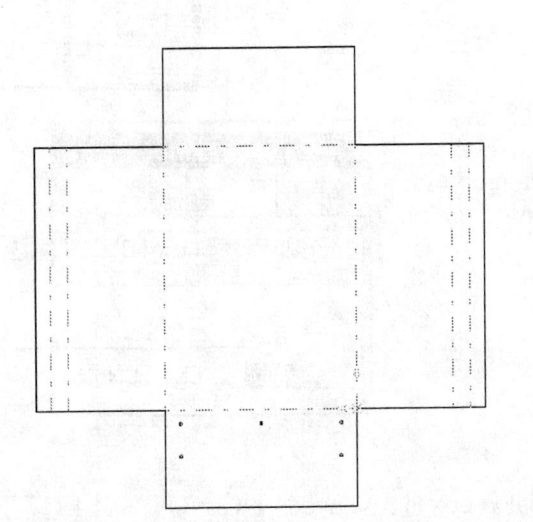

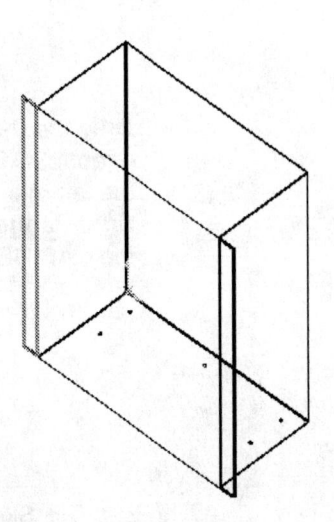

Add a Die Cutout Palette Feature

The Palette Feature directory contains examples of predefined sheet metal shapes.

Create a die cut on the right wall of the BASE-CABINET. This is for a data cable. The team will discuss sealing issues at a later date.

Create a Palette Feature.
243) Click **Tools** from the Main menu. Click **Feature Palette**.

244) Double-click the **PaletteFeatures** file folder.

Top Down Assembly Modeling | Engineering Design with SolidWorks

245) Double-click the **Sheetmetal** file folder. The Feature Palette Sheetmetal directory is displayed.

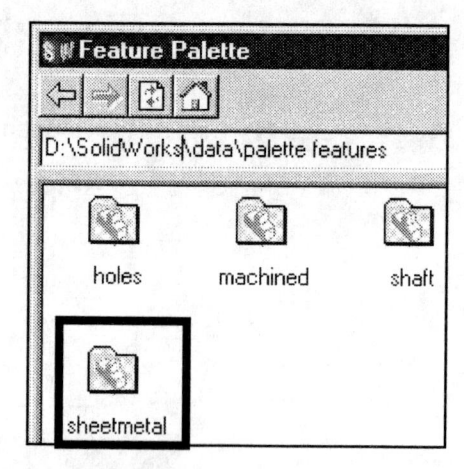

246) Drag the **d-cutout** to the Graphics window. Release the left mouse button on the **right flange outside face** of the CABINET.

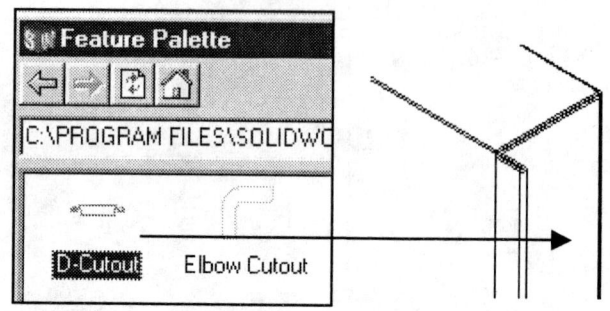

The Edit This Sketch dialog box is displayed. DO NOT SELECT THE FINISH BUTTON AT THIS TIME.

> Position the feature by using the Sketch Modify tool. Then attach the dangling dimensions indicated by red handles.

247) Rotate the d-cutout. Click **Tools** from the Main menu. Click **Sketch Tools, Modify**. The Modify Sketch dialog box is displayed.

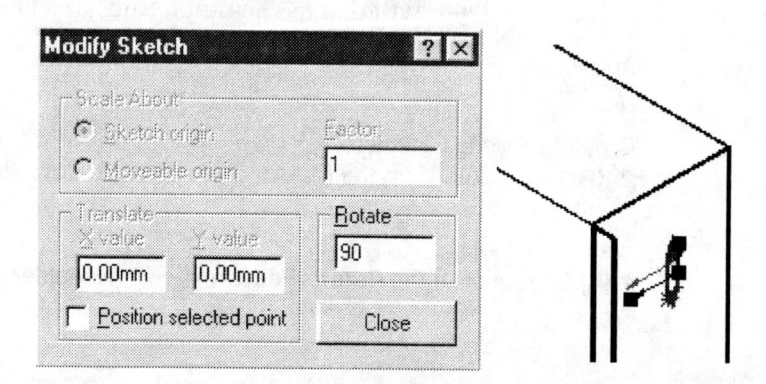

PAGE 6 - 48

Engineering Design with SolidWorks · Top Down Assembly Modeling

248) The mouse pointer displays the modify move/rotate icon. Enter **90** in the Rotate text box.

249) Press the **Enter** key.

250) Click **Close**.

251) Dimension the d-cutout. Display the Right view. Click **Right**.

252) Click **Dimension**. Create a vertical dimension. Click the **mid point** of the d-cutout. Click the CABINET **Origin** in the lower right corner. Enter **300**.

253) Create a horizontal dimension. Click the **mid point** of the d-cutout. Click the CABINET **Origin**. Enter **50**. Click **Next** from the Edit This Sketch dialog box.

Note: With thin sheet metal parts, select the dimension references with care. Use the Zoom and Rotate commands to view the correct edge.

Use reference planes or the Origin to create dimensions. The Origin and planes do not change during the flat and formed states.

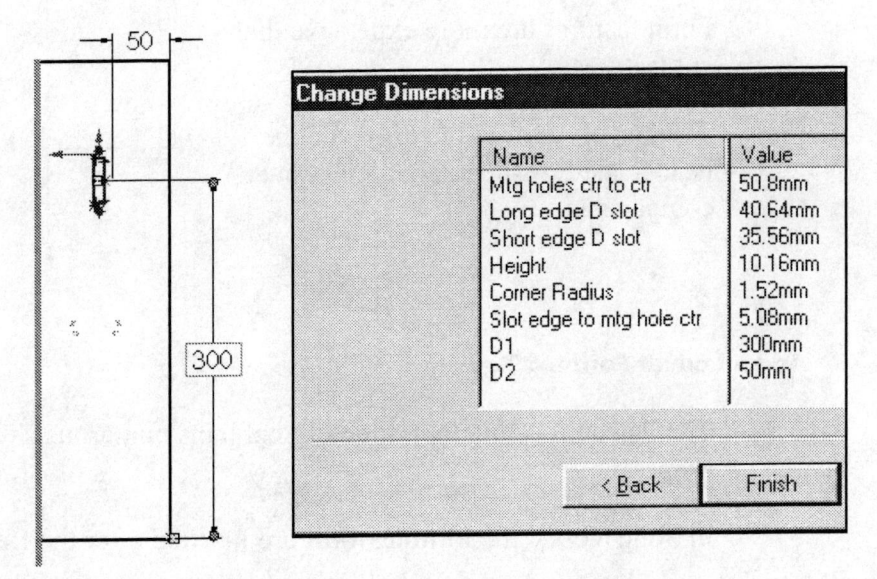

Name	Value
Mtg holes ctr to ctr	50.8mm
Long edge D slot	40.64mm
Short edge D slot	35.56mm
Height	10.16mm
Corner Radius	1.52mm
Slot edge to mtg hole ctr	5.08mm
D1	300mm
D2	50mm

254) Accept the default dimension for the d-cutout. Click **Apply**.

255) Display the d-cutout. Click **Finish**.

256) Click **Isometric**. The d-cutout goes through the right and left side. Through All is the current Type option.

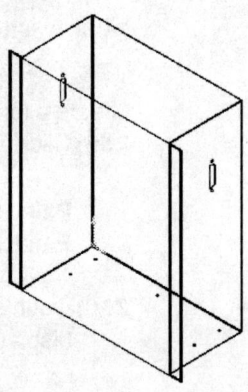

PAGE 6 - 49

Top Down Assembly Modeling   Engineering Design with SolidWorks

257) Change the D-Cutout1 to enter through the right flange. Right-click on the **D-Cutout1** in the FeatureManager. Click **Edit Definition**. Click **Up To Next** from the End Condition drop down list.

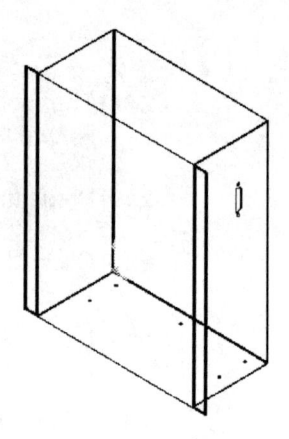

258) Click **OK**. The d-cutout feature is positioned before the Flat Pattern1 icon in the FeatureManager.

The d-cutout is incorporated into the Flat Pattern. It is cost effective to perform cuts and holes in the flat state.

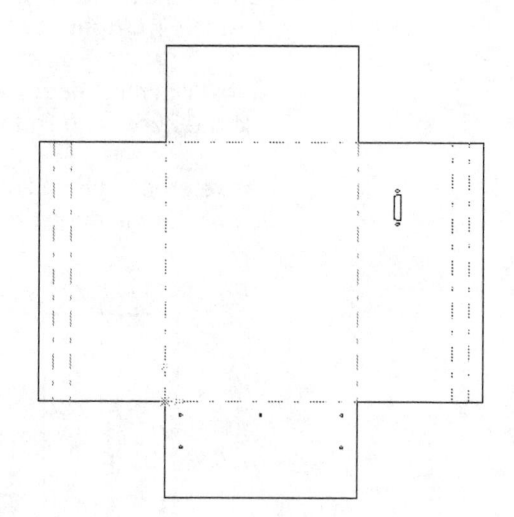

Form features, such as louvers are added in the formed state.

Form features are more expensive than cut features.

The louvers are used to dissipate the heat created by the internal electronic components.

Add a Louver Forming Tool

The Palette forming tool directory contains numerous sheet metal forming shapes.

In SolidWorks, the forming tools are inserted after the Bends are processed.

Suppress forming tools in the Flat Pattern.

Create a formed feature.
259) Click **Tools** from the Main menu. Click **Feature Palette**. Double-click the **PaletteFormingTools** folder.

260) Double-click the **Louvers** folder. The louver is displayed.

PAGE 6 - 50

Engineering Design with SolidWorks **Top Down Assembly Modeling**

261) Press the right **Arrow key** 5 times to display the inside right flange.

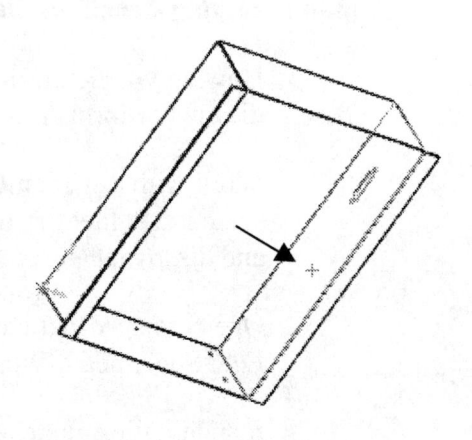

262) Drag the **louver** to the inside right flange of the CABINET. The Position form feature dialog box is displayed. DO NOT SELECT THE FINISH BUTTON AT THIS TIME.

263) Rotate the Louver. Click **Tools**, **Sketch Tools**, **Modify**. Enter **270** for Rotate.

264) Press the **Enter** key.

265) Click **Close**.

266) Display the Right view. Click **Right**.

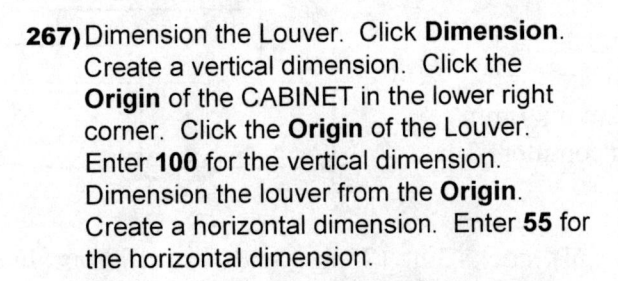

267) Dimension the Louver. Click **Dimension**. Create a vertical dimension. Click the **Origin** of the CABINET in the lower right corner. Click the **Origin** of the Louver. Enter **100** for the vertical dimension. Dimension the louver from the **Origin**. Create a horizontal dimension. Enter **55** for the horizontal dimension.

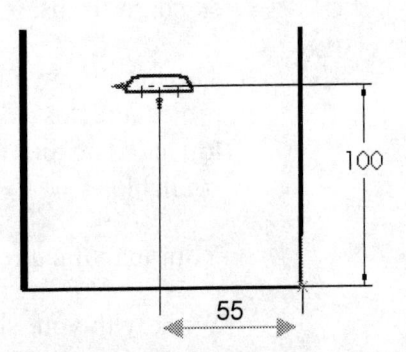

268) Display the Louver. Click **Finish** Position Form Feature dialog box.

269) **Close** the FeaturePallete dialog box.

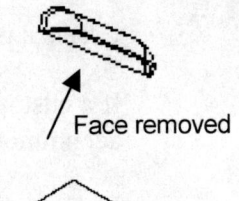

270) **Rotate** the view. The Louver form removes a face on the right flange.

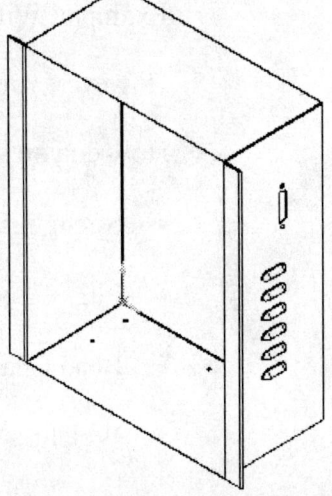

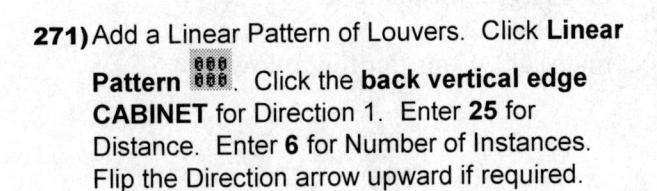

Face removed

271) Add a Linear Pattern of Louvers. Click **Linear Pattern**. Click the **back vertical edge CABINET** for Direction 1. Enter **25** for Distance. Enter **6** for Number of Instances. Flip the Direction arrow upward if required.

272) Click **OK**.

273) Save the CABINET. Click **Save**.

PAGE 6 - 51

Manufacturing Considerations

How do you determine the size and shape of the louver form? Are additional die cuts or forms required in the project?

Work with a sheet metal manufacturer. Ask questions. What are the standards? Identify the type of tooling in stock? Inquire on form advantages and disadvantages.

One company that has taken design and form information to the Internet is Lehi Sheetmetal, Westboro, MA. (www.lehisheetmetal.com).

Standard dies, punches and manufacturing equipment such as brakes and turrets are listed in the Engineering helpers section.

Dimensions are provided for their standard forms.

The form tool you used for this project creates a 32mm x 6mm louver. The tool is commercially available.

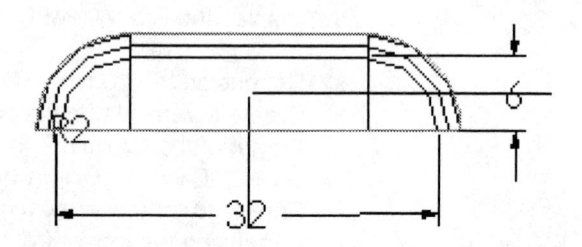

Your manufacturer only stocks 3in., (75mm) and 4in., (100mm) louvers.

Work with your sheet metal manufacturer to obtain a cost effective alternative.

Create a pattern of standard square cuts to dissipate the heat.

If a custom form is required, most custom sheet metal manufacturers can accommodate your requirement.

Example: Wilson Tool, Great Bear Lake, MN (www.wilsontool.com).

However, you will be charged for the tool.

How do you select material? Consider the following:

- Strength.

- Fit.

- Bend Characteristics.

- Weight.

- Cost.

All of these factors influence material selection. Raw aluminum is a commodity.

Large manufacturers such as Alcoa and Reynolds sell to a material supplier, such as *Pierce Aluminum. Pierce Aluminum in turn, sells material of different sizes and shapes to distributors and other manufacturers.

Material is sold in sheets or cut to size in rolls, Figure 6.1.

U.S. Sheet metal manufacturers work with standard 8ft – 12ft stock sheets.

For larger quantities, the material is usually supplied in rolls. For a few cents more per pound, sheet metal manufacturers request the supplier to shear the material to a custom size.

Check Dual Dimensioning from Tools, Options Document Properties to display both Metric and English units.

Do not waste raw material.

**ALUMINUM SHEET
NON HEAT TREATABLE, 1100-0
QQ-A-250/1 ASTM B 209**

All thickness and widths available from coil for custom blanks

SIZE IN INCHES	WGT/SHEET
.032 x 36 x 96	11.05
.040 x 36 x 96	13.82
.040 x 48 x 144	27.65
.050 x 36 x 96	17.28
.050 x 48 x 144	34.56
.063 x 36 x 96	21.77
.063 x 48 x 144	43.55
.080 x 36 x 96	27.65
.080 x 48 x 144	55.30
.090 x 36 x 144	46.66
.090 x 48 x 144	62.26
.100 x 36 x 96	34.56
.125 x 48 x 144	86.40
.125 x 60 x 144	108.00
.190 x 36 x 144	131.33

Figure 6.1
*Courtesy of Piece Aluminum Co, Inc.
Canton, MA

Optimize flat pattern layout by knowing your sheet metal manufacturers equipment.

Discuss options for large cabinets that require multiple panels and welds.

In Project 1, you were required to be cognizant of the manufacturing process for machined parts.

In Project 4 and 5 you created and purchased parts.

Whether a sheet metal part is produced in or out of house, knowledge of the materials, forms and layout provides the best cost effective design to the customer.

You require both a protective and cosmetic finish for the Aluminum BOX.

Review options with the sheet metal vendor. Parts are anodized. Black and clear anodized finishes are the most common. In harsh environments, parts are covered with a protective coating such as Teflon® or Halon®.

The finish adds thickness to the material. A few thousandths could cause problems in the assembly.

Think about the finish before the design begins.

Suppress the Louver.
274) Right-click **Louver1** from the FeatureManager. Click **Suppress**. The Linear Pattern is a child of the Louver. The Linear Pattern is suppressed.

Create a sketched pattern of Cuts to dissipate the generated inside heat.

275) Display the Right view. Click **Right**.

276) Click the **CABINET right face** in the Graphics window. Click **Sketch**.

277) Create the first square. Click **Rectangle**. Sketch a **10**mm square in the lower left corner.

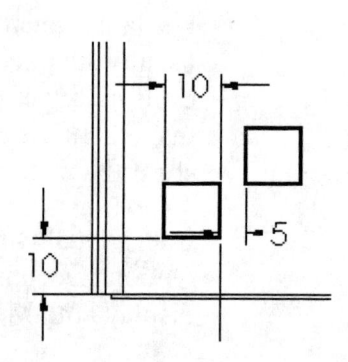

278) Create a horizontal dimension. Select the **Origin** for the dimension reference. Select the **right vertical line** of the square. Enter **80**.

279) Create a vertical dimension. Select the **Origin**. Select the **bottom line** of the square. Enter **10**.

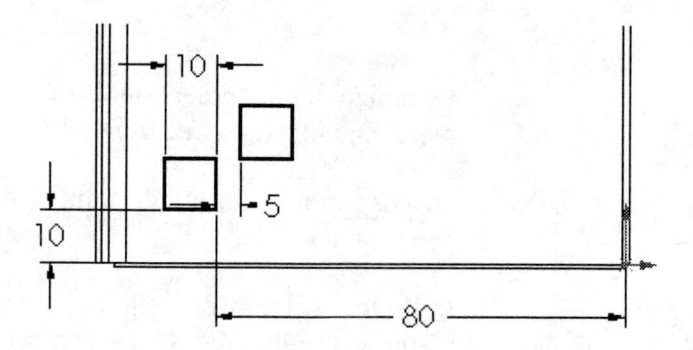

280) Create a second square. Click **Rectangle**. Sketch a rectangle to the right of the first square.

281) Click **Add Relations** from the Sketch Relations toolbar. Click the **top line** of the second square. Click the **left line** of the second square. Click **Equal**.

282) Click **OK**.

283) Click the **top horizontal line** of the first square.

284) Click Add Relations.

285) Click the **top horizontal line** of the second square. Click **Equal**.

286) Click **OK**.

287) Click **Dimension**. Enter **5** between the squares.

288) Click Add Relations.

289) Click the **top horizontal line** of the first square

290) Click the **bottom horizontal line** of the second square.

291) Click **Collinear**.

292) Click **OK**.

293) Create the sketch pattern. Click **Linear Step & Repeat** from the Sketch Tools toolbar. Select all **eight lines** of the squares.

294) Enter **2** for Number for Direction1.

295) Enter **30** for Spacing for Direction1. Press the **Preview** button.

296) Enter **13** for Number for Direction2.

297) Enter **20** for Spacing for Direction2.

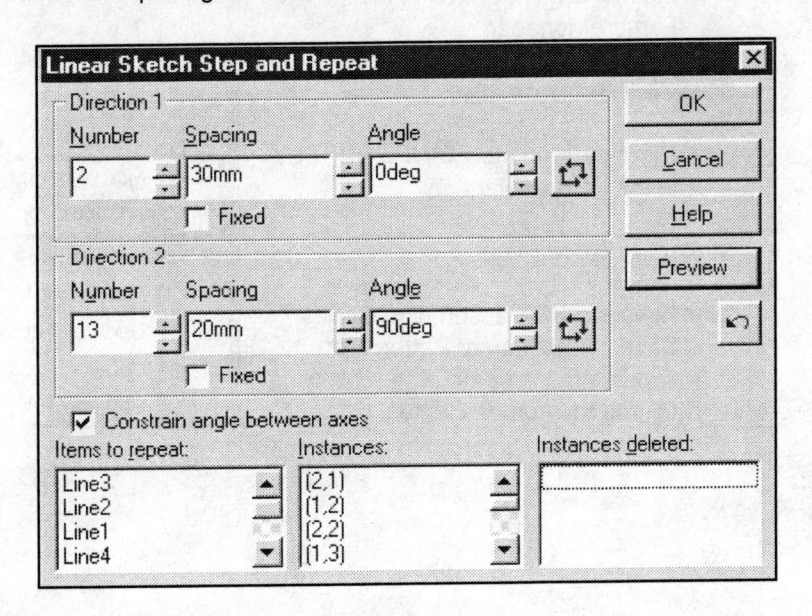

298) Click **OK**.

299) Extrude the sketch. Click **Extruded Cut**.

300) Check Link to thickness.

301) Click **OK**.

302) Click Isometric.

303) Save the CABINET in its formed state. Click **Save**.

304) Return to the BOX assembly. Click **Window**, **BOX**.

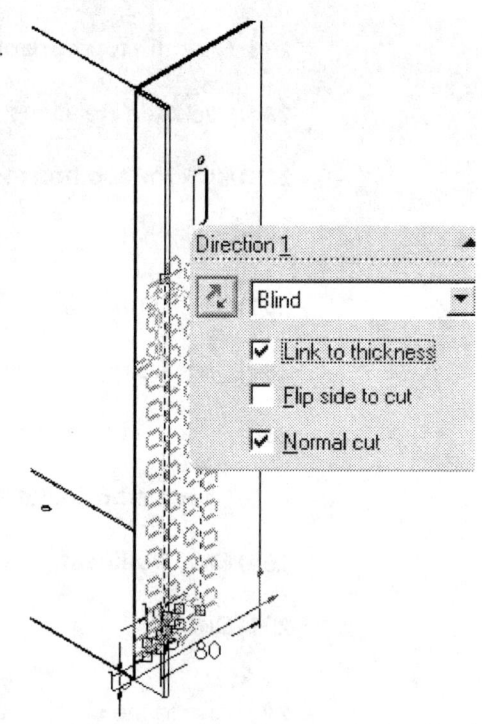

305) The MOTHERBOARD and POWERSUPPLY are suppressed. Restore the MOTHERBOARD and POWERSUPPLY. Right-click on the **MOTHERBOARD** in the FeatureManager.

306) Click Set to Resolve.

307) Right-click on the **POWERSUPPLY** in the FeatureManager.

308) Click Set to Resolve.

309) Save the BOX assembly. Click **Save**. Note: Linear Patterns are suppressed in some of the following figures to improve clarity.

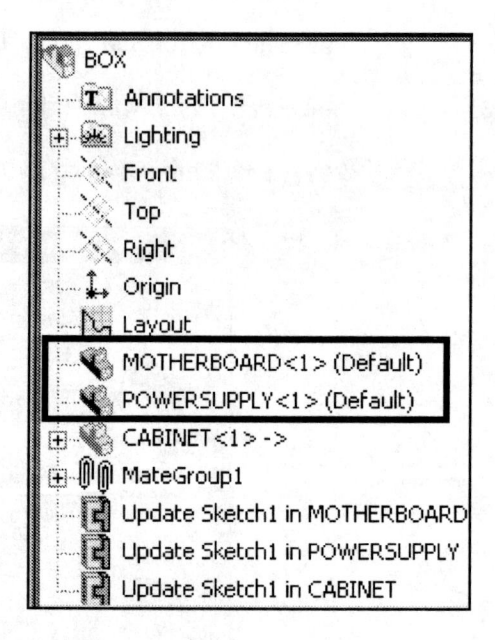

Engineering Design with SolidWorks — Top Down Assembly Modeling

PEM Inserts

The BOX assembly contains an additional support BRACKET that is fastened to the CABINET. To accommodate the new BRACKET, the Linear Pattern of 5 holes is added to the CABINET.

How do you fasten the BRACKET to the CABINET?

Answer: Utilize efficient and cost effective self-clinching fasteners.

PEM Fastening Systems, Danboro, PA USA manufactures self-clinching fasteners, called PEM® Fasteners.

PEM® Fasteners, Examples

PEM® Fasteners are used in numerous applications and are provided in a variety of shapes and sizes.

IGES standard PEM® models are available at (www.pemnet.com) and are used in the following example.

Answer the following questions in order to select the correct PEM® Fastener for the assembly:

QUESTION:	ANSWER:
What is the material type?	Aluminum
What is the material thickness of the CABINET?	1mm
What is the material thickness of the BRACKET?	19mm
What standard size PEM® Fastener is available?	FHA-M5-30
What is the hole diameter required for the CABINET?	5mm ± 0.08
What is the hole diameter required for the BRACKET?	5.6mm maximum
Total Thickness = CABINET Thickness + BRACKET Thickness + NUT Thickness Total Thickness = 1mm + 19mm + 5mm = 25mm (minimum)	25mm (minimum)

The PEM® Fastener for the project selected is FHA-M5-30.

- FH designates the stud type.
- A designates the material type, Aluminum.
- M5 designates the thread size.
- 30 designates the overall length, 30mm.

The overall dimensions of the FH PEM® Fasteners series are the same for various materials.

Do not contain PEM® Fasteners as a feature in the Flatten State of a component.

In the sheet metal manufacturing process, PEM® Fasteners are added after the material pattern has been fabricated.

Proper PEM® Fasteners material selection is critical. There are numerous grades of aluminum and stainless steel. Test the installation of the material fastener before final specification.

Note: When the fastener is pressed into ductile host metal, it displaces the host material around the mounting hole, causing metal to cold flow into a designed annular recess in the shank.

Ribs prevent the fastener from rotating in the host material once it has been properly inserted. The fasteners become a permanent part of the panel in which they are installed.

During the manufacturing process, studs are installed by placing the fastener in a punched or drilled hole and squeezing the stud into place with a press.

The squeezing action embeds the head of the stud flush into the sheet metal.

Obtain additional manufacturing information that directly affects your design.

Manufacturing Information:	FH-M4:	FH-M5:
Hole Size in sheet metal (CABINET).	4.0mm ± 0.8	5.0mm ± 0.8
Maximum Hole in attached parts (BRACKET).	4.6mm	5.6mm
Centerline to edge minimum distance. (from center point of hole to edge of CABINET)	7.2mm	7.2mm

Engineering Design with SolidWorks — Top Down Assembly Modeling

Obtain the fastener.

310) **Download** the sheet metal IGES zip file from the URL, www.pemnet.com.

311) Login in. Enter **user name** and **password**.

312) Register for a **user name** and **password** if you do not have one. They will be emailed to you.

313) Download two different metric studs. Select the **IGES library** and the **Search Metric Fasteners**.

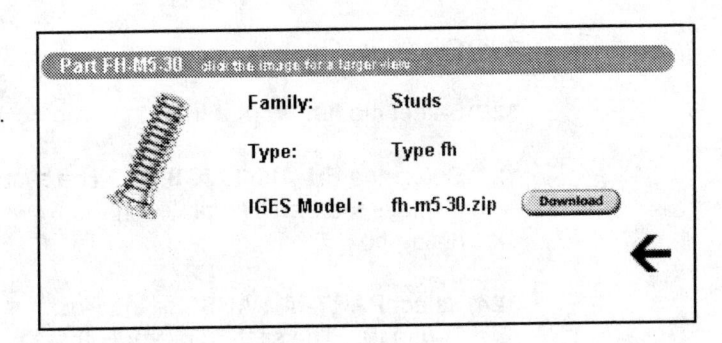

314) Enter **FH-M5-30** for the part number. A preview of the part is displayed.

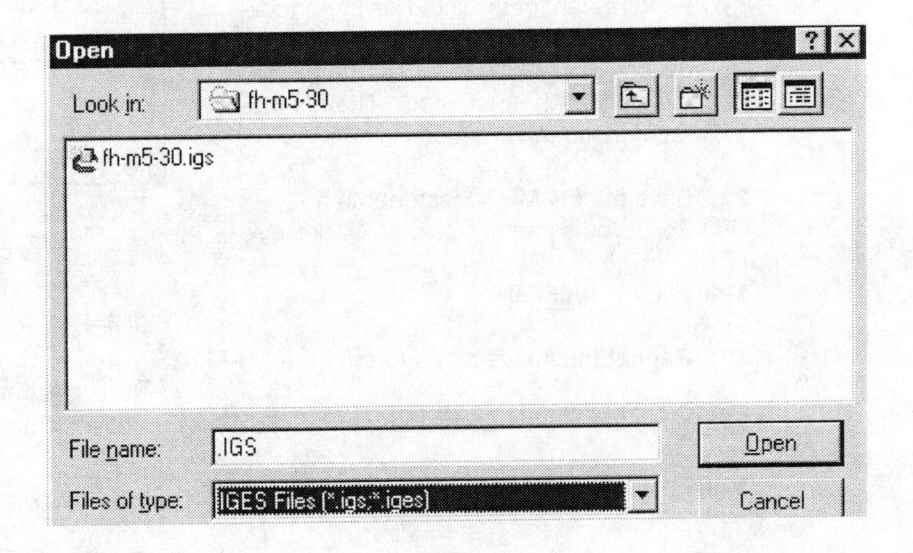

315) Click the **Download** button. Enter **FH-M4-25**.

316) Click the **Download** button. The part geometry for Aluminum and Stainless Steel fasteners is the same. The "A" is not required for download.

317) **Unzip** the zipped files, **FH-M5-30** and **FH-M4-25**.

318) Select the **ENGDESIGN-W-SOLIDWORKS\VENDOR-COMPONENTS** file folder to save the IGES files. The IGES files, FH-M5-30.IGS and FH-M4-25.IGS are created. You will use the FH-M4-25.IGS file again. Note: You cannot insert an IGES component directly into the BOX assembly.

319) Open the IGES model in SolidWorks.

320) Save the model as a SolidWorks part file.

321) Click **File**, **Open**.

322) Select the **IGS** format from the drop down list.

323) Select the **FH–M5-30.IGS** file. The system displays the Part Templates in the New dialog box.

324) Select **PART-MM-ANSI** for the Part Template. The system prompts that the IGES surfaces are being converted. The fastener is displayed in the Graphics window. The first feature in the FeatureManager is listed as Imported. Reference Imported geometry with dimensions. Convert faces and edges in the sketch when required. The FH-M5-30 requires no modification.

325) Save the FH-M5-30 fastener in a SolidWorks part file format. Click **Save**.

326) Enter **FH-M5-30**.

327) Repeat the above procedure for the FH-M4-25 fastener.

328) Save the FH-M4-25 fastener in a SolidWorks part file format. Click **Save**.

329) Enter FH-M4-25.

Insert the FH-M5-30 fastener into the BOX assembly.

330) Click Insert, Component, From File.

331) Double-click the part **FH-M5-30**.

332) Click inside the **Graphics window** in the inside bottom left corner near the seed feature of the Linear Pattern of Holes.

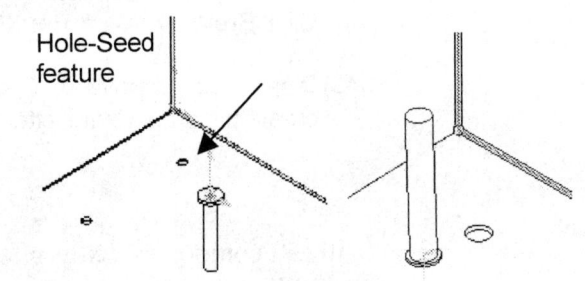

333) Zoom in on the Hole and fastener.

334) Rotate the **FH-M5-30** component. Click **Rotate Component** from the Assembly toolbar.

335) Rotate the **FH-M5-30** component until the head faces downward. Click **OK**.

336) Mate the FH-M5-30 component. Click **Mate**.

337) Create the Concentric mate. Click the cylindrical **shaft** of the FH-M5-30 fastener. Click the cylindrical **face** of the back left **Hole**. Click **Concentric**.

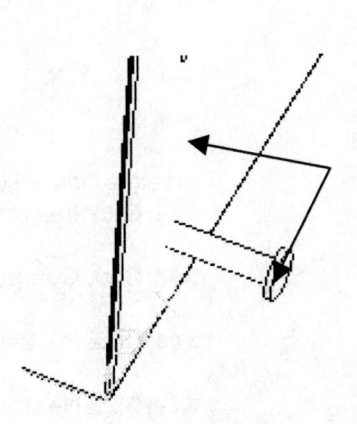

338) Click **OK**.

339) Rotate the view. **Zoom** in on the bottom face of the FH-M5-30 fastener.

340) Click **Mate**.

341) Create a Coincident mate. Click the **head face** of FH-M5-30 fastener. Click the **outside bottom face** of the CABINET. Click **Coincident**. Click **OK**.

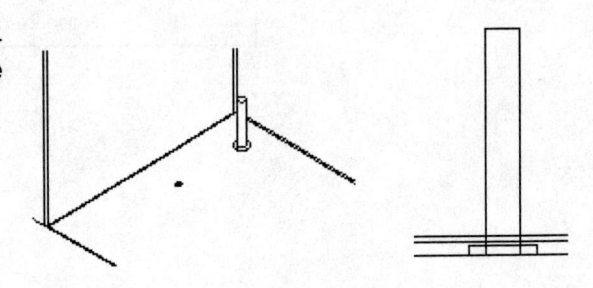

342) Fit to the Graphics window. Press the **f** key.

343) Display an **Isometric** view.

344) Create a Parallel mate. Click **Mate**. Click the **Right** plane of the FH-M5-30 fastener from the FeatureManager. Click the **Right** plane of the CABINET. Click **Preview**. Click **Parallel**. Click **OK**.

345) Display the Front view. Click **Front**. The FH-M5-30 component is used to create a Component Pattern.

Utilize a component pattern to create multiple copies of a component in an assembly.

The FH-M5-30 references the Hole1 seed feature. The Component Pattern displays the five FH-M5-30 fasteners.

Create a Derived Component Pattern.
346) Click **Insert** from the Main menu.

347) Click **Component Pattern**. The Pattern Type dialog box is displayed.

348) Click the Use an existing feature pattern (Derived) check box.

349) Click **Next**.

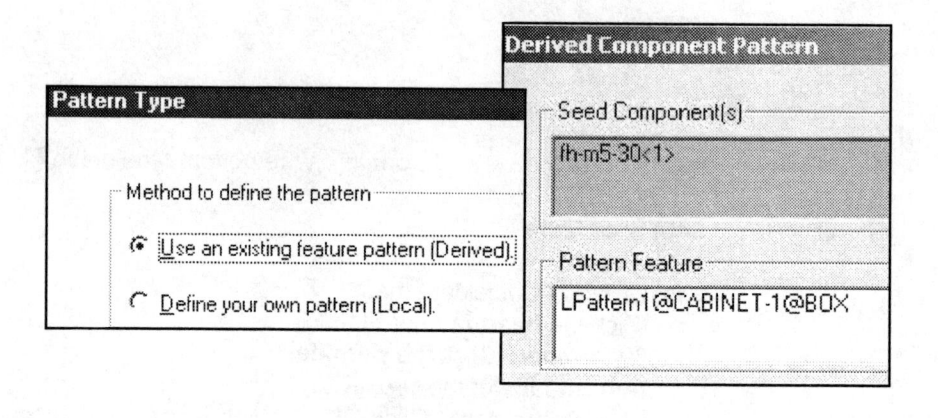

Engineering Design with SolidWorks *Top Down Assembly Modeling*

350) The Derived Component Pattern dialog box is displayed. Click **FH-M5-30** from the FeatureManager for the Seed Component text box.

351) Click **inside** the Pattern Feature. Click the **Lpattern1** under the CABINET part for the Pattern Feature.

352) Display the Derived LPattern. Click **Finish**. The additional FH-M5-30 instances are located under the Derived LPattern1 in the Feature Manager.

353) Expand the Derived LPattern1. Click the **Plus** ⊞ icon to the left of Derived LPattern1 in the FeatureManager.

354) Save the BOX. Click **Save**.

355) Click **Yes** to save the document.

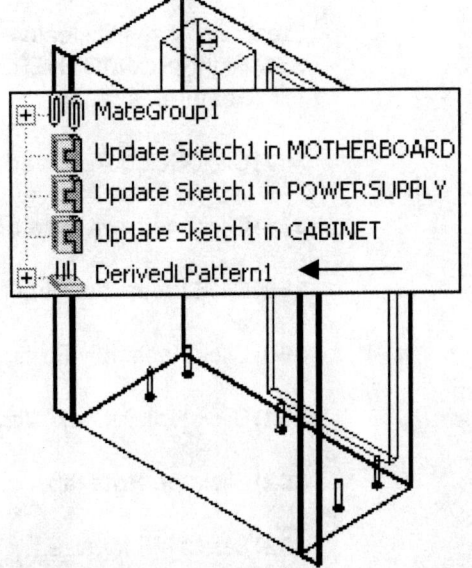

The purchasing manager of your company determines that the FH-M5-30 fastener has a longer lead-time than the FH-M4-25 fastener.

The FH-M4-25 fastener is in stock.

Time is critical.

You ask the engineer creating the corresponding BRACKET if the FH-M4-25 fastener is a reliable substitute.

You place a phone call. You get voice mail. You send email. You wait for the engineer's response and create the next assembly feature.

Add an Assembly Hole Feature

Additional holes are created through multiple components in the BOX assembly. A through Hole is required from the front of the MOTHERBOARD to the back of the CABINET.

Use the Hole Wizard as an Assembly feature.

Create an Assembly feature with the Hole Wizard.
356) Right-click **MOTHERBOARD** from the FeatureManager.

357) Click Set to Resolved.

358) Click the **front face** of the MOTHERBOARD.

359) Click **Insert** from the Main menu.

360) Click Assembly Feature.

361) Click **Hole**. Click **Wizard**.

362) Click the **Hole** tab.

363) Select **M5** for size.

364) Click **Through All** for Hole Type Depth.

365) Click **Next**. DO NOT SELECT FINISH AT THIS TIME.

366) Dimension the hole on the MOTHERBOARD. Click **Dimension**. Create a horizontal dimension. Click the **left edge** of the MOTHERBOARD. Click the **center point** of hole. Enter **25**.

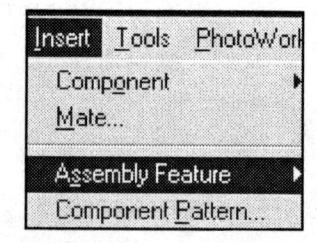

367) Create a vertical dimension. Click **Origin** of the CABINET. Click the **center point** of the hole. Enter **50**.

368) Click **Finish** from the Hole Placement dialog box. The M5 hole is added to the FeatureManager.

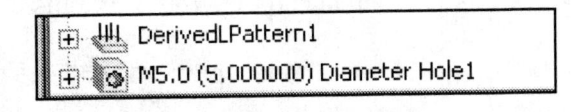

369) Save the BOX. Click **Save**. Click **Yes** to rebuild assembly.

370) Close all parts and assemblies. Click **Window**, **Close All**.

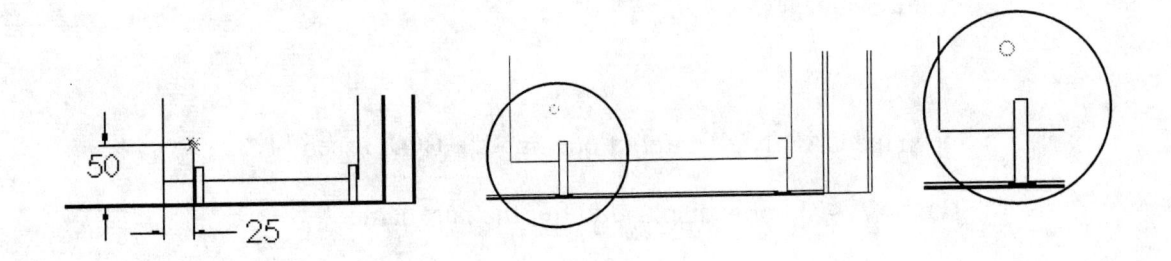

FeatureManager and the Assembly

The FeatureManager contains numerous entries in an assembly. Understanding the organization of the components and their mates is critical in creating an assembly without errors.

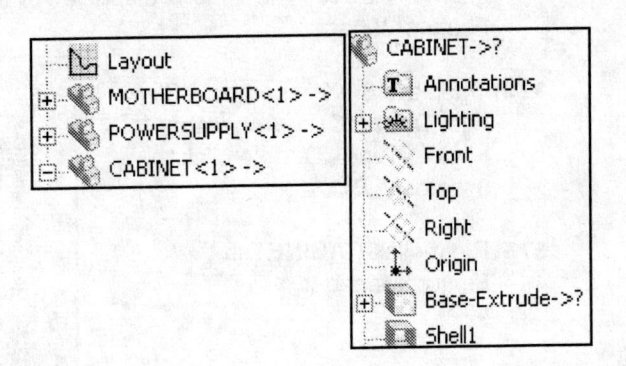

Errors occur within features such as a radius that is too large to create a Fillet feature. When an error occurs in the assembly, the feature or component is labeled in red.

Mate errors occur when you select conflicting geometry such as a Coincident Mate and a Distance Mate with the same faces.

In the initial mating, undo a mate that causes an error. Mate errors occur later on in the design process when you suppress required components and features. Plan ahead to avoid problems.

Use reference planes for Mates that will not be suppressed.

When a component displays a '->' symbol it references geometry from the assembly or from another component.

When the system does not locate reference geometry from part name displays a '->?' symbol to the right of the component name in the FeatureManager.

Open the CABINET without opening the BOX assembly.

The '->?' symbol is displayed after the part name.

Open the CABINET part.

371) Click File, Open. Browse to ENGDESIGN-W-SOLIDWORKS\PROJECTS.

372) Click **Part** for Files of Type.

373) Double-click the part, **CABINET**. The CABINET cannot locate the referenced geometry BOX. The "->?" is displayed to the right of the CABINET name in the FeatureManager.

374) Open the BOX assembly. Click **File**, **Open**. Click **Assembly** for Files of Type. Double-click the assembly, **BOX**.

375) Right-click **CABINET** in the FeatureManager.

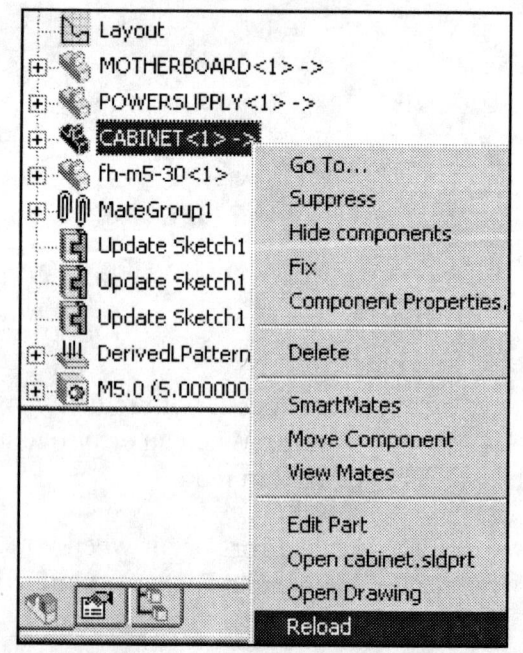

Engineering Design with SolidWorks   **Top Down Assembly Modeling**

376) Click **Reload**. Reload the CABINET.

377) Click **OK**. The CABINET references the BOX assembly.

378) Return to the BOX assembly. Click **Window**, **BOX**.

The BRACKET engineer finally contacts you. You proceed to replace the FH-M5-30 fastener with the FH-M4-25 fastener.

Use Replace to exchange the FH-M5-30 fastener with tFH-M4-25 fastener in the BOX assembly.

Edit the CABINET and change the Hole size of the seed feature from M5 to M4.

Replace the component.
379) Right-click **FH-M5-30** from the FeatureManager.

380) Click **Replace**.   Replace six FH-M5-30 with FH-M4-25 by replacing the Seed Feature.

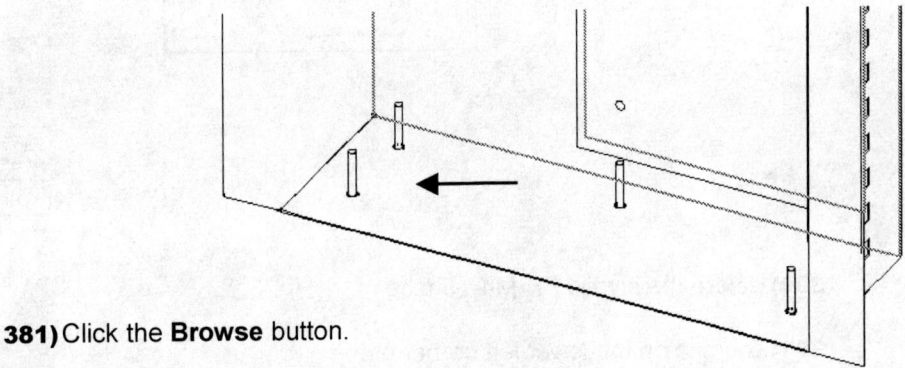

381) Click the **Browse** button.

382) Double-click **FH-M4-25**.

383) Click **OK**.

384) Click **Yes** to the Internal ID question.

385) The five FH-M5-30 are replaced with the five FH-M4-25. Click **OK**.

Two Mate Entity warning errors occur. A red X is displayed on the Mated Entities dialog box.

Redefine the Concentric Mate and Coincident Mate.

Top Down Assembly Modeling   **Engineering Design with SolidWorks**

Edit the Mates.
386) **Close** the Error message box.

387) Expand **Mates** in FeatureManager.

388) Right-click **Concentric1**.

389) Click **Edit Definition**.

The Selection text box is displayed.

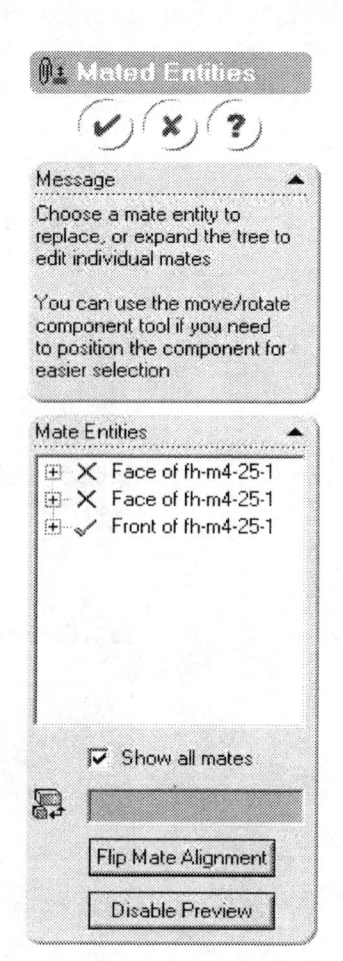

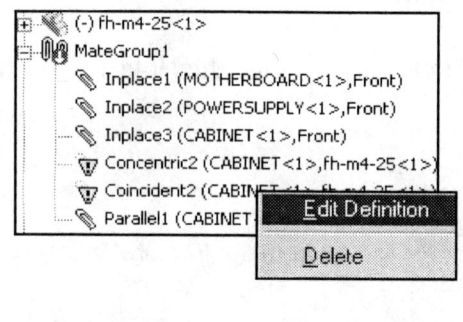

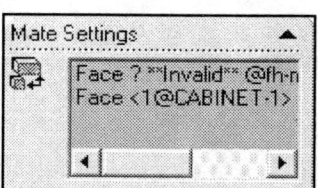

390) **Delete** the invalid FH-M4-25 face.

391) Zoom in on the lower left corner of the box. Press the **Left arrow** key. Press the **Down arrow** key.

392) Select the **cylindrical face** of the back left FH-M4-25 fastener.

393) Click **Preview**.

394) Click **OK**.

395) Click **Close** to the Rebuild error message.

396) Click **OK**.

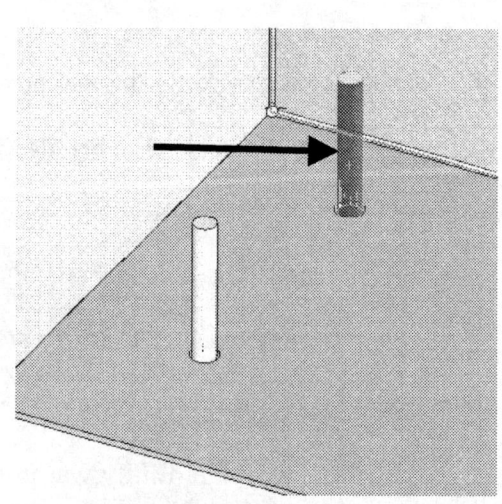

PAGE 6 - 68

Engineering Design with SolidWorks   **Top Down Assembly Modeling**

397) Right-click **Coincident1** in the FeatureManager.

398) Click **Edit Definition**. The Selection text box is displayed.

399) **Delete** the invalid FH-M4-25 face.

400) Click the **Bottom** view.

401) Select the **head** of the back left on the FH-M4-25.

402) Click **Preview**.

403) Click **OK**.

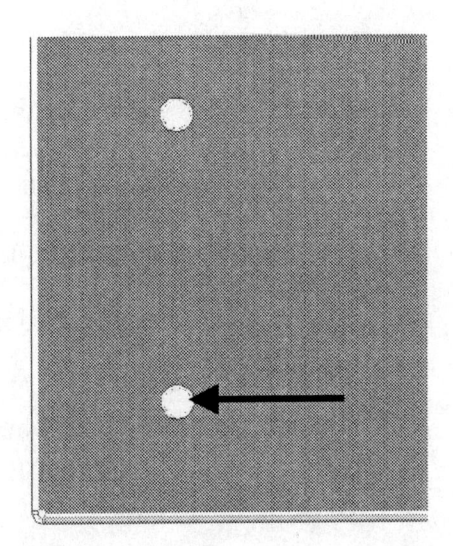

Edit the M5 Hole to an M4 Hole.
404) Right-click **CABINET** from the FeatureManager. Click **Edit Part**.

405) Expand CABINET in the FeatureManager. Right-click the **M5 Hole**. Click **Edit Definition**. The Hole Definition dialog box is displayed.

406) Select **M4** for Size from the Hole dialog box. Click **Next**.

407) Click **Finish**. The M4 Hole is displayed in the FeatureManager.

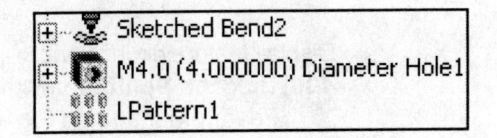

408) Return to the BOX assembly. Click **Edit Part** from the Assembly toolbar.

The Hole of the CABINET requires a M4 hole 4mm ± 0.8. Add the limit dimensions to the drawing.

Equations

How do you insure that the fasteners remain symmetrical with the bottom of the BOX when the box-width dimension varies?

Answer: With an Equation.

Equations use shared names to control dimensions.

Use Equations to connect values from sketches, features, patterns and various parts in an assembly.

Create an Equation.

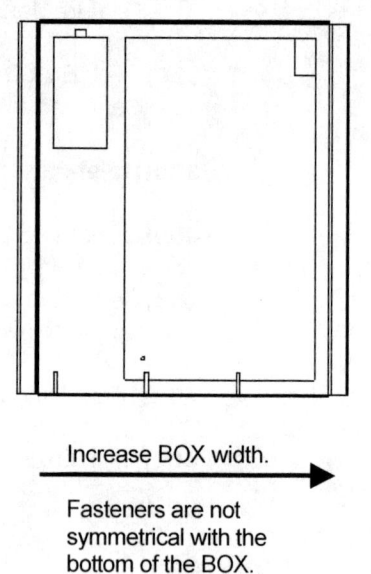

Increase BOX width.

Fasteners are not symmetrical with the bottom of the BOX.

Each dimension has a unique variable name. The names are used as Equation variables. The default names are based on the Sketch, Feature or Part.

Feature names do not have to be changed. However, when creating equations, rename variables for clarity.

Display feature dimensions.
409) Drag the **Splitbar** downward to display the BOX FeatureManager in the upper half. Display the CABINET in the lower half.

410) Right-click the BOX **Annotations** folder.

411) Click Show Feature Dimensions.

412) Display the **Front** view.

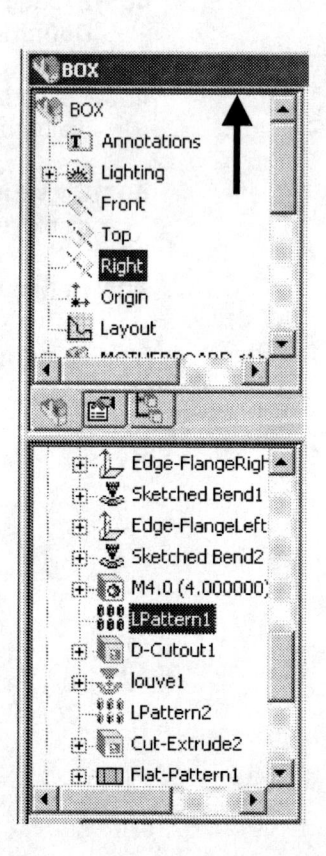

413) Edit the dimension Name in the Dimension Properties. Double-click **Lpattern1** in CABINET. Right-click on the Linear Pattern1 of Cabinet horizontal dimension **125**. Click **Properties**.

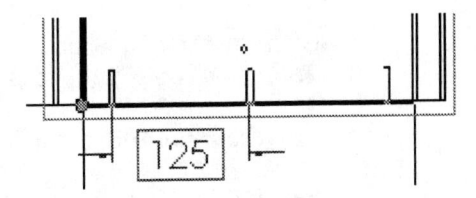

414) Rename the D3 dimension Name. Enter **lpattern-bottom**. Click **OK**.

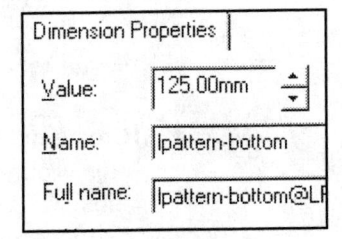

415) Click **OK** from the Dimension Property Manager.

416) Double-click **M4.0** in the FeatureManager. Right-click on the M4.0 horizontal dimension **25** in the lower left corner. Click **Properties**.

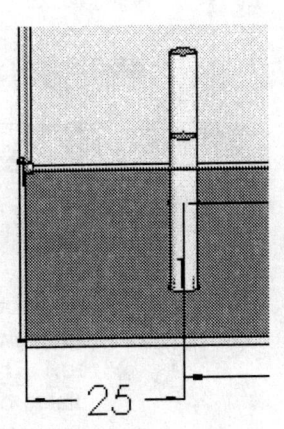

417) Rename the D1 dimension Name. Enter **hole-bottom**.

418) Click **OK**. Click **OK**.

419) Right-click **Annotations** below the CABINET entry in the FeatureManager. Click **Show feature dimension**. Click the **135** dimension.

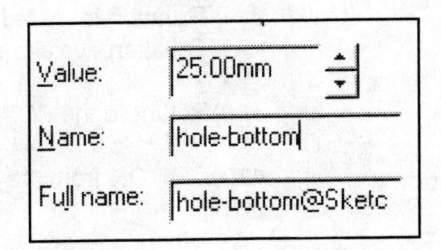

Top Down Assembly Modeling   Engineering Design with SolidWorks

Create the first Equation.
420) Click **Tools, Equations**.

421) Click the **Add** button from the Equations dialog box.

422) Create the first half of equation1. Select the lpattern-bottom horizontal dimension, **125**. The variable "lpattern-bottom@Lpattern1@cabinet.Part" = is added to the New Equation text box.

> "lpattern-bottom@LPattern1@CABINET.Part"=

423) Create the second half of equation1. Enter **0.5* (** from the keypad. Select the box_width horizontal dimension, **300 from the Graphics window.** The variable "box_width@layout" is added to the equation text box.

> "lpattern-bottom@LPattern1@CABINET.Part"=0.5*("BOX_WIDTH@Layout"

424) Enter **–2*** from the keypad.

425) Select the M4 hole-bottom dimension, **25**. The variable "hole-bottom@Sketch12@cabinet.Part" is add to the equation text box.

426) Enter **)** from the keypad.

> "lpattern-bottom@LPattern1@CABINET.Part"=0.5*("BOX_WIDTH@Layout"-2*"hole-bottom@Sketch27@CABINET.Part")

427) Display the first equation. Click **OK** from the New Equation dialog box. The Equation dialog box contains the complete equation. A green check mark placed in the first column indicates that the Equation is solved. The Equation evaluates to 125mm.

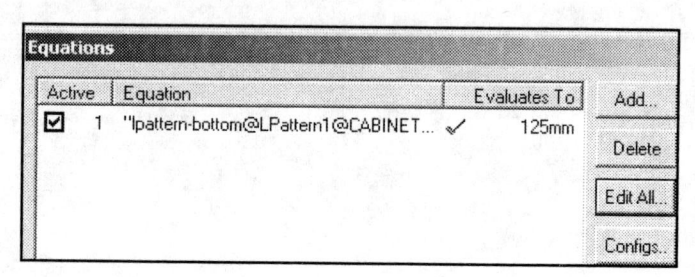

Active	Equation	Evaluates To
✓	1 "lpattern-bottom@LPattern1@CABINET... ✓	125mm

Add...
Delete
Edit All...
Configs..

428) Return to the BOX assembly. Click **OK**.

429) Click **OK** from the Dimension PropertyManager.

Engineering Design with SolidWorks   Top Down Assembly Modeling

430) Save the BOX assembly. Click **Save**.

431) Click **Yes** to save referenced documents.

Verify the Equation.
432) Display all suppressed components. Right-click **POWERSUPPLY** from the FeatureManager.

433) Click Set to Resolved.

434) Modify the dimensions. Double-click the horizontal dimension box-width, **300**. Enter **400**.

435) Double-click the dimension box-height, **400**. Enter **500**.

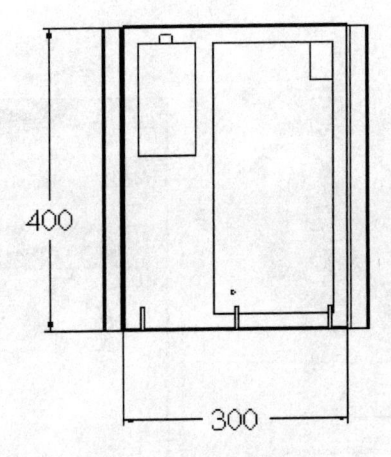

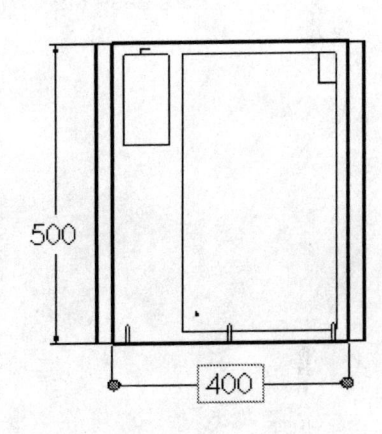

436) Click **Rebuild**.

437) Modify the dimensions. Double-click on the horizontal dimension, **400**. Enter **300**. Double-click the vertical dimension, **500**. Enter **400**. **Rebuild**.

438) Hide the dimensions. Right-click the CABINET **Annotations** folder.

439) Uncheck Show Feature Dimension.

440) Right-click the BOX **Annotations** folder.

441) Uncheck Show Feature Dimension.

442) Save the BOX. Click **Save**.

443) Click **Yes** to update the referenced models.

PAGE 6 - 73

Design Tables

A Design Table is an Excel spreadsheet used to create multiple configurations in a part or assembly.

Utilize the Design Table to create three configurations of the BOX Assembly:

- Small.
- Medium.
- Large.

Create the Design Table.
444) Click Insert, Design Table.

445) Accept the defaults. Click **OK**.

446) Select the input dimension. Hold the **Ctrl** key down. Click the **box_width@Layout** dimension. Click the **box_height@Layout** dimension. Release the **Ctrl** key.

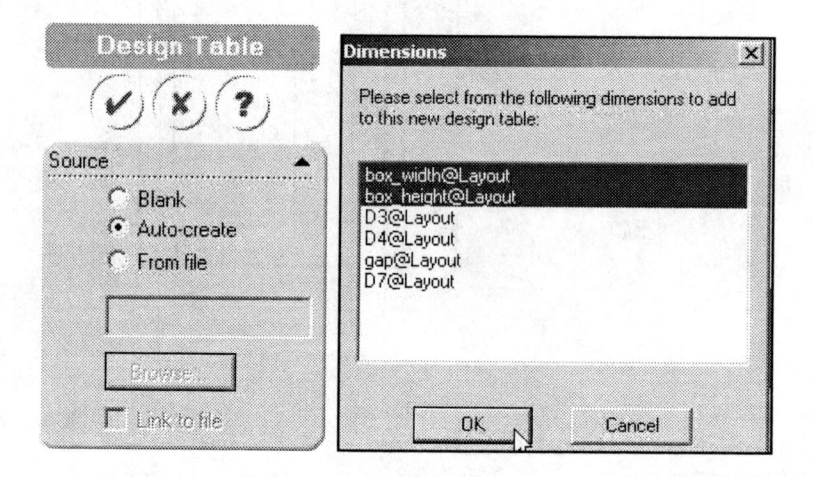

447) Click **OK**. The input dimension names and default values are automatically entered into the Design Table. The value Default is entered in Cell A3. The values 400 and 300 are entered in Cell B3 and Cell C3, respectfully.

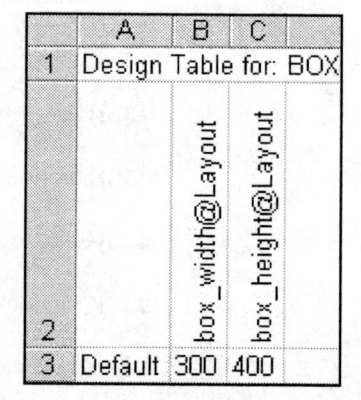

Enter the three configuration names.
448) Click Cell **A4**. Enter **Small**.

449) Click Cell **A5**. Enter **Medium**.

450) Click Cell **A6**. Enter **Large**.

451) Enter the dimension values for the Small configuration. Click Cell **B4**. Enter **300**. Click Cell **C4**. Enter **400**.

452) Enter the dimension values for the Medium configuration. Click Cell **B5**. Enter **400**. Click Cell **C5**. Enter **500**.

453) Enter the dimension values for the Large configuration. Click Cell **B6**. Enter **500**. Click Cell **C6**. Enter **650**.

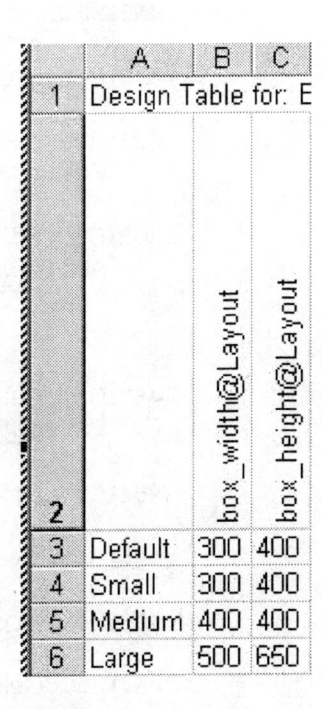

	A	B	C
1	Design Table for: E		
2		box_width@Layout	box_height@Layout
3	Default	300	400
4	Small	300	400
5	Medium	400	400
6	Large	500	650

454) Build the three configurations. Click a **position** outside the EXCEL Design Table in the FeatureManager. Click **OK** to generate the configurations. The Design Table icon is displayed in the BOX FeatureManager.

455) Display the configurations. Click the **Configuration Manager** icon.

456) Double-click **Small**.

457) Double-click **Medium**.

458) Double-click **Large**.

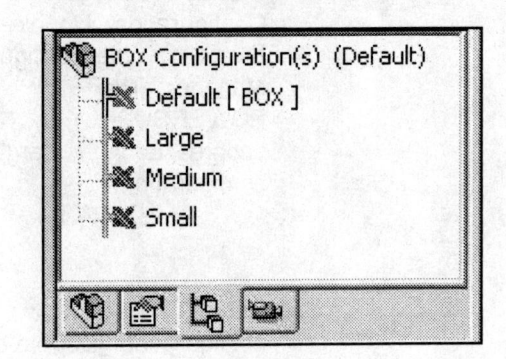

BOX Configuration(s) (Default)
- Default [BOX]
- Large
- Medium
- Small

Top Down Assembly Modeling   **Engineering Design with SolidWorks**

459) Fit the model to the Graphics window. Press the **f** key.

460) Return to the Default configuration. Double-click **Default**.

461) Return to the Design Table. Right-click the **Design Table** from the BOX FeatureManager.

462) Click **Edit Table**. Click **OK** from the Add Rows and Columns dialog box.

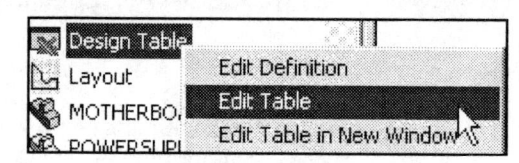

463) Click Cell **D2**. Enter **$State@POWERSUPPLY**<1>.

464) Click Cell **D3**. Enter **R** for Resolved.

465) Copy Cell **D3**.

466) Select Cell **D4** and Cell **D5**. Click **Paste**.

467) Click Cell **D6**. Enter **S** for Suppressed.

468) Update the configurations. Click a **position** outside the EXCEL Design Table.

	A	B	C	D
1	Design Table for: BOX			
2		box_width@Layout	box_height@Layout	$state@POWERSUPPLY<1>
3	Default	300	400	R
4	Small	300	400	R
5	Medium	400	400	R
6	Large	500	650	S

469) Display the Large Configuration. Double-click **Large** from the Configuration Manager. The POWERSUPPLY is suppressed in the Large configuration.

470) Return to the Default configuration. Double-click **Default**.

471) Fit the model to the Graphics window. Press the **f** key.

472) Save the BOX Assembly. Click **Save**.

473) Click **Yes** to save the referenced documents.

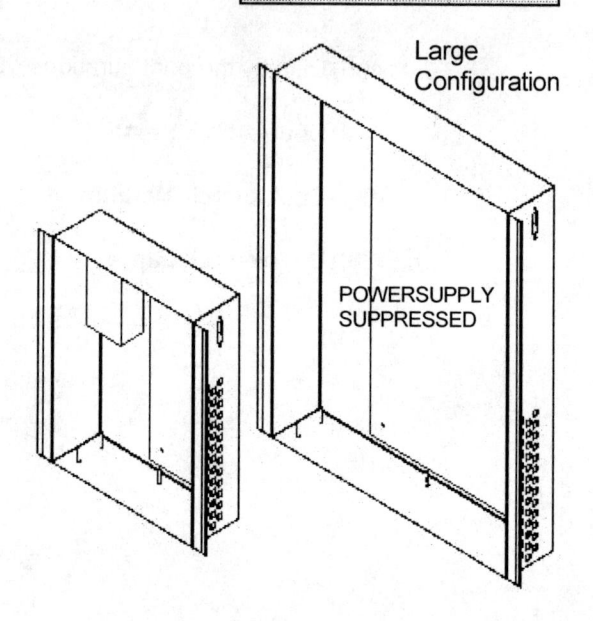

Dimensions, Color, Configurations, Custom Properties and other variables are controlled through Design Tables.

For additional information on part and assembly configurations, see On Line Help and

- Assembly Modeling with SolidWorks, SDC Publications.

- Drawing and Detailing with SolidWorks, SDC Publications.

You receive the 2D populated board from the PCB manufacturer and import the file into CircuitWorks.

Save the CircuitWorks file as MOTHERBOARD.sldprt.

Open the part from CircuitWorks into SolidWorks BOX assembly.

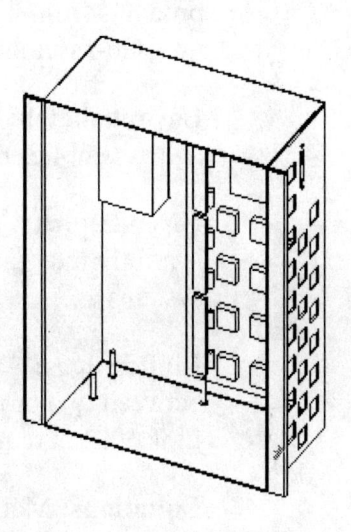

Project Summary

You created three different BOX sizes utilizing the Top down assembly modeling approach. The Top down design approach is a conceptual approach used to develop products from within the assembly.

You obtained an understanding of sheet metal fabrication. You created a sheet metal part from a solid extruded box and exploded sheet metal Flange and Hem features.

You modeled sheet metal components in their 3D formed state and will have them manufactured in their 2D flatten state.

Project Terminology

Top Down design approach: A conceptual approach used to develop products from within the assembly. Major design requirements are translated into sub-assemblies or individual components and key relationships

Layout sketch: Specifies the location of the key components. Components and assemblies reference the Layout Sketch.

CircuitWorks: CircuitWorks from Zeal Solutions (www.circuitworks.co.uk) is a fully integrated data interface between SolidWorks and PCB Design systems.

Link Values: Are used to define equal relations. Create an equal relation between two or more sketched dimensions and or features with a Link Value. Link Values require a shared name. Use Link Values within the same part.

Equations: Mathematical expressions that define relationships between parameters and or dimensions are called Equations. Equations use shared names to control dimensions. Use Equations to connect values from sketches, features, patterns and various parts in an assembly. Use Equations in different parts and assemblies.

Design Table: A Design Table is an Excel spreadsheet used to create multiple configurations in a part or assembly. Utilize a Design Table to create the small, medium and large BOX configurations.

Bend Allowance (Radius): The arc length of the bend as measured along the neutral axis of the material.

Flat Pattern: The flat manufactured state of a sheet metal part. It contains no bends.

Project Features

Extruded Boss/Base: Add material to the part. Utilize to create the CABINET.

Shell: Creates a constant wall thickness for the CABINET Extruded Base. The Shell feature represents the thickness of the sheet metal utilized for the cabinet.

Rib: Adds material between contours of existing geometry. Use Ribs to add structural integrity to a part.

Rip: A Rib feature is a cut of 0 thickness. Utilize a Rib feature on the edges of a shelled box before inserting sheet metal bends

Insert Bends: Converts a solid part into a sheet metal part. Requires a fixed face or edge and a bend radius.

Flange: Adds a flange to a selected edge of a sheet metal part.

Hem: Adds a hem to a selected edge of a sheet metal part.

Flattened: Creates the flat pattern for a sheet metal part. The Flattened feature toggles between the flat and formed state. A Flat configuration is created. Utilize the Flat configuration in the drawing to dimension the flat pattern.

Suppressed: A feature that is not displayed. Hide features to improve clarity. Suppressed features to improve model Rebuild time.

Pattern: A Pattern creates one or more instances of a feature or group of features. A Linear Pattern requires a seed feature and an edge reference for Direction1 and or Direction 2.

Derived Pattern: Pattern created inside the assembly. A Derived Pattern utilizes a component from the assembly. There are two methods to reference a Derived Pattern. Method 1 - Derived Pattern references a Linear or Circular feature pattern. Method 2 – Derived Pattern references a local pattern created in the assembly.

Reload: Refreshes shared documents.

Replace: Substitutes one or more instances of a component with a closed different component in an assembly.

Assembly Feature: Features created in the assembly such as a hole or extrude. The Hole assembly feature was utilized to create a hole through 2 different components in the BOX assembly.

Questions

1. What is a Top Down approach?

2. When do you create a Layout Sketch?

3. How do you create a new component in the context of the assembly?

4. What is the difference between a Link Value and an Equation?

5. Name three characteristics unique to sheet metal parts.

6. Identify the indicator for a part in an edit state.

7. Where should you position the Layout Sketch?

8. For a solid extruded block, what features do you add to create a sheet metal part?

9. Name the two primary states of a sheet metal part.

10. How do you insert a formed sheet metal feature such as a dimple or louver?

11. Identify the type of information that a sheet metal manufacturer provides.

12. What is an Assembly Feature?

13. What features are required before you create the Component Pattern in the assembly?

14. Define a Design Table.

Exercises

Exercise 6.1: L-BRACKET.

Create the L-BRACKET sheet metal parts.

Create a family of sheet metal L-BRACKETS, Figure E6.1a. L-BRACKETS are used in the construction industry.

Standardize your design. Create an eight hole L-BRACKET, Figure E6.1b. Use Equations to control the hole spacing.

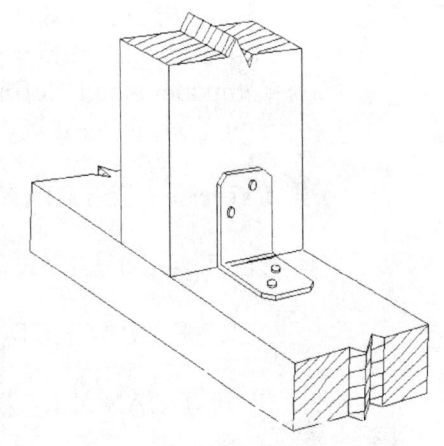

Figure E6.1a

Stong-Tie Reinforcing Brackets
Courtesy of Simpson Strong Tie
Corporation of California

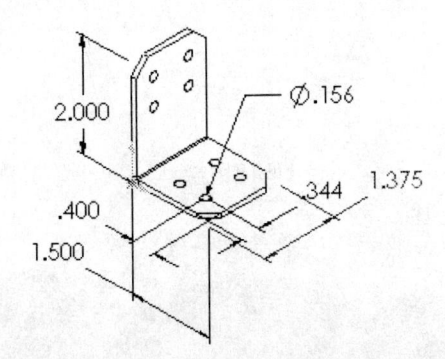

Figure E6.1b

Review Design Tables in On-line Help and Getting Started. Create a Design Table for the L-Bracket. Rename the dimensions of the L-Bracket.

	A	B	C	D	E
1	Design Table for: lbracket				
2		height@Sketch1	width@Sketch1	depth@Base-Extrude-Thin	hole1dia@Sketch2
3	First Instance	2	1.5	1.375	0.156
4	Small2x1.5x1	2	1.5	1	0.156
5	Medium3x2x2	3	2	2	0.156
6	Large4x3x4	4	3	4	0.25

L-BRACKET FAMILY TABLE

Exercise 6.2: LAYOUT SKETCH.

Create a Layout Sketch in the Assembly.

Create a Layout Sketch for a Top down design in a new appliance or consumer electronics (refrigerator, stereo CD player). Identify the major components that define the assembly. Define important relationships with Equations and Link Values.

Exercise 6.3: TRIANGULAR SUPPORT ASSEMBLY.

The TRIANGULAR SUPPORT Assembly consists of the following parts:

ANGLE-BRACKET-13HOLE

SINE-TRIANGLE

STANDOFF

MACHINE SCREW

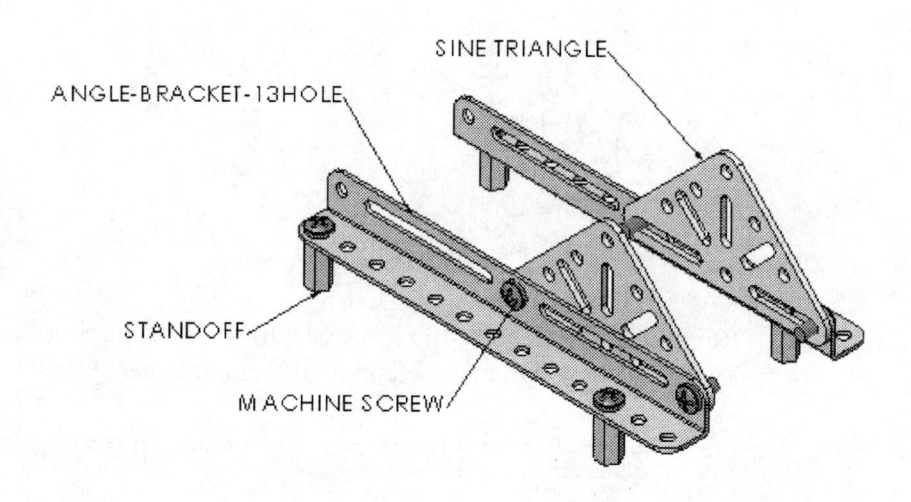

TRIANGULAR SUPPORT Assembly
Courtesy of Gears Educational Systems

Exercise 6.3a: ANGLE-BRACKET-13HOLE.

Create the ANGLE-BRACKET-13HOLE. Note the ANGLE-BRACKET with 7 holes was created in Exercise 4.1c.

The ANGLE-BRACKET-13HOLE is manufactured from .060 [1.5mm] Stainless Steel. Insert Sheet Metal Bends to flatten the part.

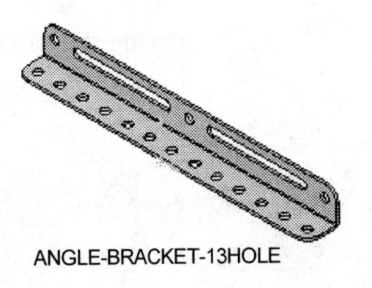

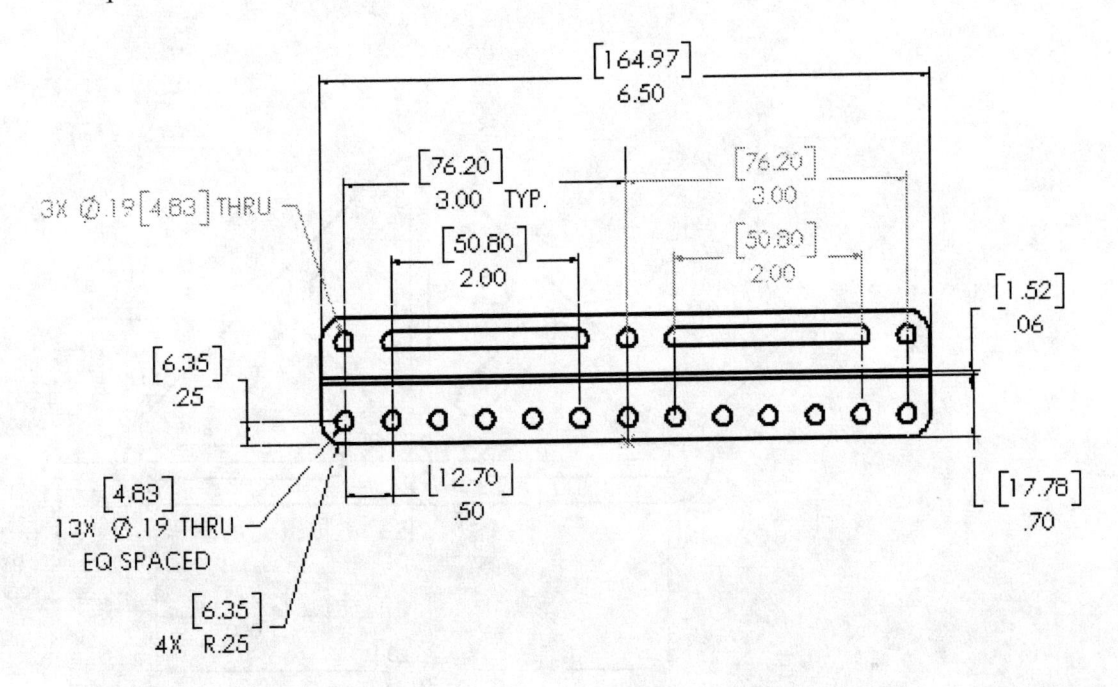

ANGLE-BRACKET-13HOLE

Save the ANGLE-BRACKET-13HOLE in the 3D formed state.

Exercise 6.3b: SINE-TRIANGLE.

Create the SINE-TRIANGLE on the Front plane. The SINE-TRIANGLE is manufactured from .060 [1.5mm] Stainless Steel.

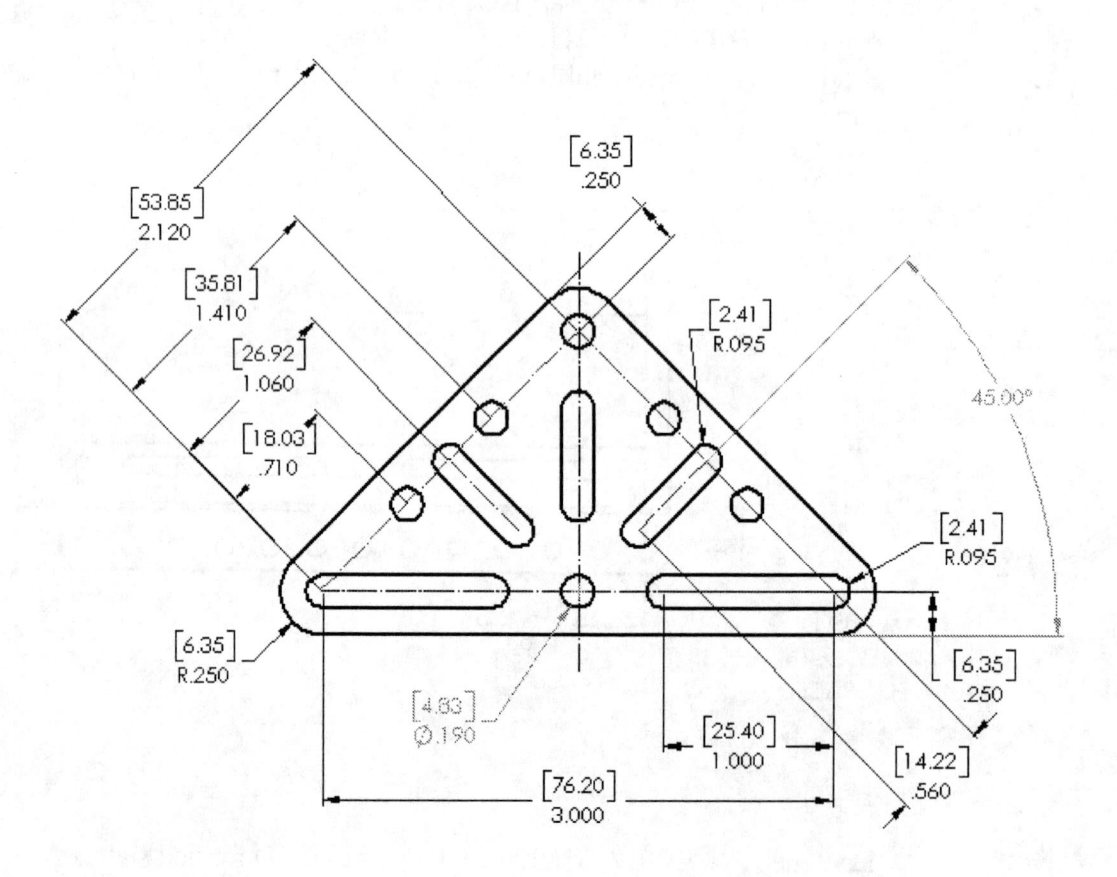

Exercise 6.3c: STANDOFF and MACHINE SCREW.

Create the hex profile STANDOFF. Utilize a 10-24 Tapped Hole. Utilize a 10-24 MACHINE SCREW. Note the MACHINE SCREW was created in Exercise 4.1d.

For Metric sizes utilize an M5 Tapped Hole and an M5 Machine Screw.

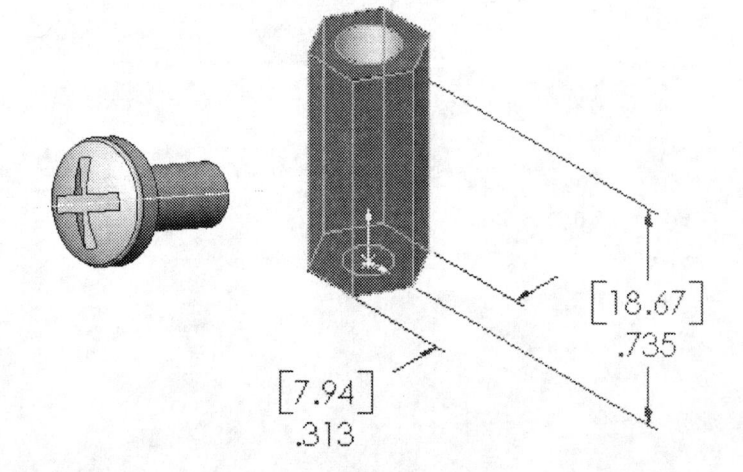

Exercise 6.3d: TRIANGULAR-SUPPORT ASSEMBLY.

Create the first half of the TRIANGLE SUPPORT assembly.

Insert the ANGLE-BRACKET-13HOLE part. The ANGLE-BRACKET-13HOLE is fixed to the Origin.

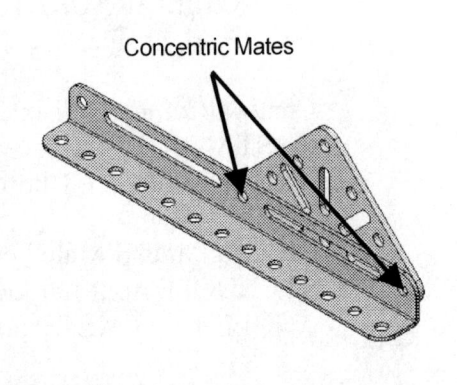

Insert the SINE-PLATE part behind the ANGLE-BRACKET-13HOLE. Create a Concentric Mate between the SINE-PLATE bottom right hole and the ANGLE-BRACKET-13HOLE right hole.

Create a Concentric Mate between the SINE-PLATE bottom left hole and the ANGLE-BRACKET-13HOLE center hole.

Create a Coincident Mate between the SINE-PLATE front face and the ANGLE-BRACKET-13HOLE back face.

Create a Create a new plane, Plane1. Plane1 is offset 1.75[44.45mm] from the front face of the ANGLE-BRACKET-13HOLE.

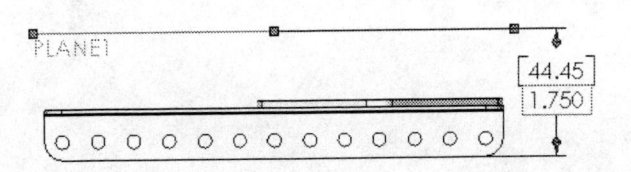

Insert the first STANDOFF part.

Create a Concentric Mate between the STANDOFF Tapped Hole and the ANGLE-BRACKET left hole.

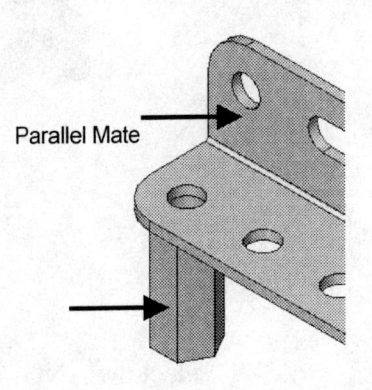

Create a Coincident Mate between the STANDOFF top face and the ANGLE-BRACKET bottom face.

Create a Parallel Mate between the STANDOFF front flat face and the ANGLE-BRACKET front face.

Insert the second STANDOFF. Create a Concentric Mate between the STANDOFF Tapped Hole and the ANGLE-BRACKET third hole from the right side.

Create a Coincident Mate between the STANDOFF top face and the ANGLE-BRACKET bottom face.

Create a Parallel Mate between the STANDOFF front flat face and the ANGLE-BRACKET front face.

Insert the MACHINE SCREW part. There are 4 MACHINE SCREWS. Create a Concentric Mate and Coincident Mate for each MACHINE SCREW.

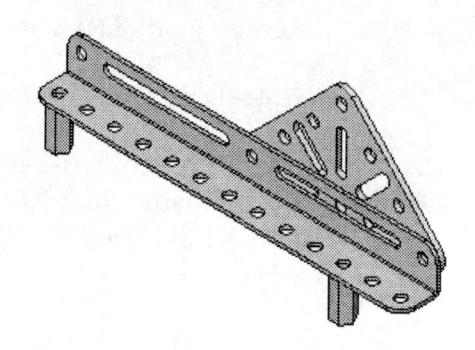

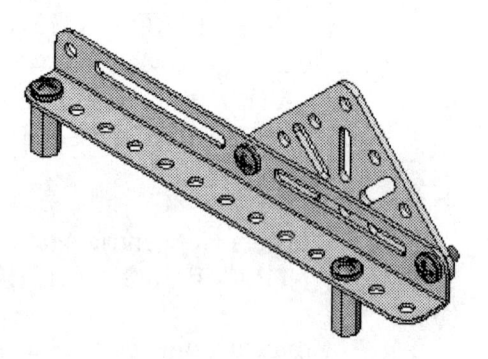

Mirror all components about Plane1. Select Insert, Mirror Components. PLANE1 is the Mirror Plane.

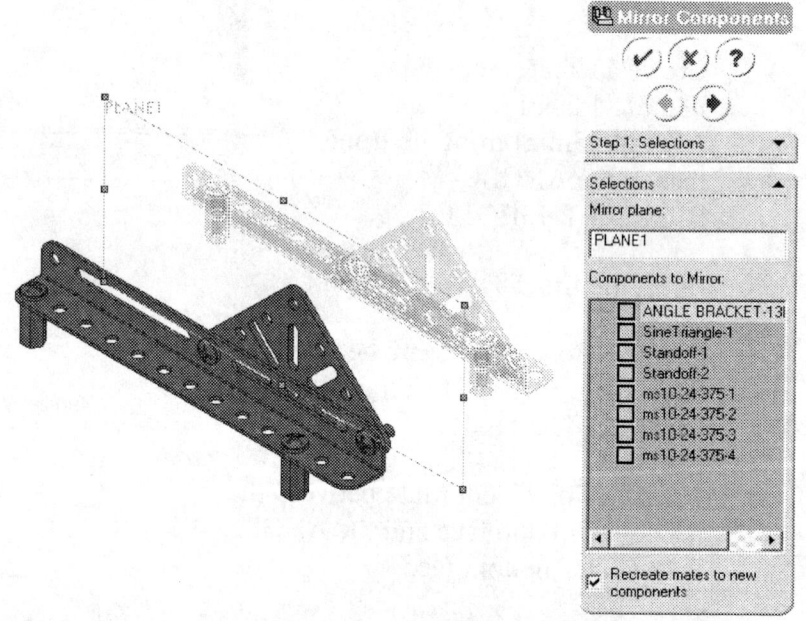

Select the ANGLE-BRACKET-13HOLE, SINE-TRIANGLE, STANDOFF<1>, STANDOFF<2>, and 4-MACHINE SCREWS from the FeatureManager.

Check the Recreate mates to new components.

If the mirror components are upside down, click the blue arrow. Select the Reorient Components button.

Organize components in the FeatureManager. Group the MACHINE SCREWS. Drag each MACHINE SCREW to the bottom of FeatureManager. The 8-MACHINE SCREWs are listed sequentially. Select the eight MACHINE SCREWS. Right-click Add to New Folder. Enter hardware for folder name. All the MACHINE SCREWS are listed below the hardware folder.

Group the STANDOFFs. Drag STANDOFF<1> above STANDOFF<3>. Drag STANDOFF<2> below STANDOFF<1>. The four STANDOFFs are grouped together.

Select the four STANDOFFs. Right-click Add to New Folder. Enter standoffs for folder name.

All the STANDOFFs are listed below the standoffs folder.

Suppress the hardware folder from the FeatureManager.

The TRIANGLE-SUPPORT Assembly is complete.

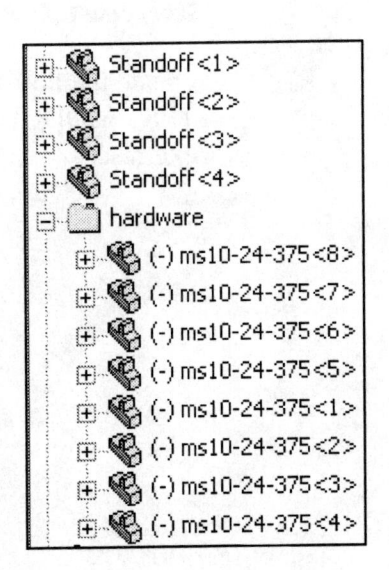

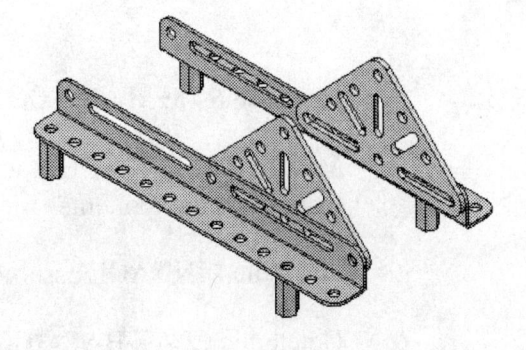

Exercise 6.4: PNEUMATIC TEST MODULE Assembly.

Create the PNEUMATIC TEST MODULE Assembly. The PNEUMATIC TEST MODULE Assembly utilizes 4 subassemblies created in previous exercises.

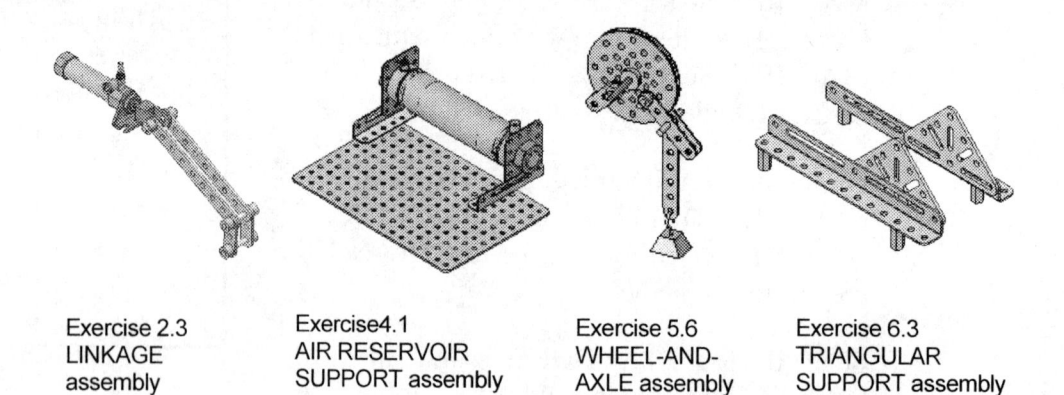

Exercise 2.3
LINKAGE
assembly

Exercise 4.1
AIR RESERVOIR
SUPPORT assembly

Exercise 5.6
WHEEL-AND-
AXLE assembly

Exercise 6.3
TRIANGULAR
SUPPORT assembly

Courtesy of SMC Corporation of America & Gears Educational Systems

Exercise 6.4a: Modify the LINKAGE assembly.

Modify the LINKAGE assembly. Delete components no longer required. Insert new components.

Open the LINKAGE assembly.

Delete the FLAT-BAR-3HOLE<1> and FLAT-BAR-3HOLE<2>.

Delete AXLE<3>. Delete the SHAFT-COLLAR<5> and SHAFT-COLLAR<6>.

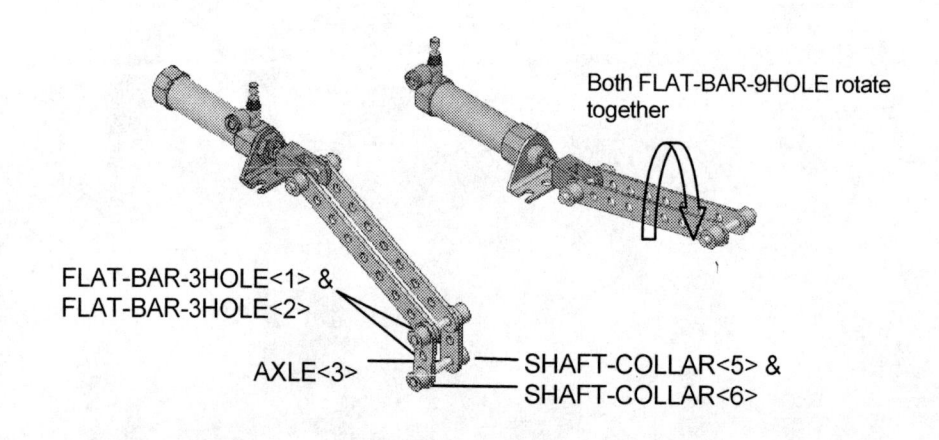

Both FLAT-BAR-9HOLE rotate together

FLAT-BAR-3HOLE<1> &
FLAT-BAR-3HOLE<2>

AXLE<3>

SHAFT-COLLAR<5> &
SHAFT-COLLAR<6>

Insert the first STANDOFF part.

Create a Concentric Mate between the Temporary Axis of the STANDOFF Tapped Hole and the Axis of the half Slot Cut.

Create a Coincident Mate between the STANDOFF top face and the BRACKET bottom face.

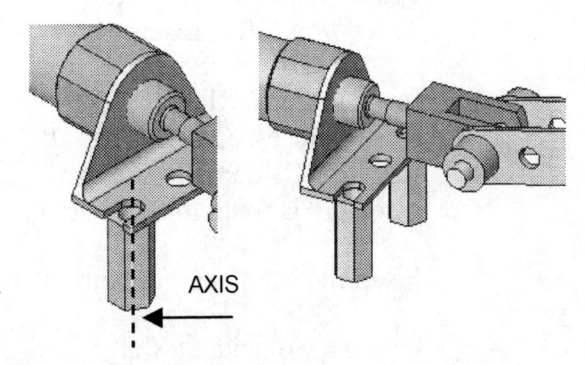

Create a Parallel Mate between the STANDOFF front face and the BRACKET front face.

Insert the second STANDOFF part. Create a Concentric, Coincident and Parallel Mate.

Save the LINKAGE assembly.

Exercise 6.4b: Modify the WHEEL-AND-AXLE assembly.

Open the WHEEL-AND-AXLE assembly.

The WHEEL-AND-AXLE rotates in the PNEUMATIC TEST MODULE assembly.

Expand the Mates. Suppress the Parallel Mate for the LINK-AND-HOOK and WHEEL-AND-AXLE Top Plane. The LINK-AND-HOOK is free to rotate.

Suppress the WHEEL-STATIC Angle Mate between the WHEEL Top Plane and the WHEEL-AND-AXLE Top Plane. The WHEEL is free to rotate.

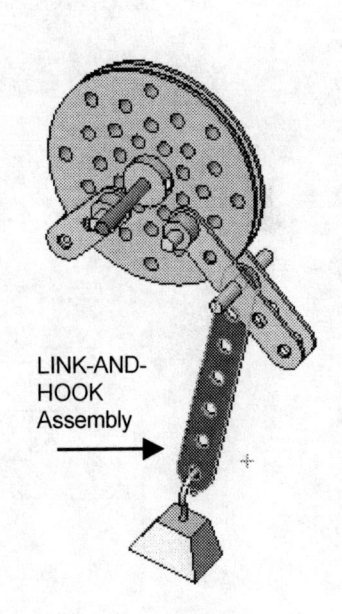

Save the WHEEL-AND-AXLE assembly.

Exercise 6.4c: PNEUMATIC TEST MODULE Assembly.

Create the PNEUMATIC TEST MODULE assembly.

Insert the AIR RESERVOIR SUPPORT assembly. The AIR RESERVOIR SUPPORT assembly is fixed to the Origin.

Concentric Mate Hole Position 5th row, 4th column.

Insert the LINKAGE assembly.

Create a Coincident Mate between the front STANDOFF Tapped Hole and the 5th row, 4th column in the FLAT PLATE part.

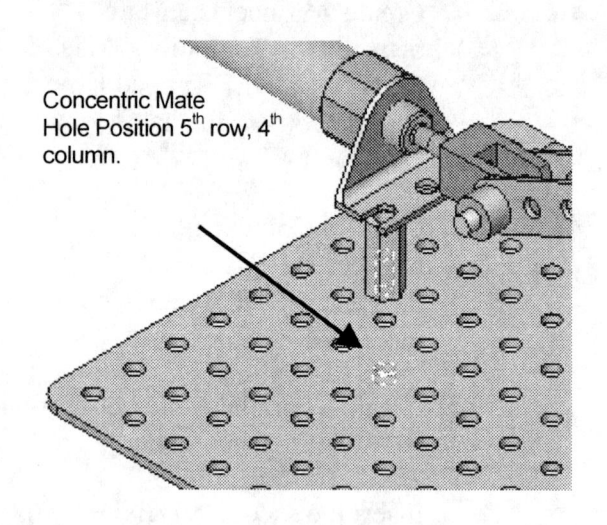

Create a Coincident Mate between the STANDOFF bottom face and the FLAT PLATE top face.

Create a Parallel Mate between the LINKAGE Front plane and the PNEUMATIC TEST MODULE Front plane.

The LINKAGE assembly is fully defined and centered on the FLAT PLATE.

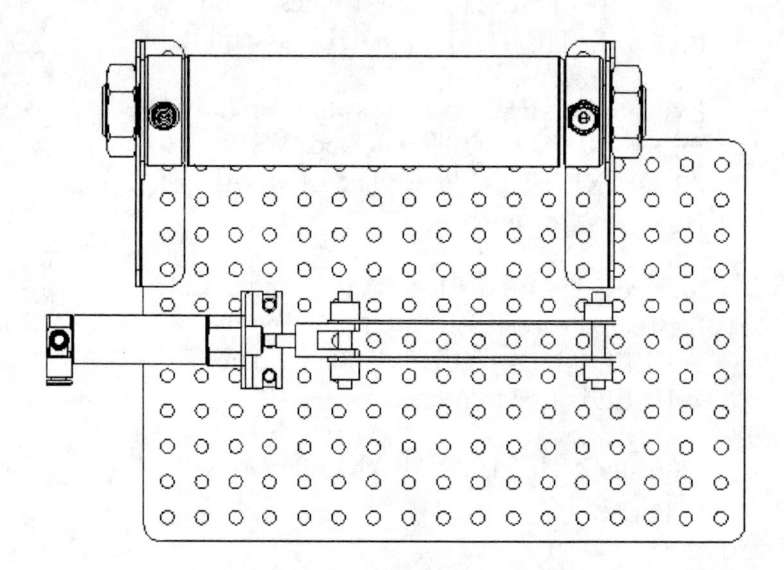

Components do not translate or rotate after insertion into the assembly. The LINKAGE assembly FLAT BARs do not rotate after insertion into the PNEUMATIC TEST MODULE assembly. The LINKAGE assembly is in the Rigid state.

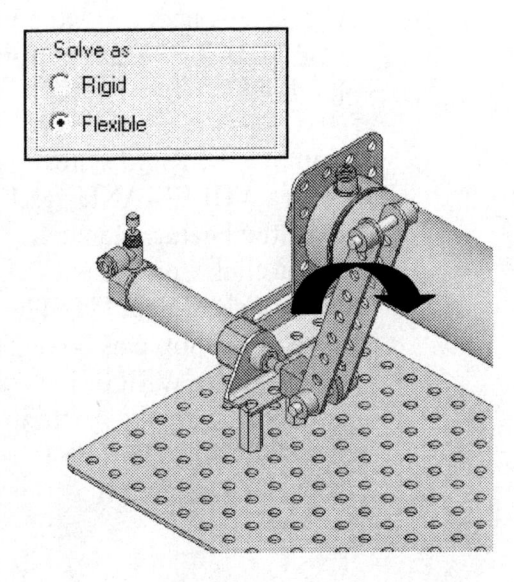

Remove the Rigid state. Select LINKAGE from the FeatureManager. Right-click Component Properties. Select Flexible for state in the Solve as box. The two FLATBAR-9HOLE in the LINKAGE assembly are free to rotate in the PNEUMATIC TEST MODULE assembly.

Insert the WHEEL-AND-AXLE assembly.

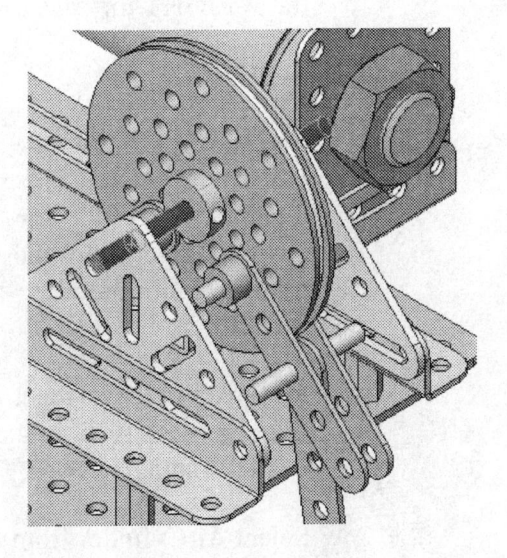

Create a Coincident Mate between the WHEEL-AND-AXLE Front plane and the LINKAGE assembly Front plane.

Create as Concentric Mate between the WHEEL AXLE-3000 and the SINE-TRIANGLE top hole.

Move the two FLAT-BAR-7HOLE downward.

Create a Concentric Mate between the LINKAGE right AXLE-1375 and the FLAT-BAR-3HOLE bottom left hole.

Remove the Rigid state. Select WHEEL-AND-AXLE from the FeatureManager. Right-click Component Properties. Select Flexible for state in the Solve as box. The LINK-AND-WEIGHT assembly is free to rotate in the PNEUMATIC TEST MODULE assembly.

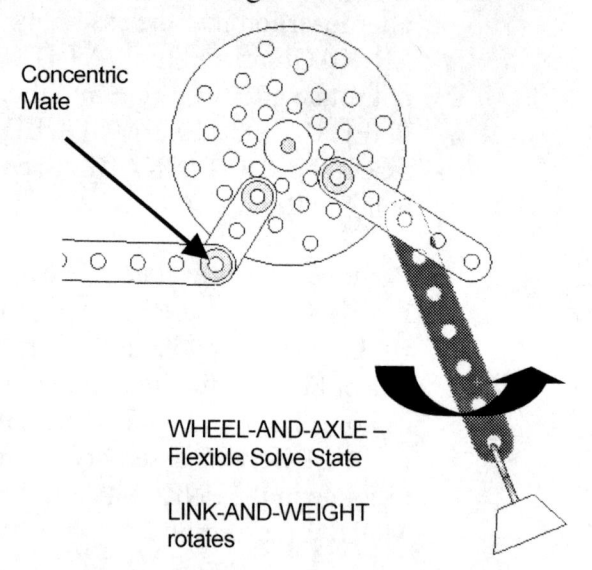

Concentric Mate

WHEEL-AND-AXLE – Flexible Solve State

LINK-AND-WEIGHT rotates

Create a Parallel Mate. Select the WEIGHT top face. Select the PNEUMATIC TEST MODULE Top plane.

The WHEEL does not translate or rotate. The linear motion of the Air Cylinder determines the rotational motion of the WHEEL.

Remove the Rigid state. Expand LINKAGE in the FeatureManager.

Select AirCylinder from the FeatureManager. Right-click Component Properties. Select Flexible for state in the Solve as box.

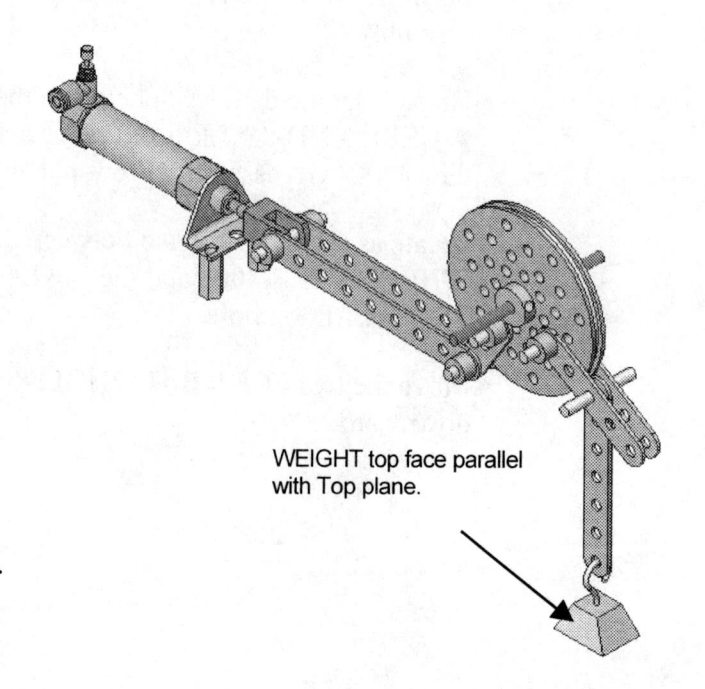

WEIGHT top face parallel with Top plane.

Engineering Design with SolidWorks **Top Down Assembly Modeling**

Expand NCJ22005Z-MP assembly.
Expand the MateGroup1 folder.
Click Stroke-Distance Mate.
Right-click Suppress.

Move the LINKAGE assembly.
Right-click the ROD CLEVIS in
the Graphics window. Click Move
Component. Select By Delta XYZ.
Enter 1.000 [25.4mm] for ΔX.

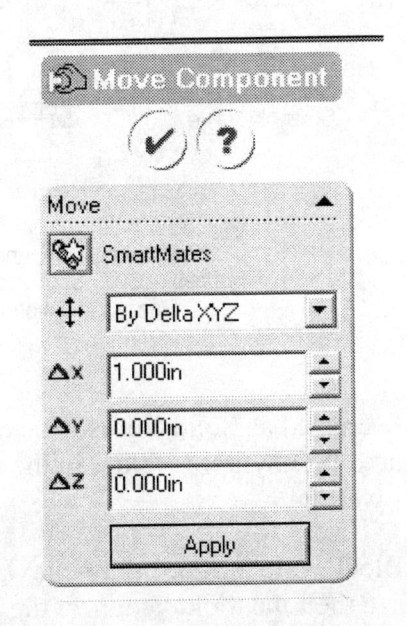

The AirCylinder Rod extends to the right. The WHEEL rotates and lifts the WEIGHT in a vertical direction.

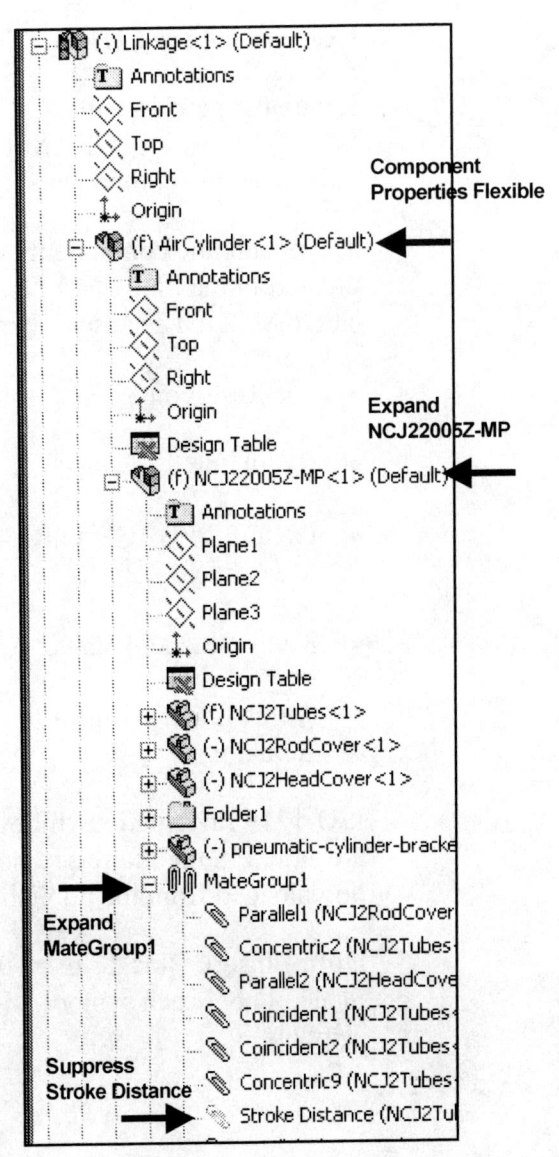

Save the PNEUMATIC TEST MODULE assembly.

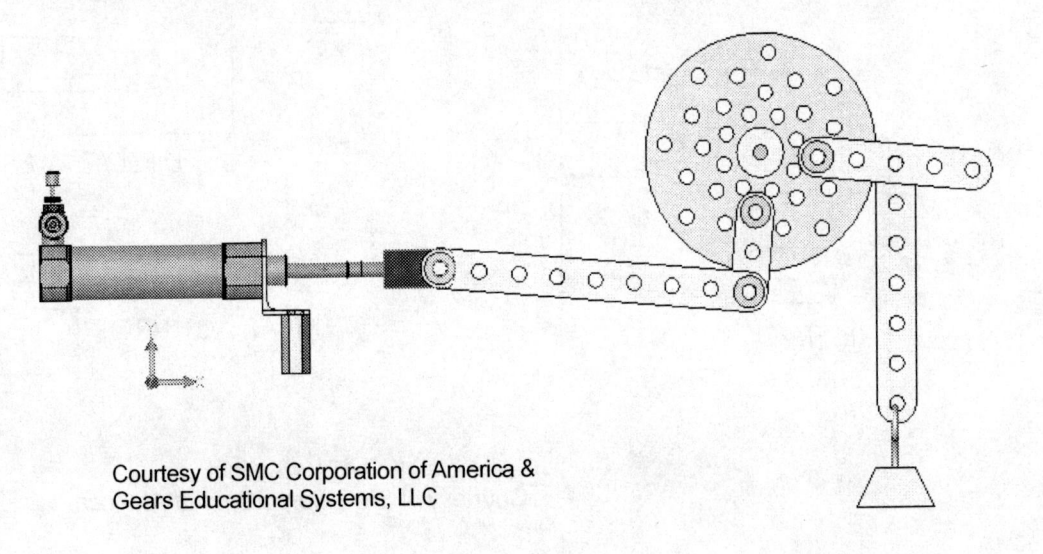

Courtesy of SMC Corporation of America &
Gears Educational Systems, LLC

PAGE 6 - 93

Exercise 6.4d: SCHEMATIC DIAGRAM

Create the new drawing, SCHEMATIC DIAGRAM for the pneumatic components.

The pneumatic components utilized in the PNEUMATIC TEST MODULE Assembly are:

- Air Reservoir
- Regulator
- ON/OFF/PURGE Valve – Mechanical 2/2
- 3Way Solenoid Valve
- Air Cylinder – Linear Actuator

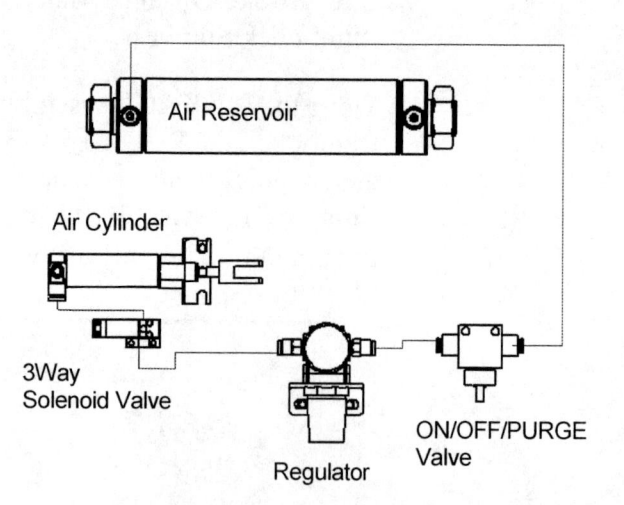

Pneumatic Components Diagram
Courtesy of SMC Corporation of America and Gears Educational Systems.

ISO-1219 Pneumatic Symbols are created as SolidWorks Blocks. The Blocks are stored in the Exercise Pneumatic ISO Symbols folder in the files downloaded from the publisher's website.

Utilize Insert, Block. Insert the Blocks into a B-size drawing. Enter 0.1 for Scale. Label each symbol. Utilize the Line tool to connect the pneumatic symbols.

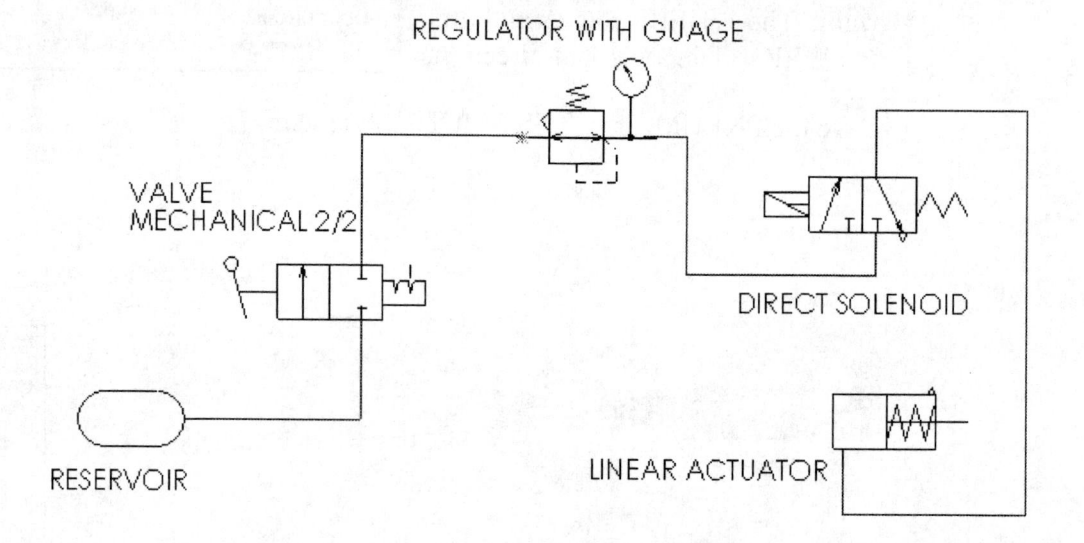

ISO-1219 Symbols
Courtesy of SMC Corporation of America

Exercise 6-4e: PNEUMATIC TEST MODULE Assembly.

The PNEUMATIC TEST MODULE Assembly was partially completed in Exercise 6.4. Additional pneumatic components are required.

The Regulator utilizes a plastic knob to control the pressure from the Air Reservoir.

The Knob controls the Pressure at P2 by adjusting the screw loading on the setting Spring. The Main valve is held open, allowing flow from the inlet, P1 to the outlet, P2. When the air consumption rate drops, the force at P2 increases. This increase in force causes the Diaphragm to drop maintaining the constant pressure through the valve.

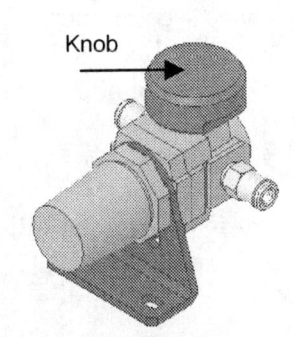

Regulator Assembly
Courtesy of SMC Corporation of America

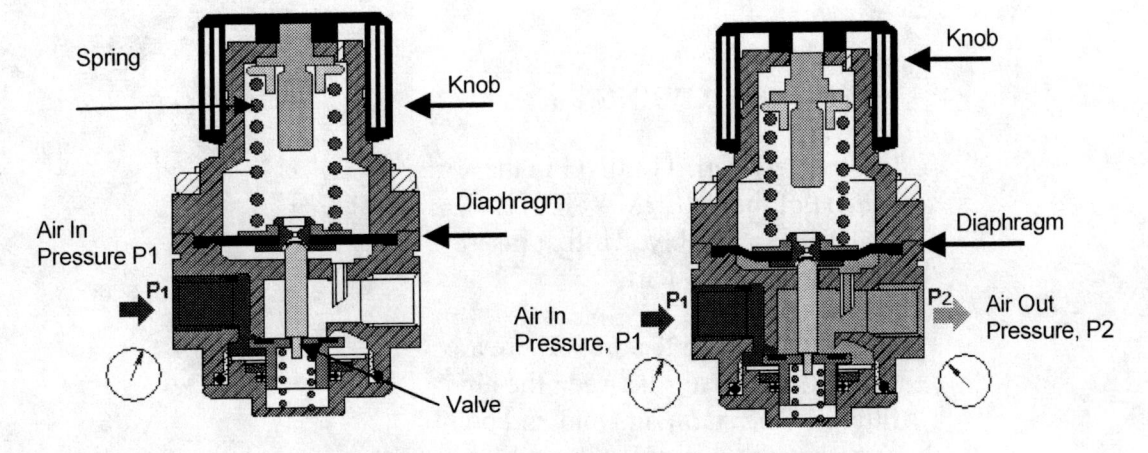

Images Courtesy of SMC Corporation of America

Plastic components are utilized in a variety of applications. The ON/OFF/PURGE value utilizes plastic components for the Knob and Inlet and Outlet ports.

The ON/OFF/PURGE valve controls the airflow from the Regulator. See Exercise 2.4 for valve operation.

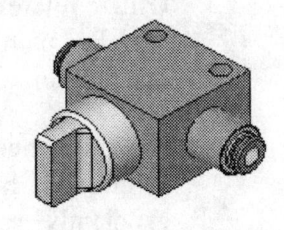

ON/OFF/PURGE Valve
Courtesy of SMC Corporation of America

The 3Way Solenoid value utilizes a plastic housing to protect the internal electronic components. The 3Way Solenoid value controls the electrical operation of to the Air Cylinder. The Solenoid acts like a switch.

Insert the following components into PNEUMATIC TEST MODULE Assembly:

- ON/OFF/PURGE VALVE (Exercise 2.4)

- REGULATOR

- SOLENOID

3WAY SOLENOID VALVE
Courtesy of SMC Corporation of America

Exercise 6.4f: TUBING Part.

Insert a new part, TUBING in the context of the PNEUMATIC TEST MODULE assembly. Utilize Insert, Component, New Part.

Utilize the Sweep feature. There are four different paths between the air fittings. The start point and end point of each path references the air fittings between pneumatic components. The profile cross section is a 5/32[M5] circle.

Utilize reference planes to create sketched paths. Utilize on line help for information to create 3D sketches.

Note: Engineers utilize the SolidWorks Piping Add-In software application to efficiently create tubing, wiring and piping in assemblies.

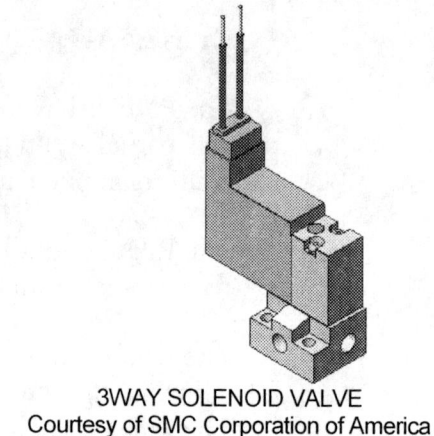

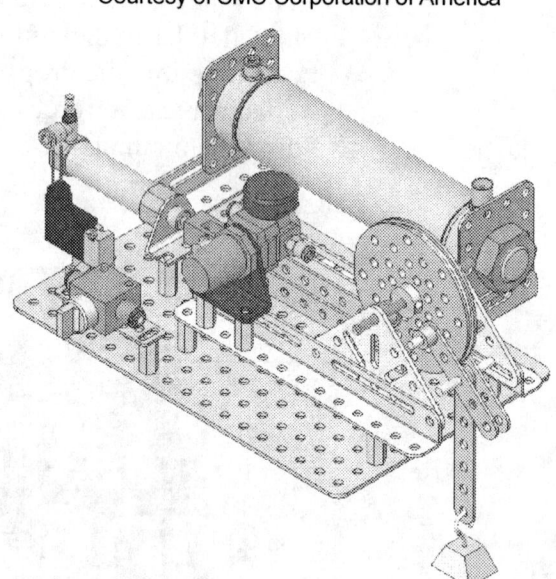

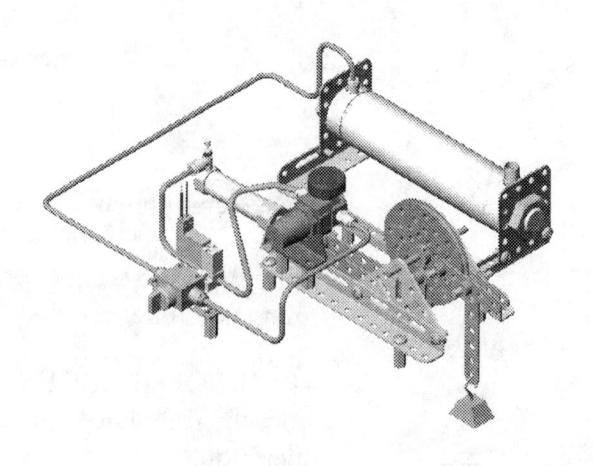

Engineering Design with SolidWorks   Top Down Assembly Modeling

Exercise 6.5: Industry Collaborative Exercise.

Create a new SolidWorks part from imported 2D geometry.

In Exercise 1.11 through 1.13, you created profiles from the company 80/20, Inc. (www.8020.net). Hundreds of .dwg(AutoCAD Drawing) and .dxf files (Drawing Exchange Format) exist on their web site. Many companies support .dwg or .dxf file format.

Download the 2D .dwg library from www.8020.net (8Mb) or open the 40-4325 drawing from the files downloaded from the publisher's website.

Open the part, 40-4325.dwg. Click Open. Select .dwg for File type. Click Import to Part. Click Next. Click Millimeters for Units. Click Finish. Select a Part metric template from the New dialog box.

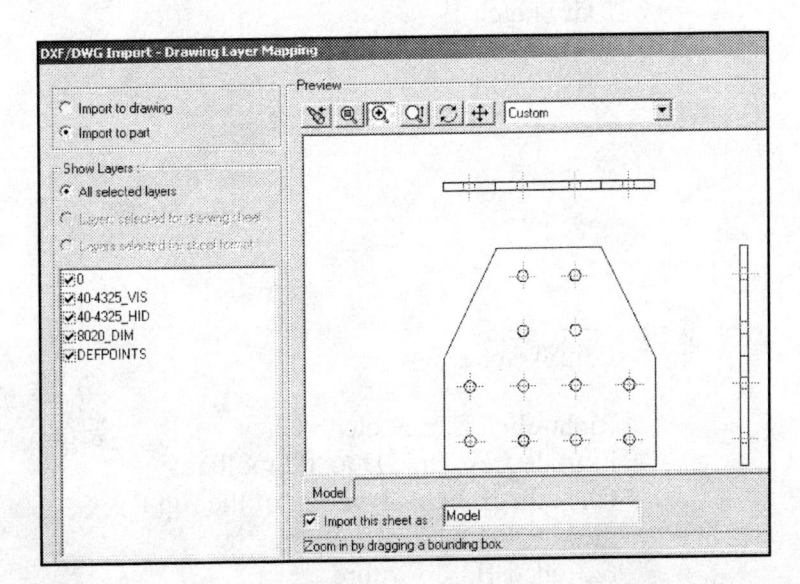

A New part is opened and the 2D Autocad geometry is inserted into Sketch1. The 2D to 3D Toolbar is displayed.

Window Select the Top view. Delete the geometry.

Define the views in Sketch1. Window Select the Front view. Click the Front icon from the 2D to 3D Toolbar. Sketch2 is displayed in the FeatureManager. Do not exit Sketch1.

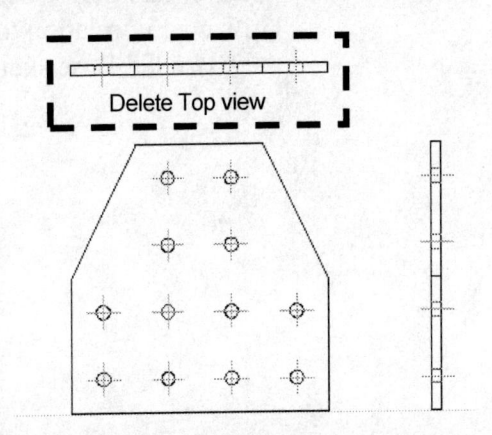

Model Courtesy of 80/20, Inc., Colombia City, IN USA.

Top Down Assembly Modeling **Engineering Design with SolidWorks**

Window Select the Right view. Click the Right ⬜ icon from the 2D to 3D Toolbar. Sketch3 is displayed in the FeatureManager. The Right view rotates 90 degrees.

Align the Right view. Display the Isometric view. Click the lower left corner of the right view sketch. Click Align Sketch ⬜ icon from the 2D to 3D toolbar. The right view is aligned to the Front plane.

Exit Sketch 1.

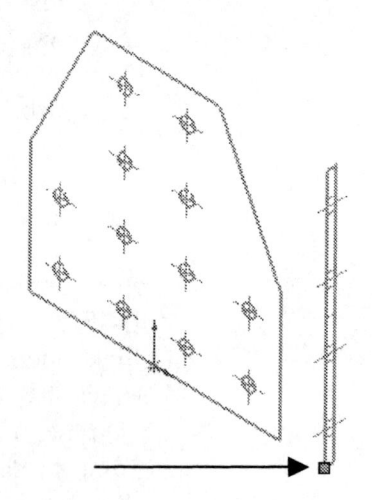

Edit Sketch2.

Right-click Edit Sketch. Click Extrude from the 2D to 3D toolbar. Click the right back vertex of the right view. Click OK to complete the Extruded-Base-Feature.

Sketch2 contains no dimensions and is under defined. The part geometry was generated from the .dwg file.

Right-click Extrude1 in the FeatureManager. Click Edit Sketch. Utilize a Fix relation to fully define the Extruded Base sketch.

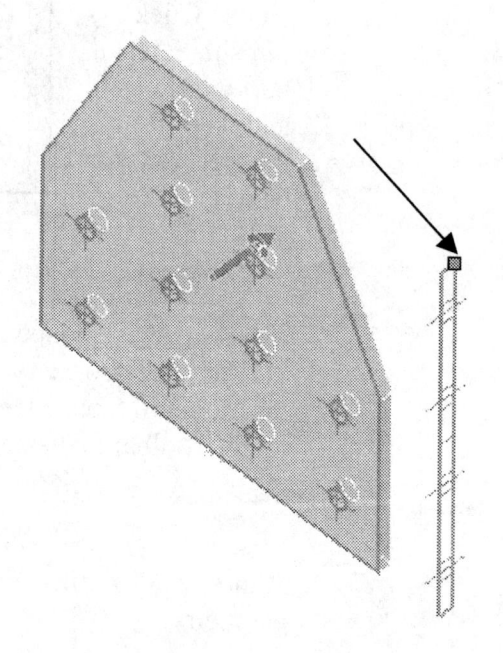

Exercise 6.6: Project Data Management (PDM).

Three months have passed. The BOX now contains hundreds of components. Many engineers and designers work to complete the project on time. How do you manage the parts, drawings and assemblies? PDM/Works is a project data management software application that runs inside SolidWorks.

The application is comprised of four major components:

- Automated Revision Control.

- Check In/Check Out of Vaulted Area for Security.

- Support of Concurrent Engineering Activities.

- Maintaining a History of Design Activities.

Use the World Wide Web to research PDM four major components. How would PDM be used in the concurrent engineering application with the BOX assembly? Identify other departments in your company that would utilize a PDM system.

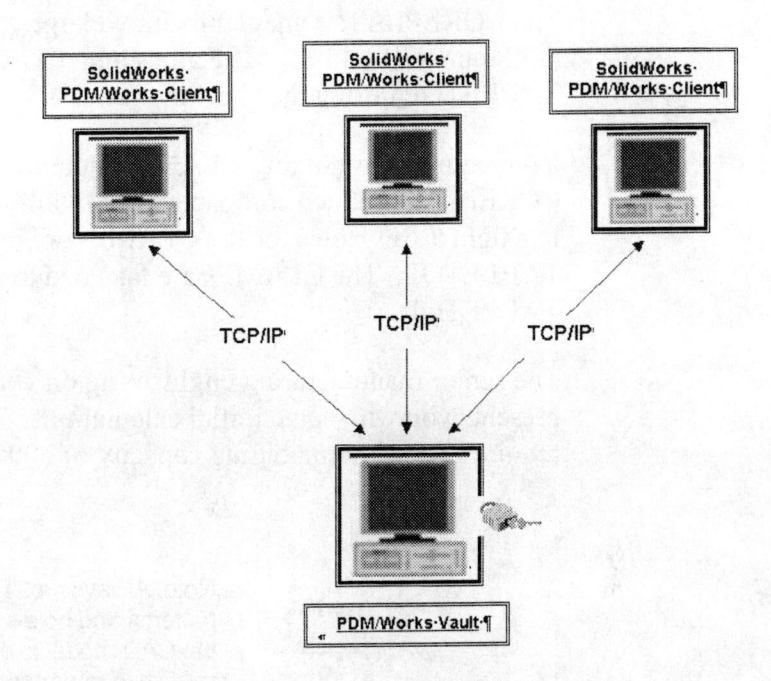

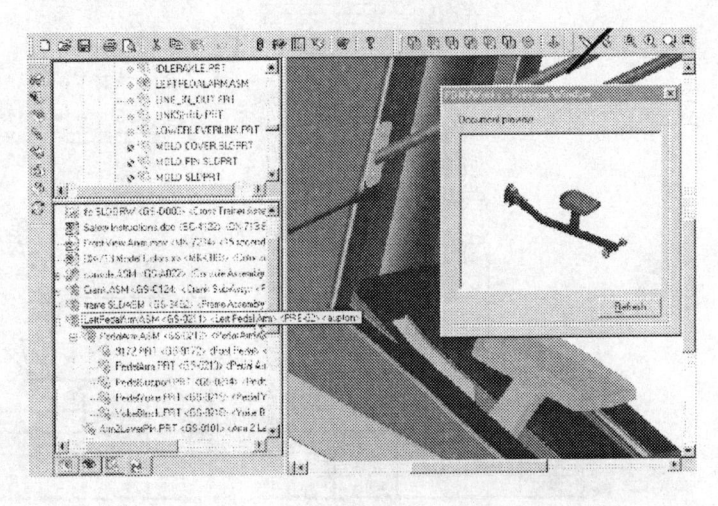

Exercise 6.7: Industry Collaborative Exercise.

Create a quality inspection CHECKING STATION assembly. Work with the team of manufacturing engineers. The quality inspection CHECKING STATION is comprised of four aluminum COREPLATEs. Each COREPLATE measures 20in. x 14in., (508mm x 355.6mm).

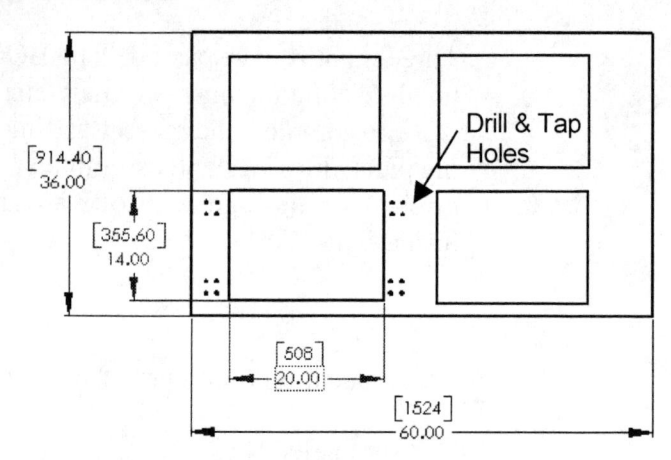

The COREPLATEs are clamped to a large steel TABLETOP 60in. x 36in., (1524mm x 914.4mm). The aluminum COREPLATEs are clamped to the TABLETOP during the checking process.

You require four holding CLAMPs mounted at the corners of the COREPLATE. Each compact CLAMP allows the quality inspector to operate in a tight area. Holes for the CLAMPs are drilled and tapped on the TABLETOP. The CLAMPs are fastened to the TABLETOP.

The senior manufacturing engineering on your team presents you with some initial calculations. The CLAMP requires a minimum holding capacity of 500 lb (340 daN).

Product Catalogs
▼Industrial Products Group
　▼World of Clamping
　　▼Horizontal Hold-Down
　　　▼Model 2371-U
　　　　2371-U
　　　▶Model 515
　　　▶Series 205
　　　▶Series 206-SS
　　　▶Series 213, 217, 227, 245
　　　▶Series 215
　　　▶Series 225
　　　▶Series 235
　　　▶Series 305, 307, 309

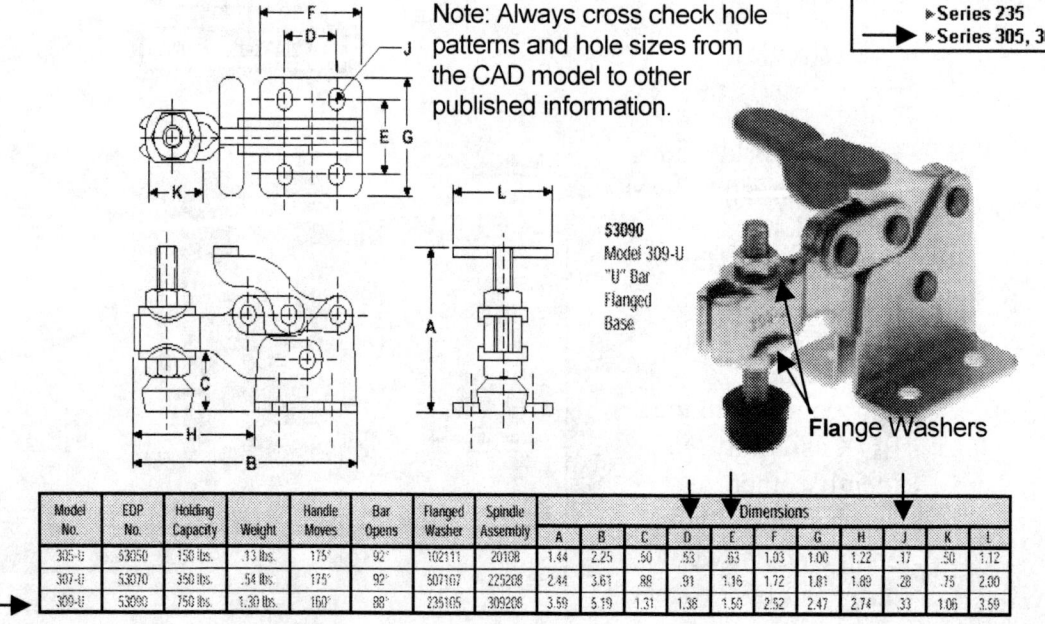

Note: Always cross check hole patterns and hole sizes from the CAD model to other published information.

53090
Model 309-U
"U" Bar Flanged Base

Flange Washers

Model No.	EDP No.	Holding Capacity	Weight	Handle Moves	Bar Opens	Flanged Washer	Spindle Assembly	A	B	C	D	E	F	G	H	J	K	L
305-U	53050	150 lbs.	.13 lbs.	175°	92°	102111	20108	1.44	2.25	.50	.53	.63	1.03	1.00	1.22	.17	.50	1.12
307-U	53070	350 lbs.	.54 lbs.	175°	92°	507167	225208	2.44	3.61	.88	.91	1.16	1.72	1.81	1.89	.28	.75	2.00
309-U	53090	750 lbs.	1.30 lbs.	160°	88°	235105	309208	3.59	5.19	1.31	1.38	1.50	2.52	2.47	2.74	.33	1.06	3.59

Engineering Design with SolidWorks | Top Down Assembly Modeling

Select the Horizontal Hold Down Clamp, part number 309-U manufactured by DE-STA-CO Industries: Madison Heights, MI USA. Visit the DE-STA-CO web site (www.destaco.com). Click the Product Catalogs button. Click Industrial Products Group, World of Clamping, Horizontal Hold-Down, Series 305, 307, 309. The Model No. 309-U row contains a Flanged Washer and a Spindle Assembly part numbers. Record the part numbers and hole dimensions.

Download the components required to create the CLAMP-WASHER-SPINDLE sub assembly.

Use the On-line Help to review the commands: Linear Pattern, Mirror, Mirror All, Component Pattern, Mirror Component.

Manually sketch a plan of your assembly. Develop the Layout Sketch in the CHECKING STATION assembly. Create the components TABLETOP and COREPLATE from the Layout Sketch. Add the CLAMP-WASHER-SPINDLE sub assembly.

Product Images
Courtesy of DE-STA-CO Industries
Madison Heights, MI USA

Insert 4 socket head cap SCREWs for each CLAMP. Create an assembly Drawing and a Bill of Materials for the CHECKING STATION.

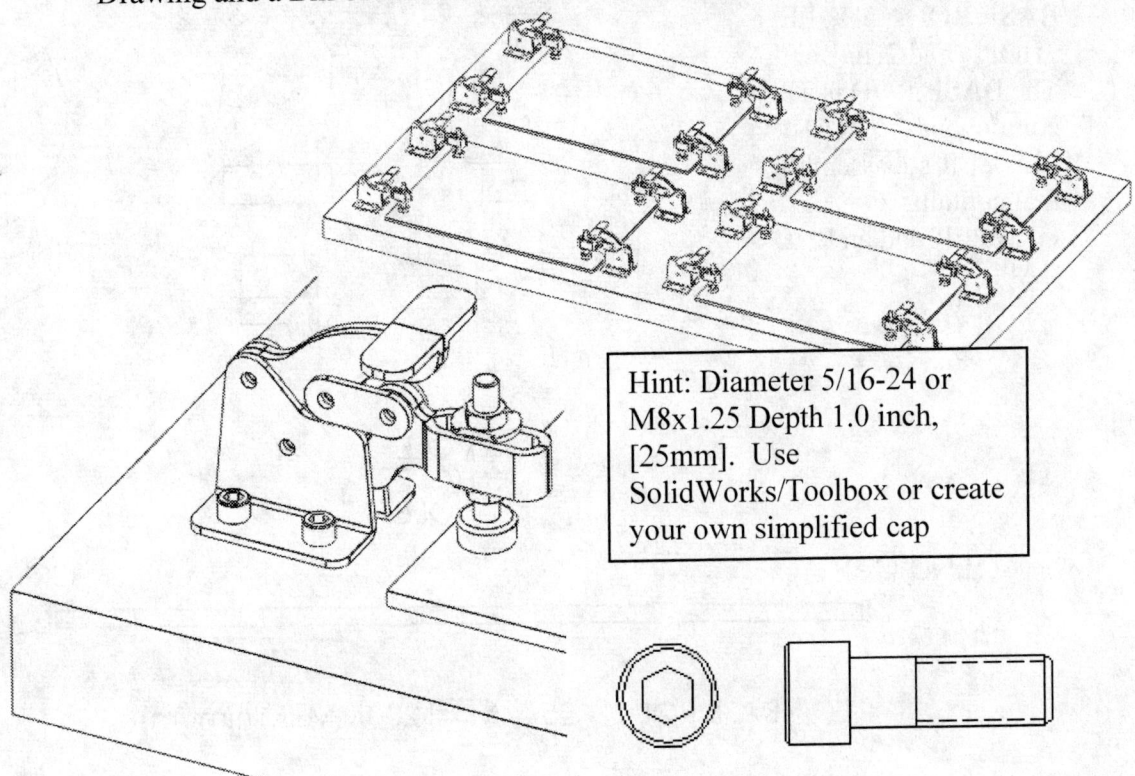

Hint: Diameter 5/16-24 or M8x1.25 Depth 1.0 inch, [25mm]. Use SolidWorks/Toolbox or create your own simplified cap

PAGE 6 - 101

Exercise 6.8: Industry Collaborative Exercise.

You are part of a team to develop a SOCCER PLAY TABLE Assembly. Partial dimensions have been provided. Develop the assembly based upon these design constraints. Research the World Wide Wide. Utilize Tools, 3D Content Central and www.globalspec.com to find components such as handles, rods, caps and bushings. Answers will vary. Read all requirements first.

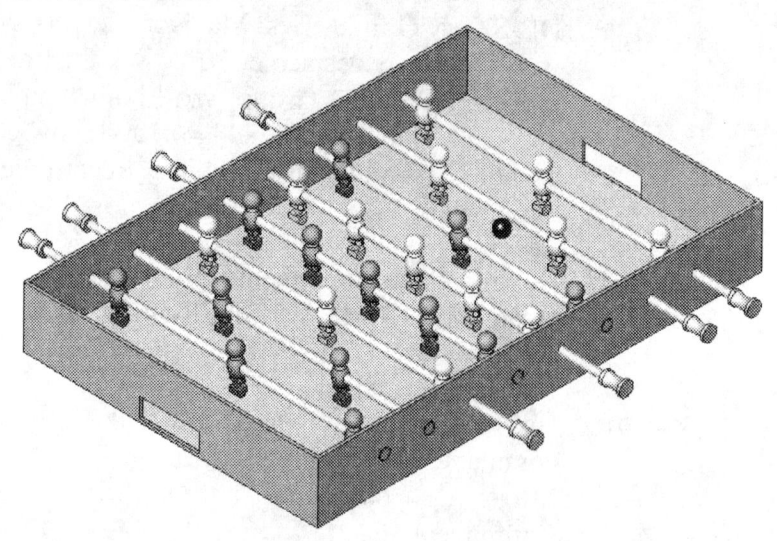

The SOCCERPLAYER is one part. The maximum dimensions are 50mm x 50mm x 165 mm.

The BASE ROD Thru Hole is 25mm. The Screw Blind Hole is 12mm.

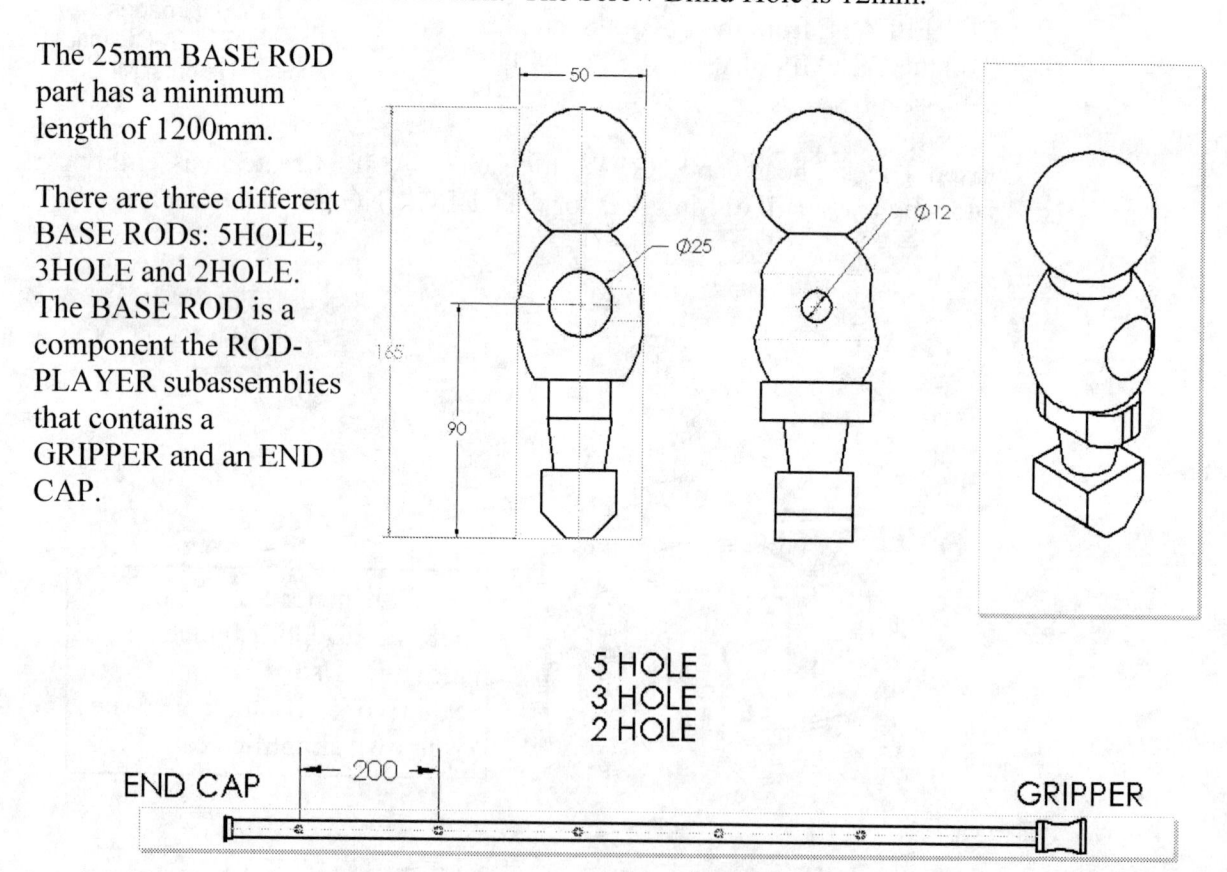

The 25mm BASE ROD part has a minimum length of 1200mm.

There are three different BASE RODs: 5HOLE, 3HOLE and 2HOLE. The BASE ROD is a component the ROD-PLAYER subassemblies that contains a GRIPPER and an END CAP.

5 HOLE
3 HOLE
2 HOLE

END CAP — 200 — GRIPPER

BASE ROD ⌀ 25 MM x 1200MM (Minimum)

There are six subassemblies.

5PLAYER-HOME	5PLAYER-AWAY
3PLAYER-HOME	3PLAYER-AWAY
2PLAYER-HOME	2PLAYER-AWAY

The TABLE size is 1000mmx1500mmx200mm. The Thickness of the TABLE is 10mm. The spacing between the six subassemblies is 150mm. The Thru Holes in the TABLE contain a Bushing. The Thru Hole height is based upon the PLAYER size.

Plan the parts and assemblies before starting this project. Utilize individual parts and assemblies or create configurations with Design Tables.

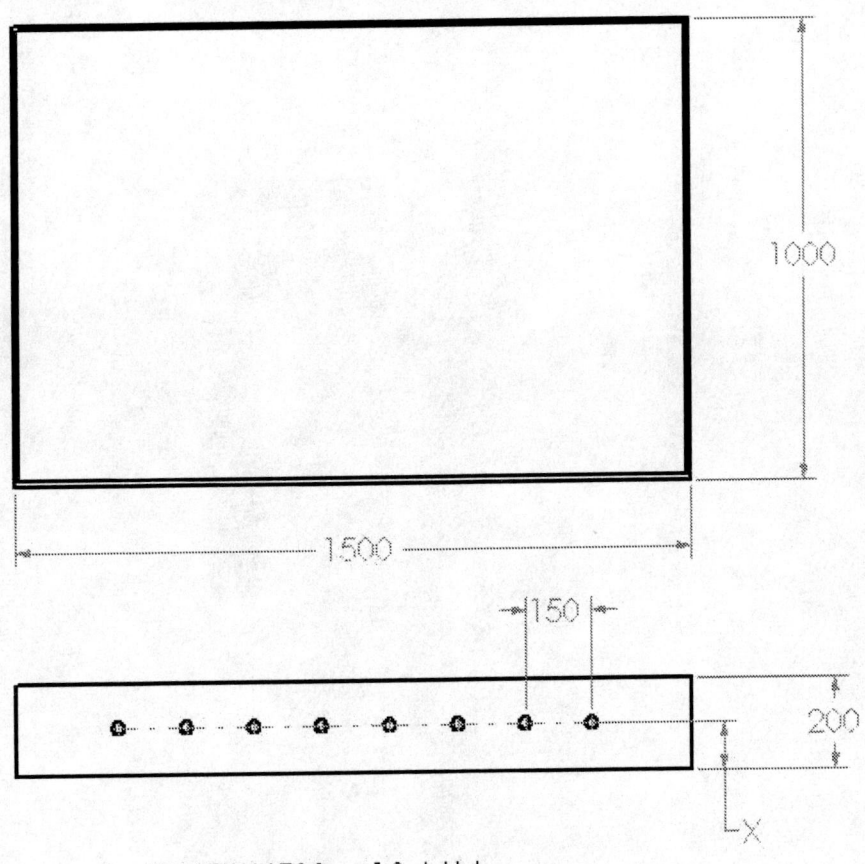

THICKNESS = 10 MM

Notes:

Appendix

Engineering Changer Order (ECO)

D&M Engineering Change Order				ECO # _____ Page 1 of __	
Product Line	☐ Hardware ☐ Software ☐ Quality ☐ Tech Pubs			Author / Date / Authorized Mgr. / Date	
Change Tested By					
Reason for ECO (Describe the existing problem, symptom and impact on field)					
D&M Part No.	Rev From/To	Part Description	Description		Owner

ECO Implementation/Class		Departments	Approvals	Date
All in Field	☐	Engineering		
All in Test	☐	Manufacturing		
All in Assembly	☐	Technical Support		
All in Stock	☐	Marketing		
All on Order	☐	DOC Control		
All Future	☐			
Material Disposition		ECO Cost		
Rework	☐	DO NOT WRITE BELOW THIS LINE (ECO BOARD ONLY)		
Scrap	☐	Effective Date		
Use as is	☐	Incorporated Date		
None	☐	Board Approval		
See Attached	☐	Board Date		

Appendix

Cursor Feedback

Cursor Feedback provides information about SolidWorks geometry. The following tables summarize cursor feedback. The tables were developed by support engineers from Computer Aided Products, Inc. Peabody, MA. Used with permission.

Sketch Tools			
	Line		Rectangle
	Circle		Ellipse
	Arc (Centerpoint, Tangent, 3 Point)		Ellipse
	Parabola		Spline
	Polygon		Point
	Trim		Extend
	Split line (not possible)		Split line (here)
	Linear step and repeat		Circular step and repeat
	Modify sketch tool		Modify Sketch (Rotate only)
	Modify Sketch (Move / Flip Y-axis)		Modify Sketch (Move / Flip X-axis)
	Move Origin of Sketch / Flip both axes		

Cursor Feedback Symbols

Courtesy of Computer Aided Products, Inc. Peabody, MA USA

Engineering Design with SolidWorks — Appendix

Sketching relationships

Icon	Relationship	Icon	Relationship
	Horizontal		Vertical
	Parallel		Perpendicular
	Tangent		Intersection
	Coincident to axis		Midpoint
	Quarter arc		Half arc
	3 quarter arc		Quadrant of arc
	Wake up line/edge		Wake up point
	Coincident to line/edge		Coincident to point
	3D sketch		3D sketch
	3D sketch		3D sketch
	3D sketch		3D sketch

Dimensions

Icon	Dimension Type	Icon	Dimension Type
	Dimension		Radial or diameter dimension
	Horizontal dimension		Vertical dimension
	Vertical ordinate dimension		Ordinate dimensioning
	Horizontal ordinate dimension		Baseline dimensioning

Cursor Feedback Symbols

Courtesy of Computer Aided Products, Inc. Peabody, MA USA

Selection

Icon	Description	Icon	Description
	Line, edge		Axis
	Select Face		Select Plane
	Surface body		Select Point
	Select Vertex		Select Endpoint
	Select Midpoint		Select arc centerpoint
	Select Annotation		Select surface finish
	Select geometric tolerance		Datum Target
	Multi jog leader		Select multi jog leader control point
	Select Datum Feature		Select balloon
	Select text reference point	This Field Left Blank	Cursors for other selections with a reference point look similar
	Dimensions		Dimension arrow
	Cosmetic Thread		Stacked Balloons
	Hole Callout		Place Center Mark
	Select Center Mark		Block
	Select Silhouette edge		Select other
	Filter is switched on		

Cursor Feedback Symbols

Courtesy of Computer Aided Products, Inc. Peabody, MA USA

Engineering Design with SolidWorks — Appendix

Assemblies			
	Choose reference plane (insert new component/envelope)		Insert Component from File
	Insert Component (fixed to origin)		Insert Component to Feature Manager
	Lightweight component		Rotate component
	Move component / Smartmate select mode		Select 2nd component for smartmate
	Simulation mode running		
Cursor feedback with Smartmates			
	Mate - Coincident Linear Edges		Mate - Coincident Planar Faces
	Mate - Concentric Axes/Conical Faces		Mate - Coincident Vertices
	Mate - Coincident/Concentric Circular Edges or Conical Faces		

Feature manager			
	Move component or feature in tree		Copy component or feature in tree
	Move feature below a folder in tree		Move/copy not permitted
	Invalid location for item		Move component in/out of sub assembly

Cursor Feedback Symbols

Courtesy of Computer Aided Products, Inc. Peabody, MA USA

Drawings

	Drawing sheet		Drawing view
	Move drawing view		Auxiliary view arrow
	Change view size horizontally		Change view size vertically
	Change view size diagonally		Change view size diagonally
	Align Drawing View		Select detail circle
	Block		Select Datum Feature Symbol
	Insert/Select Weld Symbol		Select Center Mark
	Select Section View		Section view and points of section arrow
	Select Silhouette edge		Hide/Show Dimensions

Standard Tools

	Selection tool		Please wait (thinking)
	Rotate view		Pan view
	Invalid selection/location		Measure tool
	Zoom to area		Zoom in/out
	Accept option		

Cursor Feedback Symbols

Courtesy of Computer Aided Products, Inc. Peabody, MA USA

Helpful On-Line Information

The SolidWorks URL: http://www.solidworks.com contains information on local resellers, Gold Partners, Solutions Partners and SolidWorks users groups.

The SolidWorks URL: http://www.3DpartStream.net and http://www.3DContentCentral.com contain additional engineering electronic catalog information.

The SolidWorks web site provides links to sample designs, frequently asked questions, the independent News Group (comp.cad.solidworks) and Users Groups.

Helpful on-line SolidWorks information is available from the following URLs:

http://www.mechengineer.com/snug/

- News group access and local user group information.

http://www.nhcad.com

- Configuration information and other tips and tricks.

http://www.solidworktips.com

- Helpful tips, tricks on SolidWorks and API.

http://www.doubleswx.com

- <u>How to Double Your SolidWorks Productivity</u> by Malcolm Stephens. This book is based on efficiency-increasing production engineering techniques utilizing SolidWorks.

Certified SolidWorks Professionals (CSWP) URLs provide additional helpful on-line information:

http://www.scottjbaugh.com	Scott J. Baugh
http://www.3-ddesignsolutions.com	Devon Sowell
http://www.zxys.com	Paul Salvador
http://www.mikejwilson.com	Mike J. Wilson
http://www.frontiernet.net/~mlombard	Matt Lombard

Notes:

INDEX

Engineering Design with SolidWorks provides this index to be utilized to revisit information. Do not skip steps in this tutorial-based project book. Our goal is to show how multiple design situations and steps are combined to produce a successful project. Use SolidWorks On-line help and glossary for additional options and commands.

$PRP"Number" 3-52
$PRP"Revision" 3-52
2D to 3D Toolbar 6-97
3 POINT Arc 4-38
3DPartStream 2-44
3Way Solenoid 6-94
80/20, Inc. 1-86

A

Add Relations 1-83, 2-72, 4-29
Addressing Design Issues 5-93
Advanced (Transparency) 4-49
Air Cylinder 2-75
AIR RESERVOIR SUPPORT Assembly 4-64, 4-73
Aligned dimension 4-19
Alt key 2-28
Analysis 2-66
ANGLE BRACKET 4-67, 4-68
ANGLE-BRACKET-13HOLE 6-82
Angular dimension 4-20
Animation 2-82
Annotations 6-16
Annotations Font 3-11
ANSI 3-10
ANSI B 18.3.1M-1986 standard 2-28
Arrow Keys 1-10
Arrows 3-11
ASME 14.5M Standard 3-40
ASMEY14.5M Decimal dimension display 4-11
Assembly 2-70
Assembly Hole Feature 6-64
Assembly Modeling Approach 2-6
Assembly Techniques 5-66
Assembly Template 1-18, 5-67
Associative Part, Assembly, Drawing 3-65
Autocad .dwg file format 6-97
Auxiliary view 3-26
AXLE 1-88, 3-75
AXLE3000 5-106

B

Back 1-12
Balloon 3-59
B-ANSI-MM Drawing Template 3-20
Base feature 1-82
BATTERY 4-8
BATTERYANDPLATE sub-assembly 5-78
BATTERYPLATE 4-23
Bi-directional 4-47
Bill of Materials 3-55
Boston Gear 5-118
Bottom Up assembly design approach 2-7, 2-70
BOX Assembly Overview 6-8
Boyle's Law 4-71
BULB 4-50
By Delta XYZ 2-49

C

CAPANDLENS sub-assembly 5-82
Center Marks 3-38
Centerline 1-82, 4-13
Centerpoint Arc 4-44
Chamfer Feature 1-59, 2-30
Circle 1-46
CircuitWorks 6-19, 6-77
Circular Pattern 4-56, 4-59, 5-23, 5-26
Clear All Filters 2-21
Clearance fit 2-6, 3-43
Collapse 2-38
Collision Detection 2-21
Common Units 4-70
Company Logo 3-16
Component 2-71
Component Pattern 6-62
Component Pointer 6-21
Component State 2-13
Component syntax 5-73
Computer Aided Manufacturing 2-86
Concentric Mate 2-14
ConfigurationManager 3-57
Configurations from Design Table 6-75
Convert Entities 4-31
Copy a Component 2-26
Copy and Paste 1-83
Copy Confirmation 1-63
Copy/Paste Function 1-63
Copy/Paste Pictures 3-18
Copy/Paste Sketch Threadsection 5-47

Copy/Paste Text 3-48
COSMOSXpress 2-53 2-81 2-87, 4-71
COSMOSXpress Wizard 2-58
Curve, Helix 5-46
Custom Properties 3-51
Customizing Toolbars 4-60

D

Default Templates 6-14
Delete and Edit Features 4-25
Derived Pattern 6-79
Design State 6-31
Design Table 6-74
DE-STA-CO Industries 6-101
Detail view 3-28, 3-26
Developed Length 6-32
Diameter symbol 3-37
Die Cutout Palette Feature 6-47
Dimension 1-35, 1-83
Dimension Holes 3-36
Dimensioning Standard 1-20
Display Modes 1-41
Document Properties 1-19, 3-10
Dome Feature 4-56, 5-16
Draft Feature 5-44
Drawing 3-85
Drawing Template 1-18, 3-7, 3-20

E

Edge Fillets 4-34
Edit Component Dimension 2-22
Edit Definition 1-65, 1-67, 6-68
Edit Part Color 1-70
Edit Sheet 3-6 3-19
Edit Sheet Format 3-6 3-15
Edit Sketch 1-69
Edit Sketch Plane 1-68
Edit Table 6-76
eDrawings 5-97
Elastic Modulus 2-55, 2-56
Elastic Range 2-56
Emerson-EPT 2-91
Emhart Fastening Teknologies 3-81
Enerpac 4-78
Engineering Analysis 2-81
Equal Relation 1-47, 4-14
Equation 6-16, 6-70
E-TECH 2-45

Exploded View 2-35, 2-82, 3-55
Export Files 5-95
Extruded Base Feature 1-24, 1-26, 1-32, 1-37
Extruded Base/Boss Feature 2-30, 4-15,4-18, 4-26 4-28
Extruded Base/Boss Feature 4-41, 4-47, 4-48,5-20,5-36,5-41,5-43
Extruded Cut Feature 1-24, 1-45,1-83, 2-31, 4-16, 5-20, 5-53

F

Face 1-45
Factor of Safety 2-67
Fasteners 1-43
Feature Palette 2-23, 6-50
FeatureManager Syntax 2-12, 6-65
Features 1-82
File Locations 1-17
File Management 1-11
File Name 3-50
Fillet Feature 1-24 1-50 1-83, 4-15
Fillet Rebuild error 4-18
Finite Element Analysis (FEA) 2-53
Finite Element Analysis (FEA) stiffness matrix 2-55
First angle projection 1-29
Fits 2-71
Fixed 5-71
Flange Bolt 2-23
Flange Walls 6-40
FLASHLIGHT Assembly 5-64
FLAT BAR - 3 HOLE 1-87, 3-73
FLAT BAR - 9 HOLE 1-88, 3-74
FLAT PLATE 4-65
Flat State 6-39
FLATBAR-5HOLE Part 5-107
FLATBAR-7HOLE Part 5-107
Flatten 6-39
Flip side to cut 1-66
Float 5-71
Fully defined 1-34

G

Gap between profile and extension line 3-32
GE Plastics 5-5
General Notes 3-46
Geometric Relations 1-47, 1-83, 2-71, 4-29
Geometric Tolerancing 1-28
Geometric Tolerancing Symbols 3-79
Globalspec.com 4-75, 4-77
Gold Peak Industries 4-77
Grid/Snap 1-21

GUIDE Drawing 3-8, 3-21
GUIDE Feature Overview 1-71
GUIDE Part 1-72, 3-22
GUIDE-ROD Assembly 2-8
GUIDE-ROD Assembly Drawing 3-56

H
Hardware folder 2-23
Helical/Spiral path 5-30
Heli-Coil Screw Thread Inserts 3-81
Help 1-16
Hem 6-41
HEX ADAPTER Part 5-109
Hidden Geometry 2-71
Hidden Lines Removed 1-41
Hidden Lines Visible, 1-41, 5-21
Hide Components 2-32
Hole Callout 3-36
HoleWizard 1-24, 1-52, 1-79, 1-83, 4-41, 6-44
HOOK 5-101
HOUSING 5-34

I
IGES 6-57, 6-60
IM15-MOUNT Part 4-66
i-MARK 1-Hartford 1-92
In Place Mate 6-22
In-Context 6-22
Insert Bends 6-38
Insert Component 2-9, 5-70, 6-26, 6-34, 6-61
Insert Dimensions from the Part 3-29
Instances to Skip 6-46
Interference Detection 5-94
Interference fit 2-6
Intersection 5-61
ISO 4-10
ISO-1219 Pneumatic Symbols 6-94

K
Keep Visible 5-92
Keyboard Shortcuts 1-10

L
Layer Properties 3-12
Layout Sketch 6-11
Lehi Sheetmetal 6-52
LENS 4-35
LENSANDBULB sub-assembly 5-69

LENSCAP 5-17
Light Weight Parts 2-42
Line 1-45 1-60
Linear Center Mark 3-39
Linear Motion 2-7
Linear Pattern 1-80, 5-57, 6-45, 6-51
Linear Step & Repeat 6-55
Link Properties to Drawing Notes 3-52
Link to Property 3-50
Link Values 6-16
Linkage Assembly 1-87, 2-75, 3-73, 3-77
LINK-AND-HOOK assembly 5-113
Loads 2-62
Loft 5-11 5-15, 5-37, 5-40

M
MACHINE SCREW 4-69
Make this Dimension Driven 3-42
Make-Buy Decision 2-43
Manufacturing Considerations 6-52
Mate Types 2-14
MateGroup1 2-21
Material Thickness 6-31
Mates 2-15, 2-70
Max (Arc dimension) 4-33
Maximize 1-14
Maximum variation 3-44
Measure 4-21
Menus 1-83
Midpoint 4-13
Mirror Feature 1-77
Mirror Feature 1-77, 5-62
Model Items 3-29
Modify 1-37, 1-38
Modify the Dimension Scheme 3-40
Modify the LINKAGE assembly 6-88
Modify the WHEEL-AND-AXLE assembly. 6-89
Move Component 2-13
Move Dimensions in the Same View 3-31
Move Dimensions to a Different View 3-34
Move Views. 3-24

N
Named view 1-61, 3-25
Neutral Bend Line 6-31
New Assembly 2-8
New Drawing 3-8
New Folder 1-11, 1-12

New Part 1-13
New Step 2-35
New View 1-62
Note 3-53

O

Offset Entities 4-16 4-47
Offset Sketch plane 5-12
ON/OFF/PURGE Value 6-95
Open a part within a drawing 3-47
Open a part within an assembly 2-58, 6-23
Options 1-17
O-RING 5-7
Orthographic Projection 1-28 1-29
Over defined 1-34

P

Palette Forming Tools 6-50
Pan 1-42
Parallel Mate 2-15
Parallel Plane at Point 5-59
Part Color 1-70
Part Number and Part Name 3-48
Part Template 1-18
Part Templates *.prtdot 1-22
Partial Auxiliary View - Crop View 3-33
PART-IN-ANSI Template 4-10, 4-11
PART-MM-ISO 4-11
Parts Library 2-22
PEM Fastening Systems 6-57
Perpendicular Relation 4-58
PhotoWorks 1-93
Physical Simulation 2-78
Pierce Aluminum 6-53
Pierce Relation 5-9, 5-32
Plane 1-82
Plastic Range 2-56
PLATE 1-32
Pneumatic On-Off-Purge Valve Assembly 2-83
PNEUMATIC TEST MODULE Assembly 6-88, 6-90
Point 1-45
Polygon 2-31
Project Data Management 6-99
Properties 1-47, 3-47
Push Pin 1-62

R

Rebuild 1-83, 2-72
Rebuild Error 1-67
Record Simulation 2-80
Recover from Rebuild Errors 1-67
Rectangle 1-16, 1-33
Reference Planes 1-28
Reference Triad 1-32
References 2-52
Reid Tool Supply Company 1-85
Relief 6-34
Reload 6-67
Remove 3-37
Rename feature 1-40
Rename sketch 1-40
Replace 6-67
Replay 2-80
RESERVOIR Assembly 4-70
Resize 1-60
Restraints 2-61
Revolve Boss/Base Feature 4-39, 4-46, 4-51, 4-53
Revolve Thin Feature 4-44
Revolved Cut Thin Feature 4-54, 5-22
Rib Feature 5-54, 5-55, 5-59, 5-61
Rip Feature 6-37
ROD Part 1-56
Rollback 1-65, 1-67
Rotary Motor 2-79
Rotate 1-42, 5-21
Rotate Component 2-19, 2-49, 5-91
Rotate Component by Delta XYZ 2-49
Rotational Motion 2-7
Run 2-65

S

Save As 2-52, 4-24
SCHEMATIC DIAGRAM 6-94
Screw fastener 3-80
Section View 2-39, 3-26, 3-27, 4-43
Selection Filter 2-19, 4-17
Shaded 1-41
SHAFT COLLAR 1-88, 3-76
SHAFT COLLAR-1875 5-108
SHAFT-COLLAR-5000 5-108
Sheet Format 3-7, 3-8
Sheet Metal Overview 6-30
Shell Feature 4-40, 5-21, 5-42
Shift key 2-28

Thread with Sweep Feature 5-45
Tile Horizontally 2-9, 3-22
Title Block 3-14
Tolerance 2-71
Tolerance and Fit 2-6
Toolbars 1-15 1-83
Tools 1-17
Top Down assembly design approach 2-70, 6-6
Transition fit 2-6
Transparency slider 4-49
Transparent feature 4-48
TRIANGULAR SUPPORT ASSEMBLY 6-85
Trim 4-58
TUBING 6-96

U
Ultimate tensile strength 2-57
Under defined 1-34
Units 1-20, 3-11
User Specified Name 3-62

V
Valid geometry for Mates 2-14
Velocity Slide bar 2-79
Vertex 1-45
View 1-15
View Modes 1-41
View Orientation 1-45, 4-22
Viewing Exploded State 2-38
von Mises stress 2-57

W
WEIGHT 5-102
WHEEL Part 5-110
WHEEL-AND-AXLE assembly is 5-105
Wilson Tool 6-52
Windows 2000/NT Terminology 1-7
Wireframe 1-41

Y
Yield Point 2-56

Z
Zoom In/Out 1-42
Zoom to Area 1-42
Zoom to Fit 1-42
Zoom to Selection 1-42

Show (Planes) 4-37
Show component 2-34
Show Feature Dimensions 6-16
Show Me 1-16
Show Sketch 4-30
Silhouette edge 4-45, 4-54
SINE-TRIANGLE 6-84
Sketch 1-82, 2-71
Sketch Fillet 5-39
Sketch Mirror 1-73, 1-82, 4-31 4-58, 5-38
Sketch Tools, Modify 6-48
Sketch Trim 5-38
SmartMates 2-24 2-71
SMC Corporation of America 1-89, 2-45
Socket Head Cap Screw 2-28
SolidWorks Animator 2-82, 2-87
SolidWorks SmartFastener 2-44
SolidWorks Toolbox/SE 2-44
Spin Box Arrows 1-36
Spin Box Increments 4-10
Spline 4-53
Split bar 2-38
Split the FeatureManager 2-38
Standard Views 1-15, 3-22
STANDOFF 6-84
Start 1-11 1-13 1-23
Status of a Sketch 1-82
Step Editing Tools 2-38
STL Files (*.stl) 5-96
Stop at Collision 2-21
Stop Simulation 2-80
Sub-assemblies 2-70
Sub-assembly component layout 5-64
Suppressed components 2-42, 2-72
Suppressed feature 5-27
Surface Reference plane 5-24
Sweep 5-10, 5-27, 5-49
SWITCH is 5-11
Symbols 3-37
System Feedback 1-83
System Options 1-17

T

Tangent Arc 1-75
Template 4-10
Temporary Axis 3-34, 4-43, 5-26
Third Angle Projection 1-29, 3-8
Thread 5-27